Hominid Evolution

Lifestyles and Survival Strategies

Hominid Evolution

Lifestyles and Survival Strategies

Edited by

Herbert Ullrich

εκδοσις ε α αρχαια
EDITION ARCHAEA®

1999

Layout: Elmar-Björn Krause

Production: EDITION ARCHAEA®

Printed with support of the „Universitätsklinikum Charité, Medizinische Fakultät der Humboldt-Universität zu Berlin“ and the „Deutsche Forschungsgemeinschaft“.

Printed on aging-resistant paper. / Gedruckt auf alterungsbeständigem Papier.

Printed in the Federal Republic of Germany.

ISBN 3-929439-07-7

Contents

Introduction to „Hominid Evolution - Lifestyles and Survival Strategies“

Herbert Ullrich

Lifestyles and survival strategies in hominid evolution is a subject eliciting more and more interest and gaining increasing importance in present-day human evolutionary research, especially from archaeological perspectives. Archaeologists have focused not only on the cultural remains of Palaeolithic humans but also on the reconstruction of their lifestyles, and their ways of life during the Palaeolithic period. There are many publications reconstructing Palaeolithic lifestyles. The knowledge of lifestyles is prerequisite for modeling life and survival strategies in hominid evolution. Pliocene and Pleistocene hominid groups had to find out the optimal way of life in a given natural environment and under given cultural environments on the basis of their biological equipment. Some groups experienced this with more, others with less success and some groups became extinct. The search for the optimal way of life enabled hominids to develop life and survival strategies and on the basis of these strategies to interact with their natural environments. Survival strategies might have been developed by Pliocene and Pleistocene hominid groups due to habitat preference, land use, settlement, resource use, foraging, hunting, diet, tool manufacture, co-operation, but also as a result of protection against predators, protection of territory and of fire, in breeding etc. The reconstruction of hominid lifestyles, life and survival strategies as well as their changes under special natural and cultural environments in different areas occupied by early hominids and humans should, in our opinion, become the main topic of the current and future interdisciplinary research on hominids and their environments in the Pliocene and Pleistocene.

Modeling life and survival strategies in hominid evolution is a subject which has only very recently been researched. Ten years ago Tooby & DeVore (1987) published a pilot article on "The reconstruction of hominid behavioral evolution through strategic modeling", but only a few other publications on this subject appeared during the nineteen-eighties. These mainly deal with regional variation in hunter-gatherer strategies in the Upper Pleistocene of Europe (Gamble, 1984), the reconstruction of Upper Palaeolithic hunter strategies (Klein, 1987; Olsen, 1989), early hominid subsistence strategies in seasonal habitats (Speth, 1987) and cortical motor asymmetry and hominid feeding strategies (Wundram, 1986). Neither in the "Encyclopedia of human evolution and prehistory" (Tattersall et al., 1988) and "The Cambridge encyclopedia of human evolution" (Jones et al., 1992) nor in other recent textbooks (e.g. Henke & Rothe, 1994) are strategies in human evolution dealt with. Only in very recent publications have authors focused more intensively on the modeling of strategies - life strategies and survival strategies - in hominid evolution. At the "XIII International Congress of Prehistoric and Protohistoric Sciences" held in Forli (Italy) in September 1996 about 16 papers on hominid strategies, mainly from archaeological perspectives, were presented in different sections and workshops. Land-use and settlement strategies in the Middle Palaeolithic of Italy, France, Greece, Spain and the Near East as well as during the Palaeolithic in Portugal have been studied by Alhaique et al. (1996), Auguste (1996), Bailey et al. (1996), Bar-Yosef (1996), Canals i Salomó et al. (1996), Carbonell i Roura et al. (1996), Ferreira Bicho (1996) and Shea (1996). Mobility strategies in the Middle-Upper Palaeolithic transition in south-west Germany (Beck, 1996) as well as raw material and tool production strategies (Bon & Dartiguepeyrou, 1996; DeBie, 1996; Demars, 1996; Wilson, 1996), hunting strategies in the Palaeolithic (Gardeisen, 1996a,b; Olsen, 1996) and female proto-symbolic strategies (Power, 1996) were also subjects of papers at the congress in Forli. A few other papers focusing on early hominid foraging and land-use strategies (Blumenschine, 1994; Bunn, 1994; Peters & Blumenschine, 1995; Kroll, 1995; Potts, 1994; Sikes, 1994; Stern, 1994), hunting strategies (Bratlund, 1996; Shea, 1997), core reduction strategies in the Middle Palaeolithic (Chabey & Sitlivy, 1993) and economical strategies in the Upper Palaeolithic (Bernaldo de Quirós & Cabrera, 1996) have been published in journals. Anthropological contributions are almost completely absent. Only Ishida (1991) published a paper on long distance walking strategies in earliest hominids.

Human Evolution International Interdisciplinary Project

The 40 papers presented in this book were selected from contributions given at the International Interdisciplinary Symposium on **"Lifestyles and survival strategies in Pliocene and Pleistocene hominids"** held in Weimar, May 4-10, 1997. This symposium was the second meeting of the HUMAN EVOLUTION INTERNATIONAL INTERDISCIPLINARY PROJECT *"Hominids in their environments"*. The idea of this project was conceived in 1990, when it was suggested that an interdisciplinary panel should be established to discuss the global problems of hominid evolution in relation to changes in the natural and cultural environmental. The first interdisciplinary symposium *"Man and environment in the Palaeolithic"* was held by the Institut für Anthropologie der Humboldt-Universität zu Berlin in co-operation with the Forschungsbereich Altsteinzeit des Römisch-Germanischen Zentralmusueums at Schloß Monrepos in Neuwied (Germany) in May 1993 (Ullrich, 1995b). The proceedings of the symposium were published (H. Ullrich, 1995a). At this symposium suggestions arose during discussions to hold a second meeting to discuss lifestyles and survival strategies developed by Pliocene and Pleistocene hominids and to publish a book on this topic.

The HUMAN EVOLUTION INTERNATIONAL INTERDISCIPLINARY PROJECT *"Hominids in their environments"* is a continuation of the author's research on human evolution. In 1977 an interdisciplinary working group "Probleme der Menschwerdung" (Problems of the origins and evolution of humans) was founded at the former "Zentralinstitut für Alte Geschichte und Archäologie" at the Akademie der Wissenschaften in Berlin and headed by the author. About 30 specialists from various disciplines (physical anthropology, archaeology, biology, geophysics, history, palaeontology, philosophy, psychology etc.), all from Eastern Germany, started co-operating in this project. In more than 25 colloquia and symposia, general and theoretical problems of the origins and evolution of humans were discussed under interdisciplinary aspects (Ullrich, 1992, 1993). In 1981 an interdisciplinary conference *"Anthroposoziogenese - biotischer und gesellschaftlicher Entwicklungsprozeß"* (Anthroposociogenesis - the biotic and cultural developmental process of humankind) was held in Weimar. The proceedings of this conference were published (Herrmann & Ullrich, 1985). Another international congress *"Kontinuität und Diskontinuität in der Evolution des Menschen und der menschlichen Gesellschaft"* (Continuity and discontinuity in the evolution of humans and their society) was organized in Lutherstadt Wittenberg in 1988. The main objective of the interdisciplinary working group was, however, to prepare manuscripts for the book *"Menschwerdung"* (Herrmann & Ullrich, 1991), an overall view of the hominization process from multiple and interdisciplinary perspectives. Twenty-one authors, representing eleven disciplines, have written the papers which cover a broad spectrum varying from primate evolution and hominid evolution to archaeological cultures up to the end of the Upper Palaeolithic and the social evolution of humankind.

Symposium "Lifestyles and survival strategies in Pliocene and Pleistocene hominids"

The Interdisciplinary Symposium on *"Lifestyles and survival strategies in Pliocene and Pleistocene hominids"* in Weimar, held by the Institut für Anthropologie der Medizinischen Fakultät (Charité) der Humboldt-Universität zu Berlin and sponsored by the Deutsche Forschungsgemeinschaft, was organized by the author (Ullrich, 1997a,b). 75 physical and social anthropologists, archaeologists, evolutionary biologists, botanists, ethologists, physicists, Quaternary palaeontologists, zoologists and colleagues from other disciplines from 19 European, African, American and Asian countries participated. 54 papers (all in English) were presented covering the following topics: (1) Primate models and early hominid lifestyles, (2) Ways of life and survival strategies in early hominids, (3) Human adaptation strategies, (4) Lifestyle adaptations in Palaeolithic humans, (5) Cultural adaptations, (6) Hunting and subsistence strategies and (7) Social survival strategies. The papers were followed by very lively and intensive interdisciplinary discussions. An evening session was devoted to introductions to the excursions. During the symposium the participants had the possibility to visit the exhibition of the Museum für Ur- und Frühgeschichte Thüringens as well as the Neanderthal site Weimar-Ehringsdorf, guided by R.-D. Kahlke. A full day excursion was organized to the *Homo erectus* site Bilzingsleben, where excavations have been carried out by D. Mania from the Friedrich-Schiller-Universität Jena. The Abteilung Quartärpaläontologie des Instituts für Geologie der Friedrich-Schiller-Universität in Weimar presented a video on the Lower Pleistocene mammal site Untermaßfeld (Thuringia, Germany) and a poster series on the preparation of animal skulls from this site. The participants of the symposium voted to prepare and send letters of support for the excavations of the Lower Palaeolithic sites Schöningen and Bilzingsleben. Letters were sent to the prime ministers of Niedersachsen, Sachsen-Anhalt and Thuringia asking for support for the ongoing excavations at Schöningen and to enable the Bilzingsleben finds (now in Halle, Bilzingsleben and Jena) to be stored together. In conclusion many of the symposium participants supported the idea of a continuation of interdisciplinary dialogues on the global topic "Hominids in their environments".

Unfortunately it is impossible to publish all the symposium papers in a single proceeding. The only way to publish all of the 51 papers submitted is to have two separate representative publications: this book and a special issue of the journal "Anthropologie" (Brno) dedicated to *'Lifestyles and survival strategies in Pliocene and Pleistocene hominids'* (vol. XXXVI, 1-2, 1998, 1-175). The introductions to both publications will refer to the other one.

The 11 contributions to the issue of the journal "Anthropologie" include three anthropological and eight archaeological papers. Two papers are focusing on hominid bipedalism. Vos, Sondaar & Reumer have compared the hominid and Hipparion trails from Laetoli and concluded that they share similar evolutionary pattern and that *Australopithecus* (including the species *habilis*), had a more all round gait, when compared to the striding gait of *Homo*. Czarnetzki, Dollhoff & Vacca have analysed features of human bipedalism on the femur and tibia in a pilot study resulting in a relatively clear differentiation between human bipedalism and non-human bipedalism on corresponding fossil finds of early representatives of the Hominoidea. The question whenever Neanderthals are a cold-adapted European Middle Pleistocene population is put in the paper of Condemi. In her opinion Neanderthal populations should primarily be interpreted as the result of geographic isolation, where this isolation must be related to the unique geographic position of Europe and to particular climate conditions prevailing during the Middle Pleistocene.

The paper of Soffer, Adovasio, Hyland, Klíma & Svoboda reports on numerous fiber, cordage and textile impressions on fired clay from the Eastern Gravettian sites at Pavlov and Dolní Věstonice in Moravia and demonstrate that the use of plant fibers in the manufacture of textiles, basketry, cordage and perhaps netting is evidence of a new innovation. The results of their investigations on perishable technologies enable a new look inside the hunting technology and concomitant change in the organization of hunt during the Gravettian and give a new basis to the reconstruction and characterization of the lifeway of the Gravettians. The comparison of three sites of the early handaxe culture from the Sahara, Near East and India and their aspects of behaviour by Fiedler has brought to light many common traits and has shown that the early handaxe culture was the mode of life in the human world over a long period of one million years, in which Europe played only a late and marginal role. The following two papers report new geo-stratigraphical, palaeontological and palaeoethnological results from the early Middle Pleistocene site Notarchirico in southern Italy (Piperno, Lefèvre, Raynal & Tagliacozzo) and preliminary results about the archaeology of the Middle Pleistocene site Beeches Pit in Suffolk (UK) in a European context (Gowlett, Chambers, Hallos & Pumphrey).

Four papers dealing with new important archaeological results from Asian sites. Derevianko, Petrin & Taimagambetov present methods and informative evidences from the investigation of Palaeolithic surface sites in the arid zone of Eurasia with examples from Kazakhstan and Mongolia. The cultural evolution and chronostratigraphical background in the Middle and Upper Pleistocene loess palaeosoil sections of South Tadjikistan is demonstrated for two sites (Obi-Mazar and Khonako) by Schäfer, Ranov & Sosin. According to Derevianko, Petrin, Taimagambetov & Otte Palaeolithic sites occurring in travertine and recorded in the Piedmont zone of Central Asia as well as in some other regions of Eurasia probably represent a specific adaptation of the inhabitants to the local environment. The paper of Wang Youping from Beijing is based on the results from several hundreds Palaeolithic localities focusing on human adaptations and Pleistocene environments in South China. Stone tool industries and settlement patterns indicate that human adaptations in the tropics and sub-tropics of East Asia seem to be much more complex than the bamboo hypothesis is illustrating.

Lifestyles, ways of life, life strategies and survival strategies

The manuscripts in this book, consisting of symposium papers updated to the level of ongoing discussions, and extended by many authors, were written by physical and social anthropologists, archaeologists, biologists, ethologists, historians, palaeontologists and physicists. They have been grouped in seven sections: (1) Primate models and early hominid lifestyles, (2) Ways of life and survival strategies in early hominids, (3) Human adaptation strategies, (4) Lifestyle adaptations in Palaeolithic humans, (5) Cultural adaptations, (6) Hunting and subsistence strategies and (7) Social survival strategies. These aspects are of course the main aspects addressed when studying lifestyles, ways of life, life and survival strategies but they do not include all the aspects of this topic.

Primate models and early hominid lifestyles

In his paper on lifestyles and survival strategies at the primate level Tembrock has stressed the importance of communication in the strategical and tactical organization as well as the socio-cultural institutionalization of interactions with the environment and the shifting of communication to the symbolic levels in primates for clearing the way to human evolution. The question if whether the chimpanzee or the pygmy chimpanzee represents a better ape model of incipient hominid lifestyles is put by Kortlandt and analysed by behavioural, ecological and geographical factors. His conclusion is that it is still premature to raise this question and that it is necessary to undertake more analyses of behavioural patterns. Coping with social tension and models of restoring social equilibrium after conflicts with other group members in primate societies and the inferences for peacemaking in early hominid communities is the topic of Butovskaya's paper. It is suggested that reconciliation strategies in early hominids might have been based on sexual behaviour (as in bonobos and modern man), meat sharing, friendly body contacts, ritualized gestures and utterings.

Ways of life and survival strategies in early hominids

This section includes five contributions - two are referring to australopithecine and early *Homo* finds, one on meat-eating and carcass procurement and two describe the cultural behaviour of early hominids.

According to Tobias recent studies have shown dramatic changes in our knowledge about the biological equipment of the australopithecines as well as their environment and survival: *Australopithecus* retained many primitve plesimorphous traits sharing with living apes and was more at home in a well-wooded terrain than in open savanna. It seems that survival of the australopithecines depended on their biological equipment adapted to partial arboreal life and to bipedal gait on earth. But survival was not easy for the small-brained australopithecines as can be shown by palaeodemographic studies. Only after the appearance of *Homo erectus* did human mean life expectancy increase.

The remains of the robust australopithecine and early *Homo* discovered by Schrenk in the Malawi Rift and discussed in his paper represent two eastern African endemic taxa (*Paranthropus boisei* and *Homo rudolfensis*) with different lifestyles and survival strategies which are linked to climate and habitat changes as well as on biogeographic variation and dispersal.

Opposing views of hominid strategies of carcass procurement (hunting versus scavenging) and exploitation (meat and/or marrow consumption) as the most debated topics in Plio-Pleistocene East African archaeology are analysed by Domínguez-Rodrigo. The results of his studies on medium-sized carcasses, the most dominant carcass type at early sites, support the hypothesis that hominids were not only prime accumulators of bones, but also the primary agents in the exploitation of animals by hunting or primary scavenging of undisturbed carcasses.

Cultural behaviour in Plio-Pleistocene hominids is stressed in the papers of Santangelo and Vishnyatsky from different point of views. Santangelo argues that the production of chopping tools was a "useful possession" leading to selective pressure on bigger brain size connected with the ability for culture in early hominid evolution. Only early hominids who experienced a mental commerce on tools were able to project theirs skills into the future. In Vishnyatsky's opinion the bipedalism of early hominids was a disadvantage rendering the chance for survival low in changing environments. To compensate the negative consequences of bipedalism an initial development of cultural behaviour was necessary. Culture played more and more an important role in early hominid groups. Survival in hominids is related according to Vishnyatsky to their cultural capacities.

Human adaptation strategies

Adaptation in human evolution has been a process that primarily acted on the body building and on special traits of the human body via the adequately adapted morphological (and physiological) structures. Biological adaptation acted very slowly and over a very long period. In contrast to biological (physical) adaptation, the cultural and social adaptation to natural environmental conditions acts more or less quickly and over a relatively short time.

Homo erectus was the first widely dispersed member of the genus *Homo* to leave tropical Africa and disperse into temperate regions of the Old World. Cachel & Harris focus on the adaptive zone of *Homo erectus* from an African perspective. Their conclusions show that pronounced arid adaptations of sub-Saharan Africa and larger body size document the first occurrence of *Homo erectus* in the pronounced open country and that the successful hominid migration from sub-Saharan Africa into Eurasia was not necessarily dependent upon material culture. It is suggested that the invasion of novel environments by *Homo erectus* could have lead to character release which might have increased variability in the morphological and behavioural phenotypes and that *Homo erectus* thrived on environmental disruption and habitat fragmentation.

In discussions about the oldest *Homo erectus* fossils in Eurasia the mandible from Dmanisi (Georgia) with a supposed age of around 1.9 myrs is thought to be a good candidate. Henke, Rothe & Alt clearly demonstrate that the presence of apomorphic traits in the mandible is a strong argument against affinities to an early *Homo erectus* sample and cast reasonable doubts on the claimed chronology. The authors also discuss the theories on the emergence and dispersal of early *Homo erectus* and point to the unsufficient definition of this species.

Vlček's paper recalls that two parallel human population groups can be recognized in Central Europe during the Holstein period: an advanced *Homo erectus* group represented by the human remains from Bilzingsleben, associated with pebble tools and small-dimensional industries, and an archaic *Homo sapiens* group represented by the Weimar-Ehringsdorf hominids who are morphologically closer to modern *sapiens* than to the Neanderthal forms and whose remains are associated with early Mousterian culture.

The nasal morphology of the Neanderthals is generally interpreted as an adaptation to cold and dry climate. Schwartz, Tattersall & Laitman present another view based on new studies of the nasal morphology. They see no evidence for a direct adaptation to cold and dry conditions. They suppose that Neanderthals with their apparent stability of their morphology over a long period compensated for the climatic and environmental changes they encountered by cultural means.

Changes in the patterns of adaptive strategies in the body of humans during the transition from Upper Palaeolithic to Neolithic populations in central Europe are reported by Piontek. These changes (e.g. in body size, long bone proportions, sexual dimorphism, earlier maturity), based on changes in the structure of the socio-economic system, resulted from new possibilities in the operation of natural selection (controlled by cultural factors shaping a new fertility pattern).

Leney & Foley have focused on the interaction between hominids and carnivores and have studied micro-adaptations in populations of Upper Pleistocene hyenas related to ecological changes which may be directly related to hominids, too. It is suggested that the nature and timing of variation in hominid ecologies should, under certain circumstances, be reflected by a concomitant suite of micro-adaptations in the carnivore community.

Lifestyle adaptations in Palaeolithic humans

This section included five papers referring to general aspects of lifestyle adaptations in Palaeolithic humans and comprising a broad spectrum of aspects. Hominid expansion as a function of specific populations dynamics depending on the trophic behaviour of animals and the anthropogenetic extinctions of megaherbivores are discussed by Schüle (†) & Schuster, demonstrating the close connection of human expansion and the extinction of megaherbivores in hominid evolution. To overcome ecological disturbances hominids had to develop new technologies and ecological ethics. The simultaneous evolution of large brains as well as social and ecological ethics are considered an iteration process within the complex adaptive system in human evolution.

Survival and adaptation of Pleistocene humans on islands is another topic in this section. Sondaar & de Vos conclude in their contribution that oversea dispersals of humans to islands resulted in a dramatic faunal turn-over (extinction of endemic megafaunal elements), which can be demonstrated e.g. for the islands of Flores and Sardinia about 700,000 B.P. The early colonizers of Flores and Sardinia (*Homo erectus*) had to develop new survival and adaptation strategies after the faunal turn-over.

Otte opposes the retrospective vision in the archaeological literature that on account of alimentary deficiencies during the Mousterian the Neanderthals had to struggle for "survival" (in the sense of to "remain alive"). It is emphasized that the modes of life and Mousterian technology were well adapted to the temperate conditions of interglacials and glacials. The extraordinary success of Palaeolithic populations to expand and to progress everywhere during hundreds of thousands of years demonstrates the exceptional human adaptation.

The paper of Sigmon presents a new model, different from previous explanations of hominid reproductive behaviour, suggesting that the reproductive pattern of early hominids was intimately related to the environment they inhabited. In her view early hominids fit into the evolutionary continuum of Old Worlds monkeys and apes with peak fertility periods in spite of being able to copulate throughout the year. Peak fertility periods were times during which conditions in the environment and specific habitat were most favourable for health and a general feeling of fitness.

The question if disinhibition of anti-cultural behaviour has an adaptive meaning in hominid evolution is posed by Kozintsev & Butovskaya. They have analysed laughter, weeping and yawning in primates as well as in archaic societies and suggest that the primary adaptive meaning of laughter, and possible of crying and yawning, was to maintain group cohesion by providing relief from cultural stress and thus to protect early humans from social tension and neuroses. These displays acquired new motivations after the emergence of language.

Modern models of the past are demonstrated in the life of the Eipo in the highlands of West-New Guinea. Their environment, subsistence strategies and knowledge of nature as well as their physical anthropology, medicine and culture of thought are presented by Schiefenhövel, summarizing that despite their limited material inventory the Eipo have a rich mental culture with a complicated and expressive language, a good knowledge of plants and animals, myths and poetry.

Cultural adaptations

Cultural adaptations to natural environmental conditions did not actively influence the natural environments of Palaeolithic humans but enabled life and survival strategies for the living population to be developed. Cultural continuity and changes in hominid behaviour in Africa during the Middle to Upper Palaeolithic transition is outlined by Clark. This period is characterized as a time of movements, readjustments and innovative technology with more intensive uses of food resources and of material that extended the possibilities through new and more efficient equipment - the time of preparation and experimentation preceding the appearance of modern humans and their complex Upper Palaeolithic technology.

One of the most important Middle Pleistocene sites in Europe with archaeological horizons which are appropriate for the reconstruction of the ecosphere, culture and lifestyles of a *Homo erectus* population is Bilzingsleben in Germany. The paper of D. Mania, U. Mania & Vlček analyses the geology, stratigraphy, palaeo-ecology and culture of the camp site as well as the hominid remains discovered so far and characterize the camp site as a home base occupied over a considerable period of time during the warm-temperate maximum of an interglacial about 400,000 years ago.

Evidence from technology, subsistence, settlement systems and hominid socio-ecology is given by Rolland for the Middle Palaeolithic as a developmental stage in the cultural evolution of humans. It is proposed, as a provisional hypothesis, that the emergence of home-base residences, mortuary practices, long-term home-range expansion trends, exploring and exploiting new habitats, gradual specialization of basic food staples, seasonal flux and more extensive socio-demographic spacing and networks were linked causally.

The influence of the last glacial maximum on changes in subsistence strategies and lifestyles between the Gravettian and Epigravettian in central and south-eastern Europe is evaluated in Kozlowski's paper. Different hunting strategies, social organization, ways of occupation and exploitation of territories, lithic raw material procurement and different spheres of spiritual and symbolic culture are pointed out.

Hunting and subsistence strategies

Many papers are contributed to this section. Guleç, Howell & White report on a newly discovered Lower Pleistocene faunal and artefact-bearing locality (so far the oldest one in Anatolia) in southern Anatolia showing traces of human workmanship, human transport of clastics and incisions on a metatarsus of a large bird obviously an evidence of butchering although the site itself is not a butchery site. The problems and difficulties of interpreting hominid subsistence strategies at Lower Palaeolithic sites are discussed by Turner in a case-study of the interglacial faunal remains, in association with some lithic artefacts, from Miesenheim, where not only hominids, but also carnivores and the location of the site played a role in the deposition and modification of the animal bones. Almost as old as Miesenheim is the site at Schöningen where Thieme has discovered more than a half a dozen of the world's oldest spears and other wooden implements, together with butchered remains of at least 15 horses, showing that hunting with spears originated much more earlier, at least 400,000 years ago, in human evolution.

Using lithic transport data Gamble & Steele estimated the home range patterns and dietary strategies of a Lower (Arago) and Middle Palaeolithic hominid group (Vaufrey). The calculated home ranges are, in comparison with modern data, consistent with the use of local groups of a few families focusing on meat diets. A reconstruction of Neanderthal diet using bone collagen and nitrogen stable isotopes - an new approach to investigate palaeo-diet in human evolution - by Bocherens points in the same direction: animal meat was an important part of Neanderthal diet. But also gathering fruits was a way of food supply during Palaeolithic times. Jäger & Schäfer give evidence of fruit gathering shown by the palaeobotanical examination of charred putamina of corne from the travertine site at Weimar-Ehringsdorf, where hunting activities have also been attested. Critical comments on the osteological criteria for hunting and scavenging in the Middle Palaeolithic and their economic interpretations by Bratlund show that possibilities for a positive demonstration of scavenging activities are limited and that we are much better equipped to accumulate arguments for hunting from an archaeological faunal assemblage, e.g. for megafaunal hunting by Neanderthals from Taubach near Weimar. The hunting and raw material economy during the Middle Palaeolithic in Moravia is discussed by Valoch. The animal remains from three caves used as a camp site (Kůlna), hyena den (Švédův stůl) and a cave bear site (Šipka) have been analysed with respect to hunting activity strategies. Different strategies of obtaining mineral supplies are also visible between these three caves.

Jelínek's survey of behaviour and survival strategy of early Gravettians in Moravia is based on the analysis of large accumulations of mammoth bones which demonstrate the variable character of these man made accumulations according to local conditions, needs and traditions of the site inhabitants, climatic and environmental conditions and illustrates the mode of life and survival strategy of Gravettian people. Fresh bones from hunted mammoths as well as older collected mammoth bones were used in periglacial Moravia. Strategies for survival in the Eastern Gravettian modeled on the base of resource exploitation patterns by Pichler show considerable differences between central and eastern European human groups, both regionally and diachronically. Different degress of the exploitation of local resources and different spectra of exploited animals became manifest.

Dietary shifts in the origins of the European Upper Palaeolithic is the title of Cachel's paper. She postulates that subtle dietary shifts may have affected Pleistocene human biology and social behaviour, with selection pressure exercised through population growth. The geographic dispersion and permanent settlement of anatomically modern humans may have been initiated by population growth and concomitant social complexity.

In the aceramic Neolithic of the island Cyprus around 7,500 B.C. all edible plants and animals had to be transported from the mainland. It is argued by Ronen, analysing the food animals, that the absence of *Bos* reflects a cultural induced negative attitude against this animal and that the aceramic Cypriots developed their own ideology and isolated themselves using mental and physical boundaries.

Social survival strategies

Planning capacity and symbolism, both essential and typical aspects of human culture have, according to Facchini, generated increasingly advanced abilities to control the environment, favoured adaptation to it, led humans to a more complex social organization and to better lifestyles and improved their dispersal all over the world. Planning and symbolism represent real survival strategies.

The paper of Power models strategies available to archaic *Homo sapiens* females for extracting increased investment from males. It is suggested that collective, deceptive and amplified use of red pigments as body paint confused information available to males about the female reproductive status and effectively formed a preadaptation to ritual, a female proto-symbolic strategy.

Mortuary practices and reflections on life and death in the Palaeolithic are the subject of Ullrich's paper. Two different strategies were manifested in mortuary practices: practices (1) based on disarticulated human bones resulting from manipulations on human corpses and (2) with entire intact corpses of the dead. Palaeolithic mortuary rites reflect the many unsolved problems and contradictions between life and death and the struggle for a universal conception of life and death and influenced lifestyles and survival strategies of Palaeolithic humans intensively.

Ackowledgments

I would very much like to thank the authors for preparing their manuscripts and for their inspiration in assembling this book as well as for their participation in the Weimar Symposium.

I would like to express my deep gratitute to Björn Krause, managing director of the Verlag EDITION ARCHAEA, for his efforts to publish the book in the current scope and design.

I am deeply indebted to Professor Jan Jelínek, the founder and former editor of the journal "Anthropologie" (Brno) as well as to the present editor, Dr. Marta Dočkalová, for making a special issue of this journal available for the publication of a selection of the papers from the Weimar symposium. I would like to express my deepest thanks to both of them.

I gratefully acknowledge the financial support for printing this book by the Universitätsklinikum Charité, Medizinische Fakultät der Humboldt-Universität zu Berlin and the Deutsche Forschungsgemeinschaft, Bonn.

References

ALHAIQUE, F., BIETI, A., GRIMALDI, S., KOTSAKIS, A.,KUHN, S., LEMORINI, C., MANZI, G., RECCHI, A., SCHWARCZ, H.P. & STINER, M.C. (1996): Settlement patterns and subsistence strategies in the late Mousterian site of Grotta Breuil. In: *Abstracts of the Workshops and the Posters of the XIII International Congress of Prehistoric and Protohistoric Sciences, Forli (Italia) 8-14 September 1996*, 33. Forli.

AUGUSTE, P. (1996): Les stratégies alimentaires des hominidés en France septentrionale durant le paléolithique moyen. In: *Abstracts of the Sections of the XIII International Congress of Prehistoric and Protohistoric Sciences, Forli (Italia) 8-14 September 1996*, 139. Forli.

BAILEY, G. KING, G. & STURDY, D. (1993): Active tectonics and land-use strategies: A Palaeolithic example from northwest Greece. *Antiquity* 67, 292-312.

BAR-YOSEF, O. (1996): Middle Palaeolithic settlement patterns in the Near East. In: *Abstracts of the Workshops and the Posters of the XIII International Congress of Prehistoric and Protohistoric Sciences, Forli (Italia) 8-14 September 1996*, 33-34. Forli.

BECK, R.F. (1996): Mobility strategies in the Middle-Upper Palaeolithic transitionn of southwest Germany. In: *Abstracts of the Sections of the XIII International Congress of Prehistoric and Protohistoric Sciences, Forli (Italia) 8-14 September 1996*, 171. Forli.

BERNALDO de QUIRÓS, F. & CABRERA, V. (1996): Economical strategies in the Upper Palaeolithic in the Cantabrian region. *Human Evolution* 11, 121-128.

BLUMENSCHINE, R.J., CAVALLO, J.A. & CAPALDO, S.D. (1994): Competition for carcass and early hominid behavioral ecology: A case study and conceptual framework. *Journal of Human Evolution* 27, 197-213.

BLUMENSCHINE, R.J. & CAVALLO, J.A. (1995): Frühe Hominiden - Aasfresser. In: B. Streit (Hrsg.), *Evolution des Menschen,* 68-75. Heidelberg.

BON, F. & DARTIGUEPEYROU, S. (1996): Stratégies d'approvionnement et traitement des matiéres premières lithiques: comparaison des compartements aurignaciens et gravettiens des Pyrénées centrales et occidentales françaises. In: *Abstracts of the Workshops and the Posters of the XIII International Congress of Prehistoric and Protohistoric Sciences, Forli (Italia) 8-14 September 1996*, 113. Forli.

BRATLUND, B.E. (1996): Hunting strategies in the late glacial of northern Europe. A survey of faunal evidence. *Journal of World Prehistory* 10 (1), 1-48.

BUNN, H.T. (1994): Early Pleistocene hominid foraging strategies along the ancestral Omo River at Koobi Fora, Kenya. *Journal of Human Evolution* 27, 247-266.

CANALS I SALOMÓ, A., SALA I RAMOS, R., VAQUERO RODRIGUEZ, M. & CARBONELL I ROURA, E. (1996): Occupational strategies in the Abric Romaní Middle Palaeolithic site (Barcelona, Spain). In: *Abstracts of the Workshops and the Posters of the XIII International Congress of Prehistoric and Protohistoric Sciences, Forli (Italia) 8-14 September 1996*, 245. Forli.

CARBONELL I ROURA, E., AÏMENE M., ALLUÉ, E., CÁCERES, I., CANALS, A., PASTÓ, I., SALADIÉ, P., VAQUERO, M. & VALLVERDÚ, J. (1996): Occupational strategies in Middle Palaeolithic level at Abric Romani site (Capellades, Barcelona, Spain). In: *Abstracts of the Workshops and the Posters of the XIII International Congress of Prehistoric and Protohistoric Sciences, Forli (Italia) 8-14 September 1996*, 246. Forli.

CHABAY, V. & SITLIVY, V. (1993): The periodization of ocre reduction strategies of the ancient, Lower and Middle Palaeolithic. *Préhistoire Européenne* 5, 9-88.

De BIE, M. (1996). Late Palaeolithic tool production strategies: technological evidence from Rekem (Belgium). In: *Abstracts of the Workshops and the Posters of the XIII International Congress of Prehistoric and Protohistoric Sciences, Forli (Italia) 8-14 September 1996*, 117-118.. Forli.

DEMARS, P.-Y. (1996): Les stratégies dans les choix de silex au paléolithique supérieur en Périgord: indice laminiare de l'outilage et fréquence du silex du bergeracois. In: *Abstracts of the Workshops and the Posters of the XIII International Congress of Prehistoric and Protohistoric Sciences, Forli (Italia) 8-14 September 1996*, 114-115. Forli.

FERREIRA BICHO, N. (1996): Caves, rock shelters and open-air sites: land-use during the end of the Paleolithic in central Portugal. In. *Abstracts of the Sections of the XIII International Congress of Prehistoric and Protohistoric Sciences, Forli (Italia) 8-14 September 1996*, 187. Forli.

GAMBLE, C. (1984): Regional variation in hunter-gatherer strategy in the Upper Pleistocene of Europe. In: R. Foley (ed.), *Hominid evolution and community ecology*, 237-260. Oxford.

GARDEISEN, A. (1996A): Stratégies de chasse dans le paléolithique moyen pyrénéen (France). In: *Abstracts of the Sections of the XIII International Congress of Prehistoric and Protohistoric Sciences, Forli (Italia) 8-14 September 1996*, 140. Forli.

GARDEISEN, A. (1996b): Faunal remains and hunting strategies at the Upper Pleistocene cave Le Portel (Ariège, France). In: *Abstracts of the Workshops and the Posters of the XIII International Congress of Prehistoric and Protohistoric Sciences, Forli (Italia) 8-14 September 1996,* 267. Forli.

HENKE, W. & ROTHE, H. (1994): *Paläoanthropologie.* Berlin, Heidelberg, New York.

HERRMANN, J. & ULLRICH, H. (Hrsg.) (1985): *Menschwerdung - biotischer und gesellschaftlicher Entwicklungsprozeß.* Berlin. 277 pp.

HERRMANN, J. & ULLRICH, H. (Hrsg.) (1991): *Menschwerdung. Millionen Jahre Menschheitsentwicklung - natur- und geisteswissenschaftliche Ergebnisse. Eine Gesamtdarstellung.* Berlin. 778 pp.

ISHIDA, H. (1991): A strategy for long distance walking in the earliest hominids: Effect of posture on energy expenditure during bipedal walking. In: Y. Coppens & B. Senut (dir.), *Origine(s) de la bipédie chez les hominidés,* 9-15. Paris.

JOCHIM, M.A. (1981): *Strategies for survival.* New York.

JONES, S., MARTIN, R. & PILBEAM, D. (eds.) (1992): *The Cambridge encyclopedia of human evolution.* Cambridge.

KLEIN, R.G. (1987): Reconstructing how early people exploited animals: Problems and prospects. In: M.H. Nitecki & D.V. Nitecki (eds.), *The evolution of human hunting,* 11-45. New York, London.

KROLL, E.M. (1994): Behavioural implications of Plio-Pleistocene archaeological site-structure. *Journal of Human Evolution* 27, 107-138.

OLSEN, S.L. (1989): Solutré: A theoretical approach to the reconstruction of Upper Palaeolithic hunting strategies. *Journal of Human Evolution* 18, 295-327.

OLSEN, S. (1996): Horse hunting strategies in the Palaeolithic. In: *Abstracts of the Workshops and the Posters of the XIII International Congress of Prehistoric and Protohistoric Sciences, Forli (Italia) 8-14 September 1996,* 12-13. Forli.

PETERS, C.R. & BLUMENSCHINE, R.J. (1995): Landscape perspectives on possible land use patterns for early Pleistocene hominids in the Olduvai Basin, Tanzania. *Journal of Human Evolution* 29, 321-362.

POTTS, R. (1994): Variables versus models of early hominid land use. *Journal of Human Evolution* 27, 7-24.

POWER, C. (1996): Female proto-symbolic strategies. In: *Abstracts of the Sections of the XIII International Congress of Prehistoric and Protohistoric Sciences, Forli (Italia) 8-14 September 1996,* 112. Forli.

SHEA, J.J. (1996): Settlement dynamics and lithic variability in the Levantine Mousterian. In: *Abstracts of the Workshops and the Posters of the XIII International Congress of Prehistoric and Protohistoric Sciences, Forli (Italia) 8-14 September 1996,* 41. Forli.

SHEA, J.J. (1997): Hunting strategies, weapon design, and Levantine Mousterian technological organization. Abstract. *Journal of Human Evolution* 32, A19.

SIKES, N.E. (1994): Early hominid habitat preferences in East Africa: Paleosoil carbon isotopic evidence. *Journal of Human Evolution* 27, 25-45.

SPETH, J.D. (1987): Early hominid subsistence strategies in seasonal habitats. *Journal of Archaeological Science* 14, 13-29.

STERN, N. (1994): The implications of time-averaging for reconstructing the land-use patterns of early tool-using hominids. *Journal of Human Evolution* 27, 89-105.

TATTERSALL, I., DELSON, E. & VAN COUVERING, J. (eds.) (1988): *Encyclopedia of human evolution and prehistory.* New York.

TOOBY, J. & DeVORE, I. (1987): The reconstruction of behavioral evolutionn through strategic modeling. In: W.G. Kinzey (ed.), *The evolution of human behavior: Primate models,* 183-237. New York.

ULLRICH, H. (1992): Human evolutionary research in Eastern Germany. *Human Evolution* 7, 31-39.

ULLRICH, H. (1993): Interdisciplinary studies in the project "Menschwerdung" (Origins and evolution of humans). In: B. Sigmon (ed.), *Before the wall fell. The science of man in socialist Europe,* 123-139. Toronto.

ULLRICH, H. (ed.) (1995a): *Man and environment in the Palaeolithic.* Liège (ERAUL 62). 378 pp.

ULLRICH, H. (1995b): Introduction to "Man and environment in the Palaeolithic. In: H. Ullrich (ed.), *Man and environment in the Palaeolithic,* 1-8. Liège (ERAUL 62).

ULLRICH, H. (ed.) (1997a): *Programme and Abstract Book of the International Interdisciplinary Symposium "Lifestyles and survival strategies in Pliocene and Pleistocene hominids", Weimar (Germany), May 4-10, 1997.* Weimar. 158 pp.

ULLRICH. H. (1997b): Lifestyles and survival strategies in Pliocene and Pleistocene hominids. Internationales interdisziplinäres Symposium vom 4.-10. Mai 1997 in Weimar. *Ethnographisch-Archäologische Zeitschrift* 38, 253-258.

WILSON, L. (1996): Mousterian raw material strategies in regional context in southern France. In: *Abstracts of the Workshops and the Posters of the XIII International Congress of Prehistoric and Protohistoric Sciences, Forli (Italia) 8-14 September 1996,* 112. Forli.

WUNDRAM, I.J. (1986): Cortical motor asymmetry and hominid feeding strategies. *Human Evolution* 1, 183-188.

Lifestyles and Survival Strategies at the Primate Level

Günter Tembrock

Abstract

The evolution of primates is marked by the special forms of their relations with the environment and their strategies of survival. The organization of behaviour in those living species closest to wo/man is characterized by „open programs" which allow for adaptations modified by learning and for survival strategies. Ethoecology (in nonhuman primates mainly as socioecology) investigates the patterns of interactions and the network of their causes at the proximate level, whereas sociobiology aims at finding concepts of evolutionary mechanisms at the ultimate level.

Fitness is differentiated into 4 aspects: (1) individual fitness, (2) ecological fitness, (3) reproductive (social) fitness, and (4) inclusive fitness. Socioecological studies concentrate especially on questions which deal with dyadic and triadic interactions, with cooperation and analyses of cost and benefit. The concept of competition is of particular significance with scramble and contest competition constituting important socioecological mechanisms (in this case supplemented by the principle which deals with territoriality: „bourgeois competition"). We advocate the hypothesis that maximizing inclusive fitness originally operates at the level of genotypes (as inclusive fitness is defined), however, during the evolution of primates and especially that of wo/man it shifted to new levels which are tradigenetically determined. Thus „learning" becomes a dynamic factor of evolution which makes changes of the environment possible. This type of evolution is linked to specific interactions with the environment which are nowadays called „strategies" and which during their implementation may encompass a set of alternatives labelled „tactics". Of particular importance are strategies of using resources, of social and reproductive behaviour. We suppose that during the evolutionary process mentioned earlier tactics first shifted to the phenotypical level, i.e. were extended by learning, however, on the way to wo/man strategies also developed a new quality of interactions with the environment partly through learning. Communication had a special part in the strategical and tactical organization of social relationships. The socio-cultural institutionalization (through conventions) of interactions with the environment and the shifting of communication to the symbolic level (informational messages) cleared the way for human evolution.

The evolution of primates is marked by the special forms of their relationships with the environment and their survival strategies. This applies to the ecosystem as well as to the social environment. However, learning may modify the species specific organisation of behaviour to a significant extent. Ethoecology and, in particular, socioecology investigate patterns of interactions and the structure of their causes mainly at the proximate level, whereas sociobiology is concerned with concepts at the ultimate level. In doing so, ontogeny takes an intermediate position: Micro geneses of behaviour are linked with the ecological and social environment at the proximate level, whereas the changes and phases of ontogeny are species specific and determined genetically (ultimate factors); this results in an intermediate level for the ontogeny of the organisation of behaviour (phylogenetically and micro genetically determined). The sociobiological interpretation of survival strategies of primates differentiates essentially between two classes of fitness, the

Social strategy and mating strategy (Gouzoules, 1984)	
Solitary	Polygyny, promiscuity
Family groups	Monogamy
One-male units	Polygyny
One-male units within multi-male groups	Polygyny, promiscuity
Multi-male, multi-female groups	Promiscuity, polyandry

Table 1 - An attempt at classifying social and mating strategies.

individual fitness (= Darwinian fitness = reproductive fitness) and the inclusive fitness. We prefer four classes of fitness for the analysis of behaviour: (1) the individual fitness as a general status of an individual's performance (physical fitness); (2) the ecological fitness as an adaptation to the ecological niche; (3) the reproductive (social) fitness, and (4) the inclusive fitness (contribution to the maintenance of one's own genes over generations). Socioecological studies are especially orientated towards questions dealing with dyadic and triadic interactions, with cooperation, alliances as well as analyses of cost and benefit. The concept of competition is of particular significance with scramble and contest competition constituting important socioecological mechanisms which we would like to supplement with the principle that deals with territoriality: „bourgeois competition". With such a background in the evolution of primates regarding wo/man, the sexual dimorphism referring to parameters of behaviour gains special importance (see Andersson, 1994; Short & Balaban, 1994; Jarman & Kruuk, 1996). There are three areas of body structure relevant to behaviour: body size, the size of canines, and secondary sexual characteristics. In evolutionary terms, these properties are supported by species specific particularities. Foraging strategies of vervets, langurs, baboons, and macaques are influenced by the presence of pouches which enable the monkeys to collect larger amounts of food quickly, whereas colobines (Colobinae) have a stomach partitioned into three parts to store and break down leaves with the help of bacteria. In this case the specialists predominate, in other cercopithecoids the generalists preponderate with regard to the range of food. This is also influenced by the absolute body size: Gorillas have no competitors for food bigger than themselves, and thus they can afford to be much more specialized than the two chimpanzee species. The same might have applied to human evolution when a higher capacity of ecological adaptation might have been developed.

Two environmental variables influence the strong development of canines of primate males - though to a varying extent - first, the competition for females in polygynous or promiscuous reproductive strategies, second, the defense of the social group against predators. Thus, this property also correlates with body size. Therefore, for a number of primate species there is a positive correlation between sexual dimorphism (bigger males) and polygyny; the mainly monogamous gibbon species do not have this difference in size. In these species both sexes reach maturity approximately at the same time, whereas for species with obvious sexual dimorphism and increasing difference in body size males take progressively longer to reach maturity than females. Gaining in size and power costs time until reaching the age of reproductivity. However, there are evolutionary strategic differences in solving this problem of cost and benefit as Leigh (1992) showed: Talapoin

Rang defence	Sociality	Primates
Defended by female	Forages solitarily	23 %
Defended by group	Forages solitarily	2 %
Defended by group	Foragegs socially	22 %
Undefended range	Forages solitarily	8 %
Undefended range	Forages in ephemeral group	2 %
Undefended range	Forages in persistent group	43 %

Table 2 - Female lifestyles in mammals (Jarman & Kruuk, 1996)

males (*Miopithecus talapoin*) manage to reach full body size only by prolonging growth, the males of diademed guenons (*Cercopithecus mitis*), Celebes crested macaques (*Macaca nigra*), drills (*Mandrillus leucophaeus*), and Douc langurs (*Pygathrix nemaeus*) make sure that they remain bigger by growing faster than females. Here, the concept of „growth bimaturism" is relevant: Reproductivity can already begin before full growth has been reached and for various reasons this seems to be true more often for females than for males. An extreme example is the diademed gueneon. The body size ratio of male/female is about 1.72: Females mature sexually before the males, but reach their final, disproportionately smaller size later than the males!

Gouzoules (1984) classifies the correlation between reproductive and social strategies in primates as follows: It has been established that sexual dimorphism cannot be explained simply by polygyny or male promiscuity. Foraging mechanisms also play a role: Differences in body size can be linked with varying foraging strategies and thus reduce intraspecific and, in particular, intrasocial competition. Moreover, lactating females need special protection from predators, which in evolutionary terms may also favour males' bigger body size and stronger canines.

Jarman & Kruuk (1996) distinguish between six lifestyles of mammalian females regarding spatial relationships:According to the data available on the parameters in question primates display the greatest diversity among all mammal orders. In no other order one could find all types, in seven of the sixteen orders of eutheria only one type exists.

Behind all this there are genetic and epigenetic preconditions for the organisation of behaviour which have a phylogenetic background as ultimate level, whereas ontogeny constitutes an intermediate or mixed level, and micro geneses implement parameter dependent tactics or conditional strategies at the proximate level. Dunbar (1988) prefers the convention, *„which views tactics as a set of choices specific to a particular strategic option. This leaves us free to use the terms in a strictly relative sense so that what at one level of analysis is a tactic within a particular strategy with its own array of tactical options at a lower level of analysis."* This assumption can also be realized in the model shown here at the ontogenetic level. We presume that during the „orientating phase" in which no actual information is available to help making a decision a particular motivation to display one type of behaviour preconditions conceptual behaviour that may be genetically predisposed and - in the sense of the definition - of rather strategic character. Only when external information constitutes the „orientating phase" of motivated behaviour and thus creates a „proximity field", decisions are required which allow to a greater extent for individual experiences and also for parameter dependent tactics. This applies to functional areas within the ecological as well as the social context of interactions with the environment which very often appear combined, espe-

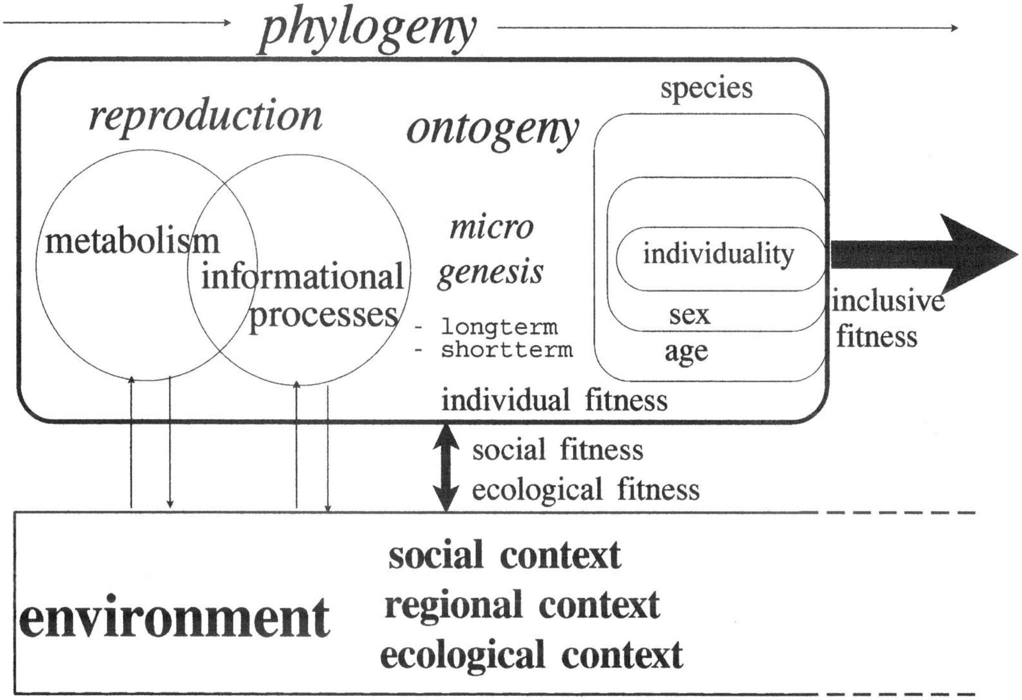

Figure 1 - A schematic description of relationships between organisms and environment. Phylogeny forms the ultimate level, microgenesis (German: Aktualgenese) the proximate level, whereas ontogeny constitutes the mixed level. Species, gender, age, individuality and reproduction go together with constitutional properties which are linked via metabolism and informational processes with the environment. Survival strategies are combined with vectors of fitness.

cially in higher primates: The social status or the social situation may take an influence on the use of resources and thus on the foraging strategy, success in finding food resources as well as spotting predators may influence the social strategy or the actual situation within the group. Putting motivation into action in the „contact field" with the functional object will finally supply the weighting for these strategies and tactics, and again, this applies to the ecological and social context. Thus one could ascribe the „tactics" displayed by primates during micro genesis to the „surface structure" of social systems sensu Hinde (1987), whereas the „survival strategies" that are often species specifically determined could be interpreted as functions of the „underlying structure" of Hinde. The first requires observations, the second further hypotheses which are to be validated.

Besides, it is obvious that „social knowledge" in primates is more differentiated than „ecological knowledge", it relates to the identity of conspecifics, their degree of relatedness, individual properties of behaviour which are also linked with the social status, relationships between individuals of the social unit and, of course, also with the bearer of knowledge in question who implements this knowledge (see Quiatt & Reynolds, 1993). This has also been shown in studies of acoustic communication, for example by Gouzoules (1984) in rhesus monkeys on Cayo Santiago (Puerto Rico), as well as by Cheney & Seyfarth (1988). Apparently, there are signal parameters in use, between which human observers cannot differentiate so that the signals are informing individuals who may benefit in their fitness. There are indications (crab-eating macaques; Dasser, 1988) that

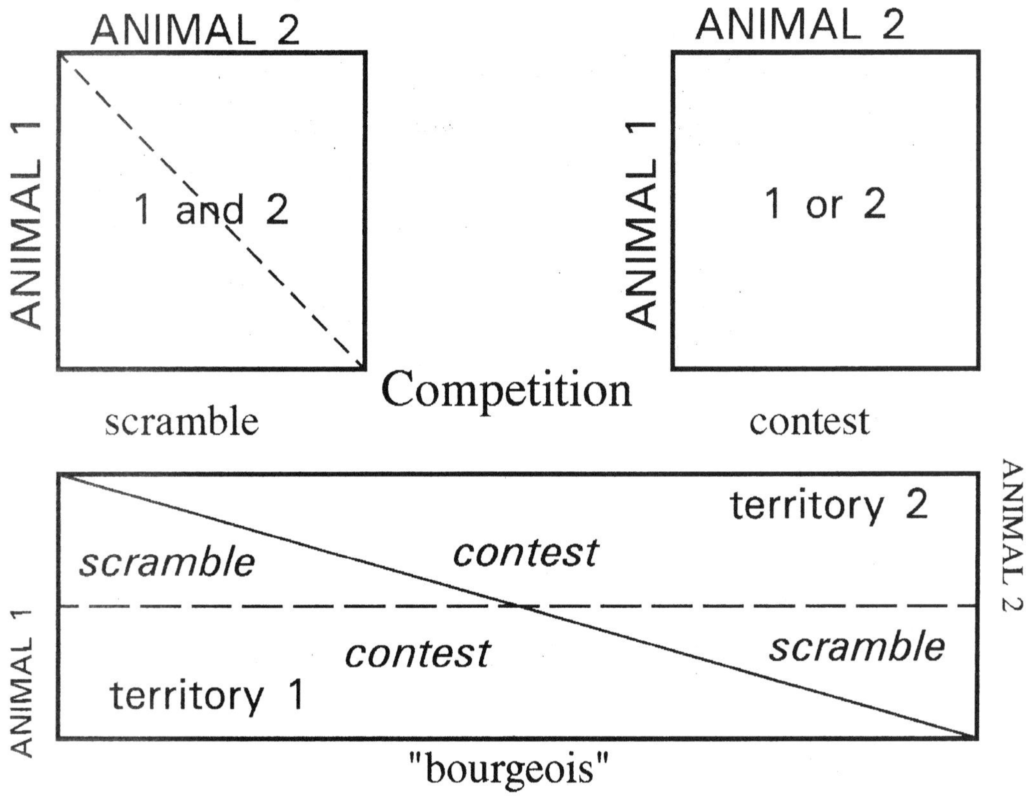

Figure 2 - The three basic rules of competition.

primates may recognize kinship, in particular mother-daughter relationships between other individuals to whom they are not related themselves. This appears to be the case especially in social units which are organized in matrilines. Such social knowledge is of particular importance for the formation of coalitions. Thus, vervets may threaten relatives of those individuals who had previously threatened their own relatives (see Harcourt, 1989). The eminent importance of social knowledge can be put into the formula: Bonobos are especially interested in bonobos. At the same time it is true that survival strategies of primates are organized socioecologically which becomes apparent in the social behaviour of chimpanzee females in Arnheim Zoo (de Waal, 1991) compared to that of females at Gombe (Goodall, 1986). At Gombe, female alliances (except for mother-daughter coalitions) could not be observed, whereas in Arnheim they were formed very quickly as a protection against attacks from the males and to influence fights for status among the males. The ecological difference cannot be overlooked: Arnheim offers plenty of food, the natural environment severe competition for food.

Foraging within the natural environment asks for a high daily time budget from primates, decisions of cost and benefit are required combined with goal-directed locomotion which demands special economies from moving tree-dwellers, in particular during fly-jumping. Apes who do not have pouches to store food, use the ground more often where they can keep their hands free for the transport of food. Mastering complex goal-directed interactions with the environment enhance

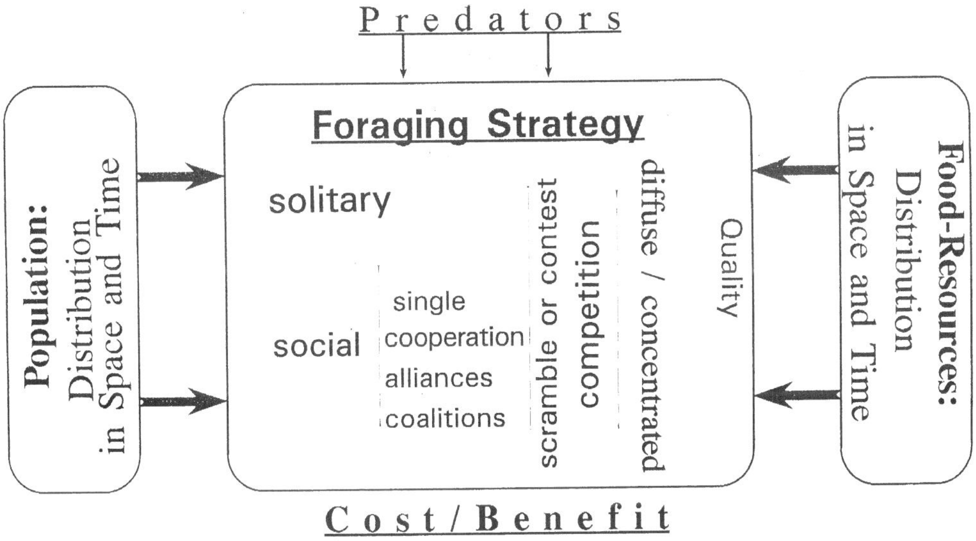

Figure 3 - Elementary factors which may influence foraging strategies.

in evolution at all levels (phylogeny, ontogeny and micro genesis) the cognitive organisation of behaviour with high competence of individually acquired knowledge and capabilities and their passing on through matrilines. This is linked with implementing communicative syntax and semantics. Furthermore, validating decisions between immediately perceivable goals (= proximity field) and those not present in time and space (distant field) also belongs to this phenomenon within the concept of socioecologically orientated organisation of behaviour. Past and future are thus integrated into the cognitive structuring of motivated behaviour. Moreover, these are also essential for the internal representation of knowledge which determines behaviour and constitutes a new form of consciousness. De Waal (1986) illustrates this with an impressive example: A male chimpanzee pursues a female, she disappears behind a tree on its right side while he is just about to run to the left, apparently he has anticipated this and throws a stone after her while running so that she again turns: He keeps his direction and thus they meet behind the tree. His tactics include five hypothetical steps: (1) I move to the left, (2) target will move to the right, (3) I throw a stone, (4) target will leap back, and (5) I reach my goal (see Quiatt & Reynolds, 1993). Keeping back of information and pretending not to be interested in situations of competition also belongs to the social strategies and tactics as well as the use of social tools to reach one's goals, already described by Wolfgang Köhler about a chimpanzee who pulled the experimenter under a high-hanging goal to reach it from the latter's shoulder. Menzel (1974) showed how a subordinate chimpanzee deceived his group about the location of a hidden food pile only known to himself by first leading them into the wrong direction, and then sneaking back to eat it on his own.

Quiatt & Kelso (1985) point out that during the evolution of primates on the way to wo/man a shift within the socioecological activities in favour of socioenergetic activities has taken place. The technoenergetic activities began with the use of objects by apes, yet took on a new quality and dimension with the use of tools by wo/man accompanied by a simultaneous reduction of ecoenergetic interactions (and strategies). This is in close relationship with our own concept:

In lesser primates the biogenetic strategies and tactics dominated, then they were superimposed by tradigenetic and finally ratiogenetic ones which became decisive for human evolution with the „Neolithic revolution".

References

ANDERSSON, M. (1994): *Sexual selection.* Princeton.

CHENEY, D.L. & SEYFARTH, R.M: Assessment of meaning and the detection of unreliable signals by vervet monkeys. *Animal Behaviour* 36, 477-486.

DASSER, V. (1988): A social concept in Java monkeys. *Animal Behaviour* 36, 225-230.

DUNBAR, R.I.M. (1988): *Primate social systems.* Ithaca.

GOODALL, J. (1986): *The chimpanzees of Gombe: Patterns of behavior.* Cambridge, Mass.

GOUZOULES, S. (1984): Primate mating systems, kin associations, and cooperative behavior: Evidence for kin recognition. *Yearbook of Physical Anthropology* 27, 99-134.

HARCOURT, A.H. (1989): Social influences on competitive ability: Alliances and their consequences. In: V. Standen & R.A. Foley (eds.), *Comparative socioecology: The behavioural ecology of humans and other mammals,* 223-242. Oxford.

HINDE, R.A. (1987): *Individuals, relationships and culture: Links between ethology and social sciences.* New York.

JARMAN, P.J. & KRUUK, H. (1996): Phylogeny and spatial organisation in mammals. In: D.B. Croft & U. Ganslosser (eds.), *Comparison of marsupial and placental behaviour,* 80-101. Fürth.

LEIGH, S.R. (1992): Patterns of variation in the ontogeny of primate body size dimorphism. *Journal of Human Evolution* 23, 27-50.

MENZEL, E.W. Jr. (1974): A group of young chimpanzees in a one-acre field. In: A. Schrier & F. Stollnitz (eds.), *Behavior of non-human primates,* vol. 5, 83-151. New York.

QUIATT, D. & KELSO, J. (1985): Household economics and hominid origins. *Current Anthropology* 26, 207-222.

QUIATT, D. & REYNOLDS, V. (1993): *Primate behaviour.* Cambridge.

SHORT, R.V. & BALABAN, E. (eds.) (1994): *The differences between the sexes.* Cambridge.

WAAL, F. de (1986): Deception in the natural communication of chimpanzees. In: R.W. Mitchell & N.S. Thompson, *Perspectives on human and nonhuman deceit,* 271-292. Albany.

WAAL, F. de (1991): *Wilde Diplomaten. Versöhnung und Entspannungspolitik bei Affen und Menschen.* München.

Ape Models of Incipient Hominid Lifestyles: Chimpanzee or Pygmy Chimpanzee (Bonobo)?

Adriaan Kortlandt

Abstract

Some authors have proposed that the pygmy chimpanzee or bonobo is more primitive and closer to man than the (common) chimpanzee. Others consider it a secondarily specialised animal adapted to the watery environments of the Central Congo Basin. In this paper the behavioural, ecological and geographical factors for supporting either of these positions are reviewed and evaluated.

Anatomically the pygmy chimpanzee may be characterised as a slender and pre-pubescent ("paedomorph") form of chimpanzee, with relatively long arms and legs but with an adult brain and adult reproductive organs. Taxonomically speaking the two species should be renamed the gracile and the robust chimpanzees in accordance with the principle of binomiality - an analogous situation exists among the australopithecines. The more agile locomotion of the gracile species indicates a more arboreal way of life, although the patterns of locomotion include several features of terrestrial origin. The more frequent bipedal gait with greater stretching of the legs in the gracile species may represent an adaptation to seasonally flooded and peat swamp habitats. Their social friendliness is a juvenile feature and indicates the virtual impossibility of bullying in the tangle of branches of the canopy. Their frequent, diversified, polymorphic and paedomorphic "sexual" behaviour is to a large extent both functionally and from the point of view of motivation, social. Calling such sexually derived patterns in a social context „sexual" is ethologically misleading.

The functional analysis of 22 eco-behavioural features shows that several of these are primitive in the one chimpanzee species and derived in the other, while for some other features the opposite applies. Some aspects are open to ambiguous or even contradictory interpretations. With yet other features, we do not understand their biological meaning. Much more meticulous descriptions and analyses of behaviour patterns are badly needed. This all means that it is still premature to raise the question of which chimpanzee represents a better model for the ape-human common ancestor and for the earliest stages of hominid evolution.

1. Introduction

Some authors have proposed on anatomical grounds that the pygmy chimpanzee or bonobo (*Pan paniscus*) is more primitive and closer to man than the (common) chimpanzee *(Pan troglodytes*) (e.g. Coolidge, 1933; Zihlman, 1984). Their arguments are in line with Haeckel's (1873) rule that ontogeny recapitulates phylogeny and Cope's (1887) rule that evolution leads to increasing size. These ideas were reinforced by the "almost human" youthful affection and social intelligence of the pygmy chimpanzee in captivity (e.g. Yerkes, 1925), eventually culminating in Savage-Rumbaugh's work (1994) with Kanzi. On the other hand, some authors have argued that *Pan paniscus* is evolutionary a secondary specialisation, derived from an ancestral form of *Pan troglodytes* and adapted to survive only (or mainly) in the evergreen, non-seasonal types of rain forest, flooded forest and swamp forest that occur in the Central Basin of the former Belgian Congo, successively called the Democratic Republic of Congo, Zaïre, and again the Democratic

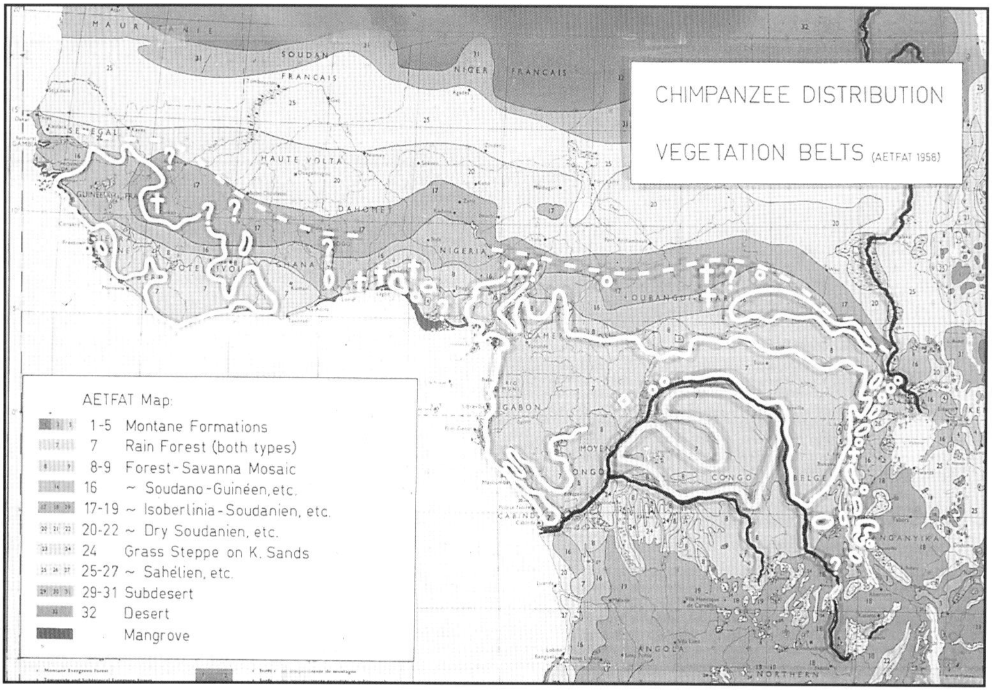

Figure 1 - *Under present day climatic conditions and disregarding human influence, about 60 % of the natural geographical range of the "common" chimpanzee consists of rain and montane forests and about 40 % forest-savanna mosaic landscapes, open woodlands and, in the Sudanian and Zambezian floral zones, dry forests. The pygmy chimpanzee or bonobo ranges almost exclusively in rain forests, seasonally flooded forests and probably (true) swamp forests, and in some areas mosaic landscapes. The faunal boundaries between the ranges of the two species are large rivers since apes cannot swim. AETFAT and A. Kortlandt.*

Republic of Congo. The latter interpretation has been proposed by Johnson (1981), Kortlandt (1972, 1995) and MacKinnon (1978). These authors' views have been inspired by, among other things, the watery and swampy environments in the Central Basin which do not represent at first sight the types of habitat where one would expect apes. Supporting (but not conclusive) anatomical, physiological and behavioural evidence has been put forward by various authors (e.g. Stanyon et al., 1986).

The two species are separated by wide water barriers, i.e. the Congo-Lualaba river and the Kasaï-Sankurunzee river (Figure 1). Apes cannot cross these barriers because they cannot swim, owing their anatomy. Thus the question of which chimpanzee is more primitive (and therefore the more human-like from an evolutionary viewpoint) leads to the question, with some exaggeration: Are we humans descended from a dryland ape of a variety of forested and mosaic landscapes, or from a wetland ape of an exclusively forested environment? The two evolutionary concepts are not, however, mutually exclusive: Some features that distinguish the two ape species may be primitive while others may be derived. The aim of the present paper is to review and evaluate the evidence and arguments from an ecological viewpoint.

The most conspicuous taxonomic difference between the two species is that the pygmy chimpanzee or bonobo is more gracile, and somewhat smaller than the „common" chimpanzee.

When fully grown, the gracile species has a pre-pubescent („paedomorphic") body and a juvenile voice, but an adult brain, adult reproductive organs and, importantly, relative to its body size longer arms and longer and straighter legs than its congeneric (Figure 2). The small size of the musuem specimens compared with those from the other side of the river inspired the name *Pan paniscus* which means little, lesser or pygmy chimpanzee (Schwarz, 1929), translated as the official English name pygmy chimpanzee (Coolidge, 1933) and the legal Belgian and Congolese-Zaïrean name *chimpanzé nain*. I personally prefer the name „robust" and „gracile" chimpanzee proposed by White (1996), firstly because they are descriptively more adequate, secondly because they are in line with the Linnean rule of binomiality (which is also standard practice in vernacular terminology) and thirdly because the word 'bonobo' has a scientifically fraudulent background [compare Tratz & Heck (1954) with Heck (1939), see also Kortlandt (submitted).] In addition the proposal draws an analogy with the robust and gracile australopithecines. For stylistic convenience the official name pygmy chimpanzee will also be used.

In the following sections I shall make an itemised comparison of the similarities and the differences between the two species with regard to size, locomotor patterns, ecology, behaviour and sociology.

2. Size

Generally, among the large mammals at the genus, species and subspecies level, forest-dwellers tend to be smaller than their relatives in more open habitats, e.g. the species or subspecies of buffalos, elephants, rhinos and various antelopes, furthermore pygmy hippos compared with hippos, okapis compared with giraffes, leopards compared with lions, guenons compared with baboons (Kingdon, 1997), and Pygmy peoples compared with the rest of us. These examples appear to represent adaptations to the tangle of vegetation in forest habitats (see also Dubost, 1979). The somewhat smaller size of the pygmy chimpanzee is therefore not surprising. However, there are some notable exceptions, e.g. the giant forest hog (*Hylochoerus*), the drill and mandrill, and the gorilla. In some instances the size variation is an expression of differences in evolutionary history and Cope's rule, e.g. among the Asian rhinos. [More palaeontological documentation in Piveteau (1957), Romer (1966), Simons (1972).] Among the primates the evolutionary increases in size can often be attributed to the Mio-Plio-Pleistocene changes of vegetation caused by climatic desiccation. These cases illustrate the effect of vegetation density but do not represent Cope's rule as such. At any rate the small size of human Pygmies is generally considered to be a case of secondary adaptation to the rain forest. As to the australopithecines, the opinion that the graciles lived in more open habitats than the robusts has been thrown increasingly into doubt. The size of *Pan paniscus* is, therefore, not an unambiguous or valid argument in favour of its alleged primitiveness.

3. Locomotion

(1) The locomotor patterns of the two species of chimpanzee are substantially the same. Both are predominantly long-distance quadrupedal ground-walkers which range mainly on the forest floor from one fruit-bearing tree to the next which they then climb. The distances covered in this way are up to 3 kms as observed in the robust species and up to 2 perhaps 3 kms in the graciles (de Bournonville, 1967; Kano, 1992). Both species now and then walk short distances bipedally, either with a waddling gait as described by anatomists, or less often with a striding gait which has been ignored by anatomists (Figures 3 a, b). They do so either to carry fruit, or to have a better view, or apparently just for a change. On branches and even on a rope they can balance, walk and run bipedally much more adroitly than any human (Rahm, 1967; Susman, 1984; Fig. 4). To move

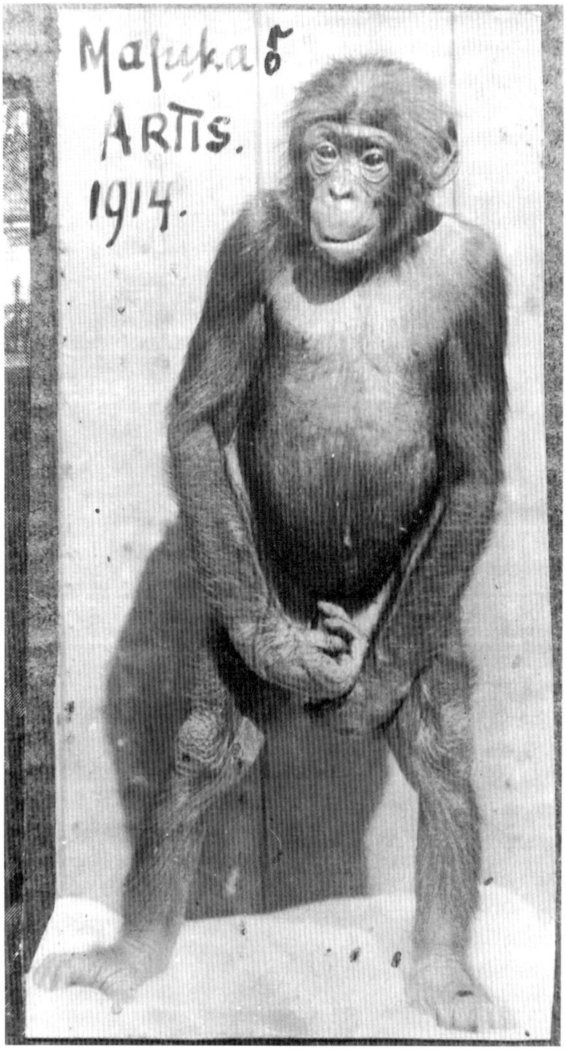

Figure 2 - The pygmy chimpanzee or bonobo has a more slender body build and somewhat longer arms and legs relative to its body size than the "common" chimpanzee. Mafuka (who lived in the Amsterdam Zoo Artis from 1911 to 1916) was the first of its kind to be recognised as a new species, but was not described scientifically and named at the time. The Amsterdam Zoo "Natura Artis Magistra".

from one tree crown to an adjoining one they may climb quadrumanously/quadrupedally, but the opportunities to do so are somewhat restricted owing to the irregularity of most rain forest canopies. Brachiation (armswinging) occurs relatively rarely among the adults. It is often used in male displays rather than for locomotion for foraging purposes, contrary to what was formerly inferred from captive observations and anatomical studies (e.g. Le Gros Clark, 1959, still reprinted unchanged in the 1980s). Descending is normally performed by embracing the tree trunk and sliding down, partly slipping and partly walking backwards. In addition, both species can leap either quadrupedally or bipedally from a high tree to dive down and land quadrupedally on a much lower tree, or even on the ground from almost unbelievable heights. Probably both can cling-and-leap quadrupedally and more or less horizontally from one tree trunk to the next like a tarsier or an indri, though this seems to have been observed only in the robust species (my observation in a papaya plantation). Altogether there appear to be no categorical differences between the two species as to the various types of locomotion they employ.

(2) There are, however, marked differences in the frequency and versatility with which the various locomotor patterns are employed by the two species. All available descriptions indicate that the adult graciles are more arboreal, agile, acrobatic, daring and accomplished than the adult robust when they move around for foraging purposes in comparable types of rain forest canopy and also in captivity (Jordan, 1977; Kano, 1979; MacKinnon, 1978; Susman et al., 1980). On the other hand, observers of the robust species, or at least of the eastern subspecies *Pan troglodytes schweinfurthii*, have expressed their surprise that these apes climb very cautiously uptree and even behave hardly like true arborealists (van Lawick-Goodall, 1968; Kortlandt, 1962; Pitman, 1931, 1942, Reynolds & Reynolds, 1965). Interestingly, the observations by Boesch & Boesch (1981) show that the western subspecies *Pan troglodytes verus* is much more accomplished arboreally than the eastern. Thus the achievement level may partly be a cultural phenomenon. Even so, *Pan paniscus* appears to be definitely more arboreally adapted than Boesch' apes. One might even guess that this difference can be explained in part allometrically because of the difference in body weight, but the behavioural difference seems to be too great for that. It does, however, suggest that the gracile species is not only anatomically but also in its locomotion a paedomorphic manifestation of the robust: the younger a *Pan troglodytes* is the more versatile and audacious it is in the trees. If so, this does not yet reveal which one of the two chimpanzee species represents the primitive and which one the derived form. Horn (1980), Jordan (1977) and Kortlandt (1972) have proposed that the more arboreal habits and greater locomotor versatility of the pygmy chimpanzee are adaptations to the frequent flooding of many of its habitats because such flooding would require more frequent moving through the canopy than on the ground. Much more precise and detailed descriptions of locomotor patterns in different types of tree architecture (e.g. Letouzey, 1986; Pieters, 1977; Tailfer, 1989) are required to clarify the issue.

(3) Observations under artificial but semi-natural conditions in zoos can often provide clues to the natural history and the evolutionary past of animal species. When zoo chimpanzees are frightened by a sudden unexpected event or noise they usually flee to the ground, especially when they are in a new environment. In the wild they tend to do the same, even though in the rain forest the trees are safer because in tall trees there are no leopards. In savannas and more or less open woodlands, however, leopards, lions and hyenas can surround a tree and wait until the prey comes down. It is safer, therefore, to pre-empt this and quickly run away on the ground. Pygmy chimpanzees, when disturbed in the wild, also flee mostly to the ground but in other cases conceal themselves in the foliage of a tree crown. In zoo cages, however, they tend to flee upwards. Orang utans on the other hand always flee to the ceiling of the cage in zoos and to the canopy in the wild. So the gracile chimpanzee must have passed through an evolutionary phase when it partly inhabited landscapes where trees stood isolated from one another. Its present-day frequent fleeing on the ground appears to be either an ancient adaptation that dates back to a time when its presumed ancestors lived in rather dry woodlands with scattered trees, or it evolved in marshy and swampy habitats which have no continuous canopy and which, moreover, are difficult hunting grounds for large carnivores.

(4) Comparative anatomy and ethology often reveal the evolutionary past. To illustrate the principle: Quadrupedalism in vertebrates dates back to the time when fishes needed fore- and hind-fins for vertical steering, just like submarines. (Hexapedalism would have been more functional, as is shown in insects.) Similarly, so-called „knuckle-walking" in the great apes is usually considered to be a secondary adaptation of an arboreal grasping hand to terrestrial quadrupedal walking. Actually, in chimpanzees, it should be called more precisely proximal-digit-walking. This pattern may also be interpreted as a primary adaptation to lengthen the forelegs in order to

get a higher viewpoint and to look over a grass and herb layer by stretching the wrist and the proximal digits, i.e. somewhat analogous to lengthening the legs by toe-walking in hoofed mammals. (Just try it yourself and notice the advantage.) Significantly, the orang utan is on a flat surface a fist-walker and occasional wrist-walker, and many platyrrhines show on a flat surface frequently or incidentally a "knuckle-palmigrade" manner of walking, i.e. with flexed fingers but with the internal side of the wrist resting on the ground (my observation). An interesting feature now is that both chimpanzee species regularly „knuckle-walk" (on their proximal digits) on more or less horizontal branches, unless such branches are too thin, whereas orang utans use their hands always in the palmigrade way even on the thickest branches. This difference indicates a semi-terrestrial origin of both chimpanzee species. While some data suggest that *Pan paniscus* „knuckle-walks" less frequently on branches (or perhaps only on thicker branches?) than its relative (Susman, 1984), other data do not support this (Doran & Hunt, 1994). Again, more precise observations of locomotor habits in various habitats and in various types of tree architecture are needed to clarify this evolutionary problem.[1] (The standardized terminolgy for locomotor patterns proposed by Hunt et al. (1996) should be recommended for this purpose.)

(5) An intriguing question is whether and why the gracile chimpanzee has such long legs, as occasionally mentioned in the literature. By comparison, the Asian orang utan, siamangs and gibbons are the most arboreally specialised, have relative to body size the longest arms and the shortest legs, brachiate the most frequently and walk bipedally with the most bent legs (Intermembral indexes 126-147 - Fleagle, 1988). In the order of decreasing arboreal specialisation and frequency of brachiation, the two chimpanzee species come after the Asian apes, having indexes of 102-106 but surprisingly the indexes of the gorillas are 116. Apparently the alleged long legs of *Pan paniscus* compared with *Pan troglodytes* are a visual illusion caused by its general body proportions, its slender build, its more frequent upright walking and, particularly, its less bent knees in bipedal gait. Even so, one wonders why it walks more bipedally and has more stretched legs. This is clearly disadvantageous when climbing from the ground along a tree trunk up to the crown (Figures 5 a, b). One might suggest that long and/or stretched legs could be useful when moving through a continuous canopy by means of bipedal walking on more or less horizontal branches, and/or when picking fruit that are hanging above (Hunt, 1994). This, however, cannot explain the difference between the two species. The mixed forest characterised by *Scorodophloeus* which is the optimum habitat in the Central Congo Basin, is almost as irregular and discontinuous as any optimum forest habitat of the common chimpanzee. Besides, there are more opportunities to pick overhead fruit when walking on the ground in the open and low types of arboreal vegetation which occur in greater abundance in the *Pan troglodytes* range (Kortlandt, 1983, 1995).

(6) The western part of the Central Congo Basin is mainly a very large delta with lakes, rivers, floodplains, seasonally flooded forests and various types of swamps. Water levels fluctuate annually or biannually, in some years up to 4 m. The general physiographic aspect is much like Holland's before dykes were built. The middle part of the Basin includes wide valleys with similar features. In such habitats upright walking with long legs and/or stretched knees would be greatly advantageous (Kortlandt, 1995; de Waal & Lanting, 1997). Long legs would facilitate crossing shallow rivers, foraging in flooded forests, fleeing when waters rise and catching fish in pools when waters go down, similar to the case with wading birds, herons, storks, etc. Even more important, in true swamps and bogs (i.e. with a predominantly organic soil) upright walking is life-saving for short-necked animals when they sink through the vegetation mat into the underlying mire. A bipedalist can pull himself out (Figure 6). I can testify to this from my field work in equally treacherous swamps in Holland. This suggests that the more upright walking of the gracile

Figure 3 - *The waddling gait with knees bent, generally at slow speeds (left). The striding gait with stretched legs, mostly used at faster speeds (right). A. Kortlandt.*

In snake-infested terrain covered by low grass or herbs chimpanzees may employ a peculiar 'hopscotch' gait, using one leg and then the other. When crossing a stream (1½ or 2 metres wide), they perform a froglike hopping jump with the legs moving simultaneously.

chimpanzee is a secondary adaptation to the special circumstances in the Congo Basin. However, there is no factual evidence of bipedal swamp walking. Both chimpanzee species have been sighted in seasonally flooded forests, but with a solid, predominantly mineral and sometimes slippery soil (Fay et al., 1989; J. M. Fay pers. comm.). Under such conditions, bipedal walking would be useless and even uncomfortable when the water level is low. Susman (1984) has observed footprints without knuckle prints along the banks of streams which suggested that the apes may cross such streams bipedally. Badrian & Badrian (1977) saw prints on an island which indicated that the apes had walked about 100 m through shallow water and Kano (1992) noticed prints that crossed a river 3 m wide and 60 cm deep. Several field workers have reported walking in rivulets, fishing in very shallow swamps with a solid bottom for minuscule animals and eating them (crabs, insects or worms?), and catching fishes in pools, playing with them, and presumably eating them. In all these cases, however, the water level apparently was too low to elicit bipedalism. Possibly the apes have been watched travelling in such true swamps as depicted in Figure 6, but if so it has not been recorded because the observers did not distinguish between true swamps and flooded solid soils. At any rate, common chimpanzees do occur frequently in the Molindi Swamp in Uganda where they apparently avoid falling into the soft, semi-fluid mud at some places by jumping bipedally with a standing start from one small but rooted tussock to the next, as I had to do myself (Kortlandt, 1972). Captivity tests with pygmy chimpanzees on various submerged substrates may clarify this issue.

4. Geographical and ecological range

(1) The ecological limits of the robust chimpanzee (apart from the water barriers) are: (a) in the West of its northern range the borderline between the *Isoberlinia soudanien* and the drier type of

soudanien, (b) in the East of its northern range the borderline between the wetter type of *soudanien* and the *Isoberlinia soudanien*,[2] (c) in the Southeast the borderline between the wetter and drier types of Zambesian woodlands, and (d) on mountains in the West about 2300 m and in the East about 3000 m. Its foraging habitats include evergreen and semi-deciduous rain forests, swamp and seasonally flooded forests, mountain forests, soudanien dry forests, and various woodlands and savannas with dispersed or scattered trees (Kortlandt, 1983). Thus it is one of the five most eurytopic large mammals of Africa, together with the elephant, the buffalo, the leopard and man. The gracile species, on the other hand, lives and forages only in evergreen and semi-deciduous rain forests, seasonally flooded forests, probably (true) swamp forests and some forest-savanna mosaics (Kortlandt, 1995; Thompson, 1997). Its ecological limit in the SE is unknown. Generally in zoology, the ubiquitous species is the most primitive among its taxon: it is the least specialised and has the widest adaptive zone, unless geographical irregularities interfere with the overall picture. From this viewpoint *Pan paniscus* is a secondary side-branch of the *Pan troglodytes* lineage.

(2) Since Mayr (1959) we are aware that the bifurcation of an evolutionary lineage requires a geographical/ecological barrier that causes a degree of genetic isolation sufficient to split a homogeneous population into two distinct daughter populations. This makes it possible to reconstruct evolutionary trees and date the divergences on geological, geomorphological and palaeoecological grounds. It is a third way to reconstruct such trees in conjunction with palaeontology and DNA taxonomy. The most plausible barriers in an evolutionary explanation of the divergences of the African Hominoidea are: (a) the Western Rift Valley with the Nile and Zambezi river systems separating the apes from the hominids at some time probably between 10 and 7 million years ago; (b) the Niger-Tilemsi-Azaouak and Cross-Benue-Chad river systems of younger but unknown geological age, which should have separated the chimpanzee from the gorilla, and which explains the difference between the eurytopic adaptations to fluctuating Plio-Pleistocene climatic zones of the chimpanzee in the West and around Central Africa as opposed to the exclusively forest adaptations of the gorilla in equatorial Africa; and the Congo-Lualaba; (c) the Congo-Lualaba and Kasaï-Sankuru river systems, together with either the Lukuga or the extensive swamps around 9° S during a wet Pleistocene phase (or a combination of both), which should have separated the two chimpanzee species (Kortlandt, 1968, 1972, 1977/80, 1995; Thompson, 1997). The last connection between the two chimpanzees should have been in the SE of the Congo Basin. If this hypothesis is valid the pygmy chimpanzee would be a late secondary side branch of the robust species.[3]

(3) During dry Pleistocene periods the Central Congo Basin, with its many rivers converging from more humid and mountainous regions, continued to contain deltaic and riverine forests. These created refuges and centres of endemism for many monkey species and subspecies (Colyn, 1987). It seems conceivable that this isolated and watery environment was the area where *Pan paniscus* began to diverge from its *proto-troglodytes* ancestral lineage as a secondary adaptation to semi-deciduous forests rooted in humid soils but in a relatively dry climate (either contrary to, or partly in line with Kortlandt's hypothesis). It is, however, equally conceivable that this watered area offered a refuge to a primitive and relatively arboreal ancestral form of chimpanzee, while its sister taxon adapted secondarily to the gradually desiccating and alternating climatic conditions during the Ice Ages elsewhere in Africa. For the time being there is no good evidence allowing us to choose between these two different (but perhaps reconcilable) hypotheses. More in-depth research on the ecology and behaviour of the two species from a palaeo-viewpoint may perhaps clarify this issue.

(4) Johnson (1981) has proposed a parallel between the (supposed) small size of *Pan paniscus* and the dwarfism often found in mammal fossils on remote islands, compared to their mainland relatives. Such secondary "insular dwarfism" is, however, caused by the fact that carnivores are generally less good swimmers than herbivores and consequently have not reached these islands (Sondaar, 1976). In such conditions becoming smaller was adaptive. This explanation cannot apply to the gracile chimpanzee because in the Congo Basin there are leopards, crocodiles and large constrictor snakes.

(5) Recently, Wrangham & Peterson (1996) have hypothesised that originally chimpanzees and gorillas lived on both sides of the Congo-Lualaba river system, but that during a Pleistocene period of drought the gorilla became extinct in the Central Basin south of the river. Consequently more herbaceous food became available to the then ancestor of the pygmy chimpanzee in that region. This should have led to its secondary evolution resulting in the present-day species. The hypothesis is ingenious, but it contains several inconsistencies:

(a) The eco-geographical conditions causing the bifurcation of the chimpanzee and gorilla lineages (before the chimpanzee-hominid splitting?) are ignored and remain unexplained (cf. Kortlandt, 1972, 1980, 1995).

(b) Even during the driest Pleistocene periods the western part of the Central Congo Basin south of the Congo-Lualaba river system still contained habitats where (proto-) gorillas could have survived and even evolved into a *paniscus* type of descendant (Colyn, 1987; Colyn et al., 1991; cf. Fay et al., 1989).

(c) The absence of the gorilla in West Africa, in spite of several mountain forest refuges during the driest Pleistocene periods (e.g. Adam, 1971-83; Hamilton 1981; Lamotte, 1942) is not explained.

(d) The remarkable ambiguous and eurytopic adaptation to forest-, woodland- and savanna-habitats of the common chimpanzee is not explained and hardly mentioned (cf. Kortlandt, many papers).

(e) Owing to the absence of gorillas in West Africa, according to the hypothesis the West African sub-species of chimpanzee should have evolved into a *paniscus*-resembling creature because there are plenty of herbaceous foods in the area.

(f) If it is true that pygmy chimpanzees live in a land of plenty and have a birth interval of only 4½ years on average (compared with 6 years in common chimpanzees) one wonders how these sexy apes have avoided Malthusian problems without waging intertribal wars, and why they have not achieved a population density higher than that of the common chimpanzee in optimum habitats (Kortlandt, 1984).

(g) The authors present their hypothesis almost as proven facts without discussing alternatives. For the rest they underline the secondary specialisation of the gracile species.

5. Food and food competition

(1) There are differences between the two chimpanzee species as to the choice of fructivorous and herbivorous foods. However, these differences appear to be largely determined by the availability of various food types at the observation sites and according to the seasons (reviewed by de Waal & Lanting, 1997). Even so, the greater sensitivity to unusual foods shown by the graciles according to zoo experiences might indicate that they are more specialised evolutionarily in their digestive tract.

(2) Probably all primates eat insects and nearly all of them hunt and eat small vertebrates, either incidentally or frequently, both in captivity and in the wild (Harding & Teleki, 1981; Kortlandt & Kooij, 1963; and many field reports by others world-wide). Apparently they need certain proteins,

Figure 4 - Chimpanzees are skilled tightrope walkers, using a striding gait even on a slack rope. A. Kortlandt.

vitamin B12, etc. which are not sufficiently available in herbs and fruit. In this respect the gracile chimpanzee is quite normal: It will eat worms. insects, probably small aquatic animals including fishes and, rarely, small reptiles and mammals as the occasion permits, but without systematically hunting them (reviewed by Lenglet, 1987 and Sabater-Pit & Veà, 1994). The robust chimpanzee is, however, quite different. These apes spurn birds and small mammals which they could easily catch, but regularly and co-operatively catch, kill and eat monkeys especially (including young baboons) and less frequently other medium-sized prey, including duiker antelopes and occasionally infant bushpigs (Lenglet, 1987 and Sabater-Pi & Veà, 1994). Furthermore, often they kill and partly eat conspecifics of neighbouring communities. Chimp-watchers call such actions towards monkeys "predatory" and "carnivorous" behaviour, but towards conspecifics "warfare" and "cannibalism". Actually, the phenomenology of both sets of behaviour clearly shows that they are largely the same. Naming them differently conceals their true nature. From a functional viewpoint, killing monkeys and outsider conspecifics is primarily extermination of food competitors, while the production of meat food is secondary because many other potential prey is spurned. From a motivational viewpoint, the killing technique and the response of the onlookers [as shown, e.g., in a picture and described on p. 64 by van Lawick-Goodall (1968)] show clearly that this is primarily an intimidation-redirection display aiming to impress anyone it may concern, while the meat is only a by-product. The resemblance with the manner of killing of Jewish children in concentration camps is indeed striking. Furthermore, the sharing of the meat after the violence has the characteristics of displacement and reassurance behaviour. Altogether, this is a unique (but all-too-human) combination and integration of heroism against outsiders, showing-off for insiders and meat production for socialisation (Kortlandt, 1972, 1974, but ignored by de Waal, 1996 and Wrangham & Peterson, 1997). Nothing of this kind has been reported for the pygmy chimpanzee. Normally, of course, the evolution of predatory behaviour proceeds from the cat-and-mouse to the lion-and-buffalo type. From this viewpoint the robust chimpanzee is highly advanced on its way toward humanisation while the gracile species is still as primitive as the arboreal monkeys - certainly not a model of hominid evolution.

6. Defence against carnivores

Among the larger primates, some species or genera use objects as intimidation tools or even as effective weapons against their natural enemies and other intruders in their range (reviewed by Kortlandt & Kooij, 1963). Two types of such defensive tool use can be distinguished. There are no intermediate forms. Predominantly ground-dwelling monkeys, mainly baboons and macaques in areas where they are not persecuted, occasionally scoop, throw or (on a slope) roll down sand, gravel, stones or other objects at predators or human intruders by means of stamping with a hand or (rarely) by underarm throwing. On the other hand, several species of exclusively tree-top-dwelling monkeys and the orang utan, when similarly disturbed, break off fruit, twigs or large dead branches and drop or throw them down on the intruder with an overarm movement. Robust chimpanzees show both patterns in both types of environment, but (in wildlife experiments with a stuffed leopard) the savanna-dwellers among them perform the first pattern in a more accomplished manner. Among forest-dwellers the second pattern is more perfected and is also performed as intimidation behaviour when they are on the ground near the test object. Furthermore, savanna-dwelling chimpanzees use big sticks as hard-hitting clubbing weapons against the model whereas forest-dwellers use them only as intimidation tools, though with some non-functional residual elements of weapon use and even weapon making. All this reflects the ambiguous adaptation of the species (Kortlandt, 1972; Kortlandt & Bresser, 1989; Kortlandt & Kooij ,1963 and n.d.; Kortlandt & Trevor, 1986; Kortlandt et al., 1981). Surprisingly, nothing of this kind seems ever to have been observed in the gracile chimpanzee. Possibly they are too shy. To ascertain the facts and evolutionary background, the stuffed leopard test should be conducted with them. The test object is available.

7. Social and sexual behaviour

(1) Pygmy chimpanzee society is a maternocracy. Female coalitions developed around the senior mothers form the leadership. They control the resources.[4] The males, on the other hand, play a rather peripheral role and are loners among their gender group. Each male remains for life bonded to his mother and under her supervision. When he gets into social trouble his mother will come to the rescue. Conversely, if he causes trouble, a group of females will call him to order. Such a kindergarten type of society is unique among primates. It can, therefore, only have evolved and derived from an earlier and differently organised type of society, or possibly even from a non-social or seasonally organised way of life.

(2) Another conspicuous characteristic of gracile chimpanzee society is its friendliness and peacefulness in combination with substantial male-female equality (e.g. de Waal & Lanting, 1997). This feature is in itself not very surprising. In a closed-canopy rain forest, bullying does not pay because a smaller and weaker individual can nearly always flee to a thinner branch beyond reach of a bigger and stronger group member. Ostracising an individual does not pay either because in the canopy are few predators. Among savanna and open woodland dwellers, on the other hand, the strong ones are as a rule the defenders without whom the group could not survive because leopards, lions and hyaenas could be lurking. Long ago the American primatologist Harlow (1951) regretted that we humans descended from the aggressive Old World monkeys rather than from the friendly New World ones with which he was also familiar in his laboratory. He apparently overlooked the fact that his rhesus monkeys were terrestrial and his capuchin monkeys arboreal. Similar ideas are nowadays in fashion with regard to the "make love not war" kind of chimpanzee.

A better guide to understanding the human model of evolution is Bigelow's „The dawn warriors" (1969). However, this does not answer the question of which one of the two chimpanzees was the primary form and nearer the ape-hominid common ancestor in social behaviour.

(3) The above is a functional (ultimate) interpretation of the friendliness in paniscus society. Alternatively, or parallel to it, a causal (proximate) explanation may be proposed. Generally, young mammals are much more tolerant, amicable and playful among themselves than adults. So it might be that the friendliness and peacefulness among the adult graciles is more or less an accidental byproduct of their anatomical and behavioural paedomorphism or juvenilisation of the species as discussed in Sections above. One could imagine that there are some common hormonal factors causing overall paedomorphism in anatomy as well as behaviour. Even so, that would in itself not elucidate whether these phenomena are evolutionarily primitive or derived.

(4) By and large the vocalisations of the adult gracile chimpanzee are homologous with those of the adult robust, but at a much higher pitch, i.e. as before puberty. This again demonstrates the overall pre-pubescent nature of the species. It also indicates that members of these ape communities are adapted to living and foraging at much closer distances to one another than their relatives: higher-pitched sounds carry less far, particularly in a forest environment. Apparently they do not need a long-distance locator call. Compare this with the songs of gibbons, the „long call" of the orang utan, the roaring of howler monkeys, the screeching of tree hyraxes, and, of course, the pant-hooting of the common chimpanzee. In the latter case the long-distance calling expresses the much larger sizes of foraging ranges in the open landscapes where common chimpanzees often occur (Kortlandt, 1984).

(5) Another conspicuous feature of *Pan paniscus* behaviour is its exuberant sexuality. It starts already before one year old, includes almost everything described in Kinsey-style books and is the predominant leisure pastime among youngsters and adults alike. Most remarkable is its stunning diversity of patterns. However, it relatively rarely aims at and leads up to copulation (e.g. de Waal & Lanting, 1997). Rather it is what Freud (1905, Engl. transl. 1953) has called the „polymorph perverse" non-integrated patterns which are commonly observed (or psycho-analytically diagnosed) in children and adolescents, and which in the course of behavioural ontogeny are - or should become - the building blocks of mature human sexual behaviour resulting from ascending integration. In his terminology these building blocks are called „sexual" because they will - or should - become a part of mature sexuality, i.e. in much the same way as an apple fruit and an apple tree are both called „apples" by the grocer and the gardener respectively. This ontogenetic principle of isolated and polymorphic development followed by ascending integration is widespread among mammals and birds (reviewed by Kortlandt, 1955; described in common chimpanzees by Plooij, 1984). The continuation (and even intensification?) of polymorph pre-pubescent sexual patterns during adulthood in the gracile species is, therefore, again a striking example of its general paedomorphism. Obviously this is a case of secondary evolution, but its adaptive value is still enigmatic.

(6) There is reason to doubt whether the term „sexual" is really appropriate for such allegedly sexual activities. In ethology, one should distinguish between the „morphology" (in space and time) of activities, their function and their motivation (e.g. Tinbergen, 1952 on derived activities, which in my opinion is his most important writing, though outdated in some respects). Nowadays, unfortunately, that part of ethology is virtually dead in primate research. Let me therefore illustrate the principle with a human example. Look at kissing in human courtship:

(a) Morphologically it does not resemble or approximate anything copulatory. Instead it is phenomenologically (i.e. morphologically) a phylogenetic derivative of oral feeding of babies,

Figure 5 - The long, muscular arms with large hands and relatively short legs enable apes to climb the thick tree trunks with smooth bark which predominate in the rain forest. Long legs are impractical for such climbing. J.C.J. van Zon (above), A. Kortlandt (below).

and of cleaning them by licking - a deparentalised derivative of parental behaviour that has been re-integrated into courtship behaviour.

(b) Functionally it is, however, both a short-term pre-copulatory stimulus and a long-term bond-forming mechanism.

(c) Motivationally it is in some cases an expression of copulatory desire, in other cases a form of social greeting, in still other situations an expression of reassuring or reconciliation, and so on.

Returning to the pygmy chimpanzee, I guess that much or most so-called sexual behaviour is *in form* a sexual derivative, but that it is both functionally and motivationally more or less similar to grooming and hugging. The evidence for this interpretation is extensive in de Waal's & Lanting's (1997) and Wrangham's & Peterson's (1996) books. The main criterion to define an activity as „sexual" in a particular context should be whether it is a part of a chain of actions aiming at and leading to coitus (functionally) and orgasm (motivationally). If my reasoning is justified, calling non-sexually motivated behaviour „sexual" is anthropomorphism of a sort, and ethologically misleading. „Sexuality-derived social" would be more appropriate. (By comparison: the forelimbs of primates are called „arms", of bats „wings", and of dolphins „flippers", rather than „legs" on comparative anatomical grounds.) A sophisticated methodology to tackle such problems of functional and motivational research has been applied to the robust chimpanzee by van Hooff (1973). A comparative study of its relative along the same line may yield interesting results.

(7) Gracile chimpanzees frequently copulate in the ventro-ventral position whereas the robust do so only rarely. Obviously this is a secondary specialisation in *Pan paniscus* because dorso-ventral is normal in the non-human primates. Orang utans often copulate in what may be called a 90° ventro-ventral position. This indicates that the peculiarity of *Pan paniscus* is probably correlated with its more pronounced suspensory climbing adaptations. It is, therefore, neither primitive nor homologous with the human „missionary position".

8. Intelligence

Long ago, those few people who had experience with both species of chimpanzee as pets or laboratory animals, generally considered the gracile one to be more intelligent than its sister species (e.g. Yerkes, 1925 and European laboratory personnel in the Belgian Congo in 1960). This was, however, a deceptive social illusion caused by the friendly temperament of *Pan paniscus*. [Even under strictly controlled laboratory conditions, many or most animals (including octopuses!) achieve higher scores of „intelligence" when their caretakers and experimenters feel sympathy and respect towards them.] In fact, „intelligence" is not a workable concept because there appears to be no general intelligence among animals. There are only more or less specialised intelligences (plural) which may remain dormant until they are activated by environmental conditions. The few data available on the two species of chimpanzee suggest that, in their evolutionary past, there must have been distinct selection processes favouring social vs. technological types of intelligence. The ecological background of these differences has, however, not yet been investigated. They remain, therefore, beyond the scope of this paper.

9. Conclusion

When Schwarz (1929) and Coolidge (1933) discovered and described a new subspecies or species of chimpanzee, it was not only „curieusement petit" but also had „alle Merkmale der Jugend" (all the characteristics of youth) and was anatomically „paedomorphic", i.e. juvenile at an adult age. Hence it was called *paniscus* which means little or pygmy chimpanzee, analogous to the Pygmy humans in the Congolese rain forest. Coolidge, however, went a step further. Being a believer in Haeckel's (1873) rule he assumed that „the more primitive or juvenile forms approach

Figure 6 - Bipedal walking is a life-saving type of locomotion in peat swamps covered with vegetation. A. Kortlandt.

more closely to an ancestral state" and consequently *Pan paniscus* „may approach more closely to the common ancestor of chimpanzees and man." Thus he opened the way to confusing Haeckel's rule and Cope's rule with secondary adaptation to an exclusively forested ecosystem. (The human Pygmies, after all, are smaller but not primitive).

In the present review an attempt has been made to distinguish these three viewpoints with regard to the ecology and behaviour of both species. Twenty-two aspects have been interpreted comparatively. Altogether the indications from these data suggest and confirm that the gracile species may be summed up, not only anatomically but also behaviourally, as a juvenile form of chimpanzee but with an adult brain, adult reproductive organs and abnormally long legs (in comparison with the other apes whose locomotion is suspensory or semi-suspensory). Five or six of the listed features may be interpreted as peculiarities indicating a secondary juvenilisation of *Pan paniscus* in the course of evolution (Nos. 3-2; 6; 7-1,3,4,5). The adaptive value of this ecological, behavioural and sociological juvenilisation is, however, not yet clear: The interpretations in the current literature are few and not very convincing. Eight or nine features (partly the same) may be seen as indications of either primary or secondary adaptations and specialisations in various types of almost exclusive forest environments (Nos. 3-2, 6; 4-1,2; 5-1,2; 7-2,4,7). Two features appear to be retentions from a life in more open seasonal or drier habitats in the past (Nos. 3-3, 4). Four other features, on the contrary, may or seem to indicate a primitive condition as a canopy arborealist (Nos. 3-2; 5-2; 6; 7-2). Three issues are open to ambiguous or even contradictory interpretations (Nos. 2, 3-2, 4-3). Finally, eight issues (partly included above) concern phenomena whose biological meaning we do not yet understand (Nos. 3-1,5; 4-4,5; 7-1,4,5,6).

Altogether this review yields a collection of diverse evolutionary viewpoints. The two species of chimpanzee apparently represent different mixtures of primitive and derived eco-behavioural characteristics. In several instances it is impossible to guess whether a feature is primitive or derived. The difference of opinion between, on the one hand, Coolidge (1933) and Zihlman (1984), and, on the other hand, Kortlandt (1972), MacKinnon (1978) and Johnson (1981) cannot yet be solved. Thus there are no ponderous arguments to consider either the robust or the gracile chimpanzee as a better model of the ape-human common ancestor and of the earliest stages of hominid evolution.

Furthermore, there is no reason to assume *a priori* that elements or aspects of behaviour that occur in the two chimpanzees and in humans should be homologous. For instance, what may be called „rightist politics" (inspired by de Waal) can be observed not only in robust chimpanzees but also in baboons, and „leftist politics" not only in gracile chimpanzees but also in elephants, with wolves in the centre. Falling in love and biparental care can be studied in humans, some primates and many birds, but not in chimpanzees. The use of objects as effective weapons occurs in humans, robust chimpanzees, baboons, certain monkeys and even ant-lions, but has not (yet) been seen in gracile chimpanzees. And so on. So when should we postulate homology between chimpanzee and human behaviour? Only when a very meticulous phenomenological analysis of the motor pattern in space and time justifies the assumptions. Raising the question of whether the robust or the gracile chimpanzee is a better model of an incipient hominid lifestyle is still somewhat premature.

Two problems emerge from this review. Firstly, why has *Pan paniscus* such long legs compared to the other, more suspensory moving apes? Secondly, why is its anatomy, behaviour and sociology juvenilised? The first problem is in principle not very difficult. One should watch them in swamps and in deeper water than has been done until now. Furthermore, one should study comparatively not only the locomotor patterns but also the tree architecture of the various types of forest in which the two species of chimpanzee occur.

The second problem is, in a way, already more than 70 years old. In 1925 the Dutch anatomist Louis Bolk proposed that the anatomical evolution from monkey to ape to man was determined by changes in the endocrine system which caused retardations in the various growth processes and eventually the persistence of juvenile or even fetal characteristics in adulthood (e.g. no coat of hair on large parts of the body, occasional absence of the M3 teeth, etc). He called this the foetalisation theory. This simplistic theory has been abandoned, of course. Still, one may wonder whether there is a genetic link between anatomical and behavioural paedomorphism in the gracile chimpanzee. Whatever the case, the problem of the proximate and ultimate causes of juvenilisation (which is also known in research on domesticated animals) deserves more multidisciplinary attention.

References

ADAM, J.-G. (1971-1983): *Flore descriptive des Monts Nimba.* Paris (Mém.Mus.Natl.Hist.Nat., 5e part.). 6 Vol.

BADRIAN, A. & BADRIAN, N. (1977): Pygmy chimpanzees. *Oryx* 13, 463-468.

BADRIAN, N. & MALENKY, R.K. (1984): Feeding ecology of *Pan paniscus* in the Lomoko Forest, Zaire. In: R.L. Susman (ed.), *The pygmy chimpanzee,* 275-299. New York.

BIGELOW, R. (1969): *The dawn warriors.* Boston.

BOESCH, C. & BOESCH, H. (1981): Sex differences in the use of natural hammers by wild chimpanzees: A preliminary report. *Journal of Human Evolution* 10, 585-593.

BOLK, L. (1925): On the problem of anthropogenesis. *Proc. Kon. Ned. Akad. Wetensch.* 29, 465-475.

BOURNONVILLE, D. de (1967): Contribution à l'étude du chimpanzé en République de Guinée. *Bulletin de l'Institut Fondamental d'Afrique Noire* 29 A, 1188-1269.

CLARK, W.E. Le Gros (1959): *The antecedents of man.* Edinburgh.

COLYN, M.M. (1987): Les primates des forêts ombrophiles de la cuvette du Zaïre: Interprétations zoogéographiques des modèles de distribution. *Revue de Zoologie Africaine* 101, 183-196.

COLYN, N.M., GAUTIER-HION, A. & VERHEYEN, W. (1991): A re-appraisal of palaeoenvironmental history in Central Africa: Evidence for a major fluviatil refuge in the Zaire Basin. *Journal of Biogeography* 18, 403-407.

COOLIDGE, H.J. (1933): *Pan paniscus*, pygmy chimpanzee from the Congo River. *American Journal of Physical Anthropology* 18,1-59.

COPE, E.D. (1887): *The origin of the fittest.* New York.

COPPENS, Y. (1994): East side story: The origin of mankind. *Scientific American* 270, 5, 62-69; 271, 4, 5.

DORAN, D.M. & HUNT, K.D. (1994): Comparative locomotor behavior of chimpanzees and bonobos. In: R.W. Wrangham, W.C. McGrew, F.B.M. de Waal & P.G. Heltene (eds.), *Chimpanzee cultures,* 93-128. Cambridge/ Mass.

DUBOST, G. (1979): The size of African forest artiodactyls as determined by the vegetation structure. *African Journal of Ecology* 17, 1-17.

FAY, J.M., AGNAGNA, M., MOORE, J. & OKO, R. (1989): Gorillas *(Gorilla gorilla gorilla)* in the Likouala swamp forests of North Central Congo: Preliminary data on populations and ecology. *International Journal of Primatology* 10, 477-486.

FLEAGLE, J.G. (1988): *Primate adaptation and evolution.* San Diego.

FREUD, S. (1905): *Drei Abhandlungen zur Sexualtheorie.* Wien (English translation: Standard edition of the complete psychological works, VII, 125-245. London).

HAECKEL, E. (1873): *Natürliche Schöpfungsgeschichte.* Berlin. 4. Aufl.

HAMILTON, A.C. (1981): The Quaternary history of African forests: Its relevance to conservation. *African Journal of Ecology* 19, 1-6.

HARDING, R.S.O. & TELEKI, G. (eds.) (1981): *Omnivorous primates.* New York.

HARLOW, H.F. (1951): Primate learning. In: C.P. Stone (ed.), *Comparative psychology,* 183-238. New York. 3rd ed.

HECK, H. (1939): Die Bonobos. *Tier & Wir* Januar/Februar 1939, 20-27.

HOOFF, J.A.R.A.M. van (1973): A structural analysis of the social behaviour of a semi-captive group of chimpanzees. In: M. von Cranach & I. Vine (eds.), *Social communication and movement,* 75-162. London.

HORN, A.D. (1980): Some observations on the ecology of the Bonobo chimpanzee (*Pan paniscus* Schwarz 1929) near the Lake Tumba, Zaire. *Folia Primatologia* 34, 145-169.

HUNT, K.D. (1994): The evolution of human bipedality: Ecology and functional morphology. *Journal of Human Evolution* 26, 183-202.

HUNT, K.D., CANT, J.G.H., GEBO, D.L., ROSE, M.D., WALKER, S.E. & YOULATOS, D. (1996): Standardized descriptions of primate locomotor and postural modes. *Primates* 37, 363-387.

JOHNSON, S.C. (1981): Bonobos: Generalized hominid proto-types or specialized insular dwarfs? *Current Anthropology* 22, 363-375.

JORDAN, C. (1977): *Das Verhalten zoolebender Zwergschimpansen (Pan paniscus Schwarz 1929).* Diss. Frankfurt a.M.

KANO, T. (1979): A pilot study on the ecology of pygmy chimpanzees, *Pan paniscus.* In: D.A. Hamburg & E.R. McCown (eds.), *The great apes,*123-135. Menlo Park.

KANO, T. (1992): *The last apes: Pygmy chimpanzees behaviour and ecology.* Stanford.

KINGDON, J. (1997): *The Kingdon field guide to African mammals.* San Diego.

KORTLAND, A. (1955): Aspects and prospects of the concept of instinct (Vicissitudes of the hierarchy theory). *Archives Néeerlandaises de Zoologie* 11, 155-284.

KORTLAND, A. (1962): Chimpanzees in the wild. *Scientific American* 206, 128-138.

KORTLAND, A. (1968): Handgebrauch bei freilebenden Schimpansen. In: B. Rensch (Hrsg.), *Handgebrauch und Verständigung bei Affen und Frühmenschen,* 39-102. Bern.

KORTLAND, A. (1972): *New perspectives on ape and human evolution.* Amsterdam.

KORTLANDT, A. (1974): Das sog. Raubtierverhalten der Schimpansen: Beuteschlagen oder Kleinkrieg? *Lecture at the 4th Meeting of Ethologists, Mainz.*

KORTLAND, A. (1977/80): The ecosystem in which the incipient hominines could have evolved. In: R.E. Leakey & B.A. Ogot (eds.), *Proceedings of the 8th Panafrican Congress of Prehistory and Quaternary Studies,* 133-136. Nairobi.

KORTLAND, A. (1980): How might early hominids have defended themselves against large predators and food competitors? *Journal of Human Evolution* 9, 79-112.

KORTLAND, A. (1983): Marginal habitats of chimpanzees. *Journal of Human Evolution* 12, 231-278.

KORTLAND, A. (1984): Habitat richness, foraging range and diet in chimpanzees and some other primates. In: D.J. Chivers, B.A. Wood & A. Bilsborough (eds.), *Food acquisition and processing in primates,*119-159. New York.

KORTLAND, A. (1994): *The true history of the Rift hypothesis of African ape-hominid divergence.* (Available from the author).

KORTLAND, A. (1995): A survey of the geographical range, habitats and conservation of the pygmy chimpanzee *(Pan paniscus)*: An ecological perspective. *Primate Conservation* 16, 21-36.

KORTLAND, A. (submitted): Pygmy chimpanzee, bonobo, or gracile chimpanzee: What's in a name?

KORTLAND, A. & BRESSER, E.G. (1963/89): *Experimentation with forest-dwelling chimpanzees in the Congo, 1963.* (16 mm film, revised video edition 1989). Amsterdam (Audio-Visueel Centrum, Universiteit).

KORTLAND, A. & KOOIJ, M. (1963): Protohominid behaviour in primates. *Symposium of the Zoological Society London* 10, 61-68.

KORTLAND, A. KOOIJ, M. (n.d.): *Protohominid behaviour in primates (Tests in captivity, 1953-1965).* (16 mm film, revised video edition). Amsterdam (Audio-Visueel Centrum, Universiteit).

KORTLANDT, A., ORSHOVEN, J. van, PFEIJFFERS, R. & ZON, J.C.L. van (1966-67/81): *Chimpanzees in the wild, Guinea 1966-1967.* (16 mm film, revised video edition 1981). Amsterdam (Audio-Visueel Centrum, Universiteit).

KORTLANDT, A. & TREVOR, S. (1964/86): *Experimentation with forest-dwelling chimpanzees in the Congo, 1964.* (16 mm film, revised video edition 1986). Amsterdam (Audio-Visueel Centrum, Universiteit).

LAMOTTE, M. (1942): La faune mammalogique du Mont Nimba. *Mammalia* 6, 114-119.

LAWICK-GOODALL, J. van (1967): *My friends the wild chimpanzees.* Washington.

LAWICK-GOODALL, J. van (1968): The behaviour of free-living chimpanzees in the Gombe Stream Reserve. *Animal Behaviour Monograph* 1, 161-311.

LENGLET;, G.L. (1987): Animals eaten by chimpanzees: A bibliographical review. *Revue de Zoologie Africaine* 101, 197-219.

LETOUZEY, R. (1986): *Manual of forest botany. Tropical Africa 1.* Nogent-sur-Marne.

MacKINNON, J. (1978): *The ape within us.* London.

MAYR, E.. (1959): Darwin and the evolutionary theory in biology. In: B.J. Meggers (ed.), *Evolution and anthropology,* 1-10. Washington.

NISHIDA, T. (1972): Preliminary information of the pygmy chimpanzee *(Pan paniscus)* of the Congo Basin. *Primates* 13, 415-425.

PIETERS, A. (1977): *Essence forestières du Zaïre.* Gent.

PITMAN, C.R.S. (1931): *A game warden among his charges.* London.

PITMAN, C.R.S. (1942): *A game warden takes stock.* London.

PIVETEAU, J. (1957): *Traité de paléontologie.* Vol. 7. *Primates, paléontologie humaine.* Paris.

PLOOIJ, F.X. (1984): *The behavioral development of free-living chimpanzee babies and infants.* Norwood.

RAHM, U. (1967): Observations during chimpanzee captures in the Congo. In: D. Starck, R. Schneider & H.-J. Kuhn (Hrsg.), *Neue Ergebnisse der Primatologie - Progress in primatology,* 195-207. Stuttgart.

REYNOLDS, V. & REYNOLDS, F. (1965): Chimpanzees of the Budongo Forest. In: I. DeVore (ed.), *Primate Behavior,* 368-424. New York.

ROMER, A.S. (1966): *Vertebrate paleontology.* Chicago. 3rd ed.

SABATER PI, J. & VEÀ, J.J. (1994): Comparative inventory of food consumed by the wild pygmy chimpanzee (*Pan paniscus;* Mammalia) in the Lilungu-Lokofe region of the Republic of Zaïre. *Journal of African Zoology* 108, 381-396.

SAVAGE-RUMBAUGH, S. & LEWIN, R. (1994): *Kanzi.* London.

SCHWARZ, E. (1929): Das Vorkommen des Schimpansen auf dem linken Kongo-Ufer. *Revue de Zoologie et de Botanique Africaines* 16, 425-426.

SIMONS, E.L. (1972): *Primate evolution.* New York.

SONDAAR, P.Y. (1976): Insularity and its effect on mammalian evolution. In: M.K. Hecht, P.C. Goody & B.M. Hecht (eds.), *Major patterns in vertebrate evolution,* 671-707. New York.

STANYON, R. CHIARELLI, B., GOTTLIEB, K. & PATTON, W.H. (1986): The phylogenetic and taxonomic status of *Pan paniscus:* A chromosomal perspective. *American Journal of Physical Anthropology* 69, 489-498.

SUSMAN, R.L. (1984): The locomotor behavior of *Pan paniscus* in the Lomako Forest. In: R.L. Susman (ed.), *The pygmy chimpanzee,* 369-393. New York.

SUSMAN, R.L., BADRIAN, N.L. & BADRIAN, A.J. (1980): Locomotor behavior of *Pan paniscus* in Zaire. *American Journal of Physical Anthropology* 53, 69-80.

TAILFER, Y. (1989): *La forêt dense d'Afrique Centrale.* Wageningen.

THOMPSON, J.A.M. (1997): *The history, taxonomy and ecology of the bonobo (Pan paniscus Schwarz), with a first description of a wild population living in a forest/savanna habitat.* PhD Thesis Oxford.

TINBERGEN, N. (1952): „Derived" activities, their causation, biological significance, origin, and emancipation during evolution. *Quaterly Review of Biology* 27, 1-32.

TRATZ, E. & HECK, H. (1954): Der afrikanische Anthropoide „Bonobo", eine neue Menschenaffengattung. *Säugetierkundliche Mitteilungen* 2, 97-101.

VRBA, E.S. (1980): The significance of bovid remains as indicators of environment and predation patterns. In: A.K. Behrensmeyer & A.P. Hill (eds.), *Fossils in the making: Vertebrate taphonomy and paleoecology,* 247-271. Chicago.

WAAL, F. de & LANTING, F. (1997): *Bonobo: The forgotten ape.* Berkeley.

WHITE, F. (1983): *The vegetation of Africa.* Paris.

WHITE, F. (1993): The AETFAT chorological classification of Africa: History, methods and applications. *Bulletin de Jardin Botanique National de Belgique* 62, 225-281.

WHITE, F.J. (1996): *Pan paniscus* 1973-1996: Twenty-three years of field research. *Evolutionary Anthropology* 5, 11-17.

WRANGHAM, R., & PETERSON, D. (1996): *Demonic males.* London.

YERKES, R.M. (1925): *Ancient human.* New York.

ZIHLMAN, A.L. (1984): Body build and tissue composition in *Pan paniscus* and *Pan troglodytes,* with comparisons to other hominoids. In: R.L. Susman (ed.), *The pygmy chimpanzee,* 179-200. New York.

Notes

[1] For other examples in which the methodology of comparative and evolutionary functional analysis is applied to locomotor patterns of chimpanzees, see Kortlandt et al. (1981); on incipient weapon use and carnivory in higher primates in general Kortlandt & Kooij (1963) and Kortlandt (1972); on hominid defense against carnivores Kortlandt (1980); and on antelopes Vrba (1980).

[2] I prefer the French technical terms because they are defined much more precisely than their English semi-equivalents and because White's (1983, 1993) concepts are defined floristically rather than physiognomically. Chimpanzee research requires chimpomorphic concepts.

[3] Coppens (1994) claims that he has independently conceived issue (a) of the above hypothesis. This is demonstrably incorrect (Kortlandt, 1994).

[4] From the current literature it is not clear to what extent there is a difference between the senior mothers, the mothers in general and the childless females as to their social role and influence, because the authors rarely distinguish these categories. Such a difference is important in other maternocracies, e.g. among deer and elephants, and in certain situations also among female common chimpanzees.

Coping with Social Tension in Primate Societies: Strategic Modelling of Early Hominid Lifestyles

Marina Butovskaya

Abstract

Because anatomically the early hominids displayed a mosaic of adaptations to bipedality and arboreality, it may be expected that their behaviour, too, possibly including their social life, resembled that of their closest primate relatives to some extent, but differed in having a number of unique characteristics. None of the extant nonhuman primates are close to the early members of the human clade. However, taking into consideration the entire range of behavioural features in cercopithecines and African apes with regard to between-species differences in habitat and lifestyle and within-species variation in social structure and behavioural strategies it is possible to delineate the most likely directions of social adaptations of early hominids in the tropical forest and in the savanna. Our basic assumption is that the rise of intelligence level in the course of primate evolution made social relationships progressively more complex. Models of male-female relationships will be suggested based on the concepts of Machiavellian intelligence and self-awareness and including deception and self-control. Also, we will discuss possible strategies aimed at maintaining group stability and reducing tension between the sexes (state alliances based on personal preferences in females and unstable alliances aimed at enhancing or maintaining social status in males). Models of restoring social equilibrium after conflicts with other group members belonging to the same or to other age-sex categories include factors like increasing role of vocal (resp. verbal) signals and exchange of gifts between aggressors and victims as a means of reconciliation, intrusion of higher-ranking individuals aimed at preventing conflicts, and the development of consolation will be discussed. It appears likely that the most recent common ancestors of humans and African apes lived in groups based on strong ties between females. Common chimps tend to develop such ties in captivity, while bonobos are strongly female-bonded in their natural habit.

Although social behaviour is undocumented by the fossil record, more and more writers are trying to reconstruct it on the basis of indirect evidence (Tooby & DeVore, 1987; Wrangham, 1987; Foley & Lee, 1989; Butovskaya, 1995; Zihlman, 1996). This paper focuses on possible models of social bonding and maintaining social stability in early hominid societies, and it uses data of primate socio-ecology, human ethology, and paleoecology.

The contribution of primate socio-ecology and the modern evolutionary theory, specifically inclusive fitness and differential reproduction paradigms, is especially relevant for the evolution of primate social systems. While body size was found to be correlated with diet, the quality and pattern of the spatial distribution of food resources makes it possible to predict certain aspects of social relationships within and between primate groups (Clutton-Brock & Harvey, 1977; Wrangham, 1980). Also, correlation has been established between terrestriality, food preferences and group size (Clutton-Brock & Harvey, 1977; van Schaik, 1989).

According to Dunbar (1988), terrestrial frugivores and generalized omnivores tend to form large foraging groups with differentiated competitive relationships among females. Group size can also depend on the environment. Observations of chimpanzees in different environmental settings have revealed a tendency to a higher level of cohesiveness in dry and open environments (Mt. Assirik, Senegal, see Tutin et al., 1983). Larger, mixed parties containing adult males are reportedly more common in nonforested habitats than are solitary individuals or parties without males. The variation in group size seems to be related not only to the nature of food distribution over the homerange, but also to predation pressure (van Schaik, 1989).

Paleoenvironment and the grouping patterns of early hominids

Recent paleoanthropological finds have extended the hominid record from 3.2 million years (Hadar, Ethiopia) to 3.9 (Awash, Ethiopia), 4.4 (*Australopithecus ramidus*, Aramis, Ethiopia, see White et al. 1994), or even 5 million years (Lothagam, Tabarin, see Hill & Ward, 1988, 1992). The period from 6 to 4.3 mya in East Africa was characterized by a moderately humid climate. During the next period, 4.4-2.8 mya, humidity increased, and 2.5 mya the climate became considerably more arid (Prentice & Denton, 1988; Coppens, 1991). So when the early australopithecines emerged, vast territories of East Africa were covered by gallery forests, woodlands, and savannas, and bipedy must have been an adaptation to open terrains. *Australopithecus afarensis* lived in a mosaic environment (Hadar, Omo, East Turkana), and fairly dry savanna (Laetoli) (White & Suwa, 1987; Bonnefille et al., 19887; Radosevitch et al., 1992). According to some writers, *Australopithecus afarensis* was yet unable to use bipedal gait for walking over long distances or running (Schmid, 1991).

The degree of sexual dimorphism in australopithecines is a matter of debate. While some writers believe that dimorphism in *A. afarensis* is more expressed than in chimpanzees or bonobos (Kimbel & White, 1988; McHenry, 1991), dental data suggest that *A. ramidus* was very similar to bonobos, implying that the similarity extended to the degree of sexual dimorphism (Zihlman, 1996).

The Pleistocene mammalian fauna of East Africa resembled the modern Ethiopian fauna but was more variable. It included saber-toothed cats (*Homotherium, Megantheron*), large cats (*Dinofelis*), large hyenas (*Chasmaporthetes*), cheetahs, and panthers (Coppens, 1985). Although they hunted mainly herbivores, hominids, too, must have been endangered. Neither large body size nor even tools would have sufficed to protect them from predators. Neither would their diurnal activity, since cheetahs and leopards can hunt during the daytime. The idea that the sociality of primates provides a protection against carnivores (van Schaik, 1989) evidently applies to early hominids in this context.

While seasonal differences in food availability can determine subgroup size in early hominids, the extent to which resources can be monopolized and the amount of food per patch influences the level of within-group competition, as demonstrated in chimpanzees, spider monkeys and long-tailed macaques (van Schaik, 1989; Chapman et al., 1995). Mass migration of large parties from one part of the homerange to another, especially during the dry season, may have been common.

Grouping patterns of African apes and humans provide us with some data concerning the stability of community social networks (gorillas of Karisoke: Fossey, 1983; chimanzees of Kasoge and Gombe: Nishida, 1979; Goodall, 1986; bonobos of Wamba and Lomako: Kano, 1980; Kano & Mulavwa, 1984; Badrian & Badrian, 1984; humans worldwide: Eibl-Eibesfeldt, 1989). Community members are well acquainted with each other. In gorillas, within-group relationships are highly stable (Harcourt, 1979; Fossey, 1983) (Table 1). In chimpanzees, bonobos, and humans the community is composed of parties which break up and form again very frequently (Wrangham, 1987). The same grouping patterns may have been common in early hominids. It is unlikely that females could travel alone or with infants or survive outside the cohesive social groups in open habitats because of predation pressure. Because bipedy had made early hominids more vulnerable to predators in the open terrain, they may have been urged to form larger travel parties. Bonds between females in early hominid societies tend to be neglected since males are regarded as the resident sex. Bipedy, however, might have stimulated longterm friendly bonds between females (Figure 1).

Species	Bonds	Cont. aggression		Dominance		Alliances	
		Males	Females	Males	Females	Males	Females
Gorilla	M-F	H	L	D	S	CA	N
Chimpanzee	M-M, F-F*	H	L	D	S	CA	FrA
Bonobo	M-M, F-F	L	L	D*	D*	CA	FrA
Man	M-M, M-F, F-F	H	L	D	S	CA	FrA
Early hominids	M-M, M-F*, F-F	?	L	D	S	CA	FrA

Table 1 - *Grouping patterns of African apes, man an early hominids.*
M-F - male-female; M-M - male-male; F-F - female-female; F-F - close friendly bonds between females are possible under some conditions; H - high; L - low; D - males are dominant over females; S - females are subordinate to males; D* - males and females could be dominant under different conditions; CA - competitive alliances; FrA - friendly preferences; N- none; ? - unpredictable.*

Data on gorillas are cited after Harcourt (1979) and Fossey (1983); those on chimpanzees after de Waal (1982) and Goodall (1986); those on bonobo after Kano (1980), Badrian & Badrian (1984) and Kuroda (1984); those on humans after Irons (1983) and our personal observations.

Hunting strategies and social relationships between and within the sexes in early hominids

The transition to a more protein-rich food has traditionally been regarded as a pivotal event in human evolution. Although recent finds from Plio-Pleistocene sites seem to support the idea of a scavenging stage (Shipman, 1986), observations on hunting strategies in chimpanzees (Wrangham & Bergmann-Riss, 1990; Stanford, 1996) suggest that early hominids did practice hunting on a larger scale than many writers tend to believe (Figure 1).

Indeed, chimpanzees not merely hunt but do it regularly. Observations collected in Tai, Mahale, and Gombe demonstrate that chimpanzees are the main predators of colobus monkeys, and chimpanzee predation is the principal factor limiting population size in the red colobus in Gombe. According to some estimates (Wrangham & Bergmann-Riss, 1990; Stanford, 1996), a chimpanzee community consisting of ca. 45 individuals can consume 600 kg of meat in carcass weight per year. They utilize virtually everything including bones, and thus no archaeological record of kills is being left. Given that most animals hunted by early hominids were small or medium-sized, archaeologists do not have much to deal with. The other possible reason behind the lack of records of early hominid hunting became evident from data on Hadza hunters-gatherers: smaller prey is either carried away by humans or its remains are quickly consumed by scavenging animals (O'Connell et al., 1992). The most reasonable assumption is that hominids practiced both hunting and scavenging (Isaac & Crader, 1981) and that their relative importance varied across the seasons.

Comparing data on chimpanzees in Gombe, !Kung hunter-gatherers, and Hadza, reveals similarities in the seasonality of diet, the peak of hunting and scavenging coinciding with the dry season (Lee, 1982; O'Connell et al., 1988; Stanford 1996). Hunting, then, may be viewed as an

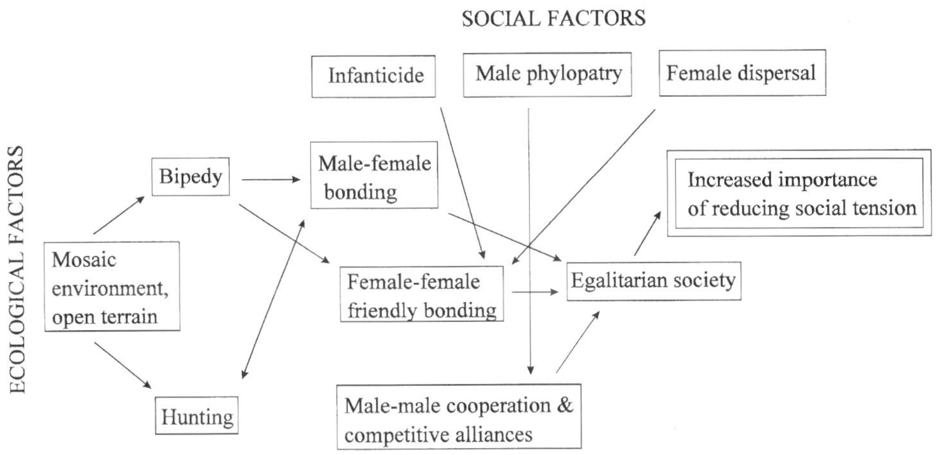

Figure 1 - Evolution of social bonding and tension reduction in early hominids.

ecological adaptation to food shortages during the dry season. Besides, hunting in chimpanzees is an important social event which is universally correlated with the increase in the size of the foraging parties and the presence of females with genital swelling. The direct effect of the presence of receptive females on the intensity of hunting in males has been demonstrated by Stanford et al. (1994). It has even been concluded that „females might be concidered the driving force behind chimpanzee hunting" (Stanford, 1996, p.106). The evolutionary roots of the male control over females, and the manipulation of females by males and vice versa, which is highly typical of humans, may well be based on a relationship between reproduction and food provisioning (Figure 1). Concealed ovulation might have made females permanently attractive thus stimulating intensive hunting and making male-female bonds more stable (Lovejoy, 1981).

The positive correlation between the number of males and the efficiency of hunting, established in chimpanzees, suggests that as hunting was becoming more and more important for the early hominids, the hunting parties were increasing in size, and cooperation was playing a more and more prominent role in the life of males. Male cooperation in hunting might be a form of competitive alliances and power control, thus increasing the reproductive chances of hunters. Maximal individual success depended on social intelligence (ability to manipulate other male group members) and instrumental intelligence, related to successful planning and the assessment of the prey's behaviour.

Evolution of male-female pair-bonding, female dispersal and infanticide

There are reasons to believe that the system of relationships in early hominids was male-resident (Foley & Lee, 1989), based on male competitive alliances, female-bonded, and tolerant, although not female-resident (Table 1). In chimpanzees, most females are not related, yet their coalitions are known to be stable and selective; kinship and friendly bonds overlap, and „the distinction between friend and foe seems infinitely sharper for females" (de Waal, 1990, p.53). Friendly female-bonding is even more common among bonobos (de Waal, 1987). Modern human females definitely follow the same patterns since their ties are highly stable and have little to do with competition for a dominance status. In many traditional societies, women who have moved to their husbands' homes establish close relationships, involving both household activities and child rearing, with their female in-laws (Table 1).

Another largely neglected factor that promoted the development of friendly ties between females is the risk that males might commit infanticide (as will be shown below, such adaptations exist in several primate species) (Figure 1). As the recent studies suggest, infanticide is an important reproductive strategy in primates (Angst & Tommen, 1977; Daly & Wilson, 1988; Hiraiwa-Hasegawa, 1992). According to the inclusive fitness paradigm, it can be advantageous for a male to invest in his own offspring and not to invest in situations where his paternity is problematic. The functional explanation of the reproductive advantages of infanticide in the latter case is that females would conceive again sooner and would invest more energy in the next infant, biologically related to the present male (Hrdy, 1979).

Perhaps infanticide is an efficient reproductive strategy of males in modern society as well. As shown by Schiefenhövel (1989), infants born in adultereous consortships or those from previous marriages run a risk of being killed in 15 out of 39 traditional societies practicing this custom. Data on Arche, the modern hunters-gatherers of Paraguay, suggest that fatherless children are 15 times more likely to be killed at the age of 2 to 15 than are those who have fathers (Hill & Kaplan, 1988). In Western societies, stepchildren are reported to be 65 times more likely to die within the first two years of life than are those who live with both biological parents (Daly & Wilson, 1988).

The female's inclusive fitness is definitely reduced due to infanticide. No wonder female primates have developed strategies aimed at preventing infanticide. While in some taxa, like macaques and vervets, females are cohesive and jointly defend infants against new males, others, such as colobuses and langurs, practice emigration and group fissioning.

The general socio-ecological model predicts that predation risk forces females to live in groups. Females compete for resources and, depending on the nature of their competition (within or between-group contest), form their social systems, which can be nonfemale-bonded, female bonded, tolerant female-bonded, or female-resident (Wrangham, 1980). This model has certain limitations, since it can neither predict the effects which the behaviour of the males (paternal investment strategies, male competitive alliances, etc.) can have on the relationships between females, nor explain male-female associations. The infanticidal models proposed by van Schaik and Dunbar (1990) and Sterck (1995) provide a necessary complement. Friendly preferences in females may be regarded as defence alliances aimed at protecting infants against infanticidal males and a system of mutual aid uniting females with newborns. Preconditions to such a system are found in nonhuman primates in the form of an increased attention females pay to nonrelated newborns, and their readiness to nurse (or even feed) and protect these infants. Given the rise of intelligence in early hominids, orientation towards female-bonding as an anti-infanticidal strategy is likely (Figure1).

Infanticide could be one of the important factors stimulating longlasting pairbonding (van Schaik & Dunbar, 1990) (Figure 1). The high sexual activity of early hominid females throughout the ovarial-menstrual cycle made them permanentlyy attractive for the males. Also, females may have developed various tactics of deception, such as concealed ovulation, aimed at making paternity recognition more difficult for the males.

Peacemaking in early hominid communities

Within- and between-group contests are a normal feature of social life in gregareous animals. Under the conditions of anegalitarian social organization and the rise in social intelligence in early hominids including mutual deception, maintaining social stability must have become highly important. Members of the same group need to reduce tension and restore relationships between former opponents. The best option is reconciliation, which has been demonstrated in many primate

Species	Reconciliation	Specific reconciliation	Initiator of reconciliation	Redirected aggression	Mediation	Friendly selectiveness	Consolation	Conflict termination	Cooperative alliances
Gorilla	Y	Y*f	V	Y*ff	N	L mf	Y*m	Y*m	R
Chimpanzee	Y	Y	A	Y	Y	H*ff	Y*	Y*	Y*mm
Bonobo	Y	Y	A	Y	Y?	H*ff	Y	Y	Y
Man	Y	Y	A	Y	Y	H*ff	Y	Y*m	Y*mm
Early hominids	Y	Y	A	Y	Y?	H*ff	Y	Y?	Y?

Table 2 - *Post conflict strategies in African apes, man and early hominids.*
Specific reconciliation patterns - special reconciliatory patterns; Initiator of reconciliation. - more frequent initiator of reconciliation.
*Y- the trait is present; Y*m - present in males' behaviour towards females; Y*f - present in the interactions of females with males; Y*mm - more typical of male-male interactions; Y*ff - more typical of female-female interactions; Y* typical of males' behaviour towards other group members; Y? - pattern is present but its expression in different sexes is unknown. Lmf - low level of selectiveness in individual preferences in both males and females; H*ff- high level of selective preferences in female-female relations; A- aggressor; V - victim.*
Data on gorillas are cited after Watts (1995a,b); those on chimpanzees, after de Waal & Roosmalen (1979), Sackin & Thelen (1984), and Goodall (1986); those on bonobos, after Kitamura (1983), Kuroda (1984), de Waal (1987, 1993); those on humans, after Sackin & Thelen (1984), Eibl-Eibesfeldt (1989), and our personal observations.

species, such as lemurs, cercopithecines, colobines, all big African apes (common chimanzees, bonobos, gorillas), and humans (Kappeler & van Schaik, 1992; de Waal & Roosmalen, 1979; de Waal, 1987; Watts, 1995 a,b; and my observations on children) (Table 2).

In all the species mentioned above, reconciliation is reported to take place in the first few minutes after the conflict. Because early hominids were gregarious, it may be expected that the same applied to them (Table 2). The use of special affiliative behaviours (aimed at appeasement, reassuring, etc.) has been documented for African apes and humans. In gorillas, reconciliation is limited to male-female interactions. Female gorillas use embracing, touching, or grumbling, as means of reconciling with males (Watts, 1995a). In common chimpanzees, reconciliation was registered in all age-sex combinations (most frequently in male-male interactions), kissing and embracing being the most typical conciliatory behaviours (de Waal & Roosmalen, 1979). The conciliatory tendency is extremely high in bonobos (de Waal, 1987) who use special post-conflict patterns of sexual behaviour, such as genital rubbing in females, mutual penis thrusting in males, ventro-ventral and ventro-dorsal matings between males and females, and genital manipulation (de Waal, 1993).

In humans, reconciliation is typical of both same-sex and heterosexual interactions. Special patterns include actual or symbolic offering of gifts, special bond-restoring rhymes, apologies, kissing, embracing, hand-holding, and touching (Sackin & Thelen, 1984). Gift offering is a common means of reconciliation both at the within-group and between-group level in most traditional societies (Eibl-Eibesfeldt, 1989). Apparently, sexual behaviour is an important means of restoring bonds between men and women.

Observations on monkeys and apes suggest that the level of reconciliation can depend on the dominance style typical of the species. In macaques, species with rigid dominance relationships (*Macaca mulatta, M. fascicularis*) display a much weaker conciliatory tendency than do egalitarian species (*M. arctoides, M. tonkeana*) (de Waal & Ren, 1988; Aureli et al., 1989). While in rhesus and longtailed macaques reconciliation is mostly initiated by aggressors, this is more often done by victims in *M. arctoides*. In the former case, victims are evidently afraid to approach dominant individuals. In African apes, the situation is different. In gorillas, who have a despotic dominance style, females who were attacked by males always initiate reconciliation with them (Watts, 1995a) (Talbe 2). In common chimpanzees, who practice egalitarian relationships, aggressors initiated almost one half of all reconciliations in the Arnhem colony, and in bonobos, who are highly egalitarian, nearly two thirds of reconciliations are initiated by aggressors (de Waal, 1987).

As our data on 6-7-year-old schoolchildren demonstrate, their social relationships may be described as relaxed and egalitarian, and reconciliation on the aggressors' initiative occurred in 66 % of the cases observed in the first minute after conflict. Admittedly, data on children may not apply to modern adults. However, it appears that they would be more applicable to early hominids (Table 2). Indeed, while children at that age are already able to form stable social bonds and make a distinction between friends, neutral group members, and enemies, their motivation to reconcile appears to be largely subconscious rather than learned. Amazingly, as revealed by individual interviewing, although most children stated that it was necessary to initiate reconciliation, they would not be too eager to reconcile; in fact, most preferred to be approached by the former opponent. As the intellectual level gets higher and higher, these tendencies may mask initial orientations and suppress the subconscious desire to reconcile in such a way as to secure personal profit.

If social relations in early hominid communities were similar to those in chimpanzee and bonobo groups, that is, relaxed and egalitarian (Butovskaya, 1995), it may be expected that aggressors were highly active in restoring social equilibrium. The fact that children, reconcile shortly after the conflict (like all the African apes), but declare that they would wait for an hour or even a day to apologize, suggests that the delay of reconciliation in modern adults is due to an intentional modification of innate behavioural predispositions which played an important role at the early stages of human evolution.

There is little doubt, then, that reconciliation was practiced by early hominids. But did its likelihood depend on the sex of the participants? It may be supposed that early hominids lived in multi-male and multi-female communities. Even if these were patrilocal (as it is most often hypothesized), it is most probable that males and females alike tended to reconcile. Males must have been interested in restoring social bonds with other group males because cooperation was necessary for the protection of territory, resources, and mating partners (Table 2). These factors are crucial for common chimpanzees communities (and possibly for bonobos). The coalition nature of male-male relations makes it impossible to predict who will be needed next as a partner in the power game. The same applies to modern human males as well. To be successful, a chimpanzee male must be able to „compromise, forgive, and forget" (de Waal, 1990, p.53). Cooperation in hunting or scavenging in hominids made such strategies even more important.

Cooperation between females was equally important for survival since all females alike needed help (especially in child rearing) (Table 2). Our analysis of postconflict interactions in children has failed to reveal any effect of sex on the reconciliation tendency.

Reconciliation in mixed-sex dyads was evidently no less important for the early hominids than was that between males or between females because of the increased probability of longlasting pairbonding. It may be suggested that in heterosexual post-conflict interactions each of the sexes had its own conciliatory strategy: for the males, it was food sharing, and for the females, sexual initiation. Concealed ovulation and permanent sexual attractiveness of females could have made the latter strategy especiallly efficient.

Redirected aggression has been described in African apes as well as in humans (Table 2). In gorillas, it was mostly restricted to female-female and juvenile post-conflict interactions (Watts, 1995b). Redirection is typical of chimpanzees (de Waal, 1982). It is probably less common in bonobos due to lower intensity of conflicts in general, lower arousal level in victims, or higher level of reconciliation compared to that in common chimpanzees (de Waal, 1993). According to our unpublished data, redirection is practiced by children in post-conflict interactions and is inversely related to reconciliation (redirection is less likely if reconciliation has occurred). The custom of blood vengeance directed against the offender's relatives (Davie, 1929; Turney-High, 1949), described in many hunter-gatherer societies, may also be regarded as a kind of redirected aggression.

Consolation (calming down the victim by a third party) is related to the species-specific cognitive level, longterm memory, ability to discriminate between ingroup members, and the level of empathy (de Waal, 1990). Consolation as a postconflict behaviour is practiced by chimpanzees, bonobos, gorillas (directed by the male leader to the females) and humans. Observations on children reveal a strong innate tendency for consoling individuals who exhibit symptoms of stress or frustration following the conflict. Perhaps early hominids were no exception (Table 2).

In gorillas, chimpanzees and humans, conflict termination is the prerogative of males probably because males generally dominanate females in these species (Table 2). In bonobos, the situation is different since females dominate males, especially in conflicts over food resources. Given that at the early stages of human evolution the role of males in group protection and hunting was increasing, it may be suggested that evolution was favouring males who intervened in conflicts and prevented violence.

Mediation of the conflict by a third party has been described in chimpanzees of the Arnhem colony (de Waal, 1993) and in human societies (Table 2). In situations where male rivals were unable to settle the conflict by their own, a female would approach one of the males, groom him for a while and then lead him to the former opponent, thus making reconciliation possible. We have witnessed almost similar situations in children: a child who had not been involved in the conflict would approach one of the opponents, calm him/her down (if it was a victim) or shame him/her (if it was an aggressor) and then lead him/her to the former opponent. Finally, the mediator would join the rivals' hands and make them recite a conciliatory rhyme (Table 2).

Conclusion

Socio-ecological theory makes it possible to assess the effect of ecological factors (quality and distribution of food resources, seasonality, and predation pressure), and social parameters (spatial distribution of female reproductive partners, and probability of infanticide) on the emergence of various types of social structures in primates. Based on the comparison of models of social bonding in extant monkeys, apes, and humans, some hypotheses can be made concerning group structure and the relationships between age and sex classes (Butovskaya, 1995).

There are reasons to believe that in terms of size, early hominid communities resembled those of chimpanzees and bonobos. In the forest, a community may have consisted of parties each of which included over a dozen adult individuals (both males and females) and their progeny (Radosevitch et al., 1992). In open areas, the parties increased in size and became more stable in composition.

Early hominid communities were likely male-resident, as in chimpanzees, bonobos and most human groups (Foley & Lee, 1989). Because most males were related and well familiar with each other since birth, they could coexist within a group. Also, they were likely to cooperate and were interested in jointly protecting females and their progeny from alien males, and their competition for females were nonviolent. Males competed with each other by means of alliances and rise in status. An indirect strategy was to manipulate other males during hunting and to make hunting more productive. This enhanced the male's chances of gaining a female and monopolizing her. Because of infanticide, the formation of stable male-female pairs may have been profitable for females as well.

Bonds between female primates are apparently more important than previously believed. Data on wild and captive chimpanzees, bonobos, and humans demonstrate that these bonds may be very stable, in fact lifelong. Unlike the situation in males, partnership between females is based on intimate affection rather than competitive benefits. This partnership protected infants from the males who tended to infanticide, and provided a basis for a collective nursing system, thus offering infants greater chances to survive.

Because social bonds within a community were highly important, it was necessary to control social stability. Reconciliation strategies might have been based on sexual behaviour (as in bonobos and modern man), meat sharing (offering gifts as a means of reconciliation is practiced in many human societies), friendly bodily contacts, ritualized gestures and utterings.

Acknowledgements

We are grateful to Russian Foundation for Fundamental Research, grant # 96-06-80405 for financial support of this study and we thank Dr. Herbert Ullrich for invitation to participate the Interdisciplinary project „Hominids in their environment“ and for inviting us to the Symposium „Lifestyles and survival satrategies in Pliocene and Pleistocene hominids“.

References

ANGST, W. & TOMMEN, D. (1977): New data and discussion of infant killing in Old World monkeys and apes. *Folia Primatologica* 27, 198-229.

AURELI, F., SCHAIK, C. P. van & HOOFF, J.A. van. (1989): Functional aspects of reconciliation among captive long-tailed macaques (*Macaca fascicularis*). *American Journal of Primatolology* 19, 39-51.

BADRIAN, A. & BADRIAN, N. (1984): Social organization of *Pan paniscus* in the Lomako Forest, Zaire. In: R. Susman (ed.), *The pygmy chimpanzee,* 325-346. New York.

BALDWIN, P.J., McGREW, W.C. & TUTIN, C.E. (1982): Wide-ranging chimpanzees at Mt. Assirik, Senegal. *International Journal of Primatology,* 3, 367-385.

BLUMENSCHINE, R.J. (1987): Characteristics of an early hominid scavenging niche. *Current Anthropology,* 28, 383-407.

BONNEFILLE, R., VINCENS, A. & BUCHET, G. (1987): Polinology, stratigraphy and paleoenvironment of a Pliocene site (2.9-3.3 MY) at Hadar, Ethiopia. *Paleogeography, Paleoclimatology and Paleoecology* 60, 249-281.

BUTOVSKAYA, M.L. (1995): Some aspects of hominid socioecology according to primatological data. In: H. Ullrich (ed.), *Man and environment in the Palaeolithic,* 309-317. Liège.

CHAPMAN, C.A., WRANGHAM, R.W. & CAPMAN, L.J. (1995): Ecological constraints on group size: An analysis of spider monkey and chimpanzee supgroups. *Behavioural Ecology and Sociobiology,* 36, 59-70.

CLUTTON-BROCK, T.H. & HARVEY, P.H. (1977): Species differences in feeding and ranging behaviour in primates. In: T.H. Clutton-Brock (ed.), *Primate ecology: Studies of feeding behaviour in lemurs, monkeys and apes,* 557-579. London.

COPPENS, Y. (1985): *L'environment des hominides au Plio-Pleistocene.* Paris.

COPPENS, Y. (1991): L'evolution des hominides, de leur locomotion et de leur environment. In: Y.Coppens & B. Senut (eds.), *Origines de la bipedie chez les hominides,* 295-301. Paris.

DALY, M & WILSON, M. (1988): Evolutionary social psychology and family homicide. *Science* 242, 519-524.

DAVIE, M.R. (1929): *The evolution of war: a study of its role in early societies.* New Haven.

DUNBAR, R.I.M. (1988): *Primate social systems.* London.

EIBL-EIBESFELDT, I. (1989): *Human ethology.* 1989. New York.

FOLEY, R.A. & LEE, P.C. (1989): Finite social space, evolutionary pathways, and reconstruction of hominid behavior. *Science,* 243, 901-906.

FOSSEY, D. (1983): *Gorillas in the mist.* Boston.

GOODALL, J. (1986): *The chimpanzees of Gombe: Patterns of behavior.* Cambridge.

HARCOUIRT, A.H. (1979): Social relationships between adult male and female mountain gorillas in the wild. *Animal Behaviour* 27, 323-342.

HILL, K & KAPLAN, H. (1988): Trade offs in male and female reproductive strategies among the Arche: part 2. In: L.Betzig, M.Borgerhoff Mulder & P.Turke (eds.), *Human reproductive behaviour,* 291-306. Cambridge.

HILL, A. & WARD, S. (1988): Origin of hominidae: The record of African large hominoid evolution between 14 MY and 4 MY. *Yearbook of Physical Anthropology* 31, 49-83.

HILL, A., WARD, S. & BROWN, B. (1992): Anatomy and age of the Lothagam mandible. *Journal of Human Evolution* 22, 439-452.

HIRAIWA-HASEGAWA, M. (1992): Cannibalism among non-human primates. In: Elgar M.A. & Crespi B.J. (eds.), *Cannibalism. Ecology and evolution among diverse taxa,* 323-338. Oxford.

HRDY, S.B. (1979): Infanticide among animals: A review, classification and examination of the implications for the reproductive strategies of females. *Ethology and Sociobiology* 1, 13-40.

IRONS, W. (1983): Human female reproductive strategies. In: S.K. Wasser (ed.), *Social behavior of female vertebrates,* 169-213. New York.

ISAAC, G.L. & CRADER, D.C. (1981): To what extent were early hominids carnivorous? An archaeological perspective. In: R.S.Harding & G. Teleki (eds.), *Omnivorous primates: Gathering and hunting in human evolution,* 37-103. New York.

KANO, T. (1980): Social behavior of wild pygmy chimpanzees (*Pan paniscus*) of Wamba: A preliminary report. *Journal of Human Evolution* 9, 243-260.

KANO, T. & MULAVWA, M. (1984): Feeding ecology of pygmy chimpanzees (*Pan paniscus*) of Wamba. In: R. Susman (ed.), *The pygmy chimpanzee,* 233-274. New York.

KAPPELER, P.M. & SCHAIK, C.P. van. (1992): Methodological and evolutionary aspects of reconciliation among primates. *Ethology* 92, 51-69.

KIMBEL, W.H. & WHITE, T.D. (1988): Variation, sexual dimorphism, and the taxonomy of *Australopithecus.* In: F.E. Grine (ed.), *Evolutionary history of the „robust" australopithecines,* 175-192. New York.

KITAMURA, K. (1983): Pygmy chimpanzee association patterns and ranging. *Primates* 24, 1-12.

KURODA, S. (1984): Interaction over food among pygmy chimpanzees. In: R. Susman (ed.), *The pygmy chimpanzee,* 301-324. New York.

LEE, R.B. (1982): Politics, sexual and non-sexual, in an egalitarian society. In: E.R. Leacock & R.B. Lee (eds.), *Politics and history in band societies,* 37-59. Cambridge.

LOVEJOY, C.O. (1981): The origin of man. *Science* 21, 341-350.

McHENRY, H.M. (1991): Sexual dimorphism in *Australopithecus afarensis. Journal of Human Evolution* 20, 21-32.

NISHIDA, T. (1979): The social structure of chimpanzees of the Mahale Mountains. In: D. A.Hamburg, & E.R. McCown (eds.), *The great apes,* 73-121. Menlo Park.

O'CONNELL, J.F., HAWKES, K. & BLURTON-JONES, N.G. (1988): Hadza scavenging: Implications for Plio-Pleistocene hominid subsistence. *Current Anthropology* 29, 356-363.

O'CONNELL, J.F., HAWKES, K. & BLURTON-JONES, N.G.(1992): Patterns in the distribution, site structure and assemblage composition of Hadza kill-butchering sites. *Journal of Archaeological Science* 19, 319-345.

PRENTICE, M.L. & DENTON, G.H. (1988): The deep-sea oxygen isotope record, the global ice sheet system and hominid evolution. In: F.E. Grine (ed.), *Evolutionary history of the „robust" australopithecines,* 383-403. New York.

RADOSEVICH, S.C., RELALLACK, G.J. & TAIEB, M. (1992): Reassessment of the paleoenvironment and prevention of hominid fossils from Hadar, Ethiopia. *American Journal of Physical Anthropology* 87, 15-28.

SACKIN, S. & THELEN, E. (1984): An ethological study of peaceful associative outcomes to conflict in preschool children. *Child Development* 55, 1098-1102.

SCHAIK, C.P. van (1989): The ecology of social relationships amongst females primates. In: V. Standen & R. Foley (eds.), *Comparative socioecology. The behavioural ecology of humans and other mammals,* 195-218. Oxford.

SCHAIK, C.P. van & DUNBAR, I.M. (1990): The evolution of monogamy in large primates: a new hypothesis and some crucial tests. *Behaviour* 115, 30-62.

SCHIEFENHÖVEL, W. (1989): Reproduction and sex-ration manipulation through preferential female infanticide among the Eipo, in the highlands of West New Guinea. In: A.E. Rasa, C. Vogel & E. Voland (eds.), *The sociobiology of sexual and reproductive strategies*, 170-193. London.

SCHMID, P. (1991): The trunk of the aurtralopithecines. In: Y.Coppens & B. Senut (eds.), *Origines de la bipedie chez les hominides,* 225-234. Paris.

SHIPMAN, P. (1986): Scavenging or hanting in early hominids: Theoretical framework and tests. *American Anthropologist* 88, 27-43.

STANFORD, C.B. (1995): The influence of chimpanzee predation on the group size and anti-predator behaviour in red colobus monkeys. *Animal Behaviour* 49, 577-587.

STANFORD, C.B. (1996): The hunting ecology of wild chimpanzees: Implications for the evolutionary ecology of Pliocene hominids. *American Anthropologist* 98, 96-113.

STANFORD, C.B., WALLIS, J., MPONGO, E. & GOODALL, J. (1994): Hunting decisions in wild chimpanzees. *Behaviour* 131, 1-20.

STERCK, L.(1995): *Females, foods and fights. A socioecological comparison of the sympatric Thomas langur and long-tailed macaque.* Utrecht.

TOOBY, J. & De VORE, I. (1987): The reconstruction of hominid behavioral evolution through strategic modeling. In: W.G. Kinzey (ed.), *The evolution of human behavior: Primate models*, 183-238. Albany.

TURNEY-HIGH, H.H. (1949/1971): *Primitive war: Its practice and concepts.* Columbia.

TUTIN, C.E., McGREW, W.C. & BALDWIN, P.J. (1983): Social organization of savanna-dwelling chimpanzees, *Pan troglodytes verus*, at Mt.Assirik, Senegal. *Primates* 24, 154-173.

WAAL, F.B. de. (1982): *Chimpanzee politics.* London.

WAAL, F.B. de. (1987): Tension regulation and nonreproductive function of sex in captive bonobos (*Pan paniscus*). *National Geographic Research* 3, 318-335.

WAAL, F.B. de. (1990): *Peacemaking among primates.* Cambridge.

WAAL, F.B. de. (1993): Reconciliation among primates: A review of empirical evidence and unresolved issues. In: W.A. Mason & S.P. Mendoza (eds.), *Primate social conflict*, 111-144. New York.

WAAL, F.B. de & REN, R.M. (1988): Comparison of the reconciliation behavior of stumptail and rhesus macaques. *Ethology* 78, 129-142.

WAAL, F.B. de & ROOSMALEN, A. van (1979): Reconciliation and concolation among chimpanzees. *Behaviour, Ecology and Sociobiology* 5, 55-66.

WATTS, D.P. (1995a): Post-conflict social events in wild mountain gorillas (Mammalia, Hominoidea). I. Social interactions between opponents. *Ethology* 100, 139-157.

WATTS, D.P. (1995b): Post-conflict social events in wild mountain gorillas. II. Redirection, side directin, and consolation. *Ethology* 100, 158-174.

WHITE, T.D. & SUWA, G. (1987): Hominid footprints at Laetoli: Facts and interpretations. *American Journal of Physical Anthropology* 72, 485-514.

WHITE, T.D., SUWA, G. & ASFAW, B. (1994): *Australopithecus ramidus*, a new species of early hominid from Aramis, Ethiopia. *Nature* 371, 306-312.

WRANGHAM, R.W. (1980): An ecological model of female-bonded primate groups. *Behaviour* 75, 262-300.

WRANGHAM, R.W. (1987): The significance of African apes for reconstructing human evolution. In: W.G. Kinzey (ed.), *The evolution of human behavior: Primate models,* 51-71. New York.

WRANGHAM, R.W. & BERMAN-RISS, E. van. (1990): Rates of predation on mammals by Gombe chimpanzees, 1972-1975. *Primates* 31, 157-170.

ZIHLMAN, A.(1996): Reconstructions reconcidered: chimpanzee models and human evolution. In: W.C. McGrew, L.F.Marchant & T. Nishida (eds), *Great ape societies*, 293-304. Cambridge.

Biological Equipment, Environment and Survival in the Australopithecine World

Phillip V. Tobias FRS

Abstract

Of the several species of Australopithecus, neither the brain, represented by endocranial casts, nor the masticatory apparatus - teeth, jaws, facial supporting structures - seem to account for the extraordinary success of the Australopithecinae. Their endocasts scarcely distinguish most members of the subfamily from chimpanzees and gorillas. The dentition lacks the large, canine teeth of the extant apes: if anything, the dental apparatus of the apemen is at a disadvantage and of less survival value, than those of the apes. Two other lines of investigation provide a possible key to the puzzle. These are furnished by the postcranial bones, and, secondly, the nature of the environment in which the apemen lived.

Recent studies have shown that Australopithecus retained many primitive or plesiomorphous traits which it shared with living apes. Some features of the shoulder girdle suggest overarm support activities, as in an animal adapted for arboreal activities. Secondly, and thirdly, the hip-joint and the knee-joint of some australopithecines are decidedly ape-like. Fourthly, the first articulated partial foot of a Sterkfontein Australopithecus has a divergent, mobile great toe and seems to be an apish prehensile foot. The spinal column, also, is markedly different from that of humans and shows some apish resemblances.

This striking array of features shared with apes betrays a creature which would be more at home in a well-wooded terrain than on open savanna. Yet, for decades, the prevailing paradigm on the environment in which Australopithecus emerged with erect bipedalism, was a savanna hypothesis. This view goes back to Dart's 1925 article in which he announced to the world the discovery of the type specimen of Australopithecus africanus. Taung being on the edge of the Kalahari Desert, Dart inferred that it had lived in the open plains. The features of the postcranial skeleton which were later to be revealed would clearly not have been advantageous in such a terrain.

However, there has been a change of paradigm. New evidence shows that the floral and faunal fossils contemporary with the apemen bespeak a forested or wooded ecology, which in places was interspersed with savanna. It seems thus that the survival of the apemen depended on that part of their biological equipment which adapted them for partial arboreal life, whilst other features equipped them for bipedal gait on earth as well as on branches of trees.

Nevertheless, although small-brained australopithecine bipeds survived for over 3 million years, if we judge by the palaeodemograhpy of the deposits, survival was not easy for the australopithecines. Of 63 specimens of Australopithecus africanus to which individual ages can be assigned, 35 % represent individuals which died before they reached anatomical maturity. Of 119 such specimens Australopithecus (Paranthropus) robustus, 60.5 % were from anatomically immature creatures. These variably high figures are related to the prevailing climate and to the degree of encephalisation among each group of early hominids.

Introduction

There are four concepts implicit in the title of this article. They are „the australopithecine world", „biological equipment", „environment" and „survival". The term, australopithecine world, has temporal, geographical and systematic (taxonomy) connotations. I shall be referring to the time period within which the australopithecines are known to have lived. The most ancient australopithecine has been assigned to the species *Australopithecus anamensis* (M.G. Leakey et al., 1995) and the oldest examples were recovered at Kanapoi, Kenya, in 1996 and dated between 4.17 and 4.12 Myr (M.G. Leakey & Ward, 1997). The latest surviving australopithecines are those classified as *A. (Paranthropus) robustus* and *A. (Paranthropus) boisei,* the latest signs of which in the fossil record are dated between 1.5 and 1.2 Myr. Thus, the genus *Australopithecus* has a duration of approximately 3 Myr, from about 4.2 to about 1.2 Myr. In arriving at these datings, I am excluding the somewhat older remains recovered at Aramis in Ethiopia and which have been dated to 4.4 Myr. Originally they were assigned to a new species in the genus *Australopithecus,* namely *A. ramidus* (White et al., 1994), but subsequently, on the recovery of more complete remains, this new species was placed in a new genus, *Ardipithecus ramidus* (White et al., 1995), although insufficient evidence has been published so far to enable other scientists to evaluate this new attribution. The proposed change of generic status has even led some colleagues to question whether the new finds from Aramis are hominids (in the classical sense - or hominins, on the basis of one of the new classifications of the Hominoidea). For purposes of this analysis, I have retained the robust and hyper-robust small-brained hominids within the genus *Australopithecus,* albeit as members of a distinct sub-genus, *(Paranthropus)* (Figures 1-3).

As defined, the genus *Australopithecus* has been found only in various parts of the African continent - namely, South Africa, Malawi, Tanzania, Kenya, Ethiopia and Chad. Although claims have been made at various times for the presence of australopithecine remains in Java and in China, these claims have not gained any major support and I omit them from this discussion (Figure 4).

Survival in the australopithecine world

While the genus *Australopithecus* spans some 3.0 Myr, on the evidence available at present, none of the species within that genus has a temporal duration longer than 1.0 to 1.5 Myr. Thus, *A. boisei* has a span of about 1.2 Myr and *A. africanus* of about 1.0 Myr [if, for purposes of this analysis, we attribute the Little Foot bones, Stw 573, to *A. africanus,* on which point Clarke and Tobias (1995) did not commit themselves, and if the foot is correctly assigned a dating of close to 3.5 Myr (Tobias & Clarke, 1996).]

Although the small-brained australopithecine bipeds survived for at least 3 Myr, the palaeodemography of the deposits suggests that the survival of individuals was not easy. A study of the demographic profiles portrayed by samples of fossil hominids is useful in this regard.

It is well known that the living and recent populations of humankind are subject to marked variations in demographic profiles (Acsádi & Nemeskéri, 1970; Franz & Winkler, 1936; Nougier, 1954; Ferembach, 1962). A high frequency of juvenile deaths and a lower mean expectation of life are found in some human populations: Such a demographic pattern is a „high risk" one. In other populations, there are a lower incidence of juvenile deaths and a higher mean life-expectancy: This is a „low risk" demographic pattern. When one studies the statistics of living populations, one finds that many factors may play a part in determining whether any population shows a high risk or a low risk demographic pattern. Among these factors may be mentioned the spectra of morbidity and of mortality, the state of nutrition, and the social, cultural and medical environment. Previous studies on fossil hominids have indicated also that, among ancient populations, the environment in which they lived was relevant to the demographic pattern of the population (Tobias, 1991, 1996).

When one considers populations of fossil hominids represented for the period from the late Pliocene to the Middle Pleistocene, four species or taxa are represented by a sufficient number of

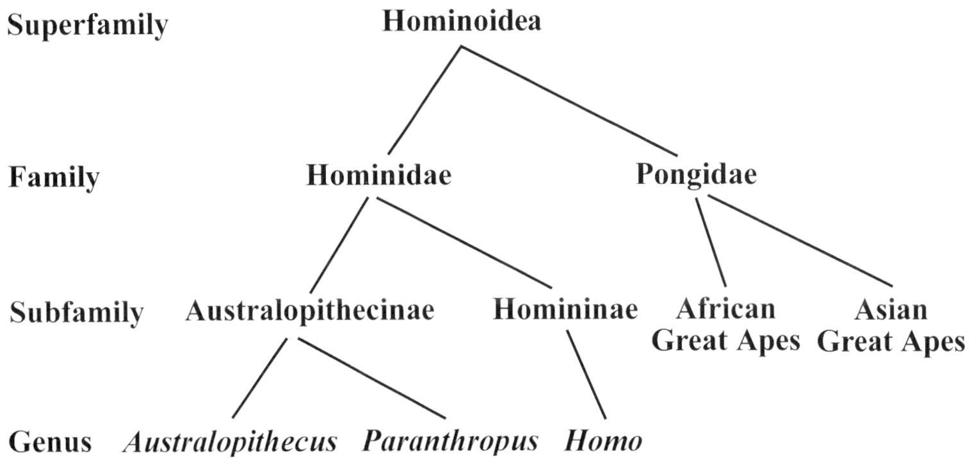

Figure 1 - A classification of hominoids that was in vogue from about the middlle of the 20th century. At this time, before molecular evolution had left its mark on palaeo-anthropology, the African great apes were still classified with the Asian great apes as pongids or members of the Pongidae. The Australopithecinae had been accepted as hominids, not pongids as G.G. Simpson in 1945 had classified them.

specimens to enable a palaeodemographic analysis to be made. These are *A. africanus* from Taung, Sterkfontein and Makapansgat in South Africa; *A. (Paranthropus) robustus* from Swartkrans and Kromdraai in South Africa; *H. habilis* from Olduvai in Tanzania; and *H. erectus pekinensis* from Zhoukoudian in China. It is likely that, when the data are made available and are collated, there may be sufficient specimens available for palaeodemographic analyses to be made of *A. boisei* from East Africa and of *A. afarensis* from Tanzania and Ethiopia (if the remains that have been assigned to this species do all belong to a single taxon, which some investigators doubt).

Both Mann (1975) and Tobias (1968, 1974, 1991) have made studies on the palaeodemography of the australopithecines. These two authors used slightly different sample sizes for *A. africanus* and *A. robustus*. For the present analysis, reliance is placed upon the data for the larger sample in each case: Thus Tobias's figures for 63 *A. africanus* specimens and Mann's for 119 specimens of *A. robustus* are used. For each fossil specimen which is age-determinable, an individual age or age-category is assigned, based on modern human standards. Whilst Mann for the most part assigned individual ages at death, Tobias grouped specimens into various categories. Ultimately, all specimens were placed in one of two categories, „Anatomically Immature" (which is equivalent in modern human terms to approximately 20 years and younger) and „Anatomically Mature" (approximately over 20 years). Another system of grouping that was used was into two groups, Younger than Childbearing Age and Of Childbearing Age. As one does not need wisdom teeth to procreate, the cut-off point between these two groups was much younger than for the prior categorisation.

It has proved instructive not only to compare the data for the two australopithecine taxa with one another, but also to compare both of these two sets of data with corresponding values for two early *Homo* taxa, namely 22 specimens of Olduvai *H. habilis* and 22 specimens of „Peking Man" *(H. erectus pekinensis)*. The results are conveyed in Table 1.

Overall, a large percentage of early hominid specimens fall into the category „Anatomically Immature". Thus, of 226 specimens pooled from all four taxa, no fewer than 126 age-determinable

Series	Anatomically Immature	Anatomically Mature	References
A. africanus (n = 63)	35	65	After Tobias (1968, 1974)
A. robustus (n = 119)	60,5	39,5	Mann (1975)
H. habilis / Olduvai (n = 22)	73	27	After Tobias (1991)
H. erectus pekinensis (n = 22)	73	27	Weidenreich (1943)

Table 1 - Percentage distributions of ages at death of anatomically immature (approximately 20 years and younger) and mature (approximately over 20 years) age-determinable specimens of various hominid series.

specimens or 56 % fall into the Anatomically Immature class. However. the percentages of immature individuals vary appreciably from a minimum value of 35 % in *A. africanus,* to much higher values in the other three taxa (60.5-73.0 %), the highest value (73 %) being attained in the two taxa of *Homo.* These results may seem to be paradoxical when one considers the geological dating of the four sets of specimens. *A. africanus* lived from about 3.5 to 2.5 Myr ago and is thus the earliest of the four taxa listed in Table 1. *A. robustus* was more recent in dating, the available samples spanning the period from about 1.8 to about 1.5 Myr ago. The dating for *H. habilis* was about 2.3 to 1.6 Myr, and that for *H. erectus pekinensis* about 0.5 Myr.

These figures negate any preconceived notion that the evolution of the Plio-Pleistocene hominids might have been accompanied by an improvement in the demographic pattern from high risk to low risk. On the contrary, the available evidence suggests that the evolution of the hominids from the late Pliocene to the earlier and Middle Pleistocene was apparently accompanied by a diminution in the chances of survival to adulthood (i.e. anatomical maturity).

These percentages of the hominid individuals who are represented by the respective series of fossils and who died before or after the attainment of anatomical maturity may cautiously be taken as evidence of the average expectation of life in each taxon; we are then led to the surprising and provocative conclusion that the larger-brained *H. habilis* and *H. erectus* populations had shorter average life spans than the small-brained *A. africanus.*

Before we seek the possible causes of this apparently paradoxical situation, it is important to address the question whether the age distributions reported here apply not only to the samples, but to the living populations from which they were drawn.

On the representativeness of the fossil populations

Several investigators have assumed or taken for granted that the demographic composition of fossil samples reflects that of the living populations (e.g. Vallois, 1937, 1961; McKinley, 1971). This was queried by Mann (1975, p. 79) who stated emphatically: „... it must be noted again that the age distributions presented here apply to the sample, and not to living australopithecine populations. Recent discussions...that have transferred these estimates to the population have little objective basis.“ Mann (1975) drew attention to the fact that among the rather large samples of the two australopithecine species he studied (totalling 174 age-determinable specimens in all), there was a complete lack of infants below the age of 2 years and few specimens to which he was able to assign an age of over 40 years. He cited a comment of Weiss (1973) that there is a tendency

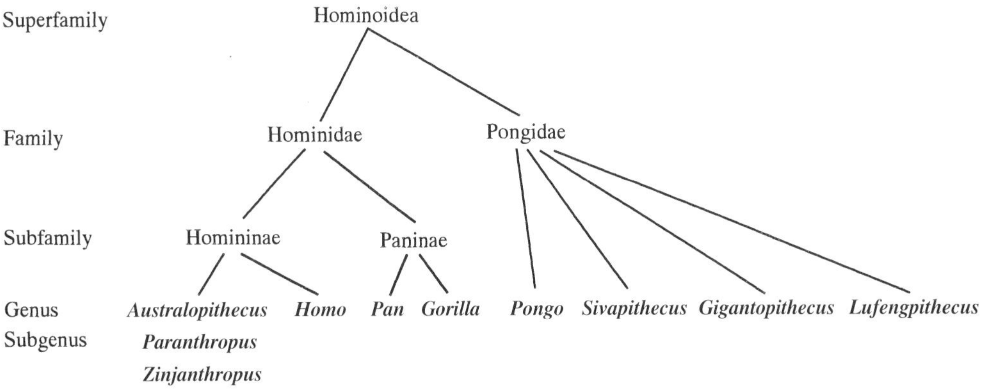

Figure 2 - *From the 1960s onwards, molecular biologists were producing evidence that separated the African great apes (chimpanzee and gorilla) from the Asian pongids and, instead, grouped them within the hominids, albeit as a separate subfamily Paninae. The previously proposed genera,* Paranthropus *and* Zinjanthropus, *were lumped into* Australopithecus *as subgenera.*

in most skeletal aging studies for „this group" (? the older individuals) to be consistently under-represented.

It is possible that differing taphonomic agencies might have been responsible for the accumulation of hominid bones in various cave depositories. Such taphonomic differences might account for differences in the demographic composition of site-samples. For example, Berger and Clarke (1995) proposed that the small animal bones which predominate in the Taung fossil deposit, including the skull of the Taung hominid child, might have resulted from the activities of aerial raptors, such as the crowned eagle, in contradistinction with the Sterkfontein and Swartkrans deposits which have numerous larger mammal bones and where terrestrial predators were probably the main bone-accumulating agencies (Brain, 1981).

Thus, if the fossil samples from two sites differ in their age composition, any of the following conditions may be applicable:

(1) The taphonomic agencies responsible for the accumulation of the bones in the two depositories might have differed, in such a way as to favour younger individuals in one and older individuals in the other. Vrba (1996) draws attention to the possibility that changes in general taphonomic conditions may themselves be „climate-driven „ .

(2) The differences between the two assemblages may be real, and may reflect differences in the conditions of life when bones from the two groups were being accumulated in their respective caves.

(3) While the environment might have been similar in the case of the two samples, evolutionary differences between the two populations might have been such as to make differing demands on the food resources of the environment. It has previously been suggested that the degree of encephalisation of a population may be one such evolutionary variable greater encephalisation making heavier demands on the resource base, and lesser encephalisation having lighter nutritional requirements (Tobias, 1996, 1997).

(4) The extent and the nature of violent deaths at the hands - or teeth and talons - of predators might have introduced differences into the age composition of prey samples in fossil depositories. Acsádi and Nemeskéri (1970: 181) suggested that death by such violent agencies, as well as from accidents, played a definite part in the taphonomy of fossil deposits and probably was important

not only in the australopithecines but among later hominids of the genus *Homo.* If there were a preponderance of large-sized predators, the age distribution of hominid prey is not likely to have been especially affected and neither anatomically immature nor mature individuals would have been especially favoured. On the other hand, if smaller predators prevailed, it might be expected that smaller - that is, younger - hominid subjects would have been favoured. As against this factor, predators that were small in body size but hunted in packs are more likely to have been equally successful against hominid adult and juveniles. In spite of Brain's claims that the agency of predators was the critical factor in the deaths of Plio-Pleistocene creatures, Newman (1993), from a study of the hominid remains from Swartkrans, concluded that signs of death from the agency of predators were evident in only a small percentage of the hominid remains in that cave.

(5) The causes of death. in general, might have differed between two fossil assemblages in which differing age compositions were manifest. Acsádi and Nemeskéri (1970: 181) concluded, in their great treatise on human life span and mortality, that, in the mortality pattern of earlier hominids, communicable diseases were less important as causes of death than they were to become later in larger populations. Instead, they asserted that: „The decisive importance lay in injuries due to accidents and violence and in diseases resulting from the factors of the natural environment or their changes. Atrophies or accumulating pathological conditions determined by regression with age also counted. The causes mentioned in the first place endangered mainly the lives of children... Consequently the child mortality rate was so high that only few of them could reach adolescent age. But those who attained it had probably sufficient resistance to environmental influences, and acquired all the knowledge that was needed to maintain life up to old age. yet adult life was not devoid of the possibilities of injuries and disasters. . . Among disasters we make special mention of the reduction or exhaustion of sources of nutrition, the resulting famines, the possibility of deficiency and nutritional diseases in general. Such circumstances must have had tragic consequences for small isolates where no reserves were available if they were unable to create better conditions in due time.“ Acsádi and Nemeskéri developed this line of thinking especially for *H. erectus* and *H. sapiens neanderthalensis.* I showed that it was almost certainly applicable as well to earlier hominids, such as *A. africanus, A. robustus* and *H. habilis* (Tobias, 1991).

Any combination of the foregoing sets of factors might have applied within any one assemblage, or in a comparison between any two sets of remains.

In view of all of the above delineated variables, environmental, taphonomic and bio-evolutionary, it is difficult to assert that the demographic pattern shown by any fossil hominid assemblage faithfully and totally reflects the demographic pattern of that population when it was living. For one thing, while it is most likely that there would have been a high infant mortality, we have nevertheless no infants of 2 years and younger. Similarly, we have no old adults in the sense of over 40 years, by modern human standards (Mann, 1975). If infant remains had been located and identified in the fossil deposits, it would follow that the percentage of Anatomically Immature individuals would have been even greater for the various early hominid assemblages. On the other hand, the lack of elderly subjects from the sample may either be taphonomic (for some reason such as neglect of them by the predators), or, more likely, reflect a demographic reality - that the creatures were simply not living to as old an age as over 40 years. Vallois (1961) has shown that „there was no gerontocracy“ right up until relatively recent Upper Pleistocene and even Holocene human populations. Hence it may be concluded that the absence of the elderly from early hominid assemblages was probably a demographic reality, reflecting the situation in nature. On the other hand, the fragility of infantile remains militates against their accumulation and preservation: Thus, their absence from most early hominid depositories until relatively recent

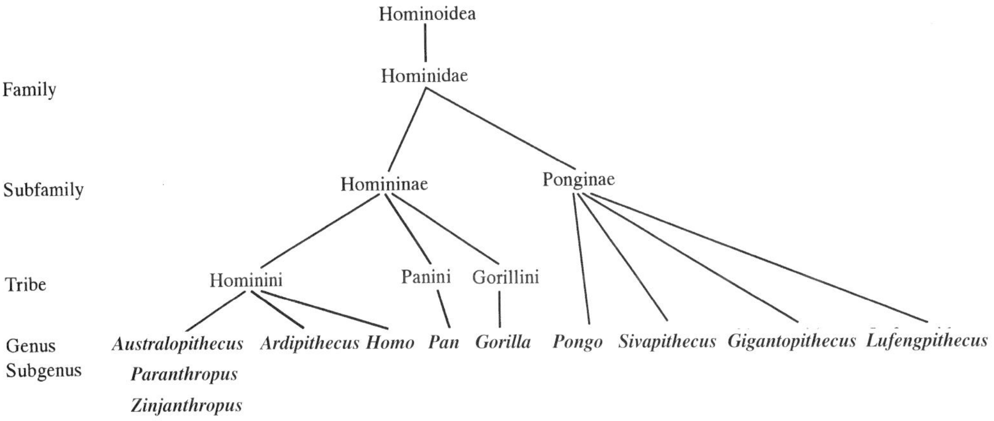

Figure 3 - From the late 1980s, an even closer association between the great apes and the other hominids was proposed and began to gain acceptance. The Asian great apes were admitted as a subfamily Ponginae in the family Hominidae, while subfamily Homininae comprised the African great apes and the human-like hominids, these being distinguished at the level of zoological tribes. On this classification, promoted by Colin Groves, what has been spoken of colloquially as „hominids" for over a century would be known as „hominins".

times may be taken as essentially taphonomic. With this in mind, we may accept that the general trend revealed by all four of our fossil hominid series, namely that they include an inordinately high percentage of Anatomically Immature individuals, is a conservative reflection of a general trend in the living populations from which those fossils were drawn.

We may next enquire what it means when we find that one fossil hominid asemblage has a far smaller percentage of Anatomically Immature individuals than others - as is the case with *A. africanus* (35 %), in contrast with the three later hominid taxa (60.5-73 %). To make a valid comparison, we must assume that roughly the same kinds and degrees of taphonomic biases have affected the various fossil series.

The African environment of the early hominid populations

For much of the 20th century, the savanna has been seen as the most probable ecological setting in which the early bipedal hominids lived. As early as Dart's (1925) article, in which he announced the discovery of the type specimen of *A. africanus* from Taung, South Africa, he claimed that „open veldt country" was the likely environment „for the production of man".

Dart (1925) contrasted the conditions of the Taung area with the luxuriant forests of the tropical belts: „For the production of man a different apprenticeship was needed to sharpen the wits and quicken the higher manifestations of intellect - a more open veldt country where competition was keener between swiftness and stealth, and where adroitness of thinking and movement played a preponderating role in the preservation of the species. Darwin has said, 'no country in the world abounds in a greater degree with dangerous beasts than Southern Africa', and, in my opinion, Southern Africa, by providing a vast open country with occasional wooded belts and a relative scarcity of water, together with a fierce and bitter mammalian competition, furnished a laboratory such as was essential to this penultimate phase of human evolution."

In making this assertion, Dart was assuming that „climatic conditions appear to have fluctuated little since Cretaceous times". Thus, for him, the relatively, arid conditions of the present-day savanna of southern Africa would have applied as well at the time of *Australopithecus.*

Broom (1933: 141), from his study of the hominid-associated fauna of Taung, supported this view: „From the remains of the animals associated with this ape *(A. africanus)* we can safely infer that the rainfall was then, as now, scanty, and that there were no forests in that region, only grassy and bushy plains from which the hills and krantzes [precipitous walls of rock] arose."

The paradigm of a savanna-bound early hominid prevailed almost up to the present. As late as 1985, Y. Coppens, E.S. Vrba and the present author pinned their faith on the savanna-bound life of the earliest hominids. At the Conference to mark the 60th anniversary of the announcement of the Taung discovery in 1985, Vrba (1985a: 199) claimed, „... the Family of Man was probably a 'founder member' of the African savanna biota.". According to Coppens (1985: 21): „The era of *Australopithecus africanus* is that of the wooded savanna and high grasses with water everywhere. That of *Australopithecus boisei* and *Australopithecus robustus* - due to climatic changes in both East and South Africa - is the age of the open savanna and prairie, sometimes steppe and even semi-desert, traversed by water sources of a reduced flow. These regions were bordered, from place to place, by the green of the gallery-forest."

The present author was no stranger to this concept. My chapter in *„Homo: Journey to the origins of man's history"* (Peretto, 1985) was entitled, „The conquest of the savannah and the attaining of erect bipedalism" (Tobias, 1985a). Speaking of the dramatic changes of climate, flora and fauna that beset Africa in the late Miocene, I reasoned: „The spread of lighter woodland and savannah and the retreat of the margins of the primaeval forests could well have created conditions in which the tendency to uprightness and bipedalism was favoured. The ability to run across the high grass cover of the savannah, perhaps from one woodland-girt stream to another, might have held advantages for those apes which could 'walk tall'. Uprightness gave its possessors a chance to see over the tall grass and to watch out for predatory enemies like the lions and sabre-toothed big cats. Seemingly it was under just such a set of conditions that the Hominidae made their appearance upon the face of the earth."

In short, the challenge for the emergence of the hominids could well have emanated from the striking environmental modifications of the late Miocene. These circumstances should not be seen as confined to cooling and its direct effects on the group with which we are primarily concerned, namely the last common ancestors of the hominids and the African great apes (Tobias,1985b). Rather, we have to deal with a much more complex skeinwork of causal mechanisms. Behind the cooling were global trends (Boaz & Burckle, 1984), Antarctic events (Flohn, 1984) and intra-African upheavals in the form of a somewhat earlier, major continental uplift (Partridge, 1985). Each of these, in turn, affected Africa secondarily in diverse ways, such as through faunal changes, indigenous and by immigration; through geomorphic responses such as the modification of drainage systems and the creation and demise of lakes; through deforestation and the spread of more open, less protective floral conditions; and through other forms of ecological diversification. With new immigrant animals, freshly-evolved endemic (or autochtonic) species, spreading Savannah, and prevailing cooler, more arid conditions, environmental challenges to survival were more exacting.

„From this African late Miocene maelstrom, one of the continent's small-brained, little specialised ape populations came forth as tottering bipeds with emancipated hands and fingers: they were the founding fathers - and mothers - of the family Hominidae, genus *Australopithecus.* Audaciously, they slipped the stagnating shackles and the cosseting constraints of their former forest retreat.

Figure 4 - *The most important sites of early (Plio-Pleistocene) hominid discoveries in Africa. The sites indicated are deposits or formations from which there have emerged specimens of Australopithecus, Ardipithecus and early Homo. All of the East and North African sites shown lie between the Tropics of Cancer and Capricorn. The South African sites lie in the temperate zone, just south of the Tropic of Capricorn.*

With curiosity, enterprise and opportunism, they ventured on to the plains of the high, inland plateau of Africa - and in time they conquered the Savannah." (Tobias 1985a: 43-44).

These statements by Vrba, Coppens and Tobias represent some of the latest pronouncements in favour of the savanna hypothesis in its varying shades of meaning. Since the middle 1980s, however, there has been a dramatic change and a shift of the paradigm.

From South and East Africa has come evidence of the Pliocene australopithecine hominids having lived in wetter, more wooded and even forested conditions. The first such evidence was forthcoming from Makapansgat in the Northern Province of South Africa (Rayner et al., 1993) and subsequent evidence came to light at Sterkfontein Member 4. Both fauna and flora provided the indications. Thus, the bovid fauna comprised browsers, as well as some grazers. At Sterkfontein, we have the presence of *Dichapetalum,* a liana or vine of a kind which in Africa today hangs from tall trees: This was found and identified by Bamford (1994) from the same stratum, Member 4, as that which has proved to be the world's richest source of fossils of *A. africanus.* These South African remains relate to deposits containing *A. africanus* (3.1 - 2.6 Myr). In a stable carbon isotope study of the habitat and diet of *A. africanus* at Makapansgat, Sponheimer and Lee-Thorp (1997) have found relatively depleted isotopic signatures throughout the foodweb at Makapansgat, which are consistent with the interpretations of a woodland habitat. Older deposits from East Africa, such as Kanapoi and Allia Bay (4.2-3.9 Myr) and Aramis (4.4 Myr), also bespeak wooded conditions, though variable and interspersed with patches of rather open grassland and of forest (Wolde-Gabriel et al., 1994; M.G. Leakey et al., 1995).

Apart from the faunal, floral and carbon isotope evidence, an indirect line of evidence has been furnished by recent studies on the foot-bones of early hominids. Stw 573, „Little Foot" from Member 2, Sterkfontein, shows morphological features which Clarke and Tobias (1995) have interpreted as human-like in some respects and ape-like and arboreally-adapted in other respects. Similarly, Deloison (1997), in a study of foot-bones from Hadar, regarded as *A. afarensis,* and of the footprints at Laetoli, Tanzania, has found anatomical evidence consistent with bipedal and occasional arboreal locomotion. Both the French and South African investigators consider that these early hominids were confirmed bipeds, but spent part of each day in the trees. Their conclusion poses the question: Were there trees large enough and with branches of adequate strength to sustain the body mass of the australopithecines? The new South and East African evidence strongly indicates that wooded and even forested conditions were available to the australopithecines. Under the old savanna model, arboreal activities by the australopithecines would not have been conceivable - or at the most would have been limited to some trees in strips of gallery forest along river courses. Thus, the morphological evidence of the feet, to which should be added Berger's (1994) evidence that the shoulder of *A. africanus* was consistent with overarm supportive activities, is in keeping with the new paradigm of a well wooded habitat for the early australopithecines.

The African environment of the robust australopithecines

It is now clear that critical changes occurred in the climate and the physical geography of Africa, between the time of *A. africanus* (about 3.5 to 2.6 Myr) and that of the later, at least partly synchronic hominids, *A. robustus* (1.8 - 1.5 Myr) and *H. habilis* (2.3 - 1.6 Myr). *A. africanus* lived at a time when Africa was more mesic, with more wooded and forested conditions. Even the temperate zone where *A. africanus* lived in South Africa, several degrees south of the Tropic of Capricorn, was wetter, more forested and frankly more tropical than it is today (Rayner et al., 1993; Bamford, 1994; Sponheimer & Lee-Thorp, 1997).

Later, from about 2.5 Myr, about the time when the *A. africanus* lineage was giving way to the line or lines of the robust australopithecines in South Africa and the co-existing line or lines of early *Homo*, there is evidence of a wave of tectonic uplift of the African continent, with cooling and desiccation, retreat of the wet forest, opening of the woodland and spread of the savanna (Shackleton & Kennett, 1975; Vrba, 1985a, 1985b; Partridge, 1985; Partridge et al., 1991).

With these changes, environmental challenges to survival were far more demanding. Not only were the climate and physical geography changing, but the fossil record reveals that conspicuous changes were occurring in the flora and the non-hominid fauna of Africa. It was a time of vigorous biological evolution and extinctions, conditions which, when added to the physical worsening, must have enhanced the competitive element in the struggle for survival.

In other words, *A. africanus* lived in more temperate times, *A. robustus* and *H. habilis* in more arid and more challenging times. We have already seen that the sample of *A. africanus* included a

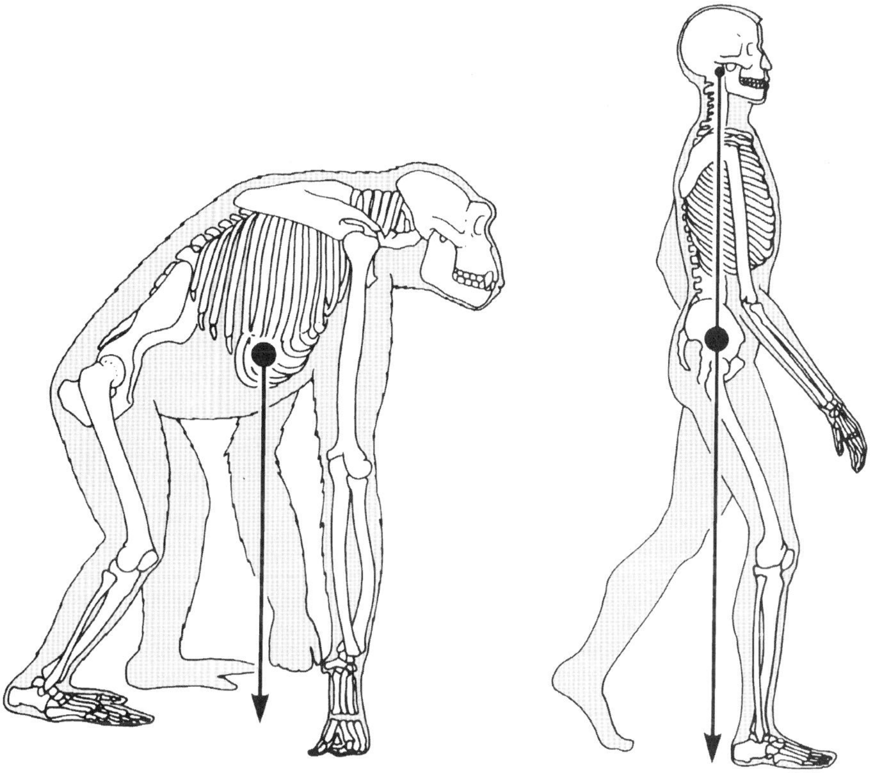

Figure 5 - The modern gorilla (left), with its oblique quadrupedal stance and gait, may be contrasted with upright standing and bipedally walking man. In the ape, the axis of the body mass (or the line of the „centre of gravity") falls between the fore-limbs and the hind-limbs. In the modern human, the weight-line passes from the occipital condyles on the basicranium, close to the spinal column, through the hipjoints on either side and so to the tripodial feet. Although various forms of bipedalism are encountered in Primates, the peculiar, modern human form of bipedalism (the striding gait) is based upon a series of distinctive anatomical adjustments affecting many parts of the skeleton from the cranial base to the feet. Some critical components of this suite of adjustments were acquired early in the process of hominisation and the acquisition of some form of bipedalism is regarded as one of the earliest criteria for the recognition of hominids (or hominins on one of the newest classifications).

larger percentage of individuals who were Anatomically Mature and earlier analyses had shown that the same sample was characterised by a larger percentage of individuals who had survived to probable procreative age (Tobias, 1974, 1991); whereas the samples of *A. robustus* and *H. habilis* were characterised by a smaller percentage (less than half) of individuals who were Anatomically Mature.

In an earlier study of the dependence of demographic patterns on prevailing environmental conditions (Tobias, 1991), I hypothesised that, when conditions were tolerably genial, or more mesic, the factors precipitating juvenile deaths would have been less in evidence, or less compelling, and a larger percentage of individuals would be expected to have survived to procreative age and even to adulthood. Contrariwise, when conditions were unfavourable, subject to cooling and aridification, the struggle to survive would have been fiercer, food shortages more debilitating, and competition between species, especially those with a similar resource base, more exacting; a larger percentage of individuals would be expected to have succumbed, as juveniles, to the ravages of the deteriorating and more competitive environment.

The only data we have bearing on the palaeodemography of the early hominids are the data on fossil hominid samples. We may test these two hypotheses against the age composition of the samples, considered against the background of the palaeo-ecological conditions pertaining to the various samples. We find that the sample collected from a population which was living under favourable conditions (*A. africanus*) is indeed marked by a smaller incidence of juveniles, while the samples of populations which were living under more unfavourable circumstances (*A. robustus* and *H. habilis*) are particularised by larger numbers of juvenile specimens. These results do not refute the hypotheses and, they provide indirect evidence that the demography of the fossil samples (save for the absent infant skeletons) reflects the demography of the living populations from which the samples were drawn.

In other words, the „lower risk" demographic profile manifested by *A. africanus* depended upon the more moderate conditions prevailing in Africa at the time, rather than on taphonomic biases. The „higher risk" demographic profiles shown by *A. robustus* and by *H. habilis* were determined by the deteriorating climatic, physical and biotic environment of the terminal Pliocene and early Pleistocene, rather than by taphonomic distortions.

This dependence of demographic patterns on the relative favourableness or adversity of the environment has a further interesting and testable corollary. If the conditions of life to which a population is subject undergo change, it would be a reasonable deduction that the demographic profiles would change, in accordance with the direction of the environmental change. Hence, if conditions changed from unfavourable to more propitious, a secular trend from a „high risk" to a „low risk" demographic pattern should take place. An environmental shift from more mesic to more malign circumstances, on the other hand, would be expected to lead from a „low risk" *A. africanus*-like pattern, to a „high risk" *H. habilis*-like demographic mode.

There is a substantial body of opinion that the taxon *A. africanus* is most likely to have been ancestral to the early *Homo* species, *H. habilis*. If for purposes of this discussion we accept this, it would follow that, round about 2.5 Myr, late-surviving members of *A. africanus* evolved into *H. habilis*, probably by a cladogenetic splitting. This happened about the time that the severe deterioration of Africa's climate and milieu took place. Under these straitened circumstances, perhaps through an evolutionary bottle-neck, *A. africanus* with its low risk demographic pattern evolved into the two derivative lineages, *A. robustus* and *H. habilis*, with their high risk patterns.

My hypothesis that changes in the demographic pattern depend upon changes in the prevailing environment is thus confirmed by the palaeo-ecological data of the fossil record.

Biological equipment and survival

The australopithecines survived over the length and much of the breadth of Africa for some 2 Myr before the late-Pliocene deterioration of the continent. Were there any features in their make-up that facilitated their survival? Neither the brain alone, as represented by endocranial casts, nor the masticatory apparatus - teeth, jaws, the temporomandibular joints and facial supporting structures - seems to provide for what appears to be an extraordinary success story. In size and in detailed morphology, the endocasts of australopithecines were scarcely distinguishable from the endocasts, or brains, of chimpanzees and gorillas. The dentition lacked the large fang-like canine teeth of the extant great apes: if anything, the dental apparatus of the australopithecines was at a potential disadvantage and of lesser survival value, than those of the extant apes.

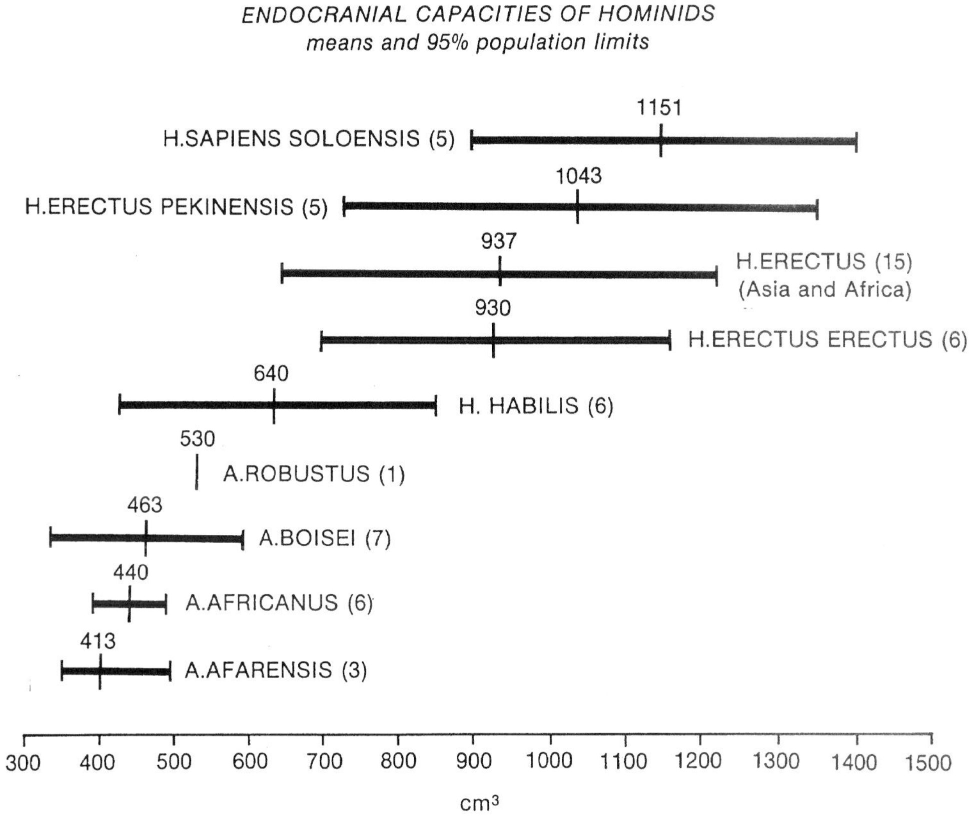

Figure 6 - *The mean values and 95 % population limits of the endocranial capacities of a number of hominid (hominin) species and subspecies.*
'A.' = Australopithecus, 'H' = Homo.
The numbers in parantheses are sample sizes.

On the other hand, the postcranial skeleton of the australopithecines provides evidence of an evolutionary novelty which might have been the key to their success. The critical feature was the evidence of bipedalism and erectness of posture. Most important of the bony adaptations to uprightness are the features of the pelvis and, in particular, its broader and shorter ilium. In addition, there were numerous other indications of adaptation to uprightness and bipedalism, from the position craniospinal joints on the basicranium to the feet (Figure 5). Indeed, if there is any broad consensus about the early emergence of the hominids, it is that skeletal adaptations to erect posture and bipedalism were probably the earliest morpho-functional criteria of the emergence of the hominid family (or hominins). To this complex of features should be added some adaptations of the canine field of the dentition, notably the reduced size of the maxillary and mandibular canine teeth and the replacement of a caniniform deciduous mandibular first molar by a more molarised or premolariform dm_1.

It may be noted in passing that, although the studies of the australopithecines over the first 50 or 60 years were preoccupied with the establishing of their hominid features, in the last decade many studies on, especially, newly found specimens of australopithecines have revealed a preponderance of ape-like and especially chimpanzee-like features which were present, and had presumably been retained, in the australopithecines. These range from the small absolute and relative size and the convolutional pattern of the brain (endocasts) to the divergent and mobile great toe of the foot. Many of these features vary among the different species of australopithecines that have been recognised and we may next enquire whether such inter-australopithecine differences were relevant to the varying demographic profiles revealed by samples of early hominids.

Increase in brain size as a challenge in the evolution of longevity

The general decline in the African climate after 2.6 Myr might have been causally related to the change from the low risk demographic profile of *A. africanus* to the high risk pattern of *A. robustus*. But, while both *A. robustus* and its approximate contemporary, *H. habilis,* show high risk demographic patterns, that of *H. habilis,* like that of *H. erectus,* betrays an appreciably higher risk pattern than that of *A. robustus*. Since these taxa shared an equally hostile environment, we need to look for some additional factor which might have contributed to one taxon's being placed at a greater risk than the other. Dr. S.U. Dani suggested to me that my data on brain-size might provide a key to this further set of differences in demographic patterns (Figure 6).

We may therefore feed into the equation the extra demands which increasing brain size or encephalisation might be expected to have made upon the body's nutritional reserves and the population's resource base. It may be postulated that, at a time of minimal encephalisation, a population's demands on the resource base would *ceteris paribus* have been minimal; during a phase of greater encephalisation, these demands would have been enhanced, whilst a population undergoing strong encephalisation would have been subject to appreciable exacerbation of its nutritional demands. We may test this hypothesis by reference to the three early hominid species we have been examining: *A. africanus* was only slightly encephalised as compared with chimpanzee; this would have made small extra demands on the resource base. Moreover, the resource base was presumably adequate at this mesic period. Only 35 % of individuals died in the 20 and under age category. *A. robustus* was moderately more encephalised than *A. africanus* (Coefficient of Cephalisation 42 % as against 36 % of adult modern human value); it lived in the cooler, drier and more demanding environment of Africa after 2.6 Myr; so survival was at a premium. The remains of 60 % of individuals fell into the 20 and under group. *H. habilis* was appreciably more encephalised than *A. africanus* and its Coefficient of Cephalisation had passed the 50 % mark in relation to the mean value in modern humans (Figure 7); it lived in the same cooler, more arid conditions as did *A. robustus,* but its greater encephalic nutritional demands placed it at a decided disadvantage in the struggle for survival. No fewer than 73 % of the age-determinable individuals were 20 years and under.

These figures appear to show that the differential longevity among these early hominid species had two principal determinants: One was the circumambient climate which could explain part of the differences, but could not distinguish between the two later synchrony species, the robust australopithecine and *H. habilis;* the other was the intensity of encephalising pressures to which the various taxa were exposed and this appears to elucidate the difference in demographic profile between the *A. robustus* and *H. habilis* populations.

It may be concluded that, while the increased metabolic burden of a larger brain would have been a restrictive handicap under any circumstances, its presence in the face of an adverse environment would be expected to have made survival even more difficult. It might even be expected to have lowered the average expectation of life, such as our demographic data for the early hominid samples have indicated. Not only did the burst of encephalisation in *H. habilis* take them by surprise, but it happened at the worst possible time!

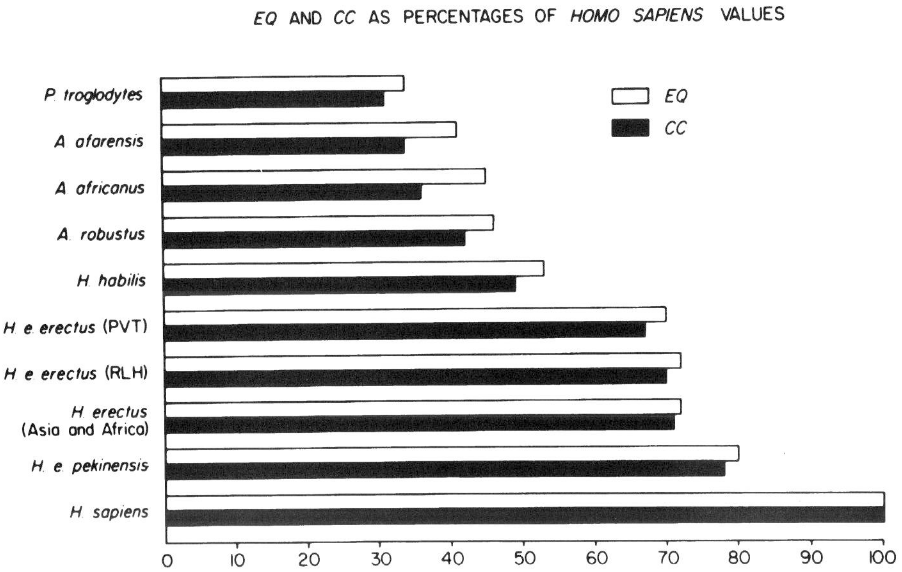

Figure 7 - *Mean values, in a selection of hominoid species and subspecies, of Jerison's Encephalisation Quotient (EQ) and Hemmer's Constant of Cephalisation (CC). The value for each taxon is expressed as a percentage of modern human mean values.*
'P.' = Pan; 'A.' = Australopithecus; 'H.' = Homo.
For Homo erectus erectus of Java, two estimates are given: one marked 'RLH' is that of R.L. Holloway; the other marked 'PVT' is that of P.V. Tobias.

On the other hand, this threat to survival and to longevity must have been more than outweighed by the selective advantages that greater encephalisation brought in its wake. The sharp change to a very high risk demographic profile went hand in hand with an enhancement of the selection pressures that favoured encephalisation.

Summary

The deterioration of Africa's ancient climate would have had a seriously deleterious effect on the chances of children surviving to procreative age, let alone to Anatomical Maturity. As a result, mean life expectancy would be inferred to have dropped and this would have been especially marked in the presence of nutrition-demanding high encephalisation. The maximum life span has probably doubled from the earliest hominids to modern *H. sapiens,* but the curve of change has not been smooth and uninterrupted. There seems to have been a decided drop in life expectancy with the emergence of *H. habilis* and this was sustained through *H. erectus.* It has been inferred from the above analysis that the critical determinants of this drop were twofold: First, the dramatic and surprising increase in brain size and encephalisation with its increased metabolic demands; secondly, the untimely occurrence of this brain burst when the climate of Africa showed, after 2.6 Myr, a marked deterioration. Only after the time of *H. erectus,* in whom encephalisation attained 70-80 % of that of modern humans, did human mean life expectancy increase. In the later stages of hominid evolution, from the middle to the later Pleistocene, increases in life expectancy and in brain size appear to have accompanied one another, to reach their zenith in the last hundred thousand years.

Acknowledgements

I should like to convey my warm appreciation to the Institut für Anthropologie, Medizinische Fakultät der Humboldt-Universität zu Berlin, the Deutsche Forschungsgemeinschaft and to Dr. Herbert Ullrich, for the generous invitation to participate in the Symposium. I am most grateful to Mrs. Heather White for preparing the typescript of my manuscript and to Mr. Terry Boraine for help with the illustrations.

References

ACSÁDI, G. & NEMESKÉRI, J. (1970): *History of human life span and mortality.* Budapest.

BAMFORD, M. (1994): In: R.J. Clarke & H.E. White (eds.), *28th Annual Report of the Palaeo-anthropology Research Unit,* 11-12. Johannesburg (University of the Witwatersrand).

BERGER, L.R. (1994): *Functional morphology of the hominoid shoulder. Past and present.* PhD. Thesis. Department of Anatomy and Human Biology, University of the Witwatersrand, Johannesburg.

BERGER, L.R. & CLARKE, R.J. (1995): Bird of prey involvement in the collection of the Taung child fauna. *Journal of Human Evolution* 29, 275-299.

BOAZ, N.T. & BURCKLE, L.H. (1984): Paleoclimatic framework for African hominid evolution. In: J.C. Vogel (ed.), *Late Cainozoic palaeoclimates of the southern hemisphere.* Rotterdam, Boston.

BRAIN, C.K. (1981): *The hunters or the hunted? An introduction to African cave taphonomy.* Chicago, London.

BROOM, R. (1933): *The coming of man: Was it accident or design?* London.

CLARKE, R.J. & TOBIAS, P.V. (1995): Sterkfontein Member 2 foot bones of the oldest South African hominid. *Science* 269, 521 -524.

COPPENS, Y. (1985): The problem of the origin of man. In: C. Peretto (ed.), *Homo: Journey to the origins of man's history,* 15-25. Venezia (Cataloghi Marsilio).

DART, R.A. (1925): *Australopithecus africanus:* The man-ape of South Africa. *Nature* 115, 195-199.

DELOISON, Y.M.L. (1997): The foot bones from Hadar, Ethiopia and the Laetoli, Tanzania footprints. Locomotion of *A. afarensis. American Journal of Physical Anthropology,* Supplement 24, 101.

FEREMBACH, D. (1962): *La necropole epipaléolithique de Taforalt. Étude des squelettes humains.* Marco Centre National de la Recherche Scientifique et de la Mission Universitaire et Culturelle Francaise au Maroc Oriental.

FLOHN, H. (1984): Climate evolution in the southern hemisphere and the equatorial region during the Late Cenozoic. In: J.C. Vogel (ed.), *Late Cainozoic palaeoclimates of the southern hemisphere.* Rotterdam, Boston.

FRANZ, L. & WINKLER, W. (1936): Die Sterblichkeit in der frühen Bronzezeit Niederösterreichs. *Zeitschrift für Rassenkunde* 4, 157-163.

LEAKEY, M.G., FEIBEL, C.S., McDOUGALL, I. & WALKER, A. (1995): New four-million-year-old hominid species from Kanapoi and Allia Bay, Kenya. *Nature* 376, 565-571.

LEAKEY, M.G. & WARD, C.V. (1997): Morphology of the *Australopithecus anamensis* lower deciduous first molar and capitate. *Journal of Human Evolution* 32, A10-11.

MANN, A. (1975): Paleodemographic aspects of the South African australopithecines. *University of Pennsylvania, Publications in Anthropology* 1, 1-171.

McKINLEY, K.R. (1971): Survivorship in gracile and robust australopithecines: A demographic comparison and a proposed birth model. *American Journal of Physical Anthropology* 34, 417-426.

NEWMAN, R. (1993): The incidence of damage marks on Swartkrans fossil bones from the 1979-1986 excavations. In: C.K. Brain (ed.), *Swartkrans: A Cave's chronicle of early man,* 217-228. Pretoria (Transvaal Museum Monograph No. 8).

NOUGIER, L.R. (1954): Essai sur le peuplement préhistoriques de la France. *Population* 9, 241-274.

PARTRIDGE, T.C. (1985): The palaeoclimatic significance of Cainozoic terrestrial stratigraphic evidence from South Africa: A review. *South African Journal of Science* 81, 245-247.

PARTRIDGE, T.C., TOBIAS, P.V. & HUGHES, A.R. (1991): Paleoecologie et affinites entre les australopithecines d'Afrique du Sud: Nouvelles données de Sterkfontein et Taung. *L'Anthropologie* 95, 363-378.

PERETTO, C. (ed.) (1985): *Homo: Journey to the origins of man's history.* Venezia (Cataloghi Marsilio).

RAYNER, R.J., MOON, B.P. & MASTERS, J.C. (1993): The Makapansgat australopithecine environment. *Journal of Human Evolution* 24, 219-231.

SHACKLETON, N.J. & KENNETT, J.P. (1975). Paleotemperature history of the Cenozoic and the initiation of Antarctic glaciation: Oxygen and carbon isotope analyses in DSDP sites 277, 279 and 281. In: J.P. Kennett, R.E. Houtz et al. (eds.), *Initial reports of the Deep Sea Drilling Project* 29, 743-755.

SPONHEIMER, M. & LEE-THORP, J. (1997): A stable carbon isotope study of environment and hominid diet at Makapansgat Limeworks, South Africa. *Journal of Human Evolution* 32, A21-22.

TOBIAS, P.V. (1968): The age of death among the australopithecines. *The Anthropologist* (University of Delhi, India), Special Volume, 23-28.

TOBIAS, P.V. (1974): Aspects of pathology and death among early hominids. *The Leech* (Johannesburg) 44, 119-124.

TOBIAS, P.V. (1985 a): The conquest of the savannah and the attaining of erect bipedalism. In: C. Peretto (ed.), *Homo: Journey to the origins of man's history,* 36-45. Venezia (Cataloghi Marsilio).

TOBIAS, P.V. (1985 b): Ten climacteric events in hominid evolution. *South African Journal of Science 81,* 271-272.

TOBIAS, P.V. (1991): The age of death of the Olduvai *Homo habilis* population and the dependence of demographic patterns on prevailing environmental conditions. In: H. Theon, J. Bourgeois, F. Vermeulen, P. Crombee & K. Verlaeckt (eds.), *Studia Archaeologica: Liber Amicorum, Jacques Nenquin,* 57-65. Gent.

TOBIAS, P.V. (1996): Changes in hominid demographic patterns, environment and encephalisation. In: L.S. Sidhu & S.P. Singh (eds.), *Human biology - Global developments,* 261-274. Ludhiana (India).

TOBIAS, P.V. (1997). Evolution of brain size, morphological restructuring and longevity in early hominids. In: S.U. Dani, A. Hori & G.F. Walter (eds.), *Principles of neural aging,* 153-174. Amsterdam.

TOBIAS, P.V. & CLARKE, R.J. (1996): Response to „Faunal evidence and Sterkfontein Member 2 Foot" by J.K. McKee. *Science,* 271, 1301-1302.

VALLOIS, H.V. (1937): La durée de la vie chez l'homme fossile. *Comptes Rend. Seances Academie des Science* 204, 60-62.

VALLOIS, H.V. (1961): The social life of early man: The evidence of skeletons. In: S.L. Washburn (ed.), *Social life of early man,* 214-235. Chicago.

VRBA, E.S. (1985 a): Early hominids in southern Africa: Updated observations on chronological and ecological background. In: P.V. Tobias (ed.), *Hominid evolution: Past, present and future,* 192-200. New York.

VRBA, E.S. (1985 b): Ecological and adaptive changes associated with early hominid evolution. In: E. Delson (ed.), *Ancestors: The hard evidence,* 63-71. New York.

VRBA, E.S. (1996): On the connections between paleoclimate and evolution. In: E.S. Vrba, G.H. Denton, T.C. Partridge & L.H. Burckle (eds.), *Paleoclimate and evolution, with emphasis on human origins,* 24-45. New Haven, London.

WEIDENREICH, F. (1943): The skull of *Sinanthropus pekinensis:* A comparative study on a primitive hominid skull. *Paleontologia Sinica* (Peking) 10, 1-484.

WEISS, K.M. (1973): *Demographic models for anthropology.* Memoirs of the Society for American Archaeology, 27. *American Antiquity,* 38.

WHITE, T.D., SUWA, G. & ASFAW, B. (1994): *Australopithecus ramidus,* a new species of early hominid from Aramis, Ethiopia. *Nature* 371, 306-312.

WHITE, T.D., SUWA, G. & ASFAW, B. (1995) Corrigendum: *Australopithecus ramidus,* a new species of early hominid from Aramis, Ethiopia. *Nature* 375, 88.

WOLDE-GABRIEL, G., WHITE, T.D., SUWA, G., RENNE, P., de HEINZELIN, J., HART, W.K. & HEIKEN, G. (1994): Ecological and temporal placement of early Pliocene hominids at Aramis, Ethiopia. *Nature* 371, 330-333.

Climate Change and Survival Strategies of Early *Homo* and *Paranthropus* in the Malawi Rift

Friedemann Schrenk & Timothy G. Bromage

Abstract

Remains of earliest Homo and Paranthropus have been recovered from two contemporaneous sites (Uraha and Malema) in the "Hominid Corridor" in Northern Malawi (Chiwondo Beds). Faunal dating suggests an age of 2.5 - 2.3 Ma for both hominids. The associated bovid and suid faunas show an overlap of a small amount of southern African with a great amount of eastern African faunal elements. Biogeographic variation in southeastern Africa may be linked to habitat change occurring due to climate change after 2.8 Ma. The hominid taxa present in the Malawi Rift, Homo rudolfensis and Paranthropus boisei, both represent eastern African endemic elements originating from the eastern African australopithecine stem group. Their different life styles and survival strategies are linked to climate and habitat change as well as biogeographic variation and dispersal. A Panafrican scenario of early hominid evolution is developed on this basis. This scenario suggests a single origin for the Paranthropus lineage but separate origins for Homo rudolfensis and Homo habilis from Australopithecus anamensis/afarensis and Australopithecus africanus ancestors respectively.

Introduction

The ancient lake deposits of northern Malawi (Figure 1) were first described by Dixey (1927). He recognised Chiwondo Beds and Chitimwe Beds which he attributed to the Pliocene and Pleistocene, respectively. Determination of the Pliocene age of the Chiwondo Beds was based on fossil vertebrates collected by him. The first major contributions to our knowledge of the palaeoecology and palaeobiogeography of the Chiwondo Beds of the northern Malawi Rift were the result of pioneering interdisciplinary explorations into the prehistory of sub-Saharan Africa by J. Desmond Clark and collegues (Clark et al., 1966, 1970; Clark & Haynes, 1970; Coryndon, 1966; Mawby, 1970). Clark led teams of researchers to conduct archaeological, palaeontological and geological work on the Chiwondo Beds in 1963, 1965 and 1967. The deposits were re-investigated by Z.M. Kaufulu and T.D. White in 1979 (Kaufulu et al., 1981). Realising the specific potential for contributions to an understanding of the Plio-Pleistocene palaeobiogeography of continental Africa, the Hominid Corridor Research Project (HCRP) began a longterm study in 1983, focusing on the regional geology and the geographic role of south-east Africa on the origin and dispersion of Plio-Pleistocene faunas including early hominids (Bromage et al., 1985; Bromage & Schrenk, 1987, Schrenk et al., 1993; Bromage & Schrenk, 1995a).

The sedimentology and stratigraphy was described by Betzler & Ring (1995), the tectonic development of the Malawi Rift was analyzed by Ring & Betzler (1992, 1995). Lake beds and fluviatile deposits of palaeolake Malawi have been subdivided into five depositional units which are bounded by unconformities reflecting sedimentary breaks (Figure 2). Prior to the formation of a major perennial lake, the depositional area was characterised by extensive fluviatile deposits mainly formed by reworked material of Jurassic red beds. Formation of a major lake after this episode is reflected in widespread lacustrine silts and sands with interbedded, mollusc-rich layers. Tectonic reorganisation of the rift finally led to localised uplift resulting in an angular unconformity and exposure of the Plio-Pleistocene deposits. The fossil localities are concentrated in the area around Malema, referred to as the "northern localities", and around Uraha near Chilumba, referred to as the "southern localities" (Figure 1).

Mammalian fauna of the Chiwondo-Beds

The sample of Plio-Pleistocene vertebrates recovered from the Chiwondo Beds in the Karonga District of Northern Malawi, includes ca. 900 identifiable fragmented remains of fish, turtles, crocodiles and large mammals and a large number of invertebrate specimens from 131 fossil localities (Bromage, Schrenk & Juwayeyi, 1995). Whereas the terrestrial vertebrate fossils are nearly always found in interbedded soft sand, the fresh water molluscs are contained in consolidated beds of carbonate cemented sandstone. Molluscan shell beds crop out as benches up to several meters thick and several hundred meters wide. Severe taphonomic processes due perhaps to the reworking of bones in beach environments and recent erosional processes led to extensive destruction of fossil material. Micro-vertebrates and carnivores are veritably unrepresented in the bone assemblage.

Three quarters of the vertebrate sample are of mammalian origin (Figure 3). Thereof ca. 150 postcranial elements have been collected, mostly high density elements - mainly fragments of long bones, vertebrae and foot bones of macro mammalian taxa. The remainder are 600 dental or cranial specimens which are taxonomically identifiable mostly to genus level. Of these the highest proportion representation show the bovids (43.3 %), followed by equids (16.5 %), hippopotamids (14.4 %), suids(11.0 %), proboscidians (8.1 %), giraffides (2.9 %), primates (1.8 %), rhinoceratids (1.0 %) and camelids (0.3 %) (Schrenk et al., 1995).

Age of the Chiwondo Beds

An estimate of the age of the Chiwondo Beds must still rely on correlation with radiometrically dated biostratigraphic units in eastern Africa. In so doing, the first definitive study by Kaufulu et al. (1981) provided dates of approximately 2.4-2.6 Ma (based on suids) and 2.5-3.0 Ma (based on bovids) for the more northerly sites and 4.0 Ma or older for the southerly Uraha sites (based on suids). Most of the Late Pliocene fossil localities now are attributed to stratigraphic Unit 3A while older Middle Pliocene localities occur in stratigraphic Unit 2 in the north and the south. An age determination of older than 4.0 Ma to about 1.5 Ma is based on the faunal assemblages described by Bromage, Schrenk & Juwayeyi (1995).

Figure 3 presents a list of mammalian taxa from the Chiwondo Beds presently recognized by us and which provide broad age constraints for the deposits. These taxa are listed together with their age ranges. The age ranges are based on radiometric age determinations of volcanic tuffs and correlations between the Koobi Fora and Shungura Formations described by Brown et al. (1985) and Brown & Feibel (1986). These ages confirm a biochronology consistent with the stratigraphy of the Chiwondo Beds as shown in Figure 2.

Mammalian biogeography

Many of the recovered faunal taxa confirm a certain degree of faunal continuity between the eastern and southern African Late Pliocene. However, our research continues also to define a unique community assemblage. Kaufulu et al. (1981) characterized the faunal discontinuity of this assemblage by noting that the predominant bovid species at Makapansgat Member 3 - *Redunca darti*, *Phenacotragus vanhoepeni* and *Makapania broomi* - do not occur in eastern African Late Pliocene sites. Furthermore, contemporaneous assemblages at Omo and Hadar, dominated by

Plio-Pleistocene Localities

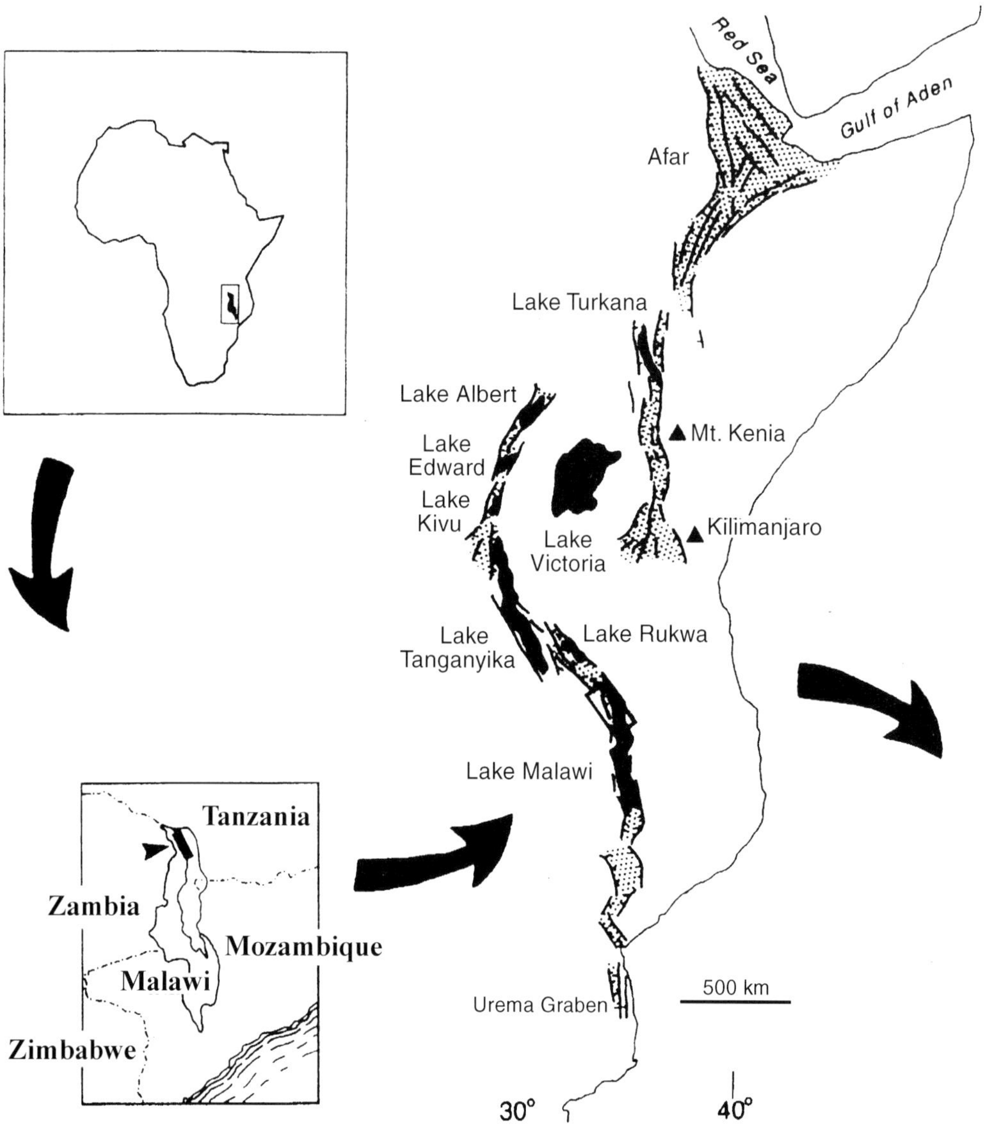

Aepyceros, reduncine genera such as *Kobus* and *Menelikia*, and tragelaphines such as *Tragelaphus nakuae,* are absent from southern African sites.

The Malawi Rift faunas are partitioned between three geographically-based groups that give clues as regards palaeoecological domains and patterns of faunal dispersion between eastern and southern Africa (Figure 3). 14 taxa taxa occur in eastern as well as southern Africa, 17 taxa are of eastern African origin and only four taxa are of southern African provenance.

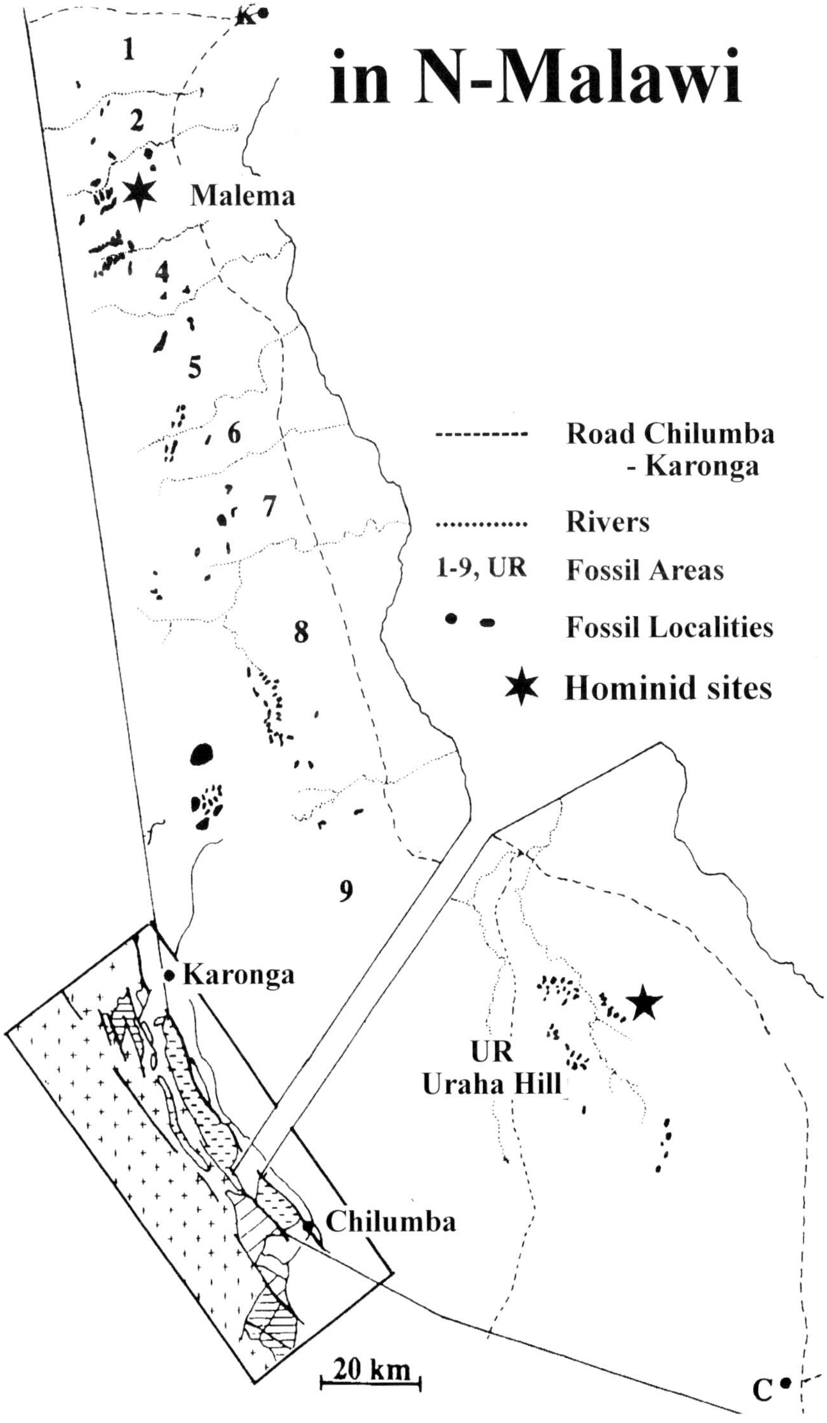

Figure 1 - Position and geology of the Malawi Rift.

The largest faunal group comprises eastern African endemics reflecting a regional African Rift Valley corridor comprised of rift "chain" montane communities and associated lowlands and grasslands shared across eastern to southeast Africa (Kingdon, 1989). Indeed, the North-South oriented structural evolution of the northern part of the African Rift Valley has been implicated as a "continuous rift corridor" related to rodent evolution and migration during the Pliocene (Denys et al., 1986; Denys, 1996). Macromammal distributions were certainly broader, however, and include East-West dispersions of savanna faunas across the Rift Valley at least as far as the Western Branch (cf. Boaz et al., 1992), while its dense forest-adapted faunas were probably distributed even further west into the Zaire Basin woodlands. Thus the impression of a 'corridor' *per se* is less one of passage along discrete faunal habitat types than it is a broad melange of interconnected habitats stretching from northeastern Africa to the Zambezian Ecozone and the temperate zone transition in southeast Africa.

The southern African endemic group is very small (Figure 2). It transpires that this is most probably due to the equatorward dispersion of faunas keeping pace with drifting vegetation belts during the aridification of global climates ca. 2.5 Ma (southern endemics dispersing to eastern Africa diminish this category while contributing to the larger shared eastern plus southern faunal group; see Figure 3). How may these geographic groupings be interpreted? The answer to this question derives from the modern synthesis of biogeography and climate in evolutionary biology (Matthew, 1939). We note that continental position and climatic conditions are responsible for the tropical *versus* temperate zonation in Africa today (Griffiths, 1984; Oliver & Crowder, 1979). These conditions are our template against which Pliocene climate change may be interpreted and faunal distributions explained.

Situated as it is between climatic regimes, the southeast African region has most probably been the meeting point for endemic faunas from the north and south since the continents have been in their current positions. Eurybiomic taxa (ie. those taxa able to acquire their resources in more than one biome; cf. Vrba, 1987 b) will have been shared between eastern and southern Africa, but taxa restricted to more specific habitats might have transgressed this region only when ecological extremes could effect latitudinal shifts in their appropriate vegetation zones. Such a latitudinal shift likely began approximately 2.8 Ma during which time cooler and dryer conditions prevailed (Bonnefille, 1976, 1980, 1995; Prentice & Denton, 1988; Vrba et al., 1985, 1989; Vrba, 1995; deMenocal & Bloemedal, 1995; Dupont & Leroy, 1995). These conditions are hypothesized to result in the equatorward shift of grassland and woodland biomes accordant with the "Habitat Theory" of Vrba (1992).

The Habitat Theory derives from seven hypotheses relating ecological conditions to faunal evolution and biogeography. Necessary features of this theory are that terrestrial vegetational habitat specificity's are heritable and thus characteristic for clades, that terrestrial mammal biomes may be distinguished by their gross vegetational physiognomy, and that distribution movements (drift) occur in the context of physical environmental change.

In order to test this proposition we have summarized the data provided by Turner & Wood (1993a) on the origin and dispersion of bovid taxa around 2,5 Ma (Table 1). *Parmularius braini*, *Tragelaphus pricei* and *Syncerus* sp. first appeared around 3 Ma. Indeed, we note that most taxa originating in southern Africa dispersed toward the equator around 2.5 Ma when aridity was increasing and biomes where shrinking towards the equator. These taxa include both dispersing open/arid and closed/dry taxa, consistent with hypothetical North-South vegetation corridors penetrating through the Zambezian Ecozone when assuming 50-60 % of present day rainfall (Cooke, 1962; see also Zinderen Bakker, 1969).

Hominid-bearing Sediments of N-Malawi

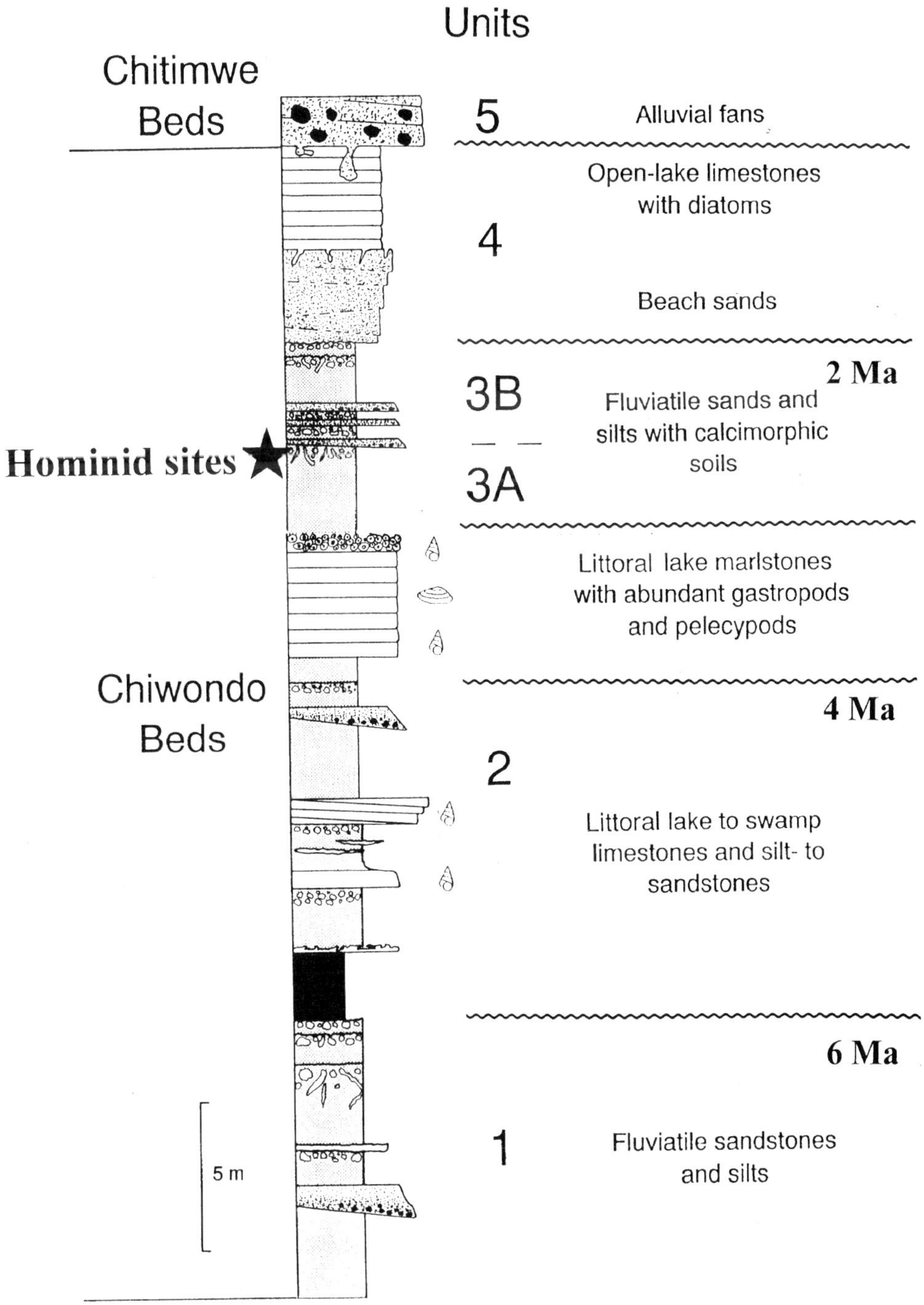

Figure 2 - Stratigraphy of the Plio-Pleistocene Chiwondo Beds.

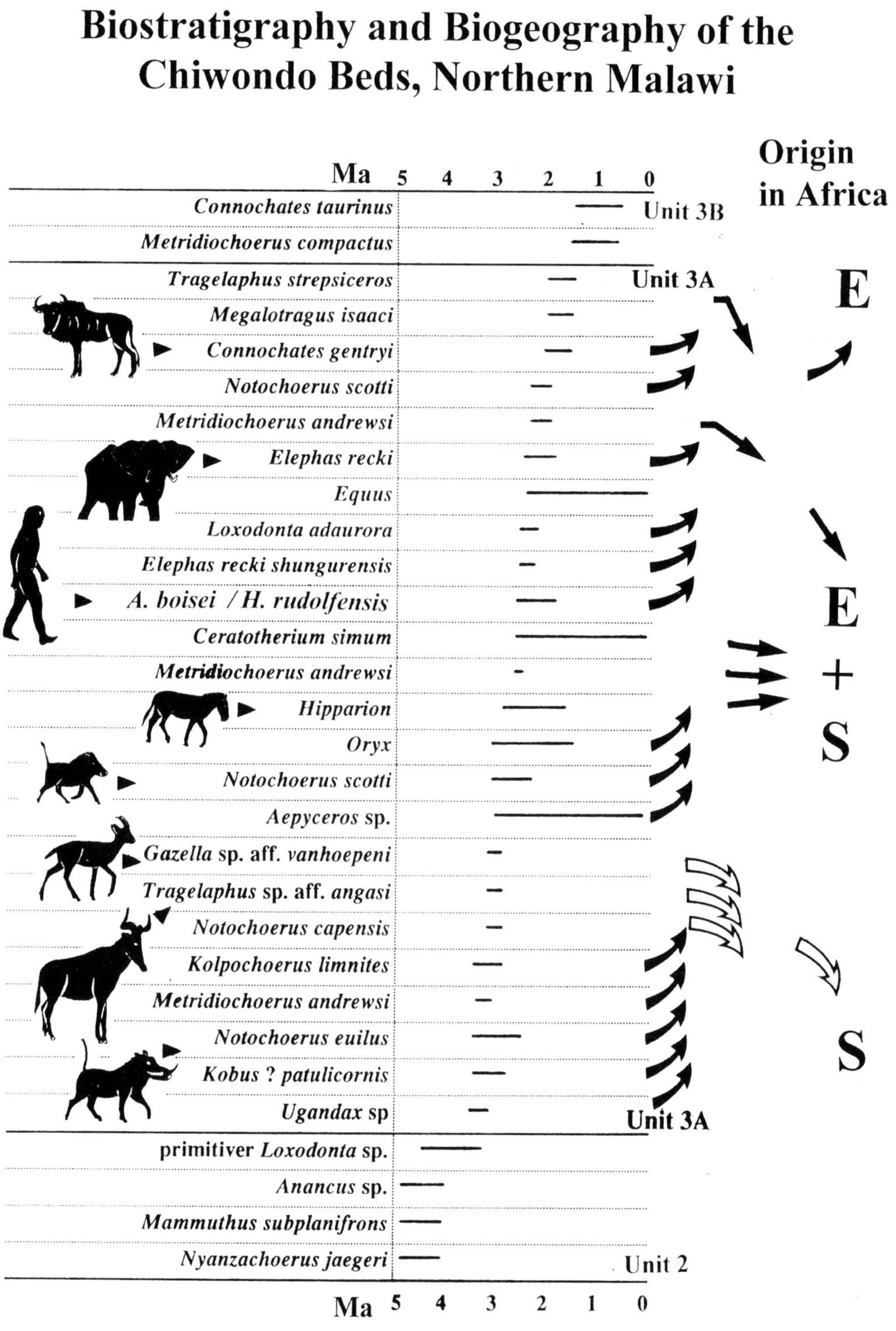

Figure 3 - Biostratigraphy and biogeography of the Chiwondo Beds.

We also note that even amongst this small group of bovids originating about 2.5 Ma there are two taxa dispersing to southern Africa at 2.0 Ma and younger (Figure 3). This may represent average environmental conditions returning to more warm and humid conditions during this time (cf. Shackleton et al., 1984). This suggests to us a model of biome expansion away from the equator sometime during the terminal Pliocene or onset of the Early Pleistocene. This is compatible with early hominid biogeography and interpretations of the origins and dispersions of *Homo* and *Paranthropus* (see below) as well as of *Australopithecus africanus* and *Homo habilis* (Bromage & Schrenk, 1995b).

Palaeoanthropology of the Malawi-Rift

During the 1991 and 1992 field seasons the Hominid Corridor Research Project has recovered a mandibular corpus, UR 501, containing third and fourth premolars and first and second molars in variable states of preservation (Figure 4) (Schrenk et al., 1993). Many absolute and relative measures respecting molar and premolar crown shape indices, some relative cusp areas, and enamel microanatomical features, as well as fissure patterns and crown morphology, are within the sample range of early *Homo*. UR 501 corresponds closely to the subset of Late Pliocene fossils from east of Lake Turkana, Kenya, that have relatively large brains and robust jaws and teeth and that have been referred to *Homo rudolfensis* Alexeev 1986, by Wood (1992b), and to which we also referred UR 501 (Bromage, Schrenk & Zonneveld, 1995).

During the fieldseason 1996 the HCRP discovered a further hominid specimen, a maxillary fragment, RC 911, at the site Malema, approximately 50 km North of Uraha (Figure 5). The specimen RC 911 represents a part of the left processus alveolaris with M^1 and a fragmentary crown of the M^2. The fragment is broken mesially just in front of the M^1 alveolus and distally of the M^2 alveolus. Crown size, shape and micropattern on the occlusal surfaces together with the height of the alveolar process suggests a determination of RC 911 as *Paranthropus boisei* (Kullmer et al., in prep.)

Southeastern African hominid biogeography

The palaeobiogeographic significance of *Homo rudolfensis* and *Paranthropus boisei* of the Malawi Rift lay in its association with an assemblage dominated by eastern African endemic faunas (Figure 3). This biogeographic pattern is most probably a reflection of the equatorward dispersion of southern faunas, keeping pace with northward-drifting vegetation belts during the aridification of global climates ca. 2.5 Ma (Bonnefille, 1976, 1980, 1995; Prentice & Denton, 1988; Vrba et al. 1985, 1989; deMenocal & Bloemendal, 1995; Dupont & Leroy, 1995). It may be surmised that *equatorial* lineages faced less the problem of latitudinal shifts in biome zonation than habitat fragmentation. Equatorial lineages having a greater diversity of habitat choices, however, remained endemic to the tropical African ecological domain and experienced patterns of extinction and speciation. Thus the palaeobiogeographic implications of the Chiwondo Beds faunal assemblages may be that *Homo rudolfensis* arose during, and partly as a result of, the ca. 2.5 Ma climatic cooling event in eastern Africa and remained endemic there in the face of prevailing equatorward dispersion tendencies in other taxa according to the Habitat Theory. This would be a demonstration of their ecosensitivity and some degree of niche-dependence in this taxon.

Homo rudolfensis was endemic to eastern Africa and conformed to the palaeobiogeographic response to environmental change exhibited by some bovids (Bromage, Schrenk & Zonneveld, 1995). This reasoning applies also to the origins of other dento-facially robust eastern African endemics during the ascendancy of cooler and dryer times between ca. 2.5 and 2.0 Ma;

Mandible from Uraha, Malawi (UR 501)

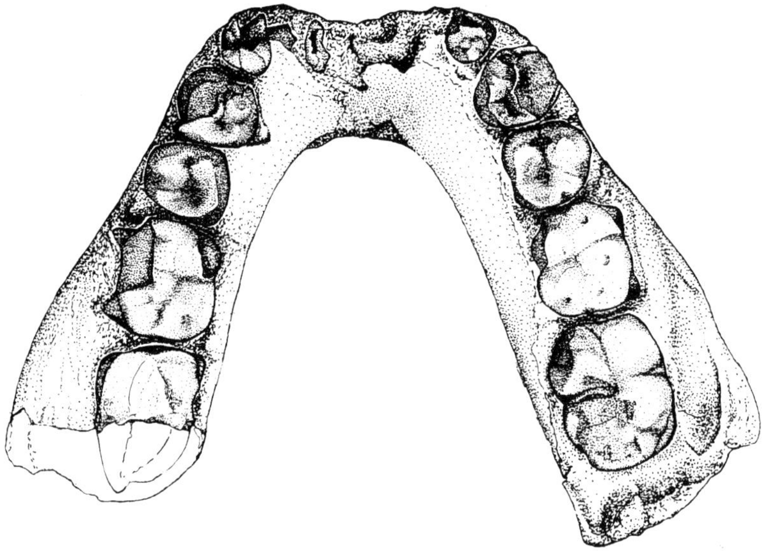

Figure 4 - Mandible UR 501 (Homo rudolfensis) from Uraha.

Paranthropus (= *Australopithecus*) *aethiopicus* (e.g. KNM-WT 17000; Walker et al., 1986) and *Paranthropus boisei*. It follows that the return to warmer and more humid conditions by 2.0 Ma would possibly have instituted or facilitated faunal dispersions away from the equator (bovid dispersions between 2.0 and 1.5 Ma did just this; see Table 1). Consistent with this view is the later age of first appearance of *Paranthropus* in southern Africa between 2.0 and 1.5 Ma at Swartkrans. We may then characterize the evolution of *Paranthropus robustus/crassidens* in southern Africa as including a moderation of the hypermasticatory trend of *Paranthropus boisei* with a diminution of the face and post-canine dentition.

The same arguments may be applied to the origin and dispersion of *Homo habilis*. If one accepts the assignment of early *Homo* specimens from Sterkfontein Member 5 (ST5) (Hughes & Tobias, 1977; Clarke, 1985) and Swarkrans member 1 (SK1) to *Homo habilis* (Brain et al., 1988), then we may attribute this taxon to the shared southern-plus-eastern African group. The earlier Late Pliocene presence of *Homo habilis* at Koobi Fora might suggest its origins in eastern Africa, increasing its distributional area into southern Africa sometime between 1.8 and 1.5 Ma during SK1 and ST5 times (Vrba, 1982).

The pattern to emerge from our palaeobiogeographic interpretations is that early hominids consistently arise in the eastern African ecological domain (which includes northern reaches of the Zambezian Ecozone in the Malawi Rift). During favorable periods, some early hominids dispersed southward beyond the Zambezian Ecozone, evolved there (perhaps due to relative isolation and/or differences associated with its temperate ecology), and became endemic there. This

Maxillary Fragment from Malema, Malawi (RC 911)

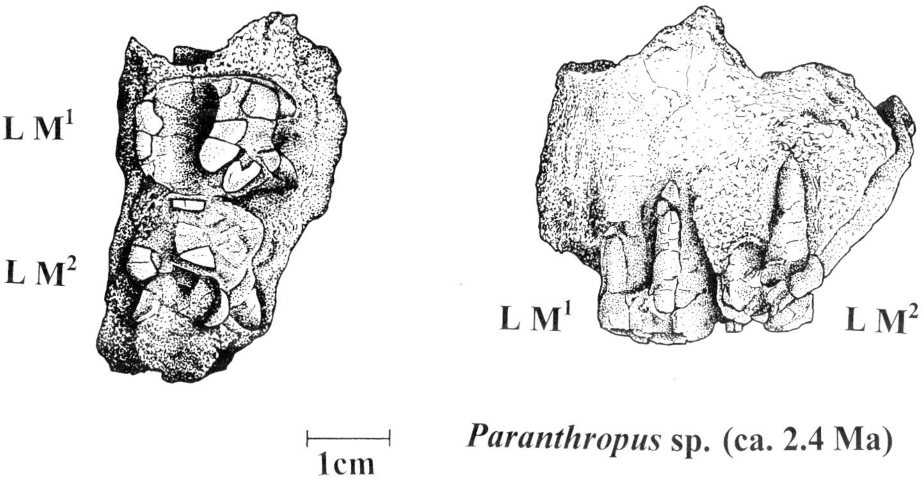

Figure 5 - Maxillary fragment RC 911 (Paranthropus sp.) from Malema.

spatio-temporal pattern is consistent with expectations that hominid diversity should be greater in tropical regions and, further, that speciation may at times be the consequence of dispersal and increasing distributional area particularly with an accompanying niche specialization (Foley, 1991).

Hominid evolutionary and survival strategies: A scenario

The geographic distribution of ancestral populations of the African ape and hominid clades was broadly constrained to tropical Africa and its lowland forests. This was a stable association until Middle-to-Late Miocene global cooling precipitated marked environmental change (cf. Brain, 1981). These climatic conditions in eastern Africa, exacerbated by the regional environmental consequences of rifting (cf. Pickford, 1991), led to a shrinking back of the forests. Indeed, significantly more open grasslands featured in eastern Africa by 8-9 Ma (Cerling et al., 1991). At the same time, Central African lowland forest refuges provided the allopatric conditions for differentiating hominoid subspecies and species (cf. Colyn et al., 1991, for a discussion of Late Pleistocene refuges).

Distribution drift of the tropical forests left more open wooded savanna in its wake (Andrews, 1981) and engendered increased habitat diversity (cf. Retallack et al., 1990). As climatic conditions became their most extreme during the terminal Miocene, some populations of arboreal hominoids found themselves peripheralized to the east along the lush riparian habitats meandering across the rain shadows of the developing African Rift Valley. Vicariance of their habitat prompted evolutionary adjustments (i.e. speciation and extinction, pulses) amongst hominoids and other mammal groups (i.e. bovids; Vrba, 1985, 1987a, 1995). The open woodland provided new niches for hominoids, but they occurred in resource patches separated by spans of non-arboreal terrain. Amongst the peripheralized populations natural selection favored those with behavioral-morphological locomotor repertoires capable of more successfully traversing this ground and taking advantage of discontinuously distributed ground-based resources. Included amongst these repertoires was the advent of bipedalism - a more energetically efficient, albeit more upright, locomotory strategy at walking speeds (Rodman & McHenry, 1980; see also Jablonski & Chaplin, 1993, regarding a hypothesized role for the selective value of bipedal displays under these new resource conditions). This uprightness had also increased thermoregulatory advantages under conditions of intense incident solar and ground level radiation in open environments (Wheeler, 1991).

As the terminal Miocene dry seasons became longer and more pronounced, ground-based foods resources included a proportionately higher diversity of underground perennial forbs while the rainy seasons would continue to supply the fruits, seeds, and pods of the woody species (cf. Peters et al., 1984). For over a million years the anatomical and behavioral complexes associated with a combined arboreal and terrestrial niche were compromised during the Mio-Pliocene to meet the demands of this mixed habitat and niche.

Australopithecus anamensis and *Australopithecus afarensis* arose in this eastern African tropical ecological domain. Their behavioural inclination was to maintain a connection to and to remain near the borders between broad riparian habitats and open woodlands during the ascendancy of more warm and humid times. Over short geological time scales, this was typically a local, non-dispersing tendency, but by approximately 4.0 Ma *Australopithecus anamensis* or *Australopithecus afarensis* successfully dispersed along the northern periphery of the tropical forests in order to give rise to *Australopithecus bahrelgazali*, known from Tchad (Brunet et al., 1996). Over longer time frames this included dispersal through the riparian "corridor" connecting eastern and southern Africa (Bromage, Schrenk & Juwayeyi, 1994) permitting population dispersal into southern Africa by 3.0 Ma. This dispersion was facilitated by the birth of palaeolake Malawi and the penetration of the Malawi Rift and its associated rift habitats into temperate Africa and the riparian corridor (Bromage, Schrenk & Juwayeyi, 1994).

This dispersing unit maintained habitat specificity's to forested environments (cf. Rayner et al., 1993) though in more environmentally temperate climes and in relative geographical isolation at the extreme distal edge of its distribution. The dispersion along changing latitudinal circumstances covaried with its transformation into, first, a geographic variant and, subsequently, into *Australopithecus africanus,* joining ranks with other southern African endemic faunas. Thus *Australopithecus afarensis* was essentially an eastern African endemic and it follows that no typical representatives are likely to be recovered from southern African deposits older than 3.0 Ma (e.g. within sub-Sterkfontein Member 4 levels).

By approximately 2.8 Ma the initiation of cooler and dryer conditions prevailed upon the African landscape, its vegetation, and its faunas, until climaxing ca. 2.5 Ma (Bonnefille, 1976, 1980, 1995; Prentice & Denton, 1988; Vrba et al., 1985, 1988; deMenocal & Bloemendal, 1995; Dupont & Leroy, 1995). During this time *Australopithecus afarensis* and *Australopithecus africanus* were each subject to unique palaeobiogeographic consequences of this global aridification, in accordance with the Habitat Theory of Vrba (1992). The rate of faunal turnover (including early hominids) in eastern Africa exceeded the southern African experience (Bromage, Schrenk & Juwayeyi, 1994) because of the relative density of "habitable" alternatives in the tropics (a consequence of greater diversity). For *Australopithecus afarensis,* then, the changing climate meant vicariance of its habitat and its distribution into more removed ecotonal riparian and closed lake margin environs. During the interim between ca. 2.8 Ma and 2.5 Ma these changing conditions engendered more extensive open habitats comprising more resistant arid-tolerant vegetation around the remaining relatively lush but narrowed "ribbons" of tree-lined riverine forest. The selective pressures of this

Habitat preference	**Origin** SA	EA	**Dispersion direction** S → E	E → S
Open / arid				
Alcelaphini				
Beatragus antiquus		EA		
Connochaetes taurinus	SA		**S → E**	
Connochaetes gentryi		EA		
Damaliscus sp.	SEA			
Parmularius braini	SA		**S → E**	
Parmularius		EA		
Hippotragini				
Oryx sp.		EA		
Antilopini				
Gazella praethomsoni		EA		
Antilope subtorta		EA		
Antidorcas recki	SA		**S → E**	
Antidorcas bondi	SA			
Closed / dry				
Tragelaphini				
Tragelaphus gaudryi		EA		
Tragelaphus		EA		E → S 2-1.5 Ma
Tragelaphus pricei	SA		**S → E**	
Aepycerotinae				
Aepyceros melampus		EA		E → S 0.5 Ma
Closed / wet				
Hippotragini				
Hippotragus equinus	SA		S → E 0.5 Ma	
Hippotragus gigas		EA		
Reduncini				
Kobus ellipsiprymnus		EA		E → S 2-1,5 Ma
Menelika lyrocera		EA		
Redunca arundinum	SA			
Bovini				
Syncerus acoelotus		EA		
Syncerus sp.		EA		E → S
Pelorovis turkanensis		EA		

Table 1 - *African fossil Bovidae first appearances approximately 2.5 Ma, their geographic origin, and their dispersion direction.*

Legend: Parmularius braini, Tragelaphus pricei, and Syncerus sp. have older origins (ca. 3.0 Ma) but dispersions around 2.5 Ma. Where the dispersion has ocurred at some date other than ca. 2.5 Ma, an approximate date has been given. Data from Turner & Wood (1993a).

habitat change resulted in the increased survival of more megadont varieties capable of feeding on tougher fruit and open woodland-open savanna food items. This was so for early hominid as well as numerous eastern and southern African large terrestrial vertebrate lineages ca. 2.5 Ma (Turner & Wood, 1993a). These pressures were likewise sufficient to result in the phyletic splitting of *Australopithecus afarensis* into *Paranthropus* and *Homo* lineages after 2.8 Ma (Vrba, 1988; Kimbel, 1995; Bobe, 1997).

The beginnings of the *Paranthropus* lineage, represented by *Paranthropus* (=*Australopithecus*) *aethiopicus* (e.g. Walker et al., 1986; Kimbel et al., 1988), maintained a reliance on fruiting resources on the riverine side of its ecotone, particularly during the dry season, but it was equally adept at grinding on the postcanine dentition those food items it required from more open habitats during more hospitable times of the year. Over time this lineage adapted itself more fully to the open habitat side of its ecotone, becoming more facially robust and megadont in order to efficiently process the tougher, more durable, vegetation of the savanna. This evolved taxon, *Paranthropus boisei,* probably never left its ancestral connection to the more closed features of its environment (cf. Shipman & Harris, 1988), however, as these habitats would continue to provide essential protection, sleeping localities, and some foods (it would be rather interesting if its enduring evolutionary history was the result of a biomic eclecticism in an otherwise so specialized a hominid).

The beginnings of the *Homo* lineage, also ca. 2.5 Ma, and represented by *Homo rudolfensis,* was an endorsement of its recency of common ancestry with *Australopithecus afarensis,* a distinction it shared with *Paranthropus aethiopicus* (Bromage, Schrenk & Zonneveld, 1994). By this token *Homo rudolfensis* then shared with *Paranthropus* a number of masticatory-related cranial and dental adaptations (Wood, 1992b, 1993) to the tougher fruit and open savanna food items sought by early hominids. It need not concern us whether these similarities between *Paranthropus* and *Homo rudolfensis* were actually shared in a temporally fleeting post-*Australopithecus afarensis* ancestor or not. The intimacy of their shared genetic heritage closely dictates their sharing, also, numerous canalized developmental pathways commanding similar responses to specific selection pressures. Thus many or all uniquely derived similarities between *Homo rudolfensis* and *Paranthropus* may be homoplasies, but this is *because* of their relative recency of ancestry. These similarities, particularly as they relate to the vegetative responses to aridifying climate change, embody our most impressive evidence to suggest that *Homo rudolfensis* remained somewhat conservative in its feeding habits and depended still, and most of all, on plant resources.

However, while *Paranthropus* was principally adapted by means of a robust masticatory system to its tough and abrasive diet, *Homo rudolfensis* exhibited an increased behavioral flexibility as its adaptation to climatic circumstances included a larger and more provoking, inquiring, and capable brain. This included a shift to proportionately less abrasive foodstuffs and more omnivorous habits (cf. Robinson, 1962).

In the Malawi Rift there is evidence for the earliest co-existence of *Homo* and *Paranthropus*. It may therefore be concluded that an alternative to the development of hyperrobust and megadont constructions existed, which was equally suitable for coping with increasingly tougher food items: this alternative was the beginning of material culture. The oldest stone tools with an age of about 2.5 Ma. are known from Ethiopia (Harris, 1986; Kimbel et al., 1996). Material culture ameliorated the effects of climate change to the degree that it enabled *Homo rudolfensis* to take advantage of other resources more efficiently than was ever possible before.

Paranthropus boisei and *Homo rudolfensis* remained endemic to tropical latitudes during this time (Bromage, Schrenk & Zonneveld, 1994). The eastern African tropical faunas, having habitable alternatives, remained within their biogeographical domain rather than brave the relative deterioration and paucity of habitats south of the African Rift Valley. It would not be for another half of a million years or so before *Paranthropus boisei,* at least, would disperse to more southern latitudes during an expansion of its biome sometime younger than 2.0 Ma.

The faunas of southern Africa were subject to a different set of environmental sequelae during the ca. 2.5 Ma cooling event. Waning of the forests and woodlands in deference to more open arid grasslands invigorated not only evolutionary adaptations to savanna life in tropical eastern Africa, but also resulted in the distribution drift northward of faunas tracking the equatorward shift of grassland and woodland biomes into eastern Africa from the south ca. 2.5 Ma (Bromage, Schrenk & Juwayeyi, 1994). Until this time the Zambezian Ecozone was an environmental wedge between the tropic and temperate zones, maintaining high degrees of southern and eastern African faunal endemism (cf. Klein, 1984; Sclater, 1896). However, the temperate zone ca. 2.5 Ma experienced more seasonal extremes, and many organisms unwittingly maintained their inherited preference for moderately seasonal climes and temperate vegetation physiognomy by moving northward with the shrinking of this biome toward the equator, effectively transgressing the Zambezian Ecozone. Amongst these migrants was *Australopithecus africanus* who, having been adapted to a modest temperate ecology, now found its suitable habitats shifted to the north toward the African Rift Valley.

During its tenure in southern Africa, *Australopithecus africanus* experienced the pressures of climate change related to increased seasonality. Some populations became peripheralized and adapted to the more open side of their woodland habitat. Indeed, selection favored masticatory trends similar to those of *Paranthropus* in eastern Africa (cf. White et al., 1981). Other populations, however, by means of a dispersing strategy not available to their contemporaries in the north, endeavored to maintain their habitat specificity's to the more closed side of their woodland habitat and so dispersed northward along the preserved riparian corridor. While dispersing toward the eastern African tropical domain, selection for increased behavioral flexibility was related to the habitat diversity of the tropics and the presence of other non-vegetative food resources available in their new region. This emerging taxon, *Homo habilis,* rapidly established itself as a categorical omnivore and found that it could buffer itself more resolutely from environmental changes. This enabled it to cross habitat boundaries more easily and also to advantage itself of more resources with its material culture. This ecogeographic perspective is consistent with one interpretation of polyphyly in *Homo* as it is currently beheld (Turner & Wood, 1993b).

By approximately 2.0 Ma Africa was rebounding from its relatively cool and dry climate to return to slightly more warm and humid conditions (cf. Shackleton et al., 1984; Bromage, Schrenk & Juwayeyi, 1994). A phase of biome expansion ensued that facilitated dispersions *away* from the equator, ending nearly one million years of relative endemism dominated by tropical equatorial speciations. *Paranthropus boisei* dispersed southward along re-established ecotonal habitats into southern Africa, varied there as a geographic variant under more temperate conditions, and evolved into *Paranthropus robustus/crassidens.* This view is consistent with a consideration of large mammal fauna biogeography and *Paranthropus* monophyly (Turner & Wood, 1993a). *Homo rudolfensis*, however, remained endemic to the eastern African tropical domain due to its preference for more open habitats around the rain shadows of the African Rift Valley.

Acknowledgements

We thank the Department of Antiquities, Lilongwe, Malawi, for giving us the permission to conduct fieldwork in the Karonga district. Figures 1-3 were prepared by Marisa Blume, Figures 4 & 5 by Claudia Schnubel. Research was funded by Deutsche Forschungsgemeinschaft (DFG), Bonn, and the National Geographic Society (NGS), Washington DC.

References

ALEXEEV, V.P. (1986): *The origin of the human race*. Moscow.

ANDREWS, P. (1981): Hominoid habitats of the Miocene. *Nature* 289, 749.

BETZLER, C. & RING, U. (1995): Sedimentology of the Malawi-Rift: Facies and stratigraphy of the Chiwondo-Beds, northern Malawi. *Journal of Human Evolution* 28, 23-36.

BOAZ, N.T., BERNOR, R.L., BROOKS, A.S., COOKE, H.B.S., DE HEINZELIN, J., DECHAMPS, R., DELSON, E., GENTRY, A.W., HARRIS, J.W.K., MEYLAN, P., PAVLAKIS, P.P., SANDERS, W.J., STEWART, K.M., VERNIERS, J., WILLIAMSON, P.G. & WINKLER, A.J. (1992): A new evaluation of the significance of the Late Neogene Lusso Beds, Upper Semliki Valley, Zaire. *Journal of Human Evolution* 22, 505-517.

BOBE, R. (1997): Pliocene environmental changes derived from faunal analysis of the Shungura Formation, Ethiopia. *Journal of Vertebrate Paleontology* 16, 23A.

BONNEFILLE, R. (1976): Palynological evidence for an important change in the vegetation of the Omo Basin between 2.5 and 2 million years ago. In: Y. Coppens, F.C. Howell, G.L.I. Isaac & R.E.F. Leakey (eds.), *Earliest man and environments in the Lake Rudolf Basin*, 421-431. Chicago.

BONNEFILLE, R. (1980): Vegetation history of savanna in East Africa during the Pleistocene. *Proceedings of the IV International Palynological Conference Lucknow* 3, 75-89.

BONNEFILLE, R. (1995): A reassessment of the Plio-Pleistocene pollen record of East Africa. In : E.S. Vrba, G.H. Denton, T.C. Partridge & L.H. Burckle (eds), *Paleoclimate and evolution with emphasis on human evolution*, 299-310. New Haven.

BRAIN, C.K., (1981): The evolution of man in Africa: Was it a consequence of Cainozoic cooling? *Geological Society of South Africa* 84, 1-19.

BRAIN, C.R., CHURCHER, C.S., CLARK, J.D., GRINE, F.E., SHIPMAN, P., SUSMAN, R.L., TURNER, A., & WATSON, V. (1988): New evidence of early hominids, their culture and environment from the Swarkrans Cave, South Africa. *South African Journal of Science* 84, 828-835.

BROMAGE, T.G. & SCHRENK, F. (1987): A cercopithecoid tooth from the Pliocene in Malawi. *Journal of Human Evolution* 15, 497-500.

BROMAGE, T.G. & SCHRENK, F. (eds.) (1995a): Evolutionary history of the Malawi Rift. *Journal of Human Evolution* 28, 1-120.

BROMAGE, T. G. & SCHRENK, F. (1995b): Biogeographic and climatic basis for a narrative of early hominid evolution. *Journal of Human Evolution* 28, 109-114.

BROMAGE, T.G., SCHRENK, F. & JUWAYEYI, Y.M. (1995): Palaeobiogeography of the Malawi-Rift: Age and vertebrate palaeontology of the Chiwondo-Beds, northern Malawi. *Journal of Human Evolution* 28, 37-58.

BROMAGE, T. G., SCHRENK, F. & KAUFULU, Z. M. (1985): Plio-Pleistocene deposits in Malawi's Hominid Corridor. *National Geographic Society Research Report* 21, 51-52.

BROMAGE, T.G., SCHRENK, F. & ZONNEVELD, F. (1995): Paleoanthropology of the Malawi-Rift: An early hominid mandible from the Chiwondo-Beds, northern Malawi. *Journal of Human Evolution* 28, 71-108.

BROWN, F.H. & FEIBEL, C.S. (1986): Revision of lithostratigraphic nomenclature in the Koobi Fora region, Kenya. *Journal of the Geological Society, London* 143, 297-310.

BROWN, F.H., McDOUGALL, I., DAVIES, T. & MAIER, R. (1985): An integrated Plio-Pleistocene chronology for the Turkana Basin. In: E. Delson (ed.), *Ancestors: The hard evidence*, 82-90. New York.

BRUNET, M., BEAUVILIAN, A., COPPENS, Y., HEINTZ, E., MOUTAYE, A.H.E. & PILBEAM, D. (1996): *Australopithecus bahrelgazali*, une nouvelle espèce d'Hominidé ancien de la région de Koro Toro (Tchad). *C.R.Acad.Sci. Paris*, t. 322, série IIa, 907-913.

CERLING, T.E., QUADE, J., AMBROSE, S.H. & SIKES, N.E. (1991): Fossil soils, grasses and carbone isotopes from Fort Ternan, Kenya: grassland or woodland. *Journal of Human Evolution* 21, 295-306.

CLARK, J.D. & HAYNES, C.V. (1970): An elephant butchery site at Mwanganda's Village, Karonga, Malawi and its relevance of Palaeolithic archaeology. *World Archaeology* 1, 390-411.

CLARK, J.D., HAYNES, C.V., MAWBY, F.E. & GAUTIER, A. (1970): Interim report on palaeo-anthropological investigations in the Lake Malawi Rift. *Quaternaria* 13, 305-354.

CLARK, J.D., STEPHENS, E.A. & CORYNDON, S.C. (1966): Pleistocene fossiliferous lake beds of the Malawi (Nyasa) Rift: A preliminary report. *American Anthropologist* 68, 46-49.

CLARKE, R.J. (1985): *Australopithecus* and early *Homo* in southern Africa. In: E. Delson (ed.), *Ancestors: The hard evidence,* 171-177. New York.

COLYN, M., GAUTIER-HION, A. & BERHEYEN, W. (1991): A re-appraisal of palaeoenvironmental history in Central Africa: Evidence for a major fluvial refuge in the Zaire Basin. *Journal of Biogeography* 18, 403-407.

COOKE, H.B.S. (1962): The Pleistocene environment in southern Africa: Hypothetical vegetation in southern Africa during the Pleistocene. *Annales of the Cape Province Museum* 2, 11-15.

CORYNDON, S.C. (1966): Preliminary report on some fossils from the Chiwondo Beds of the Karonga District, Malawi. *American Anthropologist* 68, 59-66.

DeMENOCAL, P.B. & BLOEMEDAL, J. (1995): Plio-Pleistocene climatic variability in subtropical Africa and the paleoenvironment of Hominid evolution: A combined data-model approach. In: E.S.Vrba, G.H. Denton, T.C. Partridge & L.H. Burckle (eds.), *Paleoclimate and evolution with emphasis on human evolution*, 262-288. New Haven.

DENYS, C. (1996): Olduvai rodent fauna: Palaeoecological and palaeobiogeographical affinities. A comparison between East and South African Plio-Pleistocene Faunas. *KAUPIA. Darmstädter Beiträge zur Naturgeschichte* 6, 247-261.

DENYS, C., CHOROWICZ, J. & TIERCELIN, J.J. (1986): Tectonic and environmental control on rodent diversity in the Plio-Pleistocene sediments of the African Rift System. In: L.E. Frostick (ed.), *Sedimentation in the African Rift*. Geological Society Special Publication No. 25, 363-372.

DIXEY, F. (1927): The Tertiary and post-Tertiary lacustrine sediments of the Nyasan Rift-Valley. Quart. *Journal of the Geological Society, London* 83, 432-447.

DUPONT, L.M. & LEROY S.A.G. (1995): Steps towards drier climatic conditions in northwestern Africa during the upper Pliocene. In: E.S. Vrba, G.H. Denton, T.C. Partridge & L.H. Burckle (eds.), *Paleoclimate and evolution with emphasis on human evolution*, 289-298. New Haven.

FOLEY, R.A. (1991). How many species of hominid should there be? *Journal of Human Evolution* 20,413-427.

GRIFFITHS, I.L.L. (1984): *An atlas of African affairs*. Cambridge.

HARRIS, J.W.K. (1986): Découverte de materiel archéologique oldowayen dans le rift de l'Afar. *L'Anthropologie* 90, 339-357.

HUGHES A.R. & TOBIAS, P.V. (1977): A fossil skull probably of the genus *Homo* from Sterkfontein, Transvaal. *Nature* 265, 310-312.

JABLONSKI, N.G. & CHAPLIN, G. (1993): Origin of habitual terrestrial bipedalism in the ancestor of the Hominidae. *Journal of Human Evolution* 24, 259-280.

KAUFULU, Z.M., VRBA, E.S. & WHITE, T.D. (1981): Age of the Chiwondo Beds, northern Malawi. *Annales of the Transvaal Museum* 33, 1-8.

KIMBEL, W.H. (1995): Hominid speciation and Pliocene climate change. In. E.S. Vrba, G.H. Denton, T.C. Partridge & L.H. Burckle (eds.), *Paleoclimate and evolution with emphasis on human evolution*, 425-437. New Haven.

KIMBEL, W. H., WHITE, T. D. & JOHANSON, D. C. (1988): Implications of KNM-WT 17000 for the evolution of "robust" *Australopithecus*. In: F.E. Grine (ed.), *Evolutionary history of the "robust" australopithecines*, 259-268. New York. .

KIMBEL, W.H., WALTER, R.C., JOHANSON, D.C., RED, K.E., ARONSON, J.L., ASSEFA, Z., MAREAN, C.W., ECK, G.G., BOBE, R., HOVERS, E., RAK, Y., VONDRA, C., YEMANE, T., YORK, D., CHEN, Y., EVENSEN, N.M. & SMITH, P.E.B. (1996): Late Pliocene *Homo* and Oldowan tools from the Hadar Formation (Kada Hadar Member), Ethiopia. *Journal of Human Evolution* 31, 549-562.

KINGDON, J. (1989): *Island Africa: The evolution of Africa's rare animals and plants*. Princeton.

KLEIN, R.G. (1984): The large mammals of southern Africa: Late Pliocene to recent. In: R.G. Klein (ed.), *South African prehistory and paleoenvironments*, 107-146. Rotterdam.

KULLMER, O., SANDROCK, O., ABEL, R., SCHRENK, F. & BROMAGE, T.G. (in prep.): The first *Paranthropus* from the Malawi-Rift.

MATTHEW, W.D. (1939): Climate and evolution. *Annals of the New York Academy of Science* 24, 171-318.

MAWBY, J.E. (1970): Fossil vertebrates from northern Malawi: Preliminary report. *Quaternaria* 13, 319-323.

OLIVER, R. & CROWDER, M. (1979): *The Cambridge encyclopedia of Africa*. Cambridge.

PETERS, C.R., O'BRIEN, E.M. & BOX, E.O. (1984): Plant types and seasonality of wild-plant foods, Tanzania to southwestern Africa: Resources for models of the natural environment. *Journal of Human Evolution* 13, 397-414.

PICKFORD, M. (1991): Growth of the Ruwenzoris and their impact on palaeoanthropology. In: A. l'Ehara, T. Kimura, O. Takenaka & M. Iwamoto (eds.), *Primatology today*, 513-516. Amsterdam.

PRENTICE, M.L. & DENTON, G.H. (1988): The deep-sea oxygen isotope record. The global ice sheet system and hominid evolution In: F.E. Grine (ed.), *Evolutionary history of the "robust" australopithecines*, 383-403. New York.

RAYNER, R.J., MOON, B.P. & MASTERS, J.C. (1993): The Makapansgat australopithecine environment. *Journal of Human Evolution*. 24, 219-231.

RETALLACK, G.J., DUGAS, D.P. & BESTLAND, E.A. (1990): Fossil soils and grasses of a middle Miocene East African grassland. *Science* 247, 1325-1328.

RING, U. & BETZLER, C. (1992): Normal versus strike-slip faulting during rift development in East Africa: The Malawi-Rift. *Geology* 20, 1015-1018.

RING, U. & BETZLER, C. (1995): Geology of the Malawi Rift: Kinematic and tectonosedimentary background to the Chiwondo Beds, northern Malawi. *Journal of Human Evolution* 28, 7-22.

ROBINSON, J.T. (1962): The origin and adaptive radiation of the australopithecines. In: G. Kurth (ed.), *Evolution und Hominisation,* 120-140. Stuttgart.

RODMAN, P.S. & McHENRY, H.M. (1980): Bioenergetics and the origin of hominid bipedalism. *American Journal of Physical Anthropology* 52, 103-106.

SCHRENK, F., BROMAGE, T.G., BETZLER, C.G. & RING, U. (1993): Oldest *Homo* and Pliocene biogeography of the Malawi Rift. *Nature* 365, 833-836.

SCHRENK, F., BROMAGE, T.G., GORTHNER, A. & SANDROCK, O. (1995): Paleoecology of the Malawi-Rift: Vertebrate and invertebrate faunal contexts of the Chiwondo-Beds, northern Malawi. *Journal of Human Evolution* 28, 59-70.

SHACKLETON, N.J., BACKMAN, J., ZIMMERMAN, H., KENT, D.V., HALL, M., ROBERTS, D.G., SCHNITKER, D., BALDAUF, J.G., DESPRAIRIES, A., HOMRIGHAUSEN, R., HUDDLESTUN, P., KEENE, J.B., KALTENBACH, A.J., KRUMSIECK, K.A.O., MORTON, A.C., MURRAY, J.W. & WESTBERG-SMITH, J. (1984): Oxygen isotope calibration of the onset of ice-rafting and history of glaciation in the North Atlantic region. *Nature* 307, 620-623.

SHIPMAN, P. & HARRIS, J.M. (1988): Habitat preference and palaeoecology of *Australopithecus boisei* in eastern Africa. In: F.E. Grine (ed.), *Evolutionary history of the "robust" australopithecines*, 343-381. New York.

SLATER, W.L. (1896): The geography of mammals. IV. The Ethiopian region. *Geographical Journal* 7, 282-296.

TURNER, A. & WOOD, B. (1993a): Taxonomic and geographic diversity in robust australopithecines and other African Plio-Pleistocene larger mammals. *Journal of Human Evolution* 24, 147-168.

TURNER, A. & WOOD, B. (1993b): Comparative palaeontological context for the evolution of the early hominid masticatory system. *Journal of Human Evolution* 24, 301-318.

VRBA, E.S. (1982): Biostratigraphy and chronology, based particularly on bovidae, of southern hominid-associated assemblages: Makapansgat, Sterkfontein, Taung, Kromdraai, Swartkrans; also Elandsfontein (Saldanha), Broken Hill (now Kabwe) and Cave of Hearths. In: *Pretirage, 1er Congr. Internat. Palaeont. Humaine,* 707-752. Nice.

VRBA, E.S. (1985): Ecological and adaptive changes associated with early hominid evolution. In: E. Delson, (ed.), *Ancestors: The hard evidence,* 63-71. New York.

VRBA, E.S. (1987a): A revision of the Bovini (Bovidae) and a preliminary revised checklist of Bovidae from Makapansgat. *Palaeontologia Africana* 26, 33-46.

VRBA, E.S. (1987b): Ecology in relation to speciation rates: Some case histories of Miocene-Recent mammal clades. *Evolutionary Ecology* 1, 283-300.

VRBA, E.S. (1988): Late Pliocene climatic events and human evolution. In: F.E. Grine (ed.), *Evolutionary history of the "robust" australopithecines*, 405-426. New York.

VRBA, E.S. (1992): Mammals as a key to evolutionary theory. *Journal of Mammology* 73, 1-28.

VRBA, E.S. (1995): The fossil record of African antelopes (Mammalia, Bovidae) in relation to human evolution and paleoclimate. In: E.S. Vrba, G.H. Denton, T.C. Partridge & L.H. Burckle (eds.), *Paleoclimate and evolution with emphasis on human evolution*, 385-424. New Haven.

VRBA, E.S., BURCKLE, L.H., DENTON, G.H. & PARTRIDGE, T.C. (1985): Palaeoclimate and evolution. *South African Journal of Science* 81, 224-275.

VRBA, E.S., DENTON, G.H. & PRENTICE, M.L. (1989): Climatic influence on early hominid behavior. *Ossa* 14, 127-156.

WALKER, A., LEAKEY, R.E., HARRIS, J.M. & BROWN, F.H. (1986): 2.5-Myr *Australopithecus boisei* from west of Lake Turkana, Kenya. *Nature* 322, 517-522.

WHEELER, P.E. (1991): The thermoregulatory advantages of hominid bipedialism in open equatorial environments: The contribution of increased convective heat loss and cutaneous evaporative cooling. *Journal of Human Evolution* 21, 107-115.

WHITE, T.D., JOHANSON, D.C. & KIMBEL, W.H. (1981): *Australopithecus africanus*: Its phyletic position reconsidered. *South African Journal of Science* 77, 445-470.

WOOD, B. (1992): Origin and evolution of the genus *Homo*. *Nature* 355, 783-790.

WOOD, B.A. (1993): Early *Homo*: How many species? In: W.H Kimbel. & L.B. Martin (eds.), *Species, species concepts, and primate evolution*, 485-522. New York.

ZINDEREN BAKKER, E.M. van (1969): Intimations on Quaternary palaeoecology of Africa. *Acta Botanica Neerlandica* 18, 230-239.

Meat-eating and Carcass Procurement by Hominids at the FLK Zinj 22 Site, Olduvai Gorge (Tanzania): A New Experimental Approach to the Old Hunting-versus-Scavenging Debate

Manuel Domínguez-Rodrigo

Abstract

Hominid strategies of carcass procurement (hunting vs. scavenging) and exploitation (meat and/or marrow consumption) are some of the most debated topics in East African Plio-Pleistocene archaeology. Opposite views of hominid behavior have been sustained on traditional taphonomic techniques (skeletal part representation, mortality profiles etc.), usually taking single-patterned models as referential frameworks. New taphonomic approaches (studies of conspicuous and inconspicuous bone surface modifications) recently developed, and based on multiple-patterned models, seem to be more informative about the sequence of hominid and carnivore involvement with carcasses at early sites. Experiments made by some researchers (Blumenschine, Capaldo, and Selvaggio) suggest a three-stage sequence, in which hominids are thought to have scavenged largely defleshed carcasses at felid kills.

The present study focusses on medium-sized carcasses (the dominant type at early sites) and tries to experimentally model the sequence of hominid and carnivore modification of carcasses from a new perspective. Results based on cut marks and tooth marks indicate an early access of hominids to carcasses incompatible with passive/secondary scavenging hypotheses.

Introduction

Early Plio-Pleistocene sites are the subject of an ongoing debate concerning the reconstruction of hominid feeding strategies with respect to animal carcass acquisition and consumption. The discussion of predatory (hunting) and opportunistic (scavenging) hominid behaviors has traditionally been approached zooarchaeologically through the analyses of skeletal part representation. Based on these data, some researchers have suggested that hominids were marginal scavengers (Binford, 1981 1984, 1985, 1988; Shipman, 1986), whereas others depicted them as successful hunter/scavengers, who primarily exploited carcasses for flesh rather than marrow (Bunn, 1981, 1982, 1983; Bunn & Kroll, 1986; Isaac, 1983, 1984). More recently, skeletal part profiles have also been used to argue both that hominids were transporting high-yielding meat sections from carcasses (Bunn & Kroll, 1986; Bunn & Ezzo, 1993) and, alternatively, that they were simply selecting high-yielding marrow bones that were already defleshed (Blumenschine, 1991, 1995; Blumenschine & Madrigal, 1993; Blumenschine & Marean, 1993). Bone damage has been used as a supporting argument by most of the researchers involved in the discussion. Binford (1981, 1985, 1988) used tooth marks on fossil bones to support the hypothesis that carnivores were the main consumers of carcasses. Based on the percentages and anatomical distribution of cut marks also found on some of these bones, Bunn (1981, 1982, 1983, 1991) claimed that hominids were processing fleshed carcasses. Blumenschine (1988) and Potts (1982, 1988) used, to reach different conclusions, both cut marks and tooth marks to reconstruct the sequence of intervention of hominids and carnivores in site formation, and Blumenschine (1995) further used them to hypothesize that hominids were scavenging carcasses from felid kills, from which they removed scraps of flesh, brain and the marrow contents.

One of the main problems with interpretations based on skeletal part profiles is the fact that our understanding of skeletal part representation is based on studies of the differential transport of anatomical parts by humans from kill sites to base camps, and not on what is left after their consumption, abandonment and subsequent ravaging by carnivores. Bone assemblages are the result of dynamic processes of selection and destruction, which may not reflect the initial decisions made by humans (if in fact they were the primary agents of bone accumulation). Previous studies on bone density have demonstrated that the skeletal parts composed of cancellous tissue are prone to be destroyed or taken away from their original site of accumulation by several processes: water flows, weathering, and carnivore depletion (Brain, 1981; Capaldo, 1995; Lyman, 1995). This is because these parts are less dense than other elements and usually contain significant deposits of grease, which make them attractive to carnivores.

This adds a further dimension to the discussion of the utility of skeletal part profiles for zooarchaeological analyses, because in sites where these destructive processes have been operating, especially carnivore ravaging, the distortion of anatomical representation patterns is such that we are not able to discern the original human-made patterns of bone accumulation in order to answer questions of whether they were selectively transporting certain skeletal units (e.g. limbs) or complete carcasses. The presence of both cut marks and tooth marks on archaeological faunal assemblages (Bunn, 1981, 1982, 1983; Potts & Shipman, 1981; Blumenschine, 1988, 1995) indicates that hominids and carnivores (very likely, hyenas) were involved in Plio-Pleistocene site formation. The experimentation undertaken to test the effect of carnivore ravaging on bone accumulations made by humans has led to the following conclusions (Marean et al., 1992; Blumenschine & Marean, 1993; Capaldo, 1995):

1. Axial bones (ribs, vertebrae and pelves), certain long bone epiphyses, and small compact limb bones are preferently depleted. This creates an artificial profile, dominated by some limb and cranial elements.
2. With respect to limb bones, epiphyseal fragments are more likely to be depleted than mid-shaft specimens.

Therefore, faunal assemblages at sites are the result of humans making decisions at kill sites, transporting particular bones, modifying and breaking them at camps during consumption, and subsequently abandoning them to the intervention of other agencies, either physical (e.g. water flows) and/or biological (e.g. carnivore post-ravaging). Thus, actualistic data drawn from the initial stage of this process are not useful for comparison with archaeological assemblages which have experienced post-human differential destruction.

Comparisons must be made between equivalent phenomena. This means that since sites are palimpsests - and, therefore, the result of the intervention of several agencies - single-patterned models (carnivore dens, human transport of carcasses, etc.) are not appropriate references for interpreting archaeological bone assemblages. Instead, we should be using multiple-patterned models (Blumenschine, 1988, 1995: Blumenschine & Marean, 1993; Marean et al., 1992; Selvaggio, 1994; Capaldo, 1995).

Recently, a change of focus by some researchers in the study of bone assemblages has brought new methodological and interpretative procedures that seem to be more informative than skeletal part profiles. The realization that sites were palimpsests led Blumenschine (1988) to study the timing of hominid and carnivore participation at them. He conducted several experiments on the frequency and distribution of tooth marks according to bone section in a set of assemblages first created by humans and then modified by carnivores, and in another set of assemblages in which carnivores were the primary actors. When comparing these results with the distribution of tooth marks and cut marks at Plio-Pleistocene sites, Blumenschine could determine - for the first time without ambiguous and untested speculations - the order of action of both agents at Olduvai sites: Hyenas seemed to have ravaged the bone remains accumulated by hominids (Blumenschine, 1988, 1991, 1995; Blumenschine & Marean, 1993).

The utility of this novel approach fueled other studies on bone modification as a useful source of taphonomic information, including experiments on the differential depletion of bones by hyenas (Marean et al., 1992; Capaldo, 1995), and the effect of timing of hyena access to carcasses on tooth mark and cut mark frequencies and distribution (Blumenschine & Selvaggio, 1991; Selvaggio, 1994; Capaldo, 1995; Domínguez-Rodrigo, 1997a, b). Studies on bone breakage patterns have also been made by analysing particular traces related to carnivores and humans, such as tooth marks, percussion marks (Blumenschine & Selvaggio, 1988) and bone notches (Capaldo & Blumenschine, 1994).

These experiments model site formation using a dual-patterned approach, varying hominid and carnivore access to carcasses. More recently, in order to test the hypothesis that hominids scavenged from felid kills (Blumenschine, 1991, 1995) experiments dealing with three-patterned models have been conducted (Selvaggio, 1994; Domínguez-Rodrigo, 1997a,b). The information that we are obtaining from these types of studies is potentially more useful than the traditional approaches to the study of archaeological sites (Blumenschine, 1995; Capaldo, 1995; Domínguez-Rodrigo, 1997a,b).

However, there are important differences of opinion among researchers as to how these results should be interpreted. Blumenschine (1991, 1995), Selvaggio (1994) and Capaldo (1995) claim that hominids were scavenging defleshed carcasses from felid kills, whereas Domínguez-Rodrigo (1997a,b) argues that hominids had access to fleshed medium-sized carcasses that had not been previously consumed by carnivores.

In their support of a passive scavenging scenario, Blumenschine (1995) and Capaldo (1995) dismiss the informative potential of cut marks. Although cut marks are the bone surface marks most diagnostic of hominid butchery, Blumenschine justifies his assertions by saying that we lack a referential framework on this type of bone surface modification:

„*Other potentially good proxies for biological agents of assemblage formation currently suffer from insufficient calibration by well controlled experiments and naturalistic observations in the modern world. For the Plio-Pleistocene, the cut mark data are the main case in point, despite the status of this (c.f. Bunn, 1981; Potts & Shipman, 1981) and other surface modifications as potentially the most sensitive criteria for diagnosing agency. However, there are no detailed, published control samples through which frequencies of bone cut-marked with stone tools can be used to confidently infer timing of access to carcass tissues and the amount of flesh butchered (see, however, Selvaggio, 1994). In arguing that hominids defleshed complete carcasses at the early Olduvai site of FLK Zinjanthropus, Bunn (Bunn & Kroll, 1986; Bunn & Ezzo, 1993) cites only the grosssimilarity of this assemblage's cut mark frequencies (percent of number of identified specimens) to recent archaeological occurrences. Here, Bunn bases the interpretation of one unknown (Olduvai sites) on other unknown (recent sites). While there is little doubt that Pastoral Neolithic peoples had access to fully-fleshed carcasses (Marshall, 1986), lack of precise knowledge of their butchery methods makes the assemblage a suboptimal control for evaluating Oldowan butchery. More importantly, Bunn provides no information from controlled studies that allows one to evaluate the susceptibility of the diagnosis to even small differences in the anatomical distribution and overall frequency of cut-marked bone. As a result, we do not know what specific*

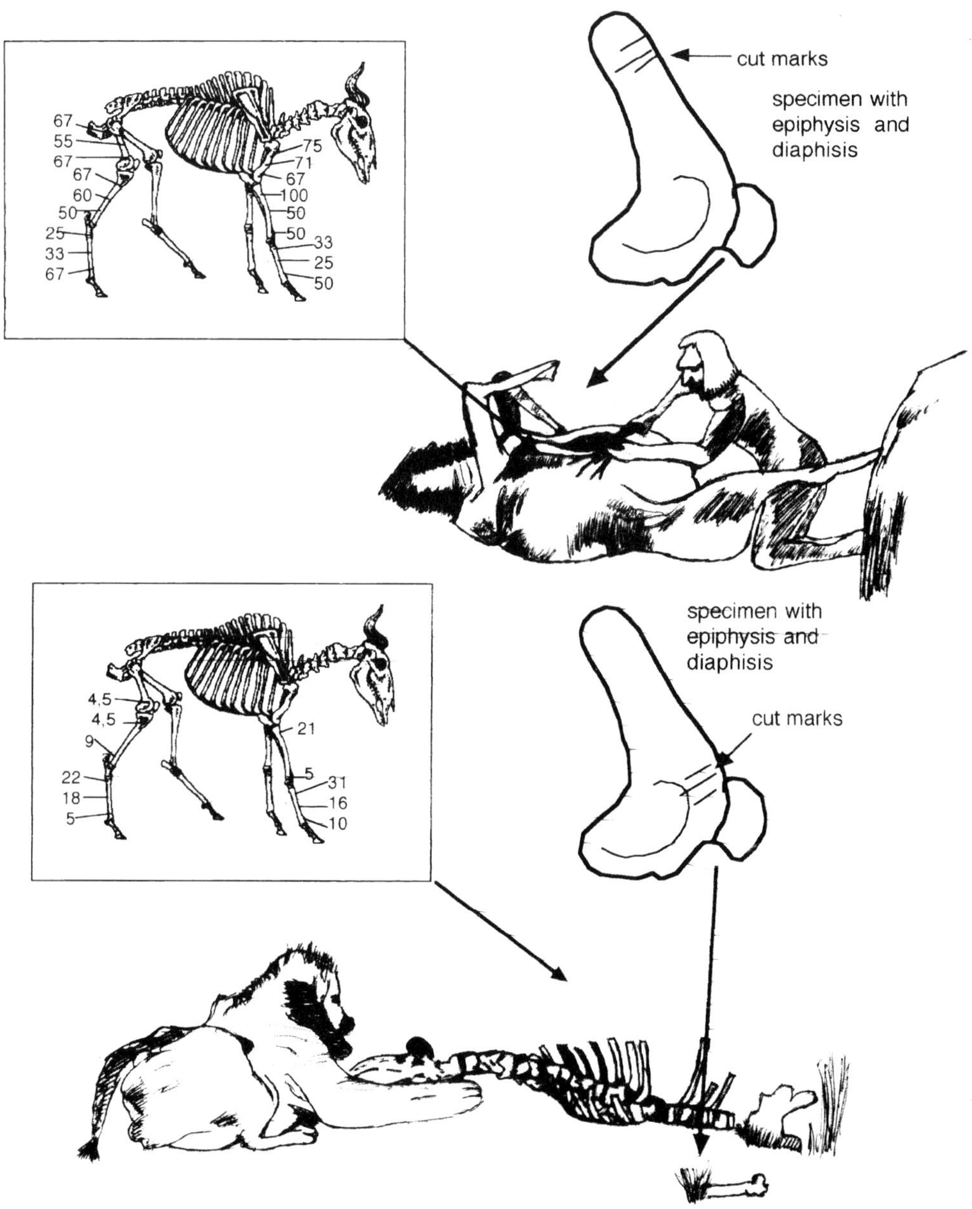

Figure 1 - Two different episodes, primary access to carcasses by hominids, and secondary access to lion kills by hominids is reflected in the distribution of cut marks on bone type and bone section. The two behaviors can be confounded using Blumenschine's, Capaldo's & Selvaggio's analyses. Cut marks will appear differentially distributed according to element type, whether carcasses were fleshed or defleshed. Likewise, the actual location of cut marks can indicate if large amounts of flesh were removed or not. The same specimen (epiphyseal fragment, according to Blumenschine's definition) can be indicative of both processes if the actual location of the cut marks are distinguished.

range of cut mark frequencies reflect bulk defleshing, as asserted for Oldowan assemblages by Bunn (Bunn & Ezzo, 1993) vs. the removal of scraps of flesh (Blumenschine, 1986) or defleshing of some bones on a carcass as opposed to all." (Blumenschine, 1995, p. 23).

Despite his experimentation, Capaldo (1995, p. 321) further claims that „*given the ambiguous nature of experimentally derived cut-mark criteria, we must avoid using them to make inferences on the amount of flesh acquired from carcasses by Plio-Pleistocene hominids, the timing of hominid carcass processing and the mode of carcass acquisition*".

With cut marks not a factor in their analyses, Blumenschine, Selvaggio and Capaldo base their 3-stage model (carnivore to hominid to carnivore) mainly on the percentage of tooth marks according to bone section, especially those occurring on mid-shaft bone specimens. As Blumenschine himself puts it:

„*Percussion mark and tooth mark frequencies, reported systematically for the first time on the assemblage from FLK Zinjanthropus, allow the rejection of the hypothesis that carnivores rather than hominids at FLK Zinjanthropus enjoyed first access to marrow in long bone cavities. This result contradicts Binford's characterization of Plio-Pleistocene hominids as marginal scavengers of carcasses heavily ravaged by bone-crunching carnivores. However, a high proportion of tooth-marked long bone midshaft fragments allows the rejection of the alternative hypothesis that carnivore access to long bones was exclusively or even primarily secondary to butchery and marrow extraction by hominids, as argued by Bunn (1986, Bunn & Kroll, 1986) and Leakey (1971). The results show that the sequence of carnivore and hominid access to long bones, and their respective carcass tissue yields, was more complex, as is consistent with a dominantly passive scavenging mode of carcass acquisition by hominids*" (Blumenschine, 1995, p. 21).

Following the same line of thought, Capaldo (1995, p. 298) argues that „*Despite the significant correlations between the bovid documented FLK 22 archeofauna and my hammerstone to carnivore experimental samples, the hominid to carnivore site formation hypothesis can be refuted*".

In direct opposition to the conclusions of Blumenschine et al. outlined above, I present the results of a new set of experiments on the timing of access of hominids and carnivores to animal carcasses that show that „passive scavenging" models are based on an overreliance on tooth marks, a faulty analytical approach to the interpretations of cut marks, and a limited consideration of all the possibilities of the Hominid-to-Carnivore experimental scenario. Comparisons of cut mark experimental results with archaeological data could be indicative of early hominid access to carcasses.

The use of cut marks as indicators of meat-eating by hominids

The results of my cut mark experiments differ from those of Selvaggio (1994) and Capaldo (1995) because we use different analytical approaches. Their studies are based on the percentage of cut marks on NISP according to bone section. Mine are based on the percentage of cut marks on NISP according to bone section and bone type.

There are two hypotheses about what cut marks on bones really reflect. Bunn & Kroll (1986) tried to prove that they are indicative of the amount of flesh attached to bones by showing that the bones with most cut marks were the highest yielding meat-bearing elements. Binford (1981, 1988), however, argued that cut marks are the result of contact between cutting tools and bones, irrespective of the amount of flesh on them. Blumenschine (1986, 1991) follows this assertion by suggesting that cut marks could result from the removal of the tiny scraps of flesh that commonly survive carnivore consumption.

Blumenschine's, Selvaggio's and Capaldo's approaches to the study of cut marks are of limited value, as the analyses of bone sections *per se* could never test, for instance, Bunn & Kroll's

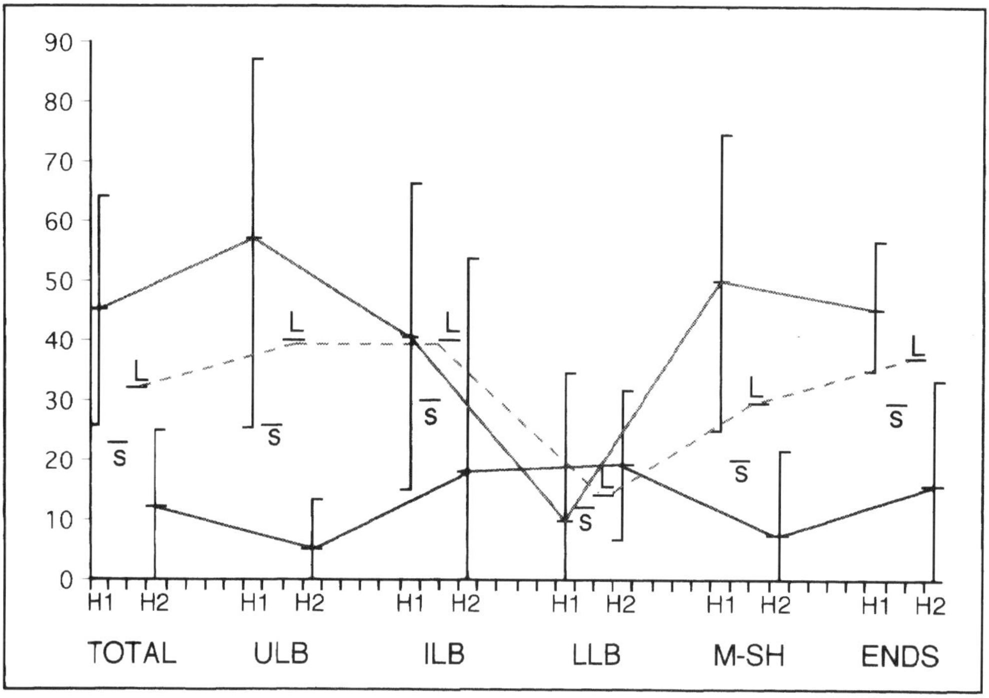

Figure 2 - Range of variation of the Hypothesis 1 (H1) & 2 (H2) experiments for cut marks and comparison with the FLK Zinj cut mark data.
For the FLK Zinj site: S = FLK smaller mammals; L = FLK larger mammals (Domínguez-Rodrigo, 1997b).

(1986) hypothesis. Flesh is anatomically distributed in a differential way according to bone element, and analyses of cut marks *per* bone section can never relate the amount of meat removed to the distribution and percentages of the resulting cut marks. For instance, two completely different behaviors may mimic each other when analysed through Blumenschine & Selvaggio & Capaldo's approach. The removal of skin and periostium from lower limb bones of carnivore-defleshed carcasses with cutting tools may leave an abundance of cut marks on all the lower-limb sections (epiphyses, near-epiphyses and mid-shaft fragments). The total percentage and sectional distribution of cut marks per carcass will be similar to the case in which hominids processed fully fleshed upper limb bones and discarded the lower limb elements. The way to differentiate these two scenarios is by quantifying the cut marks by bone type (axial, upper limb, lower limb etc.) as well as by bone section. This is important because flesh masses are diferentially distributed according to the type of element and also because the small scraps of flesh that might survive carnivore consumption also show a typical pattern of anatomical distribution according to bone type and bone section (Domínguez-Rodrigo, 1997c). This makes an excellent reference to interpret the anatomical distribution of marks on bones from archaeological sites.

However, Blumenschine & Selvaggio & Capaldo's method of analysis is problematic not only because it does not consider such a differential distribution of flesh, but also due to their definition of bone sections. A bone specimen that has a fragment of epiphysis with a section of the near-epiphysis and mid-shaft is considered by them to be an „epiphyseal specimen". Cut marks on the mid-shaft section of such fragment is classified as a cut-marked „epiphyseal fragment", which gives the incorrect impression that the cut mark is situated on the epiphysis itself (Figure 1). The high survival of epiphyseal fragments attached to shaft sections and the high percentage of marks on these large specimens will result in a relatively high number of „epiphyseal fragments" showing marks. Thus, the actual distribution of marks according to bone section is not properly taken into account. This is of real consequence when evaluating different models of hominid carcass acquisition and processing behavior. For instance, the availability of scraps of flesh from mid-shaft sections of limb bones first consumed by lions has been shown to be rather limited (Domínguez-Rodrigo, 1997c). The classification of cut marks on the mid-shaft of a specimen conserving part of the epiphysis as a cut-marked epiphyseal fragment would have quite different implications vis-a-vis a hypothesis of lion followed by hominid consumption than would its classification as a cut-marked mid-shaft fragment. I argue for a more precise classification of cut mark location that does not lump together mid-shafts and epiphyses (Figure 1).

Based on an analysis that takes more into consideration the precise location of cut marks on a particular bone and the type of bone involved, experiments by Domínguez-Rodrigo indicate that there is a clear-cut distinction between primary and secondary access to carcasses. By comparing the experimental representation of cut marks with those from the FLK Zinj site at Olduvai (Tanzania), the results suggest primary access, in which hominids were processing fleshed animals (Table 1; Figure 2).

In a set of experiments made to simulate both scenarios (primary access by hominids to fleshed carcasses and secondary access to largely defleshed carcasses from felid kills) Domínguez-Rodrigo focused mainly on medium sized (size 3) animals, because they are the predominant type of fauna at early sites (For further details on this study see Domínguez-Rodrigo, 1997a,b). With respect to the experiments simulating primary access to fully fleshed carcasses (Hypothesis 1), the results are as follows:

1. A high percentage of identified cut-marked bones with respect to total NISP.

2. Differential occurrence of cut-mark patterns. The meatiest elements (upper limb bones) are the highest ranked, with a mean percentage among all the experiments of about 60% of the total number of cut-marked bones, followed by intermediate bones with a mean close to 30%, and finally, lower limb bones (less than 10%).

3. Differential distribution of cut marks according to bone section. In the pre-ravaged human-made accumulations the mean figure of cut-marked specimens from mid-shafts (with respect to the total number of all cut-marked fragments) is about 43%. In post-ravaged assemblages, it becomes higher than 50% (Domínguez-Rodrigo, 1997a,b). This is due to the disappearance of some proximal/distal ends by density-mediated factors. Therefore, mid-shafts are the most abundant cut-marked specimens in bone assemblages that have completed the dual-patterned process.

In clear contrast, the situation observed in the experiments dealing with the hypothesis of secondary access to largely defleshed carcasses (Hypothesis 2) is significantly different:

1. A low total percentage of cut-marked bones with respect to NISP.

2. Differential occurrence of cut-mark patterns. In contrast to Hypothesis 1, the meatiest limb elements (humerus and femur) are the lowest ranked (mean=7,2% of all the cut-marked specimens). In the context of a high number of lions at a kill (such as at Maasai Mara), intermediate bones are also low-ranked (mean=10,5% of all the cut-marked specimens), with metapodials being the highest ranked elements (>60% of all the cut-marked specimens). In settings where the number of predators does not exceed 2-3 and complete consumption is delayed, cut mark representation on these bones is higher (mean=30,5%) due to the presence of larger amounts of flesh. In summary, cut marks on intermediate and lower bones are more abundant than on upper limb bones.

3. Differential distribution of cut marks according to bone section. In highly flesh-depleted carcasses only metapodials (no upper or intermediate limb bone) exhibits any cut mark on the mid-shaft. In carcasses not so heavily defleshed, this percentage is somewhat higher. Proximal/ distal ends rather than mid-shafts exhibit the highest occurrence of cut marks (mean >80%, of all the cut-marked specimens except metapodials).

The different distribution and percentages of cut-marked specimens between the two sets of experiments is explained by the differential availability of meat (see Domínguez-Rodrigo, 1997c).

If we observe the differential distribution of cut marks on bones from FLK Zinj site according to bone section, we can see that as many as 59% of the cut-marked specimens from meaty limbs of smaller animals and 73% of the cut-marked specimens from the same elements of larger animals occur on mid-shaft fragments (Bunn & Kroll, 1986). Bearing in mind that Hypothesis 1 experiments yielded a mean frequency of cut-marked mid-shaft specimens higher than 50%, the archaeological data seem to fit Hypothesis 1 better than Hypothesis 2.

With respect to the cut-mark patterns according to bone type, the data drawn from FLK Zinj show for larger animals a clear differential distribution: upper (43%) and intermediate (45%) limb bones make up most of the cut-marked specimens (88%) versus metapodials (12%). Proportionately (according to NISP) upper limbs show a greater representation of cut-marked specimens (40%) than intermediate limb bones (30%). This result is similar to that obtained in the Hypothesis 1 experiments (see Domínguez-Rodrigo, 1997a,b).

However such a comparison is somewhat incomplete. The total percentage of cut-marked specimens on the bones from FLK Zinj shows a particular pattern. However, these data is taken from Bunn & Kroll (1986) who only used bones with cut marks verified by microscopic identification. Nonetheless, there is a large number of pieces with macroscopically identified cut marks that were not used in their analysis of cut-mark patterns, even though Bunn (1981, p. 22) acknowledged that „*macroscopic examination and comparison with known types of grooves on modern bones is a sufficiently reliable basis for the identification of multiple sets of cut marks and most isolated cut marks on fossil bones.*" Therefore, I have re-analyzed the FLK Zinj data using the macroscopically identified bone sample from Bunn & Kroll's (1986) study.

Concerning the percentage of cut-marked bones with respect to NISP, it is 21,5% in smaller animals and 32% in larger animals. Cut-mark patterns are distributed preferentially on upper/ intermediate limb bones (89% in larger animals and 87% in smaller animals) with similar percentages to the Hypothesis 1 experiments: 44% on upper bones in larger animals and 31% in smaller animals, and 45% in the intermediate bones in larger animals and 56% in smaller animals. If we consider cut-marked NISP per element, this differential distribution becomes clearer. The cut-marked specimens from upper limb bones of larger animals are 57% of the total of cut-marked specimens from the meaty bonee, while intermediate bones represent only 43%. Furthermore, 70% of the cut-marked specimens in larger animals and 51% in smaller animals occur on mid-shaft specimens. Therefore, this cut-mark pattern resembles that observed in Hypothesis 1 experiments (Domínguez-Rodrigo, 1997a,b).

These patterns exhibited at FLK Zinj, whether based on the microscopically identified cut-marked specimens or on the mixed set of cut-marked bones (macroscopically and microscopically identified) are similar to those obtained in the experiments that simulated primary access of hominids to carcasses. Cut-mark percentages and distribution in upper/intermediate limb bones (especially in humerus and femur) cannot be accounted for in terms of secondary access to carcasses. Although a behavior aimed at marrow extraction could also be reflected in form of cut marks on lower limb bones - as skinning is necessary (Blumenschine, 1986) -, it would leave

HYPOTHESIS 1	TOTAL	ULB	ILB	LLB	M-SH	ENDS
n. assemblages	4	4	4	4	4	4
mean+/- S.D.	45+/-12,1	56,2+/-19,4	40,7+/-16,1	8,5+/-16,5	49,6+/-19	45,7+/-6,8
95% C.I.	(26-64)	(25,4-87)	(15,2-66,2)	(0-34,7)	(24,8-74,4)	(34,9-56,5)

HYPOTHESIS 2	TOTAL	ULB	ILB	LLB	M-SH	ENDS
n. assemblages	4	4	4	4	4	4
mean+/- S.D.	12,2+/-8,4	3,9+/-6,2	17,2+/-23,6	19,2+/-8	7,7+/-9,1	16+/-11,2
95% C.I.	(0-25,2)	(0-13,5)	(0-53,7)	(6,8-31,6)	(0-21,6)	(0-33,3)

FLK 22 Zinj	TOTAL	ULB	ILB	LLB	M-SH	ENDS
micro s	15 (51/33)	18,6 (22/118)	13,8 (23/166)	10,9 (6/55)	13,6 (35/256)	19,2 (16/83)
micro l	23 (37/159)	26 (11/42)	29,8 (20/67)	12 (6/50)	20 (21/105)	29,6 (16/54)
macro S	21.5 (73/339)	27,1 (32/118)	19,8 (33/166)	16,3 (9/55)	20,3 (52/256)	25,3 (21/83)
macro L	32 (51/159)	40 (17/42)	40,2 (27/67)	14 (7/50)	29,5 (31/105)	37 (20/54)

Table 1 - *Mean percentages of cut-marked specimens in relation to the total number of specimens from upper (ULB), intermediate (ILB), lower (LLB) limb bones, mid-shafts (MSH) and ends for the experiments (Hypotheses 1 & 2) and the FLK Zinj site.*

s = FLK smaller mammals (micro); S = FLK smaller mammals (macro); l = FLK larger mammals (micro); L = FLK larger mammals (macro).

Analysis of 95% C. I. for both sets of experiments are included. They were calculated using the t distribution, where $t_{0.025}$ is the critical value of t with n-1 degrees of freedom.

barely any scattered marks on the other bones, and surely almost none on upper limb bones and their shafts.

However, the strength of these conclusions depends on whether the range of variation of the samples of both sets of experiments do not significantly overlap. Table 1 and Figure 2 show that there is no overlap between the ratios of cut-marked specimens with respect to the total NISP of both samples. The sample of larger mammals at FLK Zinj - and this is the size of animals for which this experimentation was carried out - falls within the Hypothesis 1 range of variation. Smaller animals fall between the range of variation of both experimental samples. *The similarities between FLK Zinj and Hypothesis 1 are most striking for upper limb bones.* There is no overlap between the range of variation of Hypothesis 1 and Hypothesis 2 experiments, and both the smaller and larger mammals of FLK Zinj are situated within the range of variation of the former. For intermediate bones, although a slight overlap between both samples exists, the larger mammals of FLK Zinj are found outside this overlapping zone and well within the range of variation of the Hypothesis 1 experimental sample.

With regard to the percentages of cut-marked mid-shafts, the smaller and larger carcasses of FLK Zinj are outside the range of variation of Hypothesis 2 experiments, and the larger mammal bones fall within the range of the Hypothesis 1 experiments.

This similarity between FLK Zinj and experiments simulating early hominid access support hypotheses of hominid involvement with fleshed carcasses at sites (Domínguez-Rodrigo, 1997a,b). These results contradict Selvaggio's (1994) interpretations of the cut mark patterns of the FLK Zinj site, which could be due to the following reasons:

1. She does not differentiate between disarticulation and defleshing or meat removal from carcasses. Thus, the resulting cut mark patterns obtained by her are the result of both processes.
2. Although she used stone tools for removing the scraps of flesh on largely defleshed carcasses, she included obsidian among the raw materials utilized. Obsidian flakes are sharper and last longer than those made on basalt, lava quartz or quartzite, which are the raw materials used by Plio-Pleistocene hominids.
3. Following the analythical procedures of Blumenschine (1988), Selvaggio's analyses are based on the distribution of marks according to bone section. This type of approach does not discriminate the differential occurrence of cut marks according to bone type, nor is it useful, therefore, for testing Bunn & Kroll's (1986) hypothesis that cut marks are related to flesh extraction, which is differentially distributed on long bones (see above for discussion).
4. Several of Selvaggio's experiments are made up of single bones or sets of no more than 3 elements. Capaldo's (1995) experiments (and personal observation) suggest that carnivore post-ravaging on bone assemblages is influenced by the availability of remains. Thus, two or three bones neither realistically simulate what hyenas found at sites abandoned by hominids, nor the result of their intervention.
5. She removed the scraps of flesh and the marrow from bones herself, under the assumption that „differences in the frequencies of butchery marks are not related different butchers" (Selvaggio, 1994). I documented the variation of cut mark frequencies according to the experience of the butcher (Domínguez-Rodrigo, 1996), as was also observed by Haynes (1991). Her percentages, thus, may be inflated.
6. She does not make it clear whether she removed the periostium from bones in all carcasses. Bearing in mind that this is usually practiced by Blumenschine (1988) and Capaldo (1995), she may have done the same. This could create a higher percentage of cut marks, due to an unnecessary act, for periostium need not be removed from long bones (with the exception of metapodials) before breaking them.
7. To establish global percentages of mark distribution in the three-patterned experiments, she lumps together the data obtained from all the carcasses, irrespective of their size and the predator(s) responsible for their capture and processing. This creates a double bias. On the one side, the result of carcass consumption by solitary predators (more likely to abandon animals with higher flesh yields) is blended with that of gregarious predators (more likely to deflesh carcasses more efficiently). On the other side, and as a result thereof, the differential availability of flesh on smaller carcasses - preferentially hunted by solitary predators (Cavallo & Blumenschine, 1989) - is not differentiated from that available in larger carcasses (usually hunted and processed by gregarious predators). This creates an artificial frequency of marks that is not very useful for comparative purposes.
8. Moreover, the experimental sample used by Selvaggio (1994) is mainly made up of small carcasses. Only 4 out of the total 47 carcasses used in her study, are medium-sized animals (2 zebras, 1 wildebeest and 1 buffalo). Of these, only two were used in the three-patterned experiments under controlled circumstances. One of them consisted of 6 bones and the other of only one bone. This is obviously statistically insignificant.

Despite these methodological problems, she acknowledged that *„limbs were abandoned by lions with little or no flesh on humeri and femora. Ocassionally the skin was not completely* *consumed from the tibiae or the radius-ulnae and small scraps of flesh remained near the distal epiphyses*" (Selvaggio, 1994, p. 54-55). She further claimed that „*long bones abandoned by large groups of carnivores were usually disarticulated from the carcass and while the marrow cavity remained intact, the bones were usually encountered completely defleshed*" (Selvaggio, 1994, p. 124). Therefore, both studies (Selvaggio's and mine) lead to the same conclusion with respect to carcasses processed by lions. Upper limb bones are utterly defleshed on most ocassions. According to her, this observation could also be extended to other predators and smaller carcasses: „*rarely were scraps of flesh available on proximal long bones abandoned by carnivores in the Carnivore-Hominid sample* „ (Selvaggio, 1994, p. 122). Therefore, it is not surprising that in her experimental sample less than 5% of mid-shafts had any cut mark (Selvaggio, 1994). The fact that upper limb bones are the highest cut-marked elements in the FLK Zinj site indicates, in contrast, early access by hominids to fleshed carcasses.

Tooth marks as indicators of the timing of hominid access to carcasses

The basis of the passive scavenging model is the high rate of tooth marks on mid-shaft specimens observed at FLK Zinj. Before considering their importance, I would like to raise some methodological objections to the experiments done to interpret tooth marks on bones:

1. Some of the researchers that have performed experimental studies do not properly indicate the number of bones per experiment (e.g. Blumenschine 1988) or the time of exposure of bone assemblages to scavengers (Selvaggio, 1994) or if they do (e.g. Capaldo, 1995) it is only for one or two days. We know that the number of bones exposed to scavengers (Capaldo, 1995) and the time of exposure (Domínguez-Rodrigo & Marti Lezana, 1996) can determine the degree of bone modification and depletion at a given site.
2. The occurrences of tooth marks are recorded only according to bone section. Bone type is not considered. There is a differential incidence of tooth marks according to bone type depending on the species of carnivore involved (Domínguez-Rodrigo, 1994; Selvaggio, 1994). This is not properly addressed by some researchers.
3. They also remove the periostium from long bones (Capaldo, 1995; Blumenschine, pers. com.), prior to their exposure to scavengers, creating a distorted profile of tooth marking. All the bone flakes adhering to proximal/distal ends due to the presence of periostium are very likely to be tooth-marked by scavengers (pers. obs.), which does not exactly occur in Blumenschine's, Selvaggio's and Capaldo's experiments. I assume in my experimentation that if hominids were processing carcasses, once they broke and demarrowed the bones, they did not remove the periostium and separate the shaft flakes from the ends. Due to this, there is a bias towards a lower rate of tooth mark percentages in their Hominid-to-Carnivore experiments.
4. There are substantially more fragments of 2-3 cms. from the experimental samples than from Zinj. Bearing in mind that these bone specimens are the least likely to be marked (Blumenschine, 1988), it creates a lower percentage of tooth-marked fragments in the experiments and a higher one in FLK Zinj. If comparisons are made without including the specimens of this size, tooth mark representation on the mid-shaft specimens from the experiments are almost four times higher (almost 20%) than that documented by Blumenschine (1988) for size 3 carcasses.
5. Despite their similar methodological procedures, variation in tooth marking among the assemblages of these researchers is significant. For instance, for bones belonging to medium-sized carcasses, Blumenschine (1988, 1995) documents that 5% of mid-shaft fragments are tooth marked, whereas Capaldo (1995) obtained a mean of 15% for the same carcass size category. This casts doubt on the range of variation used by Blumenschine (1988, 1995) for tooth-marked

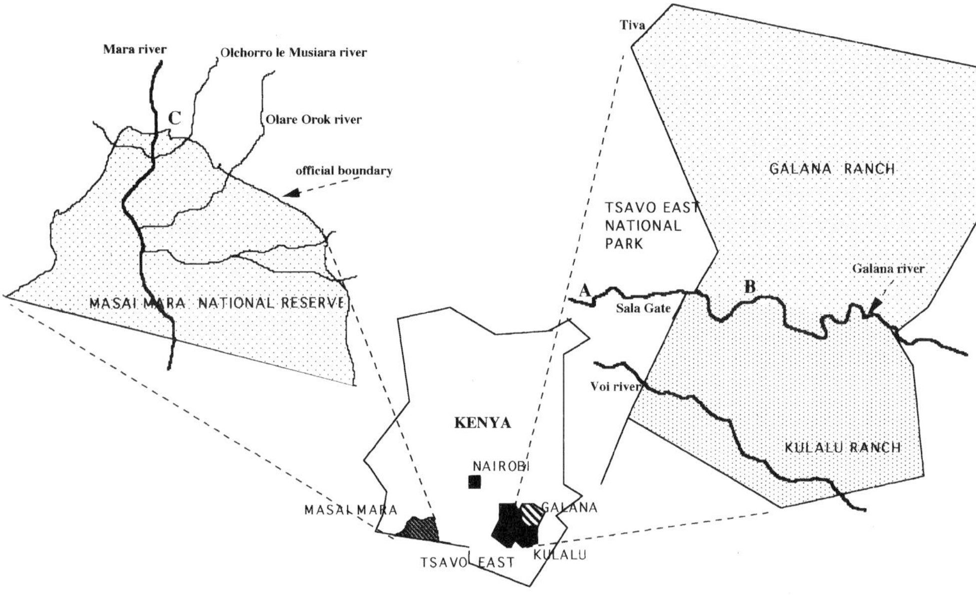

Figure 3 - Location of the places where the studies mentioned in the text were carried out. A, Tsavo East National Park; B, Galana river in Galana & Kulalu; C, Maasai Mara National Reserve, Kenya.

mid-shaft specimens (0%-10%) and suggests that the true range of variation is likely to be substantially higher (Capaldo, 1995).

In spite of these methodological and analytical shortcomings, the observation that mid-shaft specimens at FLK Zinj site are highly tooth marked (>50%) seems to be warranted (Blumenschine, 1995). Is this, however, a proof that carnivores had preceeded hominids in carcass processing? That is one of the most likely possibilities, but there are other processes that have not yet been considered.

To test the hypothesis that hominids may have been responsible for some of the tooth marks observed on fossil bones, we initiated a study with primates to document the rate of tooth marks made by them on bones (Domínguez-Rodrigo et al., n.d.). I will present here some of the preliminary results. This research was composed of two studies.

Study 1. This study was carried out using baboons of the zoo in Barcelona, Spain. Fleshed bones were given to them and after defleshing, they were broken and demarrowed. They were recovered 3 days after initial consumption (Dominguez-Rodrigo et al., n.d.). We observed that during this time the baboons were interested in the bone fragments, even though they were defleshed. Some of them still gnawed some of the bones 3 days after they had been given to them. They were regularly fed, so that this was not due to their being hungry. We observed their interest in removing the periostium with their teeth, even though the bones were broken into fragments.

Study 2. The fact that baboons might leave tooth marks on bones that were still fleshed seemed as normal as it would be for carnivores. Their interest in defleshed bones was, however, more surprising. To rule out the possibility that this interest in defleshed bones was a product of boredom or stress from confinement, a second study was carried out in Tsavo National Park (Kenya) (Figure 3). I made several experiments, but on most ocassions baboons were not interested in the bones

ZOO				
	Epiphyses	Near-epiphyses	Mid-shaft frag.	All bones
N. assemblages	3	3	3	3
X+/- S.D.	91.6+/-14.4	83.3+/-14.4	73,3+/-11,5	80,1+/-2,8
95% C.I.	(55,1-100)	(46,8-100)	(44,1-100)	(73-87,2)

TSAVO				
	Epiphyses	Near-epiphyses	Mid-shaft frag.	All bones
N. assemblages	3	3	3	3
X+/- S.D.	38,8+/-34,6	24,9+/-22	28,5+/-27,7	24,1+/-18,1
95% C.I.	(0-100)	(0-80)	(0-98,5)	(0-70)

Table 2 - Tooth mark percentages in the experiments with baboons.
Analysis of 95% C. I. for both sets of experiments. They were calculated using the t distribution, where $t_{0.025}$ is the critical value of t with n-1 degrees of freedom.

exposed. Only those cases in which baboons picked up bones or moved them from their original location were counted as actual experiments. After some unsuccessful attemps, in 3 experiments baboons moved and gnawed some of the bones. Each of these experiments consisted of 2 broken and demarrowed bones, with several bone splinters adhering to epiphyseal fragments due to the periostium. They were exposed under controlled circumstances in different places. To minimize the risk of intervention of other scavengers, 2 of them were placed near the river close to the point where baboons had repeatedly been observed to come early in the morning to drink. The other experimental assemblage was placed at a sleeping site used by the troop under observation. In none of the 3 cases did any other carnivore intervene before the retrieval of bones. Experiment 1 was monitored for 2 hours and the baboons did not pay any special attention to bones. They transported some of them a few meters from where I had left them, but did not gnaw any. Experiment 2 was controlled for 3 hours and the baboons were interested in the bones for 40 minutes. Experiment 3 (sleeping site) was monitored 4 hours. Baboons were observed to manipulate the bones at dawn and early hours in the morning.

Results. The study carried out at the zoo yielded a high rate of tooth-marked bones (x=80%) (see Table 2 and Figure 4). Epiphyseal and near-epiphyseal bones were tooth marked at a percentage higher than 80% as is observed on bones from fleshed carcasses consumed by other carnivores (Blumenschine, 1988; Selvaggio, 1994; Capaldo, 1995). Mid-shaft specimens were also highly tooth marked, with a mean slightly lower than 80% and similar to other experimental assemblages in which bones have been ravaged by carnivores. Basically, there is no difference in the number of tooth-marked bones when compared to other experimental carnivore-ravaged assemblages.

Regarding the experiments carried out in Tsavo on wild baboons and defleshed bones, although the sample is small, and therefore the ranges of variation are wide, the mean number of tooth-marked bones is high (Table 2; Figure 4). In one experiment (n1), baboons did not modify any bone at all, although they stayed around the place where bones were exposed for a while and

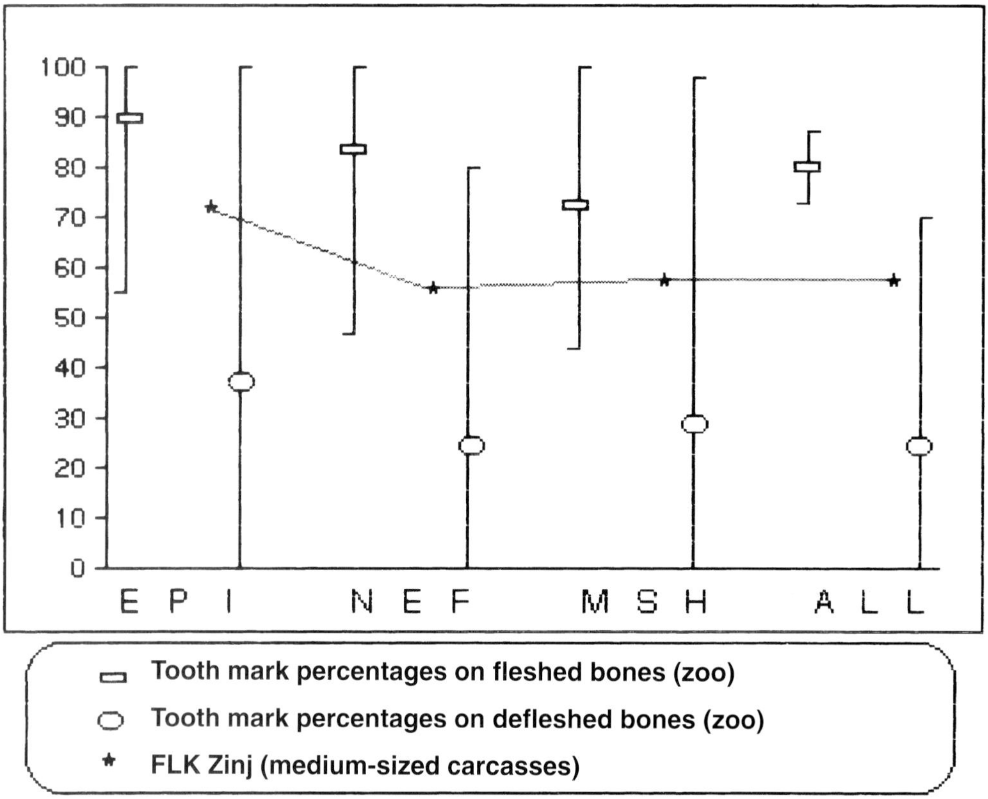

Figure 4 - Percentages of tooth marks obtained in the experiments with baboons.

transported some of them to a nearby place. This created a lower rate of tooth-marked bones than what was obtained when baboons were observed to gnaw defleshed bones. In that case, surprisingly, 28,5% of mid-shaft specimens are tooth marked, which is a higher percentage than that documented for carnivores (especially bone-crunching carnivores; namely, hyenas) in other Hominid-to-Carnivore experimental scenarios (Blumenschine, 1988; Selvaggio, 1994; Capaldo, 1995).

Although this study is at an initial stage and the percentages of tooth-marked bones might be non-representative, because of sample size, it raises three important points:

1. Baboons may be interested in defleshed bones and may leave occasionally as many tooth marks on them as bone-crunching specialized carnivores.

2. Hominids, as primates, might have been the source of some of the tooth marks observed on bones at early sites.

3. The global percentage of tooth-marked bones at a site may be the result of several agencies (e.g. hyenas, jackals, hominids and/or baboons), creating a tooth marking profile not observed in any experimental model. Our experimental models may be of limited value as long as they are single-patterned (e.g. hyenas), instead of multiple-patterned.

This study does not undermine other assertions about the meaning of the particular percentage of tooth-marked bones at FLK Zinj, but simply adds more food for thought.

The „Hominid-to-Carnivore" model revisited: A new experimental scenario

Introduction

„There are several reasons to believe that carnivores restricted the activities of hominids at the Olduvai sites. As noted previously, bones from each site exhibit tooth marks and patterns of breakage attributable to carnivores. Approximately 9 percent of the meat-bearing limb bones are unbroken and undamaged by tool cut marks. Bones unmodified by hominids could have provided edible tissues attractive to carnivores and the tooth marks of carnivores are in fact concentrated on meat-rich and marrow-rich bones. Both cut marks and tooth marks indicate that hominids and carnivores were interested in parts of the same carcasses and sometimes even in parts of the same bone" (emphasis added) (Potts, 1988, p. 343).

Potts' view of the formation of early sites, which he believes were the result of a high degree of competition between hominids and carnivores, is unsubstantiated by physical or referential evidence. The study of competition in modern riparian areas (which are the paleoecological settings where early sites occur) indicates that these habitats exhibit the lowest interaction among carnivores, being safe and suitable for prolonged activities, animal carcass preservation and bone conservation (Blumenschine, 1986, 1989; Cavallo, 1997; Domínguez-Rodrigo, 1997a,b,c). The preservation of complete bones is further proof that hominids and carnivores were not competing over the space and resources on such spots (see Capaldo, 1995). The existence of tooth marks on fossil bones from Olduvai sites simply indicate that carnivores were ravaging the remains abandoned by hominids (Blumenschine, 1988, 1995; Capaldo, 1995). There is no indication of overlap in time in the use of those places (sequential use of space) or in the unlikely case of such an overlap, no competition is suggested between both agents: nocturnal scavengers could have processed bones while hominids were sleeping at trees (Rose & Marshall, 1996; Cavallo, 1997). Such carnivores would have posed no threat to hominids as this situation is also very frequent in modern hunter-gatherer camps. Potts's (1996, p. 112) claim that *„with the exception of vultures and perhaps the cheetah, all such scavengers were potential predators of the smallish toolmakers"* is contradicted by the ethology of such animals. It is hard to believe that jackals and hyenas, which would have been the most usual scavengers attracted to the bone remains discarded by hominids, would have hunted hominids or even intimidated them at all, as these animals tend to avoid humans during daylight (Domínguez-Rodrigo, 1994).

Concerning the main evidence used by Potts, the number of complete bones, to support his claims, it should be stressed that this is not indicative of hominids processing carcasses in a hurry either. Complete bones could be due to hominids simply abandoning them at sites (if carcasses were high-yielding and hominid groups were not very large, complete processing of carcasses would not be necessary) or the result of other carnivores dragging their prey to such spots (turned into common-amenity places). In the latter case, a proportional representation of all limb bones would be expected. However, more than 50% of these complete bones are in fact metapodials (Potts, 1982, 1988). So, even if they lack any human-made modification, they could have been accumulated and abandoned by hominids, as these bones are the lowest-yielding marrow bones. The fact that they belong to animals of different sizes would be a further proof thereof. The absence of tooth marks (which has to be analysed) would go against the hypothesis that the main agent responsible for transporting and processing these bones was a carnivore (Blumenschine, 1988; Capaldo, 1995). Cavallo (1997 p. 152) suggests that *„if hominids processed carcasses in trees like leopards and left some processed and unmodified skeletal elements attached to the skin of an animal, these elements could have remained in the tree until their nutrient value had been* *depleted. After the skin of the carcass decayed, the dry bones would have fallen to the ground and become part of the archaeological record. This scenario could then explain the presence of complete long bones lacking hominid and/or carnivore modifications at the early Olduvai sites that Potts (1988) attributes to inefficient and rapid processing by hominids*". However, the abandonment of complete bones by hominids could only have been possible if the food yields obtained from carcasses were enough so that hominids did not have to exploit them thoroughly; that is, if they had been fleshed (Cavallo, 1996).

Therefore, the presence of several complete limb bones at sites could constitute one of the most important arguments undermining the passive scavenging hypothesis. Capaldo (1995) has convincingly shown that even in riparian habitats, hyenas deplete all the greasy and marrow containing bones. The fact that almost 10% of long bones at the early Olduvai sites are complete, and given that hyena post-ravaging has been documented at these sites (Blumenschine, 1988; Blumenschine & Marean, 1993), suggests not only that Cavallo might be right, but also that the original number of complete bones would have been significantly higher (Capaldo, 1995). Potts (1988, p. 343) observes that several bones exhibit „*patterns of breakage attributable carnivores*" which would not be expected if hominids had broken open all the bones. Capaldo (1995, p. 298) further claims that „*the low incidence of long bone flakes relative to other shaft fragments from FLK 22 suggests that long bones were intact when modified by carnivores*".

If hominids were aiming at marrow rather than flesh, because that is what they would have found on bones from medium-sized carcasses at lion kills (Blumenschine, 1986, 1991), it would not be expected that they had abandoned such a high number of complete limb bones. This suggests that they were exploiting carcasses not for marrow, but for flesh. This assertion is further supported by another important fact: the number of axial bones (exprerssed both in MNE and NISP) is high in most Olduvai sites. For instance, in FLK Zinj, there are more than 800 axial bone remains (the most abundant bone type after limb bones). That makes up 30% of the whole bone assemblage. Axial bones represented at early sites (both in MNE and NISP) vary from 20% to more than 40% (for instance, FLKN6) (see Capaldo, 1995). Capaldo's (1995) experiments demonstrate that axial bones at simulated sites are almost completely consumed by hyenas with a mean of 5% of survival. They are shown to be preferently deleted by these carnivores (Marean et al., 1992). Their high survivorship at Olduvai sites indicates their greater number before hyena ravaging took place at sites and also that hominids were transporting complete carcasses or sections of axial skeletons (devoid of marrow but with plentiful flesh). As Capaldo (1995, p. 350) ackowledges: „*A flesh surplus might help to explain the presence of so many axial elements, pelves and scapulae at FLK 22; that is, why would hominids transport these items to the site if they were defleshed?*" A further proof thereof is the abundance of cut marks found on scapulae, pelves and ribs (Bunn & Kroll, 1986). Capaldo (1995, p.315) further argues that his data „*initially suggest that hominids may have removed moderate to large amounts of flesh from long bones*".

There is another source of evidence that hominids might have abandoned some bones without extracting their marrow contents: the percentage of tooth marks on mid-shaft fragments at the FLK Zinj site. Blumenschine (1988) and Selvaggio (1994) have experimentally demonstrated that the defleshing of long bones from medium-sized carcasses by carnivores is usually reflected in an average of 80% of tooth-marked mid-shaft fragments. The percentage is similar in processes in which carnivores deflesh and demarrow bones, and those in which they simply deflesh carcasses and bones are broken by humans to extract the marrow (Blumenschine, 1988; Selvaggio, 1994). The percentage of tooth marks on such bone sections at the FLK Zinj site (57%) is outside the 95% confidence interval of these experiments, in which carnivores were primary agents of carcass defleshing, and those in which human defleshing and demarrowing of all the bones was followed by carnivore post-ravaging (Blumenschine, 1988; Capaldo, 1995). The only model that fits this percentage of tooth marks is the „Whole bone-to-carnivore“ scenario, in which bones defleshed by humans were demarrowed by carnivores (Capaldo, 1995).

Based on this fact, and on Potts's suggestion that several of the bones at Olduvai sites were broken by carnivores, according to the damage patterns observed on them, as well as on Capaldo's analysis of flake sizes, indicative of the same process, I tried to experimentally reproduce such a scenario. Below, I present the results of a new series of dual-patterned experiments on the „Hominid-to-Carnivore“ model, under the assumption that food yields from largely fleshed carcasses might have led to the abandonment by hominids of several marrow-bearing intact bones. This fact and the carnivore ravaging undergone by such bone accumulations might account for the particular percentage and distribution of the tooth marks found on bone surfaces at the FLK 22 Zinj site in Olduvai Gorge. Such a model such be considered as a mixture of „Whole bone-to-Carnivore“ and „Hammerstone-to-Carnivore“ experimental scenarios, using Blumenschine's (1988) and Capaldo's (1995) terminology.

Method

The study was carried out in the Maasai Mara reserve and Tsavo East National Park (Figure 3). Four different experiments were conducted: two in Maasai Mara and two in Tsavo. As in my previous experimentation, I focussed on long bones from medium-sized (size 3) carcasses, because these are predominant at early sites. Two wildebeest (Maasai Mara) and two cows (Tsavo) were used for this study. Each assemblage was composed of the 12 long limb bones from one carcass. Half of them (one front limb and one hind limb) were abandoned defleshed and demarrowed, and the other half were left defleshed but unbroken and with their marrow contents intact. Butchery was made by local people with metal knives. Then, the bones selected for demarrowing were also extensively broken by local assistants by means of quartz and basalt hammerstones. The complete bones and the fragments of each assemblage created through demarrowing were exposed (without removing the periostium) to the action of scavengers (namely, hyenas and jackals) for 13 days. Then, bones were recovered and cleaned by boiling them in a solution of water and neutral detergent (for more information about the procedures see Domínguez-Rodrigo, 1997a,b). After cleaning, each specimen was carefully examined with the aid of 16x lenses, and percussion and tooth marks were noted, taking into account the specimen on which they occurred. The definition of each bone section followed in this study is the same as those used by Blumenschine (1988): epiphyseal (bearing all or a portion of the articular surface), near-epiphyseal (lacking articular surface, but showing cancellous tissue on the medullary surface) and mid-shaft (lacking articular surfaces and cancellous tissue). The reason for using Blumenschine's nomenclature is for the sake of comparative purposes with his and others' (Capaldo's & Selvaggio's) data. Cut marks were not included in this study.

Results

The occurrence of percussion marks: Although the aim of this study was to determine whether the tooth mark percentages obtained were comparable to other alternative scenarios in which hominids played a secondary role in carcass processing, one of the main objections against it is the differential distribution of percussion marks and their lack of overlapping with other experimental results. In this regard, the mean number of bones with percussion marks is significantly lower than those obtained in other „Hominid-to-Carnivore“ experiments (Blumenschine, 1995; Capaldo, 1995). However, there is a significant overlap in all bone sections between both sets of experiments. All the percussion-marked bone sections (epiphyseal, near-epiphyseal and mid-shaft

n. assemblages	N. complete MNE	Epiphyses	Near-epiphyses	Mid-shaft frag.	Total
4	4	9	54	270	333

ALL BONES

Exp. number	Complete MNE	NISP recovered
1	-	95
2	-	93
3	1	71
4	3	74

MID-SHAFTS

Dimensions (cm)	Percentages
2-3	27 (73/270)
3-4	17 (46/270)
4-5	15 (40/270)
5-6	11,8 (32/270)
6-7	9,2 (25/270)
7-8	8,1 (22/270)
8-10	7 (19/270)
>10	4,8 (13/270)

TOOTH MARKS

	Epiphyses	Near-epiphyses	Mid-shaft frag.	All bones
N. assemblages	4	4	4	4
X+/- S.D.	80+/-17,6	49,6+/-17,2	39,1+/-14	42,2+/-12,3
95% C.I.	(52-100)	(22,1-77,1)	(16,9-61,3)	(22,6-61,8)

PERCUSSION MARKS

	Epiphyses	Near-epiphyses	Mid-shaft frag.	All bones
N. assemblages	4	4	4	4
X+/- S.D.	18,2+/-21,2	19+/-4,5	16+/-8,7	16,4+/-6,9
95% C.I.	(0-46,4)	(11,9-26,1)	(2,1-29,8)	(5,4-27,4)

Table 3 - Data of the experimental Hominid-to-Carnivore model presented in this work. Complete MNE indicate the total number of complete elements recovered after experimentation. Mean percentages of tooth-marked and percussion-marked specimens in relation to the total number of epiphyseal, near-epiphyseal and mid-shaft specimens. Analysis of 95% C. I. is included. It was calculated using the t distribution, where $t_{0.025}$ is the critical value of t with n-1 degrees of freedom.

specimens) of the FLK Zinj site appear well within the limits of the range of variation of the experimental sample of this study. It is particularly interesting to observe that although the percentage of percussion-marked bones is higher in the epiphyseal fragments, similar to the percentages obtained in the other Hominid-to-Carnivore models (Blumenschine, 1995; Capaldo, 1995), concerning near-epiphyseal specimens it is far closer to my experiments than to others. Thus, the FLK Zinj sample is shown to be within the 95% confidence interval of both experimental scenarios (Table 3; Figure 5).

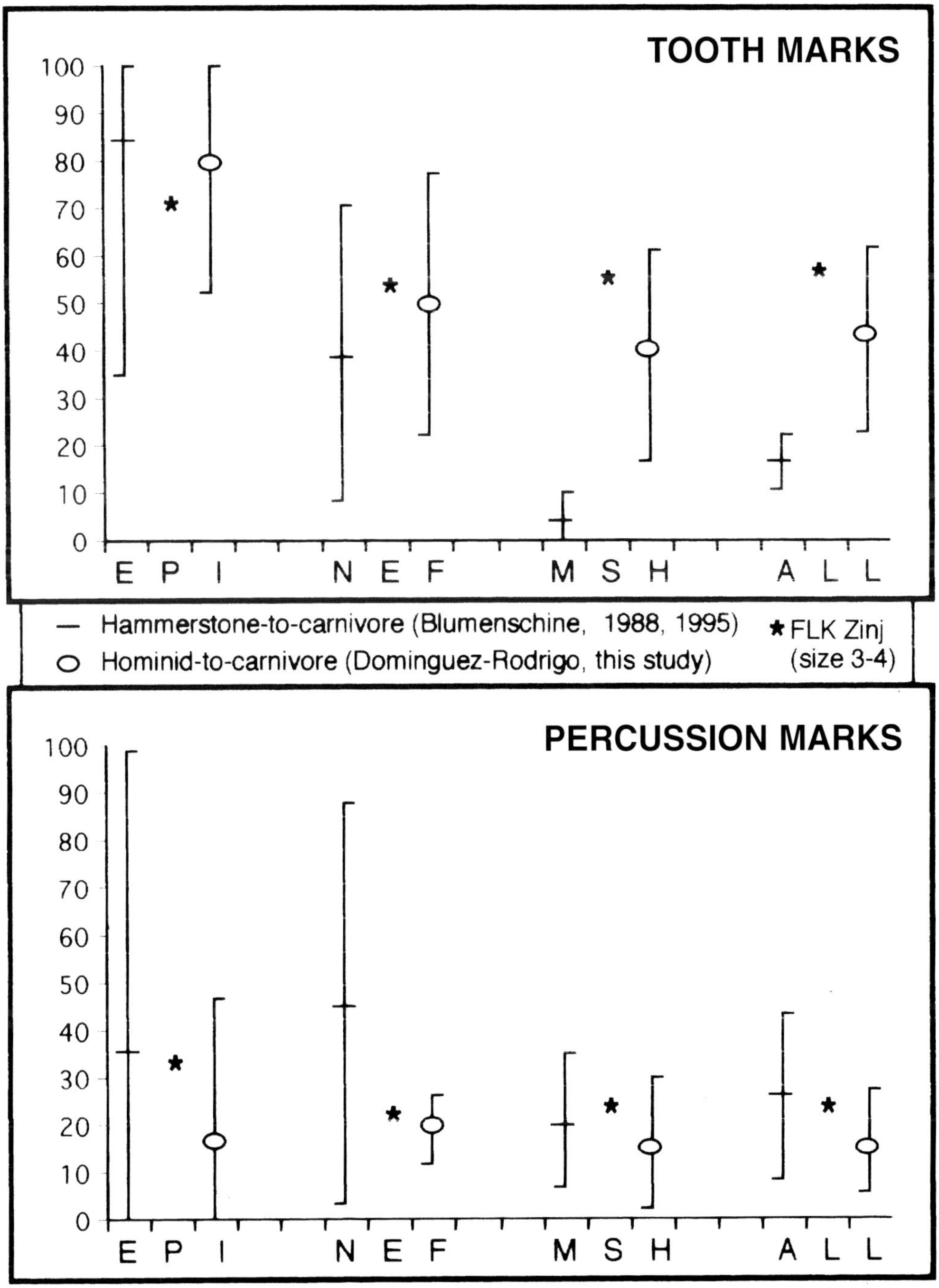

Figure 5 - *Percentages of tooth marks and percussion marks in the study simulating early access of hominids to carcasses, compared to the data from FLK Zinj and Blumenschine's model for early access by carnivores.*
See the difference in tooth marks on mid-shaft specimens for both experiments.

The occurrence of tooth marks: Tooth marks are the focus of this study. Blumenschine (1995, p. 33) observed that „*the rate of tooth marking at FLK Zinjanthropus is substantially lower than that in the carnivore only sample for all size and portion categories for which adequate samples are present. Midshaft fragments are solely responsible for the overall higher rate of tooth marking at FLK Zinjanthropus compared to that in the hammerstone-to-carnivore sample; otherwise, epiphyseal and near-epiphyseal fragments at FLK Zinjanthrpous are tooth marked at a rate predicted by the hominid-to-carnivore sample*". In no experimental study made by Blumenschine (1988, 1995) or Capaldo (1995) simulating early human access to carcasses was there observed such a high rate of tooth-marked mid-shaft bones as in the FLK Zinj site.

In the present study, I obtained a particular pattern of tooth marks according to bone section (Table 3; Figure 5). The mean of tooth-marked epiphyseal specimens is similar to that obtained in other Hominid-to-Carnivore experiments (Blumenschine, 1988, 1995; Capaldo, 1995). This is also true for near-epiphyseal fragments. The only remark that should be made is that in my experiments, the mean percentage of tooth-marked bones is more similar to the mean observed in FLK Zinj than to that obtained in other studies (Blumenschine, 1988, 1995; Capaldo, 1995).

However, the most important result of this study is that the percentage of tooth-marked mid-shaft specimens is very similar to that observed in the FLK Zinj sample. The latter falls within the range of variation of the former. In no other Hominid-to-Carnivore experiment had the FLK Zinj sample been observed to fall within such ranges of variation. Only in the „Whole bone-to-carnivore" model did Capaldo (1995) obtain a similar percentage of tooth-marked mid-shaft fragments; but in this case, the lack of percussion marks invalidates the model as a useful heuristic approach to interpreting the FLK Zinj site.

If we observe the representation of mid-shaft fragments according to their size (Table 3), we can notice there is a similar pattern to that obtained in other experimental studies and at FLK Zinj (Blumenschine, 1988, 1995; Capaldo, 1995). Thus, bias due to bone size representation is not a factor. Even the mean percentage of tooth-marked bones (irrespective of their section) is similar to that reported for the FLK Zinj bone sample.

Discussion and conclusions

I have experimentally modelled a scenario that incorporates some improvements over other Hominid-to-Carnivore studies for interpreting early Olduvai sites. With respect to bone specimen size, there is a similar pattern of representation of fragments longer than 10 cms to that observed at FLK Zinj (Table 3). It is also significantly lower than that obtained in other Hominid-to-Carnivore experiments (Blumenschine, 1988, 1995; Capaldo, 1995). The difference of specimens measuring 2-3 cms is still remarkable, although less pronounced than in other previous studies (Blumenschine, 1988, 1995; Capaldo, 1995). The pattern of bone size representation seems, thus, to better replicate that observed at FLK Zinj than any other experiment.

The most conclusive result is observed in the total number of tooth-marked mid-shaft specimens. The pattern, strikingly similar to that documented in FLK Zinj, suggests that carnivore activity at the site must not have been restricted to the ravaging of demarrowed bones broken by hominids. This should not be so surprising if we seriously consider the diagnostic framework for the 3-patterned Carnivore-to-Hominid-to-Carnivore hypothesis. In assemblages first modified by carnivores (Selvaggio, 1994) or only modified by carnivores (Blumenschine, 1988) the percentage of tooth-marked bones is much higher than that observed at FLK Zinj. If hominids had secondary access to such bone assemblages, we should expect the tooth marking percentages made by primary carnivores and those made by secondary scavengers (after the intervention of hominids) to be significantly higher than those shown at FLK Zinj.

On the other hand, only 1/6 of the complete limb bones used for this study survived carnivore consumption. This suggests, as Capaldo (1995) claims, that the original number of complete bones at FLK Zinj must have been significantly higher than that documented: about 10% of the total MNE of limb bones identified (Bunn, 1982; Potts, 1982, 1988). The tooth marks or cut marks on these complete bones still has to be carefully analysed. Some of these bones reveal, nonetheless, a significant number of modifications made by archaeologists during excavation. This circumstance and the fact that some of them have been covered with glue prevents us from doing clear identifications, as some of the marks may have been obscured by this treatement. On the other hand, tooth marks do not necessarily indicate a primary exploitation by carnivores. They could also have been made by small scavengers, such as jackals (pers. obser.), unable to break the bones. In a brief analysis of some complete elements from other Olduvai sites, I documented no clear mark that could be attributed to carnivores . This would exclude them as primary carcass consumers. Although this observation is suggestive, further analyses of these complete specimens should be carried out.

The fact that hominids may not have broken all the bones they accumulated is reflected in several other sites (e.g. FLK North 6, FxJJ 3 HAS, Gahi 5), in which complete bones from large carcasses exhibit cut marks (indicative of defleshing by hominids) but were not demarrowed. Moreover, variation among archaeological assemblages should also be properly considered. Recent examination of the bone assemblage from FxJj 50 suggest that hominids were not preceded by carnivores in the obtainment and defleshing of carcasses (Domínguez-Rodrigo, in prep.).

Thus, the available evidence supports the hypothesis that hominids were not only the prime accumulators of bones at sites, but also the primary agents to exploit fleshed carcasses. This could have been done through hunting or primary scavenging of undisturbed carcasses, but not by obtaining carcasses at other predators' kills, as suggested by passive scavenging scenarios.

Acknowledgements

I would like to thank the K. W. S. (Kenyan Wildlife Service), the A. D. C. (Agricultural Development Corporation), the National Museums of Kenya, and The Office of the President of Kenya, and especially Mr. Gishangi, Director of K.W.S. at Tsavo East National Park, for their permits and help in field research. I am also indebted to Dr. Karega-Munene, Head of the Archaeology Division in Nairobi, for his permit and help, and Jordi Serrallonga & Victoria Medina, from the Barcelona University, for their help in the study with baboons at the Barcelona zoo. I thankfully acknowledge the permit from J. Xampeny Baró, director of the Barcelona zoo. Despite the discrepancies in the interpretation of hominid behavior shown in the text, I would like to dedicate this paper to R. Blumenschine, whose efforts have shown us the way that taphonomical studies should be made, and whose knowledge and friendship I greatly appreciate.

References

BINFORD, L.R. (1981): *Bones: Ancient men, modern myths*. New York.

BINFORD, L. (1984): *Faunal remains from Klasies River Mouth*. New York.

BINFORD, L.R. (1985): Human ancestors: Changing views of their behavior. *Journal of Anthropological Archaeology* 4, 292-327.

BINFORD, L.R. (1988): Fact and fiction about the *Zinjanthropus* floor: Data, arguments and interpretations. *Current Anthropology* 29, 123-135.

BLUMENSCHINE, R.J. (1986): *Early hominid scavenging opportunities. Implications of carcass availability in the Serengeti and Ngorongoro ecosystems*. Oxford (B.A.R. International Series, 283).

BLUMENSCHINE, R.J. (1988): An experimental model of the timing of hominid and carnivore influence on archaeological bone assemblages. *Journal of Archaeological Science* 15, 483-502.

BLUMENSCHINE, R. J. (1989): A landscape taphonomic model of the scale of prehistoric scavenging opportunities. *Journal of Human Evolution* 18, 345-371.

BLUMENSCHINE, R. J. (1991): Hominid carnivory and foraging strategies, and the socio-economic function of early archaeological sites. *Philosophical Transactions of the Royal Society of London* 334, 211-221.

BLUMENSCHINE, R. J. (1995): Percussion marks, tooth marks and the experimental determinations of the timing of hominid and carnivore access to long bones at FLK *Zinjanthropus*, Olduvai Gorge, Tanzania. *Journal of Human Evolution* 29, 21-51.

BLUMENSCHINE, R. J. & MADRIGAL, T. C. (1993): Long bone marrow yields of some African ungulates. *Journal of Archaeological Science* 20, 555-587.

BLUMENSCHINE, R. J. & MAREAN, C. W. (1993): A carnivore's view of archaeological bone assemblages. In: J. Hudson (ed.), *From bones to behavior: Ethnoarchaeological and experimental contributions to the interpretations of faunal remains,* 273-300. Carbondale/Ill.

BLUMENSCHINE, R. J. & SELVAGGIO, M. M. (1988). Percussion marks on bone surfaces as a new diagnostic of hominid behavior. *Nature* 333, 763-765.

BLUMENSCHINE, R. & SELVAGGIO, M. (1991): On the marks of marrow bones by hammerstones and hyenas: Their anatomical patterning and archaeological implications. *Proceedings of the 11th Congress of the Union of International Sciences, Mainz, September 1987,* 17-32.

BRAIN, C.K. (1981): *The hunters or the hunted.* Chicago.

BUNN, H.T. (1981): Archaeological evidence for meat-eating by Plio-Pleistocene hominids from Koobi Fora, Kenya. *Nature* 291, 574-577.

BUNN, H.T. (1982): *Meat-eating and human evolution: Studies on the diet and subsistence patterns of Plio-Pleistocene hominids in East africa.* Ph. Dissertation. University of California, Berkeley.

BUNN, H.T. (1983): Evidence on the diet and subsistence patterns of Plio-Pleistocene hominids at Koobi Fora, Kenya, and at Olduvai Gorge, Tanzania. In: J. Clutton-Brock (ed.), *Animals and archaeology: Hunters and their prey,* 21-30. Oxford (B.A.R. International Series, 163).

BUNN, H.T. (1991): A taphonomic perspective on the archaeology of human origins. *Annual Revue of Anthropology* 20, 433-467.

BUNN, H.T. & KROLL, E.M. (1986): Systematic butchery by Plio-Pleistocene hominids at Olduvai Gorge, Tanzania. *Current Anthropology* 27, 431-452.

BUNN, H.T. & EZZO, J.A. (1993): Hunting and scavenging by Plio-Pleistocene hominids: Nutritional constraints, archaeological patterns, and behavioural implications. *Journal of Archaeological Science* 20, 365-398.

CAPALDO, S.D. (1995): *Inferring hominid and carnivore behavior from dual-patterned archaeological assemblages.* Ph.D. Thesis. Rutgers University, New Brunswick.

CAPALDO, S.D. & BLUMENSCHINE, R.J. (1994): A quantitative diagnosis of notches made by hammerstone percussion and carnivore ganwing in bovid long bones. *American Anthropologist* 59, 724-748.

CAVALLO, J. A. (1997): *A re-examination of Isaac's Central-Place Foraging hypothesis.* Ph.D. Thesis. Rutgers University, New Brunswick.

CAVALLO, J.A. & BLUMENSCHINE, R. J. (1989): Tree-stored leopard kills: expanding the hominid scavenging niche. *Journal of Human Evolution* 18, 393-399.

DOMÍNGUEZ-RODRIGO, M. (1994), Dinámica trófica, estrategias de consumo y alteraciones óseas en la sabana africana: Resumen de un proyecto de investigación etoarqueológico (1991-1993). *Trabajos de Prehistoria* 51, 15-37.

DOMÍNGUEZ-RODRIGO, M. (1996): Testing meat-eating in early hominids: An analysis of butchery marks on defleshed carcasses. Paper presented at the *International Congress of Paleontology and Human Evolution, Granada.*

DOMÍNGUEZ-RODRIGO, M. (1997a). Meat-eating by early hominids at the FLK 22 *Zinjanthropus* site, Olduvai Gorge, Tanzania: An experimental approach using cut-mark data. *Journal of Human Evolution* 33, 669-690.

DOMÍNGUEZ-RODRIGO, M. (1997b): A reassessment of the study of cut mark patterns to infer hominid manipulation of fleshed carcasses at the FLK *Zinj* 22 site, Olduvai Gorge, Tanzania. *Trabajos de Prehistoria* (submitted).

DOMÍNGUEZ-RODRIGO, M. (1997c): Flesh availability and bone modification in carcasses consumed by lions. *Palaeogeography, Palaeoclimatology, Palaeoecology* (submitted).

DOMÍNGUEZ-RODRIGO, M. (1997d). „Central-place" models revisited and reassessed: A study of carnivore competition in riparian and open habitats of modern savannas. *Journal of Human Evolution* (submitted).

DOMÍNGUEZ-RODRIGO, M. (in prep.): *Bone surface modifications in the FxJj 50 site, Koobi Fora (Kenya).*

DOMÍNGUEZ-RODRIGO, M. & MARTÍ LEZANA, R. (1996): Un estudio etnoarqueologico de un campamento temporal Ndorobo (Maasai) en Kulalu (Kenia). *Trabajos de Prehistoria* 53, 131-143.

DOMÍNGUEZ-RODRIGO, M., SERRALLONGA, J. & MEDINA, V. (n.d.): Disponibilidad de alimento y estrés social en papiones. Junio, 1996.

HAYNES, G. (1991): *Mammoths, mastodonts and elephants.* Cambridge.

ISAAC, G. L. (1983): Bones in contention: Competing explanations for the juxtaposition of Early Pleistocene artifacts and faunal remains. In: J. Clutton-Brock & C. Grigson (eds.), *Animals and archaeology: Hunters and their prey,* 3-19. Oxford (B.A.R. International Series, 163).

ISAAC, G. L. (1984): The archaeology of human origins: Studies of Lower Pleistocene in East Africa, 1971-1981. *Advances in World Archaeology* 3, 1-87.

LEAKEY, M. (1971): *Olduvai Gorge.* Vol. 3. *Excavations in Bed I and II, 1960-63.* Cambridge.

LYMAN, R.L. (1995): *Vertebrate taphonomy.* Cambridge.

MAREAN, C. W., SPENCER, L. M., BLUMENSCHINE, R. J. & CAPALDO, S. (1992): Captive hyaena bone choice and destruction, the Schlepp effect and Olduvai archaeofaunas. *Journal of Archaeological Science* 19, 101-121.

MARSHALL, F. (1986): Implications of bone modification in a Neolithic faunal assemblage for the study of early hominid butchery and subsistence practices. *Journal of Human Evolution* 15, 661-672.

POTTS, R. (1982): *Lower Pleistocene site formation and hominid activities at Olduvai Gorge, Tanzania.* Ph. Dissertation. Harvard University, Massachussets.

POTTS, R. (1988): *Early hominid activities at Olduvai.* New York.

POTTS, R. (1996), *Humanity's descent. The consequences of ecological instability.* New York.

POTTS, R. & SHIPMAN, P. (1981): Cutmarks made by stone tools from Olduvai Gorge, Tanzania. *Nature* 291, 577-580.

ROSE, L. & MARSHALL, F. (1996): Meat-eating, hominid sociality and home bases revisited. *Current Anthropology* 37, 307-338.

SELVAGGIO, M. M. (1994): *Identifying the timing and sequence of hominid and carnivore involvement with Plio-Pleistocene bone assemblages from carnivore tooth marks and stone-tool butchery marks on bone surfaces.* Ph. D. Dissertation, Rutgers University, New Brunswick.

SHIPMAN, P. (1986): Scavenging or hunting in early hominids: Theoretical frameworks and tests. *American Anthropologist* 88, 27-43.

Chopping Tools as "Useful Possessions": A Pliocene Tool Behaviour Leading to Bigger Brain Size Selection (and Connected Ability for Culture)

Antonio Santangelo

Abstract

The importance that must be attributed to the "usage of tools" as an operating condition is not a new thesis among the theoreticians of human evolution. This usage consequently leads to the selection of larger and more capable brains. Behaviour, under many guises, is one of the main factors of evolution. However, the statement "usage of tools" is generic; the question of the non increase in the cerebral volume of other Homininae (gorilla, chimpanzee) is still left open. The key point that tips the scales is probably not so much the instrument as such, but a new conception of the instrument.

The production of chopping tools made by some hominids lets us perceive an "interpretative" intervention in the making and in the usage of tools. Chopping tools are suitable for repetitive usage. The instrument was conceived of as "useful possession".

This acquisition must of necessity be deemed to have happened. The realisation of useful possessions infers an interpretative capacity, and finds its place within a cultural dimension. Culture must be considered to be the outcome of reflection, the processing of more than one line of thought, which concerns an object or an event that has been experienced. The mental operation during a cultural act demands a series of dramatic transformations, because the object or the event becomes the pivotal point of reference for the acception of new (= cultural) aspects of the same object or event. In the case of hominid manufacturing, the chopping tool acquired the meaning of useful possession.

Initially the brain size of <u>*Australopithecus*</u> *was not in advance of the brain size of the present-day gorilla and chimpanzees. If, however, the mental image of an employment that stretches ahead of the present moment is additionally linked to the production of the utensil, then a selective advantage is established for the species. This condition fosters an enlargement of the cerebral structure and a promotion of intellective applications. Almost certainly the crafted object, through its presence compelled advanced hominids to reflection, then to planning, then to represent future actions in their minds.*

Productions of chopping tools began at 2.7-2.5 My, maybe even earlier. Since at 2.7-2.5 My the Habilis biotype is not yet recognizable, such manufacturing must be attributed to hominid activity. With <u>*Homo habilis*</u> *(2.3-2.1 My) we face an increase of 45 % in brain volume above that of hominids. That shows an increase of the brain size, attained approximately in 0.3-0.5 My.*

The sequence of events and data given above cannot be deemed to constitute a proof, yet it is highly indicative of a powerful selective process that possibly led to the increase in brain volume. This increase was promoted by a new perception of chopping tools and by their being applied in practical day to day ecology.

Other Homininae never experienced a mental commerce on the tool, besides they are unable to project their skills into the future. Hence they did not undergo selective pressure regarding brain size. The realisation of chopping tools, cutting edges, useful possessions, had numerous and notable repercussions on the management of group life. We bring attention to the mental realisation of weapons, to the subsequent pressure for a modification of alimentary habits, hence for a sexually based division of social tasks.

Among mammals, the increase in their brain volume is mostly due to an increase in their body size that happened during their evolutionary course. Equally, an increase in brain volume can be caused by the development of sensorial systems that result more effectual for the control of new ambient niches (Jerison, 1973: 5, 15-16, 88, 345-346, 411; 1976; 1988: 417, 420; Chiarelli, 1995: 98-102). However, these modes of increase are not appropriate if we want to explain the enlargement of the brain that happened in the human lineage. Gorillas and chimps, among present day Homininae (Homininae, after Andrews classification, 1985: 21), *Australopithecus africanus* and *A. robustus* among fossil Homininae, have, or had, brains whose size remained constantly stable around an average of 400-500 cc: Although it is difficult to link the data for the evaluation of body size and brain volume in fossil species, and to compare them with those of living species (Smith, 1996: 451 on), it should be altogether realistic to evaluate the encephalisation quotient of each of these species on the basis of quite modest differences. The differences in favour of hominids are slight (Jerison, 1973: 391, 399; Tobias, 1994: 190-192, 195-197). One must therefore suppose that the cause for the increase in absolute and relative brain volume, which produced with Hominina *Homo*, must be explained with reasons that differ from those dependent on, and correlated to, the evolutionary development of the species, this species' typical condition being that of terrestrial diurnal primates.

Among theoreticians of human evolution, the thesis that attributes importance to the usage of tools is not new. As a form of behaviour, this usage is active for the development of brain size. In one of his works, Tobias likes to mention Engel's statement that "labour created man himself" (Tobias, 1989a: 146). That can be upheld in current terms on the basis of the interrelation between tool-use, cognition, and social behaviour (Gibson & Ingold, 1994; Toth & Schick, 1994: 346). On the whole, behaviour, under various forms, must be considered one of the main factors of evolution (Novák et al., 1986; Leonovičová & Novák, 1987; Piaget, 1967; 1976). However, a statement based on the "use of tools" remains generic in its appellation, and actually leaves open the question of the non increase of brain size in Homininae other than man. In fact, man was once defined the "tool maker animal" (Franklin), and yet today we know that, among Homininae, chimps too show manual abilities, are able to use natural materials and to manufacture elementary tools (Goodall, 1974; Kortlandt & Holzhaus, 1987: 473 on; Boesch, 1994: 171-181). Moreover, Boesch suggests the possibility that Homininae are able to communicate amongst themselves (Boesch, 1991: 81-90; Bodamer et al., 1994: 281 on; Tanner & Byrne, 1996: 162, 166 on). It appears that the denomination Hominina lends itself to focusing on something more than the more philogenetic relationship.

The real and advanced manipulative capacities are indisputable, but the additional fact exists that, when "under training", gorillas and chimps show themselves capable of crossing the threshold of culture. A cultural imprint is proven and evident in manifestations such as: the substitution of properties that are specific to objects (Jensvold & Fouts, 1993: 217-224), a tendency to humour (Patterson & Linden, 1984: 85-93 on, 116-117; Butovskaya & Kozintsev, 1996: 716-717), the capacity to comprehend and manage a sign language (Gardner & Gardner, 1988: 65 on; 1989 a: 337 on; 1989 b: 433 on; Patterson & Linden, 1984: 37, 39, 72, 125-126, 142). These manifestations are cultural "in the proper sense of the concept", and we insist on underlining the specific cultural character of which the species gorilla and chimpanzee are capable. Culture is considered to be the outcome of reflection, the processing of more than one line of thought, which concerns an object or an event, that has been experienced.

The mental operation during a cultural act demands a series of dramatic transformations, because the object or the event becomes the pivotal point of reference for the acception of new (= cultural) aspects of the same object or event.

Within species there is little evidence of culture. Most manifestations usually considered as cultural (because improving the intercourse of a species with nature; because meeting aspects of communication and infrasocial diffusion of information; or because they cope with practical problems) should better be considered as a form of adjustment to ambient opportunities. This statement is valid for all the species, including the other Homininae. When the events that have been experienced have not been subjected to "interpretation", there is no allowable ground for the cultural definition. Homininae different from man are able to produce culture, but their capacity is conditional upon training; in natural life gorilla and chimpanzees do not develop it, or make use of it (Toth, 1993: 791-792; Tomasello et al., 1987: 175 on; Tomasello cit. in Boesch, 1994: 171). Their cerebral abilities are employed in order to adapt to natural ambients, and to meet problems that these present at various times (McGrew, 1994: 154-156, 161; King, 1986: 361 on; Myers Thompson, 1994: 95 on). They succeed through intelligent, effectual operativity, which, however, cannot belong into the cultural definition. Additionally, their ability to produce elementary tools is limited; it does not go beyond the use of "operational concepts" (Benedetti, 1972: 268-269; Susanne, 1993: 143 on, 148). The difference between a mental capacity for "operative concepts" as in apes and monkeys, and a greater "interpretative capacity" as in humans, can be explained on the basis of the difference in brain volume and in the extension of neural networks: Extension and complication of brain network allow new functional neural states. This gap in potential was filled in human evolution through a generative development of structure capable of making a cultural capacity more and more probable (Santangelo, 1996: 21). We have insisted upon these preliminary points so that we can better face the central question of our theme, and ask how one could relate the use of tools with the increase in brain volume that has happened in Hominina *Homo*. Perhaps the key point, capable of disentangling the question, is not so much the existence of tools as such, but "a new conception" of tools. This conception appears to be cultural. In the management of life, a mark with a cultural imprint can be acknowledged only from the transition Plio-Pleistocene. It concerns the activity of lithic production made by some hominids, production which can adhere to a form of "interpretation". In particular, it concerns the production of chopping tools suitable for repetitive usage. A pebble broken by chance, found, used, and then intentionally re-produced, constitutes the late Pliocene tool. The instrument was conceived of as a useful possession, indeed with the mental perspective of being able to use it in the future (Santangelo, 1995: 45 on). This mental realisation must be considered to have logically and necessarily happened. We stress the character of "useful possession", because its realisation manifests a projection in time, and demands a quantum of conscious awareness of what has been achieved by the acting individual, who holds and keeps a stone for an operative function that has not been sensed or known earlier. This integration of intellective traits linked to a tool, goes beyond the state of "operational concepts". Operational concepts are clearly the limit within which other Homininae feel comfortable too.

The more direct way for obtaining a chopping tool is to batter a rock or a stone against a stone that acts as an anvil (Cohen, 1991: 175). The concept of a tool as a useful possession may possibly have formed primitively, as a qualitative leap into thought processing. But it is also possible that, by its very presence, a chance cutting edge may, with time, have become a useful possession. Production of chopping tools began at about 2.7-2.5 My (Tobias, 1989a: 145-146; 1990: 9; Turner, 1986: 424), maybe even earlier. Since at 2.7-2.5 My the Habilis biotype is not yet recognisable, such manufacturing must be attributed to hominid activity, that was earlier than the emergence of the human biotype.

Some populations of "advanced" *Australopithecus africanus* (Tobias, 1988: 293-303; 1989b: 123-125) may have been the makers of these artefacts, and keen observers of the "permanence" of the tool.

We point out the purport of the realisation that has happened, and have to consider it inside an evolutionary landscape with the other Homininae.

Jerison affirms that it is not necessary to possess large brains in order to make simple utensils. Chimpanzees for example have average values of cranial capacity of about 410 cc (Swindler & Wood, 1973: 92). The cerebral size of these primates appears sufficient for an ability to produce elementary tools. *Australopithecus* survived as a species for 4-5 My, and yet did not produce notable increases in brain size (Jerison, 1973: 421). However, if the image of a future usage that stretches ahead of the present is also linked to the production of the tool, then a selective advantage is created for the species that conceives of it, and this condition necessarily fosters an enlargement of the cerebral structure (Jerison, 1973: 430; Chiarelli, 1989: 207-208; Tobias, 1994: 196, 197, 200); whilst processing information, the neural network enlarges its associative capacities (hence the anatomical frame), regulating action and outcomes of processing on higher levels of function.

It is licit to infer growing of cerebral structures through a continuous change in gene frequencies, as suggested by the increase of cranial capacity, observed in the hominid sequences (Henneberg, 1989: 527). Initially the brain size of australopithecs was not in advance of the brain size of present-day anthropoids. But almost certainly the very presence of "the manufact itself" first compelled advanced hominids to reflect, to link the object into a causative sequence, then to represent future actions in their minds. We believe the cerebral structure of *Australopithecus africanus* to have been sufficient for making the tool, and then to channel it into the concept of useful possession in connection with chances of its living. In the same context, and as a consequence, Hominids had the opportunity to experiment novel ecological relationships that were brought about by their own selves as a consequence of their new "persistent" tool. A causative determinism is thus produced, which modifies the conditions of their life style.

Gorillas and chimpanzees "Homininae" on the contrary have not produced an awareness of useful possession, and are unable to project intellective and practical skill into the future (Wagener, 1987: 129; Mitchell, 1994: 224). Manufacturing remains confined to the here and now, and appears bound to the duality: "material used" and "problems to be solved" in immediate solidarity and relationship (Westergaard & Suomi, 1994: 468-469; Tomasello et al., 1987: 182; Visalberghi & Limongelli, 1991: 255). These Homininae are unable to conceive the abstraction of a manufact as an "existing object" whose links with the present situation can be severed, so that it may be stored as a tool ready for use. At the basis of this limitation there are probably a lack of the sense of time and a lack in some neurological wirement.

Gorillas and chimpanzees certainly have operative capacities, potentially cultural capacities, but there is an associated and evident incapacity for the temporal dimension. When correlating all that has been detailed above on this subject, one must give serious consideration to the thesis that a potent drive towards the selection of larger and more capacious brains followed the realisation of tool-useful possessions above that of utensils that were limited to extemporaneous usage. Realisation of useful possession, postponed employment of the tools, demand and advanced processing capacity and increasing number of neuronal links. By the same token, an explanation suggests itself for the reason why other Homininae did not meet brain expansion, in comparison with the enlargement which occurred in hominid-human groups. Gorillas and chimpanzee (higher anthropoids in general) never experienced a mental commerce on the tool, nor did they experience renewed relationships with their ambient situations, hence they did not undergo selective pressures toward increase of the brain size. In order to manage in a simply biological way, they just did not need to.

With the appearance of the Habilis biotype we meet an increase of 45 % in brain volume (from endocasts) above that of hominids (Tobias, 1986: 98-99; 1990: 5). From 450 cc average in

Australopithecus, to 643 cc in *Homo habilis* (Tobias, ibidem). The oldest Habilis biotype provide datings of 2.3-2.1 My. That shows that a great increase of the brain volume was attained approximately in a time span of 0.3-0.5 My, starting from the first production chopping tools at about 2.7-2.5 My, to the first appearances of the specimen Habilis. The sequence of events and dates that have been given cannot be considered to constitute a proof, but the new realisation in tool behaviour is extremely indicative of the possibility of a powerful selection process that led to the increase in brain volume. The increase in terms of volume and also a new modelling of the brain shape, began to manifest themselves at the Plio-Pleistocene boundary. Associative parietal and temporal areas registered the greater increase among the different areas of the cortex.

It would indeed be rash to suggest that a usage of tools that was more complex (also mentally), might have upturned the condition of substantial ambient determinism in which hominids continued to live (Butzer, 1974: 436, 451 on; Moggi Cecchi, 1994: 54-55). However, both the enlargement of the brain and its novel shaping did have consequences. They were relevant to the style of the material management of day to day life, and also to the management of social relationships. Many of the innovations that were promoted could not be scientifically proved, but they do contain an intrinsic necessity, just as today's human habits provide the very ground for their legitimisation.

The greater capacity to carry out a mental processing of practical life experiences enabled the acquisition of a psychological apperception of temporality, and hence, planning. Keeping in mind past episodes and mental anticipations of actions to be achieved, leads to realise a temporal scanning. The condensed yesterday's and today's experience manages beforehand the events of tomorrow.

This greater ability means a better mastering of the lived ambient. We could, on this subject, mention some basic "human" innovations. Let us remember home bases, retreats and the construction of shelters, butchering areas (Rose & Marshall, 1996: 307-319; Bunn & Kroll, 1986: 431-434). We point out that home bases, retreats etc. must be assessed within an integrated perspective which is indicative of a new organisation of group life and socialisation, as well as of an initial trend to modify dietary habits (Chiarelli, 1995: 106). Other Homininae are fruit and foliage eaters, and meat is only an occasional component of the nutrition of some (Teleki, 1981: 303-304 on). (And some hominid species show on their facial mass, bony crests suited for the insertion of powerful cranio-mandibular muscles).

At this point it is opportune to remember that some authors have proposed the thesis that the enlargement of the human brain should be related to the modification and improvement that their diet underwent (Leonard & Robertson, 1996: 125-129; Aiello & Wheeler, 1995: 200, 208): The assumption is implicit that early hominids had become hunters and scavengers. We do not uphold this hypothesis, even if it certainly has elements of truth. We do not share it, mainly because there is no evidence that, consuming more meat, together with the increase in brain size, there should have been a substantial auxological increment for body size and stature. An increment of body and stature has not been demonstrated in hominids (Stanyon et al., 1993: 208; Tobias, 1994: 195-196), at least till *Homo erectus*. The dietary improvement that took place with a consistent hunting may be attributed to the introduction of persistent tools, however, the enlargement of the brain was preparing at times, which were previous to *Homo erectus*.

It is better to link the promotion and selection of larger brains to the acquisition of useful possessions. Such a selective pressure is supported on the rapidity in the increasing of the cerebral volume, and on the repercussions in the application of mental capacities.

The translation of the chopping tool into the image of a cutting edge, and then into the psychological picture of a weapon, produced itself within the ambit of this multifaceted process, which is both anatomical and relatedly neuro-operative. Interpretation and re-interpretation during thought processing lead to shaping further states of „Gestalt". A weapon not only means the development of a predatory habit to be implemented for procuring sustenance, but also includes the development of new feelings within the social groups. Besides, the impact of the tools on social behaviour management has had repercussions in regards to a social division of labour.

We refer to the works by Zihlman, by Tanner, and other authors, who suggested the hypothesis that "digging sticks" might have been the first tools consistently employed by australopithecs (Tanner, 1985: 145, 149, 152, 217, 238). The first bands of hominids might have been made up of gatherers rather than hunters (Stanford & Allen, 1991: 60), and "digging sticks" suggest a basically female invention in a quite remote epoch.

The consequent later step connected with chopping tools gives origin to a sexually based division of social tasks. It is in fact the same sharing out of spheres of activity within the economy of life which, in our own days, is still basically followed by groups of hunter-gatherers. Other modifications of the psychological and factual relationships between men and women intervened as a consequence of the chopping tools, but these modifications, although certain, could in no way be specifically asserted.

The realisation of useful possessions might have been preparatory to the development of cultural capacities in general. This suggestion must not surprise. Even at lower levels in the primate scale, it has been noticed that intellective attitudes can be altered as a consequence of experimental tool using (Westergaard & Suomi, 1995: 217-222; Chiarelli, 1995: 110-111). With this, we mean that the encephalic structure which, in the hominid human line, is enabled to conceive of tool usage and of planning in time, becomes enabled at the same time as a structure capable of thoughts that are correlated to thought processing. Thus, culture is being produced. Technological ability and eminently cultural expressions are two sides of the very basic capacity to process thought. But we must immediately clarify that volume growth of the brain and hardware differentiation alone are to be considered adaptive in biological terms. They are the expression of the neural ontogenesis. The multi faceted forms in which culture may appear only signal the presence of many-sided interpretative paths.

At 2.3-2.1 My, the Habilis biotype shows that it has acquired an additional 45 % of cerebral volume, when compared with the hominid types that preceded it. Even greater cerebral increments were produced in further 0.5 My, from the Habilis biotype up to the Erectus biotype (Tobias, 1990: 5; 1994: 195; Stanyon et al., 1993: 209-211). With *Homo erectus* tools acquire new shapes and can be crafted as a scraper, a hand-axe, anvils, percussion implements, etc. (Ingold, 1994: 337 on; Ullrich, 1992: 32-33; Butovskaya, 1995: 313-314; Zihlman & Lowenstein, 1996: 696-697), according to opportunities for usage and circumstances. Besides, these manufacts were serially reproduced. This fact indicates both mental awareness and planning capacity (Chiarelli, 1989: 195-208). The voluntary action of reproducing the manufact according to criteria of symmetry, linearity, etc., is preceded by the spatial conception and "mental persistence" of the design to be executed (Chiarelli, ibidem). The conceptualisation of useful possession and its application to manufacts had consequences also for interrelations within the group, and between single individuals. A "complementation theory" (Reynolds, 1994: 407 on), which links complementary actions in the management of life is proper for humans only.

Today, there is a tendency to give some importance to forms of collaboration (for common defence, for the sharing of preys) that have been observed in the behaviour of Hominina chimpanzee, but in reality the import of these episodes is overstated, provided they are related to a true concept of collaboration (Tomasello et al., 1987: 182; Power, 1986: 251-266; Knauft, 1991: 405-407; Lancaster, cit. in King, 1986: 368).

As Köhler first pointed out, true collaboration is actually a very infrequent practice among Anthropoids, and the aims that should be shared in real collaboration are, in the intent of each individual, pursued to his own personal advantage (Köhler, 1960: 152, 154-155). On the contrary, man considers and manages ahead any form of reciprocity, and complementation, the food to be shared, the building of composite makings, and so on.

The image of the chopping tool as a weapon may have biased certain contradictory directions of human feelings. The individuals that made up the human groups constituted man's behaviour, as the utensils that united them, in their own turn, shaped their manners and their world of sociability. Next to "reciprocity behaviour" as an operating motivation of a non-contractual type, which is profoundly rooted in humans (Santangelo, 1996), exclusivity of territory was also subjoined. This behaviour sharply contrasts with the territorial indifference generally demonstrated by other Homininae (Washburn & De Vore, 1961: 92-93; Knauft, 1991: 403; Marchant, 1996: 144; Hall, 1973: 100). It is important to be fully aware of this modification, since sympatric groups of other Homininae cohabit more or less peacefully (Hall, 1973: 101; Goodall, 1974: 227, 228 on; Schaller, 1968: 253-254, 261-263, 266), while the human biotypes may not have tolerated the presence of other, sub-human groups. This is so even if the comprehensive numerosity of individuals could not have been high in relation to the vastity of the savannah. By humans, competition for "best sites" (water, butchery, etc.) must be taken into account and can explain conflictuality inter groups (Slurink, 1994: 470). Actuality of weapons may have enhanced this psychological disposition. Sometimes, conditions leading to conflictuality for food resources have been observed in chimpanzee (Lund, 1995: 225, 228-230; Power, 1986: 263; Manson & Wrangham, 1991: 369-371, 376). This remark points to the similarity in the representational capacities and psychological dispositions of chimpanzees. However, displays by chimps stop to the cognitive abilities, and no projections are produced on the cultural side.

A result of the increasing brain volume and differentiation can also be recognised in the neurological programming that governs a spoken language. There is a general agreement about connections between tool production and language. Both abilities come out from left-lateralized cerebral structures (Chiarelli, 1989; Gibson, 1994a: 190-191). However, the specific course of development of language is much discussed.

We point out that the new statement of tool production, complementation of tasks and actions, and management of life, being strictly correlated, could have functioned as pressing conditions for the development of language (Bickerton, cit. in Slurink, 1994: 469; Gibson, 1994 b: 265, 266) through the variety and reproduction of their stimuli. So realisation of the first previsional tools is thought to have contributed to the development of language.

References

AIELLO, L.C. & WHEELER, P. (1995): The expensive-tissue hypothesis. The brain and the digestive system in human and primate evolution. *Current Anthropology* 36, 199-221..

ANDREWS, P. (1985): Family group systematics and evolution among catarrhine primates. In: E. Delson (ed.), *Ancestors. The hard evidence,* 14-22. New York.

BENEDETTI, G. (1972): *Neuropsicologia. Le attività psichiche in una sintesi di psicologia sperimentale, neurofisiologia e psichiatria.* Milano.

BODAMER, M., FOUTS, D., FOUTS, R.S. & JENSVOLD, M. (1994): Functional analysis of chimpanzee (*Pan troglodytes*) private signing. *Human Evolution* 9, 281-296.

BOESCH, C. (1991): Symbolic communication in wild chimpanzees? *Human Evolution* 6, 81-89.

BOESCH, C. (1994): Aspects of transmission of tool-use in wild chimpanzees. In: K. Gibson & T. Ingold (eds.), *Tools, language and cognition in human evolution,* 171-184. Cambridge.

BUNN, H. & KROLL, E. (1986): Systematic butchery by Plio/Pleistocene hominids at Olduvai Gorge, Tanzania. *Current Anthropology* 27, 431-452.

BUTOVSKAYA, M. (1995): Some aspects of hominid socioecology according to primatological data. In: H. Ullrich (ed.), *Man and environment in the Palaeolithic*, 309-316. Liège.

BUTOVSKAYA, M. & KOZINTSEV, A. (1996): A neglected form of quasi-aggression in apes: Possible relevance for the origins of humor. *Current Anthropology* 37, 716-717.

BUTZER, K.W. (1974): *Environment and archeology. An ecological approach to prehistory.* Chicago.

CHIARELLI, B. (1989): Spatial coordination, gestural communication, the origin of language and erebral lateralization in man. *The Mankind Quarterly* 29, 195-210.

CHIARELLI, B. (1995): Le basi evolutive dell'intelligenza. *Antropologia Contemporanea* 18, 97-117.

COHEN, B. (1991): Systems thinking in the study of culture. In: *The peculiarity of man. Sociobiological perspective and other approaches*, 75-121. Poznan.

GARDNER, R. & GARDNER, B. (1988): The role of cross-fostering in sign language studies of chimpanzees. *Human Evolution* 3, 65-79.

GARDNER, R. & GARDNER, B. (1989a): Early signs of language in cross-fostered chimpanzees. *Human Evolution* 4, 337-365.

GARDNER, R. & GARDNER, B. (1989b): Prelinguistic development of children and chimpanzees. *Human Evolution* 4, 433-460.

GIBSON, K. (1994a): Overlapping neural control of language, gesture and tool-use. In: K. Gibson & T. Ingold (eds.), *Tools, language and cognition in human evolution,* 187-192. Cambridge.

GIBSON, K. (1994b): Tool use, language and social behaviour in relationship to information processing capacities. In: K. Gibson & T. Ingold (eds.), *Tools, language and cognition in human evolution,* 251-270. Cambridge

GIBSON, K. & INGOLD, T. (eds.) (1994): *Tools, language and cognition in human evolution.* Cambridge.

GOODALL, J. (1974): *In the shadow of man.* Milano (Ital. Transl.).

HALL, K.R. (1973): L'aggressività nelle società delle scimmie antropomorfe e delle scimmie in genere. In: J. Carty & F. Ebling (eds.), *The natural history of aggression*, 85-104. Milano (Ital. Transl.).

HENNEBERG, M. (1989): Morphological and geological dating of early hominid fossils compared. *Current Anthropology* 30, 527-529.

INGOLD, T. (1994): Tools, techniques and technology. In: K. Gibson & T. Ingold (eds.), *Tools, language and cognition in human evolution,* 337-345. Cambridge.

JENSVOLD, M.L. & FOUTS, R.S. (1993): Imaginary play in chimpanzees (*Pan troglodytes*). *Human Evolution* 8, 217-227.

JERISON, H.J. (1973): *Evolution of the brain and intelligence.* New York.

JERISON, H.J. (1976): Paleoneurology and the evolution of mind. *Scientific American* 234, 1, 90-101.

JERISON, H.J. (1988): Special issue of human evolution on intelligence and evolutionary biology. *Human Evolution* 3, 417-422.

KING, B.J.: (1986): Extractive foraging and the evolution of primate intelligence. *Human Evolution* 1, 361-372.

KNAUFT, B.M. (1991): Violence and sociality in human evolution. *Current Anthropology* 32, 391-428.

KÖHLER, W. (1960): *Intelligenzprüfungen an Anthropoiden.* Firenze (Ital. Transl.).

KORTLANDT, A. & HOLZHAUS, E. (1987): New data on the use of stone tools by chimpanzees in Guinea and Liberia. *Primates* 28, 4.

LEONARD, W. & ROBERTSON, M. (1996): On diet, energy metabolism, and brain size in human evolution. *Current Anthropology* 37, 125-129.

LEONOVIČOVÁ, V. & NOVÁK, V.J. (1987): *Behaviour as one of the main factors of evolution.* Praha.

LUND, M.H. (1995): Prehuman genocide. *Human Evolution* 10, 225-231.

MANSON, J. & WRANGHAM, R. (1991): Intergroup aggression in chimpanzees and humans. *Current Anthropology* 32, 369-390.

MARCHANT, L. (1996): The great apes revisited. *Current Anthropology* 37, 142-147.

McGREW, W.C.: (1994): The intelligent use of tools: Twenty propositions. In: K: Gibson & T. Ingold (eds.), *Tools, language and cognition in human volution,* 151-170. Cambridge.

MITCHELL, R.W. (1994): The evolution of primate cognition: Simulation, self-knowledge and knowledge of other minds. In: D. Quiatt & J. Itani (eds.), *Hominid culture in primate perspective*, 177-232. Niwot, Co.

MOGGI CECCHI, J. (1994): Hominid evolution: The Southern African contribution. *Antropologia Contemporanea* 17, 53-56.

MYERS THOMPSON, J. (1994): Cultural diversity in the behaviour of *Pan*. In: D. Quiatt & J. Itani (ed.): *Hominid culture in primate perspective,* 95-115. Niwot, Co.

NOVÁK, V.J., VANČATA, V. & VANČATOVÁ, M. (1986): *Adaptation, behaviour and evolution.* Praha.

PATTERSON, F. & LINDEN. E. (1984): *The education of Koko.* Milano (Ital. Transl.).

PIAGET, J. (1967): *Biologie et connaissance.* Paris.

PIAGET, J. (1976): *Le comportément, moteur de l'évolution.* Paris.

POWER, M. (1986): The foraging adaptation of chimpanzees, and the recent behaviours of the provisioned apes in Gombe and Mahale National Parks, Tanzania. *Human Evolution* 1, 251-265.

REYNOLDS, P. (1994): The complementation theory of language and tool use. In: K. Gibson & T. Ingold (eds.), *Tools, language and cognition in human evolution,* 407-428. Cambridge.

ROSE, L. & MARSHALL, F. (1996): Meat eating, hominid sociality, and home bases revisited. *Current Anthropology* 37, 307-338.

SANTANGELO, A. (1995): *Chopping tools as "useful possessions": A qualitative jump in thought processing. Brain network and cultural shape.* Milano.

SANTANGELO, A. (1996): *An outline of the building of culture.* Milano.

SCHALLER, G.B. (1968): *The year of the gorilla.* Milano (Ital. Transl.).

SLURINK, P. (1994): *Causes of our complete dependence on culture. The ethological roots of culture.* Kluwer.

SMITH, R. (1996): Biology and body size in human evolution. *Current Anthropology* 37, 451-481.

STANFORD, C. & ALLEN, J. (1991): On strategic storytelling: Current models of human behavioural evolution. *Current Anthropology* 32, 58-61.

STANYON, R., CONSIGLIERE, S. & MORESCALCHI, M.A. (1993): Cranial capacity in hominid evolution. *Human Evolution* 8, 205-216.

SUSANNE, C. (1993): Animals, a matter of definition, genetic manipulation, a matter of philosophy. *Global Bioethics* 6, 143-148.

SWINDLER, D. & WOOD, C. (1973): *An atlas of primate gross anatomy. Baboon, chimpanzee, and man.* Washington.

TANNER, J. & BYRNE, R. (1996): Representation of action through iconic gesture in a captive lowland gorilla. *Current Anthropology* 37, 162-173.

TANNER, N.M. (1985): *On becoming human.* Bologna (Ital. Transl.).

TELEKI, G. (1981): The omnivorous diet and eclectic feeding habits of chimpanzees in Gombe National Park, Tanzania. In: R. Harding & G. Teleki (eds.), *Omnivorous primates. Gathering and hunting in human evolution,* 303-343. New York.

TOBIAS, P.V. (1986): Single characters and the total morphological pattern redefined: The sorting effected by a selection of morphological features of the early hominids. In: E. Delson (ed.), *Ancestors. The hard evidence,* 94-101. New York.

TOBIAS, P.V. (1988): Numerous apparently synapomorphic features in *Australopithecus robustus, Australopithecus boisei* and *Homo habilis*: Support for the Skelton-McHenry-Drawhorn Hypothesis. In: F. Grine (ed.), *Evolutionary history of the "robust" australopithecines,* 293-308. New York.

TOBIAS, P.V. (1989a): The status of *Homo habilis* in 1987 and some outstanding problems. In: G. Giacobini (ed.), *Hominidae,* 141-150. Milano.

TOBIAS, P.V. (1989b): Hominid variability, cladistic analysis and the place of *Australopithecus africanus.* In: G. Giacobini (ed.), *Hominidae,* 119-128. Milano.

TOBIAS, P.V. (1990): Some critical steps in the evolution of the hominid brain. In: J.C. Eccles & O. Creutzfeldt (eds.), *The principles of design and operation of the brain,* 1-16. Roma.

TOBIAS, P.V. (1994): The craniocerebral interface in early hominids. Cerebral impressions, cranial thickening, paleoneurobiology, and a new hypothesis on encephalization. In: R.S. Corruccini & R.L. Ciochon (eds.), *Integrative paths to the past,* 185-203. Englewood Cliff, NJ.

TOMASELLO, M., DAVIS-DASILVA, M., CAMAK, L. & BARD, K. (1987): Observational learning of tool-use by young chimpanzees. *Human Evolution* 2, 175-183.

TOTH, N. (1993): From the hand of *Pan. Current Anthropology* 34, 791-792.

TOTH, N. & SCHICK, K. (1994): Early stone industries and references regarding language and cognition. In: K. Gibson & T. Ingold (eds.), *Tools, language and cognition in human evolution,* 346-362. Cambridge.

TURNER, A. (1986): Species, speciation and human evolution. *Human Evolution* 1, 419-430.

ULLRICH, H. (1992): Human evolutionary research in Eastern Germany. *Human Evolution* 7, 31-39.

VISALBERGHI, E. & LIMONGELLI, L. (1991): Comprehension of cause-effect relationship in tool-using capuchin monkeys (*Cebus apella*). *International Journal of Anthropology* 6, 255.

WAGENER, J.S. (1987): The evolution of man's sense of time. *Human Evolution* 2,121-133.

WASHBURN, S. & De VORE, I. (1961): *Social behaviour of baboons and early man. Social life of early man.* Chicago.

WESTERGAARD, G.C. & SUOMI, S.J. (1994): Stone-tool bone-surface modification by monkeys. *Current Anthropology* 35, 468-470.

WESTERGAARD, G.C. & SUOMI, S.J. (1995): Mirror inspection varies with age and tool-using ability in tufted capuchin monkys (*Cebus apella*). *Human Evolution* 10, 217-223.

ZIHLMAN, A. & LOWENSTEIN, J. (1996): A Spanish Olduvai? *Current Anthropology* 37, 695-697.

Why Cultural Behaviour Became a Part of Early Hominid Adaptation Strategies (On the Causes of the First „Cultural Revolution“)

Leonid B.Vishnyatsky

Abstract

While it is obvious that intellectual capabilities of chimpanzees and some other primates are quite sufficient to become a basis of rather complex forms of cultural behaviour, it is no less obvious that under natural conditions even the behavioural repertoire of the most „clever“ of non-human animals includes only single, isolated elements of culture that do not form any integral system which would have vital importance. Only in humans their survival directly depends on their „involvement“ in culture, and it was just this dependence upon a formerly disused, neglected way of behaviour that once generated quite new selective pressures, resulting in the emergence of Homo. If so, then the principal problem of human evolution may be formulated as follows: What did force our ancestors to use their cultural potentialities more actively than the progenitors of chimpanzees and other, now extinct, hominoids did? It can be hypothesized, first, that well before the hominids had to change forest habitats for open or mosaic ones they had already been anatomically pre-adapted in such a way that on the ground bipedalism was for them the most fit mode of locomotion, and, second, that originally the bipedality was their weak point not strong. Though moving on two legs better than on four, they still lost in many important respects in comparison with the quadruped apes and monkeys, which placed them at a disadvantage and, other things being equal, would render their chance for survival in changing environments rather low. Something was needed to compensate the negative consequences of bipedalism, and it was just this need that caused the initial development of culture in hominids. To survive they had to rely first of all on their cultural capacities. Performing the compensatory function, various forms of cultural activity necessarily continue to play the more and more important role in one or several groups (populations?, species?) of early hominids, until the corresponding anatomical and behavioral specializations became irreversible.

Introduction

The only substantial difference between the evolution of man and that of all the other beings is that the former process was caused by the necessity to adapt to not only and (beginning from some moment) not so natural environments as to the cultural ones, which themselves appeared and developed as means of adaptation. These artificial environments of our species is specialized to we call now culture, and this is culture humans owe both their appearance and existence. The big and complex brain, the forelimbs capable of various manipulations, as well as many other uniquely human features represent a product of natural selection guided by culture, which therefore is both man's creature and his creator. Taken in its widest sense the term culture is used to designate the totality of psychical and material phenomena which in certain respect are opposed to the natural world. Culture is a non-biological phenomenon (though its bearers are creatures of flesh and blood), it is a result of some particular behaviours which depend on and are organized by not only genes but also, and first of all, ideas. Culture in its widest sense includes all forms of behaviour based on non-genetically and, no less important, non-mechanically (i.e. through deliberate choice) acquired, accumulated and transmitted information, as well as the results of these behaviours (both material and ideal).

Accepting the above definition one cannot help acknowledging that at least some manifestations of culture can be seen in behaviours of non-human animals. Needless to say that first of all this applies to higher primates. However, while it is obvious that intellectual capabilities of the latter are quite sufficient to become a basis of rather complex forms of cultural behaviour, it is no less obvious that under natural conditions even the behavioural repertoire of the most „clever" of non-human primates includes only single, isolated elements of culture that do not form any integral system which would have vital importance. Only in humans their survival directly depends on their „involvement" in culture, and it was just this dependence upon a formerly disused, neglected way of behaviour that once generated quite new selective pressures, resulting in the divergence of evolutionary pathways of our ancestors and all the other hominoids. If so then the principal question may be formulated as follows: What did force our ancestors to use their cultural potentialities (and first of all intellectual abilities) more actively than the progenitors of chimpanzees and other, now extinct, hominoids did, who probably possessed the same or nearly the same capabilities? Why did tool use, tool manufacture and other elements of cultural behaviour cease to be something incidental for hominids and began to assume ever greater, crucial importance? Put in other words: What did cause the first „cultural revolution"? To answer these questions is to understand what the „starting device" of human evolution was and how it worked, which I consider the main problem of human evolution.

Bipedalism in the theories of human origins: A fragile link

For all their diversity the theories explaining the emergence of man rely almost invariably on two events which are considered to be of decisive importance for the beginning of the process of hominizatiion. These events are, first, the transition of some hominoids from mainly arboreal way of life in forests to mainly terrestrial existence in open or mosaic landscapes, and, second, the adoption by them of bipedal locomotion. It is generally thought that the first, having compelled the ancestors of hominids to adapt themselves to new, unwonted environments, urged them on the search of new ecological niches and thus stimulated the development of tool behaviour, sociality and so on, while the second, resulting in the emancipation of the upper limbs of the supporting and motorial functions, served as a necessary precondition of such a development. If we were able to explain: 1) What led to the change of habitats, 2) what caused the change of locomotion, and - above all - 3) why these two events have rendered the adaptation through usual biological ways impossible or insufficient, forcing the hominids to use their cultural potentiality as a last resort, then the main problem of human evolution could be considered resolved (at least in general outline). Meanwhile, only the first of the three questions can be answered more or less convincingly now. As to the causes and consequences of the transition to bipedality, the degree of clearness is here inversely proportional to the ever-increasing number of hypotheses. In theoretical constructions postulating various successions of interrelated events in human evolution this point, therefore, represents just that fragile link because of which all the chain goes to pieces. As it is impossible to dispense with this link we have to try to „restore" it.

The idea that bipedality had contributed to the development of human mind and culture was put forward by many thinkers of the 18th and 19th centuries (Helvetius, Herder, Lamarck, Wallace, Darwin, Haeckel, Engels). However, the emergence of this property itself did not find due explanation then (Bowler, 1986). Since the 30es-40es of this century predominant has become the thesis about the adaptive character of the transition to bipedality. „In the revived Darwinian climate, the challenge was to show how distinctive human traits could have arisen as adaptive responses to specific environmental conditions..." (Ingold, 1995: 381). The adherents of such an approach consider bipedalism in early hominids an adaptation to changing environments, i.e.

bipedality is thought to be a useful trait which raised the chance of its bearers for survival. All the hypotheses which answer the last criterion (let us designate them as „adaptivist") can be divided into three groups: those which explain bipedality by the necessity to relieve the upper limbs from the supporting and locomotor functions for some other ones (Darwin; Engels; Read, 1920; Washburn & Howell, 1960; Hewes, 1961, 1964; Kortlandt, 1980; Lovejoy 1981; Fifer, 1987); those which see the principal sense of the changes in the acquisition of upright posture (Hardy, 1960; Ravey, 1978); and, finally, those which lay emphasis on the energetic and/or thermoregulatory advantages secured by moving on two legs (Taylor & Rowntree, 1973; Wheeler, 1984, 1991a,b). Finally, some researchers try to solve the problem combining argumentation of several different hypotheses into a single system (Day, 1986).

The very abundance of the simultaneously existing „adaptivist" explanations alongside with the absence of widely accepted views among them are indicative of their weakness. Most of the explanations are based either on strong exaggeration of the importance of various sporadic actions (throwing, brandishing, gesticulating, transportation, sexual display, peeping over tall grass, etc.), easily performed by many apes and monkeys without any noticeable changes in ways of life and locomotion (cf. Rose, 1991: 42), or on assumptions that cannot be confirmed by any facts at all (the „aquatic" hypothesis). Another part of the hypotheses (particularly those included here into the third group) appeal to such advantages of bipedalism which could have become tangible only when the human bipedality was completely formed, but it is unlikely they were perceptible during the initial stages of its formation. As to the explanation which appeals to the tool activity one may object that in the overwhelming majority of cases both to make and to use tools is much easier in sedentary position. This is quite obvious for everybody who has observed apes or archaeologists experimenters working with wooden, stone or bone objects. Besides, judging by the available data the first bipedal hominids had appeared at least two million years earlier than the oldest known stone tools did (4.5 myr and 2.5 myr, respectively), which also impedes the proof of the direct causal link between these two events. It therefore appears very difficult (or impossible) to reveal any concrete advantages which could have been connected with the initial bipedalism in hominids. Even some proponents of the „adaptivist" hypotheses have to acknowledge that the convincing explanation „of the transition to orthograde locomotion has not yet been found" (Alexeev, 1989: 113; cf. Klein, 1989: 400).

I believe that the major cause of the general failure of the „adaptivist" explanations lies just in the initial premise which unites them. This is the logic of „adaptivism" by itself that does a bad in the present case. Usually it is taken for granted that as the bipedality arose it must have secured its bearers some selective advantages, but the latter by no means follows from the former as a matter of course. All organisms possess not only adaptive features but also neutral and even maladptive ones. Is it not possible to assume that in early hominids their bipedality was a maladaptive trait having an adverse effect on their general fitness? It can be shown that such a supposition not only conforms well to facts, but opens rather interesting ways to tackle to the problem of „starting device" of human evolution.

Bipedalism as a bad heritage of the past

As it is widely known, even in modern humans bipedalism is connected with a whole number of anatomical and physiological defects, which cause various difficulties during pregnancy and at birth, disorders in digestion, blood circulation, etc. As to the early stages of human evolution the negative aspects of bipedalism must have been felt much stronger, because the bipedality of australopithecines was, most probably, much less effective than that of *Homo sapiens*. Though some scholars continue to defend the view that the gait of australopithecines was very similar to that of modern humans, the alternative opinion seems more logical and well-founded. Indeed, „must not there have been a period of time when the human ancestor was not as good at moving on two legs as it could become, nor as good at scrambling in the trees as it once had been?“ (Susman et al., 1984: 113). As early as the middle of the 60es, having analyzed all the osteological data present at that time, Napier came to the conclusion that as compared with the walk of modern humans the walk of the first hominids (he meant *Australopithecus*) had been physiologically and energetically inefficient, „with quick, rather short steps, with their knees and hips slightly bent“, so that they could have found „long-distance bipedal travel impossible“ (Napier, 1967: 63). In Russia rather close views were expressed at that time by Yakimov (1966: 69-70). Subsequently the same or very similar opinions have been claimed by many other anthropologists (Ashton, 1981: 85; Tardieu, 1981, 1990: 494; Jungers, 1982; Jungers & Stern, 1983; Susman et al., 1984; Yurovskaya, 1989: 172-173; Clarke & Tobias, 1995; Macchiarelli et al., 1996) and confirmed by new methods of bone analysis (Spoor et al., 1994). Moreover, the study of new materials obtained in recent decades in East Africa have forced many to believe that the early hominids still retained some anatomical characteristics linked with the arboreal way of life, and, evidently, really spent not a small part of their time in the trees (Prost, 1980; Senut, 1980, 1989; Tardieu, 1981; Susman et al., 1984; Aiello, 1994: 399). Thus there is every reason to suppose that originally the bipedality of early hominids adversely affected their adaptability, placing them at a disadvantage in comparison with the close competing species which remained - like modern baboons - quadruped. Of course, such a supposition does not help to answer the question of why our ancestors became bipedal. Quite the contrary, it may seem that it complicates the problem still more. Indeed, if the bipedality was originally disadvantageous why was it let by selection? How could it have developed?

Most probably, the property under consideration was something in a kind of a „legacy“ inherited by early australopithecines from the preceding stages of their evolutionary history. Put in other words, the main morphological characteristics securing the movement on the ground without the help of forelimbs could have appeared as an adaptation to the arboreal rather than terrestrial way of life. One of the first who put forward this idea was Keith (1923: 451-454), subsequently it was sympathetically stated by some other scholars (e.g. Nesturkh, 1958: 116-117), and in recent years, thanks to the appearance of new data on the limbs of Miocene-Pliocene hominoids and hominids, it has been accepted by a whole number of anthropologists (Tuttle, 1981; Langdon, 1985; Senut, 1989; Krantz, 1991). As to the kind of arboreal locomotion that caused the formation of the anatomical prerequisites for bipedality, it is usually thought (one of recent reviews see in Rose, 1991) that it were either climbing or brachiation, which could also have combined and been supplemented with some other ways of locomotion, including those calling for erect posture and supporting role of hindlimbs (e.g. so called cruriation). In some group or groups of hominids the corresponding specializations must have gone so far, that when they would move on the ground even their rather imperfect bipedality turned out to be less inefficient and uncomfortable than quadrupedality. The former „required probably less modifications in the functional locomotor complex than re-adapt the body to a terrestrial quadrupedalism“ (Senut, 1989: 40-41).

A scenario for the first cultural revolution

All the said above allows to think, first, that well before the hominids changed forest habitats for open or mosaic ones they had already been bipedal, and, second, that originally the bipedality was their weak point not strong. Of course, both these statements by no means should be taken for granted, but at the same time they are realistic enough and do not run counter to the available data. If we accept them as working hypotheses the problem of the „starting device“ of human evolution can be solved as follows:

1) Owing to the trend to cooling and aridization characteristic of the Cenozoic (Borzenkova, 1992), during the whole Neogene the areas occupied by tropical forests were gradually decreasing and jungle landscapes were being replaced by more open ones. The environmental conditions became less stable and predictable (Hill, 1987). The decline of tropical forests and resultant decrease in amounts of vital resources rendered the competition between different populations and species of apes and monkeys more acute. Highly specialized arboreal forms, small and mobile, began to supplant hominoids (e.g. Andrews, 1981: table 2; 1986: fig.2). To survive under such conditions the latter had to master new ecological niches, and as a result in late Miocene - early Pliocene some groups of East African apes were forced to spend more and more time on the ground (in the lower strata of forest and/or the adjacent parts of savanna).

2) By that time a part of hominoids had already been anatomically pre-adapted in such a way that bipedalism turned out for them either the only possible or at least most fit mode of locomotion.

3) Though moving on two legs better than on four, they still lost in many important respects (speed, endurance, knack, etc.) in comparison with the quadruped apes and monkeys (to say nothing of predators), which placed them at a disadvantage and, other things being equal, would render their chance for survival rather low.

4) Something was needed to compensate the negative consequences of bipedalism, and this became just that need which caused the initial development of culture in hominids. They had to use their cultural capacities much more intensively than did the quadruped hominoids.

5) Especially significant climatic changes toward aridization took place in the Middle Miocene (about 6 mya) and then at the very end of Pliocene (2.5-3.0 mya), which coincides respectively with the emergence of Hominidae and *Homo* (Brain, 1981; Cronin, 1986: 85; Vrba, 1985, 1993; Reshetov, 1986: 87). Probably, the earliest australopithecines (6-3 mya), who still led a partly arboreal way of life, only slightly differed in their behaviour from present chimpanzees, though it cannot be ruled out that the former more often resorted to the help of various tools, had more complex social relations and ways of communication. As the forest habitats became less and less widespread the compensatory role of culture increased, and by the end of Pliocene the hominids of East (and may be South) Africa advanced rather far in this direction. The next cycle of aridization, which began 2,5-3,0 mya and resulted in further decline of forests and expansion of savannas, found the hominids ready to solve ecological problems at the expense of cultural development and stimulated the latter. The oldest indisputable stone tools date just from that period (Harris 1983; Semaw et al., 1997). These tools were made by either australopithecines (Susman, 1991) or earliest *Homo* (or both), whose existence was now in many respects directly dependent on culture.

6) Performing the compensatory function, various forms of cultural activity necessarily continue to play the more and more important role in one or several groups (populations?, species?) of early hominids, until the corresponding anatomical and behavioral specializations became irreversible.

References

AIELLO, L.C. (1994): Variable but singular. *Nature* 368, 399-400.

ALEKSEEV, V.P. (1989): Anthropogenesis - a resolved problem or a train of new problems? In: *Man in the system of sciences*, 111-121. Moscow.

ANDREWS, P. (1981): Species diversity and diet in monkeys and apes during the Miocene. In: C.B.Stringer (ed.), *Aspects of human evolution*, 25-61. London.

ANDREWS, P. (1986): Fossil evidence on human origins and dispersal. *Cold Spring Harbor Symposia of Quantitative Biology* 51, 419-428.

ASHTON, E.H. (1981): Primate locomotion: some problems in analysis and interpretation. *Philosophical Transactions of the Royal Society (London)* B 292, 77-87.

BORZENKOVA, I.I. (1992): *Climatic change in the Cenozoic*. St.Petersburg. (In Russian).

BOWLER, P.J. (1986): *Theories of human evolution: A century of debate, 1844-1944.* Baltimore.

BRAIN, C.K. (1981): Hominid evolution and climatic changes. *South African Journal of Science* 77, 104-105.

CLARKE, R.J. & TOBIAS, P.V. (1995): Sterkfontein member 2 foot bones of the oldest South African hominid. *Science* 269, 521-524.

CRONIN, J.E. (1986): Molecular insights into the nature and timing of ancient speciation events: Correlates with palaeoclimate and palaeogeography. *South African Journal of Science* 82, 83-85.

DAY, M.H. (1986): Bipedalism: Pressures, origins and modes. In: B.Wood, L.Martin & P.Andrews (eds.), *Major topics in primate and human evolution,* 188-202. London.

FIFER, F.C. (1987): The adoption of bipedalism by the hominids: A new hypothesis. *Human Evolution* 2, 135-147.

HARDY, A.G. (1960): Was man more aquatic in the past? *New Scientist* 7, 642-645.

HARRIS, J.W.K. (1983): Cultural beginnings: Plio-Pleistocene archaeological occurrences from the Afar, Ethiopia. *The African Archaeological Review* 1, 3-31.

HEWES, G.W. (1961): Food transport and the origin of hominid bipedalism. *American Anthropologist* 63, 687-710.

HEWES, G.W. (1964): Hominid bipedalism: Independent evidence for food-carrying theory. *Science* 146, 416-418.

HILL, A. (1987): Causes of percieved faunal change in the later Neogene of East Africa. *Journal of Human Evolution* 16, 583-596.

INGOLD, T. (1995): The hunt for human origins. *Current Anthropology* 32, 381-384.

JUNGERS, W.L. (1982): Lucy's limbs: Skeletal allometry and locomotion in *Australopithecus afarensis. Nature* 297, 676-678.

JUNGERS, W.L. & STERN, J.T. (1983): Body proportions, skeletal allometry and locomotion in the Hadar hominids: A reply to Wolpoff. *Journal of Human Evolution* 12, 673-684.

KEITH, A. (1923): Man's posture. Its evolution and disorders. *British Medical Journal* 1, 451-454, 499-502, 545-548, 587-590, 669-672.

KLEIN, R.G. (1989): *The human career.* Chicago.

KORTLANDT, A. (1980): How might early hominids have defended themselves against large predators and food competitors. *Journal of Human Evolution* 9, 19-112.

KRANTZ, G.S. (1991): Bipedalism as a brachiating adaptation. *Human Evolution* 6, 235-239.

LANGDON, J.H. (1985): Fossils and origin of bipedalism. *Journal of Human Evolution* 14, 615-635.

LOVEJOY, C.O. (1981): The origin of man. *Science* 211, 341-350.

MACCHIARELLI, R., GALICHON, V., BONDIOLI, L. & TOBIAS, P.. (1996): Hip bone trabecular architecture and locomotor behaviour in South African australopithecines. In: *Abstracts. The Sections of the XIII International Congress of Prehistoric and Protohistoric Sciences,* 109-110. Forli.

NAPIER, J. (1967): The antiquity of human walking. *Scientific American* 216, 56-66.

NESTURKH, M.F. (1958): *The origin of man.* Moscow. (In Russian).

PROST, J.H. (1980): Origin of bipedalism. *American Journal of Physical Anthropology* 52, 175-189.

RAVEY, M. (1978): Bipedalism: An early warning system for Miocene hominoids. *Science* 199, 372.

READ, C. (1920): *The origin of man and his superstitions.* Cambridge.

RESHETOV, V.Yu. (1986): *The Tertiary history of higher primates.* Moscow. (In Russian).

ROSE, M.D. (1991): The process of bipedalization in hominids. In: Y.Coppens & B.Senut (eds.), *Origene(s) de la bipedie chez les hominides,* 37-48. Paris.

SEMAV, S., RENNE, P., HARRIS, J.W.K., FEIBEL, C.S., BERNOR, R.L., FESSEHA, N. & , MOVBRAY, K. (1997): 2.5-million-year-old stone tools from Gona, Ethiopia. *Nature* 385, 333-336.

SENUT, B. (1980): New data on the humerus and its joints in Plio-Pleistocene. *Collegium Antropologicum* 4, 87-93.

SENUT, B. (1989): Climbing as a crucial preadaptation for human bipedalism. *OSSA* 14, 35-44.

SPOOR, F., WOOD, B. & ZONNEVELD, F. (1994): Implicattions of early hominid labyrinthine morphology for evolution of human bipedal locomotion. *Nature* 369, 645-648.

SUSMAN, R.L. (1991): Who made the Oldowan tools? Fossil evidence for tool behavior in Plio-Pleistocene hominids. *Journal of Anthropological Research* 47, 129-151.

SUSMAN, R.L., STERN, J.T. & JUNGERS, W.L. (1984): Arboreality and bipedality in the Hadar hominids. *Folia Primatologia* 43, 113-156.

TARDIEU, C. (1981): Morpho-functional analysis of the articular surfaces of the knee-joint in primates. In: *Primate evolutionary biology,* 68-80. Berlin.

TARDIEU, C. (1990): Des millions d'annees pour faire un bipede. *La Recherche* 220, 492-494.

TAYLOR, C.R. & ROWNTREE, V.J. (1973): Running on two legs or four: which consumes more energy? *Science* 179, 186-187.

TUTTLE, R.H. (1981): Evolution of hominid bipedalism and prehensile capabilities. *Philosophical Transactions of the Royal Society (London)* B 292, 89-94.

VRBA, E.S. (1985): Ecological and adaptive changes associated with early human evolution. In: E.Delson (ed.), *Ancestors: The hard evidence*, 663-71. New York.

VRBA, E.S. (1993): The pulse that produced us. *Natural History* 102, 47-51.

WASHBURN, S. & HOWELL, F.C. (1960): Human evolution and culture. In: S.Tax (ed.), *Evolution after Darvin.* Vol.2. *The evolution of man*, 33-56. Chicago.

WHEELER, P.E. (1984): The evolution of bipedality and loss of funtctional body hair in hominids. *Journal of Human Evolution* 13, 91-98.

WHEELER, P.E. (1991a): The thermoregulatory advantages of hominid bipedalism in open equatorial environments. *Journal of Human Evolution* 21, 107-115.

WHEELER, P.E. (1991b): The influence of bipedalism on the energy and water budgets of early hominids. *Journal of Human Evolution* 21, 117-136.

YAKIMOV, V.P. (1966). The australopithecines. In: V.V.Bunak (ed.), *Fossil hominids and the origin of man*, 43-89. Moscow. (In Russian).

YUROVSKAYA, V.Z. (1989): The evolution of hominoid locomotions. In: *Biological evolution and man*, 155-179. Moscow. (In Russian).

The Adaptive Zone of *Homo erectus* from an African Perspective

Susan Cachel & J.W.K. Harris

Abstract

Homo erectus is the first member of the genus Homo for which virtually complete postcranial anatomy can be reconstructed from fossil evidence. A number of factors can be inferred to have operated during the evolution of this species. Among these are large body size, increased heat adaptation, and dietary change as a major selection pressure. Homo erectus is also the earliest hominid taxon unequivocally thought regularly to manufacture and utilize stone artifacts. The density of the archaeological record increases with the advent of this species. Tool behavior associated with food acquisition and processing implies occupation of a niche distinctive for primates.

The first occurrence of Homo erectus in sub-Saharan Africa documents larger body size and pronounced open country, arid adaptations. Because of its novel adaptive zone, this taxon was the first widely dispersed hominid species, and its dispersal occurred at an early date. Home range size was probably larger than in other hominid species, and a wider variety of habitats were exploited. Like most large mammals, this species would have experienced very little ecological isolation. 1.8 mya marks the earliest occurrence of Homo erectus at Koobi Fora. At 1.7 mya, the archaeological record of East Africa becomes more complex. This date also marks the advent of the Karari Industry. The coincidence of these dates suggests that Homo erectus (unlike earlier or sympatric hominids) was experiencing selection pressure to exploit resources that were widely distributed across Pleistocene landscapes. This inference is supported by a variety of palaeo-environmental indicators: a major shift towards arid, open-country conditions is signaled at 1.7 mya. Archaeological survey shows that a variety of habitats were occupied after this date.

Because modern temperate species have larger home ranges and more generalized niches than tropical species, and are less affected by the existence and degree of seasonality, one might infer that similar factors were associated with the emergence of Homo erectus into temperate regions of the Old World. Because of the wide distribution of this species, niche structure probably varied with geographic range. The invasion of novel environments by Homo erectus, and the absence of hominid competitors in these environments, could lead to character release, which might increase variability in the morphological and behavioral phenotype. Phenotypic character release could explain the morphological variability which some researchers recognize by taxonomic distinctions between African, European, and Asian fossils. Behavioral character release, indicated by archaeological variability, could explain the occurrence of Acheulean variants or alternative stone tool industries.

Hominid dispersal from sub-Saharan Africa into Eurasia may have occurred in the region of the Afar Depression along a broad front, with the vector of movement then turning north into the Levant. The successful migration of Homo erectus out of the African tropics is not necessarily dependent upon material culture. We suggest that Homo erectus (perhaps like some modern baboon and macaque species) thrived on environmental disruption and habitat fragmentation, and emerged as a „weed" taxon in areas of tropical Africa where ecosystem instability was caused by Pleistocene climatic and tectonic perturbations.

Introduction

In this paper we consider the adaptive zone of the taxon *Homo erectus*, sensu latu, paying particular attention to African evidence. There is a great deal of variability within the hypodigm of this species, as broadly defined. Researchers have long argued about whether this variability reflects the existence of more than one species (e.g., Le Gros Clark,1964; Howell, 1978; Howells, 1980; Rightmire, 1990, 1996;Tattersall, 1986, 1992).Alternatively, while recognizing the strikingly polytypic nature of the fossil material, some researchers see no distinct boundary line between *Homo erectus* and *Homo sapiens*, and inveigh against the continued use of the taxon *Homo erectus* (Wolpoff et al., 1994; Wolpoff & Caspari, 1997).

Morphological variability is often considered to be merely a vexing problem that obscures the taxonomic allocation of fossil specimens, which frequently is thought to be the principal goal of palaeoanthropology. Behavioral variability is often reduced to debate about lithic typologies and the definition of stone tool industries and their boundaries. Because natural selection on morphology and behavior cannot occur without population-level variability, morphological and behavioral variability constitute primary data. In fact, the assessment of shifting variability through time may allow palaeoanthropologists to trace the activity of natural selection and other evolutionary processes. Our emphasis on evolutionary ecology leads us to recognize that morphological and behavioral variability in time and space can contribute to a better understanding of how evolution has affected fossil hominid populations.

Of all the Plio/Pleistocene hominid species of Africa, the adaptive zone of *Homo erectus* has the greatest probability of limited reconstruction.This is because the discovery of an astonishingly complete sub-adult specimen from the Kenyan site of Nariokotome allows for functional interpretations of postcranial anatomy. Detailed palaeoenvironmental reconstructions of the Lake Turkana Basin have also been performed. Archaeological survey, landscape archaeology, and lithic analysis of the archaeological record in this region have also taken place. *Homo erectus* was the first widely dispersed hominid species, and its dispersal occurred at an early date. The reconstruction of its adaptive zone is of immense theoretical importance to palaeoanthropologists, because this adaptive zone was unique among Plio/Pleistocene hominids, although it heralds the dispersal properties of later hominid species.

Palaeoenvironmental evidence

Palaeoenvironmental evidence from a number of sources indicates that more open woodland and open-country grassland habitats began to spread in Africa at the expense of tropical rainforest as early as 7 mya. This trend was probably exaggerated by the Messinian salinity crisis at 5.5 mya, caused by global cooling and a simultaneous marine regression of 40-50 meters, which was exaggerated by local tectonic effects. The Messinian salinity crisis caused remarkable desiccation of the Mediterranean Sea, the Persian Gulf, and the Red Sea (Williams et al., 1993). Mediterranean desiccation resulted in the deposition of a million cubic kilometers of evaporites, dated to three episodes occurring between 5.7-5.1 mya (Frakes et al., 1992). This catastrophic Messinian event was apparently triggered by an expansion of the Antarctic ice sheet, which lowered global sea level, and by plate tectonic activity that closed the Straits of Gibraltar. The last Messinian desiccation episode certainly appears to have been initiated by a fall in global sea level caused by Antarctic glaciation (Shackleton, 1995). The Messinian crisis or its antecedents may have launched the hominid radiation, which was based on the evolution of bipedality, if one assumes that bipedal adaptations were selected for by the spread of open-country habitats. Increasing aridity at two subsequent time periods (2.8 and 1.7 mya) may also have influenced later hominid evolution in Africa (deMenocal, 1995; deMenocal & Bloemendal, 1995).

Anatomical evidence of behavior

A remarkably complete sub-adult specimen of *Homo erectus* (KNM-WT 15000) from the Kenyan site of Nariokotome in the western Lake Turkana Basin is dated at 1.53 mya. This specimen has postcranial anatomy that is significantly different from the australopithecine condition (Walker & Leakey, 1993). The horizontal orientation of the clavicle and glenoid cavity argue for an absence of arboreal suspensory locomotion, and hence indirectly argue for increased terrestriality.

A major feature of the specimen is its modern limb proportions: the lower limb has increased in length, so that the relative proportion of upper to lower limbs is like that found in modern hominids. As exemplified by KNM-WT 15000, the evolution of relatively long lower limbs in *Homo erectus* was driven by selection pressures either for speed or for heat adaptation. A recent survey of lower limb length in hominids reveals that femur length (and therefore probable total lower limb length) increased dramatically from the australopithecines and early genus *Homo* to *Homo erectus* (Webb, 1996: Table 1, Fig. 3 & 4). Experimental analysis of walking speeds in modern humans demonstrates that maximum walking speed may be affected by lower limb length, but maximum comfortable walking speed is not. Long-distance travel is associated with comfortable walking speed. Increased lower limb length in *Homo erectus* is therefore probably not caused by selection pressure for increasing the efficiency of bipedal locomotion (Webb, 1996). The assumption that lower limb length affects bipedal efficiency has also been independently questioned by researchers studying the energetic cost of locomotion in mammals generally (Steudel, 1995, 1996; Steudel & Beattie, 1995). These researchers have concluded that body size was the single most important factor determining locomotor efficiency and economy in early hominids.

Thus, heat adaptation was probably a major selection factor in the increasing relative length of the Nariokotome lower limb, particularly given the long distal limb segments and exaggerated ectomorphy of the whole specimen (Ruff & Walker, 1993). Ruff (1991, 1994) argues that narrow pelvic bi-iliac breadth in all Plio/Pleistocene specimens complete enough to examine indicates the existence of a fundamental hominid adaptation to heat, but the greater stature and long limbs of the Nariokotome specimen indicate further adaptation to open, arid, hot environments. Heat adaptation affects the ability to range and forage in equatorial environments, and particularly affects mid-day activity levels (Wheeler, 1993). Given access to abundant, potable water, the ancestors of *Homo erectus* in East Africa may have reacted to seasonality and competition from sympatric hominid species by evolving additional morphological adaptations to dissipate heat by increasing stature, augmenting the relative length of the lower limb, and developing ectomorphy. Foraging efficiency would increase as a result of these changes. Earlier postcranial evidence is sparse; we have no knowledge of body form and limb proportions in *Homo erectus* prior to 1.53 mya. We assume that the spread of *Homo erectus* in the Old World was associated with the postcranial anatomy of the Nariokotome specimen, but no comparable North African, European, or Javanese specimen has been discovered.

Increasing aridity and the spread of grasslands in East Africa may have been important in the development of heat adaptation, because these factors increased the patchiness of hominid food resources, and necessitated more foraging time and higher activity-levels (Cachel & Harris, 1995, 1996, in press). A recent comparative analysis of primate energetics estimates that total energy expenditure in African *Homo erectus* was at least 40-45% more than for the australopithecines, and perhaps as much as 80-85% greater, depending on the type of modern model used for the estimation (Leonard & Robertson, 1997). This increase in total energy expenditure was principally caused by a markedly enlarged body size in both sexes (63 kg for adult males, 52.3 kg for adult females). Larger home ranges and a subsequent expansion in resource foraging activities would also increase total energy expenditure. Diet in these hominids must have been substantially augmented by high-quality food such as protein and fat, probably supplied from vertebrate animals.

Archaeological evidence of behavior

The archaeological record in East Africa becomes increasingly complex just after the earliest occurrence of *Homo erectus* at 1.8 mya (KNM-ER 3733 from Koobi Fora). The distributional pattern of archaeological occurrences shows a pronounced shift in hominid occupation at approximately 1.8-1.6 mya. This coincides with an arid oscillation in the paleoclimate of Africa at 1.7 mya (deMenocal, 1995; deMenocal & Bloemendal, 1995). Local tectonic activity and volcanism also changed the physical relief, the nature of local water resources, and vegetation patterns on emergent landscapes, as seen in the sedimentary basins of Olduvai and Lake Turkana (Hay, 1976, 1990; Feibel et al., 1991; Feibel, 1995). *Homo erectus* may have been experiencing selection pressure to exploit resources that were widely distributed across early Pleistocene landscapes.

By 1.7-1.5 mya, archaeological traces are found within a variety of habitats within the Lake Turkana Basin (Rogers et al., 1994). Many archaeological occurrences are found along the banks of braided river channels and in adjacent floodplain settings. Archaeological traces are found for the first time in drier, upland parts of the basin. A similar shift in hominid foraging behavior through time is seen at Olduvai Gorge, Tanzania, where sites in Bed II are also located along the banks of stream channels at higher elevations in the basin.

The earliest known stone artifacts, dating between 2.6-2.5 mya, have recently been discovered in the Gona River deposits of the Afar region of Ethiopia. The sophistication of these tools at such an early date implies an abrupt threshold for the origin of stone tool behavior, rather than a gradual, long-term continuum of change (Semaw, 1997; Semaw et al., 1997). This abrupt threshold accentuates the fact that hominids engaging in tool behavior were occupying a niche that was unique in the primate order. During the time period which we are considering here, the continuum of change from the Oldowan to the Acheulean indicates that behavioral innovations in ranging and foraging, rather than major innovations in lithic technology, were taking place.

Because of larger body size, home range size was probably larger in *Homo erectus* than in earlier hominid species. The wider variety of habitats exploited by this taxon indicates larger home range size in any case. Landscape archaeology demonstrates that patterned foraging behavior occurs in the Okote Member, dated at 1.6 mya in the Koobi Fora region of East Turkana, Kenya (Rogers, 1997). The unpatterned foraging behavior typical of modern non-human primates („routed foraging") does not occur. This indicates a sophistication in hominid ranging and foraging behaviors that does not exist at the non-human primate level. These behaviors took place in arid, open landscapes, as indicated by extensive palaeoenvironmental analysis of the Okote Member.

Phenotypic variability and ecological release

Investigation of colonizing organisms reveals that a species' entry into a new habitat with no or fewer competitors than in the original home range can create an increase in phenotypic variability - a phenomenon known as ecological or phenotypic character release (Van Valen, 1965; Grant, 1986: 340-345). Character release occurs with allopatry, because sympatric competition affects niche structure and niche breadth-morphology and behavior may change as a consequence. Phenotypic divergence between sympatric species, and the genetic bases for such ecological and morphological divergence, has been explored under laboratory conditions, as well as under field conditions (Hatfield, 1997). Adaptive differences between sympatric species can be polygenic in nature, or caused by the activity of a small number of genes with large effects. This has obvious implications for discussions of sympatric Plio/Pleistocene hominids and speciation events. However, the existence of character release per se has important implications for questions of hominid dispersion and biogeography. In particular, character release ensures that a widely dispersed species like *Homo erectus* would demonstrate great morphological and behavioral variability once released from the constraints of competition with closely related sympatric species (Cachel & Harris, 1995, 1996, in press).

Penetration of new habitats without sympatric hominid competitors might account for morphological differences within the *Homo erectus* hypodigm. Some researchers draw taxonomic distinctions at the species level between African, European, and Asian specimens. Character release may explain these differences without invoking speciation events. It is also possible that variability thought to reflect species distinctions has been exaggerated by taxonomists: A new reconstruction of hominid skull morphology from Zhoukoudian based on recent discoveries of facial, mandibular, and dental specimens indicates that facial anatomy in *Homo erectus* from the Far East was not as derived as it has generally been considered to be since publication of the 1937 Weidenreich/Swan reconstruction (Tattersall & Sawyer, 1996). Instead, the new reconstruction accords with *Homo erectus* and *Homo ergaster* material known from other regions. Similarly, the juvenile calvaria from Modjokerto, Java (Perning 1), which has had a number of very different taxonomic allocations, has recently been described as exhibiting traits characteristic of Asian *Homo erectus* (Anton, 1997). The Modjokerto specimen, dated at 1.8 mya, is currently the oldest known hominid specimen from the East Asia (Swisher et al., 1994). This date is contemporary with the earliest *Homo erectus* specimen from Koobi Fora, Kenya. The date thus confirms a very early hominid penetration of the Far East, as well as long-term morphological continuity in this region.

Technological variability across wide geographic ranges has often been analyzed in terms of limits to migration and cultural diffusion. Studies of lithic typology traditionally emphasize major behavioral differences in different regions, and have focused on the absence of the Acheulean industry in the Far East: Hand-axes and cleavers are not present, Mode 1 technology is employed, and the Levallois method does not occur. A recent review concludes that the „Movius Line" is a true phenomenon, and that its existence has remained a puzzle for over 50 years (Schick, 1994). East Asia is interpreted as being a cultural backwater; failure of the Acheulean industry and the Levallois method to penetrate to the Far East is explained either by physical barriers to human migration and cultural diffusion, or by very narrow population bottlenecks that impeded gene flow and cultural transmission (Schick, 1994: 593). According to this traditional interpretation, archaeological evidence suggests that hominids in the Far East constituted marginal populations which developed endemic behavioral traits. This interpretation has been criticized by researchers who emphasize the use of bamboo in the Far East as an alternative raw material, as well as other factors (Pope, 1995: 501-502). We argue here that, if migrating hominids were subject to behavioral character release, the existence of Mode 1 tool industries in habitats far from tropical Africa might be explained without viewing the Far East as an evolutionary cul-de-sac with an archaeologically aberrant record. Mode 1 occurrences in other areas (e.g., the Buda, Tayacian, and Clactonian industries, artifacts from Atapuerca) might be similarly explained by character release.

The discovery of artifacts dated to more than 260,000 yrs BP at Diring Yuriakh in Central Siberia attest to the wide geographic dispersion of hominids in Eurasia in spite of climatic extremes (Waters et al., 1997). Although it is possible for catarrhine primates to penetrate higher latitude areas without benefit of culture, as seen by the presence of living macaque and Pleistocene macaque species (*Macaca* spp.) in high latitudes, persistent hominid Lower Palaeolithic occupation of high latitudes would seem to indicate the technological ability to control fire.

General factors affecting dispersion

There are several general factors affecting dispersion (Cachel & Harris, 1995, 1996, in press). We consider these general factors here without extensive commentary before we discuss dispersion itself.

First, tectonic activity or Pleistocene climatic fluctuations fragment habitats, and consequently create regional faunas. This contributes to species diversity. Second, because dispersal ability is directly proportional to body size in land mammals, a large-bodied species like *Homo erectus* was probably not limited even by extensive habitat fragmentation. Third, high-latitude species have greater latitudinal range than low-latitude species, and are less affected by the occurrence or degree of seasonality. This is called Rapoport's Rule (Stevens, 1989). A consequence of this is that once hominids disperse from low latitudes, they are less constrained in terms of range or seasonal perturbations. Fourth, because the niche of a species may vary with geographic range, the wide distribution of the taxon *Homo erectus* would make it possible for niche structure to vary significantly across its distribution. Furthermore, wide distribution and low population size would ensure the operation of genetic drift. Lastly, as discussed above, ecological release with consequent phenotypic character release may also have been important.

The dispersion of *Homo erectus*

It has recently been estimated that effective hominid population sizes in sub-Saharan Africa continuously remained about three times larger than elsewhere in the Old World through the course of evolutionary time (Relethford & Harpending, 1995). If this is true, sub-Saharan Africa would have constantly served as a source area for dispersing hominids. Roberts (1984: 45) has proposed a palaeoenvironmental mechanism for triggering hominid dispersal from Pleistocene Africa. He argues that the alternation of wet and dry environments in the Pleistocene Sahara causes this dispersal. Wet environments draw hominid populations into the Sahara from more southern regions; Sahara desiccation then expels hominids out into circum-Mediterranean areas. Mathematical analysis confirms that persistent environmental variation or mosaicism does in fact promote the dispersal of natural populations (Holt & McPeek, 1996).

Average rates of morphological evolution during animal colonization of novel environments or following entry into a new adaptive zone are 100 times the average rate of morphological evolution documented for Holocene mammals (370 darwins versus 3.7 - Gingerich, 1983: Table 1). (Calculation of morphological evolution in haldanes - phenotypic standard deviation per generation - can give the intrinsic rate of change, and yield an estimate of the possible upper limit to rate of change - Gingerich, 1993). Furthermore, the literature on naturalized mammals yields many examples of species which are introduced by humans into new areas, where they flourish, spread rapidly, and markedly alter their diet and behavior to suit the novel environment (Lever, 1985). The diet and behavior of these alien species rapidly responds to differences in habitat and a relative lack of competition - a demonstration of phenotypic character release under artificial circumstances. Naturalized alien species may compete with and prey upon native species, or drive them to extinction. Successful species of naturalized mammals can tolerate a wide variety of habitats, and frequently thrive in habitats highly altered by humans. These species can disperse in a surprisingly rapid fashion.

Various routes have been proposed for the dispersal of Lower Pleistocene hominids from Africa. It is generally considered that the most likely migration route is the current land connection between Africa and Eurasia. Hominids and other land mammals are thought to have dispersed through the Sinai Desert region, and north through the Levant, which has been well explored by palaeoanthropologists. Arguments supporting this Levantine route (the Levantine Corridor) are well-known, and have been summarized in Bar-Yosef & Goren-Inbar (1993). However, we believe that another likely route is by way of the Afar Depression, through the shallow termination of the Red Sea and across the southwestern point of the Arabian Peninsula. During the Early and Middle Pleistocene, land today covered by the Red Sea may have been intermittently subaerially exposed to allow hominid movements between northeast Africa and the Arabian Peninsula and north into the Levant and thence to Eurasia. The available geophysical evidence supports a possible route from East Africa through southwestern Arabia. In fact, the Asal Rift segment of the Afar Depression is now subaerially exposed, because crustal thinning and subsidence is balanced by the injection of magma (de Chabalier & Avouac, 1994). This is a modern example of an emergent area in the Afar Depression created by processes that presumably operated throughout the Pleistocene. Hominid dispersal into Eurasia may well have occurred intermittently in this region along a broad front, as land areas emerge and subside.

Habitat disruptiom and the origin and dispersal of *Homo erectus*

Although the origin and dispersion of genus *Homo* have been frequently studied, researchers usually describe migration events, and do not employ a perspective generated from the study of the evolutionary ecology of other large mammal species (e.g., Larick & Ciochon, 1996). As is true of most large mammal species, *Homo erectus* would have experienced relatively little ecological isolation caused by geographic barriers. Nevertheless, habitat fragmentation would be accentuated by Late Neogene climate oscillations.

Fluctuations of province boundaries and widespread extinctions are palaeontological indicators of glacial climate modes (Frakes et al., 1992: 23). There is evidence that Pleistocene habitat fragmentation did have an affect on mammal distribution and community composition. For example, as a direct result of climatic fluctuations, Late Pleistocene environments in the United States were significantly more heterogeneous than in the Holocene. This environmental heterogeneity, in turn, generated significantly more diverse mammal faunas than in the Holocene. The composition of Late Pleistocene communities in America was constantly shifting. FAUNMAP evidence in North America documents community lability and extinctions during the late Pleistocene (Graham et al., 1996), and even more pronounced responses to seasonality might be expected if similar studies in a contemporary tropical habitat were conducted. One would expect this if Rapoport's rule is in effect (Stevens, 1989).

Volcanic activity can create barren, highly altered landscapes with characteristic sedimentary processes. Volcanic eruptions in East Africa are also distinguished by their peculiar hyperalkaline chemistry (Harris & Van Couvering, 1995). In addition, the distinctive nature of East African volcanic eruptions probably creates a false palaeoecological inference of aridity - an impression of „mock aridity“ that must be accounted for when ancient habitats are reconstructed (Harris & Van Couvering, 1995).

Palaeontology and archaeology both demonstrate that *Homo erectus* emerged in areas of tropical sub-Saharan Africa where habitat disruption, caused by Pleistocene climatic fluctuations, tectonic movements, and volcanism, was prevalent. East African volcanicity that persisted through millions of years may have had a major impact on Neogene ecosystems, not only through long-term habitat destruction and disturbance, but also through the generation of communities whose soil, water, and vegetation were affected by a continual influx of hyperalkaline sediments. We suggest that *Homo erectus* - perhaps like some modern macaque and baboon species (Richard et al. 1989; Burney 1995/1996: Figure 1) - thrived on environmental disruption, and emerged as a „weed“ taxon in areas of tropical Africa disrupted by Pleistocene climatic and tectonic perturbations (Cachel & Harris, 1995, 1996, in press).

Weed taxa profit from habitat fragmentation and ecosystem instability. Weed taxa would be likely to evolve, thrive, and spread under conditions of long-term, continuous environmental and community change. Stanley (1992, 1996) argues that climatic deterioration in Africa triggered the origins of genus *Homo* by reducing forest habitat. Potts (1996) argues that continued climatic fluctuations in Africa selected for hominids that were generalists. Our argument about Plio/ Pleistocene weed taxa emphasizes that climatic fluctuations and tectonic disturbances create continual long-term habitat disturbance and community change, perhaps especially in East Africa. Under these conditions, animal species that are adapted for environmental disturbance are more likely to survive, become abundant, and disperse throughout the habitats affected by such alteration (Cachel & Harris, 1995, 1996, in press). Such habitats have communities whose species composition is highly unstable. Our argument is therefore directed at the level of ancient communities.

Conclusions

1. Because of its novel adaptive zone, *Homo erectus* was the first widely dispersed hominid species. This is the earliest hominid species unequivocally thought regularly to manufacture and utilize stone artifacts. The density of the archaeological record increases with the advent of this taxon. This is also the first member of genus *Homo* for which virtually complete postcranial anatomy can be reconstructed from fossil evidence. A number of factors can be inferred to have operated during the evolution of *Homo erectus*. Among these are large body size (associated with larger home range size, K-selection, and dietary diversity), the influence of food as a limiting factor on population size, and dietary change as a major selection factor. Tool behavior associated with food acquisition and processing implies occupation of a niche distinctive for primates.

2. The first occurrence of *Homo erectus* in sub-Saharan Africa documents larger body size and pronounced open country, arid adaptations. Home-range size was probably larger than in other hominid species, with a wider variety of habitats being exploited. 1.8 mya marks the earliest occurrence of *Homo erectus* at Koobi Fora. At 1.7 mya, the archaeological record of East Africa becomes highly variable, with artifact occurrences that are generally referred to as Developed Oldowan. The coincidence of these dates suggests that *Homo erectus* (unlike earlier or sympatric hominids) was experiencing selection pressure to exploit resources that were becoming increasingly more widely distributed across Pleistocene landscapes.

3. Because modern temperate species have larger ranges and generalized niches than tropical species, one might infer that an increase in range size and the acquisition of a more generalized niche structure was associated with the emergence of *Homo erectus* into temperate regions of the Old World. The invasion of novel environments by *Homo erectus* could lead to character release, which might increase variability in the morphological and behavioral phenotype.

4. The successful migration of *Homo erectus* out of the African tropics is not necessarily dependent upon material culture. We suggest that *Homo erectus* - perhaps like some modern baboon and macaque species - thrived on environmental disruption, and emerged as a 'weed' taxon in areas of tropical Africa continually altered by Pleistocene climatic and tectonic perturbations.

References

ANTON, S. (1997): Developmental age and taxonomic affinity of the Mojokerto child, Java, Indonesia. *American Journal of Physical Anthropology* 102, 497-514.

BAR-YOSEF, O. & GOREN-INBAR, N. (1993): *The lithic assemblages of Ubeidiya*. Jerusalem (Monographs of the Institute of Archaeology, Hebrew University of Jerusalem).

BURNEY, D.A. (1995/1996): Historical perspectives on human-assisted biological invasions. *Evolutionary Anthropology* 4, 216-221.

CACHEL, S. & HARRIS, J.W.K. (1995): Ranging patterns, land-use and subsistence in *Homo erectus* from the perspective of evolutionary ecology. In: J.R.F. Bower & S. Sartono (eds), *Proceedings of the Pithecanthropus Centennial, 1893-1993.* Vol. I. *Palaeoanthropology: Evolution and ecology of Homo erectus*, 51-66. Leiden.

CACHEL, S. & HARRIS, J.W.K. (1996): The paleobiology of *Homo erectus*: Implications for understanding the adaptive zone of this species. In: G. Pwiti & R. Soper (eds), *Aspects of African archaeology,* 3-9. Harare (Papers from the 10th Congress of the PanAfrican Association for Prehistory and Related Studies).

CACHEL, S. & HARRIS, J.W.K. (in press): The lifeways of *Homo erectus* inferred from archaeology and evolutionary ecology: A perspective from East Africa. In: M.D. Petraglia & R. Korisettar (eds), *Lower Paleolithic settlement of the Old World.* London.

CHABALIER, J-B. de & AVOUAC, J-P. (1994): Kinematics of the Asal Rift (Djibouti) as determined from the deformation of Fieale Volcano. *Science* 265, 1677-1681.

FEIBEL, C.S. (1995): Geological context and the ecology of *Homo erectus* in East Africa. In: J.R.F. Bower & S. Sartono (eds), *Proceedings of the Pithecanthropus Centennial, 1893-1993.* Vol. I. *Palaeoanthropology: Evolution and ecology of Homo erectus,* 67-74. Leiden.

FEIBEL, C.S., HARRIS, J.M. & BROWN, F.H. (1991): Neogene paleoenvironments of the Turkana Basin. In: J.M. Harris (ed.), *Koobi Fora Research Project,* Vol 3, 321-370. Oxford.

FRAKES, L.A., FRANCIS, J.E., & SYKTUS, J.I. (1992): *Climate modes of the Phanerozoic.* Cambridge.

GINGERICH, P.D. (1983): Rates of evolution: Effects of time and temporal scaling. *Science* 222, 159-161.

GINGERICH, P.D. (1993): Quantification and comparison of evolutionary rates. *American Journal of Science* 293 A, 453-178.

GRAHAM, R.W., LUNDELIUS Jr., E.J., GRAHAM, M.A., SCHROEDER, E.K., TOOMEY III., R.S., ANDERSON, E., BARNOSKY, D., BURNS, J.A., CHURCHER, C.S., GRAYSON, D.K., GUTHRIE, R.D., HARINGTON, C.R., JEFFERSON, G.T., MARTIN, L.D., McDONALD, H.G., MORLAN, R.E., SEMKEN Jr., J.A., WEBB, S.D., WERDELIN, L. & WILSON, M.C. (The FAUNMAP Working Group) (1996): Spatial response of mammals to Late Quaternary environmental fluctuations. *Science* 272, 1601-1606.

GRANT, P.R. (1986): *Ecology and evolution of Darwin's finches.* Princeton.

HARRIS, J. & Van COUVERING, J. (1995): Mock aridity and the paleoecology of volcanically influenced ecosystems. *Geology* 23, 593-596.

HATFIELD, T. (1997): Genetic divergence in adaptive characters between sympatric species of stickleback. *American Naturalist* 149, 1009-1029.

HAY, R.L. (1976): *Geology of the Olduvai Gorge: A study of sedimentation in a semiarid basin.* Berkeley, CA.

HAY, R.L. (1990): Olduvai Gorge: A case study in the interpretation of hominid paleoenvironments in East Africa. In: L. Laporte (ed.), *Establishment of a geological framework for palaeoanthropology*, 23-37. Boulder, CO (Geological Society of America, Special Paper No. 242).

HOLT, R.D. & McPEEK, M.A. (1996): Chaotic population dynamics favors the evolution of dispersal. *American Naturalist* 148, 709-718.

HOWELL, F.C. (1978): Hominidae. In: V.J. Maglio & H.B.S. Cooke (eds), *Evolution of African mammals*, 154-248. Cambridge.

HOWELLS, W.W. (1980): *Homo erectus* - who, when and where: A survey. *Yearbook of Physical Anthropology* 23, 1-23.

LARICK, R. & CIOCHON, R.L. (1996): The African emergence and early Asian dispersals of the genus *Homo. American Scientist* 84, 538-551.

LE GROS CLARK, W.E. (1964): *The fossil evidence for human evolution.* Chicago. 2nd ed.

LEONARD, W.R. & ROBERTSON, M.L. (1997): Comparative primate energetics and hominid evolution. *American Journal of Physical Anthropology* 102, 265-281.

LEVER, C. (1985): *Naturalized mammals of the world.* London.

MENOCAL , P.B. de (1995): Plio-Pleistocene African climate. *Science* 270, 53-59.

MENOCAL, P.B. de & BLOEMENDAL, J. (1995): Plio-Pleistocene climatic variability in subtropical Africa and the palaeoenvironment of hominid evolution: A combined data-model approach. In: E.S. Vrba, G. Denton, T. Partridge & L. Burckle(eds), *Paleoclimate and evolution, with emphasis on human origins*, 262-288. New Haven.

POPE, G.G. (1995): The influence of climate and geography on the biocultural evolution of the Far Eastern hominids. In: E.S. Vrba, G. Denton, T. Partridge & L. Burckle (eds.), *Paleoclimate and evolution, with emphasis on human origins,* 493-506. New Haven.

POTTS, R. (1996): *Humanity's descent. The consequences of ecological instability.* New York.

RELETHFORD, J.H. & HARPENDING, H.C. (1995): Ancient differences in population size can mimic a recent African origin of modern humans. *Current Anthropology* 36, 667-674.

RICHARD, A.F., GOLDSTEIN, S.J. & DEWAR, R.F. (1989): Weed macaques: The evolutionary implications of macaque feeding ecology. *International Journal of Primatology* 10, 569-594.

RIGHTMIRE, G.P. (1990): *The evolution of Homo erectus*. Cambridge.

RIGHTMIRE, G.P. (1996): The human cranium from Bodo, Ethiopia: Evidence for speciation in the Middle Pleistocene? *Journal of Human Evolution* 31, 21-39.

ROBERTS, N. (1984): Pleistocene environments in time and space. In: R. Foley (ed.), *Hominid evolution and community ecology*, 25-53. Orlando, FL.

ROGERS, M.(1997): *A landscape archaeological study at East Turkana, Kenya.* Unpublished Ph.D. dissertation, Rutgers University.

ROGERS, M.J., HARRIS, J.W.K. & FEIBEL, C.S. (1994): Changing patterns of land use by Plio/Pleistocene hominids in the Lake Turkana Basin. *Journal of Human Evolution* 27, 139-158.

RUFF, C.B. (1991): Climate and body shape in hominid evolution. *Journal of Human Evolution* 21, 81-105.

RUFF, C.B. (1994): Morphological adaptation to climate in modern and fossil hominids. *Yearbook of Physical Anthropology* 37, 65-107.

RUFF, C.B. & WALKER, A. (1993): Body size and body shape. In: A. Walker & R. Leakey (eds), *The Nariokotome Homo erectus skeleton*, 234-265. Cambridge, MA.

SCHICK, K.D. (1994): The Movius Line reconsidered. Perspectives on the earlier Paleolithic of eastern Asia. In: R.S. Corruccini & R.L. Ciochon (eds.), *Integrative paths to the past*, 569-596. Englewood Cliffs.

SEMAV, S. (1997): *Late Pliocene archaeological occurrences from the Gona River deposits, Afar, Ethiopia.* Unpublished Ph.D. dissertation, Rutgers University.

SEMAV, S. et al. (1997): 2.5-million-year-old stone tools from Gona, Ethiopia. *Nature* 385, 333-336.

SHACKLETON, N.J. (1995): New data on the evolution of Pliocene climatic variability. In: E.S. Vrba, G. Denton, T. Partridge & L. Burckle (eds.),*Paleoclimate and evolution, with emphasis on human origins*, 242-248. New Haven.

STANLEY, S.M. (1992): An ecological theory for the origin of *Homo*. *Paleobiology* 18, 237-257.

STANLEY, S.M. (1996): *Children of the Ice Age*. New York.

STEUDEL, K. (1995): Locomotor energetics and hominid evolution. *Evolutionary Anthropology* 5, 42-48.

STEUDEL, K. (1996): Limb morphology, bipedal gait, and the energetics of hominid locomotion. *American Journal of Physical Anthropology* 99, 345-355.

STEUDEL, K. & BEATTIE, J. (1995): Does limb length predict the energetic cost of locomotion in mammals? *Journal of the Zoological Society of London* 235, 501-514.

STEVENS, G.C. (1989): The latitudinal gradient in geographical range: how so many species coexist in the tropics. *American Naturalist* 133, 240-256.

SWISHER III, C.C., CURTIS, G.H., JACOB, T., GETTY, A.G., SUPRIJO, A. & WIDIASMORO (1994): Age of the earliest known hominids in Java, Indonesia. *Science* 263, 1118-1121.

TATTERSALL, I. (1986): Species recognition in human paleontology. *Journal of Human Evolution* 15, 165-176.

TATTERSALL, I. (1992): Species concepts and species recognition in human evolution. *Journal of Human Evolution* 22, 341-349.

TATTERSALL, I. & SAWYER, G.J. (1996): The skull of „Sinanthropus“ from Zhoukoudian, China: A new reconstruction. *Journal of Human Evolution* 31, 311-314.

Van VALEN, L. (1965): Morphological variation and width of the ecological niche. *American Naturalist* 94, 377-390.

WALKER, A. & LEAKEY, R. (1993): The postcranial bones. In: A. Walker & R. Leakey (eds), *The Nariokotome Homo erectus skeleton*, 95-160. Cambridge, MA.

WATERS, M.R., FORMAN, S.L. & PIERSON, J.M. (1997): Diring Yuriakh: A Lower Palaeolithic site in Central Siberia. *Science* 275, 1281-1284.

WEBB, D. (1996): Maximum walking speed and lower limb length in hominids. *American Journal of Physical Anthropology* 101, 515-525.

WHEELER, P.E. (1993): The influence of stature and body form on hominid energy and water budgets: A comparison of Australopithecus and early *Homo* physiques. *Journal of Human Evolution* 24, 13-28.

WILLIAMS, M.A.J., DUNKERLEY, D.J., De DECKKER, P., KERSHAW, A.P. & STOKES, T. (1993): *Quaternary environments*. London.

WOLPOFF, M.H. & CASPARI, R. (1997): *Race and human evolution*. New York.

WOLPOFF, M.H., THORNE, A.G., JELÍNEK, J. & YINYUN, Z. (1994): The case for sinking *Homo erectus*. 100 years of Pithecanthropus is enough! *Courier Forschungs-Institut Senckenberg* 171, 341-361.

Dmanisi and the Early Eurasian Dispersal of the Genus *Homo*

Winfried Henke, Hartmut Rothe & Kurt W. Alt

Abstract

According to a long-standing palaeoanthropological hypothesis Homo erectus evolved in Africa from an earlier Homo-species. Many palaeoanthropologists regard this Homo erectus as the first hominid who left Africa and occupied Asia and Europe successively. However, current evidence suggests that the dispersal of early members of our own genus may have been much more complex than has been thought thus for.

The paper focuses on the early African emergence of the genus Homo and the later spread of our own genus to other parts of the Old World, as well as on the special information coming from the hominid mandible from Dmanisi (Georgia). This key fossil, initially assigned to Homo erectus, shows a mosaic pattern of plesiomorphic and apomorphic traits, which does not correspond to the morphological pattern of an early Homo erectus but quite the reverse, it has much more affinities to later specimens of the same taxon. Furthermore, the corpus of the mandible and the dentition demonstrate in some details apomorphic features compared to the described morphological pattern of Homo erectus. This in turn reveals a discrepancy between the hypothetical maximum age of 1.9 myrs - and very much younger ages too - and the expected features of such an old hominid. The presence of apomorphic traits in the Dmanisi mandible is a strong argument against affinities to an 'early' Homo erectus sample and therefore there are reasonable doubts on the claimed chronology.

In case a very high age should be proven we would have to revise our models on human origins substantially. Due to these uncertainties, i.e. insufficient definition of the Homo erectus hypodigm and the controversial age of the Dmanisi fossil - we should retrain from a phylogenetical interpretation of the Georgian mandible.

Introduction

Controversies in science are quite usual, and palaeoanthropology by no means is an exception. Sometimes new results on hominid evolution are not only solving long-standing old problems but concurrently create instead numerous new questions. This consequence can be demonstrated exemplarily by different interpretations of *Homo erectus* (see e.g. Frayer et al., 1993; Franzen, 1994a, b; Henke & Rothe, 1994, 1995; Wolpoff, 1996).

For a long time *Homo erectus* was regarded as the 'missing link' between an earlier species of the genus *Homo*, e.g. *Homo habilis* s.l., and the successor species, *Homo sapiens*. Its evolutionary intermediate position in the human family tree has been questioned more and more in the last decade (e.g. Rightmire, 1990, 1992; Tattersall, 1997). While some anthropologists questioned that this species existed at all (e.g. Wolpoff et al., 1994; Wolpoff, 1996), others claimed that *Homo erectus* lived only in Africa and Asia and may never have reached Europe (e.g. Rightmire, 1990, 1992). Further scientists hold the view that *Homo erectus* evolved from a hominid branch of African origin - possibly *Homo ergaster* - in Asia and remained restricted to the Far East (e.g. Stringer, 1993; Tattersall, 1997).

The recent discussion on the question, whether *Homo erectus* was an ancestor of our own species or an evolutionary side branch, got new stimuli from the interpretation of a fossil mandible found in 1991 by Antje Justus, co-worker of Gerhard Bosinski (Head of the Department of

Palaeolithic Reseach, RGZM, Mainz and Department of Archaeology, University of Colonia, Germany), at Dmanisi (Georgia), a site 80 km south-east of Tblisi (Dzaparidze et al., 1989/1992; Gabunia, 1992; Gabunia & Vekua, 1995; Bräuer et al., 1995/1997; Dean & Delson, 1995; Henke, 1995; Henke et al., 1995; Puech et al., 1995; Bräuer & Schultz, 1996; van den Boogard et al., 1996; Alt et al., 1997; Henke, in press).

According to current results by potassium argon, laser fusion $^{40}Ar/^{39}Ar$ and palaeomagnetic studies, latest probably dates to the final Pliocene or early Pleistocene as Majsuradze et al. (1989/1992), Gabunia & Vekua (1995), Bosinski et al. (1995) and Majsuradze (1995) proposed. Schmincke & van den Bogard (1996) determined a maximum age of around 1.9 myrs for the underlying basalt lava stream, which means at least that the fossil must be younger, but how much younger is still open to the discussion. There are severe doubts about that very high age of the mandible (Bräuer et al., 1995/1997; Dean & Delson, 1995; Alt et al., 1997), what does not mean that an age of around 2 myrs for the immigration of hominids to Eurasia is to exclude principally.

Theories on emergence and dispersal of early *Homo*

This paper focuses on the regional and chronological dispersal patterns of the genus *Homo*, on the possible taxonomic status of the Dmanisi mandible and discusses some aspects of the scenario of the assumed early invasion of hominids to Eurasia.

Though the first non-European fossil specimens of the genus *Homo* have been found during the nineties of the last century in South-East Asia almost of the recent results point to Africa as the cradle of mankind. For a long time palaeoanthropologists hold the view, that *Homo erectus* evolved in Africa as a descendant of an early *Homo*. The most possible direct forerunners under discussion were *Homo habilis* or *Homo rudolfensis*, but a new taxon, named *Homo ergaster* by Groves & Mazák (1975), becomes more and more accepted as the species who was the pioneering emigrant. Especially the excellent preserved more than 1.5 myr old fossil from Nariokotome (Turkana, Kenya; Walker & Leakey, 1986; Walker & Leakey, 1993), an adolescent boy, forewarded the idea of a separate evolution of *Homo erectus* in Asia. While some authors see *Homo ergaster* as the first *Homo*-species who migrated out of Africa, others still continue the view that *Homo erectus* took this part in our evolution. In case that *Homo ergaster* was the first human species that reached Eurasia, it would be very unlikely that those specimens who have been classified as African and Asian *Homo erectus* belong to the same taxon as e.g. Kennedy (1991) and Bräuer & Mbua (1992) claimed. After Bräuer (1992, 29) „besteht gegenwärtig wenig Grund, daran zu zweifeln, daß in Afrika und Asien und wohl auch in Europa dieselbe Art *H. erectus* existierte". Bräuer and other authors see no autapomorphies of these regional African and Asian samples, which is in total contradiction to Tattersall's view, who writes: „Many researchers began to realize that we are dealing with two kinds of early humans here, and the earlier Kenyan form is now increasingly placed in its own species, *H. ergaster*" (Tattersall, 1997, 73).

European specimens, e.g. from Mauer/Heidelberg, Bilzingsleben, who usually have been regarded as members of the taxon *Homo erectus* (*H. e. heidelbergensis*; *H. e. bilzingslebenensis*) were no longer accepted as belonging to this species, as Rightmire (1990) proposed. In this context Henke & Rothe (1994, 1995) asked whether the European *Homo erectus* is a valid taxon indeed, and others - as Wolpoff et al. (1994) - demanded to 'sink' *Homo erectus* at all, that means to merge *Homo erectus* in the evolutionary species *Homo sapiens*. The 'multiregionalists' see no distinct boundary between *Homo erectus* and *Homo sapiens* in time and space. Wolpoff and co-workers argue: „The few distinctions of *Homo sapiens* that are not shared with *Homo erectus* appear to be responses to, or reflections of, continuing evolutionary trends of increasing cultural complexity, increasing brain size, and the progressive substitution of technology for biology" (Wolpoff et al., 1994, 341).

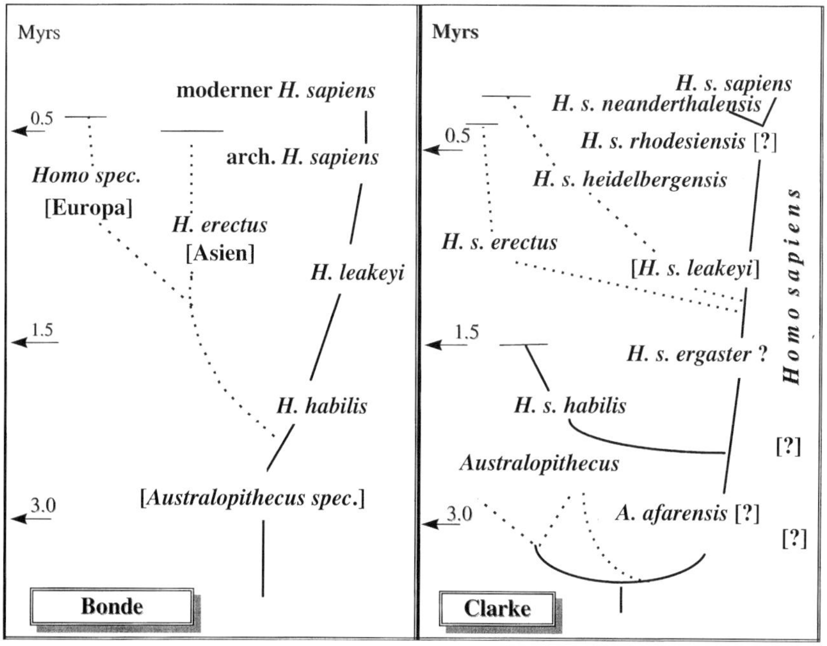

Figure 1 - Phylogenenetic model by different researchers.
a - Bonde (1989, redrawn by Henke & Rothe, 1994, 1995);
b - Clarke 1990 (reconstructed by Henke & Rothe, 1994, 1995);

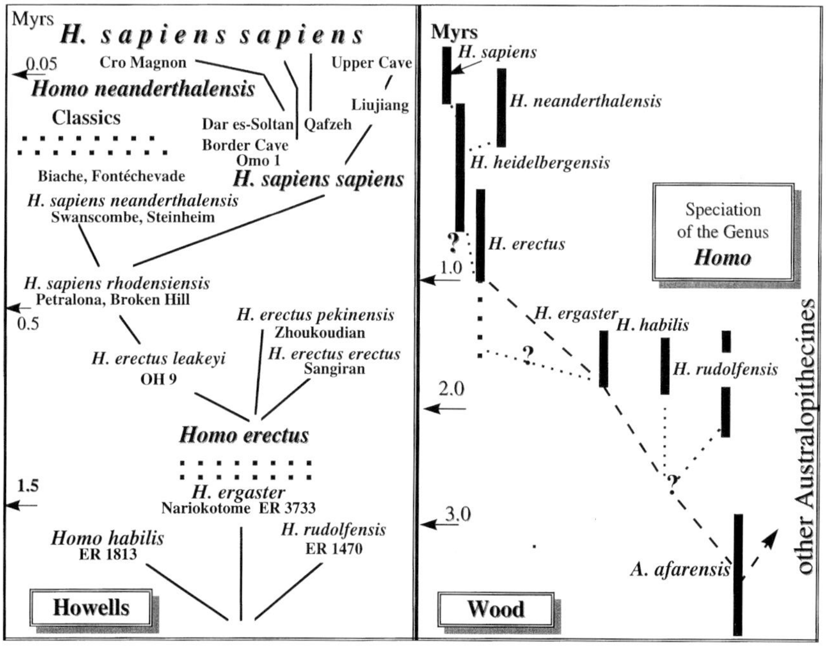

c - Howells (1990, redrawn by Henke et al., 1996, modified);
d - Wood (1996, redrawn);

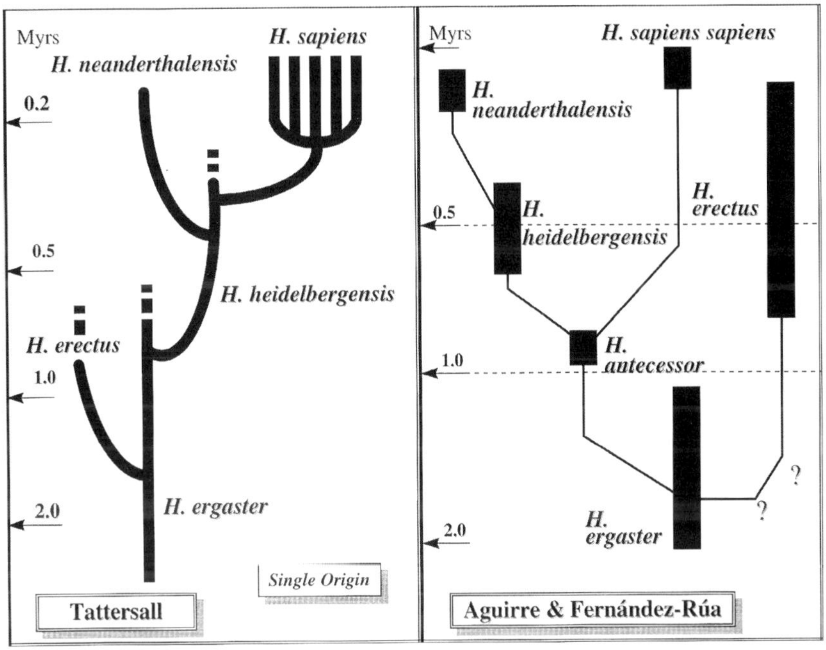

e - Tattersal (1997, redrawn);
f - Aguirre de Cárcer & Fernández-Rúa (1997, redrawn); see also Nieves (1997) with modifications concerning <u>Homo rhodesiensis</u>;

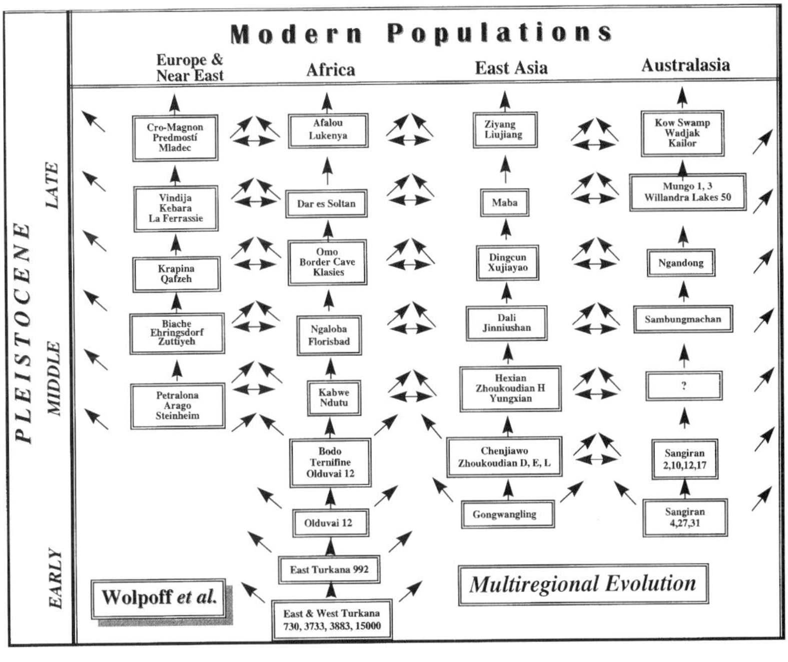

g - Frayer et al. (1993) and Wolpoff et al. (1994, redrawn).

Sites	Age (mys)	Hominid fossils and Stone-tool technology	Taxa
Hadar (Ethiopia)			
Kada Gona	**2.7-2.5**	core-flake (Omo)	
Kada Hadar	**2.3-2.2**	mandible	***Homo* sp. indet.**
		core-flake (Omo/Oldowan)	
Omo Shungura (Ethiopia)			
Member D	**2.5-2.4**	teeth	***Homo* sp. indet.**
Member E	**2.4-2.3**	teeth	***Homo* sp. indet.**
		core-flake (Omo)	
Baringo (Kenya)			
Chemeron	**2.4**	temporal bone BC 1	***Homo* sp. indet.**
Uraha (Malawi)			
Chiwondo Beds	**2.5-2.1**	mandible UR 501	***H. rudolfensis***
Nachukui (West Turkana)			
Kalochoro	**2.35-1.9**	core-flake (Omo)	
Nariokotome, Lower Natoo	**1.5**	skeleton WT 15000	***H. ergaster***
Koobi Fora (East Turkana)			
Upper Burgi	**1.90-1.88**	mandible ER 1812	***H. ergaster***
		core-flake (Omo/Oldowan)	
		occiptal bone ER 2598	***H. ergaster***
		pelvis ER 3228	***H. ergaster***
		calvarium ER 1470	***H. rudolfensis***
KBS	**1.78**	calvarium ER 3733	***H. ergaster***
Okote	**1.65-1.50**	calvarium ER 3883	***H. ergaster***
		mandible ER 992	***H. ergaster***
Swartkrans (South Africa)			
Member 1	**1.7-1.8**	calvarium SK 847	***H. ergaster***
Olduvai Gorge (Tanzania)			
Bed I	**1.8**	mandible O.H. 7	***H. habilis***
		core-flake (Oldowan)	
Lower Bed II	**1.6**	calvarium O.H. 13	***H. habilis***
		core flake (Developed Oldowan)	
Peninj (Tanzania)			
Humbu	**1.4**	biface (Acheulian)	
Olduvai Gorge (Tanzania)			
Upper Bed II	**1.2.5/1.4**	calvarium O.H. 9	***H. erectus (?)***
		biface (Acheulian)	
Konso-Gardula (Ethiopia)			
KGA 4-KGA 12	**1.4**	mandible KGA 10-1	***H. ergaster***
		biface (Acheulian)	

Table 1 - Earliest sites of the genus Homo (see also Bar-Yosef, 1994; Hill et al., 1992; Henke & Rothe, 1994; Larick & Ciochon, 1996; Schrenk et al., 1995).

Table 2 (page 143) - Early and later Eurasian dispersal sites (see also Henke & Rothe, 1994; Larick & Ciochon, 1996; Ranov et al., 1995; Rendell et al., 1987; Roe, 1995; Schmincke & van den Boogard, 1995; Vekua 1995; Wood & Turner, 1995).

SITES	AGE (MYS)	HOMINID FOSSILS AND STONE-TOOL TECHNOLOGY	TAXA *classification by single authors
Yiron (Israel)	>2,4 [?]	„tools“	
Erq-el-Ahmar (Israel)	**1.96-1.78**	Upper Erq-el-Ahmar, 2 core tools	
Ubeidiya (Israel) Ubeidiya Formation upper layers	**1.4**	biface (Acheulian)lower layers core-flake (Developed Oldowan)	
Riwat (Pakistan)	**1.96-1.78**	artifact horizon, 3 core tools	
Pabbi Hills (Pakistan)	**2.0 -0.90**	fossil horizons, core-flake (Omo)	
Longgupo (China) Middle Zone	**1.96-1.78**	mandible CV.939.1 2 core tools incisor CV.939.2	***Homo* sp. indet.** ***Homo* sp. indet.**
Sangiran (Java, Indonesia) Pucangan Formation	**1.66**	2 calvaria S 27, S 31	***H. erectus***
Modjokerto (Java, Indonesia) Pucangan Formation	**1.81**	calvarium Perning 1	***H. erectus***
Ngebung (Java, Indonesia) Kabuh Formation	**0.75-0.25**	core-flake (Sangiran Flake Industry)	
Dmanisi (Georgia) hominid locality on underlying basalt ca. 2.0	**?? - 1.9 ?**	mandible core-flake	***H. erectus* [?]**
Atapuerca (Spain) Gran Dolina (TD) 4 Gran Dolina (TD) 6	 **1.6 -0.75** **0.99-0.78**	 core-flake frontal, maxilla, mandible core-flake	 ***H. antecessor***
Orce (Spain) Fuentenueva 3	**? 1.07-0.99 ?**	temporal bone core-flake	***Homo* sp. indet.**
Isernia la Pineta (Italy) Sector I Sector II	 **0.99-0.78** **0.80-0.50**	 core-flake	
Bodo (Ethiopia) Upper Bodo Sand Unit,	**0.60**	calvarium biface (Acheulian)	***H. heidelbergensis***
Elandsfontein (South Africa) Saldana Bay	**0.70-0.40**	calvarium	***H. heidelbergensis***
Mauer (Germany)	**0.50**	mandible biface (Acheulian)	***H. heidelbergensis***
Boxgrove (England) Quarry 1	**0.50**	tibia biface (Acheulian)	***H. heidelbergensis***

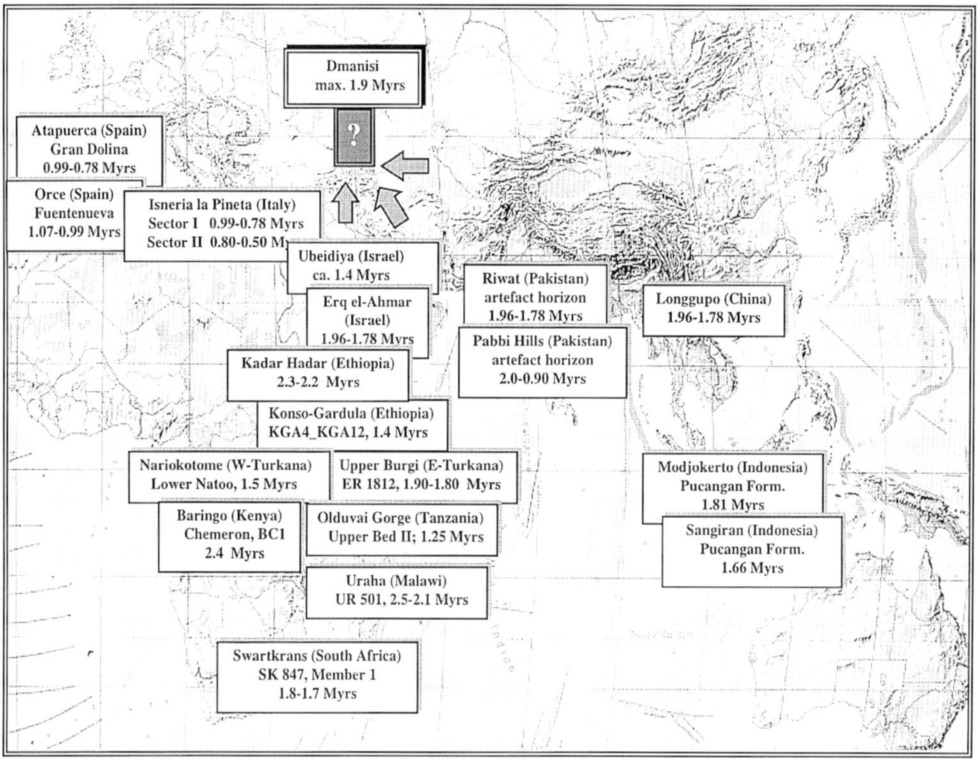

Figure 2 - Regional dispersal pattern of the genus Homo.

The above statements of palaeoanthropologists demonstrate that we seem to have many problems with our principles and methods of species recognition. Newest reports on ca. 800 000 year old fossils from Spain increase this view. The description of the Atapuerca-fossils as *Homo antecessor* by Spanish palaeoanthropologists brought up a new scenario and completed the current confusion about the phylogenetic status of *Homo erectus*.

The new pedigree of our ancestors (Aguirre de Cárcer & Fernandez-Rúa, 1997; see also Nieves, 1997) describes those European fossils, being formerly accepted as *Homo erectus*, as *Homo heidelbergensis* (see also Arsuaga et al., 1993, 1997; Carbonell et al., 1995).

This taxon is regarded as the forerunner species of the 'Neanderthals', which shall have been replaced by our direct ancestors that originated from the African taxon *Homo rhodesiensis*. The interpretation by the Spanish specialists is, that *Homo rhodesiensis* shares together with *Homo heidelbergensis* a common ancestor, which has been named *Homo antecessor* by the Spanish palaeontologists (for further information see Nieves, 1997).

The different views on the taxonomical relationships within the genus *Homo* are demonstrated in Figure 1a-g. The fact that some of the authors are referring to evolutionary species and others to chronospecies is only one cause for the obvious discrepancies, predominantly resulting from the inconsistency in the application of taxonomical principles (see Henke & Rothe, 1994, 1995; for taxonomical approaches see Ax, 1984; Hennig, 1950, 1966; Howells, 1980; Mayr, 1975; Sneath & Sokal, 1973; Tattersall, 1986).

Sites and chronology of the oldest fossils of the Genus *Homo*

Only when species - in the sense of an evolutionary species (see Ax, 1984; Wiley, 1978, 1981; Willmann, 1985) - have been adequately defined morphologically, appropriate comparisons can be made, and the distribution of character states across species be used to generate phylogenetic hypotheses. But until now, we do not have any consensus of the *Homo erectus*-hypodigm, that means we have no agreement on the question which fossils belong to the taxon that has been defined as *Homo erectus* (see Franzen, 1994 a,b; Howells, 1980, 1992; Ponce de León, 1991).

The remaining questions are:

- Is *Homo erectus* a 'good' species?
- Is *Homo erectus* the result of a cladistic event or is it a result of continuous transition?
- Is it possible to distinguish between an Asian, an African and/or an European branch of *Homo erectus*?
- Should such African and European branches, if they really exist, be called a species of their own?
- If each of these branches developed separately and continuously into *Homo sapiens*, what else is *Homo erectus* than an early or archaic *Homo sapiens*?

These questions (see also Franzen, 1994a,b; Henke & Rothe, 1995) call for the revisioning of all available materials. The first step is to describe the chronological and regional distribution of all *Homo*-fossils (Tables 1, 2 and Fig. 2). The pattern shows that the oldest fossils which are interpreted as belonging to our own genus as well as the most ancient stone tools which have been produced by *Homo* come from the Ethiopian region or nearby sites from West and East Turkana. While the Ethiopian fossils, fragmentary teeth and a mandible, are too poor to allow a valid diagnosis, the Turkana material, especially the skeleton WT 15000, the so-called „Turkana Boy", is well preserved and exhibits traits of a species of our own genus. Since many features demonstrate a much more progressive status than those fossils which have been classified as *Homo rudolfensis* (e.g. KNM-ER 1470 and UR 501) or *Homo habilis* (e.g. O.H. 7, O.H. 13).

In the following, we will not focus on the relationship of *Homo habilis, Homo rudolfensis,* and *Homo ergaster*, but concentrate on the dispersal question. As long as we regarded the oldest fossils from Eurasia to be in maximum 1 myrs old, the conclusion that *Homo erectus* has evolved in Africa afterwards left this continent seemed to be the most plausible hypothesis, but the new data from Near, Middle and Far East do not correspond with this scenario (see Table 2). While we have no fossils and only artifacts from the Palestinian 'corridor' as well as from Pakistan, there are small fossil fragments from China (Longgupo) which may indicate the presence of *Homo* sp. indet. in the Far East as many as 1.9 myrs ago.

Swisher III et al. (1994) provided further arguments for an early dispersal by calibrating an age of 1.8 myrs for the well-known fossil site of Modjokerto (Java, Indonesia), but until now we have no confirmation of this age from further research and many palaeontologists have severe doubts about this exceptional date and similar data of other places from Java (pers. comm. de Vos).

The accepted ages of the oldest fossils of Europe range up to around 0.8 myrs. The much higher age of the implements from Orce (Fontenueva) and the Gran Dolina (TD 4) (see Fig. 2) have still to be proven. This does not at all mean, however, that higher ages are impossible for palaeoecological reasons.

Within the pattern of proposed and doubted chronological data, Dmanisi is no exception (see Bosinski, 1992). The assumed maximum age of 1.9 myrs, based on archaeological and stratigraphical results, contradicts the morphological patterns, or to say it unscientifically, the morphology of the mandible and the teeth „look too young or modern". One may comment that

- parabolic dental arc, very broad; oldest *H. erectus* mandibles still with straight postcanine teeth rows;
- large teeth, relation of the teeth dimensions similar to *H. sapiens;*
- C small, in more archaic fossils with an apex protruding the occlusal level;
- horizontal abrasion of the $C_{inf.}$;
- tooth crown of the mandibular canines of similar size compared with *H. habilis*, but significantly larger than in *Australopithecus* and in anatomically modern *H. sapiens*;
- weak heteromorphy of P_{inf}.; in anatomically modern *H. sapiens* pronounced strong homomorphy;
- P_3 mesiodistally relatively long, in some cases longer than P_4, bicuspid, heteromorph, buccal cusp greater than lingual cusp;
- M small, only slightly larger than in *H. sapiens,* bucco-lingually small and taurodont, with rounded occlusal surfaces; in some cases basal cingulum present;
- Relation of the molar lengths: $M_1 \leq M_2 > M_3$; $M_1 \leq M_2$ (in older samples $< M_3$);
- Choukoudian teeth $M_1 < M_2 \geq M_3$;
- summed crown areas of P_3 - M_3 543-695 mm²; in Australopithecines ca. 861-1312 mm²; *H. habilis* 792 mm²; cf. *H. sapiens*, males 443-595 mm², females 420-562 mm².

Table 3 - Dentomorphological traits in the mandible of Homo erectus (from Henke & Rothe, 1997, tab. 16, translated).

nobody would have expected an age of 1.5 myrs for such a modern or progressive looking skeleton as the Turkana Boy exhibits before it was uncovered.

Though all possible research on the site seems to have been done (see van den Boogard et al., 1995/1997), we realize that the credibility of the proposed very high age of the Dmanisi fossil is still very low. We are of the opinion that inspite of the extensive research that has been done on the fossil itself there is still a lack of morphological analysis. For this reason we add some comparative data of dentomorphological traits for the dentition and mandible of *Homo erectus* as defined in the literature (Table 3; see Henke & Rothe, 1997). Futhermore we describe (Table 4-6) and illustrate the dentalmophological features of the Dmanisi mandible (see also Fig. 2 and 3) to allow further detailed studies.

Additional information comes from a computer tomographical research on a cast of the mandible. Special CT-software of Siemens, Germany (DENTAL CT, Somatom AR Family) used in the Department of Diagnostic Radiology, University of Freiburg, enables information of panoramic views and paraxial slices of the mandible, based on thin 1-2 mm CT scans. A very detailed comparison of the final images in multiimage display with references marking delivers much

- all mesio-distal and bucco-lingual diameters in the range of recent populations, except of the right P_3 and M_1 [= 44; 46] which are in the mesiodistal diameter somewhat larger.

Incisors:

- Incisors of medium size;
- with slight curvation at the labial side, lingually moderate concave;
- prominent basal tubercle at the basis;.
- abrasion of the crown can be interpreted as contact facette.

Canines:

- both C show accessorial cusps in the area of the distal rim;
- both canines with a pronounced distal talonid structure;
- apex of the crown relatively low, resembles in this feature those of recent populations;
- only very little contact between the C and P_3;
- incisivation of the canines which has been described by Weidenreich (1937, 1940) and Remane (1960) in the canines of *Sinanthropus* is not expressed in the same degree in the Dmanisi fossil;
- abrasion suggests that the lower canines had still a pronounced apex, a feature that does not fit in the small incisivation of „*Sinanthropus*".

Premolars:

- P_3 demonstrate only one real cusp on each side, second cusp is only a vestage; shape of the crown composed by a buccal main cusp - the protoconid - and a very low and small developed inner cusp - the metaconid;
- metaconid situated on a lingual cingulum;
- Fovea anterior smaller and less deep than Fovea posterior; cross-section of the crown mesio-distally long-oval to triangular;
- distal premolar with quadrangular to round cross-section; in comparison to P_3 size difference of the buccal and mesial cusp more balanced;
  both foveae, of which the anterior is smaller, connected on the left tooth by a furrow; in the right tooth furrow not interrupted by a rim;
- Cingulum very small.

Table 4 - Special features of the teeth of the Dmanisi mandible.

more information than usually available. Some image examples of the Dmanisi mandible are demonstrated in Fig. 4.

Finally we add some multivariate statistical results that demonstrate the heterogeneous affinities of the mandible. While the length and breadth diameters of the alveolar arch show affinities to the australopithecine and *Homo habilis/rudolfensis* patterns (see Fig. 5) there are strongest affinities to *Homo ergaster* from West Turkana (KNM ET 15000) and a so-called 'late' *Homo erectus* from Northwest Africa (Thomas Quarry 1) if the comparison is based on variables of the corpus and symphysis (see Fig 6).

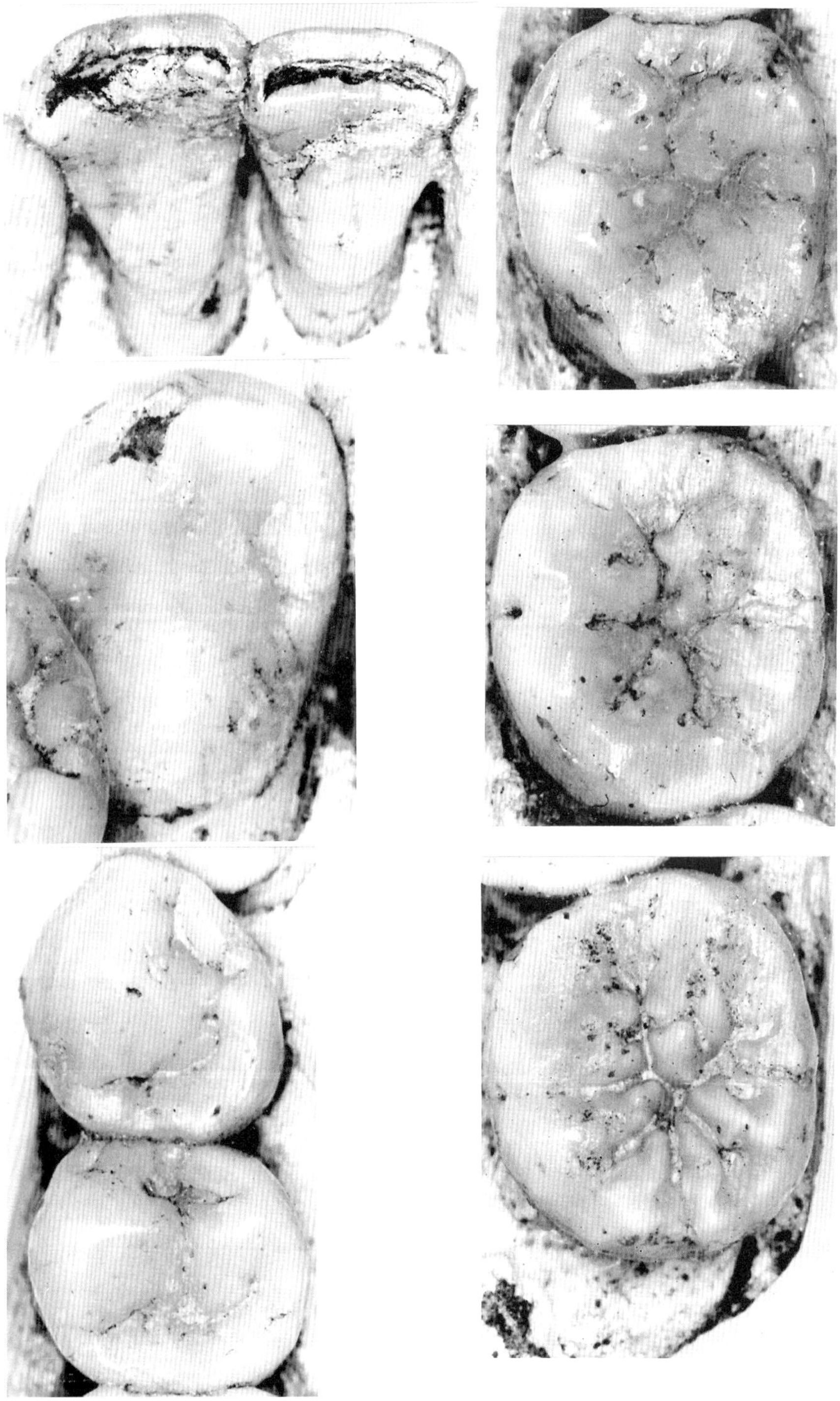

Figure 3 - Occlusal view of the left teeth of the Dmanisi mandible. Left row, down: incisivi, caninus, premolars; right row: first, second and third molar (not adjusted size).

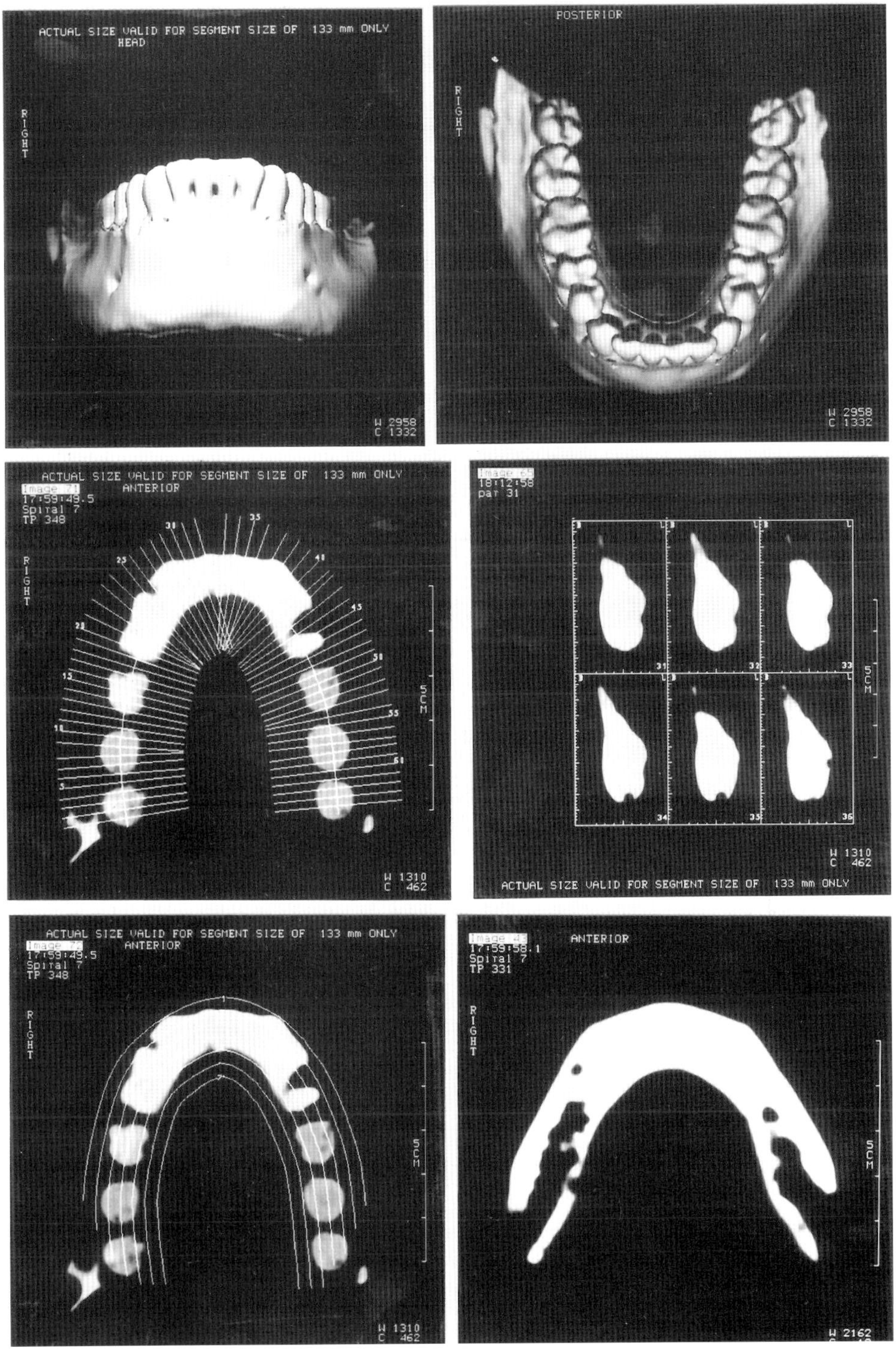

Figure 4 - *Computertomographical images of the Dmanisi mandible. (Pictures taken by C. Buitrago-Tellez and K.W. Alt, Univ. of Freiburg).*

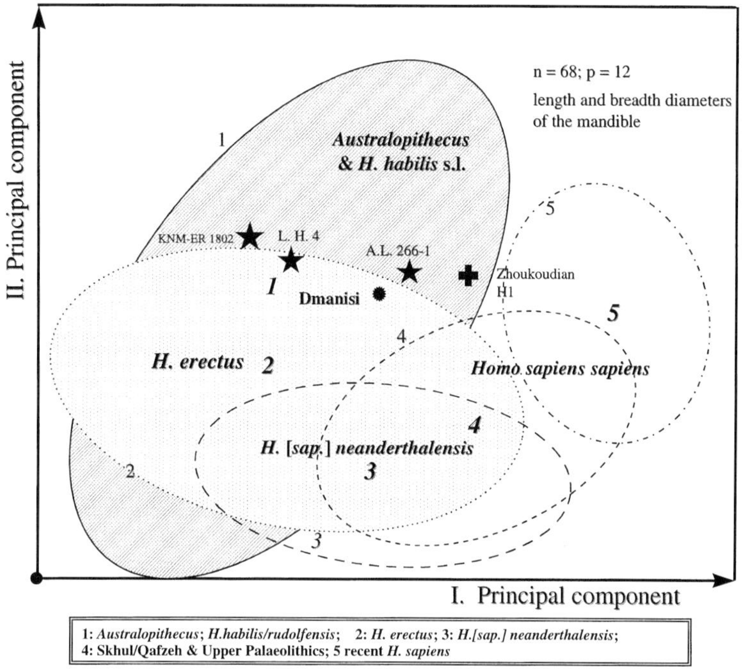

Figure 5 - *Plot-diagram and 75 %-correlation ellipses of the individual scores of a principal component analysis based on diameters of the dental arcade of 67 specimens; the scores of the Dmanisi fossil lie marginal inside the Homo erectus-ellipse and are also inside the correlation ellipse of the more ancient sample, but demonstrate distinct unsimilarities in comparison to the Homo sapiens sample (for further information see Roth, 1983; Roth & Simon, 1992; Henke et al., 1995; Alt et al., 1997).*

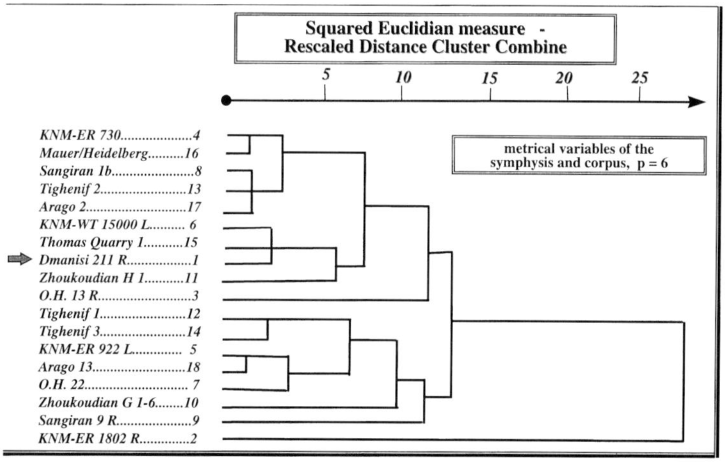

Figure 6 - *Squared Euclidian Distances based on morphological variables of the symphysis and corpus of the Dmanisi mandible.*
Nearest neighbours of Dmanisi are African fossils from Thomas Quarry and West Turkana and furthermore the Chinese mandible No. H1 from Zhoukoudian (see also Alt et al., 1997).

Molars:

- molars demonstrate a y-shape [except M_1 (46) - not to evaluate];
- size reduction from mesial to distal: M_1> M_2> M^3; this is regarded as an apomorphic trait.
- M_3 are small due to the reduction of the entoconid and hypoconid;
- there are *Foramina molaria* or *Foramina coeca* which are localised at the end of mesio-buccal fissure between the protoconid and the hypoconid (in M_1 also between the hypoconid and the hypoconulid);
- the molars have different accessorial cusps;
- M_2 and M_3 (M_1 not to evaluate) have so-called cusps-6 or a *Tuberculum sextum*, a cusp on the distal edge between the entoconid and the hypoconulid; this trait is said to be a typical feature of mongolid populations.
- all molars have a *Tuberculum intermedium* („*cusp 7 or Selenka-cusp*"); it is positioned between the metaconid and the entoconid.
- both M_1 have on the buccal side of the protoconid a *Tuberculum paramolare* (synonyma: protostylid);
- one can not exclude that these traits are *Foramina molaria*, but due to the shape, which is long and stretched, there are good arguments for the first classification; *Foramina molaria* are very frequent in mongolid populations;
- M_2 and M_3 have a *Tuberculum paracone* at the mesial border;
- M_2 and M_3 (M_1 not to evaluate) have a so-called „*deflecting wrinkle*", a medial furrow of the metaconid, which normally starts at the apex of the metaconid and runs into the direction of the protoconid, deflecting distally and in some cases showing a small cusp between protoconid and hypoconid.

Table 5 - Special features of the teeth of the Dmanisi mandible.

Progressive Features of the Dmanisi Mandible

⇒ *Mentum osseum* demonstrating a well projecting *Tuber symphyseos*
⇒ shape of the median-sagittal profil of the symphyseal region
⇒ development and inclination of the *Planum alveolare*
⇒ strongly developed *Tuberculum marginale anterius*
⇒ multiple *Foramen mentale*
⇒ area of the crown of P_3
⇒ area and shape of the crown of the P_4
⇒ area of the M_2
⇒ area and shape of the crown of the M_3
⇒ size ratio of the molars: $M_1 > M_2 > M_3$

Table 6 - Progressive features of the Dmanisi mandible in comparison to those traits that characterize Homo erectus (for further details see also Henke et al., 1995; Bräuer et al., 1995, Bräuer & Schultz, 1996; Alt et al., 1997).

Conclusions

The comparison of those dentomorphological features, that are usually accepted to characterize *Homo erectus*, with the mosaic pattern of the Dmanisi mandible reveals in some details apomorphic features of the Georgian fossil. For this reason there remain many doubts concerning its status as 'early' *Homo erectus* and even the classification as late member of this taxon does not agree with the described pattern. Contributing to this dilemma, we have to emphasize:

- that there exists a strong controversy concerning the *Homo erectus* hypodigm;
- that the frequently applied numerical taxonomical analyses are highly problematical;
- that methodologically stringent cladistic analyses to solve the *Homo erectus*-problem do not exist until yet;
- and that for these reasons we have until now no clear definition of the taxon *Homo erectus*.

Consequently, every taxonomical approach to classify the Dmanisi-mandible is at present preliminary. It can not be overlooked that the Dmanisi mandible shows some affinities to the archaic *Homo* by morphometrical and morphognostic comparisons (e.g. KNM-WT 15000; O.H. 13), but these affinities are irrelevant for phylogenetic interpretations because they are based on plesiomorphic features. Otherwise there exist numerous affinities to so-called evolved *Homo erectus* specimens (e.g. Thomas Quarry 1; Zhoukoudian H1); see e.g. Henke, 1995, 1996; Henke et al., 1995; Bräuer & Schultz, 1996; Bräuer et al., 1996; Alt et al., 1997). As the presence of apomorphies in the Dmanisi mandible seems to exclude this fossil from an early *Homo erectus* we should give a phylogenetical interpretation not until the hypodigm of *Homo erectus* has been clearly defined.

If one rejects proposal of Wolpoff et al. (1994) to include the taxon *Homo erectus* into *Homo sapiens*, one has at least to conclude that the homogeneity of a formerly accepted *Homo erectus* hypodigm has vanished by new interpretations of the relevant fossil material from the different continents of the Old World (see e.g. Tatterall, 1997).

Following the „multiregionalists" as well as those palaeoanthropologists who interprete the fossil record as multiple migration events out of Africa and successive speciation processes, there are many indications today that the former taxon *Homo erectus* (s.l.) is unvalid. While the multiregionalists assign the fossil sample under discussion to a single taxon, others see strong arguments to split the material into different taxa [e.g. *Homo erectus* (s.str.); *Homo ergaster*, *Homo antecessor*, *Homo heidelbergensis*]. As the many papers on this topic are extremely controverse we propose for the above given reasons to postpone the question of the taxonomic status of the Dmanisi mandible. We are of the opinion, that inspite of opposite views, there is still much work to be done before proposing a revised scenario of human evolution, whenever a new fossil is recovered. It should be foremost programme, to reveal the morphoclines - plesiomophic-apomorphic relationships - in primate evolution, in order to demonstrate speciation processes by differentiating autapomorphic features and to reconstruct relationship patterns by differentiating synapomorphies.

Acknowledgements

We are indebted to Dr. Herbert Ullrich who has invited us to participate in the excellently organized, very successful congress and to contribute to this volume.

The authors especially thank Dr. Antje Justus and Prof. Dr. Gerhard Bosinski (Monrepos and Colonia) for their kind permission to investigate the Dmanisi fossil and excellent cast respectively.

Finally we thank Dr. Carlos Butitrago-Tellez (Institute of Radiological Diagnostic Univ. of Freiburg/Brsg.) for his support and advice concerning the CT-analysis of the mandible.

References

ALT, K.W., HENKE, W. & ROTHE, H. (1997): Taxonomische Marginalien zum Dmanisi-Unterkiefer aufgrund dentalmorphologischer Vergleichsanalysen. *Beiträge zur Archäozoologie und Prähistorischen Anthropologie* 1, 124-133.

ALT, K.W. & TÜRP, J.C. (Hrsg.) (1996): *Die Evolution der Zähne. Phylogenie - Ontogenie - Variation*. Berlin.

AGUIRRE de CÁRCER, A. & FERNÁNDEZ-RUÁ, J.M. (1997): El equipo dientífico de Atapuerca presentó al „*Homo antecessor*", el primer europeo. *ABC de la ciencia, Viernes 30-5-97*, pág. 84.

ARSUAGA, J.L. MARTINEZ, I., GRACIA, A., CARRETERO, J.M. & CARBONELL, E. (1993): Three new skulls from the Sima de los Huesos site in Sierra de Atapuerca, Spain. *Nature* 362, 534-537.

ARSUAGA, J.-L., MARTINEZ, I., GRACIA, A. & LORENZO, C. (1997): The Sima de los Huesos crania (Sierra de Atapuerca, Spain). A comparative study. *Journal of Human Evolution* 33, 219-281.

AX P. (1984): *Das Phylogenetische System*. Stuttgart, New York.

BAR-YOSEF, O. (1994): The Lower Paleolithic of the Near East. *Journal of World Prehistory* 8, 211-265.

BOGAARD, P. van den, BRÄUER, G., GOGCAJSVILI, A., HENKE, W., MAJSURADZE, G., MUSCHELISVILI, A., PAVLENISVILI, E., SCHMINCKE, H.-U., SCHULTZ, M., SOLOGASVILI, D. & VEKUA, A. (1995, published in 1997): Der altpaläolithische Fundplatz Dmanisi (Georgien, Kaukasus). *Jahrbuch des Römisch-germanischen Zentralmuseum Mainz* 42, 21-203.

BONDE, N. (1989): *Erectus* and *neanderthalensis* as species or subspecies of *Homo* - with a model of speciation in hominids. In: Giacobini, G. (ed.), *Hominidae*, 205-208. Mailand.

BOSINSKI, G. (1992): Die ersten Menschen in Eurasien. *Jahrbuch des Römisch-germanischen Zentralmuseum Mainz* 39, 131-181.

BOSINSKI, G., GABUNIA, L.K., JUSTUS, A. & VEKUA, A. (1995): Le site de Dmanisi (Georgie, Caucase). Abstract. In: *Congreso Internacional de Paleontologia Humana, Orce*.

BRÄUER, G. (1992): Der Mensch. In: *Anthropologie heute*, 3-43. Tübingen (Deutsches Institut für Fernstudien an der Universität Tübingen).

BRÄUER, G. & MBUA, E. (1992): *Homo erectus* features used in cladistics and their variability in Asian and African hominids. *Journal of Human Evolution* 22, 79-108.

BRÄUER, G. & SCHULTZ, M. (1996): The morphological affinities of the Plio-Pleistocene mandible from Dmanisi, Georgia. *Journal of Human Evolution* 30, 445-482.

BRÄUER, G., HENKE, W. & SCHULTZ, M. (1995, published in 1997): Der hominide Unterkiefer von Dmanisi: Morphologie, Pathologie und Analysen zur Klassifikation. *Jahrbuch des Römisch-germanischen Zentralmuseum Mainz* 42, 183-203.

CARBONELL, E., BERMÚDEZ de CASTRO, J. M., ARSUAGA, J. L., DIEZ, J. C., CUENCA-BESCÓS, G., SALA, R., MOSQUERA, M. & RODRÍGUEZ, X.P. (1995): Lower Pleistocene hominids and artifacts from Atapuerca-TD6 (Spain). *Science* 269, 826-830.

CLARKE, R.J. (1990): The Ndutu cranium and the origin of *Homo sapiens*. *Journal of Human Evolution* 19, 699-736.

DEAN, D. & DELSON, E. (1995): *Homo* at the gates of Europe. *Nature* 373, 472-473.

DENNELL, R.W., RENDELL, H.M., HURCOMBE, L. & HAILWOOD, E.A. (1994): Archaeological evidence for hominids in Pakistan before one million years ago. *Courier Forschungs-Institut Senckenberg* 171, 151-155.

DZAPARIDZE, V., BOSINSKI, G., BUGIANSVILI, T., GABUNIA, L., JUSTUS, A., KLOPOTOVASKAJA, N., KVAVADTE, E., LORDKIPADNIDZE, D., MAJSURADZE, G., MGLADZE, N., NIORADZE, M., PAVLENISVILI, E., SCHMINCKE, H.-U., SOLOGAVILI, D., TUSAMBRAISVILI, D., TVALCRELIDZE, M. & VEKUA, A. (1989; published in1992): Der altpaläolithische Fundplatz Dmanisi in Georgien (Kaukasus). *Jahrbuch des Römisch-germanischen Zentralmuseum Mainz* 36: 67-160.

FRANZEN, J.L. (1994a): The *Homo erectus* problem. *Courier Forschungs-Institut Senckenberg* 171: 9-10.

FRANZEN, J.L. (ed.) (1994b): *100 years of Pithecanthropus. The Homo erectus problem*. Frankfurt a.M. (Courier Forsch.-Inst. Senckenberg 171).

FRAYER, D.W., WOLPOFF, M.H., THORNE, A.G., SMITH, F.H. & POPE, G.G. (1993): Theories of modern human origins: The paleontological test. *American Anthropologist* 95, 1, 14-50.

GABUNIA, L.K. (1992): Der menschliche Unterkiefer von Dmanisi (Georgien, Kaukasus). *Jahrbuch des Römisch-germanischen Zentralmuseum Mainz* 39, 185-208.

GABUNIA, L.K. & VEKUA, A. (1995): A Plio-Pleistocene hominid from Dmanisi, East Georgia, Caucasus. *Nature* 373, 509-512.

GROVES, C.P. &. MAZÁK, V (1975): An approach to the taxonomy of the Hominidae. Gracile Villafranchian hominids of Africa. *Canadian Museum of Geology* 20: 225-247.

HENKE, W. (1995, unpublished): *Exposé zur vergleichend-morphologischen Analyse des Hominiden-Unterkiefers von Dmanisi (Georgien, Kaukasus). Teil I (Text) und II (Abbildungen und Vergleichsdaten).* Mainz (Institut für Anthropologie der Universität MAinz). 252 S.

HENKE, W. (1996, in press): The Dmanisi mandible - morphological and phylogenetical remarks. In: N.I. Xirotiris (ed), *Perspectives in Anthropology of Past and Present Populations. E.M. Winkler Memorial Symposium, 29th Sept. to 5th Oct. 1995. Komotini (Thrace).* Proceedings of the Laboratory of Anthropology.

HENKE, W., KIESER, N. & SCHAUBELT, W. (1996): *Die Neanderthalerin - Botschafterin der Vorzeit.* Gelsenkirchen.

HENKE, W. & ROTHE, H. (1994): *Paläoanthropologie*. Berlin, Heidelberg, New York.

HENKE, W. & ROTHE, H. (1995): *Homo erectus* - valides Taxon der europäischen Hominiden? *Bulletin de la Societé Suisse d'Anthropologie* 1, 15-26

HENKE, W. & ROTHE, H. (1997). Zahnphylogenese der Hominiden. In: K.W. Alt & J.C. Türp (eds.), *Die Evolution der Zähne*, 279-360. Berlin.

HENKE, W., ROTH, H. & SIMON, C. (1995). Qualitative and quantitative analysis of the Dmanisi mandible. In: R.J. Radlanski & H. Renz (eds), *Proceedings of the 10th International Symposium on Dental Morphology, Berlin, 1995,* 459-471. Berlin.

HENNIG, W. (1950): *Grundzüge einer Theorie der Phylogenetischen Systematik.* Berlin.

HENNIG, W. (1966): *Phylogenetic systematics.* Urbana, Chicago, London.

HILL, A., WARD, S., DEINO, A. et al. (1992): Earliest *Homo. Nature* 355, 719-722.

HOWELLS, W.W. (1980): *Homo erectus* - who, when, and were, a survey. *Yearbook of Physical Anthropology* 23,1-23.

HOWELLS, W.W. (1992): *Getting here. The story of human evolution.* Washington, DC.

HUANG, W., CIOCHON, R., YUMIN, G., GU, Y., LARICK, R., FANG, Q., SCHWARCZ, H., YONGE, C., VOS, J. de & RINK, W. (1995): Early *Homo* and associated artifacts from Asia. *Nature* 378, 275-278.

KENNEDY, G.E. (1991): On the autapomorphic traits of *Homo erectus. Journal of Human Evolution* 20, 375-412.

LARICK, R. & CIOCHON, R.L. (1996): The African emergence and early Asian dispersal of the genus *Homo. Amererican Scientist* 84, 538-551.

MAJSURADZE, G. (1995, published in 1997): Geologisch-geomorphologische Untersuchungen im Masavera-Becken (Südost-Georgien). *Jahrbuch des Römisch-germanischen Zentralmuseum Mainz* 42, 76-100.

MAJSURADZE, G., PAVLENIŠVILI, E. SCHMINCKE, H.-U. & SOLOGAŠVILI, D. (1989, published in 1992): Paläomagnetik und Datierung der Basaltlava. *Jahrbuch des Römisch-germanischen Zentralmuseum Mainz* 36, 74-76.

MAYR, E. (1975): *Die Grundlagen der zoologischen Systematik.* Hamburg.

NIEVES, J.M. (1997): *Homo antecessor.* La descriptión de una nueva especie de homínido en Atapuerca aclara los orígines del hombre moderno. *Blanco - Negro, Semanario de ABC, 1 Junio 1997*, 12-21.

PONCE de LEÓN, T.M. (1991: *Die Stellung von Homo erectus im Rahmen der Hominiden-Evolution: Eine quantitative Untersuchung des Schädels.* Diplomarbeit, Anthropologisches Institut und Museum der Universität Zürich.

PUECH, P.-F., ALBERTINI, H., PUECH, S. & CHEVEAUX, J. (1995): *Homo erectus* many faces. In: R.J. Radlanski & H. Renz (eds), *Proceedings of the 10th International Symposium on Dental Morphology, Berlin, 1995,* 369-372. Berlin.

RANOV, V.A., CARBONELL, E. & RODRÍGUEZ, X.P. (1995): Kuldara: Earliest human occupation in Central Asia in its Afro-Asian context. *Current Anthropology* 36, 337-346.

REMANE, A. (1960): Zähne und Gebiß. In: H. Hofer, A.H. Schultz & D. Starck (Hrsg..), *Primatologia. Handbuch der Primatenkunde,* Bd. III, Teil. 2, 637-846. Basel.

RENDELL, H. M., HAILWOOD, E. & DENNELL, R.W. (1987): Paleomagnetic dating of the two-million-year-old artefact-bearing horizon at Riwat, nothern Pakistan. *Earth and Planetary Sciences Letters* 85, 488-496.

RIGHTMIRE, P. (1990): *The evolution of Homo erectus. Comparative anatomical studies of an extinct human species*. Cambridge.

RIGHTMIRE, P. (1992): *Homo erectus*: Ancestor or Evolutionary Side Branch? *Evolutionary Anthropology* 1, 43-49.

ROE, D. (1995): The Orce Basin (Andalucía, Spain) and the initial Palaeolithic of Europe. *Oxford Journal of Archaeology* 14, 1-12.

ROTH, H. (1983): *Comparaison statistique de la forme des arcades alvéolaire et dentaire des mandibules des Hominidés fossiles. Position de l'Homme de Tautavel dans l'évolution humaine*. Marseille: Thèse Université Aix-Marseille I (Vols. I-II).

ROTH, H. & SIMON, C. (1992): The relationship of Pleistocene and modern mandibles evaluated by principal component analysis of the arcade dimensions. In: M. Toussaint (ed.), *Five million years. The human adventure*, 111-124. Liège (E.R.A.U.L. 56).

SCHMINCKE, H.-U. & BOOGARD, P. van (1995, published in 1997): Die Datierung des Mašavera-Basaltlavastroms. *Jahrbuch des Römisch-germanischen Zentralmuseum Mainz* 42, 75-76.

SCHMINCKE, H.-U. & BOOGARD, P. van den (1996/1997): Die Datierung des Mašavera-Basaltlavastroms. *Jahrbuch des Römisch-germanischen Zentralmuseum Mainz* 43, 125-126.

SCHRENK, F., BROMAGE, T, BETZLER, C.G., RING, U. & JUWAYEYI, Y. (1995): Oldest *Homo* and Pliocene biogeography of the Malawi rift. *Nature* 365, 833-836.

SCHWARTZ, J.H. & TATTERSALL, I. (1996): Whose teeth? *Nature* 381, 201-202.

SNEATH, P.H.A. & SOKAL, R.R. (1973): *Numerical taxonomy. The principles and practice of numerical classification*. San Francisco.

STRINGER, C.B. (1993): Secret of the pit of the bones. *Nature* 362: 501-502.

SWISHER III, C.C., CURTIS, G.H., JACOB, T., GETTY, A.G., SUPRIJO, A., WIDIASMORO (1994): Age of earliest known hominids in Java, Indonesia. *Science* 263, 1118-1121.

TATTERSALL, I. (1986): Species recognition in human paleontology. *Journal of Human Evolution* 15, 165-175.

TATTERSALL, I. (1997): Out of Africa. Again.....and again ? *Scientific American* 276 (4), 46-53.

VEKUA, A. (1995, published in 1997): Die Wirbeltierfauna des Villafranchium von Dmanisi (Georgien, Kaukasus) und ihre biostratigraphische Bedeutung. *Jahrbuch des Römisch-germanischen Zentralmuseum Mainz* 42, 127-204.

WALKER, A. & LEAKEY, R.E.F. (1986): *Homo erectus* skeleton from West Lake Turkana. Kenya. *American Journal of Physical Anthropology* 69, 275,

WALKER, A. & LEAKEY, R.E.F. (eds.) (1993): T*he Nariokotome Homo erectus skeleton*. Berlin, Heidelberg.

WEIDENREICH, F. (1937): The dentition of *Sinanthropus pekinensis*: A comparative odontography of the hominids. *Palaeontologia Sinica*, New Series D, No. 1, 1-485.

WEIDENREICH, F. (1940): Some problems dealing with ancient man. *American Anthropologist* 42, 375-383.

WILEY, E.O. (1978): The evolutionary species concept reconsidered. *Systematic Zoology* (Washington) 27, 17-26.

WILEY, E.O. (1981)*: Phylogenetics: The theory and practice of phylogenetic systematics*. New York.

WILLMANN, R. (1985): *Die Art in Raum und Zeit. Das Artkonzept in Biologie und Paläontologie*. Hamburg.

WOLPOFF, M.H. (1996-1997): *Human evolution*. New York.

WOLPOFF, M.H., THORNE, A.G., JELÍNEK, J. & ZHANG, Y. (1994): The case for sinking *Homo erectus*. 100 years of Pithecanthropus is enough. *Courier Forschungs-Institut Senckenberg* 171, 341-361.

WOOD, B. (1996): Human evolution. *BioEssays*, 18 (12): 945-954.

WOOD, B. & TURNER, A. (1995): Out of Africa and into Asia. *Nature* 378, 230-240.

Homo finds from Bilzingsleben and Weimar-Ehringsdorf

Emanuel Vlček

Abstract

Two parallel human population groups can be recognized in Central Europe during the Holstein period: an advanced erectus and an archaic sapiens group. For the archaic sapiens form the subspecies Homo sapiens steinheimensis was already established. For the advanced erectus form the finds from Bilzingsleben are selected as a reference specimen and named Homo erectus bilzingslebensis. A detailed description of the neurocranium of the Bilzingsleben individual no. 1 and no. 2, which stand close to the finds from Arago - Vértesszölös - Petralona, is given in this paper. The new reconstruction of the skull H from Weimar-Ehringsdorf shows the modern architecture of the brain box and is morphologically closer to modern sapiens than to the Neanderthal forms.

Two populations during the Holstein period

During the Holstein period, we can follow two parallel population groups in Central Europe (Vlček, 1986, 1989b): First and foremost, the group which still displays erectoid features in the structure of the brain box, the so-called advanced *erectus*, and another kind, which may be assigned to the typical ancient sapient forms, archaic *sapiens*. Such a distinction is similarly attested to in the archaeological record, where the first group is related to the small-dimensional industries and the second, instead, to the Acheulian and Early Mousterian (Svoboda, 1987, 1989). The basic morphological differences in the structure of the brain box between the two groups are given in Table 1.

When reviewing the phylogenetic trends in development of the Middle Pleistocene hominids in Central Europe, the question arises how far it is necessary to ascribe finds bearing erectoid features solely to sapient forms. We believe that the "advanced" *erectus* forms are not identical with the "archaic" *sapiens* ones, and that they belong to independent cultural contexts. The differences identified in the structure of the neurocranium reach beyond the frame of a single subspecies. That was the main reason for distinguishing, within the Middle Pleistocene in Europe, two phylogenetic subspecies: one in the frame of the *Homo erectus* species, and the other in *Homo sapiens*.

For the archaic *sapiens* form the subspecies *Homo sapiens steinheimensis* (Berckhemmer, 1936) was already established. For the advanced *erectus* forms, the most easily datable and adequately morphologically definable find from Bilzingsleben is selected as a reference specimen and named *Homo erectus bilzingslebensis* (Vlček, 1978). The new finds from Bilzingsleben confirm the viability and justifiability of the proposed taxonomy.

In both phylogenetic subspecies, we observe, as their development progressed, the pattern of individual features. One observation is quite noteworthy: The changes in the shape of the neurocranium are retarded compared to the development of the brain. Therefore, the study of the brain development, hence the endocranium, is perceived as decisive in the evaluation of the finds. In the old archaic finds from Olduvai OH 9, Arago, and Bilzingsleben, a certain balance is discernible in the development of a small and primitive endocranium and a primitively shaped neurocranium. For instance, the frontal part of the endocranium of the aforementioned finds is on

	Homo erectus bilzingslebensis		*Homo sapiens steinheimensis*	
	Male	Female	Male	Female
Max. skull length	cca. 210	less than 200	cca. 200	184-210
Skull capacity	1,200 ccm	1,100-1,160 ccm	1,400 ccm	1,200-1,300 ccm
Torus supraorbitalis	frontally flattened	frontally flattened	—	rounded
Squama frontalis	flat slanting	flat slanting	—	tuber centrale
Linea temporalis	doubled	doubled	—	simple
Postorbital narrowing	clearly formed	clearly formed	—	weak
Occiput	broken	broken	rounded	rounded
Max. length	i = op	i = op	i under op	i under op
Torus occipitalis	above edge of curving	saddlelike on the edge of curving	below op	below op
Planum occipitale	low to medium high	low	high	high
Norma occipitalis	low, broad, max. width in the lower third	low, broad, max. width in the lower third	high, max. width in center of lower third	high, Hausform
Norma occipitalis	parietalia saddlelike	parietalia saddlelike	parietalia saddlelike	parietalia saddlelike
Planum nuchae	flat, convex	flat	convex	convex
Torus angularis	+	+	-	-
Os interparietale	+	+	-	-
Bone thickness	thick	thick	thin	thin
Rostrum orbitale	++ pointed	++ pointed	+? low	+? low

Table 1 - Chief differences in the shape of the neurocranium between the advanced Erectus and archaic Sapiens forms.

a lower level of development and is comparable with Sinanthropus. But within the same erectoid group, the finds from Petralona have shown that the primitively shaped neurocranium did contain a considerably developed brain, corresponding, in terms of development, with the Neanderthal forms. Thee same applies to the Vértesszölös 2 find. It follows that the development of individual components proceeds not evenly, but in a mosaiclike fashion, and that the decisive impetus for further changes is given by the development of the brain: the endocranium.

"Advanced" *Erectus*

Central Europe's oldest finds so far come from the Vértesszölös locality in Hungary. This is an open settlement in a travertine quarry near the village of Vértesszölös, some 50 km west of Budapest. During a systematic survey, carried out between 1963 and 1968, small-dimensional tools (Kretzoi & Vértes, 1965) were unearthed in an archaeological context in a layer corresponding

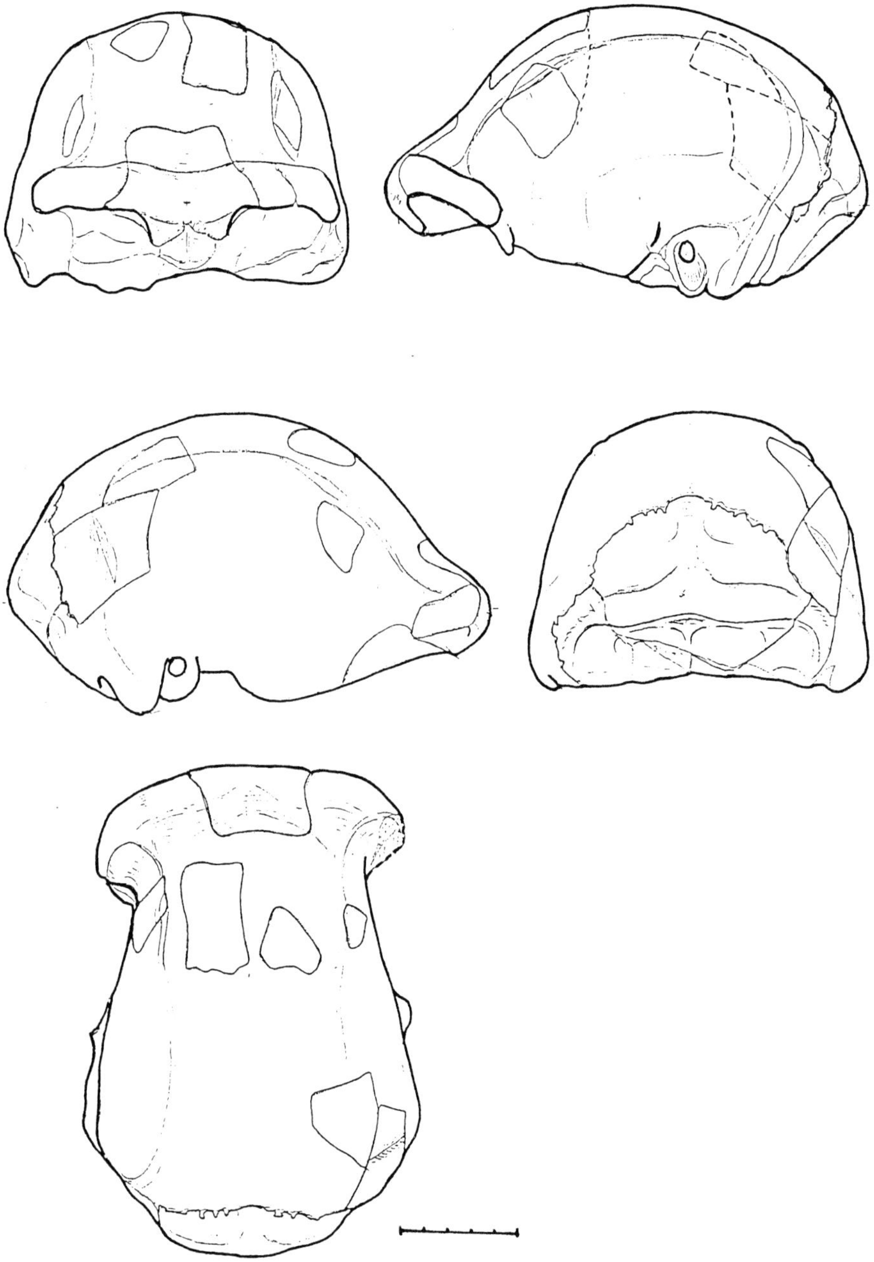

Figure 1 - *Bilzingsleben 1. Fragments of the skull 1 compared to Olduvai Hominid 9 specimen. (Orig. E. Vlček).*

with Upper Biharian (inter-Mindel), and in 1964 and 1965, modest human remains were also found: three milk molars of a seven-year-old child and fragment of a permanent molar, which were designated as Vértersszölös 1, and in the stratigraphically younger horizon, a fragment of an occipital bone of an adult male (Vértes, 1965; Thoma, 1966, 1967, 1969). Thoma (1966) classified these finds taxonomically as *Homo (erectus seu sapiens) palaeohungaricus*.

Further finds come from Bilzingslehen in Central Germany from an open settlement in travertines in the "Steinrinne" site, 35 kilometers north of Erfurt. During a systematic survey of the area on the bank of a stream flowing out of a mineral spring crater, deliberately crushed human skulls and isolated teeth have been found repeatedly since 1971, both in the settlement area and in the sediments of the adjacent brook. The locality is dated to the younger phase of the Holstein interglacial *senu lato* (stage 9, about 350,000-420,000 BP). Small dimensional tools are accompanied by Middle Pleistocene fauna and thermophile flora (Mania et al., 1980; Mania & Weber, 1986; Mania & Vlčck, 1981, 1987; Vlček 1978, 1979, 1983, 1986, 1987, 1989a).

Since the first find of a human skull fragment in 1972, remains of at least three individuals, two of adult age and one child, have been discovered so far. In morphological terms, the finds from Vértesszölös and Bilzingsleben are close to the human forms from Arago near Tautavel in France (de Lumley, 1982) and to the find from Petralona in Greece (Stringer et al., 1979; Murrill, 1980, 1981). While the settlements in Vértesszölös and Bilzingsleben were located near flowing openings of mineral springs (i.e. in travertines) the finds from Arago and from Petralona were made in caves, which causes various difficulties in dating such finds more precisely. Dating has not yet been definitely established, but the age of these finds corresponds to the periods of the antepenultimate and penultimate glaciations. The new dating of the find from Petralona is in the range of 240,000-160,000 BP (Xirotiris & Vlček, 1982).

In morphological terms the above-mentioned finds represent a group in which numerous erectoid aspects of *Homo erectus* are still present. In addition, it is possible to assign to this group the find of a mandible from Mauer near Heidelberg, whose assumed age corresponds with PK VII. The find was, however made in a secondary position in a sandpit. The original position of the fossil has not been unequivocally determined even after repeated attempts.

The morphology of this group is best represented by a partial reconstruction of the neurocranium of the man from Bilzingsleben, first in the individual No 1 (Figure. 1). As for the forehead, the most conspicuous feature is the huge torus supraorbitalis, which is not interrupted in the glabella area. Its size is remarkable. Between the glabella and the crista frontalis, the frontal bone is 25 mm thick, and in the nasion point torus, it is 21 mm high. The supraorbital torus is medium thick. The front surface of the torus was frontally flattened. The eye sockets had a horizontal position. The sulcus supraorbitalis is not formed, but a broad and smooth depressio glabellae is clearly noticeable. The pars ossis nasalis constitutes another group of features. The root of the nose is massive and broad. The squama frontalis documents an oblique course of the frontal squama without a trace of the formation of a tuber frontale. The linea temporalis is angular and doubled. The sinus frontales are formed only in the massif of torus supraorbitalis. Their shape is cauliflower-like, made up of two chambers which are not separated by other septa. The post-orbital strangulation of the skull's vertical outline is only slightly stressed. A fragment of the os parietale dx displays in section L 3 of the lamhda seam a marked thickening of the bone into the shape of the torus angularis.

The occipital bone of the Bilzingsleben man has a torus occipitalis which sits typically saddlelike (astraddle) on the edge of the occipital bone notching. In the central section, the torus is not divided but is laterally outlined through the linea nuchae superior. Then, it continues by a laterally uplifted edge up to the asterion area. The planum occipitale is low, and the l-i-o angle measures a mere 108 degrees. The planum nuchae is flattened but even and is not, therefore, concavely deepened. The thickness of the bones is striking, primarily in the section of the sutura occipito-mastoidea, where it reaches 17 mm. The opistocranion matches the inion point. Hence, the

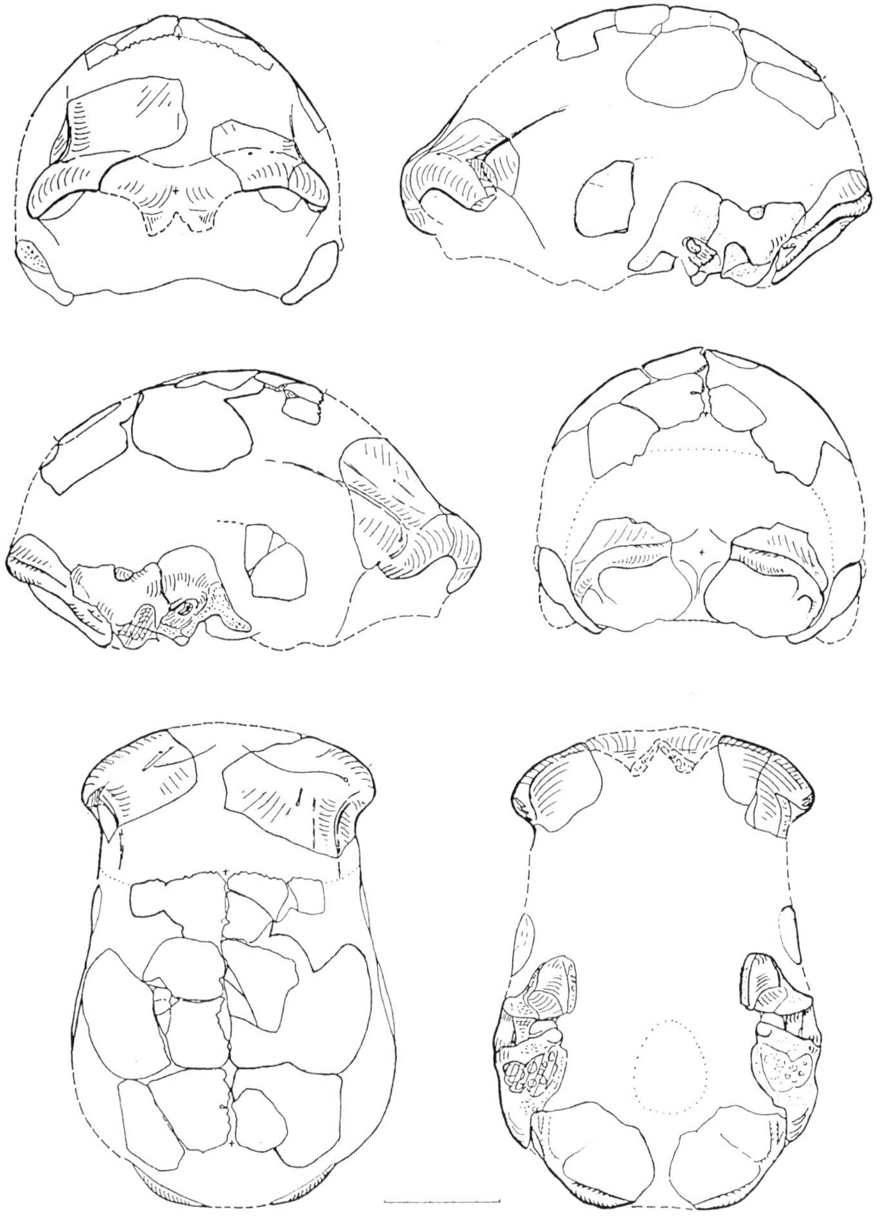

Figure 2 - Bilzingsleben 2. Neurocranium reconstruction of the individual No. 2. (After E. Vlček).

maximum length of the skull correspondends to the length g-i. The sutura lambdoidea is thickened and extends into the torus angularis described together with the parietal bone.

Further discoveries made between 1986 and 1993 (Mania et al., 1994) provide a much more accurate picture of the neurocranium of the Bilzingsleben man. It was possible to reconstruct the brain box of the second individual (Bilzingslehen 2; Figure 2). Fragments of the frontal bone B 4 and B 7 demonstrate the development of a massive torus supraorbitalis but also determine the breadth of the frontal area in the section of postorbital narrowing. Furthermore, they confirm the creation of an angular linea temporalis. The parietal parts of the skull, pieced together from several fragments, give an idea of the length and arching of both ossa parietale. The preserved fragments of the left os parietale (B 4 and D 3) prove that, in the Bilzingsleben 2 individual, the ossa interparietalia were formed in lambda just as in the finds from Arago and Petralona. As regards the Vértesszölös find, the situation in this particular area is not clear because of the incomplele state of preservation.

The find of a fragment of the right temple bone G I was a groundbreaking discovery, informing us about the formation of the auricular area, processus mastoideus and part of squama ossis temporalis. Last but not least, another fragment of the occipital bone A 3 confirms the occipital notching and the formation of a typical toruslike torus occipitalis which is more accentuated than in the Bilzingsleben 2 individual.

Thanks to the greater number of fragments of the right and left parts of the skull of the individual of Bilzingsleben 2, it is possible to complete the missing parts in a mirror fashion, thus substantially contributing to the picture of the original shape ol the skull of the Bilzingsleben hominid.

Very important features are recorded in the endocranium morphology. Particularly typical is the developed *bec encephalique*. The rostrum orbitale is narrow and high, just as in the finds from Arago and Sinanthropus (Xirotiris & Vlček, 1982).

The neurocranium of the Bilzingsleben man was long, low, and broad (Vlček, 1989a). Absolutely exact identities in terms of the shape and size of the occipital parts were discovered in Sinanthropus III (with an accuracy of 1 mm). However, the frontal parts of the Bilzingsleben hominid are more robust. Further surprising correspondences are visible between the fragments of a calvarium from Bilzingsleben and the considerably older find Olduvai OH 9, both in size and - more important - in the shape of the glabellal and occipital areas. However, the frontal part of OH 9 is, on the whole, narrower. The find was compared to the skull from Petralona as well. Striking morphological similarities have been determined in the morphology and size of the frontal area. On the other hand, dissimilarities in size have been detected in occipital parts. The skull of the Petralona individual was larger in absolute terms. Utter correspondence in the formation of the occipital parts of the Vértesszölös find can be seen in the skull from Petralona.

The described characteristics of the neurocranium of the Bilzingsleben man correspond with the erectoid human forms. Comparative studies (Vlček 1986, 1989a,b) show that, on an European scale, the find from Bilzingsleben stands closer to finds from Arago - Vértesszölös 2 - Petralona. There is no chlear-cut notion yet concerning the splanchnocranium of the Bilzingsleben finds.

"Archaic" *Sapiens*

Our attention will now be focused on the second group of finds in Europe, which are represented by the discoveries at Swanscombe, Steinheim, and Ehringsdorf (Le Gros Clark & Morant, 1938; Berckhemmer, 1933, 1936; Weinert, 1936; Breitinger, 1955). The finds from Swanscombe and Steinheim are generally dated to the Holstein period, but we do not know whether to its earlier or later stages.

In the territory under review, we should concentrate on the earlier series of finds from Ehringsdorf near Weimar, Central Germany, which belong to the warm Rügen period, between 250,000 and 100,000 years (Schwartz) or 244,000-102,000 yearss (Brunnacker). The finds were made between 1908 and 1925 in an open-air settlement during travertine quarrying in several localities. The cultural association is early Mousterian. There is an extensive literature, which is summarized primarily in the most recent studies on the cultural layers and their dating, fauna, and present flora (Behm-Blancke, 1960; Feustel, 1983).

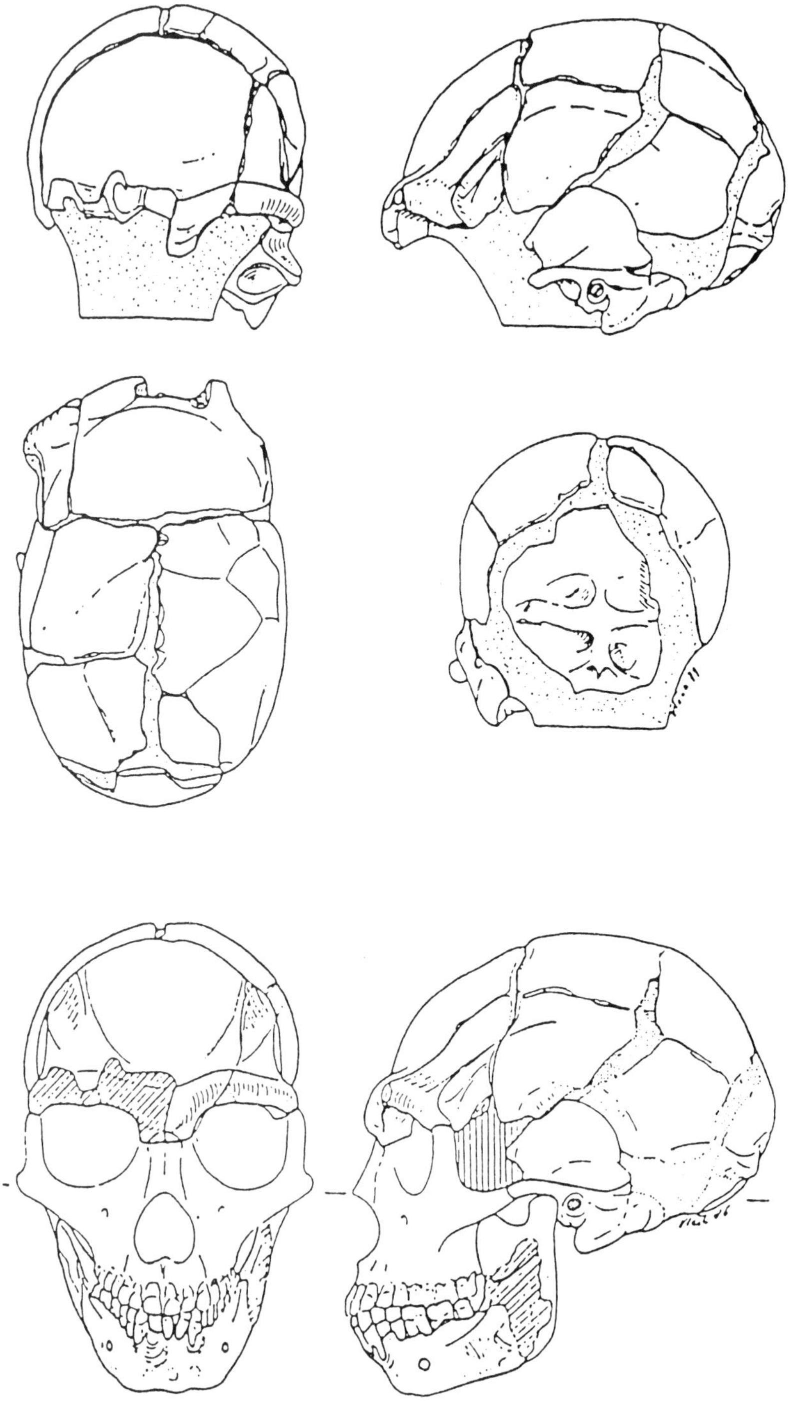

Figure 3 - Weimar-Ehringsdorf. Reconstruction of the Neurocranium E-H (above, middle) and of the skull H with mandible F (below). (Orig. E. Vlček).

As regards the available anthropological material, this includes a crushed skull of an adult individual (H) and fragments of parietal bones of another four adults (A,B,C,D). Out of the splanchnocranium, a mandibula of an adult (F) was discovered, as well as a femur fragment (E) representing the postcranial skeleton. In addition to remains of adults, there are also fragments of a child's skeleton (G). This important body of material has been studied on a number of occasions, and the skull H was reconstructed (Linding, 1934; Weidenreich, 1928; Kleinschmidt, 1931). The latest thorough revision of the material was presented by Vlček (1985, 1993).

Since the majority of information about the morphology of this group is being supplied by the Ehringsdorf find, the whole group may be characterized by the skull from Weimar-Ehringsdorf H (Vlček, 1985. 1993). The H skull is long, narrow, and lower, bearing a well-formed torus supraorbitalis, which is well separated from the frontal squama. In the glabella area, the torus is arched wavily but is not divided. The thickness of the torus is even from the glabella as far as the processus zygomaticus frontalis. On the frontal bone, the linea temporalis is formed in a crestlike manner but does not extend into the parietal bones. There is no postorbital narrowing of the skull at all. The squama ossis frontalis is steep, and the forehead is bomb-shaped, with maximum rounding in the metopion area. This arching in the middle of the frontal squama is clearly delineated so that a tuber frontale was formed.

The parietal parts are well arched, and the occipital parts are raised backward but rounded. There is no notching on the occiput. The opistocranion is found 20 mm above the inion. The planum occipitale is also rounded and bomb-shaped. The linea nuchae superior is well formed, separating the planum occipitale from the planum nuchae. The torus occipitals is indistinct. The tubera parietalia are well formed on the parietal bones. Viewed from the back, the walls of the skull's outline are positioned in parallel, creating the characteristic "Hausform Typus" of the skull. The os temporale is relatively small, with a slightly developed processus mastoideus The porus acusticus is small, and the edge of the os tympanicum is not thickened. Similarly, other fragments of parietal bones attest to the modern architecture of the brain box of the Ehringsdorf man, which is morphologically closer to modern *sapiens* than to the Neanderthal forms.

Considerable similarities to the find from Ehringsdorf are also displayed by older finds from Steinheim and Swanscombe. In these, too, the maximum width of the skull is found in the norma occipitalis in the upper half of the norm's outline, and because of its overall formation, it is reminiscent, primarily in Steinheim, of the Hausform Typus brain box. The torus angularis, or thickening in the sutura lambdoidea, is not formed. Basic differences between both groups are borne out by medisagittal and transveral CT cuts of the Steinheim skull in comparison with the skull from Petralona.

Fossil human finds of the last interglacial

A new stage of the development of Central European Palaeolithic populations is evidenced during the last (Eemian) interglacial, which corresponds to PK III. The finds from that period include an important specimen from a travertine site at Gánovce and an isolated tooth discovered in a cultural layer in travertines at Taubach.

Early glacial Neanderthals

Another stage in the development of fossil hominids dates to the early phase of the last glacial corresponding with PK II. The available finds suggest most probably a continuity to the development of Middle Pleistocene populations of *Homo sapiens steinheimensis.*

The oldest find from Königsaue, Central Germany, is dated to the Brörup interstadial. No human bone remains have been recovered there, only imprints of human papillary lines preserved on a fragment of resin binder. Another discovery of a frontal bone from Šala nad Váhom, Slovakia, comes from river sediments The accompanying fauna dates the find, with a high degree of probability, to the early glacial period.

The other finds of fossil hominids are from caves. During a systematic research in Kůlna cave in the Moravian karst, fragments of a maxilla, os parietale, and free teeth were found. Furthermore, dating to this period are the Moravian finds of a jawbone from Šipka cave near Štramberk, fragments of a jawbone, a free molar tooth, and a fragment of os parietale and os temporale from Švedův stůl cave near Ochoz, as well as the finds of two individuals from Subalyuk cave in Hungary.

References

BEHM-BLANCKE, G. (1960): Altsteinzeitliche Rastplätze im Travertingebiet Taubach, Weimar, Ehringsdorf. *Alt-Thüringen* 4, 1-246.

BERCKHEMMER, F. (1933): Ein Menschenschädel aus den diluvialen Schottern von Steinheim a.d. Murr. *Anthropologischer Anzeiger* 10, 318-321.

BERCKHEMMER, F. (1936): Der Urmenschenschädel aus den zwischeneiszeitlichen Fluss-Schottern von Steinheim an der Murr. *Forschungen und Fortschritte* 12, 349-355.

BREITINGER, E. (1955): Das Schädelfragment von Swanscombe und das "Praesapiensproblem". *Mitteilungen der Anthropologischen Gesellschaft Wien* 84/85, 1-45.

FEUSTEL, R. (1983): Zur zeitlichen und kulturellen Stellung des Paläolithikums von Weimar-Ehringsdorf. *Alt-Thüringen* 19, 16-42.

KLEINSCHMIDT, O. (1931): *Der Urmensch.* Leipzig.

KRETZOI, M. & VÉRTES, L. (1965): Upper Biharian (Intermindel) pebble-industry occupation site in Western Hungary. *Current Anthropology* 6, 74-87.

Le GROS CLARK, W.E. & MORANT, G.M. (1938): *Report on the Swanscombe skull.* London.

LINDING, K. (1934): *Der Altsteinzeitmensch des Ilmtals: Skelettreste aus dem Travertin von Weimar-Ehringsdorf.* Weimar.

LUMLEY, M,A. de (1982): L'homme de Tautavel. In: *Congrés International de Paléontologie Humaine*, 19-136. Nice. (Prétirage).

MANIA, D., MANIA, U. & VLČEK, E. (1994): Latest finds of the skull remains of *Homo erectus* from Bilzingsleben (Thuringia). *Naturwissenschaften* 81, 123-127.

MANIA, D., TOEPFER, V. & VLČEK, E. (1980): *Bilzingsleben I. Homo erectus - seine Kultur und seine Umwelt.* Berlin.

MANIA, D. & VLČEK, E. (1981): *Homo erectus* in Middle Europe: The discovery from Bilzingsleben. In: B.A. Sigmon & J.S. Cybulski (eds.), *Homo erectus papers in honour of Davidson Black,* 133-151. Toronto.

MANIA, D. & VLČEK, E. (1987): *Homo erectus* from Bilzingslebfen (GDR) - His culture and his environment. *Anthropologie* 25, 1-45.

MANIA, D. & WEBER, T. (1986): *Bilzingsleben III. Homo erectus - seine Kultur und seine Umwelt.* Berlin.

MURRILL, R. (1980): New measurements of the face of the Petralona fossil hominid skull. *Anthropos* (Athens) 7, 40-41.

MURRILL, R. (1981): *Petralona man.* Springfield.

STRINGER, C.B., HOWELL, F.C. & MELENTIS, J.K. (1979): The significance of the fossil hominid skull from Petralona, Greece. *Journal of Archaeology* 6, 235-253.

SVOBODA, J. (1987): Lithic industries of the Arago, Vértesszölös and Bilzingsleben hominids: Comparison and evolutionary interpretation. *Current Anthropology* 28, 219-227.

SVOBODA, J. (1989): Middle Pleistocene adaptations in central Europe. *Journal of World Prehistory* 3, 33-70.

SVOBODA, J., LOŽEK, V. & VLČEK, E. (1996): *Hunters between East and West, The Paleolithic of Moravia.* New York, London.

THOMA, A. (1966): L'occipital de l'*Homo* mindelien de Vérteszöllös. *L'Anthropologie* 70, 495-534.

THOMA, A. (1967): Human teeth from the Lower Paleolithic of Hungary. *Zeitschrift für Morphologie und Anthropologie* 58, 152-180.

THOMA, A. (1969): Biometrische Studie über das Occipitale von Vérteszöllös. *Zeitschrift für Morphologie und Anthropologie* 60, 229-241.

VÉRTES, L. (1965): Discovery of *Homo erectus* in Hungary. *Antiquity* 29, 303.

VLČEK, E. (1969): *Neandertaler der Tschechoslowakei.* Praha.

VLČEK, E. (1978): A new discovery of *Homo erectus* in Central Europe. *Journal of Human Evolution* 7, 239-251.

VLČEK, E. (1979): *"Homo erectus bilzingslebensis"*- Eine neue Form des mittelpleistozänen Menschen Europas. *Ethnographisch-Archäologische Zeitschrift* 20, 634-661.

VLČEK, E. (1983): Über einen weiteren Schädelrest des *Homo erecctus* von Bilzingsleben. *Ethnographisch-Archäologische Zeitschrift* 24, 321-325.

VLČEK, E. (1985): Der fossile Mensch aus Weimar-Ehringsdorf. In: J. Herrmann & H.Ullrich (Hrsg.), *Menschwerdung - Biotischer und gesellschaftlicher Entwicklungsprozeß,* 111-117. Berlin.

VLČEK, E. (1986): Les Antenéandertaliens en Europe centrale et leur comparaison avec l'Homme de Tautavel. *L'Anthropologie* 90, 143-153.

VLČEK, E. (1987): Funde von Zähnen des *Homo erectus* aus dem Travertin bei Bilzingsleben. *Jahreschrift für mitteldeutsche Vorgeschichte* 70, 83-94.

VLČEK, E. (1988): Gánovecký nález v CT-počítačové tomografii. *Slovenska Archeologia* 36, 353-362.

VLČEK, E. (1989a): Die Hominidenreste von Bilzingsleben: Über Neufunde von 1981-1987. *Ethnographisch-Archäologische Zeitschrift* 30, 270-286.

VLČEK, E. (1989b): *Homo erectus* in Europe. *Ethnographisch-Archäologische Zeitschrift*, 30, 287-305.

VLČEK, E. (1991): L'Homme fossile en Europe centrale. *L'Anthropologie* 95, 409-472.

VLČEK, E. (1993): *Fossile Menschenfunde von Weimar-Ehringsdorf.* Stuttgart.

WEIDENREICH, F. (1928): *Der Schädelfund von Weimar-Ehringsdorf.* Jena.

WEINERT, H. (1936): Der Urmenschschädel von Steinheim. *Zeitschrift für Morphologie und Anthropologie* 25, 463-518.

XIROTIRIS, N. & VLČEK, E. (1982): Arago et Petralona: Comparison de l'endocrâne. Congrés International de Paléontologie humaine, communication 21.October. Nice.

New Thoughts on Neanderthal Behavior: Evidence from Nasal Morphology

Jeffrey H. Schwartz, Ian Tattersall & Jeffrey T. Laitman

Abstract

Previously undescribed features of Neanderthal morphology that contrast significantly with configurations in other Pleistocene hominid species as well as in Homo sapiens have recently been reported (Schwartz & Tattersall, 1996a,b, in press). Among these Neanderthal apomorphies are the development of an internal nasal margin (INM) that bears basally thick, vertically oriented medial projections; an unossified cover of the lacrimal canal; and a voluminous intrusion of the wall of the maxillary sinus into the posterior nasal cavity.

We report here the presence of the INM with its medial projections in specimens from Krapina (K3/Skull C) and Subalyuk (child) and the presence of the typical mammalian horizontal conchal crests in the crania from Kabwe and Petralona. The presence of a distinct medial projection in the Subalyuk child (ca. three years old) is consistent with other Neanderthal children (Roc de Marsal, Engis, Pech de l'Aze) and demonstrates that this unique feature of the Neanderthal nasal cavity appears early in development. Inasmuch as crests that course laterally from the posterior extent of the anterior nasal spine (the spinal crests of Gower, 1923) occur in Homo sapiens and various fossil hominids, we do not include them as part of the Neanderthal apomorphy. However, because the spinal crest does not in Neanderthals become confluent with the anterior margin of an inferior nasal concha (but instead expands to become the medial projection of the INM), we suggest that Neanderthals again differ from the primitive condition in lacking a turbinal crest (Gower's terminology). The crest that continues superiorly from the medial projection we can refer to as the "superior nasal crest"; inasmuch as a superior nasal crest does not occur in any other hominid we have so far studied, this crest is also part of the derived condition.

From the functional and physiological standpoint, an internal nasal margin within the anterior nasal aperture acts to restrict the size of the opening into the nasal cavity and provides additional surface area for mucous membranes, as would the posterior swelling deeper within the cavity. The specific configurations and spatial relationships of the conchae within the Neanderthal nasal cavity are not possible to reconstruct at this time. However, in contrast to being vacuous, the large nasal cavity of Neanderthals would have been rife with projections facilitating exposed membrane surfaces. In addition, inasmuch as restricting the space through which molecules travel acts to accelerate their speed, one wonders how this would have affected Neanderthal respiration.

The presence of derived nasal structures in the approximately 130 kyr (= late Riss-Eemian boundary) Krapina Neanderthals suggests that a relationship may have existed between these structures and adaptation to a very humid and quite cold climate. Once established, these features persisted through the appreciably greater warmth of the Middle Eemian, and subsequent cyles of warm-cold periods, as is obvious from the presence of these apomorphies in, e.g. the Neanderthals from La Ferrassie, La Chapelle-aux-Saints, and St. Césaire. It has recently been suggested (Franciscus, pers. comm.) that the Neanderthal features described above are solely by-products of a large nasal aperture. In this regard, we note that an INM is not present in other fossil hominids with relatively large nasal apertures, e.g. the specimens from Kabwe, Petralona, and Sima de los Huesos. Neanderthals clearly appear to have differed from other hominids in the anatomy underlying this crucial area of their respiratory tract and, in turn, respiratory behaviors.

Introduction

Although the hominids we refer to as Neanderthals have been known for over one hundred and fifty years, and the extensive morphological differences in cranial and postcranial morphology between them and extant *Homo sapiens* also recognized, it was only in the early twentieth century that scenarios invoking environmental reasons for these morphological differences emerged in the literature. Perhaps the most influential presentation in its day, and one which certainly impacted generations of paleoanthropologists thereafter, was given by Ales Hrdlička (1927) when he was the guest Huxley Memorial Lecturer at the Royal Anthropological Institute in London. There, in direct opposition to Sir Arthur Keith, a proponent of the Neanderthal replacement theory and an attendee at the lecture, Hrdlička argued with characteristic conviction that, so far as Neanderthals displayed features that could be accommodated as natural extensions of variations purportedly observed in present-day humans, they should be recognized for what they really were in terms of human evolution: not a separate species, but merely a phase (a European Neanderthal phase) in the transformation from an archaic and Old World distributed to a modern and globally widespread form of *Homo sapiens*. The morphological argument, while greatly elaborated by Hrdlička, was, of course, not his invention alone. Others before him (e.g. Huxley, 1863; Gregory, 1920) had outlined many of this proposition's tenets.

But Hrdlička went farther than others in trying to find an explanation for how Neanderthals and extant *Homo sapiens* could be so morphologically different from one another and still be seen as being members of the same species. Given the state of knowledge at the time about the chronology of Pleistocene fossil hominids (primarily, Neanderthals) and of the glacial and interglacial periods, Hrdlička was struck by the existence of Neanderthals prior to and at the beginning of the last glacial period, when conditions were becoming more harsh and frigid. He proposed that it was change in key environmental factors that had brought about physical change through an intensification of natural selection. The result was Neanderthal morphology. Despite its dependence on a variety of rather dubious assumptions, the environmental argument for the evolution of Neanderthal morphology has remained rather seductive.

In a seminal article published a quarter-century after Hrdlička's, Howell (1952) argued at a general level that western European or so-called classic Neanderthals had evolved their morphological distinctiveness in glaciation-produced isolation, probably through the mechanism of genetic drift. And a decade later, Coon (1962), while broadly accepting an environmental argument for the evolution of much of Neanderthal morphology, emphasized the face, and the nasal region in particular, as the focus of climatically induced morphological change. Rejecting MacGregor's widely exhibited reconstruction of the La Chapelle-aux-Saints Neanderthal as being flat-nosed, Coon (1962: 533-534) concluded that western European Neanderthals, and especially those from France, must have had large noses to go along with their broad nasal apertures, while Neanderthals from "milder" climates would have had developed "less beaky" fleshy noses. Inasmuch as two major functions of the nose are to warm and moisten air (e.g. Proetz, 1953), Coon suggested that, in response to ever-intensifying glacial conditions, the Neanderthal features of large nose and nasal aperture, forward position of the nasal region, and expansive maxillary paranasal sinuses, had been cold-related adaptations whose function would have been to distance incoming cold air from the brain, which cannot withstand extremes in temperatures as can the limbs. In Neanderthals, the "large, projecting nasal 'radiator'," as Coon described the facial-nasal-sinus complex, would have warmed the frigid incoming air. Coon further felt that the large infraorbital foramen seen in various Neanderthal specimens reflected the development of another cold adaptation. The enlarged foramen would have housed a large artery (Coon did not consider the presence of the nerve), which would have provided six or seven times more warm blood to the face than in a European. Going further yet, Coon also speculated with regard to physiology that Neanderthals would have required a higher caloric intake to survive in such cold conditions.

Trinkaus (1989) has also interpreted aspects of Neanderthal nasal region morphology as being adaptations to the cold. He suggested, however, that, rather than serving to warm incoming frigid air and to keep it well away from the brain, the large and projecting nasal aperture of Neanderthals reflects an adaptation to a dry and cold glacial environment that acted both to minimize loss of moisture and to release heat generated by an activity level that these conditions would have demanded. Interestingly, among other features that had been proposed as Neanderthal adaptations to the cold, Trinkaus accepted only these hominids' low crural and brachial indices.

A somewhat different approach to interpreting the significance of the large and projecting Neanderthal nasal region has recently been proposed by Jelinek (1994), who bases much of his hypothesis on Neanderthals not buffering themselves much against their environment with furs or other exogenous trappings. If, as is widely accepted, Neanderthal body mass was relatively large (compared to *Homo sapiens*), then Neanderthals would have required a high caloric intake (probably provided by animal fat, marrow, and brains) in order to produce the energy necessary to maintain this large body mass. The essential component in the metabolic formula from which this energy would have derived is, of course, oxygen. A larger body-massed Neanderthal, the argument runs, would have needed large quantities of oxygen to produce the level of energy required not only to sustain the level of activity generated by its larger musculature, but also to maintain a higher internal muscle temperature under cold periglacial climatic conditions. If Neanderthals maintained the necessary intramuscular temperatures to combat the cold simply by staying active (even with low level activities, such as walking and/or trotting), then their oxygen and caloric needs would indeed have been great (e.g. see Åstrand & Rodahl, 1977; Shephard, 1978). In light of these assumptions, Jelinek concludes that the large Neanderthal nasal aperture was necessary for the intake of the requisite amounts of oxygen (and concomitant exhalation of CO_2), and that respiration through such a nasal apparatus would have been less heat-wasteful than breathing through the mouth. In fact, he questions "whether, given their body structure and energy needs, Neanderthals could have survived with smaller noses" (Jelinek, 1994: 78).

Given the central role of the nasal region in such scenarios of Neanderthal adaptation (i.e. to the cold), as well as of Neanderthal behavior and lifestyles (e.g. acquisition of high calorie foods, minimal processing of leather and furs for protection and insulation, maintenance of long periods of at least low levels of physical activity, such as walking or trotting), it is of paramount importance that we understand as completely as possible the morphology of this region. Since the nasal aperture is essentially just a bony boundary between the nasal cavity and that which lies outside of it, nasal aperture size constitutes only one aspect of the morphology of the region as a whole. Our studies of the morphological details of the nasal cavity in middle to late Pleistocene hominids, as well as in extant mammals, demonstrate the uniqueness of Neanderthals in this cranial region compared not only other hominids, but to terrestrial mammals in general. It is in light of these Neanderthal autapomorphies that previous inferences about Neanderthal adaptation, behavior, and lifestyle must be re-evaluated.

The nasal region

As summarized briefly above, the nasal aperture of Neanderthals is generally acknowledged as being relatively large, at least in breadth, compared to *Homo sapiens*. However, other middle to late Pleistocene fossils hominids, which do not share all of their apomorphies with Neanderthals, also have relatively large nasal apertures, most notably the Steinheim and Petralona skulls (personal observations), as well as the Sima de los Huesos specimens (Arsuaga et al, 1993). It is also worth noting that so also do some extant Asian and African populations (Williams, 1955), including the San (Franciscus, personal communication), who are amongst the smallest of *Homo sapiens* in body size. In light of recent suggestions (Franciscus, personal communication; Tobias, 1995) that the San also share with Neanderthals certain aspects of nasal cavity morphology, we shall review what is known about this region in Neanderthals, other middle to late Pleistocene hominids, and extant *Homo sapiens*.

Typical of the nasal cavity structures of adult non-marine mammals (including bats) is the presence of some number of pairs (usually four, but occasionally five) of ethmoidally derived conchae (ethmoturbinals) posteriorly, one pair of nasoturbinals superiorly, and one pair of maxilloturbinals inferiorly (see Cave, 1973; Hershkovitz, 1977). In contrast to the ethmoturbinals, which develop as pneumaticized swellings from the ethmoid sinus itself, the nasoturbinal and maxilloturbinal (the latter identified as the inferior nasal concha in humans) arise from their own centers of ossification (Hillenius, 1994). The maxilloturbinal comes to articulate with bones contributing to the internal wall of the nasal cavity along a low-lying, often relatively horizontally oriented ridge (the conchal crest) (Figure 1). In the adult, this turbinate of independent origin has usually coalesced with the conchal crest (thus partially explaining its identification as the maxilloturbinal). In desiccated specimens of neonates and juveniles, the still growing and unfused maxilloturbinal is often missing, and thus is not associated with the minimally elevated conchal crest. To the uninitiated, it therefore might seem that the maxilloturbinal develops as an enlargement of the conchal crest; this incorrect impression may be enhanced by observation of skulls in which this turbinate has coalesced with the conchal crest (Figure 2).

The structures of the nasal cavity in many primates, including *Homo sapiens*, have been detailed by Cave (1961, 1973; Cave & Haines, 1940; also see Hershkovitz, 1977). Since some prosimians and all anthropoids lack a nasoturbinal, the absence of this bone in *Homo sapiens* would appear to be due to its primitive absence in an ancestor within the primate clade. The ubiquitous presence of a maxilloturbinal in extant non-marine mammals indicates that this bone is also retained primitively in *Homo sapiens*, but from the last common ancestor of extant mammals (assuming that cetaceans are themselves autapomorphic within Mammalia). In the development of typically two pairs of ethmoidally derived turbinates (the superior and middle nasal conchae), *Homo sapiens* is similar only to the extant large-bodied hominoids amongst primates. As such, reduction to two pairs of ethmoturbinals can plausibly be seen as a shared derived feature of extant large-bodied hominoids, which do indeed appear on the basis of other potential synapomorphies to constitute a clade (e.g. see review by Schwartz, 1988). Having only superior and middle nasal conchae, therefore, would constitute a primitive retention for the members of that clade (e.g. a primitive retention in *Homo sapiens*).

In further contrast to other non-marine mammals, the conchal crest of large-bodied hominoids is consistently confined to the anterior portion of the nasal cavity (e.g. see Cave, 1961, 1973; Cave & Haines, 1940; Hershkovitz, 1977; Schwartz & Tattersall, 1996), and, concomitantly, so is the primary anterior articulation of the maxilloturbinal (inferior nasal concha) with other structures of the nasal cavity (Figure 1). The anterior portion of the inferior nasal concha of extant large-bodied hominoids is therefore characterized by a relatively horizontal premaxillary-maxillary articulation. At the point where the conchal crest-inferior nasal conchal articulation ends, a variably broad spit of bone - the lacrimal process - projects vertically. The lacrimal process articulates along its anterior margin with a thin posteriorly directed shelf from the frontal process, which constitutes a continuation into the nasal cavity of the anterior lacrimal crest. Together this shelf and the lacrimal process form a thin bony covering over the lacrimal groove, creating the lacrimal canal; in at least the human fetus, the lacrimal groove within the nasal cavity is covered by a thin

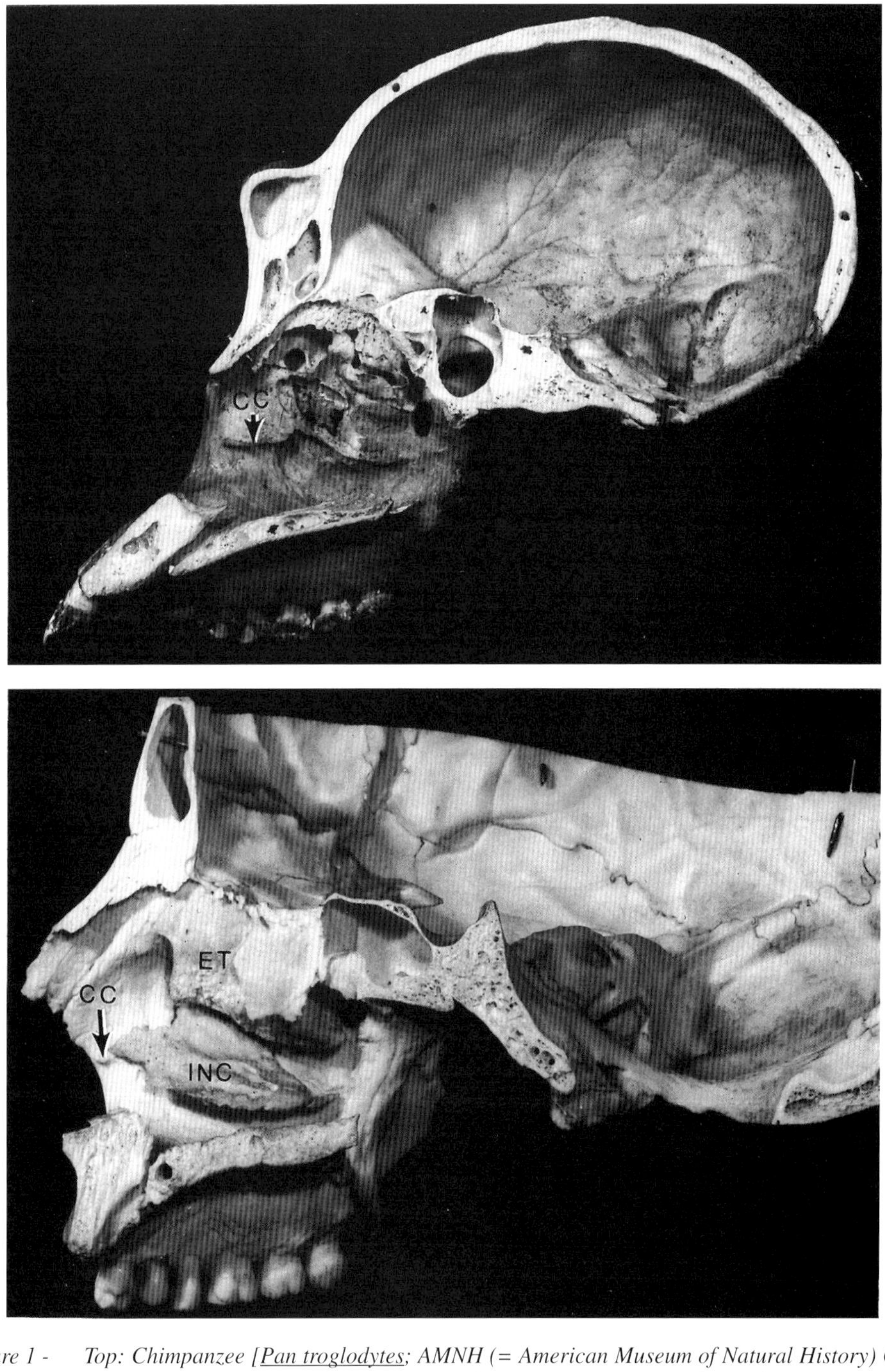

Figure 1 - *Top: Chimpanzee [Pan troglodytes; AMNH (= American Museum of Natural History) Ch.3/ l.15]. Inferior nasal conchae unfused and missing, exposing low-lying, horizontal conchal crest (CC); ethmoidal sinuses visible through damaged bone.*
Bottom: Recent Homo sapiens (Salzburg, Austria; AMNH VL/4578). Showing relationships of inferior nasal concha (INC), low-lying and horizontal conchal crest (CC) and ethmoturbinals (ET). Not to scale. © J. Schwartz.

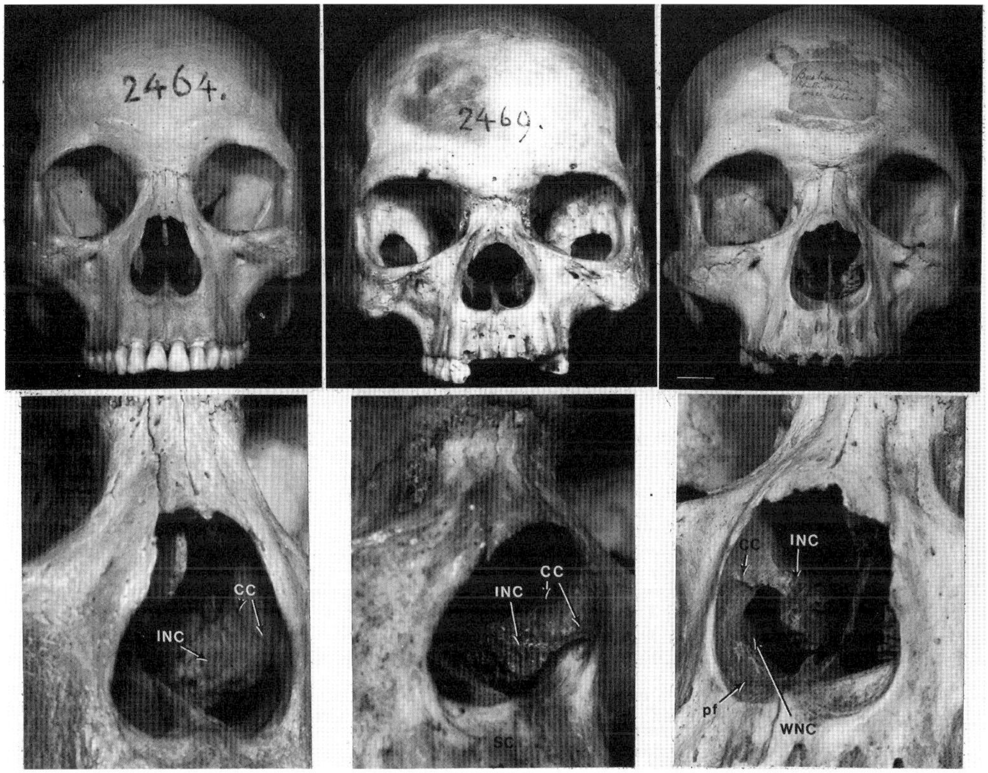

Figure 2 - *Kalahari San (Homo sapiens).*
Showing individual variation in nasal aperture size, degree of inflation of inferior nasal concha (INC), and sunkeness of wall of nasal cavity (WNC) behind the spinal/turbinal crest, inferior nasal concha, and conchal crest; in all specimens, although the inferior nasal concha is completely fused to the conchal crest (CC), traces of the latter structure can still be detected. Left (top and bottom), AMNH VL/2464; middle (top and bottom), AMNH VL/2469; right (top and bottom), AMNH 99/76.
Scale = 1 cm (applicable to all specimens in top row). © J. Schwartz.

sheet of bone (i.e. there is a lacrimal canal). Superiorly, the lacrimal process articulates with the lacrimal bone. Posteriorly, the lacrimal process grades into the superior margin of the inferior nasal concha, which forms the inferior border of the opening of the maxillary sinus. Posterior to this opening, the inferior nasal concha articulates with the ethmoid via a vertical process, posterior and inferior to which it articulates with the palatine bone. Although all anthropoid primates have some maxillary sinus development, it is only in extant large-bodied hominoids that one finds not only a true maxillary sinus (rather than a mere pocketing within the wall of the nasal cavity), but also one that drains directly into the middle meatus of the nasal cavity. Among extant large-bodied hominoids, the gorilla appears to be autapomorphic in having a dilated nasolacrimal duct that encroaches upon the maxillary sinus (see Caves & Haines, 1940).

Given the rather consistent configuration of the mammalian nasal cavity among taxa (Cave, 1973), there is little variation worthy of mention. In our studies of actual as well as of published illustrations of specimens, we noted, for example, only minor differences among specimens of

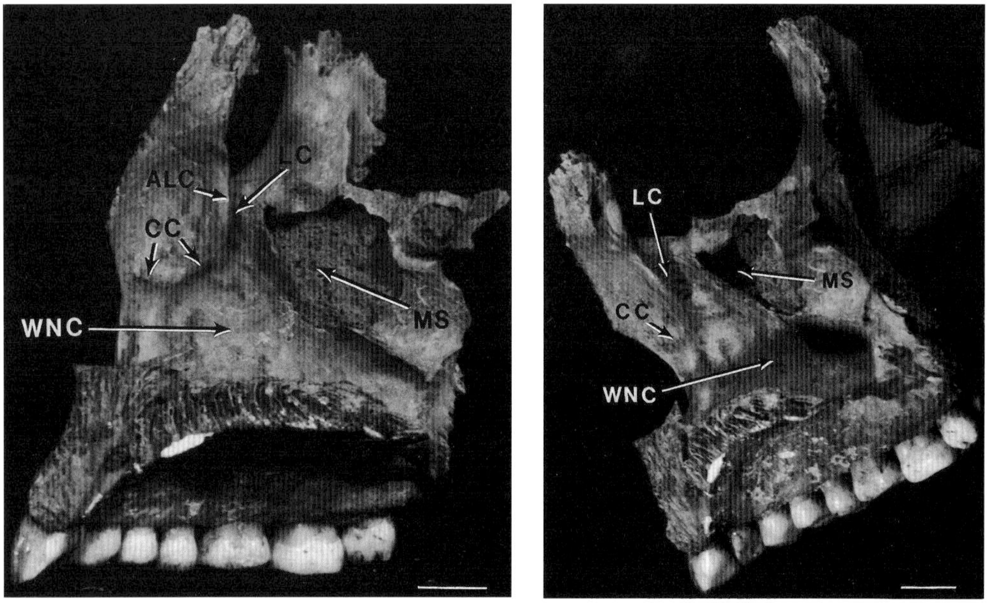

Figure 3 - *Internal aspect of the right nasal cavity of a Kalahari San (AMNH 99/8454). Showing the low-lying, rather horizontally oriented conchal crest (CC), to which the missing inferior nasal concha had not yet coalesced; the lacrimal canal, which is partially covered by a flange of bone (the anterior lacrimal crest = ALC); the sunken wall of the nasal cavity (WNC); and the reduced maxillary sinus (MS).*
Left: Lateral view; right: posterior/oblique view. Scale = 1 cm. © J. Schwartz.

each species of extant large-bodied hominoid in the lengths and/or horizontality versus verticality of the premaxillary-maxillary articulation of the inferior nasal concha and of the conchal crest. In the San sample (N = 16) we studied, one specimen (AMNH VL/2470) lacked a discernible conchal crest altogether; an isolated inferior nasal concha was lodged within the nasal cavity.

The San sample is also interesting compared to other *Homo sapiens* studied in that the lacrimal groove appears to be relatively very large for individuals of small body size; this rather voluminous lacrimal groove is more extensively covered internally by the posteriorly directed shelf of the frontal process (i.e. the platelike continuation into the nasal cavity of the anterior lacrimal crest noted within the orbit). We also observed among the San sample a more concave or depressed (i.e. sunken) lateral nasal cavity wall than one typically sees in *Homo sapiens* of larger body size (Figures 2, 3). Since the maxillary sinus lies on the other side of the lateral wall of the nasal cavity, it is reasonable to conclude that, were radiographic or tomographic studies to be done, the San would be found to have relatively less capacious maxillary sinuses. A consequence of having a sunken nasal cavity wall is that it makes it appear that the inferior nasal concha is more medially projecting than it actually is.

In contrast to the relatively stable morphology and arrangement of nasal cavity structures not only in *Homo sapiens*, but non-marine mammals in general, the configuration of the inferior margin of the nasal aperture in *Homo sapiens* varies greatly (e.g. see Gower, 1923; de Villiers, 1968). Following Gower (1923), one can identify in *Homo sapiens* the following crests (Figure 4). The lateral crest is the edge of the nasal aperture; this crest may lose distinctiveness as it nears the inferior portion of the nasal aperture, or it may retain its crispness for some extent along the

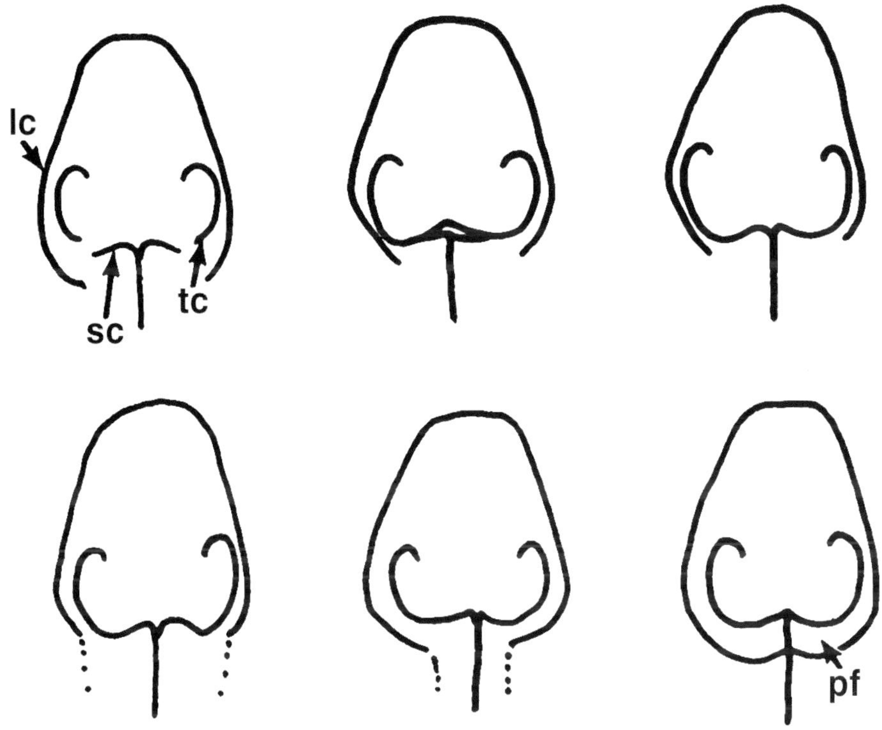

Figure 4 - *Schmatic representation of individual variation in the rim of the nasal aperture and structures just within the nasal cavity of extant Homo sapiens (after Gower, 1923). lc = lateral crest; sc = spinal crest; tc = turbinal crest; pf = prenasal fossa. © J. Schwartz.*

inferior margin, including coursing to meet the anterior nasal spine. A crest, which courses laterally from the anterior nasal spine, is the spinal crest, and a less crisply defined crest, which originates from the anterior extremity of the conchal crest (rather than, as suggested by Gower, the inferior nasal concha) and courses downward, is the turbinal crest. The spinal and turbinal crests may meet and become one, or the turbinal crest may terminate posteriorly on the anterior nasal spine and create a sulcus between it and a spinal crest that converges upon the turbinal crest somewhere along the wall within the nasal aperture. If the lateral crest courses all the way to the anterior nasal spine, and the spinal and turbinal crests become confluent, a sulcus (identified by Gower as the prenasal fossa) is formed along the inferior margin of the nasal aperture. While being, perhaps, of some interest to studies on individual variation, none of these variants - by virtue of the fact that they are variations - is of systematic valence, and thus none defines either *Homo sapiens* or any population within *Homo sapiens*.

Against this comparative background, then, we may turn to Neanderthals and other middle to late Pleistocene fossil hominids. Due to the delicate nature of all structures within the nasal cavity - not only of the turbinates, but also of the ethmoid and its perpendicular plate, the vomer, and the wall of the maxillary sinus - these structures, even in the adult skull, are easily damaged. Thus we are fortunate indeed that some fossil specimens preserve enough of the nasal region to make certain definitive observations.

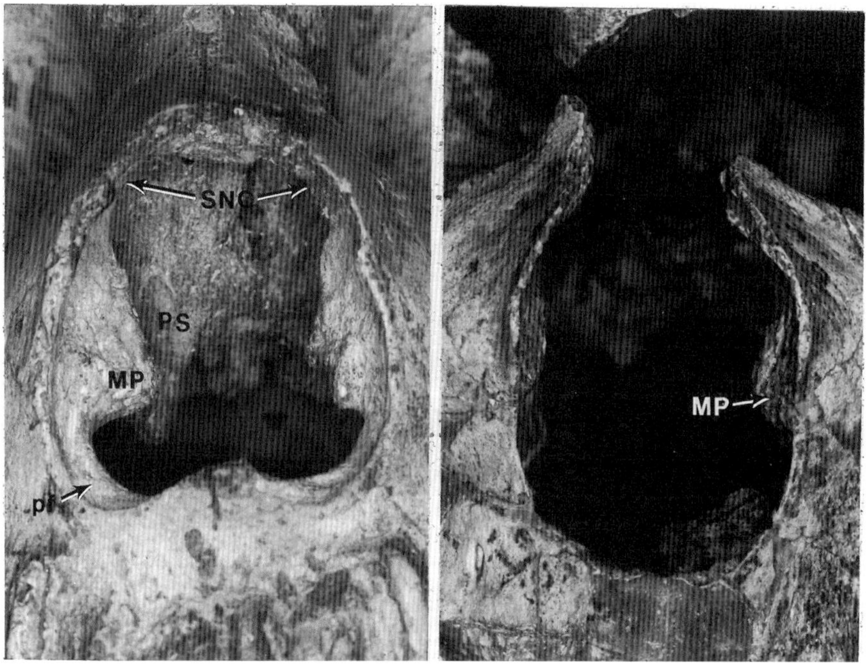

Figure 5 - *Nasal cavity structures in adult Neanderthals.*
Top left: Gibraltar 1 (Natural History Museum, London); top right: La Ferrassie 1 (Musée de l'Homme, Paris). In Gibraltar 1, the internal nasal margin (INM in the text) is essentially complete and bears a large medial projection (MP) and a strong superior nasal crest (SNC); a distinct sulcus (the prenasal fossa of Gower = pf) separates the internal from the external nasal margin; this specimen also preserves the large posterior swelling (PS) within the posterior portion of the nasal cavity. The medial projection is less protrusive in La Ferrassie.
© J. Schwartz.

As described in detail elsewhere (Schwartz & Tattersall, 1996a, in press), Neanderthals are distinguished not only from extant *Homo sapiens*, but, as far as we can ascertain from our studies and from the literature (see Negus, 1958), from all other extant non-marine mammals in important aspects of the nasal cavity. Among these potential Neanderthal apomorphies are the development of an internal nasal margin (INM) that bears basally thick, vertically oriented medial projections that are anteroposteriorly thick at the base; the absence of an ossified covering of the lacrimal canal; and a voluminous intrusion of the wall of the maxillary sinus into the posterior nasal cavity (Figure 5).

The INM and its medial projection are the most frequently preserved features of the nasal cavity of Neanderthals, both juvenile as well as adult (see Table 1). In addition to the specimens in which these features of the nasal cavity were first documented (*ibid.*), we can now add to the sample that preserves the nasal region in adequate detail another adult Neanderthal, from the Croatian site of Krapina (K3/Skull C) (Figure 6), as well as another juvenile (c. 3 years old), from the Hungarian site of Subalyuk (Figure 7). As among other adult Neanderthal specimens in which this region is preserved, the Krapina adult displays a thickened ridge originating within the nasal cavity (just posterior to the external margin of the nasal aperture = the lateral crest of Gower);

Adult	Child
Gibraltar 1	Engis
Krapina 3	Gibraltar 2
La Chapelle-aux-Saints	Roc de Marsal
La Ferrassie	Subalyuk
Shanidar 1	
St. Césaire	

Table 1 - Neanderthal specimens preserving the medial projection.

in better preserved specimens (e.g. Gibraltar 1, Figure 5) this crest (Gower's combined spinal and turbinal crests) arises from the posterior extent of the anterior nasal spine. Approximately halfway up the wall of the nasal cavity this crest expands medially into a stubby projection. A ridge continues superiorly from the medial projection into the region of the nasal bones; we identify this crest as a "superior nasal crest." In less damaged specimens (e.g. Gibraltar 1, Figure 5) the rim of the external nasal margin and the INM create a sulcus (Gower's prenasal sulcus), which is located inferiorly, extending laterally on either side of the anterior nasal spine; given the occasional presence of a similarly located sulcus in *Homo sapiens* (de Villiers, 1968; Gower, 1923), this feature is probably best considered not to be part of the apomorphic Neanderthal configuration. The stout medial projection is thick at its base and variably blunted at its tip. As appears to be the case in various speciments of Neanderthals, the medial projection does not send a bony flange or shelf over, but is instead delimited posteriorly by, the lacrimal groove; the latter feature remains, therefore, patent (rather than becoming a canal). In addition, the posterior and inferior surfaces of the medial projection may form at their confluence a horizontal margin or edge which, in tandem with the orientation of the medial projection, is lateromedially directed. Some of the morphology of the Neanderthal nasal cavity posterior to the lacrimal groove is preserved in the Gibraltar 1 skull, in which there is evidence of a balloon-like swelling of the wall of the maxillary sinus into the nasal cavity (in contrast to the more vertical or variably sunken lateral wall of the maxillary sinus in extant large-bodied hominoids). Swelling of the lateral wall of the maxillary sinus into the nasal cavity is also visible in the Kůlna Neanderthal maxillary fragment; although the bone of the lateral wall has broken off, the edge of bone remaining, which outlines the extent of the maxillary sinus, is medially arcuate.

The Subalyuk child (Figure 7) is similar to the other Neanderthal children studied so far (Roc de Marsal, Engis 2, Peche de l'Aze). As in the others, the Subalyuk child preserves a distinct medial projection. In particular, this medial projection swells out of a crestlike INM which continues superiorly beyond the medial projection as a superior nasal crest. The medial projection is very broad from front to back, and may have been pneumaticized. The lacrimal groove is very broad and deep, and was apparently not covered by bone (i.e. was not a canal). At the inferior and anterior portion of the moderately large nasal aperture is a sulcus (the prenasal fossa of Gower). This sulcus lies between the two crests (the external nasal margin or lateral crest of Gower in front, and the spinal crest behind) that emanate from the large, bifid, forward-projecting anterior nasal spines. The maxillary sinus has expanded to a point just behind the root of the deciduous upper canine. Although maxillary pneumatization in the Subalyuk child has progressed farther forward than in the other Neanderthal children, sinus expansion has not encroached upon the nasal cavity.

Obviously, the formation of the INM and its medial projections occurs early in Neanderthal development (certainly by 3-4 years), whereas expansion (via aggressive osteoclastic activity) anteriorly and laterally of the maxillary sinus begins sometime later. Clearly, in contrast not only to other large-bodied hominoids, but also to non-marine mammals in general, Neanderthals are apomorphic in these features. The unroofed lacrimal groove seen in the child is most likely due to reduced osteoblastic activity; this hypostotic condition is retained (apparently neotenically) into the adult Neanderthal. Inasmuch as a covered lacrimal groove (i.e. a lacrimal canal) has already formed at least in fetal *Homo sapiens* and at least neonatal great apes, and a lacrimal canal is present in all extant large-bodied hominoid adults, the hypostotically related absence of a covered lacrimal groove in Neanderthals (both children and adults) emerges as a developmentally based apomorphy, as does extensive pneumatization.

With regard to the features of the nasal cavity of other middle to late Pleistocene hominids that we have studied, the Steinheim skull is apomorphically similar to Neanderthals in possessing a distinct INM, complete with a superior nasal crest (Schwartz & Tattersall, 1996a, in press). The INM of Steinheim does bear a small, pointed elevation, but this slight structure lacks the bulk, depth, and prominence of the Neanderthal medial projection - the presence of which is consistent with the interpretation that the Steinheim specimen appears to be related to Neanderthals, but as a primitive sister (e.g. Schwartz & Tattersall, 1996b, in press; Stringer et al., 1984). In contrast to the Steinheim specimen and to Neanderthals, the Kabwe and Petralona skulls lack both an INM and a medial projection of any kind. Rather, these two skulls possess anteriorly within the nasal cavity a low-lying, horizontal conchal crest; they are thus primitively similar to *Homo sapiens* and other extant large-bodied hominoids in this feature.

The Neanderthal nasal region and respiration

As noted in our introduction, much attention has been paid to the relative size of the nasal aperture of Neanderthals as it relates to extrapolations about adaptation, behavior, and lifestyle. Clearly, it is not sufficient to focus on nasal aperture size alone, because individuals with relatively large apertures (Neanderthals, the Steinheim, Petralona, and Sima de los Huesos crania, as well as extant Asians and Africans) may have totally different configurations within the nasal cavity. Given the fact that wide nasal apertures are found in Asians and Africans, and that one can hardly claim that the environments occupied by subsaharan Africans and southern Asians are representative of glacial conditions while the climes of northern Europeans with narrow nasal apertures are, it is difficult to understand how the suggested correlation between nasal aperture size and climate has continued to claim the attention given to it. Furthermore, if, as has been proposed (Franciscus, personal communication), the development of the INM and medial projection, if not other features of the Neanderthal nasal cavity, is related to this hominid's development of wide nasal apertures, then we would expect to find these features in other hominids, including those individuals of *Homo sapiens* with relatively large nasal apertures. The claims sometimes heard (e.g. Tobias, 1995) that the apomorphic features of the Neanderthal nasal cavity are to be found in the Kalahari San would, if accurate (which they are not), provide further falsification of the suggestion that the Neanderthal nasal region represents an adaptation to cold climatic conditions. Nonetheless the configuration of the Neanderthal nasal cavity, as much as is clearly preserved, can indeed be discussed in its own right with regard to Neanderthal behavior and activity because Neanderthals did, after all, have to breathe.

A question that can be raised about Neanderthal respiration concerns the degree to which preserved and potentially reconstructible features of the nasal region restricted airflow. If there is a correlation between bony nasal aperture width and the width of the fleshy nose as well as of the nares (Ohki et al., 1991), then we would expect fossil hominids with *Homo sapiens*-like projecting nasal bones to exhibit the following: narrow nasal apertures with narrow fleshy noses and nares; moderately wide nasal apertures with moderately wide fleshy noses and nares; and wide nasal apertures with wide fleshy noses and nares. The wider the fleshy nose, and the wider the nares, the less resistance there will be to airflow (*ibid.*). Accordingly, Neanderthals (as well as the Steinheim, Petralona, and Sima de los Huesos fossil hominids) would have breathed with less restriction of airflow through the nares than narrow-nosed hominids. But the ease with which airflow can occur, and the velocity of the air, are two different matters.

It is a well-known rule in physics that the speed of molecules can be accelerated in one of two ways: either by heating them up, or by decreasing the volume of the structure through which they pass. Clearly, in terms of air passing through the narial openings, and lung activity apart, velocity of incoming air would be determined by nares size alone, whereas velocity of exhaled air would have the added attribute of having been warmed. It would thus seem that the possession of wide narial openings (at least wider than those of extant Europeans), as may be hypothesized for Neanderthals, would not alone contribute significantly to the acceleration of inhaled air. Considerations of exhalation will be addressed below.

A major reason that studies such as Ohki et al.'s (1991) on airflow resistance and nares width in *Homo sapiens* can be undertaken with credibility is that, minor individual variation aside, the structures of the nasal cavity (and apparently also lung volume) are the same in all *Homo sapiens*, high altitude populations included. Obviously, we cannot make this assumption with regard to Neanderthals. However, we do know that Neanderthals did not have the extraordinarily spacious and voluminous nasal cavity that has become entrenched in scenarios about this hominid's adaptation to a glacial environment, and through which air would have travelled much as it does in a typical non-marine mammal, including *Homo sapiens*. In contrast, in Neanderthals the INM, with its medial projections and the swelling of the lateral wall of the maxillary sinus farther back in the nasal cavity, would at the very least have obstructed the passage of both inhaled and exhaled air.

Although no fossil hominid specimen currently known preserves the ethmoid and its ethmoturbinals and the maxilloturbinals, we can make some predictions about how the size and spatial relations of these structures in Neanderthals would have been impacted by the INM, medial projections, and medially inflated maxillary sinuses (assuming that Neanderthals indeed developed structures we would identify as the ethmoid, ethmoturbinals, and maxilloturbinals). In a typical non-marine mammal, the first structure within the nasal cavity that incoming air encounters is the anterior end of the maxilloturbinal. In some mammals, the maxilloturbinal is strongly scrolled, but in large-bodied hominoids, for instance, it is very simple, projecting slightly into the nasal cavity and then drooping inferiorly (see, e.g. *Homo sapiens*, Figures 1, 2).

We would predict that Neanderthals had a pair of simple, rather than complexly scrolled, maxilloturbinals (Figure 7). These would not, however, have been the first nasal cavity structures encountered by incoming air. Rather, this air would have first contacted the mucous membrane-covered INM and medial projections. In Neanderthals, the pair of maxilloturbinals must have sat behind the anterioly facing pair of medial projections, so that the length of these bones was confined to the distance between the posterior surface of the medial projection and some point on the medially swollen lateral wall of the maxillary sinus. Since it is broadly analogous to the conchal crest in other non-marine mammals, the inferoposterior margin of the medial projection may have been a region of articulation of the maxilloturbinal. The configuration of the lacrimal

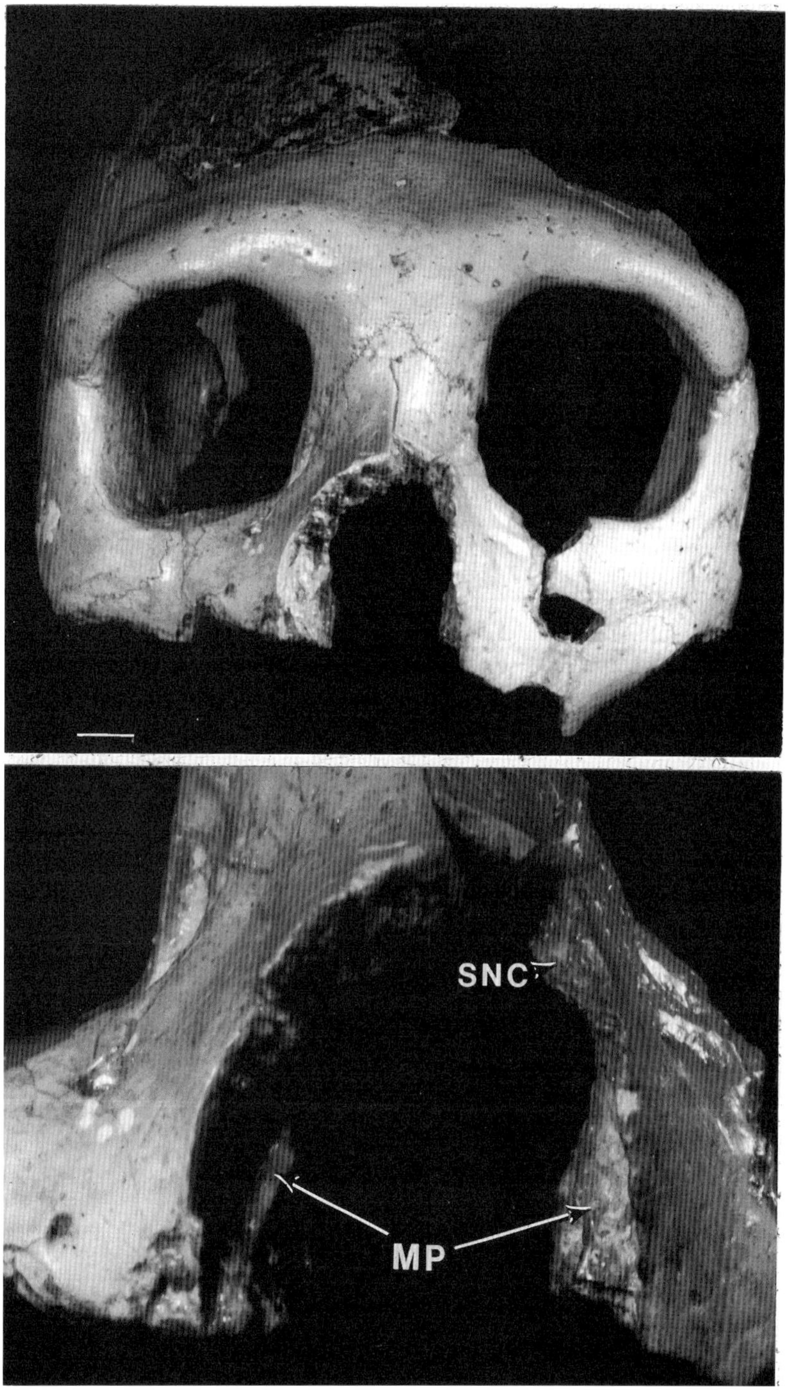

Figure 6 - *Partial cranium of Krapina Neanderthal (K3; Geološko-Paleontološki Musej, Zagreb). Although damaged, the stout base of the medial projection (MP) on each side is preserved, as is the superior continuation of the internal nasal margin [= the superior nasal crest (SNC)]. Scale for top = 1 cm; bottom not to scale. © J. Schwartz.*

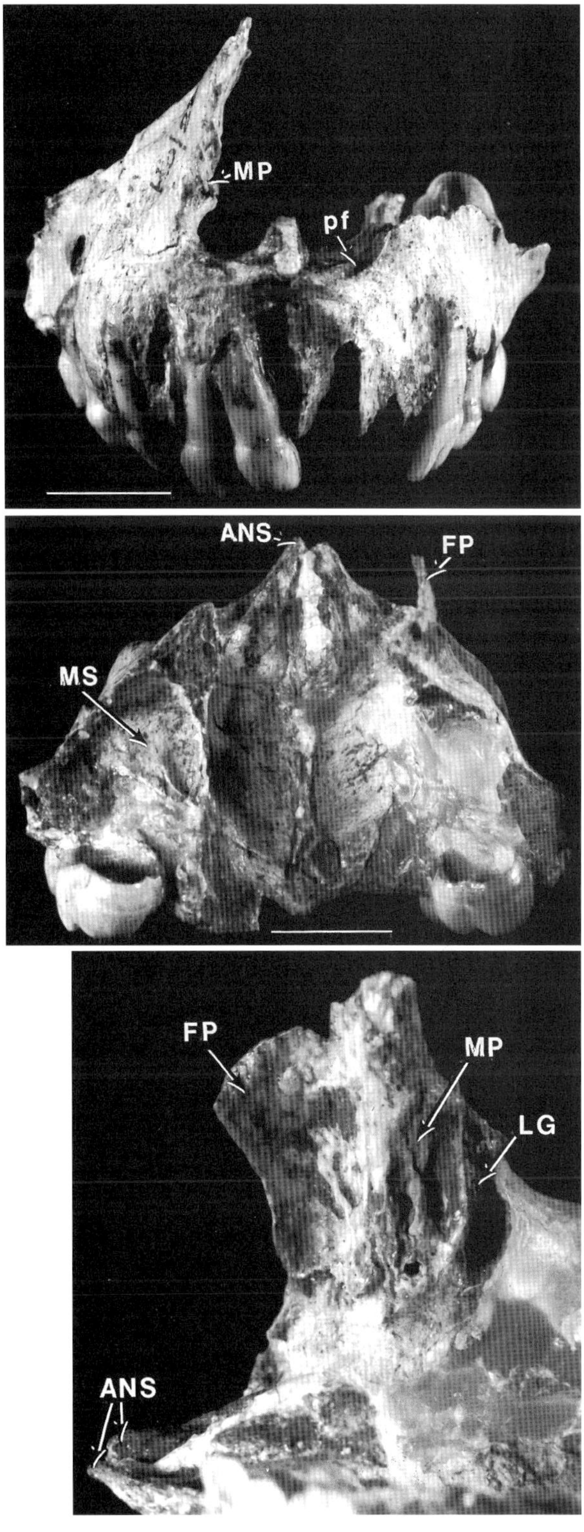

Figure 7 - Upper jaw of the Neanderthal child from Subalyuk (Temeszettudomanyi Museum, Budapest).
Top: Frontal view; although damaged, the the base of the medial projection (MP) is preserved; also note the prenasal fossa (pf), which would have separated the internal from the external nasal margin.
Middle: View from above showing large pair of anterior nasal spines (ANS, arrow points only to left spine), the still not anteriorly expanded but very wide maxillary sinus (MS), and the frontal process (FP).
Bottom: Internal view of frontal process with damaged medial projection (exposing internal structure) and lacrimal groove (LG).
Scale for top and middle = 1 cm; bottom not to scale. © J. Schwartz.

groove in Neanderthals - which lacks evidence of bone to bone contact - indicates that the hypothesized Neanderthal maxilloturbinal did not bear a lacrimal process. A conservative reconstruction of the Neanderthal maxilloturbinal, therefore, suggests that it was short, minimally or perhaps not at all flexed downward, and probably lacked processes; this bone may have extended at least as far mediolaterally as the medial projection behind which it sat and with which it probably articulated. If this reconstruction is in any way accurate, it would have provided yet another mucous membrane-bearing structure to occlude the Neanderthal nasal cavity. In addition, our reconstruction portrays a totally open lacrimal groove housing the lacrimal duct.

Reconstruction of the ethmoid and its turbinates is an even more tenuous proposition (Figure 8), which is further complicated by the fact that the region encompassing the frontal process, the medial orbital wall, and the medioinferior portion of the orbit of adult Neanderthals is also pneumaticized and inflated, apparently by the maxillary sinus. Assuming that Neanderthals did, indeed, develop a multifocular ethmoid, it is impossible to predict whether the ethmoidal air cells were primitively few in number [e.g. as in African apes and the Petralona skull (Seidler et al, in preparation)] or apomorphically more numerous [as in *Homo sapiens* (Schwartz, 1984)]. To be sure, the discovery of more and better preserved specimens of Miocene hominoids makes it increasingly likely that the last common ancestor of fossil and extant large-bodied hominoids would have developed an ethmoidal sinus or sinuses (see review by Schwartz, 1997). However, this does not preclude other aspects of the facial skeleton and paranasal sinuses from imposing themselves upon the region of the ethmoid, resulting, as in the extant orangutan, in the absence of ethmoidally derived sinuses; in the case of the orangutan, the maxillary sinuses, which are already huge and pervasive in the juvenile, invade the ethmoid, as well as the frontal, zygomatic, sphenoid, and palatine bones (Cave & Haines, 1940).

Given the extent of maxillary sinus enlargement in Neanderthals, it is possible that this hominid differed somewhat or even significantly from *Homo sapiens* in the development and adult morphology of its ethmoid. Since the orangutan, even with superposition of ethmoidal air cells by pneumatization from the maxillary sinus, still develops superior and inferior ethmoturbinals, as do all other extant large-bodied hominoids, we might expect the same of Neanderthals. Perhaps, though, given the autapomorphic expansion medially of the lateral wall of the maxillary sinus, Neanderthal ethmoturbinals would have been reduced in size relative to those of extant large-bodied hominoids. Nevertheless, any reconstruction of the Neanderthal nasal cavity that includes an ethmoid and superior and middle ethmoturbinals results in a nasal cavity that is anything but spacious and voluminous (Figures 7, 8). Rather, the Neanderthal nasal cavity would have been rife with projections, shelves, and swellings into a space that would have been divided in two by a plate consisting of the cartilaginous nasal septum, vomer, and perpendicular plate, and which would have been fully occupied in its upper half by the ethmoid and its turbinates.

Not to be overlooked in this reconstruction is the fact that, in life, all of these nasal cavity surfaces would have been covered in ciliated mucous membrane. Thus, regardless of the effect of nares width on the velocity of incoming air, it is highly probable that air travelling through the Neanderthal nasal cavity would have been accelerated both by being forced to travel through constricted passages and by being warmed as it passed over what must have been highly vascularized membranous tissues. Expired air would already be warm, but would be affected by the structurally congested nasal cavity. In this scenario, we also envision a slowing and accelerating process that recognizes upon inhalation the stepwise series of restrictions from the pair of nares, to the bisected nasal cavity that proceeds from the external nasal margin, to the INM and medial projections, then the maxilloturbinals, and, finally, to the ethmoid, ethmoturbinals, and inflated maxillary sinus; the reverse would obtain during exhalation.

On Neanderthal adaptations, behavior, and lifestyles

What, then, to make of the suggestions that Neanderthals were cold adapted hominids, at least in their nasal regions? The competing hypotheses summarized above about whether the Neanderthal nasal region functioned so as to warm and moisten incoming air, or to expel excess heat while preventing loss of moisture, are predicated on the observation that Neanderthals had large and projecting bony noses as well as on the misconception that Neanderthals had spacious nasal cavities, and that these features of the nasal region were selected for in the process of adapting to glacial conditions. If there was selection as a cold adaptation for a large and projecting nasal region, which would have borne an even farther projecting fleshy nose, then this must have come at a cost. As studies on extant *Homo* have shown, as temperature drops, skin temperature falls first at the tip of the nose, and then, just after the tip of the ear, the ear lobe and infraorbital region, the bridge of the nose; at 0° C, the areas most susceptible to frostbite - namely, the skin of the nose, ear, and chin - are also the coldest parts of the body (Roberts, 1973).

Questions of climatically induced adaptation must, of course, depend on reconstructions of the environment, in this case during Pleistocene glacial, interglacial, and periglacial periods. Apart from the fact that many Neanderthals occupied quite southerly regions, in our sample of Neanderthals that preserve the nasal cavity structures discussed above, the oldest specimen comes from the Croatian site of Krapina, which dates to approximately 130 kyr, or during the late Riss glaciation, just prior to the late Riss-Eemian boundary (the boundary between oxygen isotope stages 6 and 5e; e.g. Guiot et al., 1989); this boundary basically coincides with what Winograd et al. (1992) refer to as Termination II, a termination being the midpoint between the peaks of glacial and interglacial climates. According to data from analysis of Pleistocene pollen records in Europe, the warming trend that reached its maximum just prior to the middle of the Eemian was already well under way during the late Riss glaciation (Guiot et al., 1989). A warming trend during this period of time, which may have begun as early as 140 kyr, is also indicated by analysis of vein calcite (Winograd et al., 1992). In terms of reconstructed temperatures, the coldest it apparently got in the latter part of the late Riss (c. 135 kyr; cf. Guiot et al., 1989, and Winograd et al., 1992) was approximately and on average -2° C (with a predicted deviation around the mean of 4° C), and the warmest during the Eemian was approximately and on average 12° C (with a predicted deviation around the mean of 2° C) (see Guiot et al., 1989). Interestingly, the warmest part of the Eemian may have been warmer than some or most of the Holocene (*ibid.*). In addition to an increase in mean annual temperature, the transition from the late Riss to the Eemian would also have been characterized by glacial melting, an increase in sea levels, and an increase in mean annual precipitation. In other words, there would have been an increase in humidity.

Although the Krapina Neanderthal is the earliest specimen that preserves at least the INM and medial projections, it is not, of course, the oldest known Neanderthal. Based especially on features of the occiput (e.g. Santa Luca, 1978; Stringer et al., 1984; personal observations), the specimens from Biache (c. 160 kyr), Ehringsdorf (200 kyr +), and Swanscombe (c. 225 kyr) represent the oldest Neanderthals so far known. While the period around 150 kyr was apparently the coldest until the end of last glaciation (see Winograd et al., 1992), the time span between 200 and 240 kyr saw temperature maxima that almost reached those of the warmest part of the Eemian (*ibid.*). Since it is reasonable to conclude that, if the Swanscombe specimen is a Neanderthal, it would also have borne the suite of Neanderthal apomorphies found in the nasal region, then we are confronted with the apparent fact that these presumed cold adaptations are encountered initially during some of the warmest years of the middle to late Pleistocene. In order for the Neanderthal nasal region to be regarded as having evolved as a cold-related adaptation, then we must hypothesize the selection process to have begun and to have quickly acted during the closest preceding period of cold, i.e. 250-270 kyr (cf. Winograd et al., 1992).

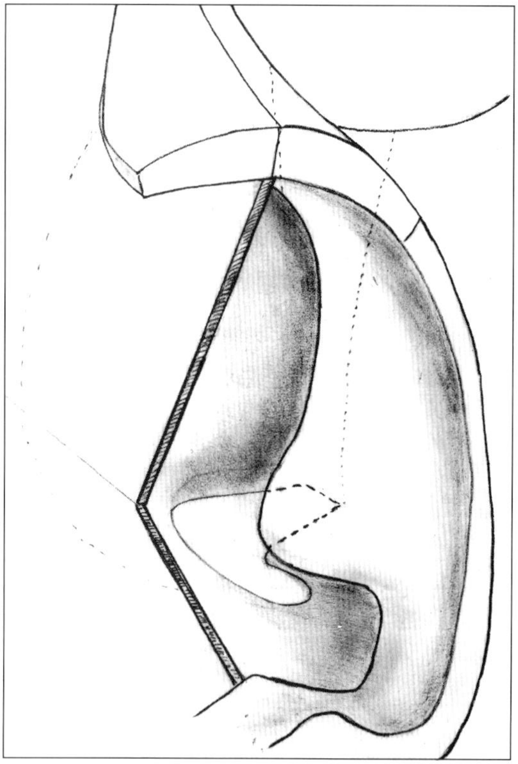

Figure 8 - *Schematic representation of the left side of the nasal cavity of an adult Neanderthal. Showing possible structure and orientation of the inferior nasal concha between the medial projection in front and the posterior swelling behind (compare to Gibraltar 1 in figure 5, and see text for discussion).* © *J. Schwartz*

If one rejects the Swanscombe and Ehringsdorf specimens as being cladistically Neanderthal (which we do not), then the Biache specimen would be the next likely candidate to have possessed the apomorphies of the Neanderthal nasal region. On the face of it, this might be appealing from a selectionist point of view inasmuch as, with occasional warming spurts, the average temperature fell between c. 190 and 150 kyr (Winograd et al., 1992). Following this line of speculation, the INM and medial projection preserved in the Krapina Neanderthal specimen would be a retention of this presumably stabilized morphology, which would then have persisted throughout subsequent major fluctuations between colder and drier and warmer and more humid periods [as is evidenced by the youngest specimens preserving this morphology, La Ferrassie 1 (40-50 kyr) and St. Césaire (36 kyr)]. If one follows the selectionist argument, one must, however, invoke some kind of mechanism for stabilizing the apomorphic Neanderthal nasal region, or for otherwise trying to explain why selection did not reject these morphologies when the environment changed to warmer and more humid conditions.

Clearly, evidence is lacking for when the distinctive features of the Neanderthal nasal region appeared. We can only surmise that their most apomorphic state was present in the first generations of this emergent hominid. But regardless of when Neanderthals first appeared, it is equally obvious that, whether or not the unique features of their nasal region represent adaptations to a cold climate, these morphologies persisted and functioned, in both cool and warm extremes, as well as during the transitional periods. Although Trinkaus (1989) and Jelinek (1994) offer seemingly opposing selectionist arguments (based on the premise of cold adaptation) for the large and projecting nasal region of Neanderthals - retaining moisture and heat from outgoing air versus humidifying and warming incoming air - it is the case that the nasal cavity mucosa actually

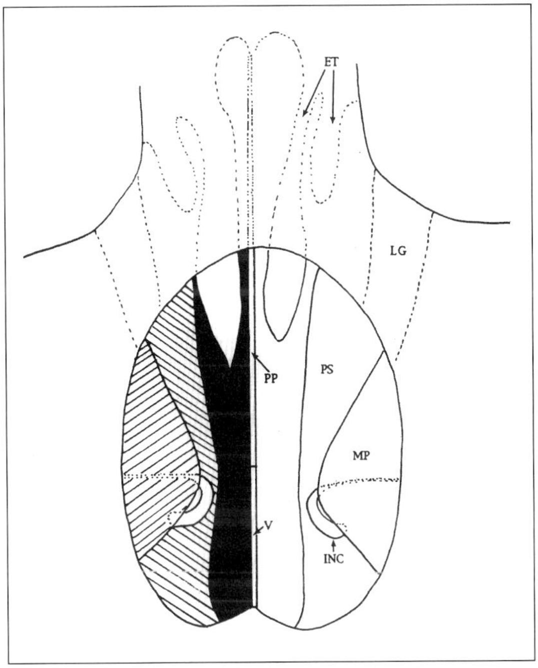

Figure 9 - Schematic representation of all possible structures within the nasal cavity of an adult Neanderthal (see text for discussion).
ET = ethmoturbinals, LG = lacrimal groove, PS = posterior swelling, MP = medial projection, INC = inferior nasal concha, PP = perpendicular plate, and V = vomer.

performs both of these functions during normal breathing, one or the other increasing in importance in synchrony with the prevailing conditions (e.g. see Roberts, 1973, for discussion of *Homo sapiens*). Thus, when temperature and humidity decrease, retention of moisture and heat increases, and vice versa (*ibid.*).

Given that Neanderthals existed for many tens of thousands of years (especially if one includes the Swanscombe specimen in the sample), and that they did so with relatively uniform cranial morphology in a diversity of locales and climatic regimes, the apparently relatively large surface area of mucous membrane exposed to incoming and outgoing air in their nasal cavity would have functioned at least adequately in both cold and dry and warm and moist climates. In periods of extreme cold, Neanderthals would probably have breathed more through their noses than orally, which, as observations on extant *Homo* demonstrate (see review by Roberts, 1973), would have served to minimize damage to pulmonary tissues from cold air.

Along with increased periods of nasal breathing, there would also probably have been an increase in the volume of inhaled particulate matter. If the surface area of mucosa within the Neanderthal nasal cavity was indeed relatively great, then so also would have been the population of cilia normally present to cleanse incoming air by moving dust and mucus forward and eventually out of the nasal cavity. Since the lacrimal duct would have drained into the nasal cavity, as would the maxillary sinus (where exactly is another question), one would predict that the role of the nasal cavity cilia would have been taxed when breathing occurred through the nose alone. With Neanderthals apparently lacking a bony roof over the lacrimal groove, there would have been room for dilation of the lacrimal duct when increased drainage was required - which, in turn, would periodically also have put demands on nasal cavity cilia.

Given all these complexities, can we actually say anything meaningful about Neanderthal behavior and/or lifestyle from study of the nasal region? If the nose would have been the first part of the body to be in danger of frostbite, then it would seem reasonable to suggest that Neanderthals exposed to glacial conditions might have tried to do something to protect this, if not other parts of the face. If glacially exposed Neanderthals did try to cover and protect the most vulnerable parts of the face, then it is reasonable to conclude that the rest of the body would have been protected in some way as well. Thus, it probably does an injustice to Neanderthals to suppose (e.g. Jelinek, 1994) not only that they were capable either mentally or neuromuscularly of only minimally covering themselves against the elements, but also that this deficiency was reflective of their lifestyle in general.

If Neanderthals did cover their bodies sufficiently to insulate themselves from the cold, as well as to maintain a proper core body temperature, then it is unnecessary to hypothesize that they required an extraordinarily high intake of high calorie food in order to sustain a high metabolic rate, which in turn requires a greater intake of oxygen, in order to exist in harsh glacial conditions (e.g. Coon, 1962; Jelinek, 1994). It would also be unnecessary to hypothesize that Neanderthals normally engaged in long, sustained periods of activity (even if only walking or trotting) in order to maintain higher body as well as intramuscular temperatures in the face of the glacial cold, as Trinkaus (1989) has done. This is inherently improbable in view of the fact that adequate core body temperature must have been maintained during sleep; but, if true, maintaining high activity levels would require a greater intake of calories, which, in turn, would require greater effort and more frequent attempts in acquiring (probably by hunting) adequate food sources. Even if Neanderthals had lacked the mental wherewithal to protect at least their extremities from frostbite, it would seem that huddling together to try to stay warm would have been a more practical solution than expending energy. In either case, Neanderthals would still have had to hunt and eat, but the critical difference would have been in the frequency with which this high energy activity took place.

Conclusion

From the foregoing it is obvious that, in spite of persistent lacunae in our knowledge of Neanderthals, both in morphology and in temporal and spatial contexts, it is nonetheless possible to address some of the age old questions regarding Neanderthals and their adaptations, behaviors, and lifestyles. Conservatively, we see no evidence in direct support of the notion that the Neanderthal nasal region was selected for as an adaptation to cold and dry conditions. It is undeniable that the uniquely configured Neanderthal nasal cavity functioned at least adequately in extreme cold, warm, and intermediate climates. We find no compelling reason to embrace the notion that under glacial conditions Neanderthals had to have maintained higher metabolic rates, which would, in turn, have required a greater intake of oxygen and more frequent consumption of high calorie food. As for being almost constantly in motion in order to maintain a certain body temperature, this hardly seems to be an energy efficient solution to dealing with the cold. What would children or the tired have done? Unless we were to imagine that all Neanderthals, presumably at a very young age, lost their fleshy noses and ears to frostbite, we must assume that they were capable of protecting themselves from the cold. If this means that they did so by exploiting furs and skins, so be it. There is abundant evidence that Neanderthals used stone implements to scrape hides, and that in certain places at least they burned fires in hearths (e.g. Bar-Yosef, 1989; Keeley & Newcomer, 1977). Certainly, if Neanderthals were capable of hunting the medium- to large-bodied mammals, or at least of butchering their frozen carcasses (e.g. Bar-Yosef, 1989; Jaubert, 1989), then they would have had these resources available to them.

The skeletons of the Neanderthals bear eloquent witness to the fact that they led hard and energetic lives. But to ascribe the necessity for such lives to cognitive, anatomical, or physiological deficiencies seems simply absurd. The Neanderthals survived for a very long and climatically unsettled time and over a vast area. Clearly, they were highly adaptable and at home in a wide variety of environments. Given the apparent stability of their morphology over an extended period of time, it seems reasonable to suppose that they compensated by cultural means for the climatic and environmental vicissitudes they encountered. To regard them as specifically "cold-adapted" hominids is to ignore the abundant evidence for their versatility. We do them scant justice by portraying them as helpless prisoners of a distinct set of presumed adaptations.

Acknowledgements

We thank Dr. H. Ullrich for inviting us to participate in his stimulating symposium, and Drs C. Stringer (London), A. Langaney (Paris), R. Orban (Brussels), E. Poty (Liège), R. Ziegler (Stuttgart), H.-E. Joachim (Bonn), J.-J. Cleyet-Merle (Les Eyzies), B. Vandermeersch (Bordeaux), J. Radovčić and M. Paunović (Zagreb), V. Dobosi and I. Pap (Budapest), , M. Dobisíková (Prague), M. Teschler-Nicola (Vienna), J. Jelínek (Brno), Y. Rak (Tel Aviv), J. Zias (Jerusalem), J. Svoboda (Dolní Věstonice), D. Pilbeam (Cambridge, MA), G. Kordos (Thessaloniki), and H. Langdon (Pittsburgh) for permission to study and photograph specimens in their charge. We also thank Drs. J.-L. Arsuaga and R. Franciscus for helpful discussion. J.H.S. was funded by the L.S.B. Leakey Foundation and the University of Pittsburgh, and I.T. and J. H. S. by the Ogden Mills Fund (AMNH).

References

ARSUAGA, J.-L., MARTÍNEZ, I., GRACIA, A., CARRETERO J.-M. & CARBONELL, E. (1993): Three new human skulls from the Sima de los Huesos Middle Pleistocene site in Sierra de Atapuerca, Spain. *Nature* 362, 534-537.

ÅSTRAND, P.-O. & RODAHL, K. (1977): *Textbook of work physiology.* New York.

BAR-YOSEF, O. (1989): Geochronology of the Levantine Middle Palaeolithic. In: P. Mellars & C. Stringer (eds.), *The human revolution: Behavioural and biological perspectives on the origin of modern humans*, 589-610. Edinburgh.

CAVE, A.J.E. (1961): The frontal sinus of the gorilla. *Proceedings of the Zoolological Society London* 136, 359-373.

CAVE, A.J.E. (1973): The primate nasal fossa. *Biological Journal of the Linnaean Society* 5, 377-387.

CAVE, A.J.E. & HAINES, R.W. (1940): The paranasal sinuses of the anthropoid apes. *Journal of Anatomy* 74, 493-523.

COON, C.S. (1962): *The origin of races.* New York.

DEAN, M.C., STRINGER, C.B. & BROMAGE, T.B. (1986): Age at death of the Neanderthal child from Devil's Tower, Gibraltar and the implications for studies of general growth and development in Neanderthals. *American Journal of Physical Anthropology* 70, 301-309.

GOWER, C.D. (1923): A contribution to the morphology of the apertura piriformis. *American Journal of Physical Anthropology* 6, 27-36.

GREGORY, W.K. (1920): On the structure and relations of *Notharctus,* an American Eocene primate. *Memoir of the American Museum of Natural History* 35: 49-243.

GUIOT, J., PONS, A., BEAULIER, J.L. de & REILLE, M. (1989): A 140,000-year continental climate reconstruction from two European pollen records. *Science* 328, 309-313.

HERSHKOVITZ, P. (1977): *Living New World monkeys (Platyrrhini), with an introduction to primates.* Volume 1. Chicago.

HILLENIUS, W.J. (1994): Turbinates in therapsids: Evidence for late Permian origins of mammalian endothermy. *Evolution* 48, 207-229.

HOWELL, F.C. (1952): Pleistocene glacial ecology and the evolution of "classic Neandertal" man. *Southwestern Journal of Anthrop*ology 8, 377-410.

HRDLIČKA, A. (1927): The Neanderthal phase of man (The Huxley Memorial Lecture for 1927). *Journal of the Royal Anthropological Institute* 57, 249-273.

HUXLEY, T. (1863): On some fossil remains of man. In: Th. Huxley, *Man's place in nature*, 157-209. New York.

JAUBERT, J., LORBLANCHET, M., LAVILLE, H., SLOTT-MOLLER, R., TURQ, A. & BRUGAL, J.-P. (1989): *Les chasseurs d'Auroch de la Borde: Un site de paléolithique moyen (Livernon, Lot).* Paris (Documents d'Archeologique Francaise, 27).

JELINEK, A.J. (1994): Hominids, energy, environment, and behavior in the late Pleistocene. In: M.H. Nitecki & D.V. Nitecki (eds.), *Origins of anatomically modern humans*, 67-92. New York.

KEELEY, L.H. & NEWCOMER, H. (1977): Microwear analysis of experimental flint tools: A test case. *Journal of Archaeological Science* 4, 29-62.

NEGUS, V. (1958): *Comparative anatomy and physiology of the nose and paranasal sinuses.* Edinburgh.

OHKI, M., NAITO, K. & COLE, P. (1991): Dimensions and resistances of the human nose: Racial differences. *Laryngoscope* 101, 276-278.

PROETZ, A.W. (1953): *Essay on the applied physiology of the nose.* St. Louis. 2nd ed.

ROBERTS, D.F. (1973): Climate and human variability. *Addison-Wesley Module in Anthropology* 34, 1-38.

ANTA LUCA, A.P. (1978): A re-examination of presumed Neandertal-like fossils. *Journal of Human Evolution* 7, 619-636.

SCHWARTZ, J.H. (1984): Hominoid evolution: A review and a reassessment. *Current Anthropology* 25, 655-672.

SCHWARTZ, J.H. (1988): History, morphology, paleontology, and evolution. In: J.H. Schwartz (ed.), *Orang-utan biology,* 69-85. New York.

SCHWARTZ, J.H. (1997): *Lufengpithecus* and hominoid phylogeny: Problems in delineating and evaluating phylogenetically relevant characters. In: D.R. Begun, C.V. Ward & M.D. Rose (eds.), *Function, phylogeny, and fossils: Miocene hominoid origins and adaptations,* 363-388. New York.

SCHWARTZ, J.H. & TATTERSALL, I. (1996a): Significance of some previously unrecognized apomorphies in the nasal region of *Homo neanderthalensis. Proceedings of the National Academy of Sciences (USA)* 93, 10852-10854.

SCHWARTZ, J.H. & TATTERSALL, I. (1996b): Toward distinguishing *Homo neanderthalensis* from *Homo sapiens*, and vice versa. *Anthropologie* 34, 129-138.

SCHWARTZ, J.H. & TATTERSALL, I. in press: Toward a definition of *Homo neanderthalensis* and H*omo sapiens*: I. The nasal region. In. J. Gilbert (ed.), *Proceedings of the Orce Conference on Human Paleontology.*

SHEPHARD, R.J. (1978): *Human hysiological work capacity.* Cambridge (International Biological Programme 15).

STRINGER, C.B., HUBLIN, J.-J. & VANDERMEERSCH, B. (1984): The origin of anatomically modern humans in western Europe. In: F. Spencer & F. Smith (eds.), *The origins of modern humans: A world survey from the fossil evidence,* 51-135. New York.

TILLIER, A.-M. (1983): Le crâne d'enfant d'Engis 2: Un exemple de distribution des caractères juvéniles, primitifs et néanderthaliens. *Bulletin de Societé Royal Belge de Anthropologie Préhistorique* 94, 51-75.

TOBIAS, P.V.T. (1995): Comment at Orce Conference, Spain (September, 1995).

TRINKAUS, E. (1989): The Upper Pleistocene transition. In: E. Trinkaus (ed.), *The emergence of modern humans: Biocultural adaptations in the later Pleistocene,* 42-66. Cambridge.

TRINKAUS, E. & HOWELLS, W.W. (1979): The Neanderthals. *Scientific American* 241, 118-133.

VILLIERS, H. de (1968): *The skull of the South African negro: A biometrical and morphological study.* Johannesburg.

WILLIAMS, R.I. (1956): The nasal index. *Annals of Otolaryngology, Rhinology, and Laryngology* 65, 171-189.

WINOGRAD, I.J., COPLEN, T.B., LANDWEHR, J.M., RIGGS, A.C., LUDWIG, K.R., SZABO, B.J., KOLESAR, P.T. & REVESZ, K.M. (1992): Continuous 500,000-year climate record from vein calcite in Devils Hole, Nevada. *Science* 258, 255-260.

ZOLLIKOFER, C.P.E., PONCE de LEÓN, M.S., MARTIN, R.D. & STUCKI, P. (1995): Neanderthal computer skulls. *Nature* 375, 283-285.

Patterns of Adaptive Strategy in the Upper Palaeolithic and post-Palaeolithic Populations: Evidence from Central Europe

Janusz Piontek

Abstract

In this paper a hypothesis has been formulated that changes in the structure of the socio-economic system between the Upper Palaeolithic and Neolithic Age resulted in new possibilities for the operation of natural selection through differential fertility. A new lifestyle which followed the transition to agricultural activity promoted a high fertility rate, and natural selection could operate through differential fertility of earlier and later maturing individuals.

The hypothesis assumes that the differences between the Upper Palaeolithic and Neolithic populations with regard to their physique (long bones, body size, proportions of long bones, sexual dimorphism, reduction in body shape) could be a result of natural selection (controlled by cultural factors shaping a new fertility pattern) which favoured the individuals who reached their maturity earlier. As a result, a new pattern of development related to earlier biological maturation of individuals was formed in the Neolithic populations.

To verify the hypothesis data concerning long bones in the Upper Palaeolithic populations, Mesolithic populations and two types of Neolithic populations: typical and non-typical agricultural society (the Linear Band Pottery Culture and Corded Ware Culture) were used.

It was proved that: (a) the described reductions of long bones in the Neolithic agricultural populations resulted from two types of biological factors: the main factors - natural selection (process of genetic adaptation to a new lifestyle - new cultural adaptive strategy), and the secondary factors - the process of adaptability changes to different biocultural situations: disease, food, work demand, settlement pattern etc.; (b) changes in morphology of long bones from the Upper Palaeolithic to Neolithic times are correlated with ecological and socio-economic changes, but from the point of view of biology they are of microevolutional character. These microevolutional changes are not caused by natural environment factors but by a new structure of the socio-economic system which formed after the transition to agriculture.

Introduction

Changes in the morphological structure of the human skeleton resulting from the transition to agriculture are most often described and analysed by physical anthropologists in the studies of skeleton populations (Anthony, 1994; Brace & Mahler, 1971; Buikstra & Cook, 1980; Collier, 1993; Formicola, 1987; Jackes, 1988; Jankauskas, 1994; Key, 1980; Larsen, 1995). On the one hand, the documentation of morphological variability of the Upper Palaeolithic and Mesolithic people (Borgonini-Tarli & Repetto, 1986; Frayer, 1981, 1984; Jacobs, 1985, 1993; Vančata, 1988, 1993) is gathered, on the other hand, various data on the influence of living conditions on the structure of the Neolithic populations skeletons are presented (Cohen & Armelagos, 1984; Huss-Ashmore et al., 1982; Larsen, 1987, 1995; Ruff et al., 1983). The comparison of both groups of the data are often the principal inspiration for the construction of models of transition to agriculture and biological effects it entails (Larsen, 1987; Hassan, 1981; Goodman et al., 1994; Wierciński, 1985).

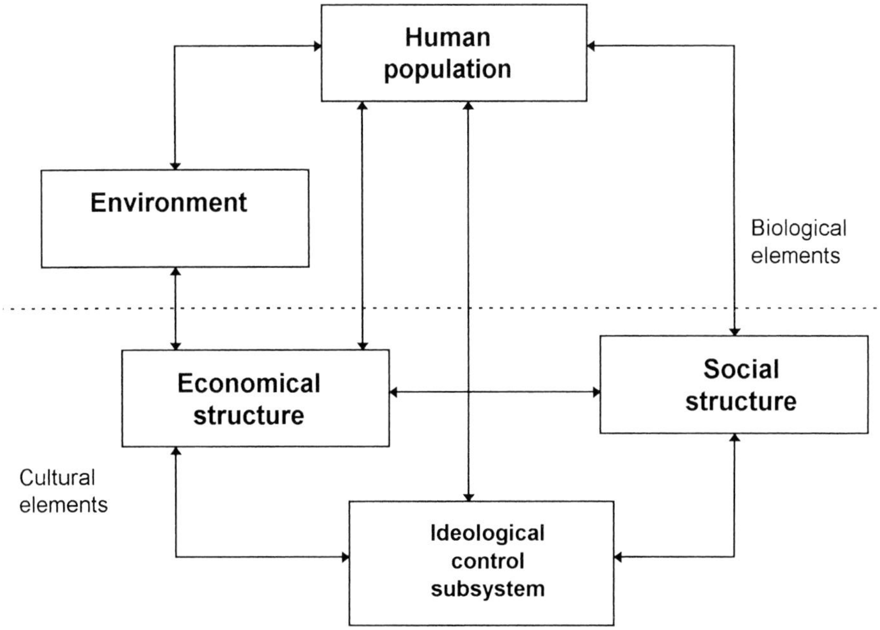

Figure 1 - The structure of the human system.

According to the model proposed by Larsen (1984) factors, such as a change in settlement pattern, change in work demand on body, change in work demand on masticatory apparatus, change in food composition, were the main factors responsible for the alterations observed in the physique of hunting/gathering/fishing populations and agricultural populations. In his latest studies Larsen (1995) extended the set of factors responsible for morphological changes of skeleton and for the health of the Neolithic population, supplementing it with food composition and nutritional quality, food processing and mastication, population growth, settlement and population distribution, work load and activity.

Hassan (1981) presented a model showing the role of the change of fertility structure resulting from the transition to agriculture. In his model factors such as work opportunities, defence requirement, work demand, opportunity for child work, abundance of baby food, reduction of competition for food among adults and children, and large populations were treated as principal factors generating higher fertility rates in the Neolithic populations. The assumptions adopted in the above described model concerning the high fertility rate in the populations of early farmers (conditioned mainly by socio-economic factors) are accepted also by other researchers (Ammerman, 1975, 1989; Carneiro & Hilse, 1966; Jackes, 1988; Larsen, 1995) and are confirmed in the research on the fertility of agricultural and non-agricultural traditional societies (Bentley et al., 1993a, 1993b).

Considering biological changes in human populations with agriculture one should - in our opinion - pose a question: whether the factors named (in quoted studies) are the factors generating the analysed biological changes, or whether they are only a description of living conditions (description of adaptation criteria) to which individuals and human populations had to adapt, either by way of genetic adaptation or by way of individual biological adaptation (adjustment or adaptability).

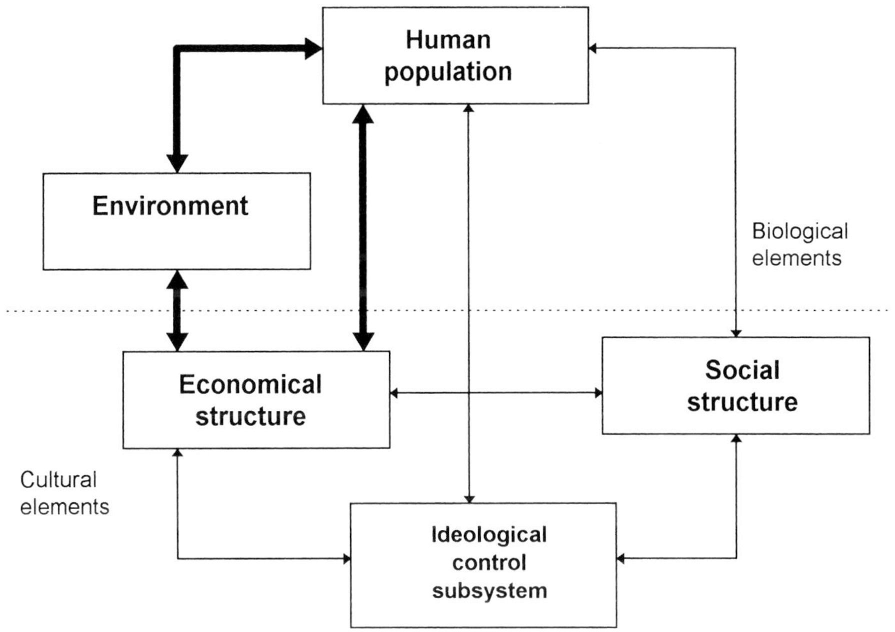

Figure 2 - The structure of the human system in the Upper Palaeolithic type strategy of adaptation.

This point of view is closer to a model of biological changes elaborated by Goodman et al. (1984). This model assumes the possibility of biological adaptation to living conditions (reaction to stress factors), or genetic adaptation (individual's death). Similar understanding of morphological reactions (of the individual and population) to the variability of living conditions was presented by Wolański (1988, 1990, 1991).

In physical anthropology certain controversies exist concerning the role of natural selection in changes of human populations. Attempts to describe the adaptive significance of different human morphological traits are still incomplete (cf. Bielicki, 1975; Foley, 1987), and the models of natural selection operation remain controversial. The difficulty in explaining the adaptation value for many human physical traits is the reason why many physical anthropologists attempt to explain biological consequences of the transition to agriculture as adaptability changes resulting from the reaction of the human system to new living conditions (Carlson & Van Gerven, 1973; Cohen & Armelagos, 1984; Goodman et al., 1984; Huss-Ashmore et al., 1982; Larsen, 1987, 1995). However, they do not try to answer the question: How was the genetic structure of the population changed after the transition to agriculture? Certain theories assume that agricultural populations inhabiting Central Europe in the Neolithic Age moved to this area from other regions, and hence were already genetically adapted to new environmental conditions (Ammerman & Cavalli-Sforza, 1984).

The structure of the human system: The model for Upper Palaeolithic and post-Palaeolithic populations

Approach

Studies on biological and cultural changes in the human species are carried out in both natural and social sciences. However, the postulates of interdisciplinary studies on biocultural evolution

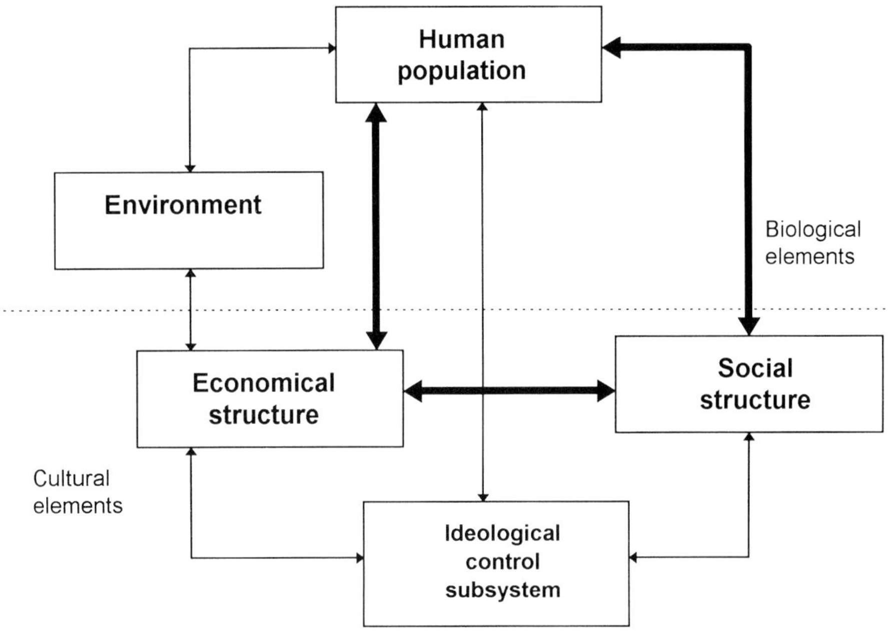

Figure 3 - The structure of the human system in the post-Palaeolithic type strategy of adaptation.

within the research programs constructed on a common theoretical and philosophical plane are not being realised.

The mono-aspectual approach is decidedly prevalent, where a given discipline in dealing with a particular aspect of biocultural evolution of human populations, distinguishes defined factors and, independently from other disciplines and factors, describes and explains a different kind of a process or phenomenon.

Generally speaking, studies on the biocultural evolution of man are conducted on the basis of miscellaneous interpretative theories, and within the frames of different cognitive orientations.

As the mono-aspectual studies do not promise any cognitive success, we would like to present an outline of multi-aspectual studies, simultaneously submitting an explanatory theory, methodology of studies and an epistemological orientation.

The proposed model of multi-aspectual studies on the biocultural evolution of man (Piontek & Weber, 1985) is characterised by: (1) the theory of evolution as a general explanatory theory; (2) the general theory of systems as a method; (3) the idealisational conception of research as an epistemological orientation; (4) the multi-aspectuality as the range of substantial studies.

Biocultural evolution is shaped under the influence of interactional factors, both biological and cultural, the activities of which are combined with the environmental conditions.

The essence of the presented, multi-aspectual model depends on the fact that in the descriptions and explanation of phenomena and processes of biocultural evolution both biological and cultural events are taken into consideration, while applying the idealisational method (Nowak, 1977).

According to this method, the revealing of the most essential determinants (factors) of any process or phenomenon (hence also of biocultural evolution) should be carried out through isolation and elimination of considerations (on the basis of a defined theory) of arbitrary, insubstantial determinants, which accompany the process of biocultural evolution.

Elements of the biocultural system	Theoretical knowledge - main processes	Main factor - side factors	Essential structure of the factors
A. Population	Theory of evolution - natural selection	A. Size of the population a. age structure b. density of population	A A, a A, a, b
B. The natural environment	Theory of evolution - adaptive dependence	B. Affluence of the environment c. utilisation of the niche d. distribution of wealth	B B, c B, c, d
C. Society	Theory of social development - adaptive dependence	C. Social structure e. distribution of groups f. the relation: group-group	C C, e C, e, f
D. Economy	Theory of social development - adaptive dependence	D. Exploitation efficiency g. exploitation,production h. production of tools and implements	D D, g D, g, h
E. Consciousness	Competitive dependence - mechanisms of ideological control	E. Ideological control subsystem i. technique of education j. ritual magical-religious interventions	E E, i E, i, j

Table 1 - A model of multi-aspectual studies on biocultural evolution.

In multi-aspectual studies on the biocultural evolution we suggest discussing not the individual, but the class of the main factors isolated in such a way, that each of them is at the same time the main factor of the distinguished aspects of studies when the idealisational method is being applied (Table 1).

Let us, therefore, consider the following model situation. If we assume that studies on biocultural evolution are carried out in five aspects (from I to V), distinguishing in each aspect one main factor (from A to E), and two side factors (from a to j), then following the idealisational conception of scientific research we can build aggregated essence structure of factors (S_{AG}) determining the biocultural evolution.

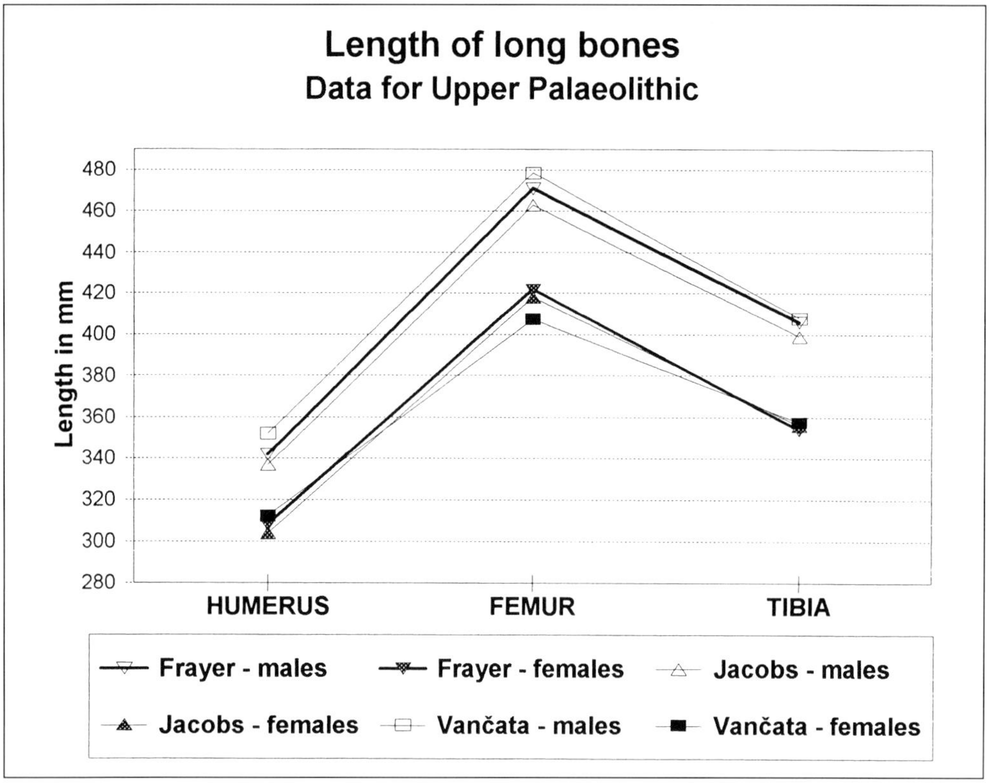

Figure 4 - Comparison of length of long bones: data for Upper Palaeolithic.

This can be presented as follows:

S_{AG}: (2) [A,B,C,D,E]

(1) [A,B,C,D,E] [a,b,c,d]

(0) [A,B,C,D,E] [a,b,c,d] [f,g,h,i,j]

In explaining the above given essential structure we would like to draw attention to the fact that index 2 means the level represented by the aggregated class of the main factors and is called the internal level, where the A,B,C,D,E factors are the main ones of each of the five aspects of studies. The other factors, marked from a to j, defined as side factors, appear on successive levels of essentiality.

Definition and structure of the human system

In every human system at least five most important elements (minimal number) can be distinguished: human population, human environment (abiotic, biotic, cultural), economic structure, social structure, ideological control subsystem. The following elements (factors) are among the most important which shape the biocultural state of a human population (Figure 1).

Factors forming biological structure of the human population operate in the environment of cultural agents and in this connection their role and significance must be estimated from the point of view of general characteristic of socio-cultural system.

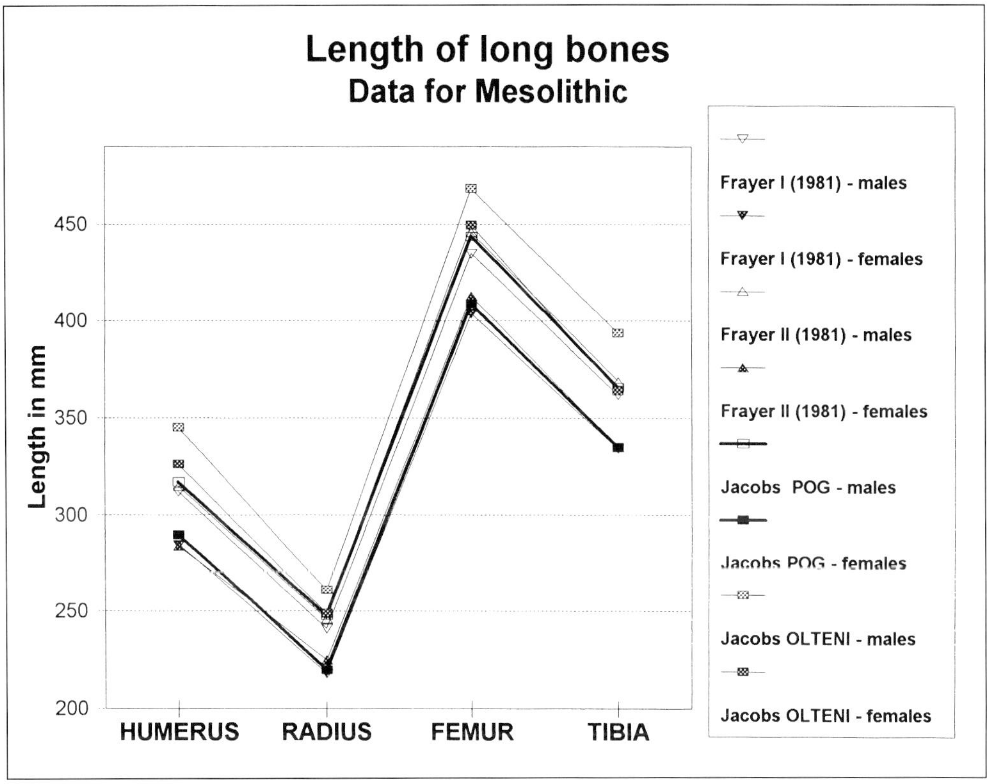

Figure 5 - Comparison of length of long bones: data for Mesolithic.

We proposed the definition of the biocultural system as a class of a certain number of important elements ensuring the realisation of human biological and cultural needs as well as duration of the system and reproduction. In the human system adaptive strategy is generated.

Adaptive strategy is understood as "a pattern of adaptation, adaptability or/and adjustment to the environment in biological and/or cultural way" (Wolański, 1990: 4).

The model of human system for Upper Palaeolithic and post-Palaeolithic populations

We advance a hypothesis that at successive stages of the biocultural development of human populations the above factors interacted on different levels of significance and that significant factors showed different essential structures (different adaptive strategies).

Thus, we assume that in the Upper Palaeolithic times (and in contemporary hunters-gatherers populations) relations between factors, such as the size of a population, the capacity of the ecological niche and the type of economic exploitation are most important (Figure 2), while in the transition to agriculture the relations among the main elements changed. Abiotic and biotic environment became the components of secondary importance. The relationship between a human population, a new type of the economic structure and social structure became the principal relation (Figure 3).

In the Upper Palaeolithic times the relationship between the elements, such as human population, human environment, economic structure, was the most important relation. In the Upper Palaeolithic

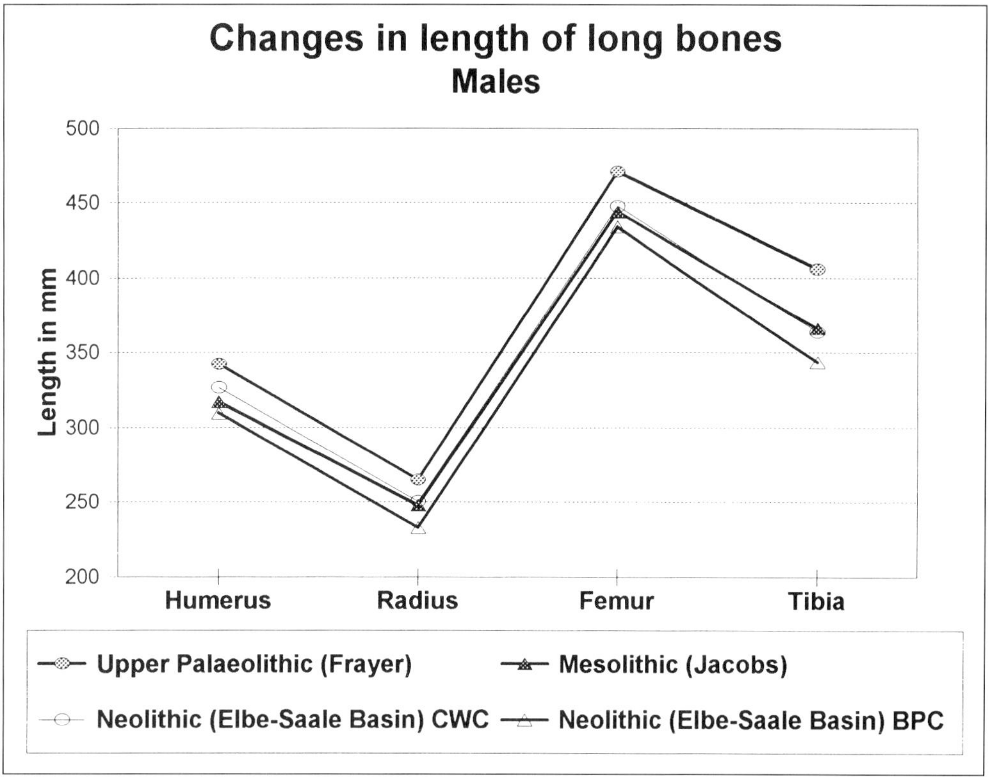

Figure 6 - Changes in length of long bones from Upper Palaeolithic to Neolithic: data for males populations.

population biological adaptation (relation between population - in biological sense - and environment, abiotic and biotic) which caused processes of rasogenesis was also important; in the transition to agriculture abiotic and biotic environment became the components of secondary importance. Relations among the population, the type of economic structure and the type of social structure became the principal relations. Relation change between the main elements in the transition to agriculture carried the substantial ecological and biological consequences.

In the transition to agriculture abiotic and biotic environment stopped to be the most essential criterion of natural selection. As a result of cultural information these populations (ideological control subsystem and social structure) could also alter the level of carrying capacity with regard to economic structure typical of them.

Hypothesis

Changes in the relations between the essential elements of the human system in the period between the Upper Palaeolithic and Neolithic times gave way to new possibilities of natural selection operation through differential fertility. A new lifestyle promoted a high general fertility rate, which was emphasised in models describing the transition to agriculture (Hassan, 1981; Ammerman, 1989). Also modern studies of the fertility rate in traditional agricultural and non-agricultural populations indicate that a higher fertility is primarily associated with the intensification of agriculture (Bentley et al., 1993a, 1993b).

Factors	Upper Palaeolithic K-selection	post-Palaeolithic r-selection
Climate	Fairly constant and/or predictable; more certain	Variable and/or unpredictable; uncertain
Mortality	More directed	Non-directed
Population size	Fairly constant in time, equilibrium; at or near carrying capacity	Variable in time, non-equilibrium; usually well below or above carrying capacity
Intragoups variability	Low	Medium to high
Intergroups variability	High	Medium to small
Opportunity for natural selection	By differential mortality	By differential fertility
Ecosensitivity	Low	High
Nutrition	Good quality protein	Poor quality protein
Leads to	Efficiency	Productivity
Selection favours	Slower development Delayed reproduction Larger body size Low maximal rate of increase	Early development Early reproduction Small body size High maximal rate of reproduction
Microevolution	Rasogenesis and/or origin of local races	Secular trends in morphological traits

Table 2 - General characteristics of the Upper Palaeolithic and post-Palaeolithic types of adaptive strategy. (Pianka's model for K-selection and r-selection. After Pianka 1970: 593 with modifications).

In the Neolithic adaptation strategy type there was a preference for high general fertility and natural selection could operate through differential fertility rate of earlier and later maturing individuals. In this situation the earlier maturing individuals (those who started to reproduce earlier) had a chance to produce a higher number of offspring than the individuals starting their

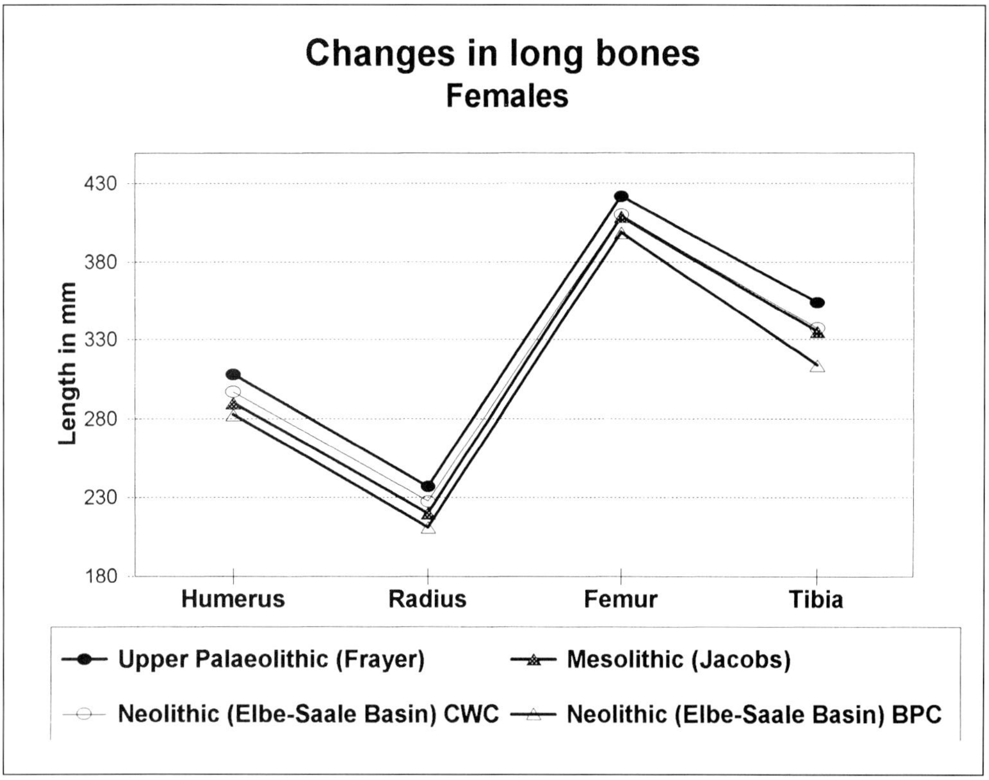

Figure 7 - Changes in length of long bones from Upper Palaeolithic to Neolithic: data for females populations.

reproduction later, while mortality rate was similar for both of these groups. This adaptation strategy feature is characteristic of the r-selection model elaborated by Pianka (1970). In this model natural selection favours early maturation of individuals, early reproduction, higher general fertility rate as well as a small body size (Table 2).

To explain the differences between the Upper Palaeolithic and Neolithic populations with regard to their physique (long bones, body size, proportion of long bones, sexual dimorphism, reduction in body shape) we advanced a hypothesis saying that in the Neolithic populations in which cultural factors shaped a new fertility type, natural selection could operate through differential fertility of earlier and later maturing individuals, favouring a new pattern of growth (early maturation).

Transition to agriculture: Biological consequences - changes in body size, long bones proportion and sexual dimorphism

To verify the above mentioned hypothesis we chose the data on the characteristics of long bones describing body size, body shape and long bones proportions. Our choice resulted from the fact that according to numerous works the study of morphology of postcranial skeleton can provide very important information for the examination of different biological and ecological traits of the human groups living in various social and economic systems (Frayer, 1980, 1981, 1984; Larsen, 1987; Henke, 1988; Vančata, 1988).

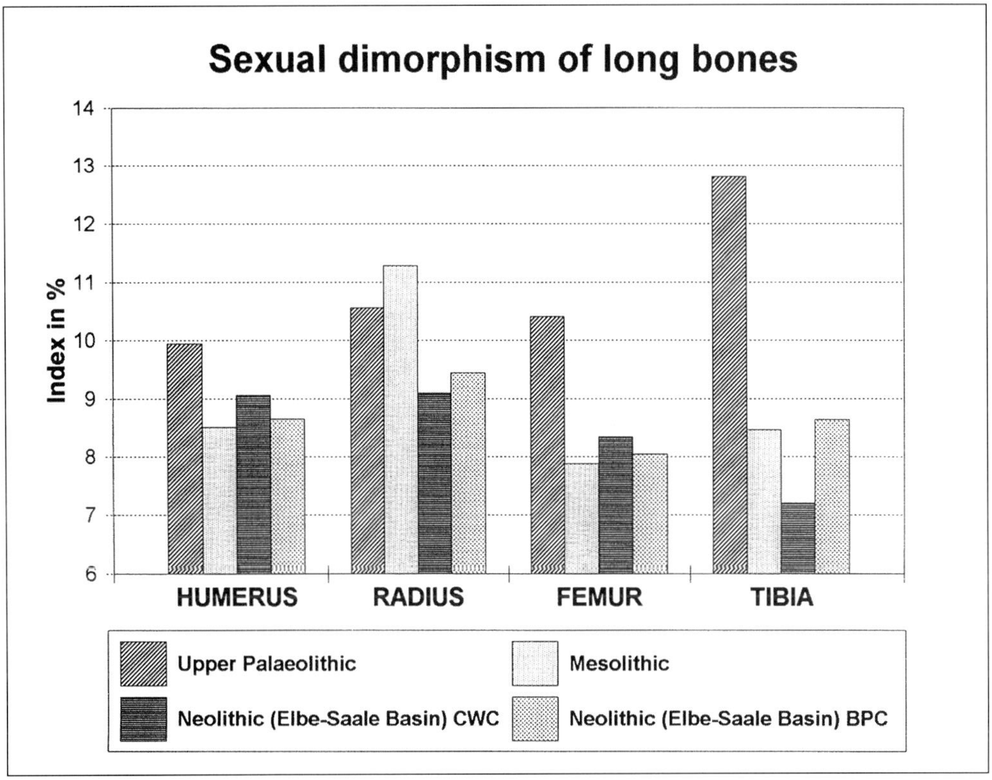

Figure 8 - Changes in sexual dimorphism of long bones from Upper Palaeolithic to Neolithic.

The size and shape of long bones as well as their sexual differences, robusticity, and temporal changes of these traits seem to be also the most important indicators of a basic adaptive and ecological pattern in human microevolution, as well as a good indicator of adaptive strategy in different cultural periods.

On the basis of the comparative analysis of variability of the long bones traits from the Upper Palaeolithic and Mesolithic archaeological sites (data collected by different authors) we selected data on the morphology of long bones in these populations and data for two Neolithic types of populations: typical of agriculture - the Linear Band Pottery Culture (BPC) from Elbe-Saale Basin and populations that are not "typical of agriculture" (mixed economy) - the Corded Ware Culture (CWC) from Elbe-Saale Basin.

The data under comparison concerning the morphology of long bones from European findings from the Upper Palaeolithic Age were published in Frayer's (1981), Jacobs's (1985) and Vančata's (1993 - individual data and arithmetical means were drawn from a data base provided by this author) works. The comparison of these three groups of data showed that Freyer's (1981) data can be located somewhat in the middle between the two others. That is why they were chosen for further comparison (Figure 4).

Data on the morphology of long bones from European findings dated back to the Mesolithic times were drawn from Frayer's (1981) and Jacobs's (1985) works. Four collective groups were

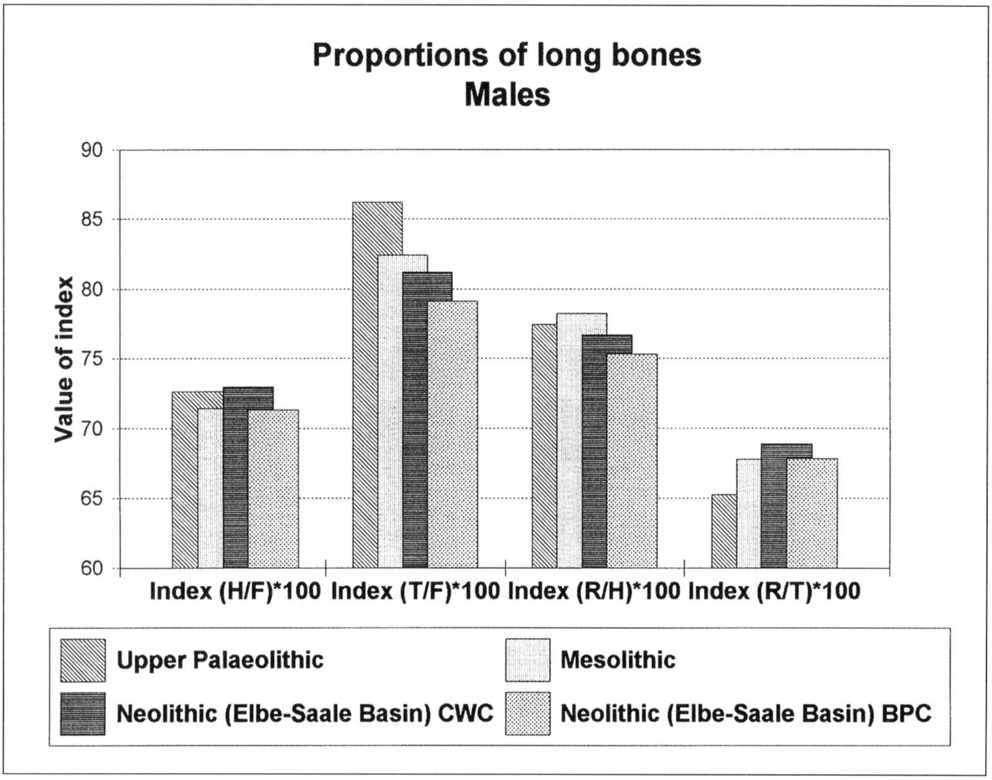

Figure 9 - Changes in proportions of long bones from Upper Palaeolithic to Neolithic: data for males populations.

compared (Figure 5). It turned out that the data collected by Jacobs (1985) compiling Post-glacial specimens (POG) occupy intermediate position in relation to other data, to the data from the Olteni burial ground in particular.

In the case of the Neolithic populations from Central Europe the data collected from various archaeological sites located in the Elbe-Saale Basin are documented most completely (numbers of skeletons, number of studied characteristics, archaeological description of findings) - (Bach et al., 1975; Bach, 1978).

The studied long bones' characteristics in the Upper Palaeolithic, Mesolithic and in Neolithic ages are shown in Figures 6-11.

On the basis of the data represented in these Figures we can draw the following conclusions:

(a) Three different morphological patterns of long bones - the Upper Palaeolithic, Mesolithic and Neolithic ones - had probably existed in the course of *Homo sapiens* microevolution in the Upper Palaeolithic/Neolithic transitional period (a similar conclusion was drawn by Vančata, 1990).

(b) Morphology of long bones of the Mesolithic populations of transitional character, and „non typical" agricultural the Corded Ware Cultures populations show some traits similar to the Mesolithic populations.

(c) Morphology of Neolithic populations from two types of archaeological cultures (LBPC and CWC) seems to be of a different character. The Corded Ware Cultures populations are in some respects similar to the Mesolithic populations, the Linear Band Pottery Culture populations show

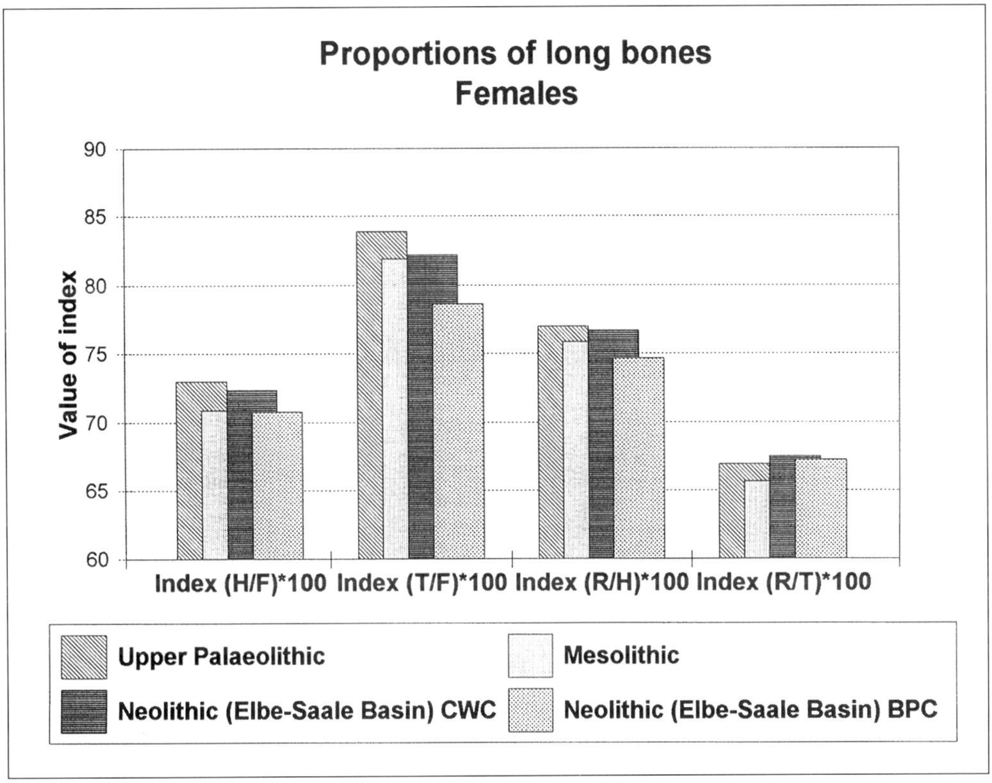

Figure 10 - Changes in proportions of long bones from Upper Palaeolithic to Neolithic: data for females populations.

an extremely different pattern morphology of long bones in relation to the Upper Palaeolithic populations.

According to the assumptions adopted in the proposed hypothesis (concerning the role of natural selection in the shaping of a new pattern of development in the Neolithic farmers' populations) the obtained results indicate that the following features can be observed in the Neolithic farmers' populations: a reduction of long bones growth, changes in size, reduction in shape, changes in proportions and changes in sexual dimorphism of long bones.

The diagram in Figure 12 shows an attempt to explain the processes causing the above mentioned changes. The diagram depicts a processual model of skeletal adaptation and changes of long bones in the transition to agriculture in which the role of a genetic process of adaptation and process of morphological adaptability changes. This model assumes that the principal factor shaping a new pattern of the morphological structure of long bones in the Neolithic farmers' populations was differential reproduction of individuals representing two different patterns of ontogenetic development: a fast pattern of growth (early maturation) and a slower pattern of growth.

According to the adopted hypothesis and the model constructed on its basis, the changes in the morphological structure of long bones are most evident in "typical" farmers' populations that is in the Linear Band Pottery Culture. These were the populations in which agriculture constituted the main area of economic activity.

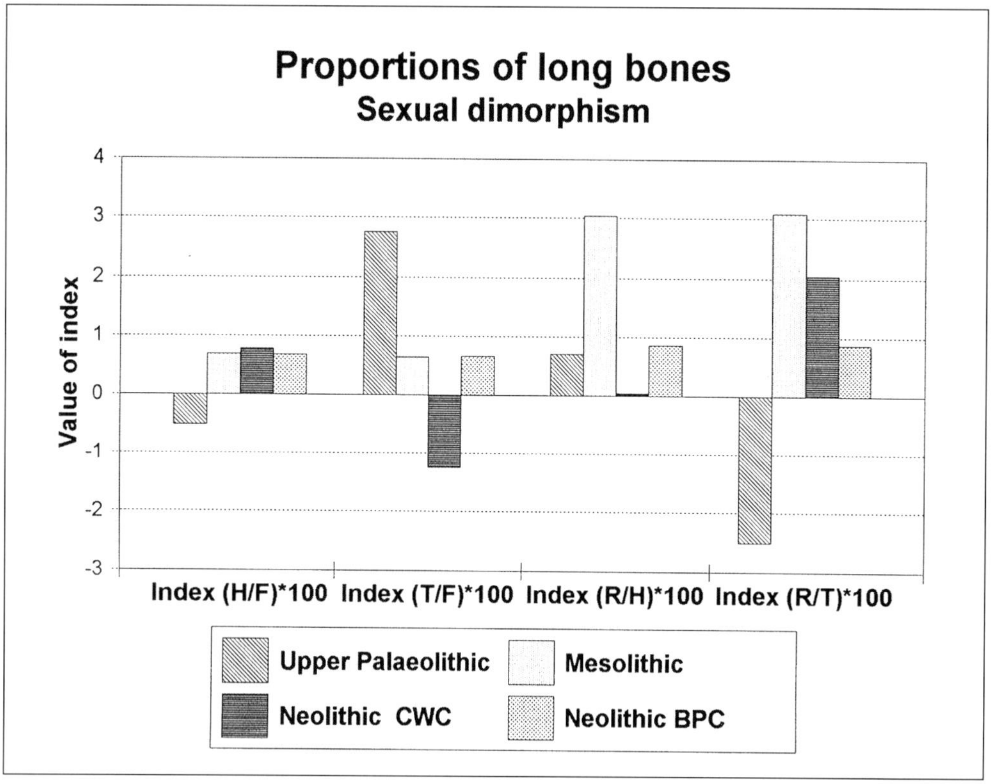

Figure 11 - Sexual dimorphism in proportions of long bones.

The Corded Ware Culture populations, whose cultural system (according to the data based on archaeological research) was much more differentiated than that of the Linear Band Pottery Culture populations, possessed a different adaptive model. In this model, we are ready to attribute decisive importance to the economic and technological factors, as far as cultural factors are concerned. These two factors, economic and technological, made it possible to differentiate economic activities to a certain degree and, at the same moment, they allowed for the exploitation of more varied parts of ecumena.

Also paleodemographic data (Piontek & Marciniak, 1990, 1992; Marciniak, 1992) indicate a different ecological and cultural situation of BPC and CWC populations. With the assumption of a high fertility rate in BPC populations (lack of biodemographic data determining the actual fertility rate in these populations) their adult life expectancy was lower than that observed in CWC populations. This could result in comparable biological dynamics of both types of the Neolithic populations. In BPC populations high fertility was probably related to a high children mortality and, as it is suggested by paleodemographic data, to a higher death rate among adults at the reproductive age. In CWC populations the overall population growth was determined by a lower fertility rate, but - even with a longer (than in BPC populations) interbirth interval - the number of individuals surviving the entire childbearing period, or its considerable part, was higher (longer life span of adult individuals).

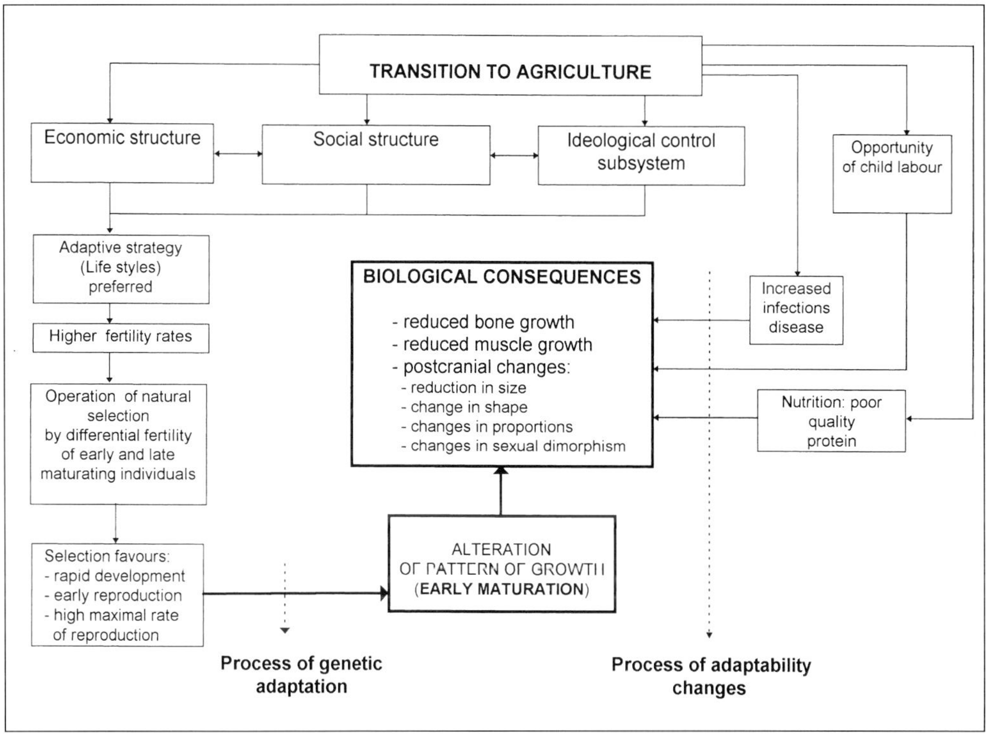

Figure 12 - Processual model of skeletal adaptation and changes of long bones in the transition to agriculture.

If the conclusions resulting from paleodemographic data are reliable and the reconstructions of the fertility structure in both types of the Neolithic populations are close to actual data (Piontek & Weber, 1990), one can state that the process of favouring by natural selection of a new pattern of development related to earlier commencement of reproduction by individuals, could concern only BPC population, since socio-economic factors characteristic of CWC societies did not have to, and probably were not, oriented toward a high fertility rate, which is confirmed by modern studies of traditional societies' fertility (Bentley et al., 1993).

In our opinion the differences as to the degree of body size reduction, long bones length and their proportions between BPC and CWC populations in relation to the Upper Palaeolithic populations may be explained with a different pattern of individual development characterising BPC populations.

If the advanced hypothesis is correct, one should expect the detection of further differences of biological features between BPC and CWC populations, since their adaptation strategies, although very similar in their general pattern, characterised various types of economic, social and ideological structure. This issue, in our opinion, requires further studies taking into consideration biological and cultural factors in the shaping of skeleton's morphological structure. Too often physical anthropologists treat skeletons from various Neolithic sites as a single set, while the transition to the Neolithic type of economy could progress according to various (at least two) variants of the forming of new adaptation strategy.

Conclusion

1. Neolithic type of adaptive strategy creates new possibilities for the operation of natural selection. The adaptive strategy of the Linear Band Pottery Cultures promoted high fertility rate, and natural selection operation by differential fertility of slower and faster development of the individuals.

2. Natural selection favours fast development, early maturation and early reproduction. Alteration of the pattern of growth in the Neolithic agricultural populations results in a new morphological pattern of long bones: very short length, different proportions (short radius and tibia), small sexual dimorphism.

3. The described reductions of the length of long bones in the Neolithic agricultural populations are caused by two types of biological factors: main factors - natural selection (process of genetic adaptation to a new lifestyle - new cultural adaptive strategy), and secondary factors - process of adaptability changes to different biocultural situations: disease, food, work demand, settlement pattern etc.

4. Changes in the morphology of long bones in the transition from Upper Palaeolithic to Neolithic times are correlated with the ecological and socio-economic changes, but from the biological point of view they are changes of microevolutional character. These changes of microevolutional character are not caused by natural environment factors but by a new structure of socio-economic structure which came to existence after the transition to agriculture.

References

AMMERMAN, A.J. (1975): Late Pleistocene population dynamics: An alternate view. *Human Ecology* 3, 219-233.

AMMERMAN, A.J. (1989): Population studies and the archaeologist. *Norway Archaeological Revue* 22, 65-87.

AMMERMAN, A. J. & CAVALLI-SFORZA, L. (1984): *The Neolithic transition and the genetics of populations in Europe*. Princeton, NJ.

ANTHONY, D. W. (1994): On subsistence change at the Mesolithic-Neolithic transition. *Current Anthropology* 35, 49-58.

BACH, A. (1978): *Neolithische Populationen im Mittelelbe-Saale-Gebiet. Zur Anthropologie des Neolithikums unter besonderer Berücksichtigung der Bandkeramiker.* Weimar.

BACH, A., BACH, H., GALL, W., FEUSTEL, R. & TEICHERT, M. (1975): Beiträge zur Kultur und Anthropologie der Mitteldeutschen Schnurkeramiker II. *Alt-Thüringen* 13, 43-107.

BENTLEY, R. G., GOLDBERG, T. & JASIENSKA, G. (1993a): The fertility of agricultural and non-agricultural traditional societies. *Population Studies* 47, 269-281.

BENTLEY, R. G., JASIENSKA, G. & GOLDBERG, T. (1993b): Is the fertility of agriculturalists higher than that of nonagriculturalists. *Current Anthropology* 34, 778-785.

BIELICKI, T. (1975): Natural selection and human morphology. In: F. M. Salzano (ed.), *The role of natural selection in human evolution*, 203-216. North-Holland Publishing Company.

BORGONINI-TARLI, S.M. & REPETTO, E. (1986): Skeletal indicators, subsistence patterns and activity regime in the Mesolithic sample from Grotta dell'Uzzo (Trapani, Sicily): A case study. *Human Evolution* 1, 331-352.

BRACE, C. L. & MAHLER, P. E. (1971): Post-Pleistocene changes in the human dentition. *American Journal of Physical Anthropology* 34, 191-204.

BUIKSTRA, J. E. & COOK, D. C. (1980): Palaeopathology: An American account. *Annual Review ofAnthropology* 9, 433-470.

CARLSON, D. S. & Van GERVEN, D. P. (1973): Masticatory function and post-Pleistocene evolution in Nubia. *American Journal of Physical Anthropology* 46, 495-506.

CARNEIRO, R. L. & HILSE, D. F. (1966): On determining the probable rate of population growth during the Neolithic. *American Anthropologist* 68, 177-180.

COHEN, M. N. & ARMELAGOS, G. J. (1984): *Paleopathology at the origins of agriculture*. Orlando, San Diego, New York, London.

COLLIER, S. (1993): Sexual dimorphism in relation to big-game hunting and economy in modern human populations. *American Journal of Physical Anthropology* 91, 485-504.

FOLEY, R. (1987): *Another unique species. Patterns in human evolutionary ecology*. Essex.

FORMICOLA, V. (1987): Neolithic transition and dental changes: the case of an Italian site. *Journal of Human Evolution* 16, 231-239.

FRAYER, D. W. (1980): Sexual dimorphism and cultural evolution in the Late Pleistocene and Holocene of Europe. *Journal of Human Evolution* 9, 399-415.

FRAYER, D. W. (1981): Body size, weapon use, and natural selection in the European Upper Paleolithic and Mesolithic. *American Anthropologist* 83, 57-73.

FRAYER, D. W. (1984): Biological and cultural change in the European Late Pleistocene and Early Holocene. In: F. H. Smith & F. Spencer (eds.), *The origins of modern humans: A world survey of the fossil evidence*, 211-250. New York.

GOODMAN, A. H., MARTIN, D. L. & ARMELAGOS, G. J. (1984): Indications of stress from bone and teeth. In: M. N. Cohen & G. J. Armelagos (eds.), *Paleopathology at the origins of agriculture*, 13-49. San Diego, New York, London.

HASSAN, F. A. (1981): *Demographic archaeology*. New York, London.

HENKE, W. (1988): Die Menschen der letzten Eiszeit. Zur Frage der Differenzierung der endpleistozaenen Hominiden Europas. *Anthropologische Anzeiger* 46, 289-316.

HUSS-ASHMORE, R., GOODMAN, A.H. & ARMELAGOS, G.J. (1982): Nutritional inference from paleopathology. *Advances in Archaeological Method and Theory* 5, 395-474.

JACKES, M. K. (1988): Demographic change at the Mesolithic-Neolithic transition: Evidence from Portugal. *Rivista di Antropologia* 66, 141-158.

JACOBS, K. (1985): Evolution in the postcranial skeleton of Late Glacial and early Postglacial European hominids. *Zeitschrift für Morphologie und Anthropologie* 75, 307-326.

JACOBS, K. (1993): Human postcranial variation in the Ukrainian Mesolithic-Neolithic. *Current Anthropology* 34, 311-324.

JANKAUSKAS, R. (1994): Lithuanian Mesolithic and Neolithic graves: Data on the transition from a foraging to food-producing economy. *Anthropologie* (Brno) 32, 165-168.

KEY, P. (1980): Evolutionary trends in femoral sexual dimorphism from the Mesolithic to the late Middle Ages in Europe. *American Journal of Physical Anthropology* 52, 244.

KOZLOWSKI, J. K. (ed.) (1973): *The Mesolithic in Europe*. Warsaw.

LARSEN, C. S. (1987): Bioarchaeological interpretations of subsistence economy and behavior from human skeletal remains. *Advances in Archaeological Method and Theory* 10, 339-445.

LARSEN, C. S. (1995): Biological changes in human populations with agriculture. *Annual Review of Anthropology* 24, 185-213.

MARCINIAK, A. (1992): Cultural adaptive strategies in the Neolithic in Central Europe within the context of palaeodemographic studies. *Journal of European Archaeology* 1, 141-151.

NOWAK, L. (1977): *Wstep do idealizacyjnej teorii nauki*. Warszawa.

PIANKA, E. R. (1970): On r- and K-Selection. *American Naturalist* 104, 592-597.

PIANKA, E. R. (1974): *Evolutionary ecology*. New York.

PIONTEK, J. (1989): Biological consequences of the "Neolithic Revolution": The case of the Middle European populations. In: I. Hershkovitz (ed.), *People and culture in change*, 367-374. Oxford.

PIONTEK, J. & MARCINIAK, A. (1990): *Biocultural perspectives on ecology of the prehistoric populations from Central Europe*. Warszawa .

PIONTEK, J. & MARCINIAK, A. (1992): Anthropological structure and culture adaptive strategies of the Neolithic populations in Central Europe. *International Journal of Anthropology* 7, 71-86.

PIONTEK, J. & WEBER, A. (1985): Studies on biocultural evolution - a multiaspectual model. *Collegium Antropologicum* 9, 215-222.

PIONTEK, J. & WEBER, A. (1990): Controversy on paleodemography. *International Journal of Anthropology* 5, 71-83.

RUFF, C. B., HAYES, W. C. & LARSEN, C. S. (1983): Changes in femoral structure with the transition to agriculture in the Georgia coast. *American Journal of Physical Anthropology* 60, 247-248.

RUFF, C. B., LARSEN, C. S. & HAYES, W. C. (1984): Structural changes in the femur with the transition to agriculture on the Georgia Coast. *American Journal of Physical Anthropology* 64, 125-136.

TRINKAUS, E. (1983): Human biocultural change in the Upper Pleistocene. In: E. Trinkaus (ed.), *The Mousterian legacy,* 165-200. Oxford.

TRINKAUS, E. & TOMPKINS, R. L. (1990): The Neanderthal life cycle: The possibility, probability, and perceptability of contrasts with recent humans. In: C. J. de Rousseau (ed.), *Primate life history and evolution*, 153-180. New York.

VANČATA, V. (1988): Ecological aspects of skeletal sexual dimorphism in microevolution of *Homo sapiens. Anthropologie* (Brno) 26, 83-92.

VANČATA, V. (1990): Microevolution of femur and tibia in the Upper Palaeolithic-Neolithic period. Paper presented at *IIIrd Symposium on Upper Palaeolithic, Mesolithic and Neolithic Populations of Europe and the Mediterranean Basin, Budapest, September 1990.*

VANČATA, V. (1993): Evolution of hominid femur and tibia: A morphometric approach to the evolutionary research in anthropology. *Human Evolution* 8, 65-79.

WIERCIŃSKI, A. (1978): The meaning and scope of anthropology. *Collegium Antropologicum* 2, 10-16.

WIERCIŃSKI, A. (1985): Bioanthropological consequences of the "neolithic revolution". *Studies in Physical Anthropology* 8, 105-110.

WOLAŃSKI, N. (1988): Ecological aspects of the growth and development of man. *Collegium Antropologicum* 12, 7-21.

WOLAŃSKI, N. (1990): *Glossary of terms and annotated bibliography for human ecology*. Warsaw.

WOLAŃSKI, N. (1991): Human ecology, medicine and anthropology. *Collegium Antropologicum* 15, 7-26.

WOLAŃSKI, N., CROGNIER, E. & BERNIS, C. (1990): International program in human ecology: Biosocial status of human populations as an indicator of environmental changes. *Collegium Antropologicum* 14, 173-182.

Competition Ecology: Pleistocene Hominids and the Carnivore Community

Mark Leney & Robert A. Foley

Abstract

Hominid evolutionary history is characterised by a series of shifts in ecology, away from an ancestral (primate) mode, towards a range of strategies that, in combination, remain an unique hominid adaptation. However individually elements of these strategies show considerable overlap with carnivore ecologies. Carnivore communities are highly structured; the fine scale adaptation of one species is shaped by coevolutionary relationships with the sympatric populations of competitor species. This suggests that the nature and timing of variation in hominid ecologies, insofar as they represent incursions into niches potentially occupied by carnivores, should be reflected by a concomitant suite of micro-adaptations in the carnivore community. Case studies presented here indicate that it may be possible to measure this. Preliminary results suggest that populations of Upper Pleistocene hyaenas (Crocuta) show significantly different microadaptations of the postcanine tooth row, possibly related to the density of sympatric hominid. Additionally, late Pleistocene bears on the European mainland show a pattern of gracilisation of limb elements correlated with the pattern of hominid occupation; this includes rare transitional forms contemporary with the earliest penetration of Upper Palaeolithic culture into the Sierras of Northern Spain.

Towards hominid palaeoecology

Foley (1987) proposed several categorisations of methods and processes that inform us about the biology of hominids. One was primarily concerned with the disciplinary boundaries and divided the field into studies of the fossil record, studies of comparative biology (qualified with the term primates appended in brackets), studies of modern human populations, and studies of the archaeological record. He then suggested that Physical Anthropology and Palaeolithic Archaeology are sub-disciplines or emergent study areas combining two or more of these components. The point here is that these are both concerned immediately with hominids: tools used by hominids and the bones of hominids and primates as models for hominids.

In a more conceptual framework, Foley later suggested that overarching biological theory coupled with middle range theory (*sensu* Binford, 1977, 1981) could provide answers to our questions about hominids in four rather different modes:

1) Firstly, inference: The use of chains of logical propositions to build theory which is then tested against independent components of the evidence. For example, one might infer that the percentage of a species in an archaeological accumulation was proportional to their importance in hominid diets. Actualistic studies can test this and might discover that it led to an overrepresentation of large species thereby leading to a modification of the original inference.

2) Secondly, there are isolating techniques where specific, otherwise minor, aspects of the total evidence can been shown to have particular and possibly categorical relevance for key questions: Did Late Middle Palaeolithic people at Grotta Guattari, in Latium, North Italy, hunt prime adult prey? Age profiles of the prey accumulations isolate a definite signal suggesting that they do (Stiner, 1994).

3) Thirdly, probabilistic techniques have proved a useful third strand where multiple processes contribute to a palimpsest record; or where specific processes have individually unpredictable results but are stochastic, such that well sampled pattern can distinguish underlying processes.

4) Finally, the fourth strand is one of comparative biology, where inter-specific patterns are extrapolated to apply to hominids. Some have criticised the casual use of such approaches, particularly loosely applied correlational techniques (Smith, 1996). However, we have argued that such concerns are not so relevant when the primary purpose of such techniques is to generate useful hypotheses which are later to be tested (Leney & Foley, 1996).

The main thrust of the rest of this paper is broadly concerned with this comparative technique, with a conceptual twist. The primary focus of the analysis is upon carnivores. The comparative framework for hominids is further divided (again by Foley, 1987), into a number of domains of ecological overlap; terrestrial primate, savannah community, large mammal, tropical environment or coloniser species for example. The particular characteristics of other members of these sets are reviewed in the light of their importance for hominid biology.

Once more it is possible to see that, in the first three modes, hominids are investigated by the immediate patterns *hominids* themselves left in the physical record; they are hominid-centred, naturally enough, but this single-minded approach does not properly acknowledge the interaction of the hominids and the environment. There is nothing wrong with anthropocentrism when studying humans. But, of all the approaches discussed, comparative biology, and in particular comparative ecology, is the area which is least concerned with human uniqueness and more focused on the place of humans within their ecological community. Thus the community should be the focus of analysis here, and not the hominids. Concentration on hominids to the exclusion of effects between and within other species is likely to overemphasise hominid ecological interactions, resulting in a limited perspective upon hominid ecology (Foley, 1987: 255). The work presented here begins to redress this. It seeks to examine quantitative factors outside the immediate record of the physical remains of hominids and their artefacts. However, it is still a potential source of information about hominid behaviour. While this information brings its own epistomological problems, it is independent of many of the biases that may be distorting our ecological interpretation of the hominid record.

Comparative palaeoecology: The palaeo-community

Where early hominids are considered as members of the wider community evolving in the African savannah biome (broadly construed), the issue of competition and community ecology is touched on in two main themes. On the one hand, character displacement is seen in the increasingly different adaptations of the robust and gracile lineages (Wolpoff, 1988), and on the other, ecological succession/replacement is put forward as a model explaining the demise of both primate competitor species, such as *Theropithecus oswaldi* (Shipman et al., 1981, 1982; Shipman, 1982), and for explaining changes in species diversity and abundance in the large carnivore guild. There are variations on these themes: Foley's own (1984), suggests that the establishment of new environmental conditions is followed initially by opportunist colonisation at low diversity. This is succeeded by a high diversity specialist assembledge as niches are partitioned, finally giving way to a third stage, where increasingly specialised lineages become vulnerable to competitive exclusions or chance extinction, thus leading to the establishment of new generalists and sub-maximum diversity. In this case, 'third-stage generalist' hominid species (i.e. *Homo erectus* sensu lato) can be seen as the cause of competitive exclusion in both hominid and carnivore guilds. Alternatively, Walker (1984) suggests more specifically that the shift in ecology of the hominids was causally responsible for the break-up of the earlier carnivore/prey community leading to the extinction of many species.

Putting aside the issue of whether the meat was alive or dead when the hominds obtained it, there is no dispute that the rise of flesh-eating played an important role in hominid evolution. The comparison of hominids to carnivores can thus be made on several levels. At the most abstract, analytical techniques can be tested on the carnivore fossil record merely to ascertain if they provide sensible and justified answers in a lineage that is phylogenetically independent of hominids but still shares certain ecological traits. Less tangental is comparison at the level of process. Similar evolutionary and ecological patterns might be expected in carnivore and hominid lineages. For processes shaping speciation or biogeography or thermoregulation or day range, carnivores provide valid null models with which to investigate hominids.

However, there is a third level at which a hominid/carnivore comparison can be construed. The inferences that we can draw from the parallels between species are not always theoretical; there are also the direct evolutionary and ecological interactions amongst the species compared. Not only has the biotic environment determined the evolutionary trajectory of hominids, but hominids themselves must have caused reciprocal evolution in their biotic environment. Hominids have removed energy from the predator/scavenger system. This has to be reflected in the ecology of that system. The challenge is to measure such changes in palaeo-communities.

Hominids and carnivores: From competition to adaptation

The interaction between hominids and carnivores is not a new topic by any means. Carnivores are seen as hominid competitors or predators and thus are considered to have shaped hominid evolutionary ecology. They are potential causes of confusion when they contribute to patterns of bone destruction, selective transport and accumulation (e.g. Blumenschine, 1986, 1987, 1988); they are models against which we can assess the activities of hominids (e.g. Stiner, 1991); and in same vein, (as ecological parallels) the richness and diversity of a carnivore community circumscribe the niche(s) that hominids occupied (Gamble, 1986, pp.305-342). As mentioned previously, Foley had touched on the turnover of carnivore species and hominid ecology, echoing the suggestions of Walker (1984), several chapters in Martin & Wright (1967), and Mosiman & Martin (1975), amongst others, although causation continues to be a point of debate. Finally, in the Upper Pleistocene of Europe, both Kurtén (1965, 1967, 1968; Kurtén & Anderson, 1980) and Turner (1981, 1986, 1992a) have dealt with the most recent spate of extinctions as well as looking at other aspects of the carnivore/hominid interaction that will be addressed below.

In competition with the hominids of the European Upper Pleistocene, there were many medium and large-bodied carnivore species (Kurtén, 1968; Stuart, 1982; Reynolds, 1911). Like the hominids, these species had enormous geographical distributions (MacDonald, 1984), they were intelligent, often highly social and, most importantly, flesh formed an important component of their diets. Yet, despite the obvious parallels, the study of Upper Pleistocene ecosystems has tended to neglect the interactions between carnivores and hominids. Most of these studies to-date have dealt with the question of extinction. The usual paradigm involves debate concerning a hypothesis that humans were or were not responsible for the extinction of one or more faunal elements either as a predator or a competitor. Whilst extinction may be a striking feature of the fossil record - for European carnivores, see Turner (1992b) - it is by its nature a stochastic process. Furthermore, it is a rare event and, taken together, these two characteristics make it hard to attribute firm causes to such events. A more subtle indicator of the inroads of one species into the niche of another may be the change population density of the various species involved in a competitive interaction. It is, however, extremely difficult to arrive at accurate estimates of this type from the fossil record. On the other hand, if the biotic interaction was responsible for modifying the selective regime under which the competing species lived, small changes in adaptation might be apparent from relatively sparse fossil evidence. This reflects the fact that the adaptation of the population as a whole is a property of individual fossils in a way that numerical information about population density is not. The strongest manifestations of such adaptive shifts should be found in characters directly related to the competitve aquisition of the resources in question.

From competition to variation

Variation *within* carnivore taxa had been considered by workers primarily concerned with systematics (Leakey, 1976; Howell & Petter, 1976; Hendey, 1974). Whilst there was often the secondary goal of examining the diversity of a palaeo-community, none of these earlier researchers (and few of the hominid ecologists) had really examined variation in carnivores at levels beyond presence or absence. Nor had many considered that microevolutionary changes in the morphology of *carnivores* might indicate changes in their ecology in the way that say, gradulaistic increases in the levels of bunodonty and enamel thickness, molarisation of the premolars and the development of megadontia within hominid lineages are taken to indicate quite subtle shifts in ecology. Such attempts to examine variation within carnivore lineages were largely concerned with body-size, its putative relationship to environmental temperature, *sensu* Bergmann's rule (Klein, 1986; Klein & Scott, 1989) and prey size (Anyonge, 1993). However, these studies had raised more complex questions when different functional complexes had generated different predictions of body-size for fossil populations leading to the reasonable hypothesis that marked discrepancies in the allometries of different functional units might reflect independent adaptation of those units. Kurtén (1955, 1967) and Turner (1981) had both made the further inference that it might be possible to study these independent components of shape variation and to draw conclusions about micro-adaptation from them, although neither had been able to implement a sufficiently general methodology to do this.

From variation to adaptation

The shape variation between species is an important paradigm for characterising the differences in adaptation between them. In carnivores, the morphology of the tooth row and the morphology of the locomotive elements are important determinants of resource partioning, defining, as they do, the fundamental adaptations of carnivores: to catch, to kill and to consume. This leads to a structuring of the carnivore guild with respect to dentition and locomotion (VanValkenburgh, 1988, 1989). The relatively consistent nature of this structure, at least partially independent of phylogeny, implies that carnivores experience strong selection with respect to niche/resource partitioning. This, in turn, suggests that competition with other carnivores plays an important part in the evolution of micro-morphology. Changes in the ecology of one species in the 'guild' should therefore be reflected in the morphological adaptations of the other species, a phenomenon known as character displacement. The empirical evidence for such an effect in extant communities is hotly debated; (see for example, Lewin, 1983, and Diamond, 1987). Critics have pointed to the statistical likelihood of species appearing to be evenly spaced along a character continuum, even without a structuring effect such as competition. This provoked an animated debate about formulating null hypotheses to test if inter-specific competition had indeed been important in structuring communities or guilds. In many ways, it is probably better to look for individual examples of such systems working dynamically, in order to demonstrate their validity, rather than relying on probablistic models. With respect to the structuring of carnivore communities, there is a certain amount of evidence along these lines.

Carnivores, competition and adaptation: The argument from extant communities

Specifically, Tamar Dayan and her co-workers (Dayan et al., 1992, and references therein) have shown that carnassial size varied between populations of carnivores with respect to the species with which they were sympatric. The absence of a larger competitor, the Wolf (*Canis lupus*), in Egypt was associated with evolution of a larger body size in the Golden Jackal (*Canis aureus*), than conspecific populations in the Middle East where historically *lupus* had been present. This type of study grew out of investigations of deviations in size due to character displacement (Hutchinson, 1959). Only later did it lead to the conclusion that it was actually the size of specific functional elements (canines or carnassials in Dayan et al., 1992) that changed the most between the populations. Therefore, when taking competitive effects into account, one must look beyond size changes. VanValkenburgh & Wayne (1994) have shown that competing sympatric carnivores may fail to manifest competitive divergence in body sizes, but that they do show *shape* divergence in sympatry.

Some of the characterisation of the interactions between carnivores (especially bears) and Upper Pleistocene hominids have been quite lurid, involving ritual cults and deliberate trapping of large carnivores (see Allsworth-Jones, 1990, pp.197-200, for a critical discussion). The earlier interpretations of the interactions between *Homo* and *Ursus* are also covered quite extensively in Erdbrink (1953, pp.472-526) but most of these are extremely fanciful and they are not reiterated here. This is not to say that we reject the notion of direct physical confrontation between carnivores and hominids. It remains likely that, of the many biotic interactions that characterised the Pleistocene ecosystems, the interaction between carnivores and humans were particularly intense, in an ecological, if not dramatic, sense. Carnivores of widely divergent body sizes and habits compete for resources at carcasses. Murie (1981) records that wolves, bears, foxes and wolverines (*Gulo gulo*) have been observed in competitive interactions over the same carcasses. Coyotes (*Canis latrans*) are also known to scavenge from unattended carcasses that also provide food to all four of the other species. Within similar ecosystems, medium and large cats (i.e. Lynx, *Felis lynx*, and Mountain Lion, *Felis concolor*), also suffer loss of prey to other carnivores as well as scavenging from the kills of others. Without exception, large and medium-sized carnivores share a considerable proportion of their resources and both inflict and endure many forms of ecological interference. Kruuk (1972, pp. 128-147) gives extensive details of the competitive interactions of Hyaenas and a wide range of other carnivorous species. Murie (1981) also details competitive interactions of *Ursus arctos*, where bears exclude other species from resources both actively (chasing other species), and passively (other species avoiding areas known to be occupied by bears). In addition, he records that species of widely different sizes such as Fox (*Vulpes vulpes* L.; *circa* 7 kg; Sheldon, 1992) and Brown Bear (*Ursus arctos* ssp; *circa* 300 kg; Bunell, 1984) compete (in this case non-confrontationally) by reciprocal raiding of food caches. It is highly probable that Palaeolithic peoples interacted in all these ways with the sympatric carnivores, both through direct confrontation and through indirect interference. The strength and nature of such interactions should reflect the nature of the strategies adopted by those people. Thus more intense competition with one species, let us say Wolves, would, all other things being equal, lead to the most marked micro-adaptive changes in the Wolf lineage and more minor or negligible changes in other species.

Diversity within the species: The species is not the unit of adaptation

Under the punctuationist orthodoxy, we assume that the majority of adaptive change takes place relatively rapidly and that 'palaeo-taxa' represent more or less uniform slices of adaptation. This is not an adequate description of the ecology of carnivores, as examination of the modern material shows for two related reasons. The carnivore 'species' is often rather difficult to define, and even then, it usually contains considerable ecological diversity. When one examines diversity in modern carnivores, the application of the species construct has been confused and fraught with contradiction (Roy et al., 1994). For example, amongst the wolf-like canids there are a group of 'species' including the Dingo, Simien jackal, Red wolf, Coyote and Gray Wolf which all show some evidence of actual or potential gene flow between them (Wayne, 1993) while on the whole they have clearly marked morphologies (Gotteli et al., 1994). On the other hand, the morphology of the Gray Wolf is by no means uniform over its whole range (Jolicoeur, 1959, 1975), there are several well marked morphologies, although, by ancestry, these do not form monophyletic clades (Wayne, 1993). The same can be said of *Ursus arctos* and Golden Jackals (*Canis aureus*), both of which have environmental tolerances and biogeographic ranges similar to those of Paleolithic hominids. Thus most species constructs miss a whole level of biological complexity in carnivores. Carnivore populations adapt to their local environments but, perhaps because heterogeneity in the environment does not generally constrain individual dispersal, thus minimizing the chances of isolation, carnivores do not readily speciate. Banfield (1953) records that a Wolf was observed as an adult over 1,000 km from the natal den in which it had been tagged, whilst a female Polar Bear (*Ursus maritimus*) was observed to have used maternity dens in both Greenland and Alaska (Durner & Amstrup, 1995), travelling a total of 5,256 km in the first year of telemetry. Conventional speciation biology suggests that outbreeding in this context can be maladaptive when it takes place with individuals from populations adapted to other environments (Rao & deBach, 1969), yet carnivores clearly do develop local adaptations despite this gene flow. More recent formulations of speciation mechanisms (e.g. Patterson, 1985) have criticised such relational concepts and have looked instead for a more circumscribed definitions of species and their immediate causation. Whatever the mechanism, the process of speciation seems to happen at a different scale in carnivores (and possibly hominids), so that the discrepancy between adaptation and speciation is more readily apparent than in most lineages.

What we now perceive as a single extant species, such as the Brown Bear (*Ursus arctos* sensu lato), contains considerable structured variation attributable to the dynamic between local selection and gene flow. It is proposed that the morphological differences between such very closely related groups can be attributed almost exclusively to adaptation. The corollary of this assumption is that such characters will not reliably indicate phylogeny; suites of morphological 'characters' can be the product of single adaptive complexes so that fairly exact morphological identity can, and will, arise as a result of parallel selection. The dynamic nature of such adaptation removes much of the permanence of contingent events. Therefore, the most likely suites of adaptation will occur again and again, independently. In the wolf-like canids, this is demonstrated very clearly by the Simien 'jackal' (*Canis simensis*). Although recently shown to be very closely related to the other wolf-like canid species, this animal is morphologically jackal-like (Gotteli et al., 1994). Its vulgar name, Simien *jackal* or even sometimes Simien *fox* is testament to this parallelism.[1] A similar effect can be seen in the cranial morphology of *Canis lupus papilles*, the Arabian Wolf and that of *Canis latrans*, the Coyote. These two widely separated populations both show the same combination of small body size, dental reduction and long thin snout associated with a diet of small prey, vegetable matter and minor carrion. This can be seen as a sort of multiregional adaptation in Wolves. It appears that a correspondence between finely resolved morphological distinctions and differences in the selective environment provides considerable scope for investigating micro-evolutionary processes in these extremely wide ranging species.

Micro-adaptation in Pleistocene populations?

Over the relatively long time period involved in the incursion of Pleistocene hominids into the carnivore guild, some carnivore lineages should show a coevolutionary trajectory, mirroring the nature and extent of the local changes in hominid ecology. Any such coevolutionary pattern amongst the carnivores will be overlaid with other patterns, due to changes in the nature of the physical environment and the other biotic interactions that mediated selection in Pleistocene populations. Some of these patterns can legitimately be seen as independent of hominid ecology, whilst others may covary with the impact of the hominids. Clearly, causality resides predominantly in the changes in the 'third factors', the glacial cycle springs to mind, but it remains true that the effect of such a third factor on either the hominids or carnivores must lead to changes in the interaction between them. To such as extent, it is meaningless to pursue the distinction between immediate and coincident causation too far. Advances in the technology utilised by hominids take place in different places at different times and so, to a certain extent, it may be possible to test the effect of behavioural advances by studying the onset of the effect at different times and places (and thus, presumably, under different climates).

Characterising micro-adaptation

The carnivore fossil sample

The material used was taken from the available collections in the Natural History Museum in London, Naturhistoriches Museum and Paläontoloigisches Institut der Universität, Wien; Zientzi ElkateaAranzadi, Donostia (San Sebastián); Muséo Nacional de Ciéncias Naturales, Madrid, and Muséum National d'Histoire Naturelle, Paris. My choice of institution was guided by the implied size of the collection and my wish to cover a reasonable geographic spread, and thus should not be taken as a complete study or an evenly sampled selection. The lack of in-depth cataloguing of non-hominid material meant that much of the material located was effectively without provenance. Nevertheless the material was mostly of last glacial or last interglacial age. Some material had better stratigraphic definition than this, but most did not. Poorly dated material was incorporated in the study to provide a fuller characterisation of the structure of variation in the Pleistocene sample. This ensured that the multivariate analyses were conducted in an empirically relevant set of spaces, i.e. those defined by the structure f variation in the wider sample. The material tended to fall into four distinct biogeographical regions, Britain, Southwest France, Cantabrian Spain and Eastern Europe. This last sample is mainly, but not exclusively, drawn from high altitude sites in Austria; it also includes a certain amount of material from mountain sites in the Balkans, justified on the grounds that the mountain ranges involved are broadly contiguous. These natural divisions are allowed to guide some of the analyses but the initial multivariate steps were not influenced by the potentially arbitrary nature of such choices. Although there are data for various elements, the shape of the tooth row (i.e. covariances between dental metrics) always seemed the best potential marker of resource mediated competition for reasons outlined above. In addition, I (Leney, 1996) wanted to examine limb elements to get some independent index of overall size and to investigate locomotory adaptation. We have used metapodials for this but there is also accessory data from the long bones.

Variables and data

We took maximum length and breadth dimensions of post-canine teeth in associated dentitions, i.e. those with the teeth more or less *in situ* in the maxilla or mandible. This was because we were principally concerned with the covariance of the measurements between the teeth. For metapodials, we decided that the mechanical properties of the metapodials as a whole were well enough represented by two measures of length, and three transverse and three dorso-ventral measures of width. These multivariate sets were partly augmented by using a number of techniques to allow the incorporation of cases with some missing data, effectively preventing the loss of cases with partial information. These methods are discussed at length in Leney (1996) and material that is currently in preparation. Where these methods are germane to the analysis they are mentioned below, but detailed treatment is beyond the scope of this paper.

The analyses

The main intention of this rather extensive set of analyses was to quantify the nature of the variation within the two lineages selected for the initial study; *Ursus arctos sensu lato* (Brown, Grizzly and Cave Bears) and *Crocuta crocuta* (Spotted and Cave Hyaenas). We did not want to allow preconceptions about the differences between various subsets of the sample to influence the way in which the variation was initially quantified. Although we were also keen to maximise the information available from the fossil material, the use of single metric variables from fragmentary remains was clearly inadequate. To this end, we only used fossil elements where a number of measurements could be made on a single individual, and subjected these data matrices to Principal Components Analyses or PCA (Pearson, 1901). We do not propose to go into this in great detail here; essentially this results in a projection and ordination of the data which seems inherently sensible. All variables and cases are initially given equal weight in the analysis, autocorrelation amongst the variables is taken into account and is in fact fundamental to the technique, whilst unusual cases are easily detected as multivariate outliers and their effect on the analysis can be determined by iteratively removing the cases in question. Although 'multivariate normal' data are of considerable use when interpreting the results, they are not essential. PCA is quite robust in this sense and can produce perfectly useful descriptions of multivariate data where it is non-normal.

Principal components

For those unfamiliar with the technique, it is best illustrated by using an initial example with just two morphometric variables. We might call them *length* and *width*. Where *length* and *width* are plotted for a sample, there is often a strong correlation between them. The bivariate data usually form something like an ellipse when projected into the arbitrary space defined by the orthogonal (right-angled) opposition of the two variables. Interpretation is relatively simple, usually the long axis of this ellipse represents the common size component of the variation in the two measurement dimensions, and the secondary axis represents the residual deviation from this underlying relationship. Where length in on the x axis, cases above the long axis of the ellipse described by the scatter of the data are '*fatter*' (persuing the length *vs* width example) in proportion to their distance from the line. What has happened here is, geometrically, merely a rotation. We have conceptually rotated the two dimensional measurement axes, to align them with the empirically determined axes inherent to the data, and then used these new axes, the 'Principal Components' (PCs), to interpret the variation in the sample. These (two) components remain orthogonal; that is to say that, in the geometry of the measurement space, they are at right angles, and furthermore that they delineate statistically independent fractions of the sample variance. It is easy to visualise an extension of this into three dimensional figure space, if a third variable, say '*depth*', were to be added. Although harder to grasp, PCA does just this for any number of measurement dimensions one might choose to use. In addition, as more variables are used, PCA helps to determine which component dimensions are important, interpretable or significant and those which are negligible, due to noise or error, by determining how much of the total sample variance each PC explains. Although this is a gross simplification of the ways in which PCs can and should be interpreted, it lies at the heart of most of the more sophisticated approaches, but this is as far as we will go here. There is an extensive body of published work discussing the interpretation of PCA (see Klingenberg, 1996 for a recent example).

Micro-adaptation: Case studies

Some of the results of two such analyses are presented below. The data are introduced, the Principal Components are briefly described, the principal component scores of various sub-sets of the sample are compared and subjected to statistical tests where appropriate; finally interpretations are offered in each case.

Bear metapodials

The first data set discussed concerns the metapodials of the fossil bears. The eight variables were entered into a PCA for each of the ten metapodials, and each of these data sets produced more or less the same pattern of components. Slightly less than 70 % of the variance is accounted for by a single axis consisting of a positive combination of all variables. Almost 20 % of the remaining variance is accounted for in a second factor that contrasts the four transverse measurement with the two estimates of 'length', something that often termed 'robusticity' in the conventional idiom. However, it is important to recognise that this robusticity is not an allometric consequence of size; it is a relative thickness or thinness of the element after accounting for the null expectations of the general relationship between length and width variables, an expectation that is empirically derived from the sample itself.[2]

As none of the individual metapodial elements departed substantively from this pattern, all the data were pooled into a single sample. This was achieved by taking a modern sample of bear metapodials where the association between the metapodial elements was undoubted (as they came from the same animal), and then using their relative proportions to scale the fossil material by element, so that each fossil bone had a series of z-score variables essentially representing the number of standard deviations seperating it from the sample of modern bears. Where the results of such an analysis speak to the overall differences in size and shape of metapodials in terms of overall between-population differences in the structure and function of the animals feet, the meaning is clear. However, where results involve differences between metapodial elements or between front and hind feet for example, it must be recalled that all the variables are standardised to the size and shape of the modern sample and thus the between-element differences must be carefully interpreted with respect to this.

This resulted in a PCA of a matrix with 256 metapodials including 30 modern cases, and all the fossil material including that usually assigned to *arctos* and *spelaeus*. Two main components accounted for 82 % of sample variation; PC1 representing size and simple allometry, and PC2 representing an independent axis of robustisity/gracility similar to that seen in the analyses of separate metapodials. There was generally a clear separation between the material assigned to the two sub-species, although there is intermediate material to which we will return to later. The bones generally assigned to *arctos* were gracile but not always smaller than the *spelaeus* material. The Oxygen Isotope Stage 6 sample of '*arctos*' from the Bear Stratum at Tornewton (Devon, Britain) was similar in its large size and extreme gracility to the modern comparative material of the 'Kodiak' bear which is as large as the largest cave bears. Other Pleistocene '*arctos*', particularly those from the terminal Pleistocene of Cantabrian Spain was very similar to smaller extant Brown Bears such as those still found in the region today. However, there was also considerable variation within the material traditionally placed in the *spelaeus* taxon, especially with respect to the second PC, that describing the robust/gracile continuum. There were significant heterogeneities with respect to metapodial elements both between populations and within-population themselves.

Differences between populations

Using the best 'population' samples available, the samples from the Upper Palaeolithic deposits at Montgaudier (Southwest France), Troskaeta and Arrikrtuz (Cantabrian Spain) and Grabowica (Eastern Europe), there is no significant between-group effect on the first PC (interpreted as size). However, on the second component, some populations are distinctly different. The Spanish material is the most robust; the Eastern European sample, on average, slightly less so, although this difference is not significant. Montgaudier is the most gracile group; weighted linear contrasts show that Troskaeta is the most robust group ($t = 5.58$, $df = 110$, corrected $p<0.005$) and that Troskaeta, Arrikrutz and Grabowica are significantly more robust than the Montgaudier sample ($t = -5.05$, $df = 110$, corrected $p<0.005$).[3]

Within sites, but between the metapodial elements, the following effects were also marked. In Spain the samples from Arrikrutz and Troskaeta both have significantly more robust front feet than hind feet (for PC2 scores, Mann-Whitney $U = 21$, $n = 12,13$, $p = 0.0012$ and $U = 7$, $n = 11,6$ $p = 0.025$ respectively). This robusticity is not seen in the Montgaudier ($p = 0.35$) sample or at Grabowica ($p = 0.11$). However, at Montgaudier the front foot is relatively small (for PC1 score $U = 58$, $n = 18,16$ $p = 0.003$).

The general pattern amongst those sites with tractable sample sizes indicated that the material traditionally referred to *spelaeus* encompasses considerable morphological variation, and that some of this variation corresponds to differences between populations. The most robust sample is found in Spain, the most gracile in Southwest France, and a modestly robust population in the Eastern European sample.[4] In addition, the differences between the scores for fore and hind feet in the earlier Spanish fossil (fore paw reinforced) support the conclusion that they were very heavily built, as the front foot carries a disproportionate fraction of the body weight. The more even reinforcement of the fore and hind feet in the Eastern European and French population suggest a lighter animal, similar in proportion to the modern bears standard. The relatively enlarged hind foot in the French material suggests greater emphasis on locomotion as opposed to support, given the disproportionate stress imposed on the hind foot by locomotion (Manter, 1938).

Interpreting the differences

This gives us some grounds to speculate about the ecological differences between these populations. The morphology of the robust Spanish bears indicates a very large slow moving animal, presumably a herbivore (possibly scavenger). The large body would have supported a low-quality, high-bulk diet, and this was probably most efficiently obtained over a relatively small range leading to relatively high population densities exploiting typical bulk bear foods such as tubers, fruit and other plant material. The Eastern European population appears to be rather similar, but the reinforcement of the metapodials is less extreme. The best sample of this material comes from Grabowica which is situated in the mountains of Bulgaria. These fossil do not however, differ from the more northerly material, mainly Austrian, with which it is grouped in this analysis. Examining the range of dietary ecologies amongst modern bears it appears that the smaller, more lightly built French Cave Bear was probably more of an active hunter. For bears, this typically means small mammals and fish, but larger prey are not unusual. This strategy would have involved more dispersed resources, wider ranges and lower population densities than those potentially supported by a more herbivorous diet.

Why should this be so? Explanations can be cast in various modes, ecological geographical and chronological and various combinations of these three. The Alpine material is never associated with human occupation. None of the cave bear sites used is associated with definite Middle or Upper Palaeolithic industries and it seems fair to assume that the humans population densities of these high altitude areas were very low throughout the potential period of accumulation (100-25K). Cantabrian Spain was sparsely occupied in the period concerned (Straus, 1990) and there are very few associations between these bear sites and any definite archaeological horizons (We return to this point in a moment). On the other hand, the richness and diversity of Middle and Early Upper Palaeolithic (MP and EUP respectively) sites in Southwest France[5] is a testament to relatively high population densities of hominids. Our tentative hypothesis would be that ecological interference from hominids was the cause of smaller body size and increased mobility amongst bear in the French sample. The inference that this might be associated with increased carnivory is currently unsupported by the evidence from dental adaptation, and thus it remains open to question. Alternative scenarios suggest that rather than driving sedentary omnivore bears into a more peripatetic and predatory lifestyle, consistent hunting patterns amongst MP and EUP people (Stiner, 1994) might actually encourage bears to give up competing with humans for prey and turn instead to a more plant-based diet. The broader pattern of bear morphology in the European Upper Pleistocene provides some support for this. Britain, which has consistently marginal hominid occupation, also has a continuous record of gracile, arctoid bears and shows very low frequencies of the more spelaeoid bears when compared to continental sites (Leney, 1996, pp. 171-180).

Chronologically, it may be that the poor sampling of the material has resulted in different times being sampled in different areas. We saw that strong differences were found between Montgaudier and Troskaeta. The bear horizons at Montgaudier are probably comtemporary with the Upper Palaeolithic deposits (perhaps Magdalenian), whilst the Troskaeta material is considerably older. The process of gracilisation might be common to both regions but is obscured by the lack of Middle Paleolithic associated bear samples in France and Early Upper Palaeolithic samples from Spain. Better samples and dates would help here.

So this leaves unresolved the question as to whether gracility at Montgaudier is attributable to local ecology (possibly hominid density or common environmental determinants of hominid and ursine ecology) or to wider chronological effects such as a change in climate. In truth, these explanations are not completely exclusive as a change in climate might merely precipitate the movement of the local ecological factor (perhaps hominids) from one region to the next.

Additional evidence: The Middle to Upper Palaeolithic transition

There is however a small vein of additional evidence. Completely gracile bears had come to occupy Northern Spain towards the end of the Pleistocene and there are a number of subfossil finds that testify to this. At the same time the Upper Palaeolithic culture(s) had penetrated into the interior of the Cantabrian Cordillera. What had happened in between? At present we have data for just one key site which might help to resolve the nature of this transition. This site is known as Lezetxiki. Lezetxiki is a cave near Garagarza in the province of Guipúzcoa in Northern Spain. This has one of the best characterised local stratigraphies and quite extensive details are given in Altuna (1972). Fortunately, the stratigraphic position of the faunal elements was recorded *on the bones themselves* which has ensured that the information has survived the vagaries of the curatorial process. The few elements of relevant material in this study comes from Level 5a which is an archaeologically sterile horizon closely associated with an overlying 'proto Aurignacian' industry in Level 4c. Immediately underlying this *Ursus* material are Levels 5b and 6 which contain advanced Mousterian horizons. Level 7 contains an older Mousterian industry. Altuna (1972, p. 435) shows the frequency of species in the various levels of Lezetxiki. Level 6 is transitional. Above Level 6 are increasing numbers of *Cervus elephas* and *Capra pyrenaica*, whilst 6 and 7 are characterised by high frequencies of *Dicerorhinus*, and *Bison priscus*. *Rangifer* appears most frequently in Levels 3a and b (where they are associated with an 'advanced Aurignacian'), suggesting that cold peaked around this point; or alternatively, that reindeer hunting was at a maximum at this point (but that there may have been further cold periods where the climate completely prevented hominid penetration into the Cordillera). *Gulo gulo*, a species associated with modern boreal zones (Bjärvall and Magoun, 1984) and Pleistocene peri-glacial conditions (Stuart, 1982, p.86; Kurtén, 1968, pp.91-93), is only recorded from Level 2 suggesting that this might represent the glacial maximum.

There is a finite radiocarbon date for Level 3a of 19,340 ± 780 (I-6144), cited in Straus (1990). The nearby site of Aitzbiarte (Altuna, 1972) has *Rangifer* only in the Magdalenian and Solutrian and not in later 'Azillian' (Neolithic/Mesolithic) or the earlier Aurignacian, suggesting that the development of the Upper Palaeolithic was well advanced here by the onset of the glacial maximum. But, the Lezetxiki Level 3a date is associated with an '*Aurignacian/Perigordian ordinaire*' industry (Straus, 1990) which would seem very late for such a 'primitive' tool kit. In his synthesis of the trans-Pyrennian areas Straus (1990) cites 31-23 Kya for the Aurignacian, and 21-20 Kya for the Gravettian/Upper Perigordian which overlaps the local 'Solutrian' with Gravettian elements persisting into the Solutrian and some earlier appearances of 'Solutrian' typology.[6]

Although there may be some difficulty in maintaining a unilineal succession of typology in the region, it is clear that a synthesis of the date, faunas and industries combines to place the *Ursus* material from Level 5a at, or around, the Middle to Upper Palaeolithic transition. The site is additionally unusual as an Upper Pleistocene hominid site, because it is located some 25 km into the Sierras which rise sharply from the coast of Northern Spain at this point, and even lies in the Mediterranean, rather than the Atlantic watershed. The lack of adjacent lowlands or substantial coastal plain clearly differentiates Cantabrian Spain from the Southwest corner of France, *contra* Gamble (1986, pp.71-73).

This Level 5a material, which includes just a handful (or perhaps a pawful) of *Ursus* metapodials is distinctly more gracile than the bulk of the Spanish spelaeus material. Despite the fact there are only four metapodials, these are significantly more gracile than either of the main Spanish *spelaeus* samples, Troskaeta and Arrikrutz. The chance of such gracile metapodials being drawn from the same population as Troskaeta and Arrikrutz is less than 1 % (Mann-Whitney U = 32, n = 4,59, p = 0.0055). That this is not a statistical artefact, such as a mixed sample of both robust and gracile bears at Lezetxiki, is demonstrated by Figure 1 where it is possible to see that these four metapodials cluster in the intermediate morphospace. These bones have a transitional morphology which is a property of the individual metapodials. This places them in very similar morphospace to the French material, such as the sample from Montgaudier. This evidence is supported by similar relationships in the distal long bones where an ulna and two tibias from Lezetxiki all appear relatively small and gracile when compared to the wider sample of Cave bear material.

Other material from Spain provides weak support for a transition from robust to gracile at, or around, the beginnings of Upper Palaeolithic colonisation. Morphologically gracile '*spelaeus*' metapodials are also found in small quantities at Cueva Salitre near Santandér. Altuna (1972) notes an association between these bear bones and the cultural levels, although it is now unclear if this was with the Early or Later Upper Palaeolithic horizons. What remains certain is that the material post-dates the arrival of the Upper Palaeolithic in Northern Spain. Cueva de la Maltraviesa also contains small quantities of a similar animal, bones which are possibly associated with a 'Magdalenian' occupation (1990). The site of Congosto also appears to have a few of these rather gracile cave bears but has no stratigraphic association. Of the sites with the robust variant, the complete absence of cultural levels at Peña de Mazo, Cubias Negra, Arrikrutz and Cueva Berriatua suggest that these sites predate the occupation of the Cantabrian interior that took place in the Upper Palaeolithic, although this is indirect, negative evidence. There is stronger evidence to suggest that Troskaeta represents a cave fill of OIS 4 through 3 (Llopis et al, 1954) which would place these before the advent of the local Upper Palaeolithic. In summary then, of the material in the Madrid and San Sebastián museums, no *robust* spelaeoid bears are associated with Upper Palaeolithic facies, whilst the more gracile samples (even where these are just single elements) are never definitely associated with levels that predate the Upper Palaeolithic.

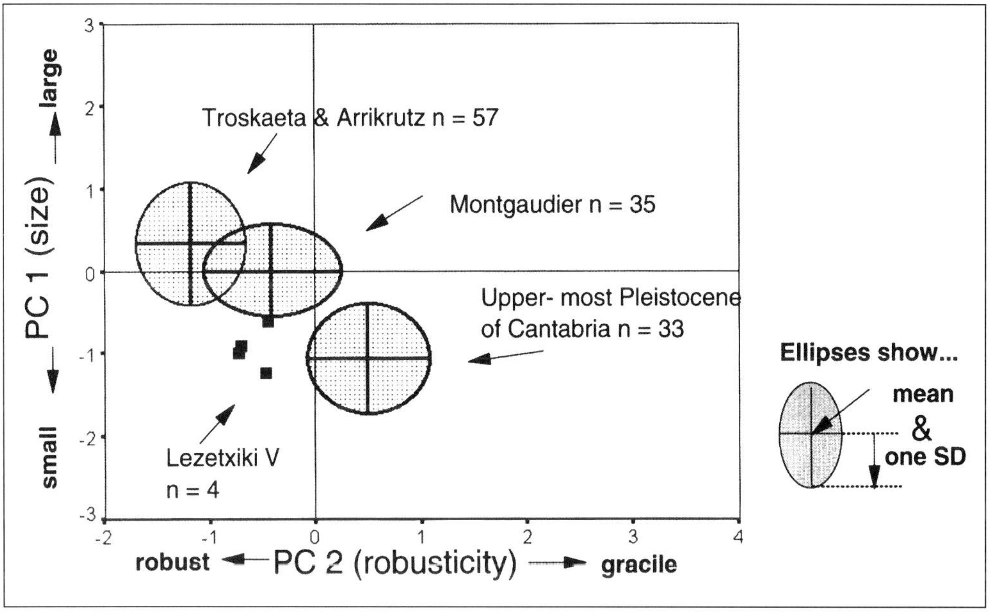

Figure 1 - *Plot of principal components of the pooled metapodial sample of fossil bears (selected sites shown). This plot of the first two principal components clearly shows the marked differences between three of the main samples of south-western European Upper Pleistocene Ursus. The four metapodials from Level 5 of Lezetxiki clearly fall into a unique portion of the morphospace, intermediate between that occupied by Ursus spelaeus (Troskaeta, Arrikrutz and Montgaudier) and Ursus arctos (Upper-most Pleistocene of Cantabrian Spain). This diversity coupled with the variation between the morphology of the two samples traditionally classified as Ursus spelaeus suggests that the "two species model" of late Pleistocene bears as an inadequate description of their ecological diversity.*

This fails to test whether the gracilisation of cave bears, and possibly the replacement of the spelaeus morphology with arctos was a product of changes in the climate or some third factor in the environment, or whether it was a product of hominid impact on the bears' ecology. Perhaps more material will come to light to resolve this. One prediction would be that, given hominid agency, where shifts in hominid ecology can be pin-pointed, the changes in bear morphology would follow after a minor lag. However, where competition from hominids was not driving the system, changes in bear morphology would be as likely to predate hominid shifts as postdate them. This seems to require remarkable precision in the dating of changes in the archaeological and palaeontological record but, with many sites there may be a statistically observable effect, using data similar to that observed at Lezetxiki, given the magnitude of the morphological shift observed in this limited sample.

Hyaena teeth

Similar problems dog the analysis of between-population variation in the Spotted Hyaena, *Crocuta crocuta*. In most of the continental European localities studied, the stratigraphic information available for non-hominid fossils is very sketchy. Nevertheless the following comparisons can be made. In Britain there are well defined OIS 5e and OIS 4-3 faunas, Ipswichian and Devensian

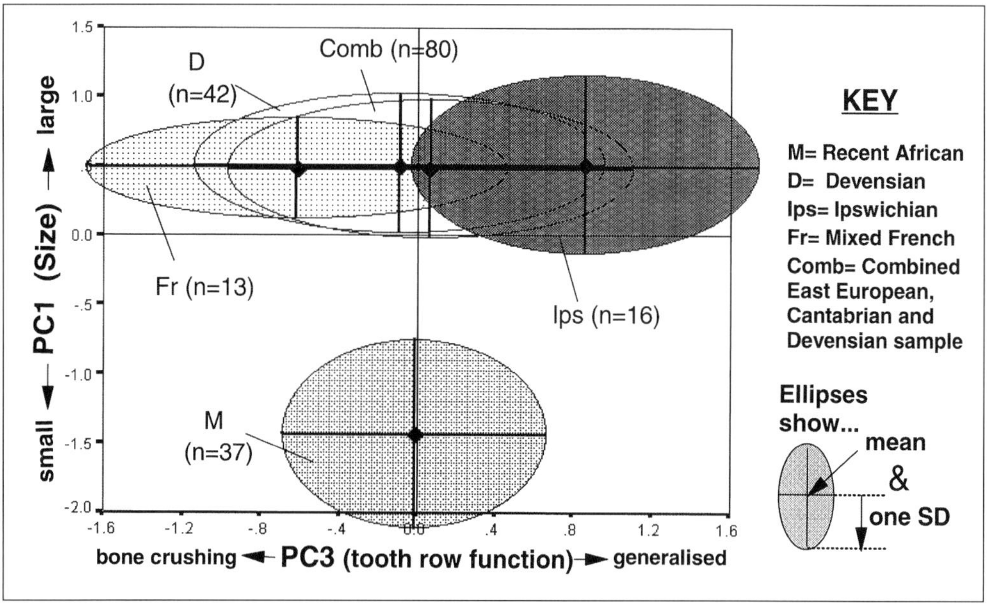

Figure 2 - Plot of principal components of the mandibular teeth of Crocuta, by population. This plot illustrates two important between-sample heterogeneities in the thrid principal component of variation: The Ipswichian Crocuta population (from Britain's last interglacial) has significantly more generalised dention than the Devensian (British last glacial). In addition, despite poor stratigraphic resolution, the French sample of Upper Pleistocene Crocuta showed a significantly greater bone-crushing adaptation than a combined sample including Upper-Pleistocene material from Cantabrian Spain, high altitude sites in East Europe, and the British Devensian (all populations associated with lower densities of hominid occupation than the French Upper Pleistocene).

respectively. The Ipswichian is characterised by a range of enivornmental markers related to the warm climate, coupled with the presence of hippopotamus, the absence of horses and minimal evidence for hominid occupation (see Stuart, 1982, p.176). On the other hand, the Devensian was much colder, but conversely shows increased evidence of hominid occupation (Sutcliffe, 1995). Samples of post-canine mandibular dentitions were taken for these chronological units, and these were augmented with material from modern (African) *Crocuta* as well as fossil material from Eastern Europe, France and Spain.

Comparing these sets of fossils, it is evident that there are between-population differences. The main contrast between the Ipswichian and Devensian emerging from the PCA is a component explaining about 17 % of the total sample variance. This effectively contrasts the relative size of the second premolar with all the other teeth (the third and fourth premolars and the first lower molar, the carnassial). There is no significant difference in size between the Ipswichian and Devensian judged by the scores of the first principal component (a weighted positive linear combination of all variables). As this size is removed in PC1, the larger second premolars in Ipswichian hyaenas are balanced by a reduction in the other dental dimensions in the analysis, which is confirmed by the small negative coefficients on all these variables. Previous analyses using only a single tooth, the carnassial (Klein & Scott, 1989), have concluded that Ipswichian

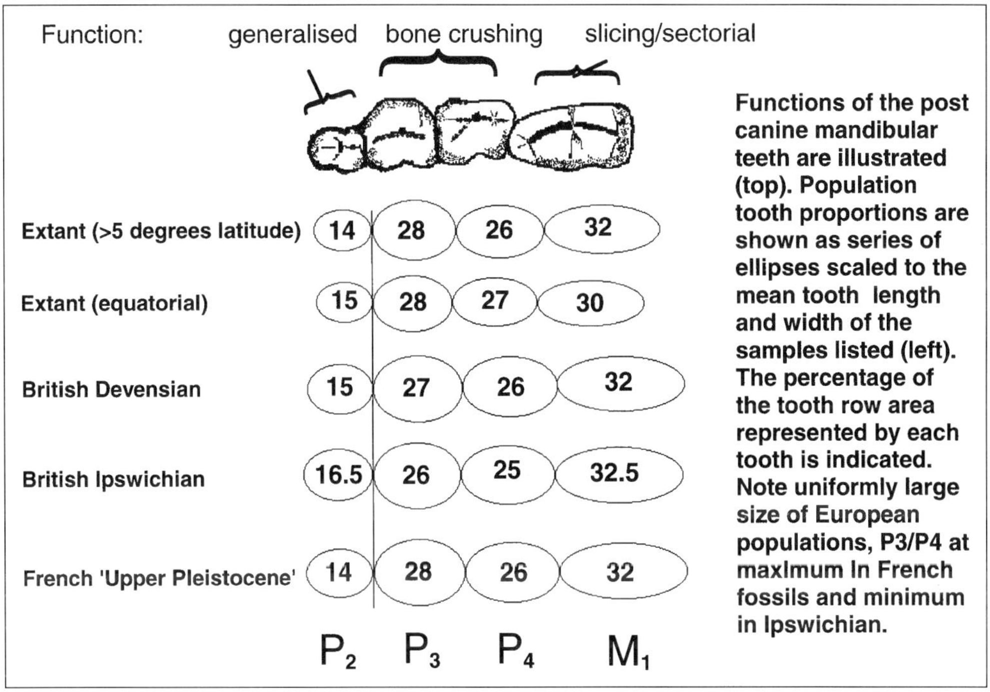

Figure 3. - Hyaena tooth proportions.

hyaenas were smaller. We believe that this study proves that, on the basis of more complete dentitions there is no significant difference in *size*, but that there is one of *shape*. In the Ipswichian, the enlarged anterior premolar and more evenly sized posterior teeth are reminiscent of a more dog-like dentition. This suggests an adaptation to a more generalist primary carnivore diet, as opposed to the specialist bone-crushing adaptation seen in modern African and Devensian population (for PC2 scores, the Ipswichian has significant relative enlargement of the lower second premolar when compared to British Devensian, Eastern European and French samples; One-way ANOVA with Scheffé correction). Figure 2 shows the distribution of the Ipswichian and Devensian cases projected into the plane defined by the first and third Principal Components.

Putting aside this chronological comparison for the time being and returning to the geographical comparisons of the material assumed to represent the last glacial (broadly defined) across the sample areas, we see a distinct pattern: The French material is rather different, significantly so when compared to the combined sample of material from all the other sites, Spanish, British (Devensian) and Eastern European. Figure 2 also shows the projections of the French and 'Combined' samples into the PC1/PC3 space. When the PCA is repeated without the Ipswichian sample, it is the third component, accounting for around 10 % of all sample variance, that is the axis expressing the inter-population differences. This component shows that the French sample has significantly smaller lower second premolar and enlarged third and fourth premolar relative to the other three populations.[7] For a linear contrast of British, Austrian and Spanish last glacial material with French last glacial, $t = 2.09$, $df = 89$, $p = 0.03$, and this despite the rather small sample size for the French population. This suggests that increase in the size of the third and fourth premolars in the French material was compensated by an additional reduction in the second premolar indicating a shift from a generalist carnivore diet towards increasing specialisation in bone crushing. This mirrors the morphology of modern East African material (after size is taken into account) suggesting that the feeding ecology of the extant animals is probably a better model for the French populations than it is for the other fossil groups.

Given what we can infer about the hominid populations contemporary with these fossil samples, it seems that the presence of hominids is associated with increasing bone crushing specialisation in the hyaenas, suggesting more secondary carcass feeding and less primary carnivory. Alternatively, this might indicate a shift towards larger prey where significant extra resources could be gleaned from bone crushing. While they are statistically significant these are very fine micro-adaptations; Figure 3 shows the projections of the population mean dimensions of the elements of the post-canine tooth row in the original measurement space where, even made explicit, the shape differences remain hard to see. The multivariate characterisation of these slight differences is the only way to properly addresses such variation if the nature of the differences in adaptation between populations is to be elucidated.

Conclusion

We hesitate to draw direct inferences about hominid ecology from either of these case-studies; the picture is too incomplete. However, we hope that this work demonstrates that it is feasible to tease out population differences in relatively subtle aspects of morphology and to relate these to feeding ecology. With more species, particularly wolves and foxes, coupled with better chronological resolution of the material, it will be possible to test the effect of hypothesised shifts in hominid ecology on the carnivore guild. The goal of this work is to draw attention to the dynamic nature of adaptation in carnivores. It suggests how this might be used to investigate ecological changes which may be directly related to hominids or may merely help in our understanding of the background processes shaping the ecology of carnivory in the late Pleistocene. The variation in the fossil record is waiting to be characterised: variation within-populations that was the raw material upon which natural selection worked, and variation between-populations that records the micro-adaptations created by that same process.

Notes

1) It should also be remembered that variation in the environment can produce ontogenetic differences from an identical genotype, but this kind of phenotypic plasticity can really be seen as an adaptation itself, *sensu* Slobodkin, (1968). Where our interest is primarily ecological, a distinction between genetic and environmental 'causes' of adaptation is not particularly relevant.

2) In fact here we have not used logged data and thus run the risk that strongly curved allometries might swamp the independent fractions of robusticity. However, this would lead to a specific and predictable artefactual distribution of cases with respect to the robusticity axis which is not seen in any part of this analysis. It therefore seems that the heterogeneity with respect to robusticity within the sample is a much stronger effect than the distortion introduced by curvilinear intraspecific allometry. Avoiding the unnecessary use of logs, eliminates artificial compression of data towards the larger end of the size specturm. As the range of sizes is very large and the curvature in the data is rather small, it was felt that the PCA of the raw data was a preferable description.

3) Ten statisitical tests were used here, six pair-wise comparisons and four contrasts. Therefore using the Bonferroni correction all probability values are inflated ten-fold. This does not alter the conclusions here that these two effects are the strongest and remain highly significant.

4) This pattern is confirmed by data from other Eastern European sites such as Mixnitz and Schottloch which contain relatively robust elements and French sites such as St. Aubin, G.d'Oiselles, Echenoz and L'Herm which contain spelaeoid material of a more gracile nature (Leney, 1996, pp.165-171).

[5] Notwithstanding potential sampling biases (Campy & Chaline, 1993), there seems to be a real concentration of sites here.

[6] Straus considers this date rather late for such a primitive Upper Paleolithic industry. In this study, the important point is really that Level 5a comes exactly between the terminal Mousterian and the first traces of the EUP. My interpretation of the late date for the Aurignacian industry at Lezetxiki 3a is that this elevated site represents a seasonally occupied hunting outpost and that the 'Aurignacian' character is more a property of the activity patterns at the site than the chronological context of the tool makers.

[7] This analysis utilised the missing data protocols. As a between-population difference was expected, cases with one missing tooth were allowed into the analysis. Each missing datum was replaced by using the overall mean for that variable using the whole sample. Any bias that this introduces tends to obscure between-group differences as the global mean is used. However, the French population remains significantly different despite this potentially homogenising procedure. This allows us to dismiss the suggestion that using only complete dentitions has biased the nature of the sample becuase of unkonwn local differences in taphonomy. In addition, without using missing data, the French material was not actually significantly different from the other populations; this demonstrates the utility of the conservative application of missing data techniques in multivariate analysis of fossils, as the cases with partial information may make an important contribution to the sample.

References

ALLWORTH-JONES, P. (1990): The Szeletian and the stratigraphic succession in Central Europe and adjacent areas: Main trends, recent results and problems for resolution. In: P. Mellars (ed.), *The emergence of modern humans: An archaeological perspective,* 160-242. Edinburgh.

ALTUNA, J. (1972): Fáuna de mamíferos de los yacimentos prehistóricos de Guipuzcoa. *Munibe* 24, 1-464.

ANYONGE, W. (1993): Body-mass in large extant and extinct carnivores. *Journal of Zoology* 231, 339-350.

BANFIELD, A.W.F. (1953): Range of individual wolves (*Canis lupus*). *Journal of Mammalogy* 34, 389-390.

BINFORD, L.R. (1977): *For theory building in archaeology.* New York.

BINFORD, L.R. (1981): *Bones: Ancient men and modern myths.* New York.

BJÄRVALL, A. & MAGOUN, A.J. (1984): Wolverine. In: D.W. MacDonald (ed.), *The encylopedia of mammals,* 120-121. Oxford.

BLUMENSCHINE, R.J. (1986): Carcass consumption sequences and the archaeological distinction of scavenging and hunting. *Journal of Human Evolution,* 15, 639-659.

BLUMENSCHINE, R.J. (1987): Characteristics of an early hominid scavenging niche. *Current Anthropology,* 28, 383-407.

BLUMENSCHINE, R.J. (1988): An experimental-model of the timing of hominid and carnivore influence on archaeological bone assemblages. *Journal of Archaeological Science* 15, 483-502.

BUNELL, F. (1984): Grizzly bear. In: D.W. MacDonald (ed.), *The encyclopedia of mammals,* 88-90. Oxford.

CAMPY, M. & CHALINE, J. (1993): Missing records and depositional breaks in French late Pleistocene cave sediments. *Quaternary Research*, 40, 318-331.

DAYAN, T., SIMBERLOFF, D., TCHERNOV, E. & YOMTOV, Y. (1992): Canine carnassials - character displacement in the wolves, jackals and foxes of Israel. *Biological Journal of the Linnean Society* 45, 315-531.

DIAMOND, J.M. (1987): Character displacement - will grebes provide the answer. *Nature* 325, 16-17.

DURNER, G.M. & AMSTRUP, S.C. (1995): Movements of a polar bear from Northern Alaska to Northern Greenland. *Arctic* 48, 338-341.

ERDBRINK, D.P. (1953): *A review of fossil and recent bears of the Old World with remarks on their phylogeny based upon their dentition.* Utrecht.

FOLEY, R.A. (1984): Early Man and the Red Queen: Tropical African community evolution and hominid adaptation. In: R.A. Foley (ed.), *Hominid evolution and community ecology: Prehistoric human adaptation in biological perspective,* 85-110. London.

FOLEY, R.A. (1987): *Another unique species.* Harlow, Essex.

GAMBLE, C. (1986): *The Palaeolithic settlement of Europe.* Cambridge.

GOTTELI, D., SILLEROZUBRII, C., APPLEBAUM, G.D., ROY, M.S., GIRMAN, D.J., GARCIAMORENO, J., OSTRANDER, E.A. & WAYNE, R.K. (1994): Molecular-genetics of the most endangered canid - The Ethiopian wolf *Canis simensis. Molecular Ecology* 34, 301-312.

HENDY, Q.B. (1974): The late Cenozoic carnivora of the South Western Cape Province. *Annals of the South African Museum* 63, 1-369.

HOWELL, F.C. & PETTER, G. (1976): Carnivora from the Omo Group Formations, Southern Ethiopia. In: Y. Coppens, F.C. Howell, G.L. Isaac & R.F. Leakey (eds.), *Earliest man and environments in the Lake Rudolph Basin,* 314-331.Chicago,

HUTCHINSON, G.E. (1959): Homage to Santa Rosalia or why are there so many kinds of animals? *American Naturalist* 93, 145-159.

JOLICOEUR, P. (1959): Mutivariate geographical variation in the Wolf *Canis lupus* L. *Evolution* 13, 283-299.

JOLICOEUR, P. (1975): Sexual dimorphism and geographical distance as factors of skull variation in the wolf *Canis lupus* L. In: M.W. Fox (ed.), *The wild canids,* 54-61. New York.

KLEIN, R.G. (1986): Carnivore size and Quaternary climate change in Southern Africa. *Quaternary Research* 32, 153-170.

KLEIN, R.G. & SCOTT, K. (1989): Glacial interglacial size variation in fossil spotted hyenas (*Crocuta crocuta*) from Britain. *Quaternary Research* 32, 88-95.

KLINGENBERG, C.P. (1996): Multivariate Aallometry. In: L.F. Marcus, M. Corti, A. Loy, G.J.P. Naylor & D.E. Slice (eds.), *Advances in morphometrics,* 23-49. London.

KRUUK, H. (1972): *The spotted hyena.* London.

KURTÉN, B. (1955) Sex dimorphism and size trends in the Cave Bear, *Ursus spelaeus* Rosenmüller and Heinroth. *Acta Zoologica Fennica* 90, 1-48.

KURTÉN, B. (1965): The carnivora of the Palastine caves. *Acta Zoologica Fennica* 107, 1-74.

KURTÉN, B. (1967): Some quantitative approaches to dental microevolution. *Journal of Dental Research* 46, 817-828.

KURTÉN, B. (1968): *Pleistocene mammals of Europe.* London.

KURTÉN, B. & ANDERSON, E. (1980): *Pleistocene mammals of North America.* New York.

LEAKEY, M.G. (1976): Carnivora of the East Rudolf succession. In: Y. Coppens, F.C. Howell, G.L. Isaac & R.F. Leakey (eds.), *Earliest man and environments in the Lake Rudolph Basin,* 302-313. Chicago.

LENEY, M.D. (1996): *Morphology and microevolution in Pleistocene carnivores.* Cambridge.

LENEY, M.D. & FOLEY, R.A. (1996): Biology and body size in human evoluition: A comment. *Current Anthropology* 37, 467-468.

LEWIN, R. (1983): Santa Rosalia was a goat. *Science* 221, 636-639.

LLOPIS LLADO, N. & ELOSEGUI, J. (1954): Sobre las características del relleno de la Sima de los Osos de Troskaeta-ko-kobea, (Ataun-Guipúzcoa). *Munibe* 6, 38-46.

MacDONALD, D.W. (1984): Carnivores. In: D.W. MacDonald (ed.), *The encyclopedia of mammals,* 18-25. Oxford.

MANTER, J.T. (1938): The dynamics of quadrapedal walking. *Journal of Experimental Biology* 15, 522-540.

MARTIN, P.S. (1967): Prehistoric overkill. In: P.S. Martin & H.E. Wright (eds.), *Pleistocene extinctions,* 75-120. New Haven.

MARTIN, P.S. & WRIGHT, H.E. (1967): *Pleistocene extinctions.* New Haven.

MOSIMAN, J.E. & MARTIN, P.S. (1975): Simulating overkill by Paleoindians. *American Scientist* 63, 304-313.

MURIE, A. (1981: *The Grizzlies of Mount McKinley.* Washington D.C.

PATTERSON, H.E.H. (1985): The recognition concept of species. In: E. Vrba (ed.), *Species and speciation,* 21-29. Pretoria.

PEARSON, K. (1901): On lines and planes of closest fit to a system of points in space. *Philosophical Magazine* 2, 557-572.

RAO, S.V. & deBACH, P. (1969): Experimental studies in hybridization and sexual isolation between some *Aphytis* species (Hymenoptera: Aphelinidae). III. The significance of reproductive isolation between interspecific hybrids and parental species. *Evolution* 23, 525-533.

REYNOLDS, S.H. (1911): *A monograph of the British Pleistocene mammalia.* Vol II. *British Pleistocene Hyaenidae, Ursidae, Canidae and Mustelidae.* London.

ROY, M.S., GEFFEN, E., SMITH, D., OSTRANDER, E.A. & WAYNE, R.K. (1994): Patterns of differentiation in North American wolflike canids revealed by analysis of microsattelite loci. *Molecular Biology and Evolution* 11, 553-570.

SHELDON, J.W. (1992): *Wild dogs: The natural history of the non domestic Canidae.* London.

SHIPMAN, P. (1982): More on theropithecus at Olorgesailie - age structure and mortality - reply. *Current Anthropology* 23, 351-352.

SHIPMAN, P., BOSLER, W. & DAVIS, K.L. (1981): Butchering of giant geladas at an Acheulian site. *Current Anthropology* 22, 257-268.

SHIPMAN, P., BOSLER, W. & DAVIS, K.L. (1982): On the arguments of butchering of giant geladas - reply. *Current Anthropology* 23, 110-111.

SLOBODKIN, L.B. (1968):Towards a predictive theory of evolution. In: R.C. Lewontin (ed.), *Population biology and evolution,* 187-205. Syracuse.

SMITH, R.J. (1996): Biology and body size in human evolution: Statistical inference misappli(ed.) *Current Anthropology* 37, 451-481.

STINER, M.C. (1991): Food procurement and transport by human and non-human predators. *Journal of Archaeological Science* 18, 455-484.

STINER, M.C. (1994): *Honor amongst thieves: A zooarchaeological study of Neandertal ecology.* Princeton.

STRAUS, L.G. (1990): The Early Upper Palaeolithic of Southwest Europe: Cro-Magnon adaptations in the Iberian Peripheries, 40,000 - 20,000 BP. In: P. Mellars (ed.), *The emergence of modern humans: An archaeological perspective,* 276-302. Edinburgh.

STUART, A.J. (1982): *Pleistocene vertebrates in the British Isles.* London.

SUTCLIFFE, A.J. (1995): Insularity of the British Isles 250,000-30,000 years ago: The mammalian, including human evidence. In: R.C. Preece (ed.), *Island Britain: A Quaternary perspective,* 127-140. London.

TURNER, A. (1981): *Aspects of the palaeoecology of large predators including man during the British Upper Pleistocene, with particular emphasis on predator-prey relationships.* Ph.D. Thesis. University of Sheffield.

TURNER, A. (1986): Correlation and causation in some carnivore and hominid evolutionary events. *South African Journal of Science* 82, 75-76.

TURNER, A. (1992a): Villafranchian-Galerian larger carnivores of Europe: Dispersions and extinctions. *Courier Forschungs-Institut Senckenberg* 153, 153-160.

TURNER, A. (1992b): Large carnivores and earliest european hominids - changing determinants of resource availability during the lower and middle pleistocene. *Journal of Human Evolution 22,* 109-126.

VanVALKENBURGH, B. (1988): Trophic diversity in past and present guilds of large predatory mammals. *Paleobiology* 14, 155-173.

VanVALKENBURGH, B. (1989): Carnivore dental adaptation and diet: A study of trophic diversity within guilds. In: J.L. Gittleman (ed.), *Carnivore behaviour, ecology and evolution,* 410-436. London.

VanVALKENBURGH, B. & WAYNE, R.K. (1994): Shape divergence associated with size convergence in sympatric East-African jackals. *Ecology* 75, 1567-1581.

WALKER, A.C. (1984): Extinction in hominid evolution. In: M.H. Nitecki (ed.), *Extinctions,* 119-152. Chicago.

WAYNE, R.K. (1993): Molecular evolution of the dog family. *Trends in Genetics* 9, 218-224.

WOLPOFF, M.H. (1988): Divergence between early hominid lineages: The roles of competition and culture. In: F.E. Grine (ed.), *History of the 'robust' australopithecines,* 485-497. New York.

Ethology and Population Dynamics: Keystones for Anthropogenic Extinctions and Hominid Expansion

- The Eutrophic Zone, its Genesis and Implications -

Wilhelm Schüle † & Sabine Schuster

Abstract

Expansion of animal (and human) populations is a function of species-specific population dynamics, which depend, above all, on the animals' trophic behaviour. To reconstruct past behaviour and population dynamics, actualistic analogies are indispensible. Behaviour is controlled by genetically fixed patterns to a high degree. Plio-Pleistocene megaherbivores were without natural enemies and had therefore lost the genetic patterns for enemy recognition and adequate reactions. Due to this loss, they were easy prey for hominids armed with thrusting spears. The improvement in the hominids' trophic basis resulted in the growth of hominid populations. Megaherbivore numbers declined as losses to hominid spears were not replaced because of low megaherbivore reproduction rates. Along the limits of their original range, hominids could kill megaherbivores immigrating from beyond. This resulted in the development of a stable Eutrophic Zone. In the course of a long co-evolution, some megaherbivores managed to increase reproduction rates and adapt to the new superpredator. These are the ancestors of extant megaherbivores in the Old World tropics. When all unwary megaherbivores had disappeared, progressive australopithecines / early hominines had to expand their range. In Eurasia, hominid progress was limited by climatic parameters. Only along this new Eutrophic Zone, which shifted with climatic oszillations, can a long co-existence of Homo sp. and megaherbivores be observed beyond the Old World tropics. In the course of the last glaciation, Homo sapiens sapiens improved techniques of thermo-insulation, which enabled him to exterminate unwary megaherbivores in the cold climates beyond the former Eutrophic Zone. South of East Beringia, access to America was blocked by the iceshield. When this barrier melted, America lost most of its fauna to invading man. In Australia, which man had reached earlier by raft or boat, the same process can be observed. Disappearance of most megaherbivores and establishment of an anthropogenic fire-regime caused ecological disturbances world-wide. In order to survive in the impoverished hinterland, the hominids had to develop new technologies and ecological ethics. The simultaneous evolution of large brains as well as social and ecological ethics are considered an iteration process within the complex adaptive system of human evolution.

Neogene African Hominoidea were omnivorous, as are most primates (Teleki, 1973; Harding & Teleki, 1981; Fleagle, 1988; Coppens, 1990). Seasonal trophic bottleneck situations in the semi-arid savannas and mosaic forests forced them into an increasingly carnivorous way of life. It is very likely that termites, small rodents, eggs and young birds soon failed to satisfy them. For larger prey, however, they needed weapons.

We will not go into the wide-spread hypothesis claiming that the hominids satisfied their need for protein by scavenging. This view corresponds neither to the behaviour of any living primate, nor is it consistent with the ecological and physiological prerequisites. It simply is based on the erroneous opinion that large animals can not be killed using primitive weapons (W. Schüle, 1990 b; W. Schüle & Schuster, 1996, 1997).

Up to the present, pongids have been observed using cudgels only when spontaneously defending their children (Kortland, 1968). Nevertheless, we claim that Mio/Pliocene "protohominids" used cudgels, both for defence and for hunting. We do so mainly for one reason: Without the neccessity to carry objects permanently, even during swift locomotion, there is no plausible explanation for the evolution of bipedalism and of the hominids' highly refined cooling system. Only weapons must be carried around permanently (W. Schüle, 1990 b). The osteological finds of *Australopithecus (Ardipithecus) ramidus* from the Lower Pliocene of Ethiopia (Wood, 1994; White et al., 1994, 1995; Woldegabriel et al., 1994) are a bit too scanty yet to allow specific statements about its bipedalism. One can, however, presume that australopithecines (or "proto-australopithecines") were - at least rudimentarily - able to hunt with cudgels by that time.

As long as the early hominids hunted small game using cudgels, ecological consequences were light and are not to be discerned palaeo-ecologically. Their new potential prey quickly learned that some of the previously harmless Hominoidea had become dangerous after they had started to both walk upright and to carry cudgels in their hands. The prey of the evolving hominids became as wary towards them as it has always been towards its other predators. Species diversity was not affected by the early australopithecines cudgels; at the most, population sizes of the prey species were reduced insignificantly. The preys reproduction rates were sufficiently high and flexible to compensate for the additional losses.

The giant species of the African tortoise may possibly be an exception. On other continents and on islands, their disappearance was always correlated with the appearance of *Homo* spec. The species of giant tortoises still existing on the Galapagos Islands and on the Seychelles have survived because of strict protective regulations only. Their strategy of reproduction is adjusted to high loss of juveniles, but not of adult individuals. These long-lived animals mature late, and if the reproducing individuals die all too soon, populations collapse. In Africa, giant tortoises seem to have disappeared by the end of the Miocene or in the Lower Pliocene already. That feeds suspicions that Mio/Pliocene Hominoidea of the strain leading to the australopithecines were already able to crack adult tortoises with large stones (Leakey, 1968), or to tear them out of their shells bit by bit by using wooden barbs. Tortoises do not migrate to evade seasonal aridity or periods of severe drought. Consequently, they helped the hominids through seasonal or occasional periods of trophic bottleneck situations, and reduced trophically caused losses.

Population dynamics of the earliest hominids were positively influenced by their new behaviour, i.e. by hunting with cudgels and by cracking of giant tortoises. Their range, however, will probably not have expanded significantly.

Right from the beginning, the hominids needed tools for their new trophic behaviour. Not only did they need wooden hunting weapons (which have not been conserved for obvious reasons), but they also had to use sharp cutting edges to cut their prey. Their dentition was not suited for that, being relatively weak if compared to that of recent pongids. In our oppinion, this is an indication for a very early start of the extracorporal processing of hard dietary components. Unlike the brachiating pongids, the hominids did not possess extremely strong frontal extremities to tear the prey apart. For that, they did not need the so-called pebble tools, but sharp flakes. Only the keenness of the cutting edge was relevant, not the shape of the stone flake. The cores were the waste, not the flakes. Cores were carried along only if one could not hope to find stones suitable for producing sharp edges in the hunting area. Possibly this also applies to the hand axes which appeared much later, but they might also have served as tools or objects of prestige. For cracking bones, any good-sized stone sufficed. As weapons, stones were nearly worthless, whether worked or not. Their successful use as projectiles would have required extraordinary force and high accuracy. The size limit of potential prey hunted this way would have been quite low. At what time slings made from cords or strips of leather were invented cannot be determined (W. Schüle & Schuster, 1996, 1997).

Different from the cudgel, the wooden spear - not to be confused with a javelin! - turned out to be an ecological time-bomb. It proved to be the instrument that both facilitated and forced the hominids expansion.

With a wooden spear of sufficient length, one could kill any animal that did not flee or attack. There was no longer an upper limit to prey size. Actual proof that such hunting methods were applied dates to a much younger epoch, though. The Eemian yew spear from Lehringen (Germany) was found in a position indicating it was rammed into the abdomen of a sub-adult *Palaeoloxodon* from beneath and behind (Jacob-Friesen, 1956, 1959). We shall refer to that later.

For earlier periods, there is only an indirect argumentation. In Africa, several species of megaherbivores, i.e. animals weighing more than about 1,000 kg, became extinct in the Upper Plio- and Lower Pleistocene without descendants or substitutes. Extinctions occured primarily among those species, which substantially exceeded this weight-limit (Maglio & Cooke, 1978; Martin, 1984). Beyond Africa, these and other species of a similar weight-class disappeared much later (Kurtén, 1968; Kurtén & Anderson, 1980; Kowalski, 1986; Alberdi et al., 1995). That disappearance roughly coincides with the arrival of *Homo* spec. in the dispersal areas of these species (Martin, 1984).

Only in the Old World tropics, the biotopic niches of extinct megaherbivores were at least partly re-occupied by surviving species. Everywhere else, all megaherbivores became extinct without being substituted.

Additionally, some very large terrestrial Cercopithecidae of the genera *Dinopithecus* and *Theropithecus* as well as the robust australopithecines disappeared in Africa in the lower Pleistocene. In Southern Asia, the (at least) gorilla-sized Hominoidea *Meganthropus* and *Giganthropus* dissapeared. These can be viewed both as prey and as direct biotope competitors of the contemporary hominines.

Had the extinctions of large terrestrial vertebrates which gradually affected the whole earth in the course of the Quaternary been caused by climatic changes, then these extinctions should have co-occured with the climatic oscillations of the Pleistocene. That is not the case. In Africa, extinctions occured near-exclusively among megaherbivores. Outside of Africa, the later *Homo* spec. appeared, the more species became extinct and the smaller extinct animals became, as extinctions also affected the medium weight class (50 kg to 1,000 kg). The sequence of extinctions was - as far as the data can be obtained with sufficient accuracy - proportional to the animals body size: the bigger the animal, the earlier it became extinct. Some large predators had probably specialized on juvenile megaherbivores as prey, especially sabertoothed cats (Machairodontidae). These also became extinct as the populations of their potential prey grew too small.

Small species with weights below 50 kg were nearly unaffected. Such animals, however, are usually climatically more sensitive than large ones. Many extremely stenothermal species survived all climatic fluctuations of the Pleistocene, many euthermal species died out.

To attribute this wave of extinctions to anthropogenic causes, parameters of ethology and population dynamics are necessary, both for the hunters and for their potential prey. That is the only way to explain why even early hominids had the ability not only to kill large animals, but also to exterminate whole species.

A number of parameters, with their internal complexity, essentially control the population dynamics of any given species. These are: the endogenous reproduction rate per generation, the exogenous loss due to predators, diseases and catastrophes, and the trophic basis of the population. Trophic behaviour, therefore, plays a key role. Territorial parameters ultimately are also trophic ones. The change of a single parameter will affect the population size, either positively or negatively. All other parameters are in turn affected through feedback, and new feedback is the result.

To elucidate behaviour and reproduction rates of animals (including man) of the past, an analysis of the fossil (including the archaeological) record alone is insufficient. Working with present-day analogies is indispensable.

Konrad Lorenz showed that animal (and human!) behaviour is controlled by genetically fixed patterns of behaviour. The cognitive stimuli triggering these patterns are genetically fixed as well. The absence of such patterns in the genetic code results in an inability to acquire specific patterns of behaviour empirically, neither can such patterns be handed down to descendants.

Within the limits of given genetic patterns, a variability both of the triggering stimuli and of the range of actions exists. In vertebrates, this variability as well as ability for empirical learning and for handing down acquired knowledge is controlled by the number of ganglia and by the structure of the central nervous system, especially of the cerebral cortex (Lorenz, 1966, 1969; Immelmann, 1983).

Even if that appears misleading at first glance, we shall initially concentrate on endemic faunas of islands. These are the key to understanding continental conditions.

Even on large islands, only small endemic terrestrial predators exist. On small islands, there are no predators at all. This is because predators require far greater territories than their potential prey. On islands, even those as big as Madagascar, existing areas are too small to permanently maintain viable populations of larger endemic predators.

Because of that, endemic island vertebrates gradually lose the genetic bases for behavioural patterns concerning predator recognition and adequate reactions, i.e. for flight or attack. Senseless flight and purposeless attack are a waste of energy and may result in accidental injuries, so they represent negative factors of selection. On islands, endemic vertebrates are genetically unafraid of terrestrial predators.

As a rule, genetically fixed patterns of behaviour are highly specific. Animals small enough to be attacked from the air are not at all unafraid of volant enemies. On very large or on offshore islands we find birds of prey; on small islands there are skuas, frigate birds and other predatory sea birds. To be genetically unafraid of birds would be lethal for small animals.

Penguins are genetically unafraid ashore and on ice, but while they are in the water, even the shadow of an orca or a leopard seal will make them panic and flee. To predatory sea birds, they only react at their nest when eggs and young are endangered. The adult penguins themselves are too big a prey.

For endemic species on islands, there are no losses due to predators. As a consequence, their reproduction strategy is genetically adapted to that, and the animals have very low reproduction rates. If this were not the case, mass reproduction would ensue and the trophic basis of the animals would collapse.

When a new predator appears and colonizes the island, the endemites cannot compensate for the high losses. Their populations decrease rapidly.

If the new predator specializes on certain prey, it soon destroys its luxuriant food supply. Distances in searches for prey increase, as does the energy needed. The population of the potentially neo-endemic predator will starve just before the last reproductive remnants of its palaeo-endemic prey have disappeared. The survivors now have the chance to build up new populations.

If, however, the new predator is a trophic generalist, it can change to other food and so exterminate its prey without having to starve.

Man is a trophic generalist, i.e. an omnivore, as are most primates. He can exterminate his preferred prey and then change to other food. Man will survive, but the genetically unafraid island fauna wont. All bigger and many smaller palaeo-endemic island species disappeared quickly as soon as *Homo* spec. appeared on formerly predator-free islands.

For that, *Homo* did not need especially efficient weapons. Even for the sub-fossil Moas of New Zealand, for Holocene giant ostriches, pigmy hippopotamus, and for the gorilla-sized lemurs of Madagascar, clubs and spears were sufficient weapons. They all became extinct as soon as *Homo* appeared on their islands (W. Schüle, 1993).

Excessive body size protects against predators nearly as efficiently as living on an island. Steller's sea cow (*Hydrodamalis [Rhytina] gigas*), a megaherbivore living in shallow water, became extinct few years after its discovery. Only boats, spears and clubs were necessary to manage that. The animals were genetically unafraid. The remnant population which became victim of the 18th century sealers lived on island coasts, but during the Pleistocene, there were *Hydrodamalis* populations everywhere in the shallow bays and offshore islands of the northern pacific, from Japan to California. They were exterminated as soon as *Homo* appeared there. Special weapons were unnecessary.

The genetically unafraid big whales were threatened only after ocean-going ships spread. Formerly, they could only be hunted near the coasts. Had they had not lost the genetic basis for predator recognition and adequate reactions, they would not be on the verge of extinction today. Even with long-range harpoon guns, whales diving off in time and re-emerging in a distant location cannot be slaughtered. Melvilles vindictive Moby Dick is a masterpiece of literature. Ethologically, however, it is an absurdity.

Terrestrial animals - if big enough - have no natural enemies as well, neither on islands nor on continents. Nevertheless, their young are still threatened. Protecting the young, however, represents a complex of genetically fixed behavioural patterns and cognitive triggers which differs from the one for self protection. Even genetically unafraid Steller's sea cows hurried in when their calfs were threatened. They tried to form a protective ring - and were themselves slaughtered without any resistance. Similar patterns of behaviour can be observed among whales confronting orcas, among musk-oxen confronting wolves and polar bears, and among elephants confronting big felines and hyenas. Confronted with man, that behaviour is suicidal, even if spears are his only weapons.

Today, megaherbivores survive only where they experienced long co-evolution with australopithecines and early hominids. In the megaherbivores of the African Pliocene, i.e. in mastodonts, in *Dinotherium* , in *Libytherium*, and in *Ancylotherium* (Fig.1), the genetic basis of enemy recognition and adequate reaction either was present rudimentarily, or it had been lost completely during the generations long increase in body size. They were genetically as unafraid as Steller's sea cow.

Such unafraid megaherbivores can easily be killed by ramming a sharpened, long hardwood spear between the ribs from beneath and behind. Those kind of spears are easy to make. Recent pongids posses amazing skills in making, using and handing down knowledge of tools, most of which are made from organic materials (Heltne & Marquardt, 1989; Lethmate, 1990; McGrew, 1992; Kano, 1992; Wrangham et al., 1994). With that in mind, one is inclined to believe that progressive australopthecines - or at least the earliest hominids - were able to make sufficiently long and sharpened spears, to use them and to hand down the necessary knowledge. Such long, sharp spears do not have to be as well-balanced as javelins. If need be, even a splinter torn off by lightning-stroke will work.

Although much younger, the Eemian yew spear from Lehringen an der Aller (Germany) was discovered among the ribs of a sub-adult male *Palaeoloxodon* (Jacob-Friesen, 1956; Thieme & Veil 1985; see also Wagner, 1988).

Figure 1 - *Spießjagd früher Hominiden. A - Dinotherium = Deinotherium, B - Libytherium = Sivatherium, C - Ancylotherium, D - Anancus. Reconstruction of animals after E. Thenius.*

Extant sub-adult elephants are often found at the periphery of a herd. They are no longer protected by the behavioural patterns for protecting the young, so they are highly endangered. If they were genetically unafraid - as one can assumed Pliocene megaherbivores were - they would be at mercy of any new predator without protection or defense.

This is different for calves. In vertebrates taking care of their offspring, the young which are not yet independent flee to their mothers, not away from danger. This reaction is probably not triggered by the danger itself, but by the mothers warning cry. Vice versa, a juveniles cry of fear is a strong trigger for the intervention of its mothers. These behavioural patterns are genetically fixed even if the adult animals themselves do not have any natural enemies and are genetically unafraid. As a consequence, young megaherbivores were less threatened than juveniles, provided their mothers classified a new enemy as being dangerous. A preference for young megaherbivores, which is sometimes postulated for our ancestors, is a fairy tale. Their mothers, or - in species with highly developed social behaviour - possibly other adult individuals of the herd, would have protected them. Saber-toothed cats (Machairondidae), for which suspicions that they hunted young megaherbivores are probably justified, must have had highly developed hunting techniques to outmanoeuvre mothers and aunts. The Quaternary palaeontologist Björn Kurtén portrayed this very nicely in his novel "The dance of the tiger". Tigers, lions, and even hyenas occasionally kill stray, ill, or over-inquisitive calves, but there is hardly any chance for them to kill well behaved, healthy young elephants.

In megaherbivores, the loss of juveniles affects population dynamics much more negatively than the loss of very young individuals. Due to their late maturity, replacing juveniles requires many years. Young calfs can be replaced much faster: the mother will soon be with young again.

Loss of infants is allowed for in the reproduction strategy even of the largest animals, but not the loss of adults.

We do not know of any spears from the African or South Asian Plio/Pleistocene, but around that time most of the megaherbivores disappeared from the Old World tropics. They were apparently unable to adapt their behaviour and reproduction rates in time to cope with the hominids changed behaviour.

Only those megaherbivores survived which were able to adapt to the new enemy. These are the direct ancestors of extant elephants, rhinos and hippos.

That adaptation was probably not due to a reactivation of old behavioural patterns or to the development of new ones. Instead, existing patterns for protecting the young probably varied (and extended), so that the megaherbivores could cope with the spear-carrying hominids.

Behaviour based on genetic patterns can be significantly varied only by highly cerebralized animals. It is a characteristic feature that the brains of surviving megaherbivores are larger and more structured than those of the extinct forms. Comparable casts, however, are so far available only for proboscidian skulls (Dechaseaux, 1958).

Prior to the adoption of hunting small game we postulate for the Upper Miocene/Pliocene, the hominids food supply had been quite sparse and subject to strong seasonal fluctuations in the semi-arid savannas of Africa. By the use of spears, it was improved nearly instantaneously, and became crisis-proof. In undisturbed areas, megaherbivores are present in the lowest numbers both of species and of individuals, but they represent most (frequently exceeding 90 % !) of the grazers biomass (Bourliere, 1963). With reproduction rates being constant, the hominids losses due to predators or to lack of food declined considerably after the invention of the spear.

Eutrophization always leads to mass reproduction of the profiting species. When the trophic basis of the hominids had been quite essentially expanded by tapping the megaherbivore food source, the first population explosion was an inevitable consequence. The reduction of losses to predators enhanced this effect. Resource over-exploitation had to follow.

There was also hardly any trophic limit to the band size of the spear-carrying hominids: One dead *Dinotherium* or *Ancylotherium* was sufficient for many co-eaters. If need be, a single spear-hunter could make the kill. His position in the hierarchy of the group - and thus the number of his descendants - certainly surpassed that of a club-hunter, who could come up with nothing more than a hare or a porcupine. Striving for social advancement is one of the most powerful drives among all higher primates - including man (de Waal *passim*).

The much-acclaimed improved weaponry of the advanced hunters does represent technical progress. Ecologically, however, it works like a boomerang, just like the simple wooden spear.

For expanding the range of *Homo* spec. and for exterminating so many genetically unafraid or inexperienced animals, technological progress which improved spear hunting was nearly meaningless. For hunting game which had become shy in the hinterland, however, it did make a difference. Contrary to a widely held opinion, the new weapons did not result in permanent enhancement, but, in the long run, in a further reduction of resources. Populations of prey animals shrank. The wider the range of long-distance weapons, the larger the flight- or attack-ranges of shy game became. Each hunt became longer and more energy-consuming.

Temporary improvements were only possible through weapons with still longer ranges. In the long run, however, these had the same negative effects. Exhaustion of the resources exploitable with any new technology forced either a further improvement of the latest invention, or the development of a completely new technology. That, in turn, resulted in a further over-exploitation of resource. Throughout the history of humankind, this cycle has been repeated incessantly (Cohen, 1977; Crutzen & Müller, 1989).

Along with the dwindling animal resources, population densities of hominids which were at least seasonally dependent on hunting also declined. Only in exceptionally favourable situations could they change to different food partially or completely.

At the same time, however, the threat posed by attacking prey and by carnivores was reduced by every additional meter which their weapons carried. Every increase of the hunter populations, though, meant an additional burdening of game stocks, i.e., a reduction of their trophic basis - unless the hunters would have developed an ecological ethics, i.e. if their had learnt from their mistakes. But that was a slow process.

The improvement of weapons - no matter at what stage - had a much bigger impact on intra-species conflicts than on profitable hunting. It thereby influenced the expansion of hominid populations, whether these were defined genetically or culturally.

The pressure on genetically unafraid megaherbivores grew synchronous with the numbers and the efficiency of the hunters. The reproduction strategy of the megaherbivores was not adapted to the loss of adult individuals. Stocks collapsed, and so did the ample trophic resources of the spear-carriers. Mathematical models clearly demonstrate that (A. Schüle, 1990, 1992).

Local sedentary game species of medium weight do have natural enemies and have therefore never lost the genetic basis of their behavioural patterns for recognition of enemies, flight, or attack. In the ancestral biotopes of the spear-carrying hominids, these animals had become timid or aggressive. Both their flight distances and critical distances (Lorenz, 1969) increased quickly. The hominids spears were unbalanced and could therefore not be thrown. The growing distances could not be overcome with these simple weapons. Hunting medium-sized game became difficult and dangerous. Hunting distances lengthened, more energy was used. It also became dangerous to hunt those megaherbivores which had adapted to the hominids behaviour in the course of a long co-evolution. The trophic basis of the spear-carriers grew smaller, while their losses due to hunting accidents and throphic bottleneck situations.

Wherever unafraid megaherbivores had disappeared and other remaining game had become timid or dangerous, hominid population densities decreased considerably.

The game of medium size not only survived the initially high losses caused by the hominids without extinction of species, but, due to its higher reproduction rates, it also regenerated quickly after having become timid or aggressive. Additionally, it also profited from the reduction of the hominid populations.

Only at the outer limits of their range were the hominids still able to kill immigrating unafraid megaherbivores and inexperienced game of medium size with their spears. At those outer limits of their area of dispersal, an eutrophic zone was formed for the hominids which permitted relatively high population densities as compared to the core areas.

If the spear-carrying hominids wanted to continue to live in the land of milk and honey by exploiting unafraid megarherbivores and inexperienced game, they had to expand their range constantly. That means they had to migrate to areas that still held both unafraid megarherbivores and inexperienced game. The hominids colonized biotopes uncharacteristic for them, first in Africa, and later also the in the tropic and subtropic areas of Eurasia - and plundered them. But only on the expansion front was there a well laid table; hominids in the hinterland were on short commons. There, population densities were low.

In some distance to the eutrophic zone, there were no immigrating megaherbivores. They could not have come that far. If, however, the hominids left an area in the hinterland for several years or decades, game populations regenerated - and just as quickly unlearned to recognize the hominids as predators and to react to them accordingly. Today, this effect can be observed in any game reserve. When the hominids returned, for some time the hunt was good and easy. Had the game become timid again, life became hard for the spear-carriers, and they did well to move on. Sedentary hunters can and could exist - if ever - under extraordinary circumstances only. Normally, they are an ethological, population-dynamical, and ecological absurdity.

Homo erectus probably reached the southern tip of Asia in Java by crossing the dry shelf during a cold period already at the beginning of the Lower Pleistocene (Swisher et al., 1994.; see also de Vos & Sondaar, 1994). In southern Asia, the eutrophic expansion zone had collapsed prior to that when the areas that were left to the unafraid megaherbivores became too small to sustain reproducing populations.

In southern Asia, both the species that became extinct and those that survived were of the same or at least of closely related genera as in Africa. Only the Hippopotamidae did not survive in southern Asia: *Hexaprotodon* disappeared.

Afterwards, the sea blocked the way to Australia until deep into the Upper Pleistocene, i.e. for more than 1.5 myr. Recent investigations indicate that *Homo sapiens sapiens* reached Australia during the last glacial cycle only - and plundered it with spear and fire (Flood, 1995).

Only off-shore islands visible from the mainland were probably reached even earlier by *Homo erectus*, e.g. Corso-Sardinia in the Mediterranean Sea, some islands in the southern Wallacea east of Java, as well as Celebes-Sulawesi and the Philippines. Non-human island hoppers had reached these islands as well, among them Rhinoceratinae (only on the Philippines) and Proboscidea. Endemic faunas on these islands, which, independent of body size, were genetically mostly unafraid, dissapeared in some places in the Middle Pleistocene already (W. Schüle, 1993). The early, Lower Pleistocene finds of *Homo* spec. in southwest Europe (Gibert, 1993; Roe, 1995; Lumley, 1990) could possibly be explained more easily this way than by their by-passing the Mediterranean in the cold north. The Straits of Gibraltar were narrower in the Lower Pleistocene than they are today; during glacials, large portions of the African shelf between Sicily and Africa fell dry. The mastodon *Anancus* seems to have disappeared significantly earlier in Africa and southern Europe than in the north.

If we presume that some human colonists might by chance have returned from their more or less willingly discovered islands to the mainland, telling others about the land of milk and honey they had left, this might have been an incentive even for *Homo erectus* to develop better water-craft. In view of recent scientific findings on the highly developed communications systems of extant bonobos (*Pan paniscus*) and chimpanzees (*Pan troglodytes*) (de Waal *passim*; Cavalieri & Singer, 1994), the idea is no longer as absurd as it might have appeared some decades ago.

We can presume that, in the exploited hinterland, the nearly inexhaustible food resources of the mud-flats and mangrove swamps of tidal coasts in South Asia played an important role for the development of water-craft, i.e. for the expansion over the open sea. One can hardly expect this source of food to have escaped a hungry *Homo erectus* notice. Shell middens which are so frequent in the Holocene (Fetten, 1991) are to be expected above the highest stage of interglacial (i.e., eustatic) transgressions only. If such shell middens survived, those of glacial times would today be submerged, those of the interglacials would have been eroded by the following transgression.

In the north, the hominids expansion was halted at their own climatic limit. But the situation in the north was completely different from that at the southern coasts.

At his climatic limit, *Homo erectus* had access to big game immigrating from the north, which was either genetically unafraid or it was inexperienced with regard to the human superpredator. Man could not plunder the northerly reproduction zones of these animals. As a consequence, a wide Eutrophic Zone was formed for *Homo erectus* and his successors - a zone which he could exploit with his spears but not exhaust. This represents the only explanation for the long co-existence of man and megafauna in Eurasia.

The Eutrophic Zone shifted permanently due to the climatic oscillations of the Pleistocene, but qualitatively it remained stable for about 1.5 myr. The situation remained essentially unchanged even when *Homo erectus* was replaced by *Homo sapiens praesapiens* and these again by *Homo sapiens neanderthalensis*. Despite the extremely unfavourable climate, population densities within Eurasias Eutrophic Zone were probably much higher than in the southern hinterland where there was no unafraid or inexperienced game. Animals that have become shy or agressive are hard to kill with simple spears. It is rather telling that in Eurasia most osteological and archaeological evidence is found along the northern limit of the range of Pleistocene man.

The extension of the Eutrophic Zone was not defined by man's own climatic needs only, but also by those of the tree species which provided the raw material for the spears. Beyond the northern limits of yew (*Taxus* spec.), ash (*Fraxinus* spec.) and walnut (*Juglans* spec.), hard and fracture-proof spears of sufficient length could be procured only by long-range transport. Without spears, however, the eutrophic zone was neither to be held nor to be expanded.

In the long run, *Homo* spec. could not survive outside the humid tropics and subtropics without profitable hunting, because vegetable food is not available year-round in sufficient amount and quality. North of the southern (or above the montane) snowfall-line, it can be procured during the vegetation period only.

For the profitable hunting of shy or aggressive game in the hinterland, long-range weapons with considerable penetration were necessary. For hunting genetically unafraid megaherbivores and inexperienced game, improved weapons were neither necessary nor useful. For the step-by-step extinction of so many species of big game during the Pleistocene, improved weapons were not necessary, although this has often been postulated. Megaherbivores and the larger mammals of the medium weight classes did not become extinct in the hinterland of the hominid expansion. The advancing Eutrophic front first destroyed them locally, and they became extinct when it reached their last reproduction zones. Few - if any - species of big game in the medium weight classes became victim of the javelins, bows, slings and traps of technologically advanced humans in the Pleistocene of Afro-Eurasia.

Only after *Homo sapiens neanderthalensis*, and later, on a larger scale, *Homo sapiens sapiens* had developed efficient techniques of thermo-isolation as huts, tents and, above all, sewn clothing in the course of the last glacial period, the last refuges of the unafraid or inexperienced game in Eurasia and east Beringia (northern Alaska) could be plundered. The Eutrophic Zone collapsed.

South and east of east Beringia (the Bering strait fell dry during cold periods), the expansion of *Homo sapiens sapiens* was temporarily stopped by the North American ice shield which covered the continent from coast to coast.

Only after the Edmonton Channel opened between the Montane and Laurentian ice shields on the Pleistocene/Holocene boundary, and only after it had become passable for humans and animals alike, both Americas could be reached and be plundered. All palaeo- and neo-endemic megaherbivores and nearly all endemic mammals weighing more than approximately 50 kg (Kurtén & Anderson, 1980) were exterminated in both Americas - more than 100 species all told. Furthermore, giant tortoises and many birds, even small ones, also dissapeared (Martin & Klein, 1984). Nearly all extant big mammals of North America, including the buffalo, are Eurasian immigrants adapted to hunting. They crossed the dwindling, foresting land bridge in the early Holocene. Only prong-horn antelope (*Antilocapra americana*), black bear (*Ursus americanus*), jaguar (*Panthera onca*) and Atlantic manatee (*Trichechus manatus*) are endemic species, possibly also musk-ox (*Ovibos*), caribou (*Rangifer*) and mountain goat (*Oreamnus*). None of the Holocene immigrants reached South America.

An insuperable climatic boundary that - like in Eurasia - could have led to the development of a permanent Eutrophic Zone with a steady influx of unafraid or inexperienced game, or that could at least have slowed *Homo sapiens sapiens* on his way to the south did not exist in America.

Martin (1984) described aspects of both the palaeontology and the population-dynamics of the first human plundering of both American continents in detail, although schematized and without the ethological explanation.

All older dates for a human presence south of the American ice shield arise from outdated methods, or they could be refuted. Anyway, it would be quite inexplicable why the genetically unafraid megaherbivores along with the associated faunas had not disappeared at any of the earlier dates already. The recently revived hypothesis that palaeo-Indian groups followed the west coast avoiding the hinterland rich in unafraid megaherbivores and inexperienced game, thus reached the southern Pacific coast of North America is, in our opinion, incompatible with human behaviour.

The continental expansion of the hominids ended at Cape Horn. The Straits of Magellan had long ceased to be an obstacle for *Homo sapiens sapiens*. The rapidly progressing Eutrophic Zone and its dependent human populations collapsed inevitably. The hot fights for the last naive giant sloths and mastodonts remaining on the conical southern tip of South America which must have occured among the crowded tribes of hunters are hardly in accordance with the romantic picture of the “gentle savages”.

The radical ecological changes of anthropogenic origin due to the plundering of the fauna and the anthropogenic frequent fires (Goldammer *passim*) have been described in detail by Pavel Putshkov for the Holarctis (Putshkov passim), and especially for North America (Putshkov, 1994).

Radical ecological changes which were mainly caused by the reduction (Old World Tropics) or the disappearance (beyond the Old World Tropics) of the ecologically important megaherbivores (Owen-Smith, 1988) and the anthropogenic fire frequency (Dechamps, 1984; James, 1989) accompanied the entire expansion of the hominids all over the world (W. Schüle, 1990 a; 1992 a,b; W. Schüle & Schuster, 1996).

The hominids were genuine palaeo-endemites in the mosaic forests and savannas of Africa only; outside of Africa, they were allochthonic invaders. Allochthonic biotic invasions (Groves & Burdon, 1986; Crosby, 1986; Lever, 1994) result in serious distortions of endemic ecosystems, even if intruders are less eco-aggressive than *Homo* spec. The radical ecological changes caused by the hominids were the more grave, the further the distance to their home biotopes and the later they arrived. Cultural evolution had outpaced their genetical evolution. Beyond Africa, hominids are not so much genetical as they are cultural neo-endemites.

In this context we shall have to return to the weapons of the hominids, and to the beginning of their expansion.

In our opinion, the transition to hunting with extra-corporal weapons (i.e., cudgels) we postulate had a decisive influence on the development of the australopithecines bipedalism (W.Schüle 1990 b; W. Schüle & Schuster 1996, 1997). It did not require an intelligence which actually exceeds that of extant chimpanzees or bonobos, however. The ecological consequences were anything but severe.

The consequences of the weapons for the intra-species relations of the early hominids must have been much more serious, however. The uncontrolled use of cudgels or even spears in intra-species conflicts for food and territories, in fights for rank in the hierarchy of the band, and in sexual rivalry, could easily have had lethal consequences.

Animals with corporal weapons which could be lethal during intra-species conflicts posses genetically fixed behavioural restraints to prevent fatal outcomes. In animals without such weapons, these restraints are either weak or not present at all.

The hominoid line leading to the hominids, especially the early and late hominids, never posessed efficient corporal weapons. One may presume that the restraints of Plio/Pleistocene hominids in intra-species behaviour were weaker than those among extant pongids. Being brachiators, these have much stronger fore-limbs, and they also posses more powerful dentitions than the terrestric australopithecines and hominids. Even the transition to hunting with cudgels or clubs must already have resulted intra-species conflicts which were easily fatal. The skull of a primate is easily shattered with a club. Still easier, an opponent can be killed with a spear, a javelin, or a long-range bow, not to mention modern weapons.

Thus, right from the beginning, even the simple cudgel but more so any further improvement of weapons lend superiority over opponents within the band. It also meant superiority of one band over others that either did not have any weapons or only less efficient ones.

The development of new, genetically fixed restraints in social behaviour would have needed many more generations than the empirical improvement of weapons and their use. Without the development of ethical, i.e., not genetically fixed, controls, any band - and, in the end, the species as well - would have exterminated itself quite soon. Even among chimpanzees, intra-band conflicts can be lethal although chimpanzees have quite efficient corporal weapons and, as a consequence, they presumably also posses more efficient, genetically fixed restraints (Goodall, 1991; de Waal, 1993).

A band that was able to develop an "ethical use of weapons" for use among themselves, but that would still use their weapons unscrupolously against other bands (just like Jane Goodall's chimpanzees use their physical and numerical superiority in "tribal wars" [Goodall, 1991]), must have gained considerable territorial and reproductive advantages over other bands as well as over closely related species. The giant terrestric primates of the Old World Tropics and *Australopithecus robustus* probably fell victim to that type of behaviour in the Lower Pleistocene already.

Ethical behaviour is very demanding to the capabilities of the central nervous system. The development of more and more efficient weapons, combined with a socio-ethical use of weapons within ones own band (or confederation of bands), but not against other bands, and the enhancement of the brain's capabilities, form a behaviourally induced loop of positive feedback, i.e. a fast iterative cycle in the complex adaptive system of the evolution of armed hominids (Prigogine & Nicolis, 1987, Gell-Man, 1994).

Once social ethics had been established in a band or group of bands, they could then be transferred to the protection of their "own" stock of game. This was an imperative requirement for their sustainable use, and for the bands own survival in the hinterland of the Eutrophic Zone. The "ecologically correct" behaviour of traditional communites was not born from the seemingly inexhaustible plenty of the Eutrophic Zone, but from the scarcity in the overexploited hinterland.

In man's genetic pattern, eco-ethical behaviour is laid down even less than socio-ethical behaviour. Only the ability is contained in the genetical information of *Homo*. If new, unexploited areas are developed, or if new, only apparently inexhaustible resources are made accessible by new technologies, then the ecological ethics are lost to a great extent and must be re-learned and handed down anew after depletion or complete exhaustion of the resources.

Eco-ethical behaviour is much more demanding on the capabilities of the brain than socio-ethical behaviour: it requires conscious or unconscious insight into processes of much higher complexity. Only by eco-ethical behaviour a band of hunting hominids is able to make sustainable use of its resources.

Such insight can, however, only be achieved by individuals. The band, the tribe, or the population, they all require traditional rules, the observance of which is controlled continuously.

For the handing down of empirically learned hunting methods or similar behavioural patterns, or of simple methods for production and use of primitive weapons and tools, imitation of the behaviour of the elder generation would suffice. This can easily be observed among lions handing down hunting techniques or among pongids handing down "types of tools" and their use (Lethmate, 1990, 1991; Wrangham et al., 1994).

For handing down socio- and eco-ethological patterns of behaviour and for control of obedience thereto, neither simple imitation nor even the concatenation of word-like utterances is sufficient. A "grammatical" combination of "words", *ergo language*, seems to be inevitable. Only a central nervous system whose abilities exceed that of a chimpanzee's brain can achieve this. The necessity to hand down such patterns of behaviour exerted considerable selective pressure on the development of the brain. The necessity to avoid the disastrous effects of efficient weapons must have affected the brain's development much stronger than their production and use. The most striking leap in the development of the human brain occurred on the borderline between *Australopithecus* and *Homo*. Prerequisite for a further enlargement of the brain, however, was an efficient cooling system able to avoid overheating of the central nervous system (Falk, 1992). The cudgle-hunting of the australopithecines laid the thermodynamical foundations for that development.

The taboos of "primitive" and the complicated rules of "complex" religions of advanced civilizations are rooted, at least partially, in the socio- and eco-ethical necessities of the early use of weapons. In the monotheistic religions, especially in atheism, eco-ethological behaviour assumes a lower priority than socio-ethological aspects, and must therefore be replaced by laws. The control of obedience to these laws by the society sadly turns out to be considerably less efficient than enforcing religious rules.

Both traditional and high-tech cultures face that self-same problems today. Nowadays, however, not only trophic, but also all other resources are affected, not only megaherbivores but the whole biosphere. The causes for this disaster are to be found in new technologies, i.e. changed forms of behaviour and their far-reaching consequences. New technologies, whether they be the invention of spears or the domestication of plants and animals, always have similar effects: the trophic basis is expanded, losses are reduced, exponential growth of reproduction rates ensues, and behavioural reactions are inadequate (Bender, 1997; Hösle, 1991).

We cannot study this problem here in more detail. To us, however, it appears quite plausible that the successive waves of genetically or culturally defined hominid forms who - beginning in Africa in the Plio-/Pleistocen, and later repeatedly expanding from Africa all over Eurasia - replaced their respective predecessors and thereby surged out over the Eutrophic Zone at the primary expansion front of the hominids, is closely connected with this problem.

The worldwide expansion of the Europeans in the historical modern periods - despite all the differences in details - represents the current analogy thereto. It was made possible only because the Europeans utilised their superior weapons technology in their newly acquired territories unscrupulously against both their respective predecessors and the game species these exploited. The more intensive, exhaustive exploitation of resources allowed the Europeans to have temporarily substantially higher reproduction rates than those of the hunting, gathering, or "primitive" but sustainably farming aboriginals. A further ecological impoverishment right down to complete desertification were and are the consequences.

It may well be that Konrad Lorenz was right when he said that man might just be too stupid to survive.

Acknowledgements

For help and inspiration, in some way or other, we would like to thank: Emiliano Aguirre, Teresa Alberdi, Meinrat O. Andreä, Andrea Bender, Paul J. Crutzen, Alfred Czarnetzki, Sophie Ehrhardt †, Josep Gibert, Johann G. Goldammer, Vittorio Hösle, Kenneth J. Hsü, Ralf-Dieter Kahlke, Heiko Kallweit, Otto Kegel, Adriaan Kortlandt, Ulrich Lehmann, Konrad Lorenz †, James Lovelock, Andreas Mahler, Christian Maise, Paul S. Martin, Ernst Mayr, Ana Mazo, Karl Joseph Narr, Julia Noack, Sandra Pichler, Pavel Putshkov, Achmed Schüle, Hansjürg Steinlin, Erich Thenius, Konrad Weidemann and Karl-Friedrich Weiß.

Initial investigations were sponsored by Deutsche Forschungsgemeinschaft and Volkswagenstiftung. Later investigation was made possible by Freiburger Institut für Paläowissenschaftliche Studien F.I.P.S.

Thanks to all of them, and to many more, not mentioned here.

References

ALBERDI, M.T., LEONE, G. & TONNI, E.P. (1995): *Evolución biológica y climática de la región pampeana durante los últimos cinco milliones de años. Un ensayo de correlación con el Mediterráneo occidental.* Madrid.

BENDER, A. (1997): *Beute unter Tabu. Traditionelles Umweltverhalten in Mikronesien und Polynesien.* Freiburg/Brsg. (Paläowissenschaftliche Studien PAST, vol.1.).

BOURLIERE, F. (1963): Observations on the ecology of some large African mammals. In: F.C.Howell & F.Bourlière (eds), *African ecology and human evolution,* 407-427. Chicago.

CAVALIERI, P. & SINGER, P. (eds), (1994): *Menschenrechte für die Großen Menschenaffen.* München.

COHEN, M.N. (1977): *Foodcrisis in prehistory.* New Haven.

COPPENS, Y. (1990): Hominoides, hominides et hommes. In: *5 millions d'années. L'aventure humaine*, 26-45. Bruxelles (Catalogue).

CROSBY, A.W. (1986): *Ecological imperialism. The biological expansion of Europe.* Cambridge.

CRUTZEN, P.J. & MÜLLER, M. (1989): *Das Ende des Blauen Planeten ?* München.

DECHAMPS, R. (1984): Evidence of bush fires during Plio-Pleistocene in Africa (Omo and Sahabi) with the aid of fossil woods. *Paleoecology of Africa* 16, 291-294.

DECHASEAUX, C. (1958): Encéphales de Proboscidiens fossiles. In: J.Piveteau, *Traité de paléontologie*, vol. 2, 296-298. Paris.

FALK, D. (1992). *Braindance.* New York.

FETTEN, F. (1991): Gibt es eine "Molluskenhaufen-Ökonomie"? In: *Festschrift für Wilhelm Schüle,* 131-144. Buch am Erlbach (Internationale Archäologie. Veröffentlichungen des Vorgeschichtlichen Seminars Marburg, Sonderband 6).

FLEAGLE, J.G. (1988): *Primate adaptation and evolution.* San Diego, New York.

FLOOD, J. (1995): *Archaeology of the dreamtime. The story of prehistoric Australia and its people.* Sydney, London.

GELL-MANN, M. (1994): *The quark and the jaguar.* New York.

GIBERT, J. (ed.) (1993): *Proyecto Orce-Cueva Victoria (1988-1992): Presencia humana en el Pleistoceno inferior de la región de Granada y Murcia.* Orce (Museo de Prehistoria Josep Gibert).

GOLDAMMER, J.G. (ed.) (1990): *Fire in the tropical biota. Ecosystem processes and global challenges.* Berlin, Heidelberg, New York (Ecological Studies. Vol. 84).

GOLDAMMER, J.G. (ed.) (1992): *Tropical forests in transition. Ecology of natural and anthropogenic disturbance processes.* Basel.

GOLDAMMER, J.G. (1994): Feuerökologie. *Spektrum der Wissenschaft* 7/94, 86-94.

GOODALL, J. (1991): *Wilde Schimpansen. Verhaltensforschung am Gombe-Strom.* Hamburg.

GROVES, R.H. & BURDON, J.J. (1986): *Ecology of biological invasion.* Cambridge.

HARDING, R.S.O. & TELECKI, G. (1981): *Omnivorous primates.* New York.

HELTNE, R. &.MARQUARDT, L.A. (eds.) (1989*): Understanding chimpanzees*. Cambridge.

HÖSLE, V. (1991): *Philosophie der ökologischen Krise*. München (Moskauer Vorträge. Beck'sche Reihe 432).

IMMELMANN, K. (1983): *Einführung in die Verhaltensforschung*. Berlin.

JACOB-FRIESEN, K.H. (1956): Eiszeitliche Elefantenjäger in der Lüneburger Heide. *Jahrbuch des Römisch-Germanischen Zentralmuseum Mainz* 3, 1-22.

JACOB-FRIESEN, K.H. (1959): *Einführung in Niedersachsens Urgeschichte*. Hildesheim.

JAMES, S.R. (1989): Hominid use of fire in the Lower and Middle Pleistocene: A review of the evidence. *Current Anthropology* 30, 1-26.

KANO, T. (1992): *The last ape: Pygmy chimpanzee behaviour and ecology*. Stanford.

KORTLANDT, A. (1968): Handgebrauch bei freilebenden Schimpansen. In: B.Rensch (ed.), *Handgebrauch bei Affen und Frühmenschen*, 59-102. Bern, Stuttgart.

KOWALSKI, K. (1986): *Die Tierwelt des Eiszeitalters*. Darmstadt.

KURTÉN, B. (1966): Pleistocene mammals of the Bering Bridge. *Commentationes Biologicae. Societas Scientarium Fennica* 19 (8), 1-7.

KURTÉN, B. (1968): *Pleistocene mammals of Europe*. London.

KURTÉN, B. & ANDERSON, E. (1980): *Pleistocene mammals of North America*. New York.

LEAKEY, L.S.B. (1968): Bone smashing by late Miocene Hominidae. *Nature* 218, 528-530.

LETHMATE, J. (1990): *Evolutionsökologie und Verhalten der Hominoiden*. Tübingen.

LETHMATE, J. (1991): Haben Schimpansen eine materielle Kultur? *Biologie in unserer Zeit* 21/3, 132-139.

LEVER, Ch. (1994): *Naturalized animals*. London.

LORENZ, K. (1966): *Evolution and modification of behavior*. London.

LORENZ, K. (1969): *Über tierisches und menschliches Verhalten*. Band I-III. München.

LUMLEY, H. de & LUMLEY, M.A. de (1990): La conquete de l'ancien monde par l'*Homo erectus*. In: *5 millions d'années. L'aventure humaine*. Bruxelles (Catalogue).

MAGLIO, V.J. & COOKE, H.B.S. (1978): *Evolution of African mammals*. Cambridge.

MARTIN, P.S. (1973): The discovery of America. *Science* 179, 969-974.

MARTIN, P.S. (1984): Prehistoric overkill: the global model. In: P.S. Martin & R. Klein (eds.), *Quaternary extinctions*, 354-403. Tucson.

MARTIN, P.S. & KLEIN, R. (eds.) (1984): *Quaternary extinctions*. Tucson. 2nd ed.1989.

McGREW, W.C. (1992): *Chimpanzee material culture. Implications for human evolution*. Cambridge.

OWEN-SMITH, N. (1989): *Megaherbivores: The influence of very large body size on ecology*. Cambridge.

PRIGOGINE, I. & NICOLIS, G. (eds.) (1987): *Die Beherrschung des Komplexen*. Vol.1. München, Zürich.

PUTSHKOV, P.V. (1989): *Uncompensated Pleistocene extinctions: Supposed mechanism of the crisis*. Kiev (Special Volume of the Schmalhausen-Institute of Zoology, Academy of Sciences of Ukraina). (Russian).

PUTSHKOV, P.V. (1991): Uncompensated Würm extinctions. Part 1. Largest herbivores: Eury- or stenobiotic forms? *Vestnik Zoologii* 1991/5, 45-53. (Russion with abstract in English).

PUTSHKOV, P.V. (1992a): Uncompensated extinctions in Würm. Part 2. Transformation of the environment by gigantic herbivores. *Vestnik Zoologii* 1992/1, 58-66. (Russian with abstract in English).

PUTSHKOV, P.V. (1992b): Uncompensated extinctions in Würm. Part 3. Overkill, underkill and other factors. *Vestnik Zoologii* 1992/4, 73-81. (Russian with abstract in English).

PUTSHKOV, P.V. (1993a): Uncompensated extinctions in Würm. Part 4. Old World: The mild variant of the crisis. *Vestnik Zoologii* 1993/1, 63-71. (Russian with abstract in English).

PUTSHKOV, P.V. (1993b): Uncompensated extinctions in Würm. Part 5. Crisis in the Palaearctic. *Vestnik Zoologii* 1993/4, 59-67. (Russian with abstract in English).

PUTSHKOV, P.V. (1994): Uncompensated Würm extinctions. Part 6. Final Communication: The crisis outside the Old World. *Vestnik Zoologii* 1994/2, 65-73. (English with abstract in Russian).

ROE, D.A. (1995): The Orce Basin (Andalucía, Spain) and the initial Palaeolithic of Europe. *Oxford Journal of Archaeology* 14/1, 1-12.

SCHÜLE, A. (1990): Simulation of populations dynamics as a means of palaeoecological research. In: D.L.Bruton & D.A.T.Harper (eds.), *Microcomputers in paleontology*, 82-105. Oslo (Contrib. Palaeontology Mus. Univ. Oslo 370).

SCHÜLE, A. (1992): A finite-element model to simulate distributions of large mammals. In: J.G. Goldammer (ed.), *Tropical forests in transition. Ecology of natural and anthropogenic disturbance processes,* 225-240. Basel.

SCHÜLE, W. (1990 a): Landscapes and climate in prehistory. Interaction of wildlife, man and fire. In: J.G. Goldammer (ed.), *Fire in the tropical biota. Ecosystem processes and global challenges,* 273-318. Basel.

SCHÜLE, W. (1990 b): Human evolution, animal behaviour, and quaternary extinctions: A paleo-ecology of hunting. *Homo* 41, 228-250.

SCHÜLE, W. (1992 a): Vegetation, megaherbivores, man and climate in the Quaternary and the genesis of closed forests. In: J.G. Goldammer (ed.),*Tropical forests in transition. Ecology of natural and anthropogenic disturbance processes,* 45-76. Basel.

SCHÜLE, W. (1992 b): Anthropogenic trigger effects on Pleistocene climate? *Global Ecology and Biogeography Letters* 2, 33-36.

SCHÜLE, W. (1993): Mammals, vegetation and the initial human settlement of the Mediterranean islands: A palaeoecological approach. *Journal of Biogeography* 20, 399-412.

SCHÜLE, W. & SCHUSTER, S. (1996): Klima, Speer und Feuer. Zur ökologischen Rolle des frühen Menschen. *Jahrbuch des Römisch-Germanischen Zentralmuseum Mainz* 42, 207-255.

SCHÜLE, W. & SCHUSTER, S. (1997): Anthropogener Urwald und natürliche Kultursavanne. Paläowissenschaftliche und andere Gedanken zu einem sinnvollen Naturschutz. In: B.Gehrken, C.Meyer (eds.), *Natur- und Kulturlandschaft* 2, 22-55. Höxter.

SWISHER C.C., CURTIS, G.H., JACOB, T., GETTY, A.G, SUPMIO, A. & WIDISMORO (1994): Age of earliest known hominids in Java, Indonesia. *Science* 263, 1118-1121.

TELEKI, G. (1973): *The hunting behaviour of wild chimpanzees.* Lewisburg.

THIEME, G, & VEIL, S. (1985): Neue Untersuchungen zum eemzeitlichen Elefanten-Jagdplatz Lehringen, Lkrs.Verden. *Die Kunde* NF 36, 11-58.

VOS, J. de & SONDAAR, P. (1994): Dating hominid findings in Indonesia. *Science* 266, 1726-1727.

WAAL, F.B.M. de (1982): *Chimpanzee politics.* London.

WAAL, F.B.M. de (1993): *Wilde Diplomaten.* München.

WAAL, F.B.M. de (1995): Die Bonobos und ihre weiblich bestimmte Gemeinschaft. *Spektrum der Wissenschaft* 5/ 95, 76-83.

WAGNER, E. (1988): Die Grabung an der Fundstelle "Bunker" und im Travertinbruch Haas in Stuttgart-Bad Cannstatt. In: *Archäologische Ausgrabungen in Baden-Württemberg 1987,* 17-19. Stuttgart.

WHITE, T.D., SUWA, G. & ASFAW, B. (1994): *Australopithecus ramidus,* a new species of early hominid from Aramis, Ethiopia. *Nature* 371, 306-312.

WHITE,T.D., SUWA, G. & ASFAW, B. (1995): *Ardipithecus ramidus,* a new species from of early hominid from Aramis, Ethiopia. *Nature* 375, 88.

WOLDEGABRIEL, G., WHITE, T.D., SUWA, G., RENNES, P., HEINZELIN, J. de, HART, W.K. & HEIKEN, G. (1994): Ecological and temporal placement of early Pliocene hominids at Aramis, Ethiopia. *Nature* 371, 330-333.

WOOD, B. (1994): The oldest hominid yet. *Nature* 371, 280-281.

WRANGHAM, W.C., McGREW, W.C., WAAL, F.B.M. de & HELTNE, P. (1994): *Chimpanzee cultures.* Cambridge.

ZISWILER, V. (1965): *Bedrohte und ausgerottete Tiere. Eine Biologie des Aussterbens und des Überlebens.* Berlin, Heidelberg, New York.

Survival and Adaptation of Pleistocene Humans on Islands

Paul Y. Sondaar & John de Vos

Abstract

In general it is considered that Homo erectus colonized about 50,000 years ago the islands. However the islands of Sardinia and Flores indicate that colonization took place much earlier, probably by Homo erectus. Survival and adaptations strategies of these early colonizers are discussed.

Introduction

Overseas dispersal is mostly considered a relative recent event in human history. In this view *Homo sapiens* was the first to cross seabarriers. *Homo sapiens* should have reached the island continent Australia about 50,000 years ago and the colonization of the islands of S.E Asia are placed in the same time interval (Bartstra et al., 1991; Allen, 1991; Jones, 1993). Arrival of humans on the Mediterranean islands is thought an even more recent event (Cherry, 1992). However this view is not tenable any more:

1. On the island of Flores (Indonesia) a lithic industry was found *in situ*, associated with fossils of the Middle Pleistocene *Stegodon trigonocephalus* (Maringer & Verhoeven, 1970). Recently, these finds were confirmed (Sondaar et al., 1994; Morwood et al., in press) and dated approximately 700,000 B.P. This implies that it was not *Homo sapiens*, who was the first to cross the sea-barrier, but *Homo erectus* instead.

2. Martini (1992) described artifacts *in situ* from Middle Pleistocene sediments at Sa Coa de sa Mua, near Perfugas on Sardinia. In this site, three different horizons with artifacts can be distinguished. This is hard evidence for a Middle Pleistocene presence of humans on the island Sardinia. However there are no indications of the presence of Pleistocene humans on other islands of the Mediterranean. This evidence points to the fact that colonization of an island is not only depending on "How to reach it" but "How to survive on it".

How to reach

Paleozoogeography and evolution of island mammals

It is evident, that the fauna on islands largely depends on which animals could reach the islands and at what time. For landmammals the most obvious obstacle is water, like channels, straits or wider seas. Simpson (1965) and Dermitzakis & Sondaar (1979) pointed out the possible dispersal routes of landmammals and their impact on the island faunal composition and evolution. The following routes are distinguished:

1. The corridor route, in which faunal interchange from one region to another is possible;
2. The filter route, where spread is probable for some animals, but improbable for others;
3. The sweepstake route, in which spread is impossible for most species, but possible under certain conditions for a small number of species;

4) The pendel route, which can be easily crossed by some mammals, while it is an insurmountable barrier for others.

The faunal evolution on an island depends highly on possible dispersal routes.

There are islands which are situated on continental shelfs and only separated from the mainland by relatively shallow water. These islands, called continental islands, are subject to being reconnected with the mainland by a relatively small lowering of sealevel.

On the other hand there are islands separated from the mainland by deep sea straits, or oceanic islands which rised from ocean depths, and are not subject to being reconnected with the mainland. These islands will mainly become populated by sweepstake dispersal and only a very few taxa will reach the island. When this colonization is successful, then an endemic fauna will evolve with very few taxa in which carnivores are lacking (unbalanced). However each island has its own geological history, which might be complex. This history might be reflected in the faunal evolution. For example unbalanced endemic islands faunas with a few taxa might be found on the present mainland. The only conclusion that can be drawn from this fact is that such areas of the mainland must once have been an island in a remote past.

The most striking change the larger mammals evolve on islands is diminution in size. The Pleistocene dwarf elephants of Cyprus, Crete and Sicily had the size of a pig (Ambrosetti, 1968; Bate, 1906; Sondaar, 1977). On the mainland, elephants and hippos did not have natural enemies, because of their larger size. On the Pleistocene islands, where there were no large predators in the fauna, size lost its significance. Small size has several advantages, such as less food needed daily, a greater mobility and a reduction of the territorial area.

Besides dwarfism of larger mammals, giant forms evolve on islands. Giant rats from the islands of S.E.Asia are a good example (Musser, 1982). The giant tortoises is an other vertebrate which might evolve on an island. The Galápagos islands provide such an example. In the past, giant tortoises were living on the Balearic Islands and the islands of S.E. Asia, like Sulawesi and the smaller Sunda islands as Flores. Though it remains uncertain whether this gigantism in island tortoises evolved on the islands, or was inherited from giant mainland ancestors (Sondaar, 1981).

Once an endemic island fauna is established, then it does not change much through the time. A striking example is the five million year faunal history of the Balearic Islands. The fauna was not affected by the global climatological changes in that time interval. On the mainland on the contrary, we find dramatic changes, such as extinctions and evolution of new taxa during that period. The arrival of humans on the Balearic Islands approximately 8,000 B.P. did change the fauna and flora, just as on the other Mediterranean islands (Sondaar et al., 1995).

Early and Middle Pleistocene dispersal routes of humans

Early Pleistocene Java

The faunal evolution of Pleistocene Java is characterized by faunal turnovers (Sondaar, 1984; de Vos et al., 1982, 1994) (Figure 1). In the Bumiaju section (West Java) and the Sangiran section (Middle Java) the same faunal turnover can be observed. The Early Pleistocene *Tetralophodon-Geochelone* fauna is replaced by the *Stegodon-Homo erectus* fauna. *Tetralophodon bumiajauensis, Hexaprotodon simplex*, at least two cervids, and a giant tortoise became extinct. This fauna, the Satir, was impoverished and did not contain carnivores, and was clearly an island fauna (Sondaar, 1984). The younger Ci Saat fauna has a mainland asiatic stamp. This fauna includes: *Hexaprotodon sivalensis, Stegodon trigonocephalus, Sus stremmi*, cervids, bovids and a *Panthera* sp. Hominids S4 and S5 from the Sangiran Formation, upper part of the "black clay's above tuff 9, might also be placed in this fauna. The Ci Saat fauna, the oldest member of the *Homo erectus-Stegodon* fauna, was still impoverished and the dispersal of this fauna was by a filter route.

By a local geological event the former island (Java) became connected by a landbridge with mainland Asia. The absolute dating by fission track of this event can be placed approximately 1.2

Figure 1 - *Catastrophic Early Pleistocene (± 1.2 Ma) faunal turnover on Java. Arrival of a mainland fauna, inclusive humans, by a land-connection causes the extinction of the island-fauna.The Satir fauna is replaced by the Trinil/Ci Saat fauna (Sondaar, 1984; de Vos et al., 1994).*

Ma (Suzuki et al., 1985). The subsequent colonization of the former island by the *Homo erectus-Stegodon* fauna (mainland), with the large carnivore *Panthera tigris* caused the extinction of the *Tetralophodon-Geochelone* fauna (island) without large predators. It might be clear that in this case it is difficult to infer which role humans, *Homo erectus*, played in the extinction catastrophe, except for the giant tortoise which must have became the victim of *Homo.* The only natural enemy of giant tortoises are humans, their extinction on island is usually related to the arrival of humans. As such they are good indicators for the arrival of man.

Middle Pleistocene Flores

Stone artifacts have been found in a fluvial sandstone layer pertaining to the Ola Bula Formation, near Mata Menge, west Flores (Figure 2). Based on the palaeomagnetic measurements the age of the artefact-bearing layer has assessed 0.7 Ma, which means that only *Homo erectus* could be the maker. The same layer contains fossil remains of a large *Stegodon*, a giant rat, teeth of a small crocodile, freshwater molluscs and plant remains (grasses). An older faunal association occurs at the base of the Ola Bula Formation. It contains a pygmy *Stegodon*, a giant tortoise and *Varanus komodoensis.* The latter was found later at another site with artifacts. This means that there had been an extinction of the dwarf *Stegodon* and the giant tortoise (Sondaar et al., 1994; van den Bergh et al., 1996) (Figure 3).

The following scenario may be reconstructed when humans arrived on the island (Sondaar, 1987): The dwarf *Stegodon* and the giant tortoise *Geochelone* were an easy prey and became extinct. The giant rats were a major source of protein for Palaeolithic humans (giant rats are still eaten by the local people), but could survive by their high reproduction rate. The settlement of human hunters on the island meant that there was a large predator on the island, and when the mainland *Stegodon trigonocephalus* from Java arrived on the island he was hunted. In consequence he did not evolve into a dwarf. The Eastern Sunda island are dry and the plateaus are the best

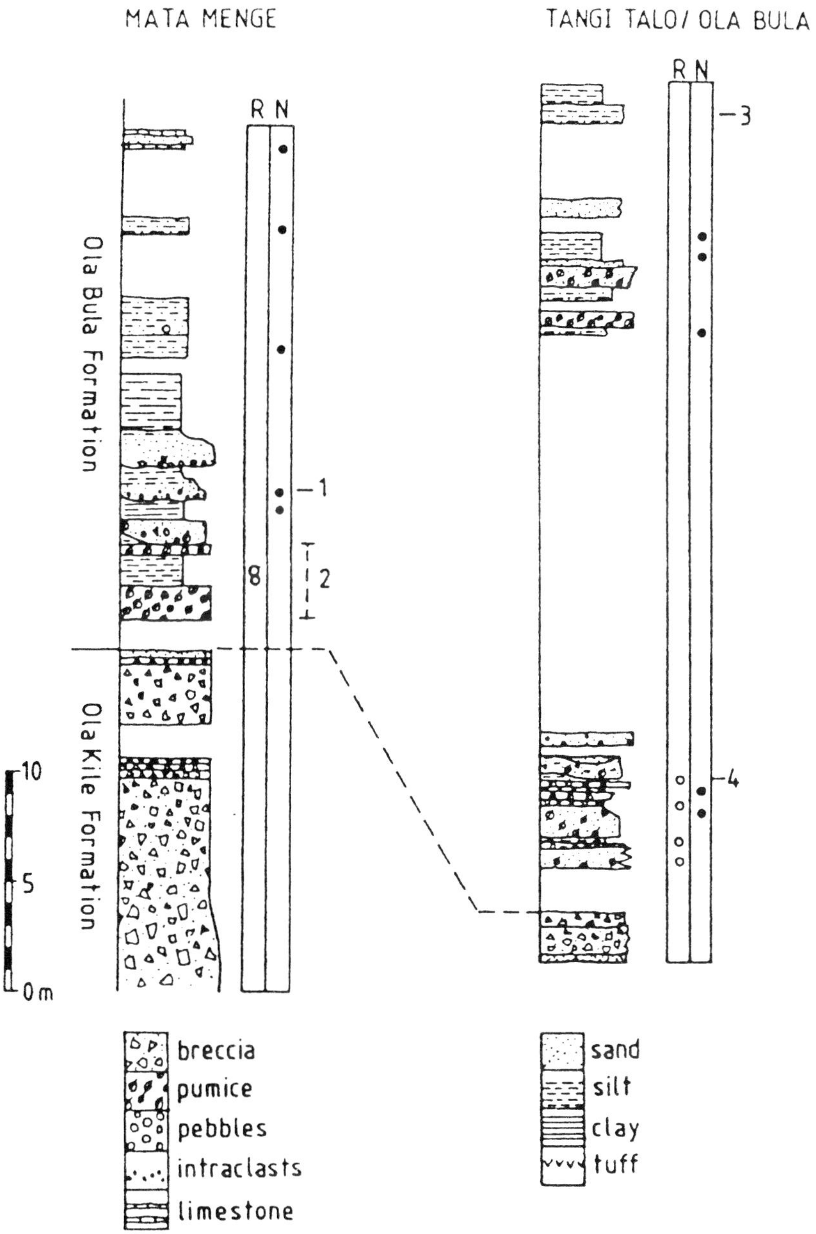

Figure 2 - *Two sections: Mata Menge (left) and Tangi Talo/Ola Bula (right). Magnetic polarity (N = normal; R = reversed) of several sampled horizons is indicated to the right of each section. 1 - Sandstone/silt layer near Mata Menge, which has yielded artifacts and fossils of Stegodon trigonocephalus florensis Hooijer; 2 - Pink/white mottled paleosoil interval; 3 - Layer near Ola Bula, excavated by Verhoeven. Fossils of Stegodon trigonocephalus florensis and Hooijeromys nusatenggara Musser have been found in this layer; 4 - Fossil-bearing layer near the hill Tangi Talo, which has yielded remains of a pygmy Stegodon, a giant tortoise (Geochelone sp.) and several teeth of Varanus komodoensis Ouwens. (After Sondaar et al, 1994).*

Figure 3 - *Extinction and recovery of Middle Pleistocene fauna in Flores (± 0.7 Ma). Arrival of Homo erectus by sweepstake dispersal on the island causes the extinction of the giant tortoise and the dwarf Stegodon. The dwarf Stegodon (Fauna A) is replaced by a Stegodon of mainland dimensions (Fauna B).*

typified as an open savanna covered with grasses. This was the same in the Pleistocene. Just as the stratigraphically older dwarf, the large *Stegodon* did evolve higher crowned teeth to chew the hard grasses (van den Bergh et al., 1996). *Homo* managed to arrive on the island by overseas dispersal and the arrival of humans on the island Flores is a clear example of the sweepstake route dispersal.

Middle Pleistocene Sardinia

The lithic industry found in Northern Sardinia (Arca et al., 1982; Martini, 1992) is evidence for a Middle Pleistocene human on the island, and the extinction of many elements of the *Nesogoral* fauna might be caused by the hunting activities of humans. Just as in Flores there was a dramatic faunal turnover in the Middle Pleistocene, and this event can be explained in a similar way. Arrival of humans on the island caused the extinction of the dwarf pig, and the endemic goat *Nesogoral*. These artiodactyls were replaced in the younger fauna by a deer with mainland proportions, *Megalocerus cazioti*. Just as the large *Stegodon* from Flores, the deer did not evolve specific island characters, like dwarfism and low gear locomotion. The parallel goes further, also on Sardinia there was a mammal, *Prolagus,* which might have been a crucial source of food for the newly arrived Pleistocene humans. This hare like mammal had a high reproduction rate and we know from burnt *Prolagus* bones in Holocene deposits that Neolithic humans did eat them.

If we suppose that humans arrived on islands like Crete or Karpathos with cervids having island characteristics, than we can imagine the following scenario. The short legged and dwarfed deer would have been an easy prey and being reduced rapidly in number. This caused a deficiency in foodsupply for the Pleistocene humans as there was no other game of interest. There was a small mammal with a high reproduction rate, the murid *Mus minotaurus*, but this mouse was predated by an endemic owl and far too small to serve as foodresource for Pleistocene humans.

We can observe extinction of the large herbivorous mammals and extinction on specific level of small herbivorous mammals like *Prolagus* a pika and *Rhagamys* a murid. In the younger fauna *Prolagus* is represented by a smaller species, while the murid species is larger than its ancestor from the older fauna. Did the Palaeolithic hunter exerted a selective pressure on the large *Prolagus* which evolved into a smaller species?

The following may be concluded on human dispersal to islands:

1. *Filter route:* If a landconnection comes into existence between an island and the mainland, humans will migrate to the former island together with its accompanying fauna, causing a catastrophic faunal turnover. The local endemic island fauna will be replaced by the mainland fauna. Only a few endemics, mostly small mammals, may survive. This is the case in Early Pleistocene Java.

2. *Sweep stake route:* Overseas dispersal of humans to the island will result in a dramatic faunal turn-over. In the fossil record this is characterized by the extinction of the endemic megafaunal elements, and the arrival of a few mainland species, who do not evolve typical island characters. Dwarfism and low gear locomotion will not evolve (Sardinia, Flores, possibly Sulawesi and the Philippines). Smaller mammals may survive, and in that case will show some evolutionary changes. The dramatic faunal turn-over, on Flores (palaeomagnetism) and Sardinia (paleontological evidence) can be placed at approximately 0.7 Ma, the Middle Pleistocene.

How to survive

Survival and adaptation

Islands with Middle Pleistocene humans are Sardinia in the Mediterranean, Flores, and most probable Timor, Sulawesi and the Philippines in S.E Asia. These island have in common a mammal, that lived on the islands when humans arrived, which had high reproduction rate and were sufficiently large in size to serve as a foodresource: the pika *Prolagus* on Sardinia and giant rats on the islands of S.E. Asia. It is known that these mammals were an important foodresourse for Neolithic humans. From this it might be concluded that only those islands would have been suitable for permanent colonization by Pleistocene humans if the exploitation of the natural resources on the island could support a viable human population over a long period of time, without exhaustion of the resources.

In this scenario, not the overseas travel was the bottle-neck for early humans to settle on islands, but the availability of food. On islands with an endemic and unbalanced fauna with a low diversity, the presence of a mammal, large enough in size, with a high rate of reproduction seems to be essential for the permanent settlement of a hunter- gatherer population. The pika *Prolagus* of Pleistocene Sardinia/Corsica and the giant rats from the islands of S.E.Asia fulfilled this requirement.

The *Homo erectus* faunas of Java, some times a continental island but connected in glacial periods, are well known and show the usual diversity seen in mainland faunas. However the environments of the oceanic islands, where Middle Pleistocene humans settled, was entirely different. On Flores there were only giant rats and a *Stegodon* with mainland proportions to hunt on; on Sardinia only *Prolagus* and deer.

It is tempting to speculate if this island-human evolved special characters as an adaptation to the island environment. To answer this question we need fossils of this human. Sofar, they have not been found on Flores, but from Sardinia we have some information. Human fossils were found in Corbeddu Cave (Oliena): a temporale and a maxilla (without teeth), dated C/14 8,750 ± 140 years B.P., about 1,000 years before Neolithic humans colonized the island. The cranial-fragments show some distinct characters, which may set them apart from contemporary mainland populations (Spoor & Sondaar, 1986).

Further a proximal part of an ulna was found out of stratigraphical context. This ulna is more fossilized and is distinct in morphology from modern - as well as fossil humans. Neolithic humans from Sardinia are not significantly different from those of the European mainland. This suggests a degree of endemism in the isolated pre-Neolithic population on Sardinia. It could be argued, that the Pleistocene humans were better adapted to the island environment of Sardinia in having a hunting strategy based on deer and the pika *Prolagus*. Neolithic man, with a completely different living strategy, destroyed the environment for the Palaeolithic humans. However the scarce human fossils found sofar are not conclusive for such a hypothesis. More human fossils need to be found to test it. Potentially there are possibilities to find human fossils on Flores. The Soa basin is very rich in fossiliferous deposits, especially the upper part of the Ola Bula Formation. A large scale prospecting and excavating campaign is therefore necessary.

Conclusions

The following may be concluded:

1. If a corridor comes into existence between an island and the mainland, humans will migrate to the former island together with its accompanying fauna, causing a catastrophal faunal turnover. The local endemic island fauna will be replaced by the mainland fauna. Only a few endemics, mostly small mammals, will survive. Examples are Early Pleistocene Java and Upper Pleistocene Sicily.

2. Overseas dispersal of humans to the island will result in a dramatic faunal turn-over. In the fossil record this is characterized by the extinction of the endemic megafaunal elements, and the arrival of a few mainland species, who do not evolve typical island characters. This dramatic faunal turn-over can be dated at approximately 700,000 B.P. on Flores (palaeomagnetism) and Sardinia (paleontological evidence). The smaller mammals may survive, and will show some evolutionary changes, on specific level, after the human arrived. Once settled on the island, the fauna will evolve different from the islands without humans. Dwarfism and low gear locomotion will not evolve (Sardinia, Flores, possible Sulawesi and the Philippines).

3. Islands with unbalanced faunas are only suitable for permanent colonization by Pleistocene Man, if exploitation of the natural resources on the island can support a viable human population over a longer period of time, without exhaustion of the resources. On islands with endemic and unbalanced fauna, with a low diversity, the presence of a mammal, large enough in size, with a high rate of reproduction seems to be essential for the permanent settlement of a hunter-gatherer population. Examples for this are the ochotonid *Prolagus* from Sardinia/Corsica and the giant rats from the islands of S.E. Asia.

References

ALLEN, H. (1991): Stegodonts and the dating of stone tool assemblages in islands of S.E. Aisa. *Asian Perspectives* 30, 243-660.

AMBROSETTI, P. (1968): The Pleistocene dwarf elephants of Spinagallo (Siracuse, Southeastern Sicily). *Geologica Romana* 7, 277-398.

ARCA, M., MARTINI, F., PITZALIS, G., TUVERI, C. & ULZEGA, A. (1982): Paleolitica dell'Anglona (Sardegna Settentrionale). Richerche 1979-1980. *Quadrani*, 12, 58, fig. 27.

BARTSTRA, G.J., KEATES, S.G., BASOEKI & KALLUPA, B., (1991): On the dispersion of *Homo sapiens* in eastern Indonesia. *Current Anthropology* 32, 317-321.

BATE, D.M.A. (1906): The pygmy *Hippopotamus* of Cyprus. *Geological Magazine* 3, 241-245

BERGH, G.D. van den, MUBROTO, B., AZIZ, F., SONDAAR, P.Y. & VOS, J. de, (1996): Did *Homo erectus* reach the island of Flores?. *Indo-Pacific Prehistory Association Bulletin* 14, 27-36.

CHERRY, J.F. (1992): Paleolithic Sardinians?. In: R. Tykot & A.T. Sheffield (eds.), *Sardinia in the Mediterranean. A footprint in the sea,* 28-39. (Studies in Sardinian Archaeology).

DERMITZAKIS, M.D. & SONDAAR, P.Y. (1979): The importance of fossil mammals in reconstructing paleogeography with special reference to the Pleistocene Aegean Archipelago. *Annales Géologique des Pays Hellénique* 29, 808-840.

JONES, R. (1993): Continental reconnessance: Some observations concerning the discovery of the Pleistocene archeology of Australia. In: M. Spriggs (ed.), *Community of cultures*, 97-122.

MARINGER, J. & VERHOEVEN, TH. (1970): Die Steinartefakte aus der Stegodon-Fossilschicht von Mengeruda auf Flores, Indonesien. *Anthropos* (Fribourg) 65, 229-247.

MARTINI, F. (1992): Early human settlements in Sardinia: The Palaeolithic industries. In: R. Tykot & A.T. Sheffield (eds), *Sardinia in the Mediterranea. A footprint in the sea*, 40-48. (Studies in Sardinian Archeology).

MORWOD, M.J., AZIZ, F., BERGH, G.D. van den, SONDAAR, P.Y. & VOS, J. de (1997): Stone artefacts from the 1994 excavation at Mata Menge, West Central Flores, Indonesia. *Australian Archaeology* 44, 26-34.

MUSSER, G.G. (1982): The Trinil rats. *Modern Quaternary Research in Southeast Asia* 7, 65-85.

SIMPSON, G.G. (1965): *The geography of evolution*. Philadelphia, New York.

SONDAAR, P.Y. (1977): Insularity and its effect on mammal evolution. In: M.N. Hecht, P.L. Goody & B.M. Hecht (eds.), *Major patterns in vertebrate evolution*, 671-707. New York.

SONDAAR, P.Y. (1981): The *Geochelone*-faunas of the Indonesian Archipelago and their paleogeographical and biostratigraphical significance. *Modern Quaternary Research in Southeast Asia* 6, 111-119.

SONDAAR, P.Y. (1984): Faunal evolution and the mammalian biostratigraphy of Java. *Courier Forschungs-Institut Senckenberg* 69, 219-235.

SONDAAR, P.Y. (1987): Pleistocene mammals and extinctions of islands endemics. *Mémoirs de la Société Géologique de France*, N.S. 150, 159-165.

SONDAAR, P.Y., BERGH, G.D. van den, MUBROTO, B., AZIZ, F., VOS, J. de & BATU, U.L. (1994): Middle Pleistocene faunal turnover and colonization of Flores (Indonesia) by *Homo erectus*. *C.R. Acad. Sci. Paris* 319, serie II, 1255-1262.

SONDAAR, P.Y., McMINN, M, SEGUI, B & ALCOVER, J.A. (1995): Paleontological interest of karstic deposits from the Gymnesic and Pityusic islands. *Endins* 20 (*Mon.Soc.Hist.Nat. Balear* 3), 155-171.

SPOOR, F. & SONDAAR, P.Y. (1986): Human fossils from the endemic island fauna from Sardinia. *Journal of Human Evolution* 15, 399-408.

SUZUKI, M., WIKARNO, BUDISANTOSO, SAEFUDIN, I. & ITIHARA, M. (1985): Fission track ages of pumice tuff layers and javites of hominid fossil bearing formations in the Sangiran area, Central Java. In: N. Watanabe & D. Kadar (eds.), *Quaternary geology of the hominid bearing formations in Java*, 309-357. Bandung (Geological Reseach and Development Centre Special Publication 4).

VOS, J. de, SARTONO, S., HARDJA-SASMITA, S. & SONDAAR, P.Y. (1982): The fauna from Trinil, type locality of *Homo erectus*. A reinterpretation. *Geologie en Mijnbouw* 61, 207-211.

VOS, J. de, SONDAAR, P.Y., BERGH, G.D. van den & AZIZ F. (1994): The *Homo* bearing deposits of Java and its ecological context. *Courier Forschungs-Institut Senckenberg* 171, 129-140.

Modes of Life in the Palaeolithic: Not Survival But Well-being

Marcel Otte

Abstract

The extraordinary success of Palaeolithic populations to expand and progress everywhere, during hundreds of thousands of years, demonstrates adaptive exceptional in the animal kingdom. Only our retrospective vision dares to put in doubt such evidence. The destined blindness is due to our fear of our own past and our real animal nature, that our mind refuses.

Introduction

This contribution puts into question the often used notion of "survival" during the European Pleistocene, prior to the Upper Palaeolithic. Curiously, the archaeological literature contains frequent allusions to the alimentary deficiencies of the Mousterians. Some persistent insinuations imply a vague difficulty of provisioning of which Neandertals would have been victims. This supposed alimentary deficiency would thus be in opposition with the abundance enjoyed by people of the Upper Palaeolithic.

This assumption appears contestable to us both on a theoretical level and in the face of empirical archaeological evidence.

Longevity

If we confine ourselves to the European Middle Palaeolithic, we have an occupation which is at the same time dense and of very long duration. Around three hundred thousand years separate the early phases with the Levallois technique from the interruption provoked by "modern" man, around forty thousand years ago. This extreme duration demonstrates, on its own, the modes of alimentary equilibrium, which were continually adapted and of high quality, given varying time periods and successive environmental conditions.

A deficient alimentary system would have led to rapid and inevitable extinction of the European population at the first climatic fluctuation. This longevity proves, in contrast, the extreme adapative flexibility of Palaeolithic methods of subsistence.

Comparatively, all later periods in the history of "man the hunter", were clearly shorter, and therefore less "successful" than that of the Early Palaeolithic. Related to this, what could be the destiny of humanity today and its chances of success, for instance, for the duration of the next millenium ?

Territories

Across space as well as through time, the adaptive power of the Mousterian is manifested by the diversity of ecological habitats occupied. Diverse assemblages of the Middle Palaeolithic are found in every latitude, every altitude and every topographic context.

Progressively, the entire continent was "colonized" by Pre-Neandertals and then by the Mousterians themselves. This original migration was concomitant with an appropriate and still more intimate and subtle, and therefore more powerful, adaptation.

The progressive increase in the number of sites, their extension across space and the diversity of habitats also evidences this adaptive "success", and therefore alimentary success as well.

The prey

Bone remains found at Mousterian sites show the diversity of prey, adaptive possibilities presented by hunting techniques and the possible specialisation of Mousterian predation (Otte & Patou-Mathis, 1992).

As much as in the mountainous sites of Cantabrian Spain (Altuna, 1992) as in the Caucasus (Lioubine, 1993) or the Polish Plain (Gautier, 1989), this diversity demonstrates a broad agreement between predation and environmental resources.

Conversely, a wide range of prey shows "opportunistic" behavior, that is to say, intimately adapted to the behavior of each animal.

Environmental determinism does not exist; rather, one observes an aptitude for all forms of hunting, butchery and sharing, infinitely more rich than for the specialised hunters of the Final Palaeolithic, for example.

Hunting

Technical elaboration, attested by the lithic tools, demonstrates the ability to produce weapons and traps. These mechanical preparations (e.g., the Levallois method) were also applied to the organization of the hunt, coordination of the group and learning of prey behavior.

Some preserved wooden weapons (Lehringen, Clacton) demonstrate the existence of technical processes applied to vegetal materials, apparently elaborate and varied.

The variety itself of butchered animals should not be interpreted in terms of restrained opportunism but rather as evidence of a range of methods of appropriate hunting.

Environment

Considered over the long-term, equivalent to the Early Palaeolithic (Lower and Middle), the European environment underwent phases of rewarming, sometimes quite intense and long-lasting.

It appears therefore likely that a good part of the diet was of vegetal origin, as certain traces discovered in the Eifel volcanos (Bosinski et al., 1988) and as the customs of living hunter-gatherers show in such favorable climatic conditions (Sahlins, 1972).

It seems required as well that a primitive diet must necessarily contain an important vegetal component in order to simply maintain a population over time (Delluc, 1995).

Finally, traces of a dietary deficiency do not appear to be present on Neandertal bones.

Interpretation

Upon reflection, we can ask ourselves upon what basis the vague idea of an alimentary deficiency during the Early Palaeolithic can be founded. It does not seem to be based on available archaeological data which show, in contrast, a strong adaptive power, longevity, and a constant demographic increase. Aptitudes reflected in the technology confirm this adaptive subtlety.

This notion of deficiency, of a fight for "survival", a quest for subsistence, is present rather as a product of our own retrospective regard on the evolution of humanity.

The notion of human "progress", which haunts all the assumptions in historical and evolutionary disciplines (natural sciences and human sciences), appears to be the underlying premise of this contestable hypothesis.

"Modern" man, to which we associate ourselves in its glorious destiny, will be found in this way distinguished more clearly from its unlucky, incapable, predecessor.

We could argue, in contrast, that a reduction in hunting opportunities, which appear with the specialisation of the Upper Palaeolithic, denote a reduction in adaptive capabilities to environmental rate changes.

Discussion

While found quite frequently in the literature, archaeological data do not support the idea of a "survival" which would simply be a means to "remain alive", like a swimmer who holds his head above water.

Unconsciously, it is perhaps a matter of establishing a difference in degree between "ancestors" and us, such that the myths present a difference relative to their lack of language or social values, for example. If it had been a matter of "survival," they would not have disappeared and we would not be here to speak of them. Instead. Paleolithic populations maintained themselves over hundreds of thousands of years, were adapting, perfecting, developing, expanding everywhere.

The reason for this "spectacular success" resides in the aptitude for all forms of hunting. This consists particularly in the killing and consumption of a being *similar to humans* in behavior. It is therefore a challenge against the world. This sacrifice is an exchange with Nature, produced by the development of the conscience. This challenge is applied to the animal because it represents the best of Nature.

This phenomenon was therefore essentially spiritual in the first stage. It was first necessary *to dare to kill,* not be need but by the desire to exist in contrast with the natural condition of humans, because they had at the same time the impression of belonging to nature and the desire to be separate from it (Franck, 1990; Otte, 1996).

This sacrifice is based then on the exchange between human life and animal death. So, it requires one to kill and consume. This weapon, the only one that is truly powerful, is thus situated first in the mind; technology was made only to follow it. But this fundamental audacity allowed populations to migrate, to leave the original forest belt and to adapt to northern latitudes (Delluc, 1995; Cachel, 1997). While migration was simple in Asia, the Mediterranean formed an obstacle which filtered the expansion toward Europe. The loss of vegetal resources was compensated by meat and animal fat. It remains to know why humans migrated, because only possibilities existed, not the necessity.

The mastery of fire participated in this conquest. It brought light to darkness, warmth to cold, changed the mechanical characteristics of material, but especially, frightened animals. In this way, it reinforced the notion of power over Nature.

Technology permitted the planning, organisation, preparation, and anticipation of future needs. Social space attests to this organization, by the presence of workshops and wooden shelters. The use of wood is also found in the production of weapons: at Lehringen where stone was used to work wood (Veil, 1990-91) and at Schöningen where wooden points intended to be thrown were produced (Thieme, 1997). The extreme intensity of working of wood is found in many traditional populations today whose stone artifacts are not more sophisticated than those of the Lower Paleolithic.

The modes of life and Mousterian technology were as well adapted to the temperate conditions of interglacials as to the cold of glacial eras. They are found in wet environments, near ponds and lakes, as well as on the very dry steppes (Caucasus), on mountains (Caucasus) as well as plains (Zwollen). There was therefore a range of adaptations to all types of environments, with equal success. This was due neither to chance nor determinism, but rather to human choice, focused, decided, assumed. Technology and modes of life were sufficiently flexible and powerful for all these forms of choice. The deterministic explanation that we could consider justifying these changes would be only a note of failure of our own aptitude for intelligence and imagination. Palaeolithic humans, themselves, "survived" calmly and quietly.

When Darwinian theory was opposed to the Bible, it was still a matter of explaining the difference between man and nature. The words changed but the story remained the same. It was to designate the inferiority of preceding man: different because anterior, anterior because inferior. We continue to want to see these men of the past with problems of existence so that we can feel ourselves at the summit. To consider ourselves at the "end of evolution," our mind reclaims this story which is only a children's tale. Again, everything happens in our mind, even today . . .

References

ALTUNA, J. (1992): Le paléolithique moyen de la région cantabrique. *L'Anthropologie* 96, 87-101.

BOSINSKI, G., van KOLFSCHOTEN T. & TURNER E. (1988): *Miesemheim I. Die Zeit des Homo erectus,* Andernach (Andernacher Beiträge 2).

CACHEL, S. (1997): Dietary shifts and the origins of the European Upper Palaeolithic. In. H. Ullrich (ed.), *Abstract Book of the International Interdisciplinary Symposium "Lifestyles and survival strategies in Pliocene and Pleistocene hominids",* 19-20. Weimar.

DELLUC, G. (1995): *La nutrition préhistorique.* Perigueux.

FRANCK, M. (1990): *Le dieu à venir.* Paris.

GAUTIER, A. (1989): Preliminary notes on the fauna of the Middle Palaeolithic site at Zwolen (Poland). In: M. Otte (ed.), *L'homme de Néandertal.* Vol. 6, *La subsistance*, 69-73. Liège (ERAUL 33).

LIOUBINE, V.P. (1993): La chronostratigraphie du paléolithique du Caucase. *L'Anthropologie* 97, 291-298.

OTTE, M. (1996): *Le paléolithique inférieur et moyen Europe.* Paris.

OTTE, M. & PATOU-MATHIS, M. (1992): Comportements de subsistance au paléolithique moyen en Europe. *Paléo* 4, 29-34.

SAHLINS, M. (1972): *Age de pierre, âge d 'abondance. Economie des sociétés primitives.* Paris.

THIEME, H. (1997): Lower Palaeolithic throwing spears and other wooden implements from Schöningen, Germany. In. H. Ullrich (ed.), *Abstract Book of the International Interdisciplinary Symposium "Lifestyles and survival strategies in Pliocene and Pleistocene hominids",* 105-107. Weimar.

VEIL, S. (1990-91): Die Nachbildung der Lanze von Lehringen. Experimente zur Holzbearbeitug im Mittelpalaeolithikum. *Die Kunde* N.F. 41-42, 9-22.

Reproductive Ecology and Early Hominids

Becky A. Sigmon

Abstract

This paper suggests that the reproductive patterns of early hominids were intimately related to the environment they inhabited. Climate, temperature, photoperiocity, amount of daily sunlight and food accessibility are known to be factors that affect the breeding patterns of mammals including nonhuman primates. These influences act as timing cues, zeitgeber, that cause the reproductive system to function at peak capacity, resulting in the animal's experiencing peak breeding seasons. More recent studies are finding that reproductive patterns of modern humans may also be affected by environmental factors and that human populations also experience peak fertility periods.

The present model differs from previous explanations of hominid reproductive behavior that have suggested that early hominids had voluntary control of their sexual behavior, that they did not have breeding seasons, that females no longer experienced estrus as the periodic occurrence of peak fertility, and that males did not experince fluctuations in their reproductive capabilities. These interpretations are being contradicted by findings from more recent studies in reproductive ecology. The earlier view implied that hominids had evolved a new reproductive pattern that led to the development of the sexual division of labor, male provisioning of female in exchange for "sexual favors" and fatherhood. Alternative explanations need to be found to explain the origin of these human behaviors.

The view taken in this paper is that early hominids fit into the evolutionary continuum of Old World monkeys and apes in having peak fertility periods in spite of being able, like their Old World primate relatives, to copulate year round. Hominid matings tended to occur more frequently and fertilization was more likely during those seasons that corresponded to the times during which conditions in the environment and specific habitat were most favorable for health and general feeling of fitness. Early hominids, females and males, experienced peak fertility periods at least annually and possibly, bi-annually, depending on the environment. This would have influenced the social structure of the hominid group.

Introduction

The purpose of this paper is to explore the question of reproductive ecology in Palaeolithic hominids. Relevant research will be reviewed on the reproductive patterns of modern species of mammals, and of the more closely related nonhuman primates and modern humans. Since early hominids fit into the evolutionary framework between the latter two groups, emphasis will be placed on data pertaining to nonhuman primates and modern humans. A working model for a new way to view early hominid reproductive ecology is proposed. This model may provide insight into how reproductive patterns could have affected the lifestyle and survival strategies of Palaeolithic hominids.

Environmental influences on mammalian reproductive ecology

Studies in mammalian reproductive ecology have demonstrated relationships between environmental factors and the animal's reproductive physiology (for example, see: Baker, 1938; Amoroso & Marshall, 1960; Gilmore & Cook, 1981; Gwinner, 1986; Prentice & Whitehead, 1987). Photoperiod is considered a primary *zeitgeber* that turns on the mating mechanisms in many animals (Gwinner, 1986). Temperature is considered by Macfarlane (1974) to be another major factor that causes seasonality in conception peaks. The role of the amount of daily light has been shown to affect ovulation in mammals (Clegg et al., 1956; Ganong et al., 1963; Menaker, 1971; Karsch & Foster, 1981) and breeding seasonality in humans (Timonen & Carpen, 1968; Takahashi, 1964; Ehrenkranz, 1983a,b; Hajek et al., 1981).

Environmental temperature has been demonstrated to play a role in mammalian ovulation (Ryle, 1961; Holmes, 1968; Allen, 1978; Michael & Zumpe, 1981) and in sperm volume and testicular activity in humans (Zorgniotti, 1991; Levine, 1994) and in other mammals (Cupps et al., 1960; Holmes, 1968; Waites & Ortevant, 1968). The importance of temperature in reproductive physiology of mammals in general has been investigated by Huntington (1938), Chang et al. (1963), Macfarlane (1970, 1974), and Roenneberg & Aschoff (1990).

Adequate nutrition is another contributing factor affecting mammalian reproductive physiology (Ivanovsky, 1923; Keyes et al., 1950; Frish, 1978a,b; Allen, 1978; Michael & Zumpe, 1981; Widdowson, 1981; Campbell & Leslie, 1995). Loudon and Racey (1987) state, in their preface: "....remarkably little is known of the physiological mechanisms whereby nutrition directly regulates fertility and inter-birth interval." Their book reviews much of the literature and studies in this subject.

In human populations, activity patterns have been shown to affect female and male reproductive physiology including individual activity patterns (Levine, 1991; Cummings et al., 1983; Campbell & Leslie, 1995) and group activity patterns (e.g. Panter-Brick & Ellison, 1994).

A variety of additional stresses that may occur in human civilization are known to affect fertility, such as effects from wars (Dean, 1949; Helweg-Larsen et al., 1951), disease and psychological stress (Campbell & Leslie 1995).

Through a review of the reproductive patterns in nonprimate mammals, nonhuman primates and humans, I will consider the question of whether or not the reproductive patterns of early hominids were influenced by environmental factors, and whether or not early hominids may have had breeding periodicity.

Seasonal breeding in mammals

Throughout the plant and animal kingdoms, there is seasonality to reproduction. It applies to mammals in general (Aschoff, 1955; Follett & Follett, 1981), and more specifically to nonhuman primates (for example, see: Hartman, 1932; Gauquelin, 1968; Hafez, 1971; Graham, 1981) and modern humans (Roenneberg & Aschoff, 1990; Lam & Miron, 1994).

"The universal existence, precision and conservation of this annual timing of reproduction reflect a high selective pressure for an appropriate seasonal niche." (Roenneberg & Aschoff, 1990: 195).

Reproductive seasonality is thought to be due to the presence in animals of endogenous annual clocks (Follett & Follett, 1981; Gwinner, 1986). These circannual rhythms seem to provide a kind of template on which optimal timing of reproduction can be based, through the body's response to various environmental factors. As indicated earlier, primary *zeitgebers* that turn on this clock are thought to be photoperiod and temperature. Other environmental factors such as nutrition are also known to affect fertility. In some environments, rainfall is a factor that is necessary before breeding seasons commence, especially in areas that have seasonal dry and wet periods (Koford, 1965). Ample rainfall causes plant growth and therefore directly or indirectly increases food supply for animals.

Because the concept of seasonality in reproduction may be misconstrued to refer to the birthing season, I want to make it quite clear that this is not the case. The high selective pressure in the timing of reproduction that is being referred to here is for *breeding season*, not for birthing season.

Endogenous annual clocks are present in animals in order to prepare the reproductive physiology of the animal for its peak fertility. The birthing period of a species is a product of the seasonal breeding time in species; it is the latter that is under stringent selective pressure.

Mammalian breeding is cyclical and is directly related to the occurrence of specific factors in the environment. The importance of a given ecological variable in affecting reproduction differs from place to place. There could be one or a combination of factors that act to initiate the peak breeding period in animals. The breeding season in mammals is referred to as the rutting season which is a period of sexual excitement in adult females and males, and it is their annual time of mating. The adult female and male at this time are at their biological *peak fertility*. The environmental factors referred to in the earlier part of this paper are the initiators of the rutting season in that they act as the "releasers" of the endogenous circannual rhythms (Gwinner, 1986).

The female mammal in a state of rut is said to be in estrus, although in both adult females and males there is a cyclical development in the endocrine and reproductive systems that prepare them for a period of peak fertility and of mating (Holmes, 1968; Follett & Follett, 1981). The male is said simply to be in a state of rut. There is an annual seasonality to the breeding period, although studies have shown that mammalian fertility can be enhanced, and mating period increased to more than one per year, through improved living conditions. The conditions that apply are enhanced modifications of the environmental factors that bring on peak fertility, including amount of daylight, food available and warm temperatures (Heape, 1900; Clegg et al., 1965; Allen, 1978; Menaker, 1971; Karsch & Foster, 1981; Loudon & Racey, 1987). Domesticated animals may breed at any time of the year (Holmes, 1968). Studies suggest that the physiology of pregnant and lactating female mammals is adapted to preserve a certain amount of their energy and fat reserves for the demanding needs of the latter stages of gestation and the following period of lactation. These physiological adaptations can take the form of cyclic starvation in the polar bear (Watts & Hansen, 1987), energy balance and delayed mobilization of fat reserves in the Svalbard reindeer (Tyler, 1987), and other methods described in various sources like Loudon and Racey (1987).

Reproductive ecology in nonhuman primates

Non-human primates tend to have reproductive patterns that are similar to those in other mammals (Hafez, 1971; Graham, 1981) including seasonal breeding and a tendency toward a rutting season. Primates are different, however, in that they have developed a further elaboration of the reproductive system which is not characteristic of nonprimate mammals. Old World primates exhibit sexual behaviour, including intercourse, at times other than their regular breeding seasons (Goodall, 1968; Loy, 1971; Hafez, 1971; Rowell, 1972; Lemmon & Allen, 1978; Savage-Rumbaugh & Wilkerson, 1978: Robinson & Goy, 1986). These copulations generally do not result in conception, and they cannot be taken as an indication of the animals' most fertile time. Rather, this copulatory behaviour probably serves other purposes than just procreation, as has been suggested by Robinson & Goy (1968) and others. This pattern of extra-procreative intercourse and sexual behaviour continues throughout the evolution of the primate lineage, including the family Hominidae, but it does not seem to be a feature of nonprimate mammals.

Researchers studying primates have tended to emphasize the female rut or estrus as the primary initiator of the breeding season. The reason for this tendency is not clear, but may be partly explained by the high degree of visibility of female sexual swellings in the perineum compared to the less observable changes in the male primate in rut. However, male primates, like other male mammals in rut, also experience physiological changes in their generative system including a rise in their plasma testosterone and androgen levels (Michael & Bonsell, 1977; Michael & Zumpe, 1978), as well as morphological changes in the perineal region (Koford, 1965). These changes in the reproductive physiology may be brought on by exogenous factors in the environment (Zorgniotti, 1991; Chik et al., 1992). The breeding season is therefore a time in which female and male are at their biological peak fertility.

In primate studies, as noted, researchers have focused on the female estrus cycle rather than on the breeding season that involves biological reproductive enhancement in both females and males. It is very likely that this focus in the primate literature has influenced researchers in hominid evolution who have proposed that the "loss of estrus", rather than the loss of the breeding season, has played a large role in early hominid evolution (for example, see: Morris, 1967; Washburn & Lancaster, 1968; Short, 1980, 1994; Hrdy, 1981; Lovejoy, 1981; Fisher, 1982;). Evidence from more recent studies in reproductive ecology of humans (Roenneberg & Aschoff, 1990; Zorgniotti, 1991; Levine, 1994; Campbell & Wood, 1994), suggests that an "estrus loss or concealment" theory is a limited, possibly incorrect, view of the physiological reproductive events involved in hominid reproductive biology.

Additional studies on higher primate females, particularly of Old World nonhuman primates, provides us with more information for comparison with hominid females. The data shows that Old World monkeys and apes have a periodic menstrual cycle of approximately one per month (Robinson & Goy, 1986; Hafez, 1971; Graham, 1981), which is similar in length to that of the lunar based ovulatory cycle of human females. Robinson & Goy (1986:66) distinguish between the occurrence of estrus and of regular monthly menstrual/ovulatory cycles in primates in that estrus is "....the time of maximum follicular growth and estrogen production...." compared to other times which presumably are less than optimal.

Studies like these and others, suggests that not all ovulatory cycles are of equal fertility. This is a significant point in that it seems to apply to all Old World primates, and perhaps all primates, including hominids. This observation is confirmed because of the presence of breeding seasons, or limited times during a year when breeding and conception is more likely to occur in monkeys, apes and modern humans. Yet the cyclical regularity of an ovulatory cycle would allow for breeding and conception to be possible during any month of the year. The fact that there are months of the year when conception is more likely leads one to the conclusion that there are periods of peak fertility in primate populations, and that some ovulatory periods are of optimal fertility while others are greatly reduced in their potential for conception.

The explanation for this variation lies in the fact that environmental factors act as the "releaser" to the reproductive physiology of the body, and that it would appear the endogenous ovulatory cycle that falls closest to the timing of optimal environmental conditions for breeding will be a period of optimal or peak fertility for the animal. Macfarlane (1974) explains the reproductive physiological changes in a species that results in optimal fertility as a neuroendocrine response (see also Martini & Ganong, 1966). The sensory input from environmental stimuli has an influence on hypothalamic neurons and these lead to neurosecretion of releasing factors. He says (1974:573): "These small polypeptide molecules of three or four amino acids move to the anterior pituitary in capillaries and release follicle stimulating hormone or luteinizing hormone from basophilic cells." This release of sexual hormones in females and males provide the stimulation for the reproductive system to become enhanced to the point that the animal experiences an optimal or peak fertile period.

Other researchers have suggested that an internal breeding rhythm enabled a species to make a quick response to environmental factors (Baker, 1938), and that endogenous biorhythms, in general, are dependent on properties of the environment that can either inhibit or release the rhythm (Pittendrigh, 1981; Gwinner, 1986). Old World primates seem to have evolved a lunar-based endogenous reproductive rhythm which requires certain factors in the environment to initiate the optimal functioning of a given cycle. What the factors are seem to vary from one to another ecosystem, depending on what environmental factor or factors lead to optimal living conditions for the habitat and the species in question. Shortly after the optimal state in the environment is reached, the adult members will experience the neuroendocrine response that triggers the enhancement of the reproductive physiology producing the peak fertile period in the species.

Human reproductive ecology

Reproductive ecology in human and nonhuman primates have a great deal in common. Like nonhuman primates, the human female has a monthly menstrual/ovulatory cycle, and the human male has seasonal fluctuations in testicular activity, sperm volume and testosterone levels (Smals et al., 1976; Lincoln, 1981; Levine, 1994). It has also been shown that there is seasonal breeding in human populations (Seiver, 1981; Roenneberg & Aschoff 1990; Lam & Miron, 1994) and that the peak breeding times are associated with photoperiodicity, hours of sunlight, nutrition, activity patterns and a variety of other stress factors in a given environment.

Data on human reproductive ecology indicates that one or two, but not all of the female's ovulatory periods result in enhanced fertility (Macfarlane, 1974; Lam & Miron 1994). It is relevant at this point to consider the distinction between fecundity and fertility. The ability to reproduce, fecundity, is not the same as actually reproducing, which is fertility. Human females have a cyclical ovulatory rhythm occurring about once per month, often referred to as a lunar cycle since it corresponds with the moon's cycle with the earth. This cycle appears to be an endogenous biorhythm. Exogenous factors such as photoperiod, nutrition and activity patterns affect the degree of fertility during any of these potentially reproductive periods (Ellison, 1991, 1994; Lam & Miron, 1994; Cummings et al., 1983). This may be nature's way of assuring reproductive success or, as Lincoln & Short (1980) suggested, of inhibiting reproduction. Endogenous reproductive rhythms provide a template for the timing for fecundity. Not every cycle will have the same gametic fertility. When the ecological setting is most favourable, the biology of the female will experience its peak fertility. This fact seems to be true of humans as well as other primates.

Human males also reach their peak fertility during times of optimal living conditions (Spira, 1984; Levine, 1994; Lam & Miron, 1994) and testicular activity, sperm volume and testosterone levels are highest during these periods. The phenomenon of estrus that occurs in females has a parallel development in males in which their own reproductive biology is independently experiencing its peak fertility. Thus, both sexes simultaneously are prepared for mating at their most fertile times. This is not to overlook the fact that adult females and males influence the sexual behaviour of each other, which of course they do in any number of ways, and using a variety of the senses, as studies on pheromones, for example, have shown (Vandenbergh, 1983).

There is some evidence for an endogenous rhythm for sperm production (in macaques in Howles et al., 1982; in humans, see Spira, 1984 and Chik et al., 1992), but very little research has been done in this area.

Early hominid reproductive ecology

As noted above, some researchers (Elkin, 1954; Morris, 1967; Washburn & Lancaster, 1968; Short, 1980, 1994; Benshoof & Thornhill, 1979; Burley, 1979; Hrdy, 1981; Lovejoy, 1981; Fisher, 1982) have suggested that the estrus phenomenon was lost in hominid females at an early stage in human evolution, and that this loss was associated with certain behavioural changes in the species including the sexual division of labour and a sexual contract in which females traded "sexual favours", as used in the contemporary sense, for male provisioning. It was suggested that a female's "sexuality" came entirely under her conscious control (Short, 1994). None of these workers have noted the possible presence or the possible loss of a male rutting season in hominid evolution. Rather, it seems assumed that the hominid male is always in a state of potential sexual excitation, and is capable of mating at any time. Does this also assume that the male hominid ability to mate and capability to reproduce puts him in a state of continuous peak fertility? Of course there are two questions here with the first being his ability to mate at any time. It has already been noted that this is a nonhuman primate characteristic, specifically occurring in Old World monkeys and apes. It did not evolve first in hominids, but rather is a trait that they share with their close Old World primate relatives. Also it was noted earlier that sexual intercourse may occur but may not result in conception.

Second, was there a time of peak fertility in early hominid males, and similarly in early hominid females? Recent studies in reproductive ecology of modern humans suggest that there are peak fertility periods in males and in females (Takahashi, 1964; Timonen & Carpen, 1968; Ehrenkranz, 1983a,b; Roenneberg & Aschoff, 1990; Levine, 1991, 1994; Lam & Miron, 1994) and therefore that not all times of the year are of equal potential for conception. These patterns would have evolved from their early hominid antecedents.

With the development of the field of reproductive ecology, there is a growing body of evidence that similar and independently arising biological fertility peaks occur in female and in male mammals. The time of estrus, at least in primates, including hominids, might now better be expanded to a more inclusive term involving adult males and females in a population, and which I refer to here as the period of adult *peak fertility*. Optimal living conditions, which are relative to any given population and their habitat, is a key to understanding the period of peak fertility in a species.

Relevance to early hominid evolution

I will now address the issue of the implications to human evolution of reproductive data on Old World primates and modern humans by proposing a new model of early hominid reproductive ecology.

The following diagram (Table 1) illustrates an alternative model of reproductive ecology in hominid evolution. It suggests that early hominids (at least australopithecines to *Homo erectus*) would have experienced peak breeding seasons. Females and males would have come into rut with the generative system of both sexes experiencing enhanced fertility at the time in their environment during which there occurred optimal living conditions, keeping in mind the relative aspect of optimum. These environmental conditions would vary from place to place, depending on latitide and thus photoperiod, temperature and other environmental conditions affecting living conditions including food sources and availability. Thus, the timing of peak breeding times would have varied in early hominid populations.

Like their nonhuman primate relatives of the Old World, early hominids were capable of sexual activity and intercourse at any time of the year. Conception from intercourse would have been most likely during their season of peak fertility.

Cyclical ovulatory periods occurred in females and the one closest to the time that her body had reached a healthy and fit level during optimal environmental conditions would become the one with peak fertility, thus the one during which conception would most likely occur.

Male hominids may also have had endogenous sexual cycles but we know very little on this subject even in extant primate species. Studies have shown that male primate peak fertility is related to exogenous factors that occur in his environment. As the hominid male also reached his upper levels of fitness and health during the times in which living conditions reached optimal levels, his reproductive system responded through increase in sperm volume and testosterone levels.

	Mammals	Old World Primates	Modern Humans	Early Humans
Annual or biannual peak breeding season	Yes	Yes	Yes	Yes
Estrus in female	Yes	Yes	The occurence of 1 or 2 circannual enhanced fertility peaks suggests the biological potential	Probably
Rut in male	Yes	Probably	for a reproductive system response similar to estrus and rut	Probably
Sexual activity possible all year round	Not normally in feral animals	Yes	Yes	Yes
Cyclical ovulatory period in female (endogenous)	Possibly but exogenous factors reduce strength of response	Yes	Yes	Yes
Cyclical increase in sperm volume and testicular activity brought on by exogenous factors (also endogenous?)	Possibly but periods not known except for rut season	Probably but periods not known except for annual peak breeding season	Yes	Yes

Table 1 - Reproductive patterns of mammals and nonmammalian primates with a new model proposed for early hominids.

This new model proposes that early hominids followed a reproductive pattern similar to that in Old World primates and modern humans in having one or two peak fertility periods a year. Living in a more balanced relationship with their natural surroundings, early hominids probably had patterns more like those of their earlier Old World primate relatives. Although sexual behaviour would have been possible at any time of the year, they would have experienced annual or bi-annual peaks in breeding. These could have been times in which the group was distracted from the everyday pattern of normal subsistence activities, but these breeding seasons in hominids would be of short duration. Subsistence patterns would soon return to normal after the period of peak fertility and until the birthing season when there would be another interruption from everyday basic activities. These two times would probably have had short term effects on the movements of the group, or its migratory activities.

The fossil record and a study of paleoenvironment may provide us with information on these two major events in the life cycle of early hominids. Figuring out the timing of the breeding period could provide us with greater insight into the effect of reproductive ecology in the life of early hominids, and thus shed more light on human evolution.

References

ALLEN, W.R. (1978): Control of oestrus and ovulation in the mare. In: D.B. Creghton, G.R. Foxcroft, N.B. Haynes & G.E. Lamming (eds.), *Control of ovulation,* 453-468. London.

AMOROSO, E.C. & MARSHALL, F.H. (1960): External factors in sexual periodicity. In: *Marshall's physiology of reproduction*, Vol. 1, part 2, 707-831. London.

ASCHOFF, J. (1955): Jahresperiodik der Fortpflanzung bei Warmblütern. *Studium Generale* 8, 742-776.

BAKER, J.R. (1938): The evolution of breeding systems. In: G.R. de Beer (ed.), *Evolution: Essays on aspects of evolutionary biology*, 161-177. London.

BENSHOOF, L. & THORNHILL, R. (1979): The evolution of monogamy and concealed ovulation in humans. *Journal of Social and Biological Structures* 2, 95-106.

BURLEY, N. (1979): The evolution of concealed ovulation. *American Naturalist* 114, 838-858.

CAMPBELL, B.C. & LESLIE, P.N. (1995): Reproductive ecology of human males. *Yearbook of Physical Anthropology* 38:1-26.

CAMPBELL, K.L. & WOOD, J.W. (eds.) (1994): *Human reproductive ecology.* New York (Annals of the New York Academy of Science, Vol. 709).

CHANG, K.S.F., CHAN, S.T., LOW, W.D. & and NG, C.K. (1963): Climate and conception rates in Hong Kong. *Human Biology* 35, 366-376.

CHIK, C.L., ALMEIDA, O.F.X., LIBRE, E.A., BOOTH, J.D., RENQUIST, D. & MERRIAM, G.R. (1992): Photoperiod-driven changes in reproductive function in male rhesus monkeys. *Journal of Clinical and Endocrinological Metabolism* 74, 1068-1074.

CLEGG, M.T., COLE, H.H. & GANONG, W.F. (1965): The role of light in the regulation of cyclical estrous activity in sheep. In: *Proceedings of the Conference on Estrous Cycle Control in Domestic Animals*, 96-103. Washington DC (USDA Misc. Publ. 1005).

CUMMINGS, D.C., QUIGLEY, M.E. & YEN, S.S.C. (1983): Acute suppression of circulating testosterone levels by cortisol in men. *Journal of Clinical and Endocrinological Metabol*ism 57, 671-673.

CUPPS, P.T., McGOWAN, B., RAHLMANN, D.F., REDDON, A.R. & WEIR, W.C. (1960): Seasonal changes in the semen of rams. *Journal of Animal Science* 19, 208-213.

DEAN, R.F.A. (1949): Women war captives in Russia. *British Medical Journal* 1, 691-695.

EHRENKRANZ, J.R.L. (1983a): Seasonal breeding in humans: birth records of the Labrador Eskimo. *Fertility and Sterility* 40, 485-489.

EHRENKRANZ, J.R.L. (1983b): A gland for all seasons. *Natural History* 6, 18-23.

ELKIN, W. (1954): Social behavior and the evolution of man's mental faculties. *American Naturalist* 8, 129-154.

ELLISON, P.T. (1991): Reproductive ecology and human fertility. In: G.W. Laster & C.G.N. Mascie-Taylor (eds.), *Application of biological anthropology to human affairs,* 14-54. Cambridge.

ELLISON, P.T. (1994): Advances in human reproductive ecology. *Annual Review of Anthropology* 23, 255-275.

FISHER, H. (1982): *The sexual contract.* New York.

FOLLETT, B.K. & FOLLETT, D.E. (eds.) (1981): *Biological clocks in seasonal reproductive cycles.* New York (Proceedings of the 32nd Symposium of the Colston Research Society, Bristol).

FRISCH, R. (1978a): Population, food intake and fertility. *Science* 199, 22-30.

FRISCH, R. (1978b): Nutrition, fatness and fertility: the effect of food intake on reproductive ability. In: W.H. Mosley (ed.), *Nutrition and human reproduction*, 91-122. New York.

FRISCH, R. (1985): Fatness, menarche and female fecundity. *Perspective of Biological Medicine* 28, 611-633.

GANONG, W.F., SHEPHERD, M.D., WALL, J.R., BRUNT, van E.E. & CLEGG, M.T. (1963): Penetration of light into the brain of mammals. *Endocrinology* 72, 962-963.

GAUQUELIN, M.F. (1968): Le cycle annuel de reproduction du macaque, *Macaca irus. Bulletin Biologique France et Belgique* 102, 261-270.

GILMORE, D. & COOK, B. (eds.) (1981): *Environmental factors in mammal reproduction.* London.

GOODALL, J. (1968): The behaviour of free-living chimpanzees in the Gombe Stream Reserve. In: *Animal Behavior Monogaphs* 1, 165-311.

GRAHAM, C.E. (1981): *Reproductive biology of the great apes: Comparative and biomedical perspectives*. New York.

GWINNER, E. (1986): *Circannual rhythms*. New York.

HAFEZ, E.S.E. (1971): Reproductive cycles. In: E.S.E. Hafez (ed.), *Comparative reproduction of nonhuman primates*, 160-204. Springfield, Ill.

HAJEK, E.R., GUTIERREZ, J.R. & ESPINOSA, G. (1981): Seasonality of conception in human populations in Chile. *International Journal of Biometeorology* 25 (4), 281-291.

HARTMAN, C.G. (1932): Studies in the reproduction of the monkey, *Macacus (Pithecus) rhesus* with special reference to menstruation and pregnancy. In: *Contributions to Embryology*, vol. 23, 1. Washington.

HEAPE, W. (1900): The "sexual season" of mammals and the relation of the "pro-oestrum" to menstruation. *Quarterly Journal of Microscopical Science* 4 (1), 1-70.

HELWEG-LARSEN, P., HOFFMEYER, H., KIELER, J., THAYSEN, E.H., RHYGESEN, P. & WULFF, W.H. (1951): Famine disease in German concentration camps: complications and sequelae. *Acta Psychiatrica et Neurologica Scandinavica,* Suppl. 83, 199-206.

HOLMES, R.L. (1968): *Reproduction and environment*. New York.

HOWLES, C.M., CRAIGON, J. & HAYNE, N.B. (1982): Long-term rhythms of testicular volume and plasma prolactic concentrations in rams reared for three years in constant photoperiod. *Journal of Reproduction and Fertility* 65, 439-446.

HRDY, S. (1981): *The woman that never evolved*. Cambridge, Mass.

HUNTINGTON, E. (1938): *Season of birth*. Chichester, England.

IVANOVSKY, A. (1923): Physical modifications of the population of Russia under famine. *American Journal of Physical Anthropology* 6, 331-353.

KARSCH, F.J. & FOSTER, D.L. (1981): Environmental control of seasonal breeding: a common final mechanism governing seasonal breeding and sexual maturation. In: D. Gilmore & B.Cook (eds.), *Environmental factors in mammal reproduction*, 30-53. London.

KEYES, A., BROZEK, J., HENSCHEL, A. & TAYLOR, H.L. (1950): *The biology of human starvation*. I. Minneapolis.

KOFORD, C.B. (1965): Population dynamics of rhesus monkeys on Cayo Santiago. In: I. DeVore (ed.), *Primate behavior*, 160-174. New York.

LAM, D.A. & MIRON, J.A. (1994): Global patterns of seasonal variation in human fertility. In: K.L. Campbell & J.W. Wood (eds.), *Human reproductive ecology: Interactions of environment, fertility and behavior*, 29-45. New York (New York Academy of Science, Vol. 709).

LEMMON, W.B. & ALLEN, M.L. (1978): Continual sexual receptivity in the female chimpanzee (*Pan troglodytes*). *Folia Primatologica* 20, 80-88.

LEVINE, R.J. (1994): Male factors contributing to the seasonality of human reproduction. In: K.L. Campbell & J.W. Wood (eds.), *Human reproductive ecology: Interactions of environment, fertility and behavior,* 29-45. New York (New York Academy of Science, Vol. 709).

LINCOLN, G.A. (1981): Seasonal aspects of the testis. In: H.Burger & D.de Krester (eds.), *The testis*, 255-302. New York.

LINCOLN, G.A. & SHORT, R.V. (1980): Seasonal breeding: Nature's contraceptive. *Recent Progress in Hormonal Research* 36, 1-52.

LOVEJOY, O. (1981): The origin of Man. *Science* 23 (4480), 341-350.

LOUDON, A.S.I. & RACEY, P.A. (eds.) (1987): *Reproductive energetics in mammals*. London (Zoological Society of London Symposium, Vol. 57).

LOY, J.D. (1971): Estrous behavior of free-ranging Rhesus monkeys (*Macaca mulatta*). *Primates* 12, 1-31.

MACFARLANE, W.V. (1970): Seasonality of conception of human populations. *Biometeorology* 4 (1), 167-182.

MACFARLANE, W.V. (1974): Seasonal cycles in human conception. In: W.W. Trump (ed.), *Progress in biometeorology*, 557-577, 711-713. Amsterdam.

MARTINI, L. & GANONG, W.F. (1966): *Neuroendocrinology*. New York.

McCLILNTOCK, M.K. (1983): Pheromonal regulation of the ovarian cycle: enhancement, suppression and synchrony. In: J.G. Vandenbergh (ed.), *Pheromones and reproduction in mammals*, 113-149. New York.

MENAKER, M. (1971): Rhythms, reproduction, and photoreception. *Biology of Reproduction* 4, 295-308.

MICHAEL, R.P. & BONSALL, R.W. (1977): A 3-year study of an annual rhythm in plasma androgen levels in male rhesus monkeys (*Macaca mulatta*) in a constant laboratory environment. *Journal of Reproduction and Fertility* 49, 129-131.

MICHAEL, R.P. & ZUMPE, D. (1981): Environmental influences on the sexual behaviour of rhesus monkeys. In: D. Gilmore & B. Cook (eds), *Environmental factors in mammal reproduction*, 77-99. London.

MORRIS, D. (1967): *The naked ape*. London.

MOSLEY, W.H. (1978): *Nutrition and human reproduction*. New York.

PANTER-BRICK, C. & ELLISON, P.T. (1994): Seasonality of workloads and ovarian function in Nepali women. In: K.L. Campbell & J.W. Wood (eds.), *Human reproductive ecology*, 234-236. New York (Annals of the New York Academy of Science, Vol. 709).

PITTENDRIGH, C.S. (1981): Circadian organization and the photoperiodic phenomena. In: B.K. Follett & D.E. Follett (eds), *Biological clocks in seasonal reproductive cycles*. New York (Proceedings of the 32nd Symposium of the Colston Research Society, Bristol).

PRENTICE, A.M. & WHITEHEAD, R.G. (1987): The energetics of human reproduction. In: A.S.I. Loudon & P.A. Racey (eds), *Reproductive energetics in mammals*, 275-304. London (Zoological Society of London Symposium, Vol. 57).

ROBINSON, J.A. & GOY, R.W. (1986): Steroid hormones and the ovarian cycle. In: W.R. Dukelow & J. Erwin (eds.), *Comparative primate biology: Reproduction and development*, Vol. 3, 63-91. New York.

ROENNEBERG, T. & ASCHOFF, J. (1990): Annual rhythm of human reproduction. I. Biology, Sociology or Both? II. Environmental Correlation. *Journal of Biological Rhythms* 5 (3), 195-216, 217-239.

ROWELL, T.E. (1972): Female reproductive cycles and social behavior in primates. *Advances in the Study of Behavior* 4, 69-105.

RYLE, M. (1961): Early reproductive failure of ewes in a hot environment. 1. Ovulation rate and embryonic mortality. *Journal of Agricultural Science* 57, 1-9.

SAVAGE-RUMBAUGH, E.S. & WILKERSON, B.J. (1978): Socio-sexual behavior in *Pan paniscus* and *Pan troglodytes*: a comparative study. *Journal of Human Evolution* 7, 327-344.

SEIVER, D. (1981): Seasonality of fertility: new evidence. *Population and Environment* 10 (4), 245-257.

SHORT, R.V. (1980): The origins of human sexuality. In: C.R. Austin & R.V. Short (eds.), *Reproduction in mammals: Human sexuality*, 1-33. Cambridge.

SHORT, R.V. (1994): Human reproduction in an evolutionary context. In: K.L. Campbell & J.W. Woods (eds.), *Human reproductive ecology*, 416-425. New York.

SMALS, A.G., KLOPPENBERG, P.W.C. & BENRAAD, T.J. (1976): Circannual cycle in plasma testosterone in man. *Journal of Clinical Endocrinology and Metabolism* 42, 979-982.

SPIRA, A. (1984): Seasonal variations in sperm characteristics. *Archives of Andrology* 12, 23-28.

TAKAHASHI, E. (1964): Seasonal variation of conception and suicide. *Tohoku Journal of Experimental Medicine* 84, 215-227.

TIMONEN, S. & CARPAN, E. (1968): Multiple pregnancies and photoperiodicity. *Annales Chirurgiae et Gynaeologiae Fenniae* 57, 135-138.

TYLER, N.J.C. (1987): Body composition and energy balance of pregnant and non-pregnant Svalbard reindeer during winter. In: A.S.I. Loudon & P.A. Racey (eds.), *Reproductive energetics in mammals*, 203-229. London (Zoological Society of London Symposium, Vol. 57).

VANDENBERGH, J.G. (1983): *Pheromones and reproduction in mammals*. New York.

VANDENBERGH, J.G. & VESSY, S. (1966): Seasonal breeding in free-ranging Rhesus monkeys and related ecological factors. *American Zoologist* 6, 207.

WAITES, G.M.H. & ORTAVANT, R. (1968): Early effects of brief elevation of testicular temperature of spermatogenesis in the ram. *Annales de Biologie Animale, Biochemie et Biophysique* 8, 323-331.

WASHBURN, S.L. & LANCASTER, J. (1968): The evolution of hunting. In: R.B. Lee & I. DeVore (eds.), *Man the hunter*. Chicago.

WATTS, P.D. & HANSEN, S.E. (1987): Cyclic starvation as a reproductive strategy in the polar bear. In: A.S.I. Loudon & P.A. Racey (eds.), *Reproductive energetics in mammals*, 305-318. London (Zoological Society of London Symposium, Vol 57).

WIDDOWSON, E.M. (1981): The role of nutrition in mammalian reproduction. In: D. Gilmore & B. Cook (eds.), *Environmental factors in mammal reproduction*, 145-159. London.

ZORGNIOTTI, A.W. (ed.) (1991): *Temperature and environmental effects on the testis*. New York.

Disinhibition of Anti-cultural Behaviour: A Means of Adaptation in Hominids ?

Alexander G. Kozintsev & Marina L. Butovskaya

Abstract

Most or all archaic societies are known to have practised reversal rites during which all cultural norms were turned upside down to demonstrate how things should not be and thus to uphold social order. The ideal model of reverse behaviour was provided by the trickster myth which, like the reversal rites, were powerful laughter-provoking stimuli and may be viewed as the deepest cultural roots of modern humour.

Ethological evidence suggests that laughter is a ritualized bite derived from the relaxed open-mouth display of nonhuman primates, a signal of nonhostile intentions during quasi-aggressive play. In modern humans, laughter, like crying (which is phylogenetically somewhat younger) and yawning (which is much older) is a stereotypic vocal display that has no obvious adaptive meaning in terms of physiology but is a means of psychological defence. All these displays resemble "displacement activities" which are normally inhibited but became disinhibited in situations of ambivalence, frustration or motivational deadlock. Laughter, crying and yawning are highly communicative and even "contagious", they disrupt speech and immobilize, suggesting that they are ritualized social releasers which evolved primarily as speech-inhibiting and possibly anti-cultural responses.

Indeed, the emergence of language, which automatically gave rise to culture as a system of norms and taboos, must have been an informational and emotional shock for the hominids. Perhaps the primary adaptive meaning of laughter (and possibly of crying and yawning, which were evidently much more social in the past than they are now) was to maintain group cohesion by providing temporary relief from cultural stress and thus to protect early humans from social tension and neuroses. Although the evolutionary age of these displays is different, they may have acquired new motivation after the emergence of language. They even seem to have partly overlapped in the course of cultural evolution. Indeed, laughter and crying appear to be opposites, yet they were interchanged in many archaic rites, since death meant rebirth.

This article focuses on the evolutionary and cultural aspects of three stereotypic human displays: laughter, weeping, and yawning. They have received strikingly different amounts of attention. While laughter has always been in the foreground, weeping was far behind, and yawning was almost completely neglected. Although all the three are quite mysterious, they seem to have little in common. Their evolutionary age is widely different, yawning being the most ancient and the other two patterns much younger. Their emotional component ranges from very high in weeping to zero in yawning. Their semantics appears to be very dissimilar since laughter and weeping are regarded as opposites and yawning as neutral. Finally, their functions in modern society are different: laughter is extremely social, weeping mostly individual, and yawning plainly anti-social. However, as we will try to demonstrate, they may prove by far more similar than is traditionally believed.

We will begin with laughter and its origins. It has been claimed that before apes are taught to use sign language they are merely playful but humorless (McGhee, 1979). As our observations on chimpanzees seem to demonstrate, captivity per se may give rise to rudimentary practical humor since playful violation of a prohibition in a way strikingly similar to certain archaic forms of festive reverse behaviour in humans, known as "symbolic inversion" (Babcock, 1978), is accompanied by facial expression obviously suggestive of pleasure (Butovskaya & Kozintsev, 1996). We have described these forms of ape behaviour as "presymbolic inversion" (Kozintsev & Butovskaya, 1994).

It is beyond doubt, however, that the ability to sign automatically generates full-fledged humor based on the production of incongruities (McGhee, 1979). The question is, why? One explanation is that playing with incongruity is only possible after symbolic communication has been mastered (McGhee, 1979). But possible does not mean necessary. Why should intellectual play be necessary for anyone except intellectuals?

Although dozens of theorists have tried to explain the meaning of laughter and humor, these phenomena remain enigmatic (see Keith-Spiegel, 1972; McGhee, 1979; Morreall, 1987, for reviews). We can probably trace their history back to the archaic festive rites where cultural norms were reversed. However little we know about early forms of symbolic inversion, it is quite evident that their meaning extended far beyond the notion of intellectual play. H. Schurtz described these rites as safety valves (*Ventilsitten*), outlets for the release of subconscious collective discontent resulting from the interiorization of cultural norms and taboos. Sometimes they are seen as an antidote to tedium. Perhaps one of their objectives was to demonstrate the way things should not be and thus to uphold social order by contrasting it with symbolic mythological chaos (Stanner, 1960; Leach, 1961; Turner, 1974, 1978; Babcock, 1978; Abramyan, 1983).

On the individual level, reverse behaviour was practised by ritual clowns or similarly marginal persons whose social role in archaic societies was highly contradictory since they were both contemptible fun-makers and awesome spiritual beings surrounded by taboos (Willeford, 1969; Makarius, 1970). Reverse behaviour was modeled by the trickster myths which, like the reversal rites, were powerful laughter-provoking stimuli and may be viewed as the deepest cultural roots of humor (Radin et al., 1956; Belmonte, 1989).

Of all the numerous theories of laughter, one that is especially relevant for us at present links laughter with so-called derived, or displacement, activities of animals. The term refers to inadequate behavioural responses which appear in situations of ambivalence wherever the adequate (autochtonous) drives are in conflict or the requisite releasing stimuli are absent. This usually happens either during fighting or during sexual activity (Tinbergen, 1952). Leyhausen (1973) was probably the first to suggest that laughter is a displacement activity similar to sand digging in sticklebacks, self-licking in cats, and head scratching in humans. Later, the idea was supported by Russell (1987, 1996), who paralleled laughter with other human "substitute responses" like floor pacing, finger tapping, cold sweat, and vomiting.

Tinbergen (1952) believed that displacement activities are mere substitutes and have no motives of their own. Later, however, it was shown that they are motivated by the same stimuli as they would be in the normal context. The only difference is that these stimuli are weaker than the principal ones, so the respective actions are inhibited under normal conditions but become disinhibited in situations of conflict between stronger drives, frustration, motivational deadlock, or absence of sensory feedback (Andrew, 1956; van Iersel & Bol, 1958; Hinde, 1970; McFarland, 1985).

The disinhibition theory was extensively used by Porshnev (1974) who linked it with Pavlov's ideas to hypothesize that human language was originally a form of displacement activity that had emerged by disinhibition and had later itself become an inhibitor (specifically, a means of inhibiting action in general and aggression in particular).

So if laughter is indeed a displacement activity, one must, first, find its own motives, and second, understand the reasons of its inhibition under normal conditions. There is no doubt that laughter is indeed strongly suppressed most of the time. It always "waits" for the moment to seize any opportunity and, when disinhibited, it spontaneously breaks out and in turn completely inhibits both speech and action. This is strange, given that the emotional component of laughter is small (Deacon, 1992).

It is evidently not enough, then, to say that laughter is merely an expression of pleasure or that people laugh because humor is associated with intellectual play (McGhee, 1979), pleasant psychological shift (Morreall, 1987) or disturbance adjustment (Russell, 1987, 1996). Moreover, certain kinds of laughter are neither caused by humor nor pleasant (Black, 1982; Pfeifer, 1994), and if they do not result from organic disorders or direct stimulation of the brain, it would be wrong to discard them as "inappropriate" or "parasitic". The same applies to the laughter of preschool children which appears to be "silly" or caused by mere agitation (McGhee, 1979).

To move a step further, one should recollect that laughter is highly stereotypic, ritualized, and contagious, and these features are typical of displays which have a signal function in animals and are known as social releasers (Tinbergen, 1959). According to Tinbergen (1952), some displacement activities have been ritualized and became releasers. He suggested that laughter is a releaser that has an appeasing function and has evolved from the defense or threat display through ritualization (Tinbergen, 1959). The same view was expressed by Lorenz (1963). Indeed, ethological evidence suggests that laughter derives from the relaxed open-mouth display, or play face, which is a ritualized bite used by nonhuman primates during quasi-aggressive play (Bolwig, 1964; van Hooff, 1972). In terms of vocalization, however, human laughter is qualitatively different from that of the chimpanzee (Provine & Bard, 1996). Apes know that the play face is a releaser and conceal it when they do not want to play (Tanner & Byrne, 1993). Human laughter, too, is highly social (Chapman, 1973), and signals an intention to play (Grammer & Eibl-Eibesfeldt, 1990). It is contagious by itself, even in the absense of any humor (Provine & Yong, 1991). Also, it has been experimentally proven that humor and laughter are social lubricants that reduce anger and hostility (Dworkin & Efran, 1967; Singer, 1968; Landy & Mettee, 1969).

But if the primary function of laughter was to prevent aggression, why is it so rudimentary in nonhuman primates and so highly developed in humans? Also, why is it most expressed in young monkeys and apes, who are not yet capable of violence, and is known to decrease with age, just the opposite of what might be expected if the aggression prevention hypothesis were true?

Before attempting to answer these questions, we will briefly address weeping and yawning, of which much less is known. Certain oft-cited anecdotal zoological examples notwithstanding, weeping is uniquely human. According to Darwin (1872, p. 154), "a habit like weeping <...> must have been acquired since the period when man branched off from the common progenitor of the genus Homo and of the non-weeping anthropomorphous apes." Weeping, like laughter, appears to be a releaser since it is very stereotypic and communicative. And, like laughter, it disrupts speech and immobilizes (Deacon, 1992).

Although in modern society weeping is suppressed by adults in most situations and is not regarded as contagious, it was evidently much more social in the past. Collective weeping rites have been described in Australian aborigines and American Indians, and they were related not only to funerary events but to occasions like reception of guests where collective weeping was regarded as an expression of friendship and social cohesion (Mauss, 1969). Friederici (1907) termed this custom "greeting with tears". In many archaic religious traditions, collective weeping and laughter were performed either simultaneously or in succession since death was associated with eating, coitus, and rebirth (Reinach, 1912; Frazer, 1923; Freidenberg, 1997).

It is often claimed that weeping is adaptive because tears lubricate the eyeball and have bactericidal properties. But again, if this physiological mechanism is so valuable, why don't apes need it? Before this question is answered we can only say that weeping in humans should be best regarded as a displacement activity and a social releaser. Yawning, which is phylogenetically very much older than laughter or weeping, also appears to be a displacement activity (Tinbergen, 1952). Its physiological meaning has become less clear than ever since the old idea that yawning is caused by the deficit of oxygen and the excess of carbon dioxide in blood was not supported by experimental data (Provine et al., 1987). This does not mean that yawning is physiologically useless in all vertebrates. This only means that natural selection might have adjusted the old display, which man had inherited from his remote ancestors, for some new purpose. Indeed, according to the principle formulated by Baerends (1958), displays are evolutionary more conservative than their motivations.

Yawning is doubtless a social releaser since it is stereotypic and contagious to such an extent that the mere mention of it makes people yawn (Provine, 1986). This has evidently nothing to do with sleepiness because observing a sleeping person does not induce yawning in the observers or make them sleepy. Paralinguistically, yawning signals lack of interest (Provine et al., 1987). Although we have no evidence of collective yawning rites (the idea itself is rather funny), yawning was doubtless less individual in the past than it is now. Otherwise it is impossible to explain its extreme contagiousness. Like laughter and weeping, it is suppressed most of the time. When disinhibited, it disrupts speech and inhibits action.

So all the three expressive patterns can be regarded as both displacement activities and social releasers. Moreover, it turns out that they are more similar than they appear at first sight. Maybe they originally had a common function? If they did, what was it? Our hypothesis is very simple. However semantically dissimilar laughter, weeping, and yawning may be now, they have had a primary common function which was precisely the same as it is at present: to inhibit speech and action in as many people as possible. Total incompatibility of speech with laughter or weeping (at least sobbing) has been well demonstrated on the brain level (Deacon, 1992). Clearly, yawning is no less speech-inhibiting. Pleasure or relief that can be individually derived from these displays are epiphenomena. Each of the three can, but none of them must, be pleasant per se. Their primary adaptive value should be sought at the social, not at the individual level.

The idea that culture is a burden is very old. Perhaps its most ardent advocate was Freud (1978). But then language, too, is a burden: first, because it automatically gives rise to culture as a system of norms and taboos imposed on people (Porshnev, 1974), and secondly, because of the informational shock resulting from the acquisition of a radically new system of symbolic communication. Porshnev (1974) noted that yawning might have provided early hominids with a means of coping with language shock. The same applies to laughter and weeping. In fact, all the three are signs of refusal. According to Freud (1959), humor is "an attitude by means of which a person refuses to suffer", and the ego "refuses to be distressed by the provocations of reality." The definition is much too narrow, of course, since any reality assimilation is tiresome at the least (McGhee, 1979). Humor, then, is a playful refusal to deal with reality in a culturally appropriate way.

In the same line, weeping means a refusal to counter reality (because it is too sad or too touching), and yawning a refusal to be interested in it. And, because assimilating symbolic information, reacting to the provocations of new cultural reality in an adequate way, and even merely being interested apparently required more mental strength and persistence than might be expected from creatures who had just descended from the apes (known to be highly negativistic), refusal, if only a temporary and imaginary one, to do so meant reversal and, at the same time, liberation (Mindess, 1971). Perhaps the rudimentary "laughter" of young monkeys and apes, too, is not so much a means of preventing aggression, as a sign of liberation, and an invitation to join in getting liberated, from the necessity to comply with social norms. This is something that most adult animals either do not need or cannot afford.

For apes, as for other mammals, the principal means of preventing stress is sleep. Another means, which is important also for maintaining social integration in primate groups, is grooming. It has been hypothesized that in the course of human evolution, social grooming was replaced by language, which had become the principal factor of within-group integration (Dunbar, 1993). If so, language had not merely introduced a new stress factor but had also deprived humans of an old stress-reducing device. In this situation, sleep alone would certainly not suffice, so new means had to be urgently developed and/or old ones had to be adjusted ad hoc.

It appears likely, then, that laughter, weeping, and yawning were psychological defense means that protected early humans from stress caused by language and culture, prevented neuroses, and at the same time, being presymbolic, ensured social cohesion on a deeper level than language could do. Paradoxically, laughter and weeping are commonly regarded as atavisms (which is correct), and yet they are almost uniquely human. It is precisely at the early stage of human history that they were especially adaptive. No wonder apes begin to joke as soon as they acquire the ability to sign; they do not begin to weep evidently for biological reasons. We do not know how often they yawn or how much they sleep.

In humans, humor is indeed known to reduce stress (Dixon, 1980; Lefcourt & Martin, 1986), and it can be seen as an alternative to neurosis (Elitzur, 1990). People who are susceptible to boredom appreciate humor more (Ruch, 1988). Although the stress-reducing function of weeping and yawning has not yet been experimentally demonstrated, there is little doubt that it does, or at least did, exist.

However, because all the three patterns are manifestations of negativism and make people mute, weak, and inactive, they could only be used during relatively short breaks between normal activities, otherwise culture would have been endangered. This explains why they are real displacement activities which are suppressed most of the time. Yet, unlike displacement activities in animals, they are strongly motivated and, when disinhibited, they violently break out and begin to function as social releasers.

Laughter, weeping and yawning are not the only patterns that interfere with speech. Respiratory adjustments such as coughing, sneezing, and nose blowing are speech-inhibiting and are sometimes used as paralinguistic means. However, they are by far less contagious (in a psychological, not medical, sense) and, although sometimes being similar to displacement activities, they can hardly be described as social releasers.

References

ABRAMYAN, L.A. (1983): *Pervobytny prazdnik i mifologiya* (Archaic festival and mythology). Yerevan.

ANDREW, R.J. (1956): Normal and irrelevant toilet behaviour in Emberiza spp. *British Journal of Animal Behaviour* 4, 85-91.

BABCOCK, B. (1978): Introduction. In: B. Babcock (ed.), *The reversible world: Symbolic inversion in art and society*, 13-36. Ithaca.

BAERENDS, G.P. (1958): Comparative methods and the concept of homology in the study of behaviour. *Archives neerlandaises de zoologie* 13, 401-417.

BELMONTE, T. (1989): The trickster and the sacred clown: Revealing the logic of the unspeakable. In: K. Barnaby & P. D'Acierno (eds.), *C.G.Jung and the humanities: Towards hermeneutics of culture,* 45-66. Princeton.

BLACK, D.W. (1982): Pathologic laughter: A review of the literature. *Journal of Nervous and Mental Diseases* 170, 67-71.

BOLWIG, N. (1964): Facial expression in primates. *Behaviour* 22, 167-192.

BUTOVSKAYA, M.L., & KOZINTSEV, A.G. (1996): A neglected form of quasi-aggression in apes: Possible relevance for the origins of humor. *Current Anthropology* 37, 216-217.

CHAPMAN, A.J. (1973): Social facilitation of laughter in children. *Journal of Experimental and Social Psychology* 9, 528-541.

DARWIN, C. (1872): *The expression of the emotions in man and animals.* London.

DEACON, T.W. (1992): The neural circuitry underlying primate calls and human language. In: J. Wind, B. Chiarelli, B. Bichakjian & A. Nocentini (eds.), *Language origin: A multidisciplinary approach,* 121-162. Dordrecht.

DIXON, N.F. (1980): Humor: A cognitive alternative to stress? In: I.G. Sarason & C.D. Spielberger (eds.), *Stress and anxiety,* vol. 7, .281-289. Washington, D.C.

DUNBAR, R.I.M. (1993): Coevolution of neocortex size, group size, and language in humans. *Behavioral and Brain Sciences* 16, 681-735.

DWORKIN, E.S. & EFRAN, J.S. (1967): The angered: Their susceptibility to varieties of humor. *Journal of Personality and Social Psychology* 6, 233-236.

ELITZUR, A.C. (1990): Biomimesis: Humor, play, and neurosis as life mimicries. *Humor* 3, 159-175.

FRAZER, J.G. (1923): *The golden bough.* London.

FREIDENBERG, O.M. (1997): *Poetika syuzheta i zhanra* (Poetics of subject and genre). Moscow (1st edition, 1936).

FREUD, S. (1959): Humor. In: *Collected Papers,* vol.5. New York.

FREUD, S. (1978): Die Zukunft einer Illusion. In: *Massenpsychologie und Ich-Analyse.* Frankfurt/Main.

FRIEDERICI, M. (1907): *Der Tränengruß der Indianer.* Leipzig.

GRAMMER, K. & EIBL-EIBESFELDT, I. (1990): The ritualization of laughter. In: W. Koch (ed.), *Natürlichkeit der Sprache und der Kultur,* 192-214. Bochum (Bochumer Beiträge zur Semiotik, vol.18)

HINDE, R.A. (1970): *Animal behaviour: A synthesis of ethology and comparative psychology.* New York.

HOOFF van, J.A.R.A.M. (1972): A comparative approach to the phylogeny of laughter and smiling. In: R.A. Hinde (ed.), *Non-verbal communication,* 209-243. Cambridge.

IERSEL van, J.J.A. & BOL, A.C.A. (1958): Preening in two tern species: A study on displacement activities. *Behaviour* 13, 1-88.

KEITH-SPIEGEL, P. (1972): Early conceptions of humor: Varieties and issues. In: J.H. Goldstein & P. McGhee (eds.), *The psychology of humor,* 4-34. New York.

KOZINTSEV, A.G. & BUTOVSKAYA, M.L. (1994): Presymbolic inversion: Language and the evolution of negative behavior. In: *10th Meeting of the Language Origin Society, Berkeley, U.S.A., University of California, July 15-18, 1994. Abstracts,* 24-25. Amsterdam.

LANDY, D. & METTEE, D. (1969): Evaluation of an aggressor as a function of exposure to cartoon humor. *Journal of Personality and Social Psychology* 12, 66-71.

LEACH, E.R. (1961): *Rethinking anthropology.* London.

LEFCOURT, H.M. & MARTIN, R.A. (1986): *Humor and life stress: Antidote to adversity.* New York.

LEYHAUSEN, P. (1973): Biologie von Ausdruck und Eindruck. In: K. Lorenz & P. Leyhausen, *Antriebe tierischen und menschlichen Verhaltens,* 297-407. München.

LORENZ, K. (1963): *Das sogenannte Böse.* Wien.

McFARLAND, D. (1985): *Animal behaviour: Psychology, ethology, and evolution.* London.

McGHEE, P. (1979): *Humor: Its origin and development.* San Francisco.

MAKARIUS, L. (1970): Ritual clowns and symbolic behavior. *Diogenes* 69, 44-73.

MAUSS, M. (1969): Expression obligatoire des sentiments (rituels funeraires australiens). In: *Oevres,* vol. 3. Paris.

MINDESS, H. (1971): *Laughter and liberation.* Los Angeles.

MORREALL, J. (ed.) (1987): *The philosophy of laughter and humor.* New York.

PFEIFER, K. (1994): Laughter and pleasure. *Humor* 7, 157-172.

PORSHNEV, B.F. (1974): *O nachale chelovecheskoy istorii* (The beginning of man's history). Moscow.

PROVINE, R.R. (1986): Yawning as a stereotyped action pattern and releasing stimulus. *Ethology* 72, 109-122.

PROVINE R.R. & BARD. K.A. (1996): Laughter as a tool to study chimpanzee vocalization and speech competence. In: *Abstracts from the XVIth Congress of the International Primatological Society and the XIXth Conference of the American Society of Primatologists,* 493. Madison.

PROVINE, R.R., HAMERNIK, H.B. & CURCHAK, B.C. (1987): Yawning: Relation to sleeping and stretching in humans. *Ethology* 76, 152-160.

PROVINE, R.R., TATE , B.C. & GELDMACHER, L.L. (1987): Yawning: No effect of 3-5% CO_2, 100% O_2, and exercise. *Behavioral and Neural Biology* 48, 382-393.

PROVINE, R.R. & YONG, Y.L. (1991): Laughter: A stereotyped human vocalization. *Ethology* 89, 115-124.

RADIN, P., KERENYI, K. & JUNG, C.G. (1956): *The trickster: A study in American Indian mythology.* New York.

REINACH, S. (1912): Le rire rituel. In: *Cultes, mythes et religions,* vol. 4, 109-129. Paris.

RUCH, W. (1988): Sensation seeking and the enjoyment of structure and content of humor: stability of findings across four samples. *Personality and Individual Differences* 9, 861-871.

RUSSELL, R.E. (1987): *Life, mind, and laughter: A theory of laughter.* Chicago.

RUSSELL, R.E. (1996): Understanding laughter in terms of basic perceptual and response patterns. *Humor* 9, 39-55.

SINGER, D.L. (1968): Aggression arousal, hostile humor, catharsis. *Journal of Personality and Social Psychology, Monograph Supplement* 8 (N 1, part 2).

STANNER, W. (1960): On aboriginal religion, III. *Oceania,* 31, 100-120.

TANNER, J.E. & BYRNE, R.W. (1993): Concealing facial evidence of mood: Perspective taking in a captive gorilla? *Primates* 34, 451-457.

TINBERGEN, N. (1952): "Derived" activities: Their causation, biological significance, origin, and emancipation during evolution. *Quarterly Review of Biology* 27, 1-32.

TINBERGEN, N. (1959): Einige Gedanken über Beschwichtigungsgebärden. *Zeitschrift fuer Tierpsychologie* 16, 651-665.

TURNER, V. (1974): *The ritual process: Structure and anti-structure.* Harmondsworth.

TURNER, V. (1978): Comments and conclusions. In: B. Babcock (ed.), *The reversible world: Symbolic inversion in art and society,* 276-296. Ithaca.

WILLEFORD, W. (1969): *The fool and his scepter: A study of clowns and their audience.* Evanston.

Modern Models of the Past: The Life of the Eipo, Highlands of West-New Guinea

Wulf Schiefenhövel

Abstract

The Eipo, members of the Mek family of language and their neighbours were studied from 1974 to 1981 within the framework of the Deutsche Forschungsgemeinschaft Priority Programme „Man, Culture and Environment in the Central Highlands of West-New Guinea" initiated by G. Koch and K. Helfrich.

The life of this group of about 800 inhabitants of the upper Eipomek valley north of the central cordillera of Irian Jaya was „traditional", not only in the material sense with tools made of stone, bone and wood, a mixed subsistence strategy combining simple digging-stick swidden horticulture and collecting/hunting but, more importantly, with a social life in small, tightly knit communities. Patrilineality was combined with clanexogamy and virilocality. Political decisions as well as control of the exosphere (including major sacred ceremonies) were taken by men who dominated village and valley based politics, including large feasts for visitors and armed conflicts. Women were usually in control of the domestic scene and provided most of the daily food and firewood. There was a marked male-female dichotomy with the sacred men's house being the male and the menstruation-birth-puerperium house being the female centre. In this, the Eipo society increased biological dimorphism by cultural means. Despite their limited material inventory they had a rich mental culture with a complicated and expressive language, natural science (e.g. knowledge concerning plants and animals), myths and poetry. Examples for these aspects of their lives, also for their pragmatism and cultural flexibility vis-à-vis the presence of the scientists and the advent of a new culture, will be given.

The Eipo, in my view, can be seen as „modern models of the past".

Introduction

If we want to know something about human existence, about how hominids or hominines and their successors lived we have a number of options. To name a few: prehistory, archaeology, neurobiology, modelling human thought and behaviour including computer models, primate models, mammalian models and the study of living humans in the sense of ethnoarchaeology or ethnoprehistory. This last approach is the topic of my contribution which will present aspects of life among the Eipo in the Daerah Jayawijaya in the Indonesian province of Irian Jaya, the Western half of New Guinea (Figure 1). They do not represent hunter-gatherers, undoubtedly the earliest form of human subsistence strategy, but digging-stick horticulturists who also gather various valuable foods from nondomesticated sources.

New Guinea is one of the most fascinating areas of the world with the second largest cordillera in the southern hemisphere after the Andes. The mountains in the vicinity of the homeland of the Eipo reach up close to 5,000 meters and the lowest pass the people can take to cross the range when they visit their southern neighbours has an altitude of 3,700 meters - 800 meters higher than our highest Alpine peak in Germany. Nature in the rugged interior of New Guinea, the abode of many but mostly very small groups of mountain Papua, is stunning and demanding at the same time.

The history of these populations is likewise extraordinary. According to archaeologists (cp. Swadling, 1981) the ancestors of these people arrived 50,000 to 60,000 years ago at what is now

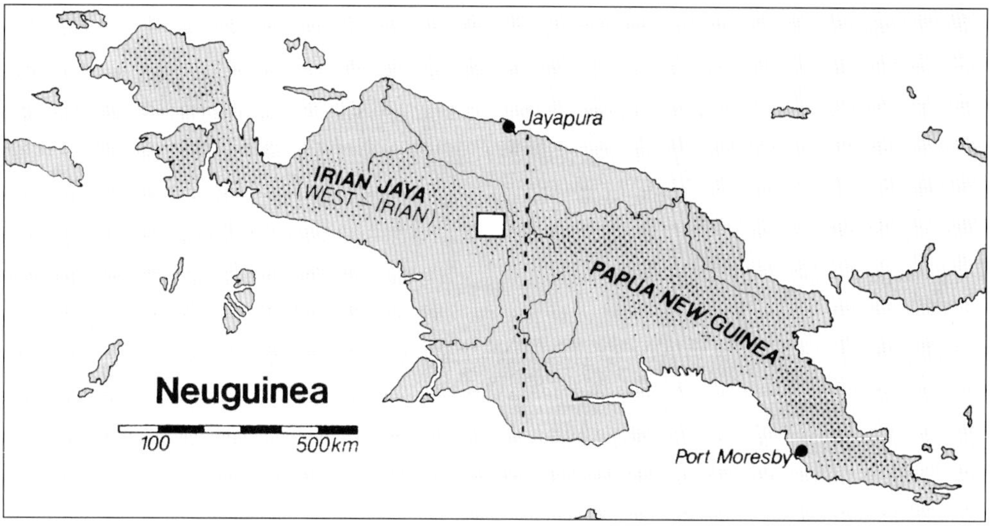

Figure 1 - Area of the Mek group of cultures and languages.

the large island of New Guinea and was then a land connected, within the Sahul shelf, to Australia. Since then they have had, until our days, a very isolated existence. This did not keep them from being among the first on our planet to discover the possibility to domesticate plants, which took place about 9,000 years ago. The highlands of New Guinea, then, are one of the early centers of world agriculture.

Parts of the nothern coast and the islands in the north and east are inhabited by Austronesian peoples who came by seagoing outrigger sailing boats several thousand years ago. The Trobriand islanders, where we also carry out fieldwork since 1982, belong to this group of cultures characterized by matrilineality, segmented societies with heritable chieftainship, sophisticated boatbuilding and navigation techniques. Melanesia, therefore, is home of two populations markedly differing in genetic stock, language and many aspects of culture.

The Eipo, their environment, subsistence techniques and knowledge of nature

"Eipodumanang", the ones living on the banks of the Eipo River, is how the people in the upper Eipomek valley call themselves. They are part of a larger group of related languages and cultures which were named "Mek", after the term for water and river (Schiefenhövel, 1976, 1991), and were studied in the framework of the interdisclipinary priority programme of the Deutsche Forschungsgemeinschaft 'Man, Culture and Environment in the Central Mountains of West-New Guinea', which was initiated by Gerd Koch and Klaus Helfrich of the Museum für Völkerkunde in Berlin.

The terrain is usually steeply incised. Anthropogenic grassland is found in a wide circle around the villages, but patches of rain forest exist in between the garden areas. The mountains, up to the tree line in about 3,500 meters, are covered by a complete layer of mountain and alpine forest.

The inventory of tools was very simple indeed: Stone adzes, stone and bamboo knives, bone awls, liana pieces and tinder to make fire, bows and barbed (but non-poisoned) arrows and various body decorations formed the bulk of the wordly possessions, easily fitting in one of the typical highland stringbags made by the women.

The villages of the Eipo and their neighbours in the Mek area had between 30 to 250 inhabitants and were usually built on spots which facilitated defense. The settlement pattern and various other facets of traditional life have partly changed due to culture change brought about mainly through the effects of a fundamentalist Christian mission. I therefore often use the past tense to describe the situation during the time of fieldwork in the seventies.

There were three functional types of houses. The family houses *(dib eik)* were the home of the nuclear plus one or more members of the extended family. The very confined space was often shared with a pig and a dog or two. Husbands not uncommonly spent much of their time and some of the nights in the more spacious men's houses *(yoek aik)* which served functions similar to an English club (no women, access only for initiated males, secret male society, talking politics) and that of the religious center of the villages with sacred relics stored there. The women's houses *(barye eik)* were situated at the fringe of the settlement and used exclusively by women during the time of menstruation, for childbirth, puerperium, times of severe disease and conflict with their husbands.

The Eipo used the slash and burn technique to make new gardens which they rotated by shifting cultivation. Primary forest was only occasionally cut down (a rather difficult task to perform with stone adzes), instead the secondary grassland of the old gardens was taken in usage again after a fallow of around 15 years. The growth, especially the circumference of a particular tree typical for this vegetation, *Trema tomentosa*, was used as bioindicator for a sufficient duration of the fallow period. Rainfall in the years 1975-76 was very high, reaching 6,000 mm *per annum*. Gardens were, therefore, often made in slopy terrain so that a natural drainiage was ensured. The techniques to grow sweet potatoe *(Ipomoea batatas)*, the staple, taro *(Colocasia esculenta)*, the archaic and sacred food, various leefy greens rich in protein, sugar cane *(Saccharum officinarum)*, possibly originating in New Guinea, banana *(Musa paradisiaca)* and other food plants were very developed. The number of cultigens for taro and sweet potatoe is impressive - a convincing sign of the agricultural ingenuity of these mountains Papua and proof for the fact that humans, under Neolithic conditions, have increased biodiversity (Table 1).

Our ethnobotanic research (Hiepko & Schiefenhövel, 1987) revealed that the Eipo, as many other ethnic groups around the globe, have an astounding body of knowledge of nature. They classified various plants as belonging together, described in the term of human kinship. It took the botanists careful study of the respective herbarium species and literature to come up with the same conclusion. This was, for example, the case with two seemingly very different kinds of plants which the Eipo informants insisted of being related and which were thought by the botany specialists in our team to be not even related on the family level. They turned out not only to be both *Saxifragaceae*, but also belonging the the same genus *(Polyosma)*. Many more examples of the very developed system of understanding nature were found during fieldresearch, also in the field of ethnozoology (Blum, 1979). It is quite obvious that humans have the capacity of being natural scientists in the double sense of the word.

Pigs *(Sus scrofa)* are domesticated by the Eipo and reared in the custody of women. But as food for people is already scarce, there is little surplus left to be fed to the pigs. They are killed in ceremonial context only and do not provide everyday protein. About half the population is barred from consuming the highly valued meat and fat of these animals, as there "totemic" ancestor was the pig. There was no difference in health between the ones occasionally eating pork and the ones who didn't.

Most of the protein for the women and their small children came from utilizing almost all (except, e.g. earth worms) possible sources: insects, spiders, larvae, eggs, reptiles, snakes etc. In this way, by collecting and consuming a few gramms of valuable animal protein per day, the women and children, the most needy of good nutrition in every society, obviously received enough essential amino acids. Also in the realm of other food resources carved out protein niches for women and children were found: after initiation, the first stage of which often took place at an age around 5, most of the many species of birds were taboo for males and thus available for women and their offspring. Except some cases with iodine deficiency goiter, they were no clinical signs of malnutrition.

Species	Malingdam	Moknerkon	Tali
Taro (*Colocasia esculenta*)	56	39	65
Sweet potatoe (*Ipomoea batatas*)	34	30	44
Banana (*Mesa x paradisiaca*)	21	20	24
Sugar cane (*Saccharum officinarum*)	17	21	21
bace - Vegetable (*Saccharum edule*)	13	16	12
touwa - Vegetable (*Abelmoschus manihot*)	13	12	13
teyang - Vegetable (*Setaria palmifolia*)	10	18	10
mula - Vegetable (*Rungia klossii*)	5	8	9

Table 1 - Cultural diversity of common Eipo food plants. Number of cultivars in 3 different villages. (From Hiepko & Schiefenhövel, 1987).

Physical anthropology and medicine

The physical anthropology of the Eipo has been described by the late Ernst Büchi (1981), who also discusses the pgymy problem and the possibly ancestry of the Eipo and other highland populations in New Guinea. Male average body length was 146.2 cm, this is very short even in New Guinea highland standards. Most skulls were mildly brachycephalic. Surprising for us, the first Europeans who lived with them for a longer period (in my case 22 months), was the general good status of health despite the fact that no modern medicine was available.

About 90 % of all diseases were infections in the wider sense of the term (bacterial, protozoic, helminthic, viral; Schiefenhövel, 1982), a classic morbidity pattern for pre-immunization, pre-antibiotic times, thereby probably for most of the history of humankind. Amazing was the physical performance of these small, but very muscular and extremely well trained people. Woman were able to carry their own body weight (approximately 40 kgs) for hours without apparent fatigue and men exhibited stunning power, speed and stamina especially in intravillage fighting and intervalley warfare. These athletic feats are the more surprising as protein supply was, as mentioned, extremely limited, i.e. about half the WHO daily minimum requirements.

Menarche, birth and early childhood

Menarche, however, was very late: around 17 years. Menstruation in societies like that of the Eipo is a generally rare event. A young women may have her period about 10-20 times or so before she becomes pregant. She will then give birth and breastfeed her baby until the age of at least two, usually three years. The number of menstruations before the onset of the next pregancy is bound to be small, also as sexual intercourse after the long post-partum sex-taboo tends to be regular. Until menopause, in the second half of the fourties, women have thus had only an estimated 40 or 50 cycles - the situation in industrialized countries, with a roughly tenfold number of menstruations, is markedly different.

Birth involved a number of birth attendants, one or two of them had midwife functions. As is typical for by far the majority of cultures the preferred body postures during delivery were vertical. The parturients had the possibilty to react to the birth process inside their bodies by choosing and changing body positions and generally adapting their behaviour (walking, resting, sitting, lying, standing squatting, knee-elbow position, screaming, sighing, eating, drinking, actively seeking or declining support etc.). The biocybernetically regulated process of birth was very little interfered with, a wise decision as all vaginal and other interal manoeuvers would only make things worse. Archaic forms of obstetric care, body contact, holding, massaging, giving support in every respect, birth in the wellknown territory of the women's house etc. seem to have been quite effective as infant mortality in the first year was, compared to other Third World countries (often more than 20 %), surprisingly low (around 5-6 %).

Early childhood was characterized by subtle intuitive parenting, breastfeeding on demand for about 2 1/2 years and having the baby in body contact most of the time and in the midst of the others. From an evolutionary point of view this is precisely what one should expect for little hominids and *Homo sapiens* to grow up: Provide all the necessary tactile, olfactory, acoustic, visual, material, social, psychological and intellectual stimuli which are typical for the society as a whole to the little being who so effectively absorbs the cues of such rich learning environment.

Late menarche corresponded with a late onset of body growth. Under the prevailing critical conditions of life - limited protein supply, high energy expenditure during work, often in the rain (no protecting dress), and the cold nights (around 13° centigrade, no cover) plus parasite load - the bodies of children suffer stunting for many years and take a long time to mature. But, this seems to be a useful adaptation: by far most children are very active and vital in play, work and mental tasks. They are "small but strong" (Obrist, 1992).

Gender antagonism and political leadership

The village plan (Figure 2) of Munggona, the central settlement of the upper Eipomek valley, indicates a typical feature of the Papuan highlands communities: marked gender antagonism. The world of women and men are separated physically and conceptually.

Women are seen as intrinsically dangerous, particularly with regard to their sexual and reproductive powers; to come into contact with menstrual blood means sure death for a man. There is, on the other hand, love (often very romantic and erotic as expressed in songs; vide below), affection, friendship and partnership. Men have, as in many societies, an option for polygyny; most of them do not manage to suceed, however, in the difficult task to diplomatically handle two spouses and their kin. In rare cases two men may be living together with one wife in a polyandrous bond. Premarital sex was accepted, extramarital affairs violated the rules of moral conduct but still occured. Patrilineality, virilocality, general patrifocality and patriarchy with regard to most political decisions were characteristics of the Eipo society as those of other highland populations.

On the basis of their intellectual, oratorical, social and physical power, big men led the communities as persons who took initiative, pursued plans and ensured the following of rules and traditions, though they also used them to their own advantage. In this protomeritocracy leadership was dependent on the actual power of the leader. In case he showed signs of weakening and loosing persuasive skill he lost his position, too. There was no institutionalized chieftainship.

Social norms were, above all, maintained through public opinion, which, in turn, was effected trough gossip, discussion of disputed issues and the use of extrahuman powers in black magic, alledgedly performed by mostly female, sometimes male witches. In a very similar way wrongdoing was believed, according to the animistic religious system, to be punished by the spirits of the dead or spirits in nature, often conceptualized to be there since the onset of times. Such sanctions were thought to be enforced *hic et nunc*: If somebody did something wrong, he or she or one of their close kin, perhaps also their pig or garden, could expect disease, death or ill fate within a foreseeable period of time. These kinds of religious systems are thus very different from the ones of Mediterranean and Asian origin which threaten to punish misbehaviour only in the afterlife. To become sick in the Eipo, as in many other traditional, society is part of a norm-enforcing system directing extrahuman executive powers at health and wellbeing.

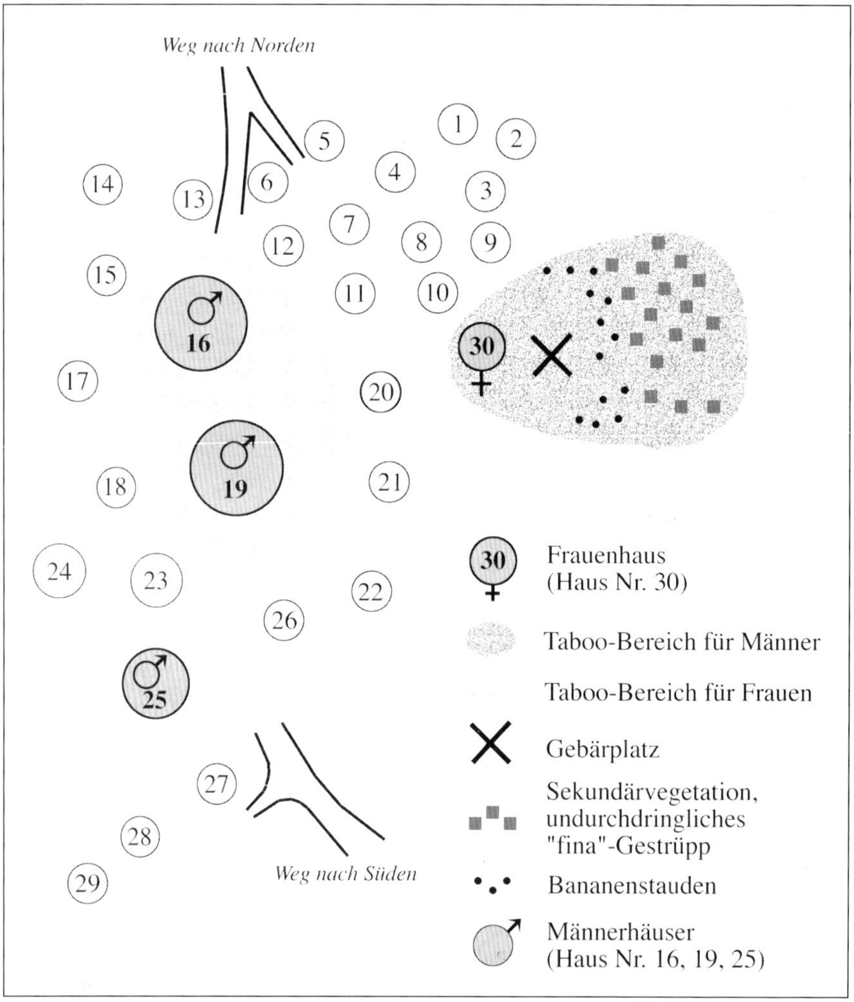

Figure 2 - The village of Munggona in the Eipomek Valley.

Aggression

Despite the fact that the Eipo were generally friendly and controlled, the potential for aggressive acts was quite high and did not need much triggering. Until recently, intraalliance fights and interalliance warfare took a toll of approximately 3-4 dead per 1,000 inhabitants per year. About 25 % of all males were likely to be killed rather than die of natural causes. Exocannibalism was sometimes practiced, explicitely to utterly destroy the body of the hated enemy, not to incorporate his or her powers or to supplement protein supply. This kind of warfare, then, seem to be motivated by a very strong sense of belonging and by building up an artificial cultural boundary between otherwise very similar populations. This process, character enhancement in evolutionary terminology, is probably one of the factors which have led to the stunning linguistic and cultural diversity in Melanesia. - In the seventies, the Eipo were just about 800 souls, but considered themselves, in the usual mechanism of ethnocentricity, as a people quite outside the other groups around them.

Culture of thought

The Eipo made very few carved or painted objects, in contrast to their neighbours in the east and west who had sacred wooden shields and wooden drums. While there is little visual art apart from decorating one's own body in often spectacular displays, oratory skills are remarkable. Men and women are very well able to voice their opinion, often their grievances, in very convincing and socially skillful ways. Sometimes these public talks are filled with subtle or ribald allusions and sometimes almost incomprehensible to the outsider because of *kunuse yupo*, the usage of code words for tabooed terms or concepts. The texts of profane and sacred songs and chants, too, are convincing proof of the artistic and intellectual power of these isolated mountain Papua. One of the love songs which were recorded during the project (Arthur Simon) and first translated by Volker Heeschen gives vivid evidence of this. It is here included in the English transliteration by the poet Fred Turner (taken from Hiepko & Schiefenhövel, 1987):

Oleto's Song to Her Lover

Ikokiye dam, kokbare dam,
na lebnuke, na nuknuke,
na ning guma, na ning anya,
na lebnuke, na nuknuke.

kidiknamne, banabnamne.
na dukiye, na kokaye
na lebnuke, na nuknuke.
se, kayape, nay kayape,
kidiknamne, banabnamne.

se, fotong o, nay atiye,
na lebnuke, na nuknuke.
table ya metebum,
like ya kwinebum,
koubuknamne, webuknamne.

lung aik dam, fan aik dam
kidikamne, banabnamne.
ya metebum, ya kotobum.
urya ya metebum wilalyam,
dibrenangkin, kwinibnangkin, delebnangkin.

fin dam sekne, nab dam sekne
cebroblyame, sekdoblyame.
ape lunge, ape fane
kwinibnangkin, delebnangkin.
kidikse, se wicape, nay kayape

kidiknamne, banabnamne.
dibrenangkin, mebdeibnangkin,
se wicape, nay kayape.
fin dam sekne, nab dam sekne
toubuke dablyam, lekuke dablyam.

subkor welen korbik welen,
ati teleb welebnamne, yanibnamne,
ape yane, ape fane.

By ferny-limb, by curly-tree,
I heard one say, I heard one say,
'What's thine is mine and is not thine',
I heard her say, I heard her say.

Lost will I be and far away.
'My curly-limb! My ferny-tree!',
I heard her say, I heard her say.
Friends and gossips! Old companions! -
Lost will I be and far away.

'Oh curly-heart, o housey-home',
I heard me say, I heard me say.
With thine axe thou sheared the slim-tree,
cut and tore the rooted wild-tree;
tending my garden I would be.

Wild weed by the threshold, wild growth by the door. Lost will I be and far away.
Thy clever axe will strike and tear;
here, come and fell the quick-elm tree!
I'd know thy flesh, and root it up, pull thee onto me.

That flesh-bracelet, pulse-fireband,
flame-vine, snap it from thine arm!
Wild heart, weed-grass, wild growth by the door,
I'll twist thee free, I'll pull thee free.
And lost, my sisters, friends, I be.

Lost will I be and far away.
I'll know thy flesh; thou'lt groan for me.
Gossips and sisters! O my friends!
The pulse-bracelet, the wirst-vine,
Bite it and burn with alchemy!

And as the fruit-vine climbs the tree,
I'll clamber, my housey-home, and stay,
long-legs, wild growth, at home in thee.

Conclusion

As Marcel Otte has said during the conference, early hominids and humans did not only "survive". They created communities, culture, a host of knowledge about themselves and their environments, they shaped nature according to their needs and technological capacities. In short, they were typically human just like ourselves who work in front of computers and enjoy fieldwork and going to scientific meetings. From an evolutionary point of view there can only be little, if any, difference in the basic emotional and cognitive structures between stone age woman and man and us today.

It remains a mystery to me why, in disciplines like philosophy, history, anthropology and sociology, cultural relativism and, in recent times, postmodern views are so en vogue. It is almost impossible to convince some colleagues of the observation that romantic love seems a universal phenomenon. Commonly held wisdom has it that this kind of erotic, sublime affection could have sprung into existence only through the various constituents of bourgeois society. "Primitive" people, so an often painted picture, have sex, often indiscriminately, and nobody knows who the father of the children is, hence matrilineality (and matriarchy) as archaic forms of human society. These kinds of concepts can't work if one includes just a grain of evolutionary biology into the perspective. I am convinced that, quite in contrast to the dictum of Bronislaw Malinowski (1929; otherwise a brilliant anthropologist), the Trobrianders and members of any other human society, either exsting or extinct, know and knew about biological fatherhood. Men and their families are very aware of this and precisely attempt to guard the one big male Achilles' heel carefully - precisely because they know: *pater semper incertus.*

Quite striking is the fact that societies like that of the Eipo have developed, completely on their own and of course without access to the writings of Carolus Linnaeus and other great thinkers, a very sophisticated system of knowing about and classifying the dead and living things in nature. Ethnobotanical research in the last decades (cp. Berlin et al., 1973) has shown that humans are natural scientists par excellence. One of the many intriguing questions arising from this kind of research is: "How, on earth, did our ancestors develop this amazingly precise knowledge and how did they manage to keep this intellectual thesaurus in their culture without loosing something and without formal training in schools, universities and without any form of written record?" One only has to think of the tenthousands of medicinal plants used around the globe. How did they find them? Surely not by conducting a huge pharmacological trial and error search programme.

I believe that these kinds of questions are important for the study of hominid and human existence. In order to understand life in Neolithic and Palaeolithic conditions we can utilize insights from many old and new fields, including evolutionary biology, human ethology, brain research, biopsychology and crosscultural anthropological research. The scenarios emerging from these thrilling attempts will, I am convinced, show that our distant forefathers were very much like ourselves.

References

BERLIN, B., BREEDLOVE, D. & RAVEN, P. (1973): General principles of classification and nomenclature in folk biology. *American Anthropologist* 75, 214-242.

BLUM, J.P.(1979): *Untersuchungen zur Tierwelt im Leben der Eipo im zentralen Bergland von Irian Jaya (West-Neuguinea), Indonesien.* Berlin.

BÜCHI, E. (1981): *Physische Anthropologie der Eipo im zentralen Bergland von Irian Jaya (West-Neuguinea), Indonesien.* Berlin.

HIEPKO, P. & SCHIEFENHÖVEL, W. (1987): *Mensch und Pflanze. Ergebnisse ethnotaxonomischer und ethnobotanischer Untersuchungen bei den Eipo, zentrales Bergland von Irian Jaya (West-Neuguinea), Indonesien.* Berlin.

MALINOWSKI, B. (1929): *The sexual life of savages in Northwestern Melanesia.* London.

OBRIST-van EEUWIJK, B. (1992): *Small but strong. Cultural concepts of (mal)nutrition among the Northern Kwanga (East Sepik Province, Papua New Guinea).* Basel.

SCHIEFENHÖVEL, W. (1976): Die Eipo-Leute des Berglands von Indonesisch-Neuguinea. *Homo* 26, 263-275.

SCHIEFENHÖVEL, W. (1982): Results of ethnomedical fieldwork among the Eipo, Daerah Jayawijaya, Irian Jaya. *Medika* (Jakarta), 11, 829-843.

SCHIEFENHÖVEL, W. (1991): Eipo. In: T. Hays (ed.), *Encyclopedia of world cultures.* Vol. II. *Oceania,* 55-59. Boston.

SWADLING, P. (1981): *Papua New Guinea's prehistory.* Port Moresby (National Museum and Art Gallery).

Cultural Continuity and Change in Hominid Behaviour in Africa During the Middle to Upper Pleistocene Transition

J. Desmond Clark

Abstract

The end of the Middle Pleistocene was a time of fluctuating climatic episodes towards a steadily drying climate that was accompanied by movements of plant, animal and human communities in and out of the Sahara which acted as a bottleneck that inhibited or stimulated repopulation as rainfall and temperatures fluctuated. During the latter times the Sahara became a highly favourable habitat for repopulation from south of the desert and for interaction between hominid populations coming from both south and north.

In the central African tropics the basin of the Congo River system became drier, the evergreen humid tropical forest was confined to refugia and was replaced in many parts of the basin by more open country, savanna woodland and grasslands. Human populations had occupied the peripheral parts of the basin in the earlier Pleistocene but human settlement in the more central parts of the basin does not appear to have become possible or been attempted until the end of Oxygen-isotope Stage 7 and Stage 6. This was a time when handaxes and cleavers of the late Acheulean techno-complex were replaced by other kinds of artifacts each of which appears to be specifically designed to perform a more limited range of tasks than were the multipurpose bifaces of the Acheulean. This large tool component of the succeeding cultural stages - the Sangoan and Lupemban techno-complex - is found not only in the Congo Basin but extends into the Lake Victoria Basin in East Africa, the West African savanna and in the southern tropics into Angola, Zambia, southern Tanzania and Malawi.

Elsewhere in northern and southern Africa and the Sahara this large tool component is absent and cultural assemblages of the Middle Stone Age exhibit varied specialisation of the small tool component. This small tool refinement and specialisation is present also with the Central African techno-complexes. Throughout the continent lithic technology can be seen to continue with very little change from the end of the Earlier into the Middle Stone Age but with growing refinements, in particular in the use of the Levallois and discoid core methods of flake production. While technology was slow to change the refinement and innovations in Levallois flake and blade production, though spatially sporadic at first, are indications of the development of new material equipment with which to carry out the different sets of activities now being performed.

This specialisation begins mostly between 200-100 kya though it is likely to have begun earlier in some regions in response to more rapid climatic change. By 40 kya, a number of regionally specialised Middle Stone Age cultural entities can be recognised and identified by their tool-kits which are based in part on a technology which, though refined, is a continuation of that found with the later Acheulean Complex.

This paper goes on to compare some of the tool-kits found and to propose that some, if not all, of these, though environmentally induced, were made possible by an improved socio-economic structuring of hominid groups that is reflected in tool forms and the ways these were used.

Technological innovation at this time is closely related to the invention of hafting or mounting in various ways the stone working parts to form the first simple composite tools. Much equipment must have remained unhafted - such as the core-axes, large blades and some small scrapers - but at the East and South African sites others, points for example, have reduced and trimmed butt ends and the tanged tools of the Aterian techno-complex in the Sahara are clear indications of hafting for the more efficient use of different lithic artifact types.

Until late in the Middle Stone Age the core-axe remained unhafted. With the Sangoan this tool is heavy and crude with two main attributes: a heavy retouched butt and a rounded, bifacially thinned distal end. Sangoan core-axes are mostly heavy and sub-triangular in plan form; the early Lupemban examples being heavy and, more often parallel-sided. In some cases both kinds have had the end intentionally removed to produce a new, sharp edge to the tool. Numbers of these resharpening spalls have been found at Kalambo Falls with the Sangoan and the earlier Lupemban techno-complex. The core-axe must have been a tool with which to make other equipment and, being heavy and suited to hard usage, was possibly employed in the preliminary working of wood. More refined wood-working may be reflected in the chisel ends of some of the parallel-sided core-axes and in the denticulate, concave and notched scraper forms as well as the proto-burin/chisels which, though present in the Acheulean, increase in numbers in the post-Acheulean assemblages.

Another cultural innovation of this time is the rubbing or grinding stone which, during the Middle Stone Age, is widely manifest in the Sahara, in Central and in South Africa and also, though very rarely, with the later Acheulean at the Kalambo Falls. This grinding activity suggests that more persistent and additional uses were being made of resources - plant and, perhaps, animal - and may also be implying perhaps expanding group size, more regular seasonal movement patterns and more intensive use of resources, all taking place simultaneously with the appearance in the continent of anatomically Modern humans.

This period - from the end of the Acheulean to the appearance of the Upper Palaeolithic (between 50 and 40 kya) - was a time of movements readjustments and innovative technology with more intensive uses of food resources and of materials that extended the possibilities through new and more efficient equipment. It was the time of essential preparation and experimentation preceding the appearance of Modern humans and their complex technology in the Upper Palaeolithic.

This paper is an attempt to use stone artefact assemblages that belong in the transitional time period between the end of the Middle and beginning of the Upper Pleistocene (>200-100 kya) to identify behavioural changes which took place in the lifeways of hominid populations in Africa. Reliable data for this time period are very sparse. Dating, climatic and environmental contexts, associated hominid fossils, and well excavated sealed archaeological assemblages are largely lacking. There are very few well documented sites with climatic information that cover this crucial period in early human social reorganisation. This period saw the end of the Lower Palaeolithic biface and chopper/flake tool kits and its replacement by others that continued, and refined one main component of these archaeological aggregates associated first with *Homo erectus* and later with archaic *Homo sapiens*. The introduction of new artefact forms implied innovative, (and presumably, more efficient) group ways of using the natural resources of the environment by populations that were on the threshold, and saw the appearance, of anatomically modern humans.

Africa is a vast land mass, highly variable in climate and biomes, and in the ingenious ways in which present-day human populations have attained an efficient equilibrium with their habitat (White, 1983; Dorst & Dandelot, 1970; Murdock, 1959). By the time of the fully developed Middle Stone Age (c. 100-40 kya) a number of broad, regional complexes can be identified (Clark, 1992: Fig. 1) on the basis of their stone tool kit, with the beginnings of this regional readjustment occurring appreciably earlier.

Perhaps the most important problem before discussing behaviour, is the lack of a reliable chronology for sites of this period beyond the range of radiocarbon and mostly too young for dating by the potassium argon method, or relating to sites from regions where K/Ar cannot be used due to the absence of volcanic sediments. U-series, TL, and ESR are not as yet fully reliable techniques but, with the improving refinement and increasing number of dates, they are being regularly used and relied upon; sometimes, unfortunately, as firm ages even where single dates are concerned. This is understandable in large part because these dates are all that are available. With this cautionary note, we will now examine the lithic assemblages, dating and the nature of the climate and environments at the time of the appearance of the earliest Middle Stone Age.

Its origins lie in the Acheulean biface techno-complex, the main cultural entity occurring in the Middle Pleistocene and is thought to have been replaced sometime between 300 and 200 kya by several new tool kits. Bifaces of Acheulean type appear to have disappeared by 240-250 kya on the evidence from East Africa (McBrearty et al., 1996), from North Africa and the Sahara at much the same time (Sahnouni et al,.1996; Wendorf & Schild, 1980), and also in South Africa (Volman, 1984). In Europe also the late Acheulean at La Micoque is dated at 300 - 350 kya (Falgueres et al., 1997).

The recent evidence from the Kapthurin Formation is especially important as it provides good stratigraphic support for the near contemporaneity of assemblages with bifaces, others based on blades, and yet others with heavy duty tools like those of the Sangoan (McBrearty et al.. 1996: 575-576). This locality in the Lake Baringo basin, Kenya, confirms the Levallois technique to have been a late Acheulean invention, but it also shows assemblages occurring without handaxes and therefore, reminiscent of the Middle Stone Age between 780-500 kya. This should be no surprise since there are numerous examples of assemblages without bifaces in the Middle Pleistocene, where retouched flake tools are closely comparable technologically to those of the early Middle Stone Age (Clark, 1964, 1969, 1975; Clark & Kurashina, 1979), although core/flake components with retouched forms are more usually associated with bifaces (Leakey, 1996). Some aggregates are mostly bifaces with a few flakes, but usually there are both bifaces and retouched flake components. Sometimes also, the flake component of the Middle Pleistocene hominid tool kit occurs by itself both in Africa and Eurasia. This variability is presumably related to various different behavioural activities.

The idea that the Acheulean consists *only* of bifaces appears to continue still for some researchers, and it is time that the considerable variability to be seen in Middle Pleistocene tool assemblages is more generally recognised. Such variability in Acheulean/Developed Oldowan aggregates was at one time thought to be the work of biologically distinct hominids, but by the Middle Pleistocene this view can no longer be sustained. The change from the Lower to Middle Palaeolithic in Africa if not also elsewhere is one of degree, not of total replacement and though large bifaces disappear, the core/flake component continues sometimes along with the Levallois technique, all of which undergoes regional modification. In this transitional period, therefore, we can see both continuity and change.

In North Africa, the Sahara, South Africa, and in much of eastern Africa also, the most significant change in lithic technology was the disappearance of bifaces, and refinement and innovation in the retouched light duty stone tool component. These were mostly open savanna regions that also saw rapid development of the Levallois method of obtaining thin flakes and blades. In most of tropical west, central, and south central Africa, however, a heavy duty tool component continued, though very different from the Acheulean biface, and evolved together with the light duty flake component, presumably because in these more closed habitats these heavy duty tools were an essential part of the subsistence strategies of the prehistoric populations there (Klein, 1989: 259-263) .

The supporting data is discussed below - chronology, followed by evidence for climatic change and the ways in which stone tools underwent change in part, perhaps, due to changing ecological conditions and, in part, due to readjustments in social organisational abilities to compensate for the overall increase in human population densities. 'Density' is, of course, a misnomer since human populations must have been very sparsely spread over the continent, sometimes migrating when major climatic change so dictated such as the desertification of the Sahara, or the expansion of tropical biomes at the end of the Pleistocene (Clark, 1964, 1992).

Chronology

There is now general agreement that the end of the Acheulean is well beyond the lower limit of radiocarbon dating, and that the few early dates associated with the Acheulean are infinite dates. In North Africa, the Acheulean had probably disappeared before 200 kya. But, based on the Haua Fteah evidence, two tip ends of bifacially made tools were found with a 'pre-Aurignacian' blade industry at the base of the cave deposits of the o/i Stage 5, and date between 80-130 kya (McBurney, 1967).

At Jebel Irhoud, the age of the late archaic *Homo sapiens* fossils and the Middle Stone Age industry belong in the o/i Stages 5 and 6 with an ESR date of 190-130 kya proposed (Grün & Stringer, 1991; Hublin, 1992). The calvarium from Salé in Morocco is believed to be nearer 400 kya, and thus to belong in the Harounian transgression and to be contemporary with an artefact assemblage of late Acheulean age (Biberson, 1961). This stratigraphically late final Acheulean Stage 8 from Casablanca though undated, probably represents the general composition of final Acheulean assemblages with blades north of the Sahara (Clark, 1992 a).

In the Sahara, there are several U-series dates associated with seven wet periods and Acheulean tools in o/i Stages 6, 7, and 8 ranging between >300-160 kya and correlated with deep sea cores from the Mediterranean and the Atlantic (Haynes et al., in press; Lezine & Casanova, 1991; McHugh et al., 1988; Szabo et al., 1989; Wendorf et al., 1991 1993). In the eastern Sahara, the Acheulean from a mound spring is dated at 350 kya by U-series and the succeeding occupations by the Middle Stone Age groups from Bir Sahara and Bir Tafawi date to the Last Interglacial and before, between 70-130 kya (Wendorf et al., 1991). A date has also recently been obtained for the 'Lower Levalloisian' from the Rufuq Pass, Kharga Oasis of 220 kya (Schwarcz 1996, pers. comm.).

South of the desert in West Africa, the Acheulean appears to be almost entirely absent except from the grassland regions of Nigeria and Ghana. The Acheulean in the Sudan remains undated. In the Horn (Ethiopia and Somalia), it is well represented with a long chronology and though the terminal stages are still undated, ages for the Middle Stone Age suggest that it had disappeared before the appearance of the Sangoan in the Kasim el Girba region in southeast Sudan (Shiner et al., 1971).

In East Africa, there is a U-series date of 260 kya for a late Acheulean from the Kapthurin Formation containing a hominid mandible in association with the proto-Levallois flaking method and blade production (Leakey, M. et al., 1969; McBrearty et al., 1996). As we have seen, ongoing work at Kapthurin suggests that around this time there was significant variability in the artefact assemblages. A Sangoan is present at Simbi in the Lake Victoria basin dated by K/Ar between 200 and 50 kya (McBrearty, 1991). In the Lake Eyasi basin, northern Tanzania, the Sangoan may date from between 300 and 200 kya (Mehlman, 1987, 1991). There is also one U-series date of 270 kya for a late Acheulean aggregate which is stratigraphically older than a Sangoan aggregate at Isimila in central Tanzania (Howell et al., 1972).

At Kalambo Falls, Zambia, where the Sangoan unconformably overlies the Acheulean, the radiocarbon dates are all infinite, and an amino acid racemisation date of >100 kya is thought to be minimal. Recently, U-series dates for wood from Sangoan and Acheulean horizons have been obtained (McKinney, 1996, pers. comm.) dating respectively from 80-50 kya and 200-100 kya.

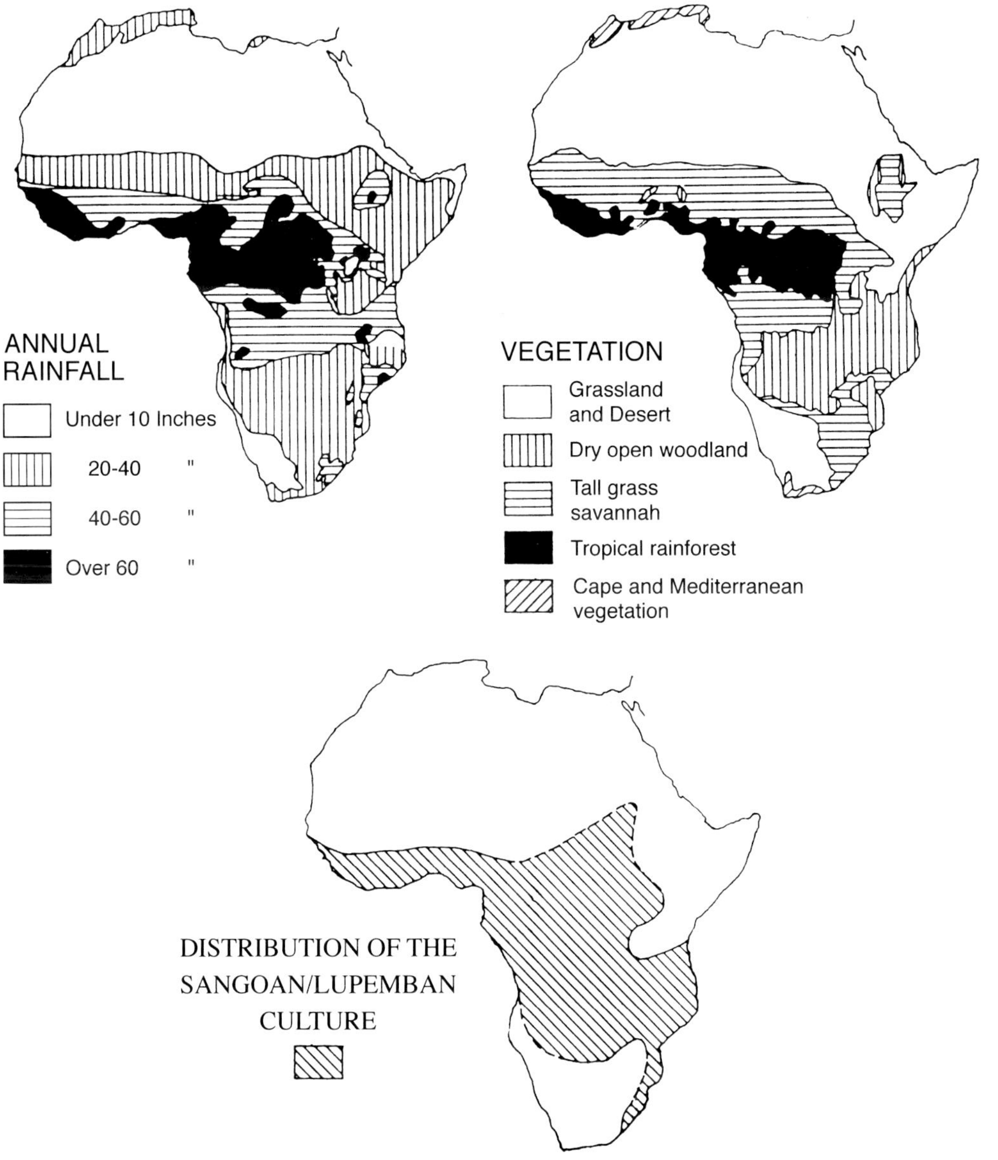

Figure 1 - Rainfall, vegetation and distribution of Sangoan/Lupemban culture.

Again, it is probable that these should be considered as minimal ages if the dates from East Africa referred to above are confirmed.

From South Africa, the Acheulean (Fauresmith) at Elandsfontein and Kathu Pan date on fauna between 700-600 kya (Klein, 1989: 211). Dates for the Early Middle Stone Age in South Africa confirm that it was already established by c. 100 kya (Deacon, 1992) at Klasies River Mouth Caves and Border Cave (Beaumont et al., 1978; Grün et al., 1990). These dates suggest that assemblages designated "early Middle Stone Age" and Sangoan appeared at much the same time in the continent, somewhere around 300 - 200 kya, but in different ecosystems.

Habitat

The climatic and environmental record for the African continent prior to 50 kya is poorly known as there are very few pollen records or other evidence, except for deep sea cores from which some correlations can be made to suggest the environmental context of terrestrial sites of the immediately post-Acheulean artefact assemblages.

In North Africa, the final Acheulean site at Sidi Zin, Tunisia, is considered to belong to a time of increased aridity (Gobert, 1950), but the Acheulean in the Sahara is found in association with streams and lakes where climatic changes from warmer and wetter, to drier and colder appear to have been fairly rapid.

Equally significant are changes that are inferred, and sometimes confirmed with respect to the first human occupation of the Congo basin and West Africa. The Acheulean appears to be confined to the peripheral edges of the basin. There is little evidence for occupation of the drier, western part of the basin and none at all for penetration of its central region. It is presumed that during much of the Middle Pleistocene, tropical evergreen forest occupied most of the Congo basin but that at some time during o/i Stages 6, 7, and 8, drier and cooler conditions resulted in retreat of the forests which were replaced by tropical deciduous woodland and grassland; the latter in particular in those southern and western regions covered by the original Tertiary expansion and later redepositions of the Kalahari Sands. This drying climate was the time when hominid populations moved into the central basin of the Congo and its tributaries and developed a very distinctive Middle Stone Age techno-complex, the Sangoan and its successor, the Lupemban (Clark, 1962; Mortelmans, 1962; van Moorsel, 1956). These techno-complexes are not confined to the Congo basin, but are very widely spread in south central Africa which today is a vegetational mosaic dominated by deciduous woodlands of the *miombo* type (*Brachystegia*) with riparian evergreen forest along its streams, montane forest, and grassland on the high plateau lands and broad, grass filled and seasonally swampy valleys (*dambos*) that meander through the woodlands (White, 1983: 86- 101) . This kind of habitat is attested also for the late Acheulean at Kalambo Falls where pollen and macro-plant remains show that there is very little difference between the vegetation associated with the Acheulean and that of the present-day (Taylor et al., in prep.; van Zinderen Bakker, 1969). Pollen evidence shows that the climate associated with the Sangoan was alternating between milder, and colder and drier episodes. This may, perhaps, be seen as the most likely pattern that prevailed when the Sangoan techno-complex replaced the Acheulean in south central Africa (Figure 1.) The evidence for grasslands at Simbi in the Lake Victoria basin (open forest mosaic) is not inconsistent with this (McBrearty, 1992a, 1 992b) .

I see this "transitional period", >200-70 kya, as one of two emerging major techno-complexes: (1) the Middle Stone Age (*sensu stricto*) with regional variants in the classic sense of light duty tools in the more open and drier tropical habitats; and, (2) the Sangoan/Lupemban complex associated with tropical and subtropical woodlands.

And let me repeat here that I am not implying and never have implied a kind of ubiquitous cover of trees nor of evergreen tropical forest as has sometimes been mistakenly suggested. This ecosystem is the mosaic referred to above, and is likely at times to have been more open, while at others, to have been modified by the ingress of montane species under cooler temperatures, or at other warmer times included lowland tropical species. These woodlands have always been associated with extensive areas of grasslands on high plateaus and plains.

Pollen data for the main part of the later Pleistocene are more numerous and more reliable (Vincens et al., 1989, 1991) but the record stops before ±50 kya, and only rare and isolated windows into earlier times are known. The pollen record from Kalambo Falls is one of these bridging the transition from Acheulean through the Sangoan, to the Middle Stone Age Lupemban Complex.

Associated hominid fossils

Before discussing the lithic assemblages associated with these two major techno-complexes, it is necessary to review briefly the hominid fossils that can be associated with this transitional period. These have all been described by a number of authors and it is only necessary here to show that both late archaic and modern human fossils are found in the period between >200-70 kya.

The Ngaloba cranium from Laetoli (Tanzania), associated with an early Middle Stone Age is dated to 120 kya (Day & Stringer, 1982; Harris & Harris, 1981). The partial crania from Lake Eyasi (Tanzania) believed to be associated with the Sangoan, are also archaic and may date to well before 130 kya (Mehlman, 1987). The skull and postcranial bones from Kabwe, Zambia most probably date to the time of the Sangoan or final Acheulean, as assemblages of both occur adjacent to the cave site (Clark, 1959; Klein, 1973) . Even though Kabwe is archaic it shows a number of modern features. Singa (Sudan) (Stringer, 1979) is another cranium said to be associated with a Sangoan tool kit (Ziegert, 1981). Jebel Irhoud in Morocco is on the verge of being modern and ancestral to the later Dar-es-Soltan fossils with Aterian tools. In the Kibish Formation in the Lower Omo Valley, Ethiopia, one cranium (Omo II) has archaic features predominating, while Omo I is fully modern. These crania are said to be from the same horizon (Member I) (dating to 130 kya) and artefacts with Omo I are identified as Middle Stone Age. The Klasies River Mouth Cave hominid fossils from South Africa show predominantly modern features and date between 100-90 kya. More fragmentary fossils combine this blending of archaic and modern characteristics in the same fossil, while others are more clearly modern or archaic (e.g. Florisbad) (Bräuer, 1984). Until precision in dating can produce temporal definition, these fossils cannot be placed in chronological order and their significance clarified.

Archaeology

It is necessary to emphasize again, the continuity in lithic technology between the late Acheulean, Sangoan, and Middle Stone Age complexes. Core and flake forms continue throughout. In particular, both unspecialised and specialised cores continue but show more elaborate and specialised preparation for flake removal, and increase in standardisation of retouched tools, and also an increased selection in the materials used. The soft hammer flaking that produced the finely made bifaces of the Acheulean has been replaced by hard hammer flaking, both direct and indirect, and retouch is mostly stepped.

Populations remained sparsely distributed, but some of these were moving into new terrain as climatic amelioration in the Sahara and in Equatoria opened up favourable biomes previously of restricted use or unavailable to them. At this threshold of readjustment it is to be expected that experimentation led to the emphasis of one or other lithic component of an assemblage as choice and habitat dictated. Such adaptation led to the appearance of various flake scraper and point forms using the Levallois and disc core methods in the Sahara, the Horn, and much of east and south Africa. In the northwest and Mediterranean belt of North Africa, blades became a feature seen already with the final Acheulean in Morocco and later with the 'pre-Aurignacian' at Haua Fteah. In East Africa, the stratigraphic evidence from Kapthurin in the Baringo basin seems to demonstrate the existence of three contemporaneous cultural entities - a final Acheulean, a Sangoan, and an industry that emphasised blades. Blades and the Levallois technique based on obsidian and Eocene chert, are present from the beginning of the Middle Stone Age in Ethiopia (+180 kya) (Wendorf & Schild, 1974) and Somalia. A blade industry also overlies the Middle Stone Age at Enkapuni-ya-Moto in the Kenya Rift before 55 kya (Ambrose, 1996 pers. comm.).

By contrast, in Equatoria and south-central Africa, the emphasis is on both light duty and heavy duty tools. The Levallois method and blades only appear later, and at first only rarely with the

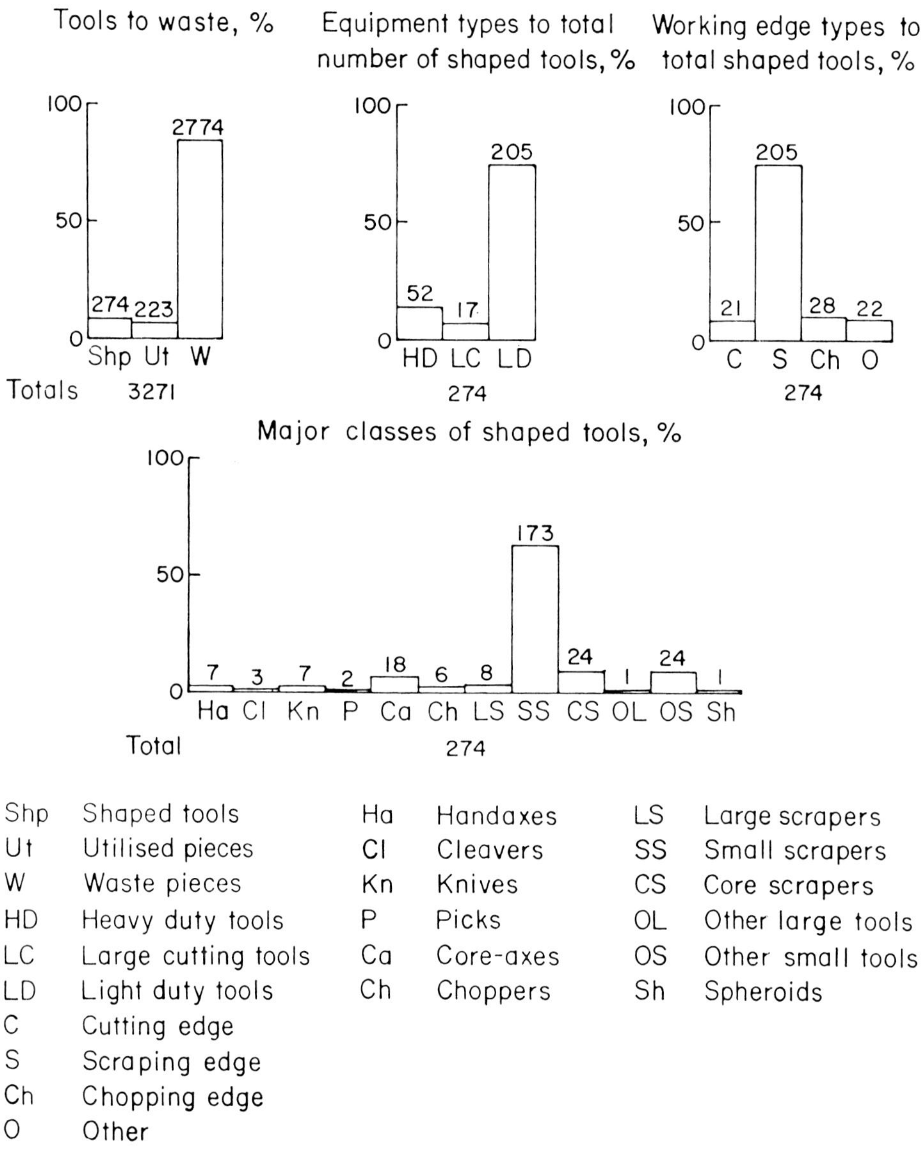

Figure 2 - Histogram Kalambo Falls Sangoan industries.

early Lupemban. In South Africa again, the emphasis is on blade technology based on quartzite (Cape and Transvaal) and indurated shale (Orange River basin and Natal).

The Sangoan is always found underlying the Middle Stone Age in the Sudan, east and south Africa; the Middle Stone Age assemblages there do not contain any heavy duty core-axes (Clark, 1988). In streams in southern and south central Africa, as well as in Uganda, the Sangoan is usually found in the basal gravels of the later Pleistocene alluvial sediments. Heavy duty tools are usually found without the light duty tool component, due no doubt to fluvial sorting. This has led some researchers to think of the Sangoan as consisting of aggregates of large picks and core-axes,

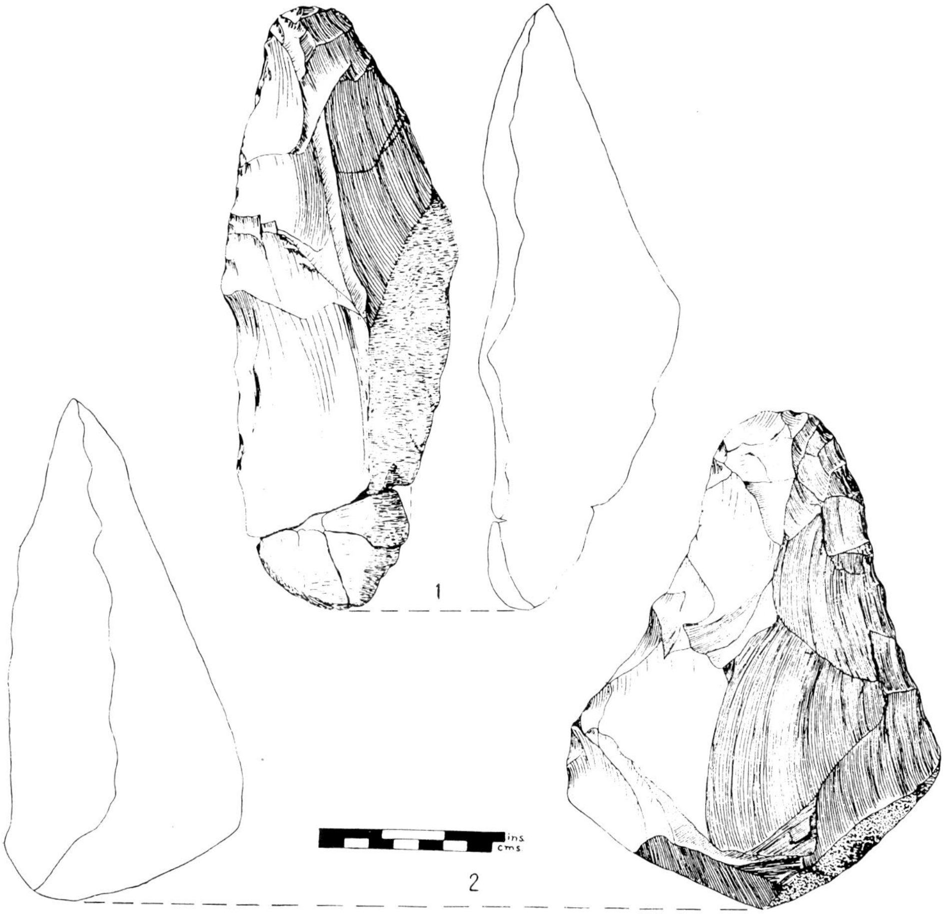

Figure 3 - *Sangoan tools from Kalambo Falls, Zambia.*
1. Bifacial, parallel sided core-axe with rounded working end made from feldspathetic quartzite. (B 1292159).
2. Bifacial, sub-triangular core-axe with accuminate edges and lingnate working end. Made from a cobble of hard quartzite. (B/214/59).

but nothing more. For this reason, the Sangoan assemblages from Kalambo Falls, which consist of *both* heavy and light duty artefacts, show that the heavy duty component is designed to make it possible to use certain resources of the biome that are either absent or no longer considered important by Middle Stone Age groups living in other ecosystems. The light duty tool component of the Sangoan does not differ significantly in retouched tools, or flake and core forms from those of the early Middle Stone Age in the Sahara, east or south Africa; except in the apparent absence of the Levallois method of flake production.

So, it is my contention that early Middle Stone Age assemblages (*sensu stricto*) exist contemporaneously with Sangoan and Lupemban in those parts of the continent where the need for heavy duty tools did not exist. This must remain contentious though a probability until greater precision in dating becomes available.

The likelihood of this contemporaneity is supported again by evidence from Kalambo Falls. Since this has only previously been reported in a paper difficult to access (Clark, 1964), and treated more generally in a recent comparative review (Sheppard & Kleindienst, 1996), it is set out again here to emphasise certain features of Sangoan technology (Clark,1964).

Figure 2 shows the breakdown by tool forms of an assemblage from a stoneline at Site B2/59 and also indicates the percentages in which individual types are present. When the debitage flakes and cores are analysed, they conform closely to those found with the Acheulean which is stratigraphically below. In addition, there is a small number of Acheulean type bifaces, but these are rare and have been replaced by core-axes. This heavy duty tool component is not nearly as important percentage-wise as is the light duty tool component; in particular, the various forms of small scrapers. (Figure 4, nos. 4-7). The Sangoan at Kalambo Falls is, therefore, a tool kit of light duty tool forms to which are added the heavy duty core-axes and picks.

Sangoan core-axes (Figure 3) differ from Acheulean bifaces in several ways. They weigh more than bifaces. The refined soft hammer flaking is absent except at the distal end and the tool is blocked out by hard hammer, primary flaking. Lateral edges converge or more rarely are parallel. Butt ends are broad, and often the cortex is retained. The distal end is the working part of the tool as is also its weight. It would seem (Figure 4, no. 1) that these tools were, indeed, designed for heavy duty work since a number have the tip ends broken off and were resharpened or re-used. (Figure 4, nos. 2 and 3). We recovered a number of these distal ends, and it is contended that these core-axes were most probably intentionally resharpened when they became blunt. Figure 2 shows the typical Sangoan convergent core-axe and the less common parallel sided core-axe which becomes dominant in the succeeding earlier Lupemban Industry. If this aggregate is typical of the Sangoan tool kit, then other Sangoan assemblages, for example, those from the lower levels of the redeposited Kalahari Sands II in northeastern Angola (Figure 5) which shows the emphasis on heavy duty tools may indicate they are special purpose assemblages, or more probably may have been distorted by natural sorting by slopewash or creep. One or other of these alternatives may also explain the hillside assemblages in the Lake Victoria basin, Uganda (Cole, 1967; O'Brien, 1939; Wayland, 1923).

Innovation and technology

The variability that makes its appearance with the earliest post-Acheulean industries (>200 kya) must reflect differences in the way resources were used and in how lifeways differed from those that prevailed during the Acheulean. While the earlier post-Acheulean industries continue the basic technology and light duty tool component of a late Acheulean tradition, the regional variability that now becomes apparent suggests that new and more efficient experimental ways of exploiting resources were being adopted.

Undoubtedly, one of the most important inventions was that of the simple composite tool in which a predetermined stone working part is mounted with other materials to produce a more efficient implement often set in various different kinds of handles or hafts that are easily replaced. If curation of equipment was important, as I believe it to have been, movement could have been greatly facilitated and light duty tools more easily transported while heavier equipment would be left behind to be used again on the next seasonal round once movements were more systematically structured. Various kinds of hafts have been suggested but probably in Africa, for hand held tools, the most common was a lump of mastic that could provide an efficient grip and could be easily modified or changed simply by heating the mastic to make it malleable by working it with the hands. Hafting is attested by the reduction of bulbs and platforms to produce a thinner and more regular proximal end for mounting. The Aterian tang, though later, is a further and incontrovertible

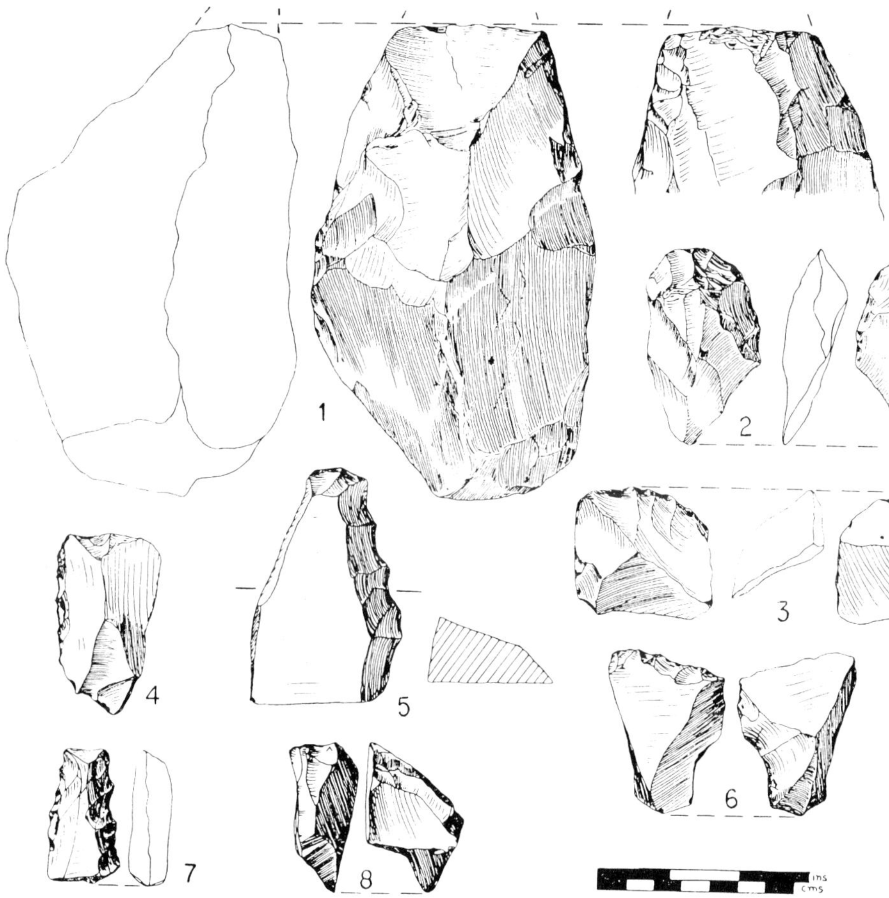

Figure 4 - *Sangoan tools from Kalambo Falls, Zambia.*

1. Thick butted, convergent edged core-axe of hard quartzite. The original working end has been removed by a blow struck on the dorsal face to form a straight chisel or core-axe edge which has later been retouched and utilised. (B2/II/59).

2. Resharpening flake from a core-axe of hard quartzite with linguate working end. The rejuvenating blow was struck close to one edge to detach the whole of the end of the tool. (B/134/59).

3. Resharpening span from a core-axe of hard quartzite removing the whole of the end of tool by striking it close to the edge of one face.(B/194/59).

4. Single edge side-scraper with straight, finely serrated edge made on a primary flake of chert. Shallow angle of retouch. (B 1/IV/Bi/56).

5. Notched end and side-scraper made on a tabular piece of chert. Blunt angle of edge retouch. (B/311/59).

6. Straight, single edge side-scraper and notch made on a chunk of feldspathic quartzite. Blunt and steep angles of edge retouch. (Bi/IV/B/56).

7. Double edge serrated side-scraper made on a thick primary flake of chert. (B1298159).

8. Proto-burin with four facets opposed to a joint plane. Made on a thick lump of chert. (A11931599).

indicator of hafting. The presence of differential patina, identified as due to protection by the hafting medium on a 54 kya Mousterian scraper from Quneitra, Israel (Friedman et al., 1994-1995) and on scrapers from southwest France (Anderson-Gerfaud, 1990) are further confirmation of hafting. Proximal reduction is not found with Kalambo Falls Acheulean small flake tools, however, and it is probable that they were hand-held without hafts as indeed many of the Middle Stone Age tools may also have been.

Shea (1991) identified Levallois points as spear points with the Levantine Mousterian and the stone blade may have been one of the most important incentives for hafting by producing a cutting, as well as a piercing, weapon that complements the pointed wooden spear. It is very impressive that pointed wooden spears, identified for throwing were already present in a Middle Pleistocene context in northern Germany, between 400 and 300 kya (Thieme, 1997) suggesting an as yet unrecognised ability and inventiveness by the biface makers who may have also made additional uses of wood and its by-products.

Various different activities have been identified on the basis of use-wear and polish on the edges of stone tools. Keeley and Toth (1981) show something of the range of materials worked; evidence that can often be supplemented by research into butchery, and skin, plant, and wood working.

The working of wood in more open and temperate ecosystems may take many forms, but Middle Palaeolithic/Middle Stone Age assemblages (*sensu stricto*) suggest it is probable that no heavy wood working was undertaken in these ecosystems. However, wood working is represented by scrapers and small adze-like stone equipment such as can be seen in the polish on the proximal ends of the cordiform bifaces of the Mousterian of Acheulean tradition (Anderson-Gerfaud, 1990).

It has been suggested that Sangoan core-axes might have been used for digging in the ground, but unless for shallow rooted plants or shallow burrowing animals, this seems impractical due to the shape and length of the tool for digging deeper holes. A digging stick is infinitely preferable. Prizing off blocks of raw material from outcrops (quarry sites) is another suggestion for the use of these tools; this can be expected to have produced much fracture and damage to the core-axes, whereas wedges and fire setting would have been much more efficient (Davies, 1976). Another suggested use that I made several years ago which has met with varied reception, is that the core-axes that supersede the Acheulean in a large part of sub-Saharan Africa, were primarily wood-working tools. Even in seemingly extensive grasslands there are still some trees, and if wood-working was an important activity, it implies not a little equipment would have been made from it. For instance, if the habitat was woodland there would have been more incentive to work wood to provide a range of useful products. It seems more probable, therefore, that this was the primary use for core-axes and proto-burins (Fig. 4, no. 8), (as wedges for splitting wood) and further processing by a range of flake and core scrapers. Few would query the use of more refined Lupemban core-axes as adze and axe tools, hand-held or hafted, for tree climbing, fruit collecting, and wood-working. The evolution of this tool type is well represented at Kalambo Falls and the Lake Victoria basin, and can be shown to have evolved from the Sangoan core-axe.

Equally as important of course for the people of the central African woodlands are the by-products, particularly bark, resin, and nutrients from the cambium or inner bark layer.

There is very good evidence at Kalambo Falls for the use of fire by Acheulean and Sangoan populations. A most efficient way of working wood without metal tools is by controlled charring and scraping as is still the case in many ethnographic contexts. Pieces of wood modified from use with evidence of charring are present with the Acheulean at Kalambo Falls. Fire setting and chopping will fell a tree and make possible further wood working into a range of tools and utensils (bowls, troughs, large and small pointed digging tools, spear shafts, and even canoes).

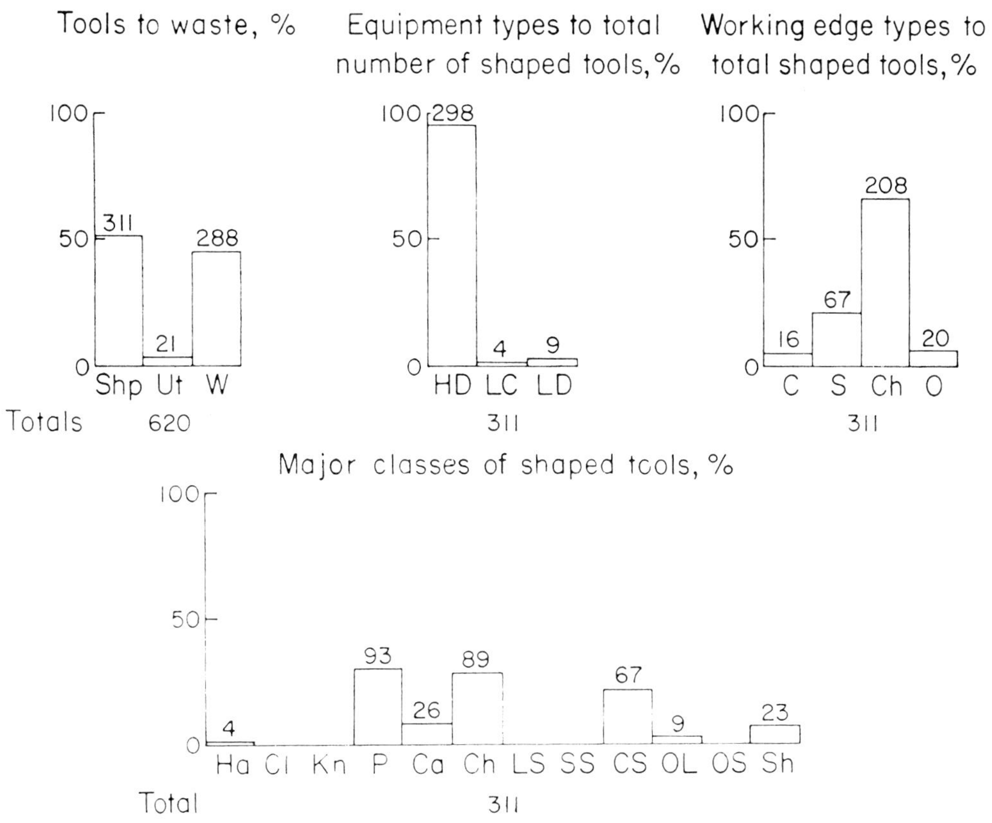

Figure 5 - *Analysis of Sangoan industries for three sites in Northeast Angola (Mussolexi, Camafufo, Catongula).*

Equally important is the availability of latex and resin for joining materials. This is regularly used today for mending cracks or holes in aluminum and clay pots. When the bark is removed from a tree, the outer part has many uses as carrying trays, constructing small sleeping platforms in trees, and ground shelters. The inner bark of some trees yields both protein and sugar and when stripped away, can be used to make rope, string and carrying bags, as well as bark cloth. How far back bark cloth was invented is not known. It was certainly present during the Later Stone Age (Deacon & Deacon, 1980). It is by no means impossible that bark cloth began to be made during the Middle Stone Age, if not earlier. It is not all that difficult a process - beating to spread the fibres, soaking in water, and further beating, layering and soaking a number of times, produces a durable covering for a dwelling structure, or as clothing or bags for carrying. The polished and glossy faces of what we have called 'rubbing stones' found with the Kalambo Falls Acheulean, Sangoan and later industries may have been used in the process of bark cloth making. The distribution of bark cloth today is to a great extent complementary to the range of Sangoan/ Lupemban techno-complexes but whether or not bark cloth began to be made at this early time, while a possibility, remains to be seen. Since the Sangoan core-axe belongs with a woodland, mosaic ecology there is a strong possibility that it is associated with trees and their by-products. A woodland economy does not necessarily imply that wooden equipment and utensils were important and numerous, but it is usually found to be the case. On the other hand, in more open ecosystems as for example, in the Sahara and Somalia today, wooden utensils or tools are also important.

The Sangoan/Lupemban techno-complex is but one example of a major and easily visible kind of variability that is present by the beginning of the Middle Stone Age in Africa south of the Sahara. It is an example also of actively expanding lifeways during the prelude to the appearance of anatomically modern humans. It belongs also in the least known time period of any part of African prehistory because of the paucity of sites, uncertainty of dating, and minimal palaeoenvironmental evidence. In some ways, the technology is a continuation. In others it shows more intensive experimentation and development of new ways of expanding control of territory and relieving intergroup stress within increasing population densities by more structured societal behaviour and by the introduction of a more regularly organised economy, as a means of establishing recognition of group identities and territories. Such innovation is also necessary to maintain cohesion in rapidly expanding communities widely distributed over the landscape in rapidly changing and sub-dividing component units, group identity must have been essential and it is possible this was achieved by more regular recognition of genetic lineages, family units, and affiliations. Mostly this scenario remains supposition with few data as yet to back it up. But, as chronological reliability increases for this transitional and early stage of the Middle Stone Age, so will the reasons and causes of these changes become more visible and identifiable.

References

ANDERSON-GERFAUD, P. (1990): Aspects of behavior in the Middle Paleolithic. Functional analysis of stone tools from Southwestern France. In: P. Mellars & C. Stringer (eds.), *The human revolution: Behavioural and biological perspectives on the origins of modern humans,* 389-418. Edinburgh.

BEAUMONT, P. B., de VILLIERS, H. & VOGEL, J.C. (1978): Modern man in sub-Saharan Africa prior to 49,000 B.P. A review and evaluation with particular reference to Border Cave. *South African Journal of Science* 74, 409-419

BIBERSON, P. (1961): *Le paléolithique inferieur du Maroc Atlantique.* Rabat (Publication Service des Antiquites du Maroc, no.17).

BRÄUER, G. (1984): A craniological approach to the origins of anatomically modern *Homo sapiens* in Africa and the implications for the appearance of modern humans. In: F. Smith & F. Spencer (eds.), *The origins of modern humans: A world survey of the fossil evidence,* 327-410. New York.

CLARK, J.D. (1959): Further excavations at Broken Hill, Northern Rhodesia. *Journal of the Royal Anthropological Institute* 89, 201-232.

CLARK, J. D. (1962): *Prehistoric cultures of Northeast Angola and their significance in tropical Africa.* Lisbon (Museo do Dundo Companhia de Diamantes de Angola. Publicaçones Culturais, no. 62).

CLARK, J. D. (1964): The Sangoan culture of Equatoria: The implications of its stone equipment. In: *Miscelanea en Homenaje al Abate Henri Breuil,* 309-325. Barcelona.

CLARK, J. D. (1969): *Kalambo Falls prehistoric site.* Volume I. *The geology, palaeoecology and detailed stratigraphy of the excavations.* Cambridge.

CLARK, J. D. (1975): A comparison of the Late Acheulian industries of Africa and the Middle East. In: K. W. Butzer & G. Ll. Isaac (eds.), *After the australopithecines,* 605-659. The Hague.

CLARK, J. D. (1988): The Middle Stone Age of East Africa and the beginnings of regional identity. *Journal of World Prehistory* 2, 235-305.

CLARK, J. D. (1992 a): The Earlier Stone Age/Lower Palaeolithic in North Africa and the Sahara. In: F. Klees & R. Kuper (eds.), *New light on the Northeast African past. Current prehistoric research,* 19-37. Köln (Africa Praehistorica 5).

CLARK, J. D. (1992 b): African and Asian perspectives on the origins of modern humans. In: M. J. Aitken, C. B. Stringer & P. A. Mellars (eds.), *The origins of modern humans and the impact of chronometric dating,* 201-215. Princeton (The Royal Society. Philosophical Transactions: Biological Sciences. 337).

CLARK, J. D. & KURASHINA, H. (1979): Hominid occupation of the east-central highlands of Ethiopia in the Plio-Pleistocene. *Nature* 282, 33-39.

COLE, G. H. (1967): The later Acheulian and Sangoan of southern Uganda. In: W. W. Bishop & J. D. Clark (eds.), *Background to evolution in Africa,* 481-526. London.

DAVIES, O. (1976): The 'Sanepan' industries. *Annals of the Natal Museum* 22, 885-911.

DAY, M.H. & STRINGER, C.B.. (1982): A reconstruction of the Omo Kibish remains and the *erectus-sapiens* transitions. In. *1er Congrès International Paléontologie Humaine. Pretirage 2,* 814-846. Nice.

DEACON, H.J. (1992): South Africa and modern human origins. In: M.J. Aitken, C.B. Stringer, & P.A. Mellars (eds.), *The origins of modern humans and the impact of chronometric dating,* 177-184. Princeton (The Royal Society. Philosophical Transactions: Biological Sciences. 337).

DEACON, H.J. & DEACON, J. (1980): The hafting, function and distribution of small convex scrapers with an example from Boomplaas Cave. *South African Archaeological Bulletin* 35, 31-37.

DORST, J. & DANDELOT, P. (1970): *A field guide to the larger mammals of Africa.* London.

FALGUERES, C., BAHAIN, J.-J. & SELEKI, H. (1997): U-Series and ESR dating of teeth from Acheulian and Mousterian levels at La Micoque (Dordogne, France). *Journal of Archaeological Science* 24, 537-545.

FRIEDMAN, E., GOREN-INBAR, N., ROSENFELD, A., BAR MARDER, O. & BURLAN, F. (1994-1995): Hafting during Mousterian times - further indications. *Journal of the Israel Prehistoric Society* 26, 8-31.

GOBERT, E.G. (1950): Le gisement paléolithique de Sidi Zin. *Karthago* 1, 1-51.

GRÜN, R., BEAUMONT, P.B. & STRINGER, C.B. (1990): ESR dating evidence for early modern humans at Border Cave in Southern Africa. *Nature* 344, 537.

GRÜN, R. & STRINGER, C.B. (1991): Electron spin resonance dating and the origin of modern humans. *Archaeometry* 33, 153-199.

HARRIS, J.W.K. & HARRIS, K. (1981): A note on the archaeology at Laetoli. *Nyame Akuma* 18, 18-21.

HAYNES, C.V., MAXWELL, T.A., El HAWARY, A., NICOLL, K.A. & STOKES, S. (in press): An Acheulian site near Bir Kiseiba in the Darb el Arba in Desert, Egypt. *Geochronology.*

HOWELL, F.C., COLE, G.H., KLEINDIENST, M.R., SZABO, B.J., & OAKLEY, K.P. (1972): Uranium series dating of bone from the Isimila prehistoric site, Tanzania. *Nature* 237, 51-52.

HUBLIN, J.J. (1992): Recent human evolution in northwest Africa. In: M. J. Aitken, C. B. Stringer, & P. A. Mellars (eds.), *The origins of modern humans and the impact of chronometric dating,* 185-191. Princeton (The Royal Society. Philosophical Transactions: Biological Sciences., no. 337).

KLEIN, R.G. (1973): Geological antiquity of Rhodesian Man. *Nature* 244, 311-312.

KLEIN, R.G. (1989): *The human career: Human biological and cultural origins.* Chicago.

KEELEY, L. & TOTH, N. (1981): Micro-wear polishes on early stone tools from Koobi Fora, Kenya. *Nature* 293, 464-465.

LEAKEY, M.D., with D. ROE (1996): *Olduvai Gorge. Excavations in Beds III, IV and the Masek Beds 1968-1971.* Cambridge.

LEAKEY, M.D., TOBIAS, P.V., MARTYN, J.E., & LEAKEY, R.E.F. (1969): An Acheulian industry with prepared core technique and the discovery of a contemporary hominid mandible at Lake Baringo, Kenya. *Proceedings of the Prehistoric Society,* 48-76.

LEZINE, A.M. & CASANOVA, J. (1991): Correlated oceanic and continental records demonstrate past climate and hydrology of North Africa (0-140 k). *Geology* 19, 307-310.

McBREARTRY, S. (1991): Recent research in western Kenya and its implications for the status of the Sangoan industry. In: J. D. Clark (ed.), *Cultural beginnings. Approaches to understanding early hominid life-ways in the African savanna,* 159-176. Bonn.

McBREARTRY, S. (1992 a): The Sangoan record at Simbi, Kenya. *Nyame Akuma* 37, 50-51.

McBREARTRY, S. (1992 b): Sangoan technology and habitat at Simbi. *Nyame Akuma* 38, 34-39.

McBREARTRY, S., BISHOP, L. & KINGSTON, J. (1996): Variability in traces of Middle Pleistocene hominid behavior in the Kapthurin Formation, Baringo, Kenya. *Journal of Human Evolution* 30, 563-580.

McBURNEY, C.B.M. (1967): *The Haua Fteah (Cvrenaica) and the Stone Age of the south-east hard terrains.* Cambridge.

McHUGH, W. P., McCAULEY, J.F., HAYNES, C.V., BREED, C.S. & SHABER, G.G. (1988): Paleorivers and geoarchaeology in the southern Egyptian Sahara. *Geoarchaeology* 3, 1-40.

MEHLMAN, M. (1987): Provenience, age and associations of archaic *Homo sapiens* crania from Lake Eyasi, Tanzania. *Journal of Archaeological Science* 14, 133-162.

MEHLMAN, M. (1991): Context for the emergence of modern man in eastern Africa: Some new Tanzanian evidence. In: J. D. Clark (ed.), *Cultural beginnings. Approaches to understanding early hominid life-ways in the African Savanna,* 177-196. Bonn.

MOORSEL, H. van (1956): Esquisse préhistorique de la plaine de Leapoldville. *Bulletin Royale des Sciences Coloniales, Brussels,* 582-595.

MORTELMANS, G. (1962): Vue d'ensemble sur la préhistoire du Congo occidental. In: G. Mortelmans & J. Nenquin (eds.), *Actes du IVème Congrès Panafricain de Préhistoire et de l'étude du Quaternaire,* 129-164. Tervuren (Musee Royale de l'Afrique centrale. Ser. in-8°. Sciences humaines,no. 40).

MURDOCK, P. (1959): *Africa: Its peoples and their culture history.* New York.

O'BRIEN, T.P. (1939): *The prehistory of the Uganda Protectorate.* Cambridge.

SAHNOUNI, M., de HEINZELIN, J., BROWN, F. & SAOUDI, Y. (1996): Récentes recherches dans le gisement Oldowan d'Ain Hanech, Algerie. *Comptes Rendus Academie des Sciences, Paris.* tome 323, Serie IIa, 639-644.

SHEA, J.J. (1991): Spear points from the Middle Paleolithic of the Levant. *Journal of Field Archaeology* 15, 441-450.

SHEPPARD, P.J., & KLEINDIENST, M.R. (1996): Technological change in the Earlier and Middle Stone Age of Kalambo Falls (Zambia). *African Archaeological Review* 13, 3, 171-196.

SHINER, J.L., MARKS, A.E., de HEINZELIN, J., CHMIELEWSKI, W. & HAYS, T.H. (1971): *The prehistory and geology of the Northern Sudan.* Parts I & II. Report to the National Science Foundation. Grant GS 1192.

STRINGER, C.B. (1979): Reconstructing recent human evolution. In: M.J. Aitken, C.B. Stringer, & P.A. Mellars (eds.), *The origin of modern humans and the impact of chronometric dating,* 217-224. Princeton (The Royal Society. Philosophical Transactions: Biological Sciences, no. 337).

SZABO, B.J., McHUGH, W.P., SHABER, G.G., HAYNES, C.V. & BREED, C.S. (1989): Uranium-series dated anthigenic carbonates and Acheulean sites in Southern Egypt. *Science* 243, 1053-1056.

TAYLOR, D.M., MARCHANT, R. & HAMILTON, A.C.: A re-analysis and interpretation of palynological data from Kalambo Falls prehistoric site. In: J.D. Clark (ed.), *Kalambo Falls prehistoric site.* Volume 3. (in preparation).

THIEME, H. (1997): Lower Palaeolithic hunting spears from Germany. *Nature* 385, 807-820.

VINCENS, A. (1989): Les forêts claires Zambéziennes du bassin Sud-Tanganyika. Évolution entre 25000 et 6000 ans B.P. *Comptes Rendus Academie des Sciences Paris,* tome 308, Serie II, 809-814.

VINCENS, A. (1991): Late Quaternary history of the South-Tanganyika basin climatic implications in south central Africa. *Palaeo. Palaeo. Palaeo.* 86, 207-226.

VOLMAN, T.P. (1984): Early prehistory of Southern Africa. In: R.G. Klein (ed.), *Southern African prehistory and paleo-environments,* 162-220. Rotterdam.

WAYLAND, E.J. (1923): *Some primitive stone implements from Uganda.* Occasional Paper No. 1, Geological Survey of Uganda.

WENDORF, F., CLOSE, A., SCHILD, R., GAUTHIER, A., SCHWARCZ, H., MILLER, G.H., KOWALSKI, K., KROLIK, H., BLUSZCZ, A., ROBINS, D., GRÜN, R. & McKINNEY, C. (1991): Chronology and stratigraphy of the Middle Paleolithic at Bir Tarfawi, Egypt. In: J.D. Clark (ed.), *Cultural beginnings: Approaches to understanding early hominid life-ways in the African savanna,* 197-208. Bonn.

WENDORF, F. & SCHILD, R. (1974): *A Middle Stone Age sequence from the Central Rift Valley. Ethiopia.* Warsaw.

WENDORF, F. & SCHILD, R. (1980): *Prehistory of the eastern Sahara.* New York.

WENDORF, F., SCHILD, R., CLOSE, A.E. et al. (1993): *Egypt during the last Interglacial: The Middle Paleolithic of Bir Tarfawi and Bir Sahara East.* New York.

WENDORF, F., SCHILD, R., CLOSE, A.E., SCHWARCZ, H.P., MILLER, G.H., GRÜN, R., BLUSZCZ, A., STOKES, S., MORAWSHA, L., HUXTABLE, J., LUNDBERG, J., HILL, C.S. & McKINNEY, C. (1994): A chronology for the middle and late Pleistocene wet episodes in the eastern Sahara. In: O. Bar-Yosef & R.S. Kra (eds.), *Late Quaternary and Paleo-climates of the Eastern Mediterranean,* 147-168. Tuscon.

WHITE, F. (1983): The vegetation of Africa and UNESCO Vegetation Map of Africa. *Nature Resource* No. 20, 86-101. Paris, U.N.E.S.C.O.

ZIEGERT, H. (1981): Abu Hugar Paleolithic site (Blue Nile Province, Sudan). A preliminary Report. *Xth U.I.S.P.P. Congress, Mexico City. 1981 Section 2 Cl.*

ZINDEREN BAKKER , E.M. van (1969): The Pleistocene vegetation and climate of the Basin. In: J.D. Clark (ed.), *Kalambo Falls prehistoric site.* Volume I. *The geology, palaeoecology and detailed stratigraphy of the excavations,* 57-84. Cambridge.

The Bilzingsleben Site - *Homo erectus*, his Culture and his Ecosphere

Dietrich Mania, Ursula Mania & Emanuel Vlček

Abstract

The Bilzingsleben site is situated on the northern border of the Thuringia Basin in the Lower Wipper Valley. Its build-up is a travertine sequence at the base of which the Lower Palaeolithic find horizon was discovered. Six terrace-travertine sequences were observed corresponding to as many climatical basic cycles (Bilzingsleben I-IV). The age of the site Bilzingsleben II is 400,000 B.P.

More than 1,300 m² of the main find horizon have been excavated so far. The finds seem to have been discarded at a place used as a home base by Homo erectus individuals for a substantial period of time during the warm-temperate maximum of the interglacial. The palaeo-ecological setting of the site will be discussed in this paper. There will also be given an overview of the archaeological and palaeoanthropological finds.

Bilzingsleben is the most important site of the Lower Palaeolithic in Central Europe. This is a camp site of *Homo erectus* with a lot of cultural remains, with living traces and structures of a settlement. The site has an age of 400,000 years B.P.

The archaeological horizon is embedded in a travertine complex near the village Bilzingsleben on the northern border of the Thuringian basin.

Geology and stratigraphy

Two sediment sequences from the region of the Saale river and the foreland of the Harz mountains yielded a new basis for the geological and archaeological chronology beginning with the Elsterian glaciation. These are the terrace-travertine sequence from the Wipper valley near Bilzingsleben (Figure 1) and the limnic sequence from the brown coal mining near Schöningen (Mania, 1997a; Thieme & Mania, 1994; Thieme & Maier, 1995).

Six large climatic cycles (Bilzingsleben I-VI, Schöningen I-VI) were recognized after the Elsterian glaciation each lasting about 80,000 to 120,000 years. They correspond to as many interglacials and glacials. Three climatic cycles alone were identified for the phase between the Elster glaciation (Elster II) and the Saale glaciation *sensu stricto* (Saalian I = Drenthian). This phase is the so called „Holsteinian complex“ (Bilzingsleben/Schöningen I-III). The Saalian complex is subdivided by an interglacial (Bilzingsleben IV, Schöningen IV, Neumark-Nord, Ehringsdorf, Langenbogener Bodenkomplex of Lengefeld-Bad Kösen). The interglacial in the Cycle V is the Eemian interglacial, in Cycle VI is the Holocene. This last climatic cycle between Eemian and Holocene (Weichselian) is subdivided by 15 smaller cycles. The base of this stratigraphy is the Aschersleben lake in the Northern Harz foreland - the most complete sequence of the last 125,000 years on the continent (Mania, 1997a-c).

The terrace-travertine sequence of Bilzingsleben was subject to a special geological mechanism which was regulated by climatic changes. It is characteristic of the river valleys of the Mittelgebirge forelands (Figure 2) and it is subdivided into four phases:

1. *Early glacial phase.* Relatively humid cold climate. Fluviatile erosion, transport and accumulation of gravels are synchroneous processes in the course of which the valley was eroded with a gravel cover developed on its bottom. Duration: 50,000- 80,000 years.

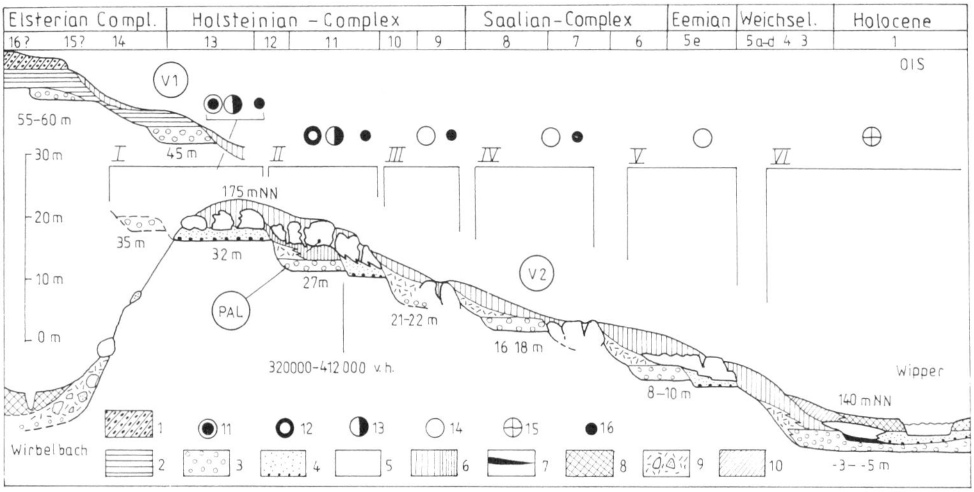

Figure 1 - *Bilzingsleben. The terraces-travertine sequence of the Wipper valley near the village. 1 - groundmoraine, 2 - glazi-limnic silts „Bändertone", 3 - cold climatic gravels, 4 - warm climatic fluviatil gravelly sands, 5 - travertine, 6 - loess (upper loess and Middle Pleistocene loess), 7 - organic sediments, 8 - loams and silts of the flood plain, 9 - slope debris and solifluction sediments, 10 - humic slope sediments, 11 - Theodoxus serratiliniformis fauna, 12 - Corbicula fluminalis fauna, 13 - Helicigona banatica fauna, 14 - Helix pomatia fauna, 15 - common molluscan fauna of woods, 16 - Celtis australis.*
V1 and V2 - the first and the second glaciation terrace, Pal - Palaeolithic find horizon. OIS - Oxygen isotopes stages.

Figure 2 (page 295) -The stratigraphy and climatic record of the last 800,000 years in the Elbe-Saale region. A, B - stratigraphy, KZ - cold stage, glazial, WZ - warm stage, interglazial; C - warm- and cold stages; D - cilmate record, 1 - arctic, 2 - subarctic, 3 - boreal, 4 - temperate; E - the Bilzingsleben sequence; F - the Schöningen sequence; G - important profiles and special sediment seqences; H - soil complexes: uBK, oBK - lower and upper Pedocomplex of Mahlis, FBK - Pedocomplex of Freyburg, RBK - Pedocomplex of Rudelsburg, LBK - Pedocomplex of Langenbogen, NBK - Pedocomplex of Naumburg, KV soil of Kösen, HB - soil of Holocene.
Faunas: Af - Archidiskodon meridionalis fauna, Mtf - Parelephas trogontherii fauna, Mtf - Mammuthus primigenius fauna, Paf - Palaeoloxodon antiquus fauna, Lf - Lagurus lagurus fauna, Wf - Holocene fauna of woods.
Molluscan faunas: 1 - Fagotia acicularis fauna, 2 - Helicigona banatica fauna of the Middle Pleistocene type, 3 - H. banatica fauna of the Upper Pleistocene type, 4 - Theodoxus serratiliniformis fauna, 5 - Corbicula fluminalis fauna, 6 - common Central European fauna , 7 - fauna of the forest steppes, 8 - fauna of the loess steppes and the tundras.

2. *Pleniglacial phase.* Arid and cold climate. The processes described above have come to a standstill. Permafrost prevails. Solifluction and aeolic events occur. Loess is formed. Duration: 10,000-20,000 years.

3. *Late glacial phase.* The climate becomes more and more humid and warm. The river is refilled with water and cuts into the gravel cover. To a lesser degree, sands and gravels are again transported and accumulated. During a period of rewarming and reforestation, however, these processes are stopped again. Duration: 5,000-10,000 years.

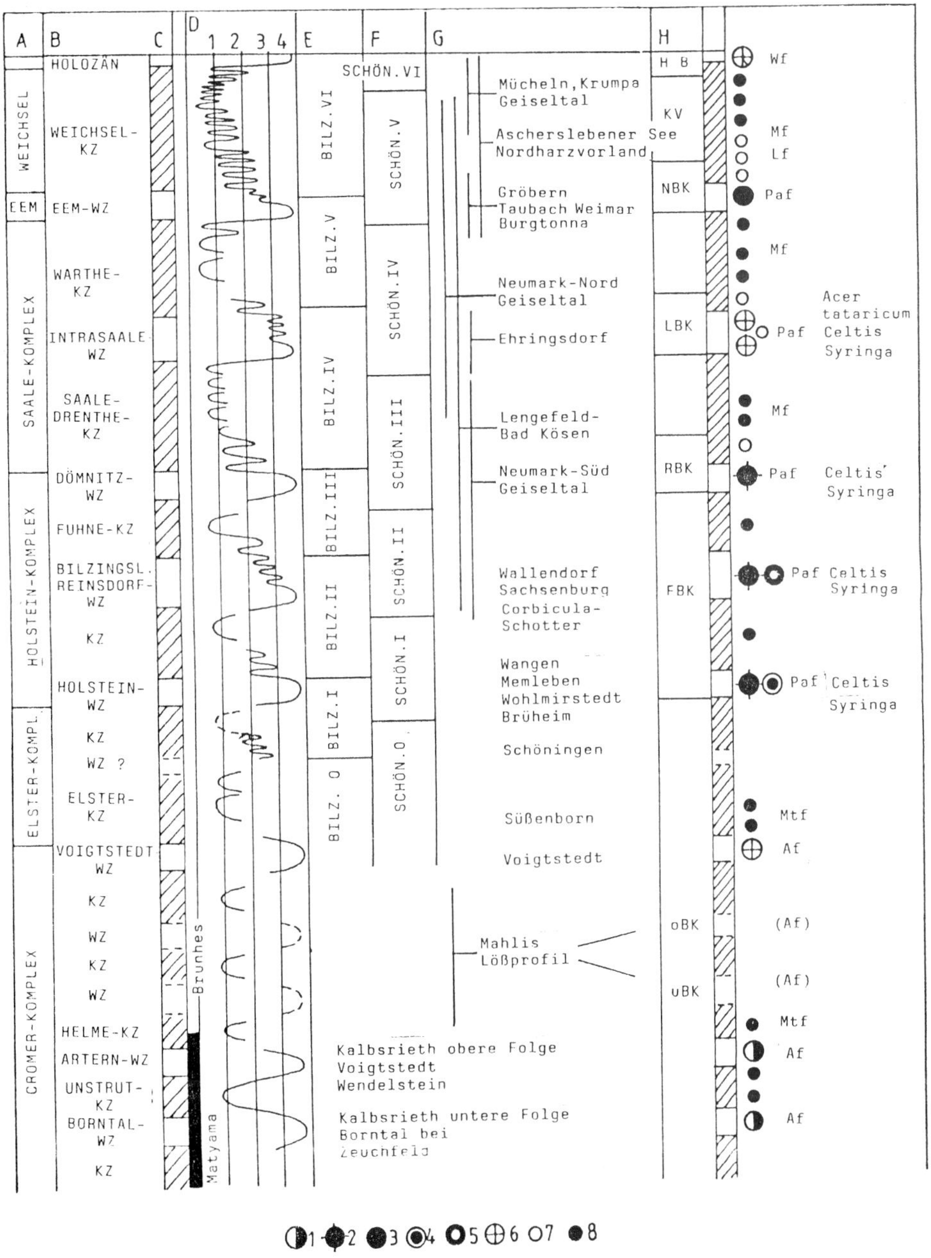

4. *Interglacial phase.* Warm temperate climate. Thick travertine covers are formed. The interglacial weathering processes favour the production of forest soils (Parabraunerde). Duration: about 10,000 years.

The terrace situated at 45 m above the recent bottom of the Wipper valley carries the glacial series of the Elster glaciation. The subsequent terrace-travertine series lies at 32-35 m above the recent bottom (Bilzingsleben I). The terraces situated at 26-27 m, 20-22 m, 16-18 m and 8-10 m contain the cycles of Bilzingsleben II-V. The base of cycle VI lies below the recent bottom. The

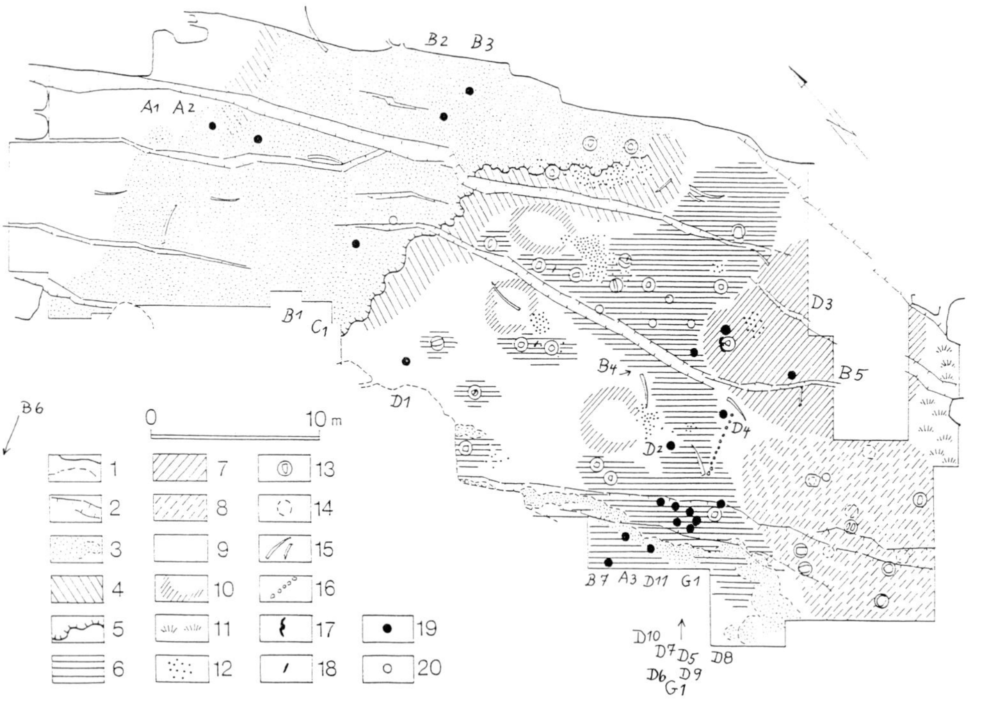

Figure 3 - *Bilzingsleben. The map of the camp site.*
1 - limits of excavation, 2 - vaults, 3 - travertine sands of the channels and the deluvial fan, 4 - activity zone near the border line, 5 - border line, 6 - activity zone: so called „Workshop zone", 7 - pavement, 8 - special workshop zone with burned blocs and large anvils, 9 - other activity zones, 10 - dwelling structures, 11 - reed on the border of the lake, 12 - fire places, 13 - small workshop with anvil, 14 - burned bloc, 15 - tusks, 16 - line of large pebbles, 17 - a special workshop with skull of a bison on the pavement, 18 - intentionell engraved bone artefact, 19 - human skull fragment, 20 - human tooth.

terrace situated at 16-18 m corresponds with the Saale glaciation *s.str.* Thus, the travertine complexes of the Bilzingsleben cycles I-III demonstrated three completely developed interglacials of the period between the Elster and the Saale glaciation. They are separated from each other by glacials with phenomenons of the permafrost and with loess formation. The climatic conditions of these interglacials are mediterraneously and subcontinentally influenced.

The oldest interglacial of Bilzingsleben I (Schöningen I) corresponds to the Holsteinian interglacial. The interglacial of Bilzingsleben II contains the Palaeolithic find horizon. The equivalent in the Schöningen sequence is the interglacial Schöningen II, so called „Reinsdorfian", as well with Palaeolithic find horizons (Thieme et al., 1993; Thieme, 1996). The interglacial of Bilzingsleben III (Schöningen III) can be correlated with the Dömnitzian (Erd, 1978; Cepek, 1986; Urban, 1997). Bilzingsleben II and III are not events of an early glacial period of the Saalian (Litt & Turner, 1993; Eißmann & Litt, 1994; Urban, 1997). They are separatly large cylcles in the Pleistocene climatic sequence. From this reason, we called this part of the Middle Pleistocene „Holsteinian complex".

Bilzingsleben I

A fluvial-limnic sequence covered by a travertine of several metres is situated on the 32 m terrace. The sandy-gravelly sediments contain a *Theodoxus serratiliniformis* fauna. This river snail is related to *Theodoxus danubialis* found in the lower course of the Danube. Apart from *Helicigona banatica*, the following Mediterranean and Southeast European species occur: *Aegopis verticillus, Pagodulina pagodula, Discus perspectivus, Pseudalinda turgida, Iphigena tumida*, as well as the Atlantic species *Azeca menkeana* and the fossil species *Acicula diluviana*.

Larger mammals, such as the straight tusked elephant *Palaeoloxodon antiquus* and the rhinoceros *Dicerorhinus kirchbergensis*, have also been recorded from the *Theodoxus* deposits.

A *Helicigona banatica* fauna has been collected from the overlying sandy-silty parts of the fluvial deposits. *Helicigona banatica* is also typical for the travertine layers in the upper part of the sequence. This travertine has yielded *Corylus avellana, Quercus* sp., *Alnus glutinosa* and *Celtis* sp.

Bilzingsleben II

The interglacial deposits consist of a travertine complex laterally interlaced with fluvial sands on top of the glacial 27 m terrace (gravels, solifluction deposits and a loess cover with a thickness of 3 m). The sequence contains at the base of the travertines the well known Palaeolithic horizon of Bilzingsleben.

In the surroundings of the site, we have found the *Corbicula fluminalis* in the interglacial fluvial deposits of this Middle Pleistocene terrace. The *Corbicula fluminalis* fauna with interglacial terrestrical species is also embedded in the fluvial facies of the travertine complex.

The *Corbicula fluminalis* fauna is a typical fluvial fauna for this Middle Pleistocene interglacial in the Elbe-Saale region. Recently, *Corbicula fluminalis* lives in the large rivers of southwestern Asia with brackish water. The influence of salty water in the interglacial Middle Pleistocene rivers of the Elbe-Saale region comes from the salinar of Trias and Permian in the underground.

The characteristical terrestrical molluscan fauna of the travertine complex II is the *Helicigona banatica* fauna with *Azeca menkeana* and *Acicula diluviana*. Species indicative of a wooded and an open environment are common in the fauna. Furthermore Mediterranean and Southeast European species are present: *Helicigona banatica, Aegopis verticillus, Discus perspectivus, Pagodulina pagodula, Iphigena tumida, Iphigena densestriata, Truncatellina claustralis, Cepaea vindobonensis*. The smaller mammals are represented by stratigraphically important species such as the watervole *Arvicola cantianus* and *Trogontherium cuvieri*. Also *Macaca sylvana* occurs. The larger mammals are represented in a *Palaeoloxodon antiquus* fauna.

A study of the abundant floral remains showed the presence of a rich mixed oak-forest flora, more specifically a Buxo-Quercetum, Buxo-Syringetum and Berberidion association with many thermophile exotics such as *Buxus sempervirens, Pyracantha coccinea, Celtis australis, Syringa josikaea, Juniperus sabina, Vitis sylvestris* and other southern species. *Potentilla fruticosa*, a species with a more continental distribution, also occurs.

Bilzingsleben III

The travertine differs distinctly from the older ones because of ist lithology and its fossil content. It contains a *Helix pomatia* fauna with some exotic elements (*Aegopis verticillus, Discus perspectivus, Iphigena densestriata, Truncatellina claustralis*) along with components of a temperate mixed forest (*Quercus* sp., *Corylus avellana, Syringa josikaea, Ulmus* sp., *Fraxinus excelsior* and *Celtis* sp.). Other south and southeast European species of molluscs are *Iphigena latestriata, Iphigena tumida, Pagodulina pagodula, Valvata naticina*.

Figure 4 - Bilzingsleben. Camp site. A dissection of the pavement area.

The limnic-telmatic depostis from the upper series of the Körbisdorf gravels in the Geisel valley (Mania & Mai, 1969) and the slope debris - soil complex of Lengefeld-Bad Kösen profile (Rudelsburg soil complex - Mania & Altermann, 1970, new investigations Mania et al., 1995) are also indicative of the „Bilzingsleben III-interglazial". A fully developed *Helicigona banatica* fauna associated with its particular *Azeca menkeana* fauna was found in the valley of the Geisel and in Lengefeld Bad-Kösen. A hornbeam-oak forest with Mediterranean and Pontic spezies (e.g. *Crataegus pentagyna, Prunus mahaleb, Acer monspessulanum, Quercus pubescens, Azolla filiculoides*) was determined in the profile of the Geisel valley. In the Rudelsburger soil-slope debris complex *Celtis* sp. occured.

The latest Bilzingsleben interglazial (Bilzingsleben III) might correspond to the Dömnitz Interglazial (Cepek, 1986; Erd, 1978), which according to pollen analysis resembles the Wacken and Schöningen interglazial (Urban et al., 1991). The middle Bilzingsleben interglazial (Bilzingsleben II) represents an older interglazial phase which does not correspond to the Holsteinian *sensu stricto* because of clear differences in the palynological record (Erd, 1997). The palynological picture shows similarities with that from the so-called Reinsdorf interglazial deposits exposed in the browncoal-pit near Schöningen, which also contain artefacts and wooden spears and clubs (Thieme & Maier, 1995; Thieme, 1997a,b). The pollen succession of the Reinsdorf interglazial differs from the Holsteinian and the Schöningen interglazial because of a high amount of *Abies* pollen and of an other course of the succession in the climatic optimum (Erd, 1997; Urban, 1997). The *Heliciogona banatica - Helix pomatia - Azeca menkeana* fauna of the three Bilzingsleben interglacials represents apparently typical Middle Pleistocene features also recognised in other interglacial deposits of the „Holsteinian complex", e.g. in the travertine of Brüheim, Schwanebeck and Osterode near the „Fallstein" (Mania, 1973, 1983). These Middle Pleistocene faunas share the occurence of the now extinct species *Acicula diluviana*.

A chronological indication was obtained by the $^{234}U/^{230}Th$ and ESR dates of the middle interglacial, the Bilzingsleben II deposits with the Palaeolithic find horizon (Schwarcz et al., 1988). The dates of 320-414 Kyr B.P. suggest a correlation of the Bilzingsleben II interglazial with OIS 11.

The Palaeolithic site and the culture

The terrace sequence of the middle complex (Bilzingsleben II) displays the following succession: Wipper gravels (1 m thick) - a gravelly silty solifluction layer (o,3 m thick) - aeolian loess with several ice wedge generations (3-4 m thick). After an erosional phase the loess was bleached by a pseudogley, then dissected by a brook. The interglacial travertine precipitation had already occured in the channels of the brook, and an up to 6 m thick travertine profile subsequently developed on top of them. At the base of this travertine the Palaeolithic horizon is present. It rests on the loess surface, which has constituted the flat and horizontal surface of a shore terrace of a shallow lake, as well as the subsoil of the basin cut by the brook channels. Adjacent to the shore terrace, an alluvial fan composed of travertine sands was deposited from the west. The fan and the shore terrace are the two types of facies containing the cultural remains, and are known in the literature on Bilzingsleben as the „Schwemmfächer" and the „Uferterrasse" respectively. The archaeological horizon is covered by a 60 cm thick „*Chara*" limestone overlain by a complex of travertine several meters thick. More than 1300 m^2 of the main find horizon with 3500 kg cultural objects have been excavated so far. The finds seem to have been discarded at a place used as a home base by early humans occupied for a substantial period of time during the warm-temperate optimum of the interglacial.

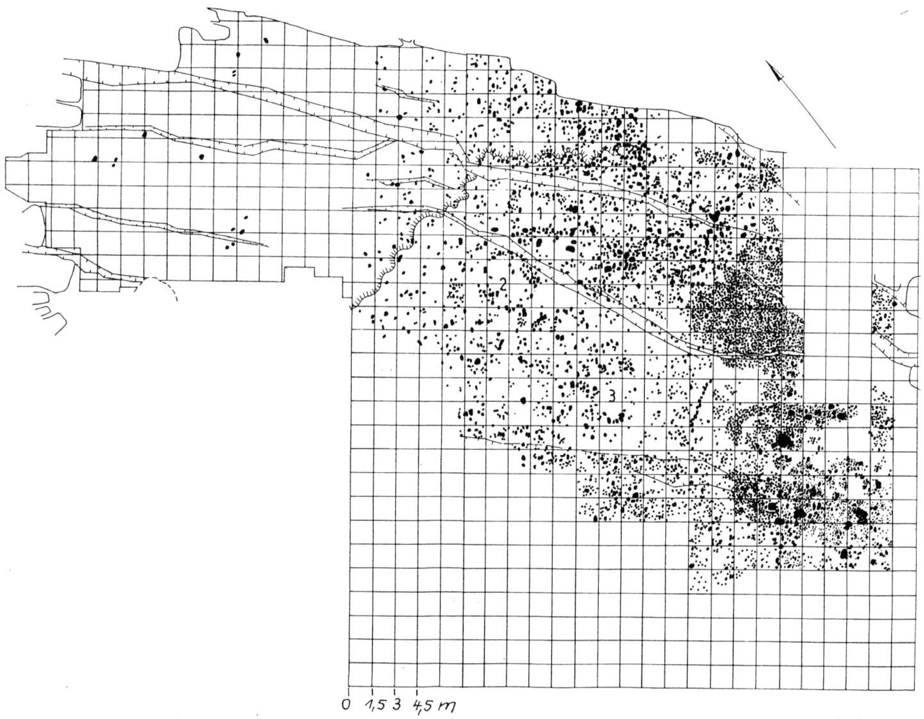

Figure 5 - *Bilzingsleben. Camp site. Distribution of the manuports (pebbles) and pebble tools. 1, 2 and 3 - dwelling structures.*

The camp site (Figure 3) was situated on a shore terrace at the edge of a shallow lake fed by an ascending karst spring. Close to the site was the outflow of the source. To the west, behind it, the slope of the valley rose, and to the northwest, east and southeast the valley lowland spread. The prevailing vegetation of this area can be inferred from the floral remains: a light, dry oak wood, predominantly interspersed by a thicket of *Buxus*. Meadows with shrubs of *Buxus sempervirens, Syringa josikaea, Pyracantha coccinea, Potentilla fruticosa, Corylus avellana, Swida sanguinea, Viburnum lantana* and other species were also determined. *Cotoneaster integerrimus* and *Juniperus sabina* shrubs grew on the slopes of the valley. Thick mixed oak forests with *Taxus baccata* occured in the narrow valley north of the site. In the valley, thicket of willow, reeds, meadows and swampy wood prevailed.

Judging from the floral and faunal remains, the climate was warm and relatively dry. The average temperature in January was -0,5° to +3 °C, in July +20° to 25 °C with an annual average of +10° to 11 °C. Seven months had average temperature of more than +10°C. The annual precipitation was approximately 800 mm. The climate was submediterranean and subcontinental influenced.

The most important types of vegetation were Buxo-Quercetum, Buxo-Syringetum and Berberidion (Mai, 1983, 1988, 1989). The contained Mediterranean and Southeast European species such as *Buxus sempervirens, Pyracantha coccinea, Celtis australis, Syringa josikaea, Juniperus sabina, Vitis sylvestris* associated with the subcontinental species *Potentilla fruticosa*. Furthermore, the flora consisted of the following species: *Marchantia* sp., *Thelypteris thelypteroides, Taxus baccata, Picea abies, Quercus robur, Alnus glutinosa, Betula pubescens, Populus tremula, Tilia*

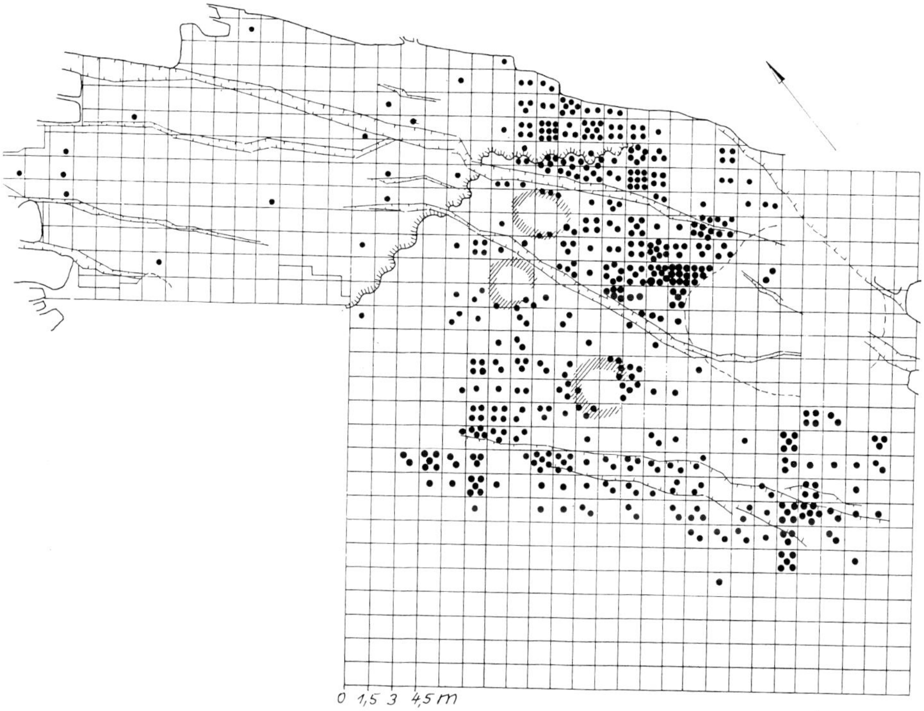

Figure 6 - *Bilzingsleben. Camp site. Distribution of the large hammerstones.*

platyphyllos, Pyrus sp., *Prunus avium, Prunus padus, Acer campestre, Acer pseudoplatanus, Fraxinus excelsior, Berberis vulgaris, Corylus avellana, Salix cinerea, Salix purpurea, Rubus* sp., *Cotoneaster integerrimus, Crataegus* sp., *Euonymus* sp., *Cornus mas, Swida sanguinea, Hedera helix, Rhamnus frangula, Viburnum lantana, Peucedanum alsaticum, Galeobdolon luteum* (Mai, 1983). The palynological analysis shows the phase with the oak mixed forests and with a maximum of the hazel and the hornbeam phase (Erd, 1997).

The characteristic association of the molluscan fauna found in the archaeological horizon is the *Helicigona banatica* fauna with in total 90 species (Mania, 1983). Thirty three of these are wood species. Besides *H. banatica* other South and Southeast European species occur: *Pagodulina pagodula, Discus perspectivus, Aegopis verticillus, Iphigena tumida.* Species of open terrain frequently occur indicating the occurence of light woods, forest steppes and open areas. Apart from common open ground species (*Pupilla muscorum, Vertigo pygmaea, Truncatellina cylindrica, Vallonia pulchella, Vallonia costata*) Southern steppe forms appear (*Truncatellina claustralis, Cepaea vindobonensis, Pupilla triplicata*). Among the aquatic fauna, *Theodoxus serratiliniformis* and *Belgrandia germanica* are characteristic elements. The wood fauna is associated with the atlantic species *Azeca menkeana* and with the fossil species *Acicula diluviana*.

As yet, twenty species of ostracods have been determined (Diebel & Pietrzeniuk, 1980). The occurence of the semi-aquatic *Microdarwinula zimmeri*, today in regions near the Equator as well as a salt-water species, is noteworthy.

The vertebrate species identified are mostly from the archaeological horizon. The great number of skeletal remains are considered to be food refuse left by humans:

Figure 7 - *Bilzingsleben. Camp site. Distribution of the choppers and chopping tools from pebbles.*

Pisces: *Silurus glanis, Tinca tinca* (Hebig, 1983).

Amphibia, Reptilia: *Bufo bufo, Natrix natrix* (Böhme, 1997).

Aves: *Anas platyrhynchos, Bucephala clangula, Cygnus olor, Haliaeetus albicillus, Strix aluco, Turdus* sp. (Fischer, 1997).

Micromammalia: *Talpa* sp., *Sorex araneus, Spermophilus* sp., *Glis glis, Apodemus flavicollis, Clethrionomys glareolus, Arvicola cantianus, Microtus oeconomus, M. subterraneus, M. arvalis-agrestis, Trogontherium cuvieri, Castor fiber* (Heinrich, 1994, 1997a,b).

Larger mammals: *Canis lupus, Vulpes vulpes, Meles meles, Martes* sp., *Felis silvestris, Panthera (Leo) spelaea, Ursus deningeri-spelaeus, Macaca sylvana, Palaeoloxodon antiquus, Equus mosbachensis-taubachensis, Dicerorhinus kirchbergensis, Dicerorhinus hemitoechus, Sus scrofa, Cervus elaphus, Dama* sp., *Capreolus capreolus, Bos primigenius, Bison priscus* (Fischer, 1991, 1997; Guenther, 1991; Musil, 1991a,b; Toepfer, 1983; Turner, 1997).

The game, which is hunted by man, has following frequence of individuals: elephants 12.1 %, rhinos 26.4 %, wild cattles 5.0 %, wild horses 2.7 %, bears 11.3 %, lions 1.1 %, deers 12.8 %, wild pig 1.9 %, roe deers 1.9 %, wolf 0.3 %, beavers 14.4 %, Trogontherium 4.2 %, wild cat 0.3 %, smaller animals 4.5 %, *Silurus glanis* 0.3 % (U. Mania, 1997).

Favourable conditions of preservation have preserved occupation structures and activity zones at the site. Occupied with detailed taphonomic analyses of the living floor we have already distinguished several patterns within the find material such as the foundations of three simple dwelling structures. In front of them, there were hearths and two or three workshops with anvils of stone or bone of elephants, with artefacts and fractured bones. Farther away from the dwellings we have discerned a workshop zone resembling the workshops found in front of the shelters.

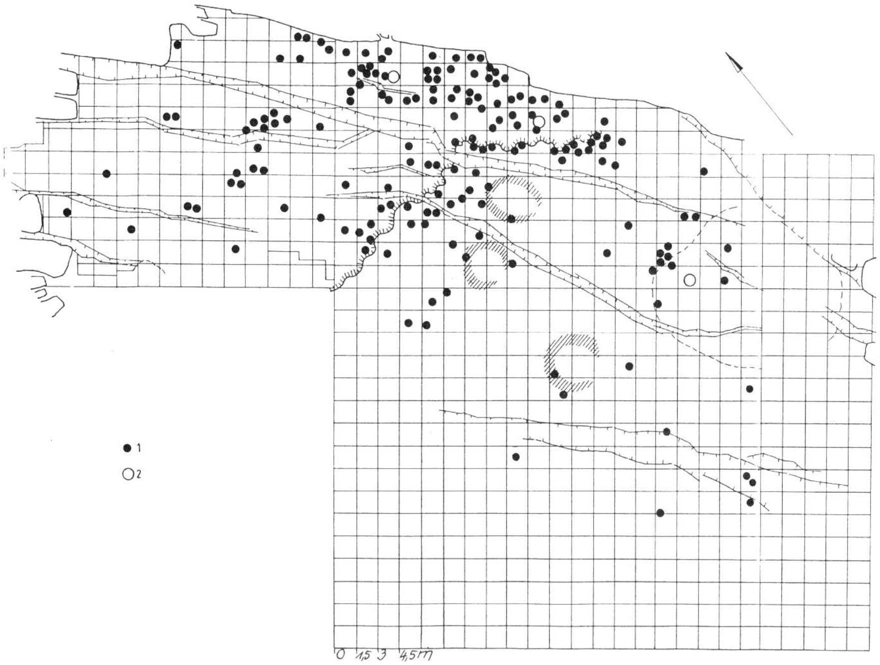

Figure 8 - Bilzingsleben. Camp site. Distribution of the mattoks and cudgels from antlers of roe deer.

Within this zone, there is an oval to circular paved area (Figure 4) with a diameter of 9 m. It is composed of pebbles and bone fragments of nut- to fist-size and with larger dimensions. There were no objects found lying on its surface. According to the archaeological evidence, special cultural activities may have been carried out there.

The distribution of specialised forms of tools, of debris related to the manufacture of the tools as well as bone, antler, ivory and wooden artefacts (Figures 5-12) is indicative of the existence of other areas of specific activities. The alluvial fan reaching to the shore-lines seems to contain the waste dump of the camp site. The latter is assumed to have been the home base of a group of humans for some time. With this situation of the Bilzingsleben site we have the evidence for the existence of a socio-cultural environment of the man before 400,000 years.

The artefacts can be differentiated according to their functions and their specifical material. Large pebble tools were made of quartzite, limestone and crystalline rocks. These are choppers, chopping tools and large hammerstones. With small hammerstones from quartz pebbles remarkably small specialized flint tools were made. The flint tools (Figure 13) are 8 to 100 mm long. Their edges are retouched, but bifacial and unifacial retouched tools also occur. The standard forms are knives, backed knives, scrapers, hand-axe like points, Tayac points, Quinson points, denticulates and notches, borers and special core-like tools.

Another complex of large tools consists of compacta of the long bones of elephants. There are large scrapers with grip-like parts and retouched edges. They are 15 to 80 cm long. Other tools of bone are cisel-like and chopper-like tools. Work support were made of shoulder blades or pelvis bones of elephant or rhino. Picks and club-like tools were manufactured from deer antlers.

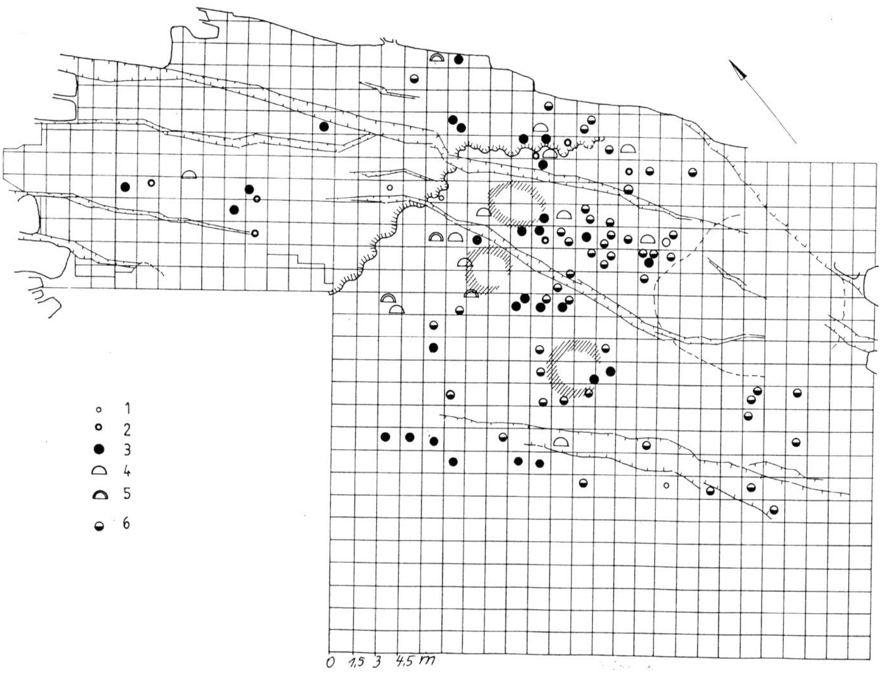

Figure 9 - *Bilzingsleben. Camp site. Distribution of the scrapers from compact bone. 1 - very little scrapers (< 10 cm), 2 - little scrapers (11-16 cm), 3 - medium large scrapers (17-30 cm), 4 - large scrapers (31-45 cm), 5 - very large scrapers (46-72 cm), 6 - scraper-like tools.*

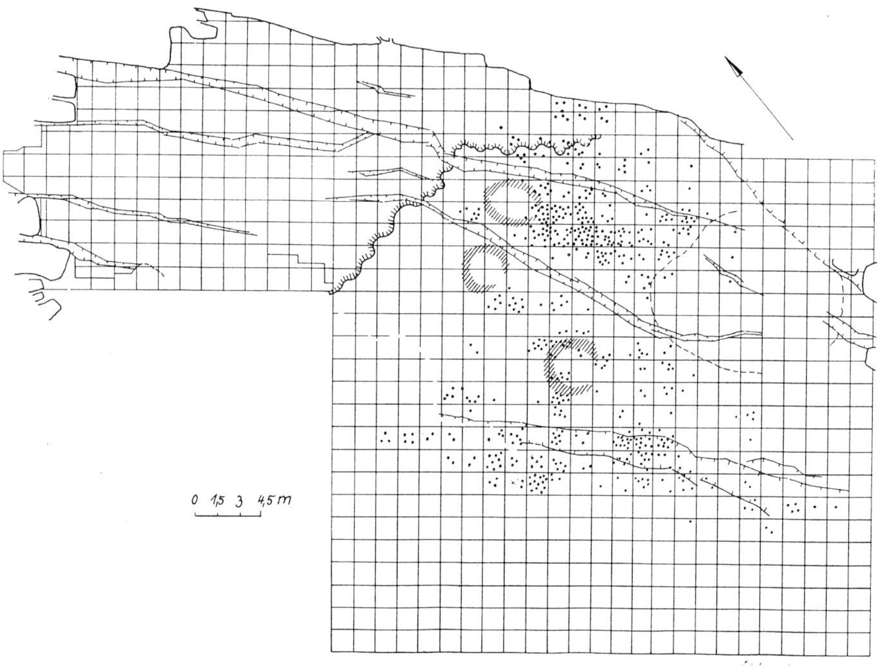

Figure 10 - *Bilzingsleben. Camp site. Distribution of bone flakes.*

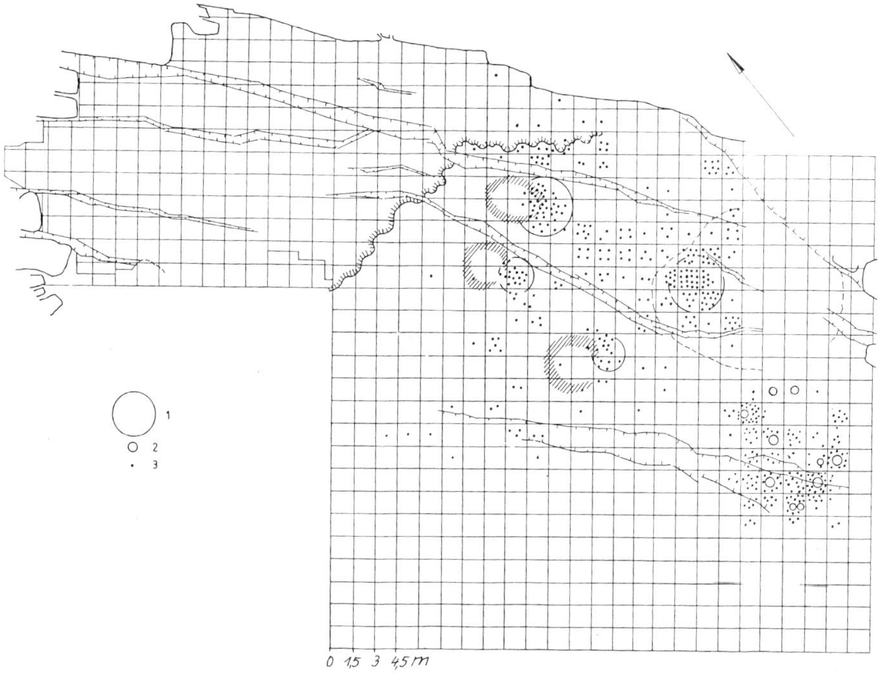

Figure 11 - *Bilzingsleben. Camp site. Distribution of the heated respectively burned stones (3), blocs (2) and fire places (1).*

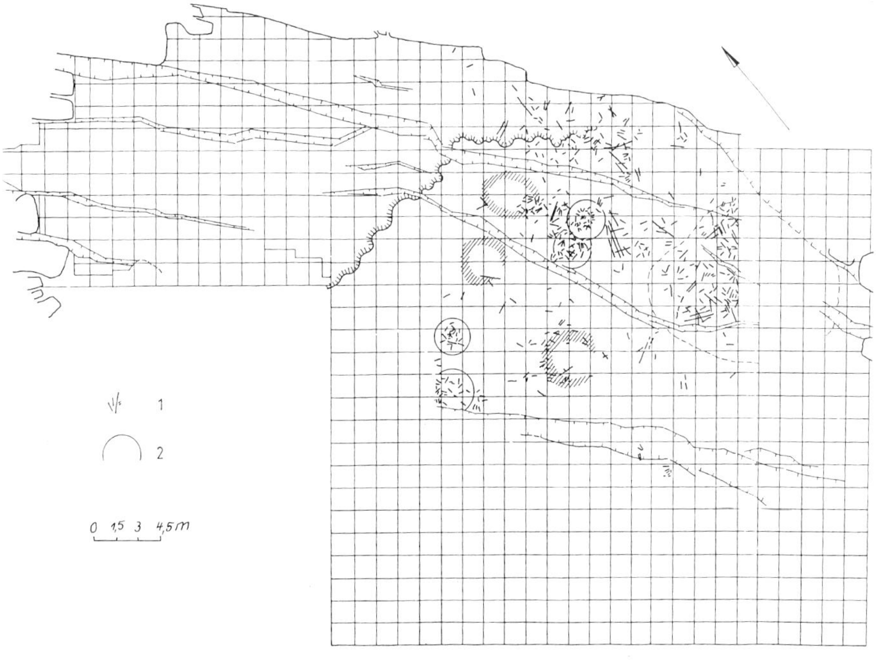

Figure 12 - *Bilzingsleben. Camp site. Distribution of wood remains (1) and workshops for the working of wood (2).*

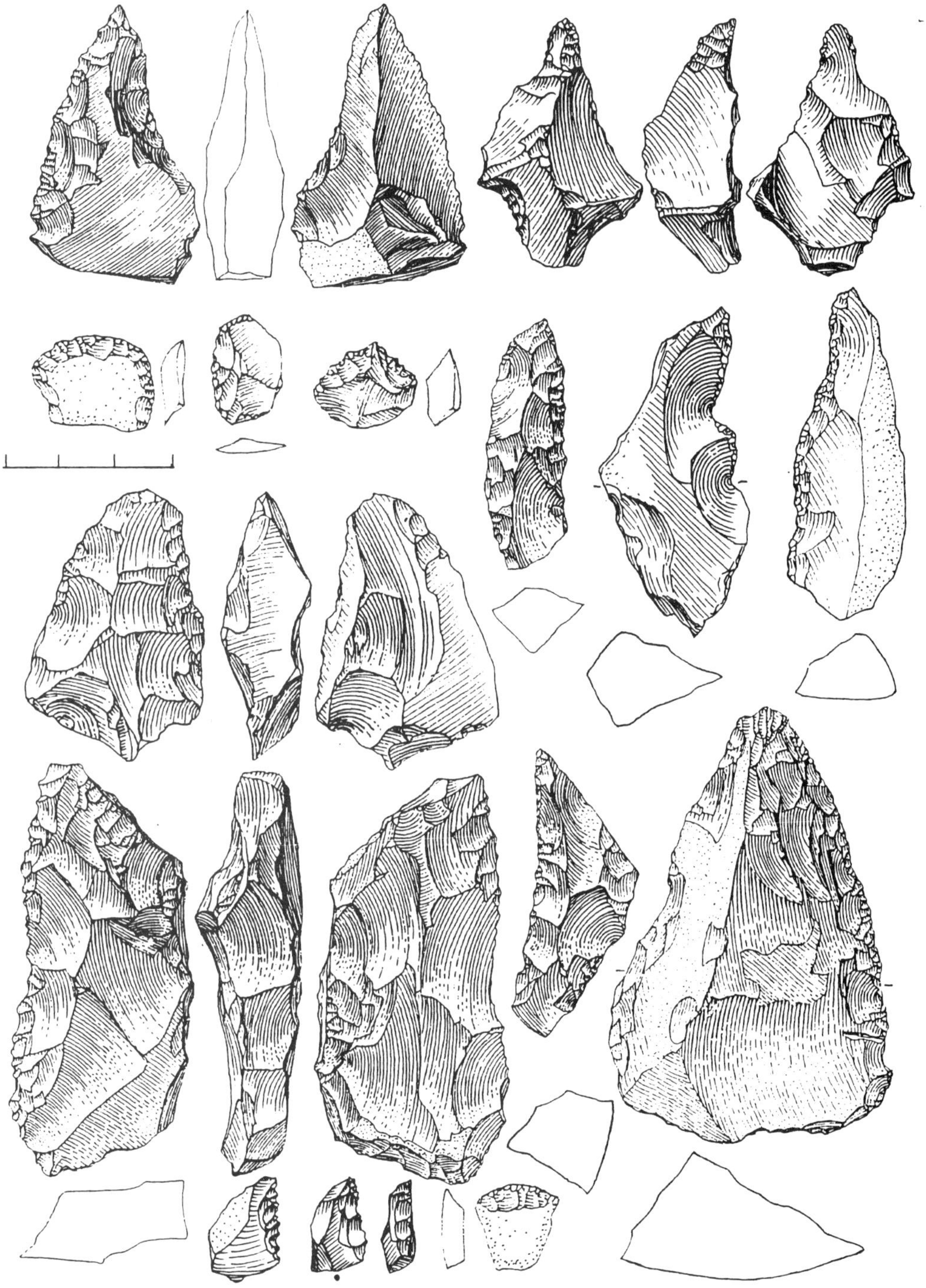

Figure 13 - Bilzingsleben. Some tools from flint.

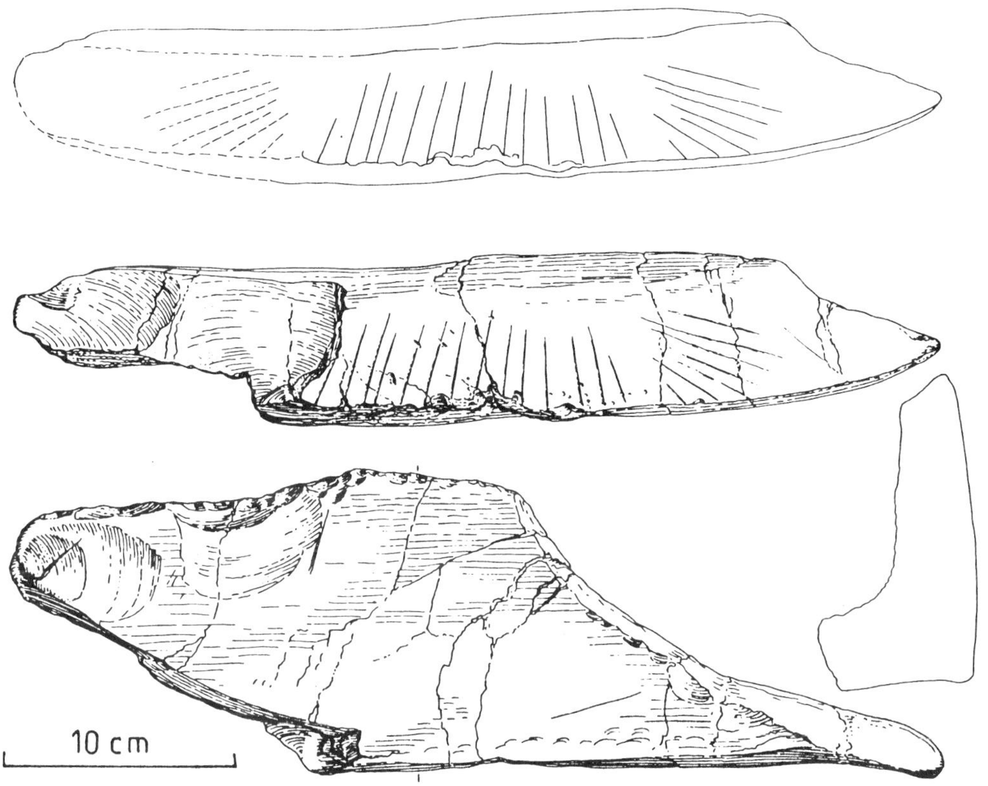

Figure 14 - Bilzingsleben. Artefact from a splinter of tibia of an elephant with intentionell engravings on the flat surface.

A great number of wooden remains may represent artefacts, too: rod-like, hook- and spade-shaped forms were found. Straight objects with a length of 180 to 230 cm are wooden spears. Some of these spears were situated near a workshop with an anvil from a tibia of elephant and choppers from quartzite for working of wood.

Some bone artefacts (Figure 14) exhibit deliberately engraved sequences of lines (Mania & Mania, 1988). They are the evidence for the ability of abstract thinking and for the existence of a language.

Palaeoanthropology of the Bilzingsleben site

Friedrich von Schlotheim mentioned the find of a human skull in 1818 and in 1820, which had originated from the travertine of Bilzingsleben. Unfortunately, this find, as well as a molar tooth, which was discovered in 1927, have disappeared. Our excavation carried out since 1969. In this time we have found more then 20 fragments of human skulls (Figures 15-18) and 6 teeth:

Bilzingsleben A1 - Middle fragment of Os occipitale (22.10.1972)

A 2 - Fragment from the left side of the Os occipitale (6.8.1974). A1 and A2 are from the same bone.

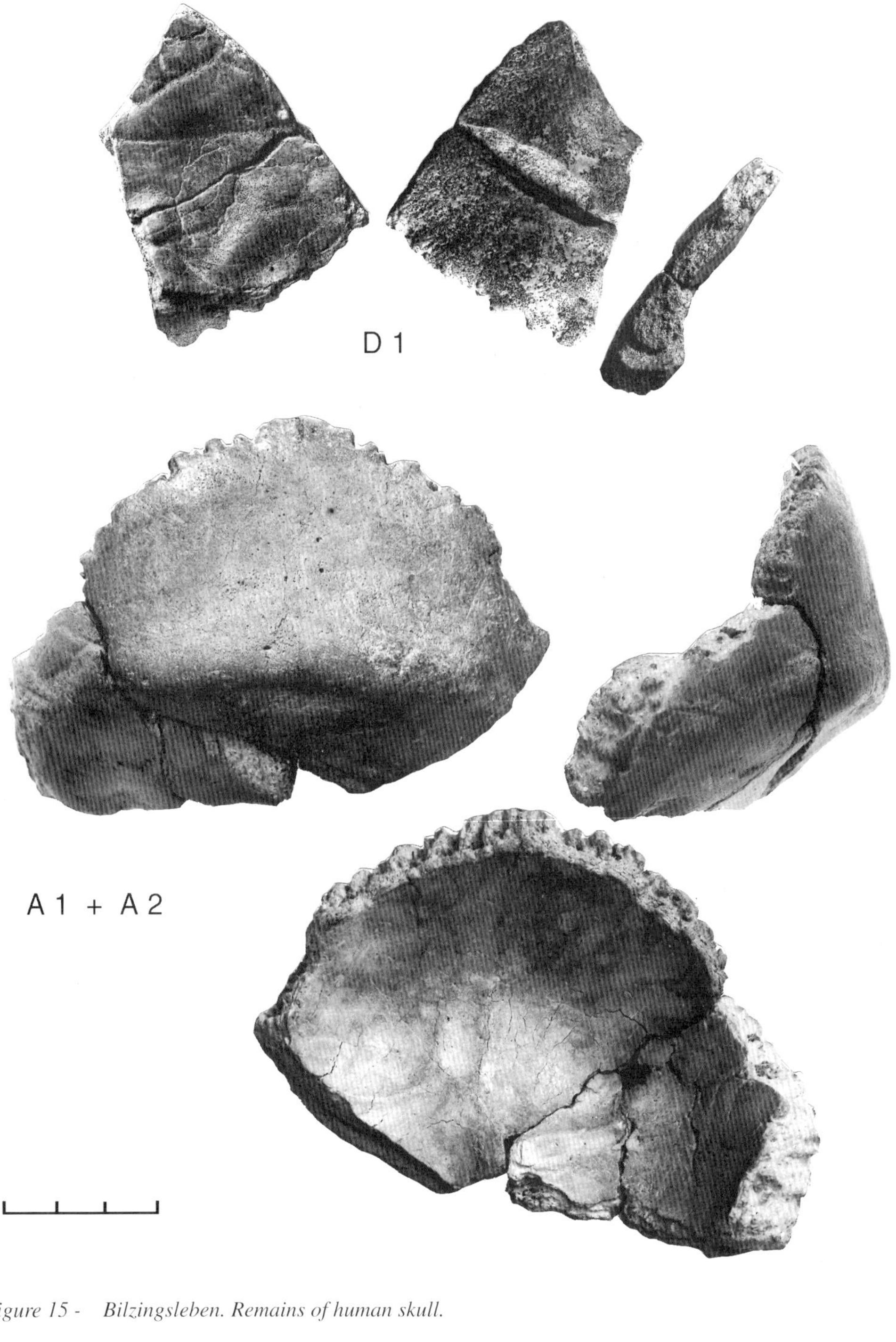

Figure 15 - Bilzingsleben. Remains of human skull.

A3 - Fragment of an Os occipitale, left side (14.5.1993)
B 1 - Middle fragment of Os frontale (15.7.1975)
B 2 - Middle fragment of the squama of Os frontale (15.8.1976)
B 3 - Fragment from the left side of squama of Os frontale (26.7.1979)
B 4 - Left fragment of the orbital region of Os frontale (2.4.1986)
B 5 - Upper fragment of the squama of Os frontale (26.8.1989)
B 6 - Fragment from the right side of the squama of Os frontale (29.8.1991)
B 7 - Right fragment of Os frontale (27.7.1993)
C 1 - Molar (M 1/2) of the right maxilla (3.7.1976)
D 1 - Posterior fragment of the right Os parietale (5.7.1977)
D 2 - Upper anterior fragment of the right Os parietale (14.8.1984)
D 3 - Posterior fragment of the left Os parietale (8.8.1986)
D 4 - Posterior fragment of the left Os parietale (14.7.1986)
D 5 - Posterior fragment of the right Os parietale (30.7.1992)
D 6 - Upper anterior fragment of the right Os parietale (25.9.1992)
D 7 - Anterior fragment of the right Os parietale (3.10.1992)
D 8 - Anterior fragment of the right Os parietale (30.10.1992)
D 9 - Middle fragment of the right Os parietale (31.7.1992)
D 10 - Large middle fragment (3 pieces) of the left Os parietale (25./28.9.1992)
D 11 - Upper posterior fragment of the right Os parietale (16.6.1993)
E 1 - Molar (M 2) of the right mandibula (30.6.1982)
E 2 - Molar (M 1) of the left mandibula (11.6.1983)
E 3 - Molar (M 3) of the left mandibula (27.7.1985)
E 4 - Prämolar (12.8.1994)
F 1 - Milkmolar (m 1/2) of the left mandibula (31.7.1984)
G 1 - Right Os temporale (4 pieces) (31.7.1992)
(Mania et al., 1994, Mania, 1997 b).

With the help of the skull remains it is possible to reconstruct two skull individuals („Bilzingsleben 1" and „Bilzingsleben 2" - E.Vlček, Prag) (Vlček, 1996 - see also the paper of Vlček in this volume).

The individual 1 consists from A1, A2, B1, B2, B3, B 5, D1 and D5. The investigations conducted by Vlček (1978, 1980, 1991) demonstrate the great morphological similarity of the finds of Olduvai Hominid 9, Pithecanthropus VIII and Sinanthropus III. Consequently, the skull 1 from Bilzingsleben must be assigned to the Middle Pleistocene group of *Homo erectus* („*Homo erectus bilzingslebenensis*" Vlček 1978, 1991).

The skull 2 consists from A3, B4, B7, D2, D3, D4, D6, D7, D8, D9, D10, D11, G1. The reconstruction (E.Vlček) is better possible than with skull 1. Bilzingsleben 2 is the prototype of the advanced European *Homo erectus*. Perhaps, the fragment B 6 demonstrate a third individual.

Acknowledgments

We are grateful to the Volkswagen Foundation, Hannover, the Leakey Foundation, Oakland (U.S.A.) and the Deutsche Forschungsgemeinschaft, Bonn for the financial and technical support of our Research Project Bilzingsleben (Friedrich Schiller-University Jena).

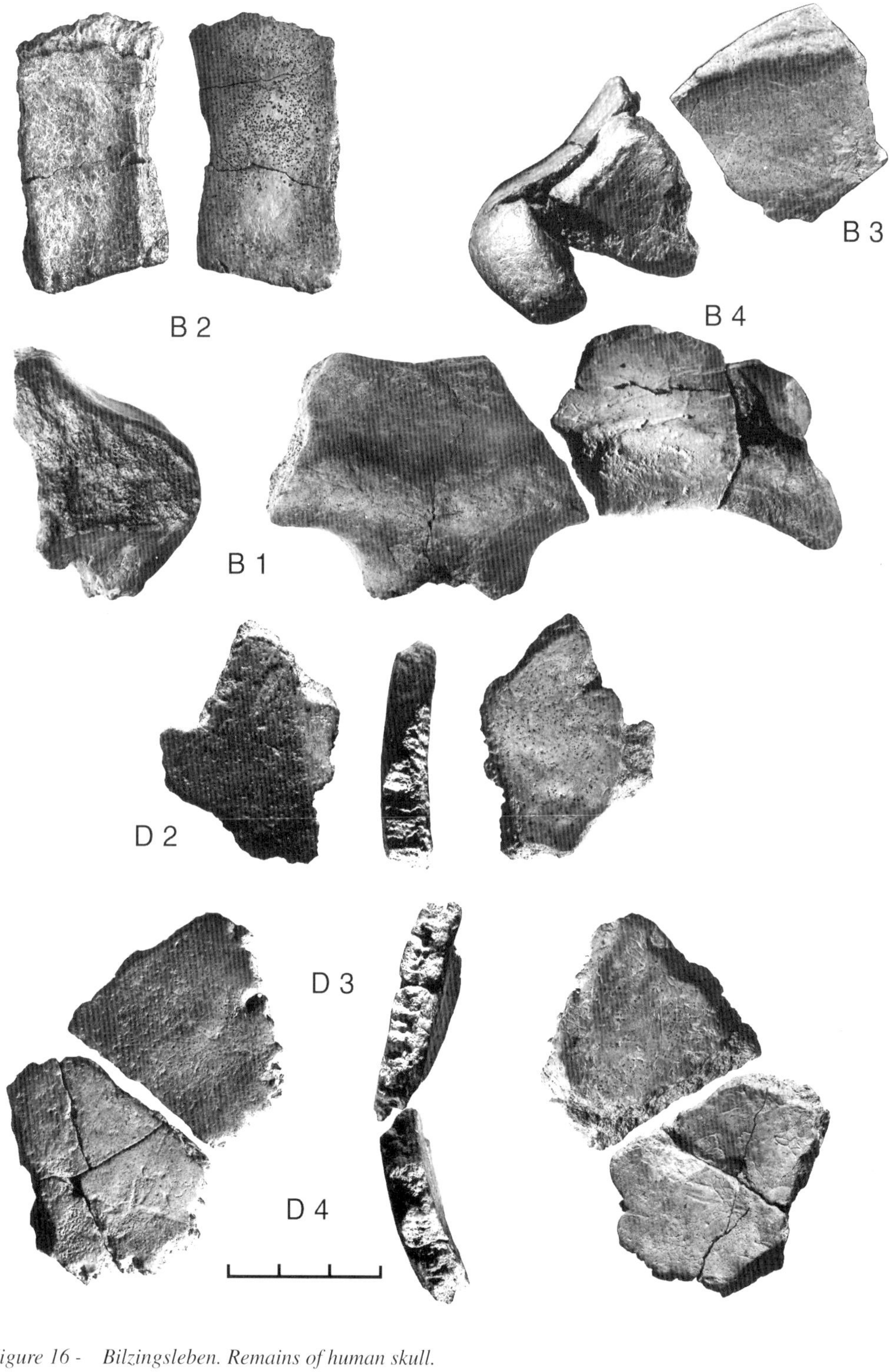

Figure 16 - Bilzingsleben. Remains of human skull.

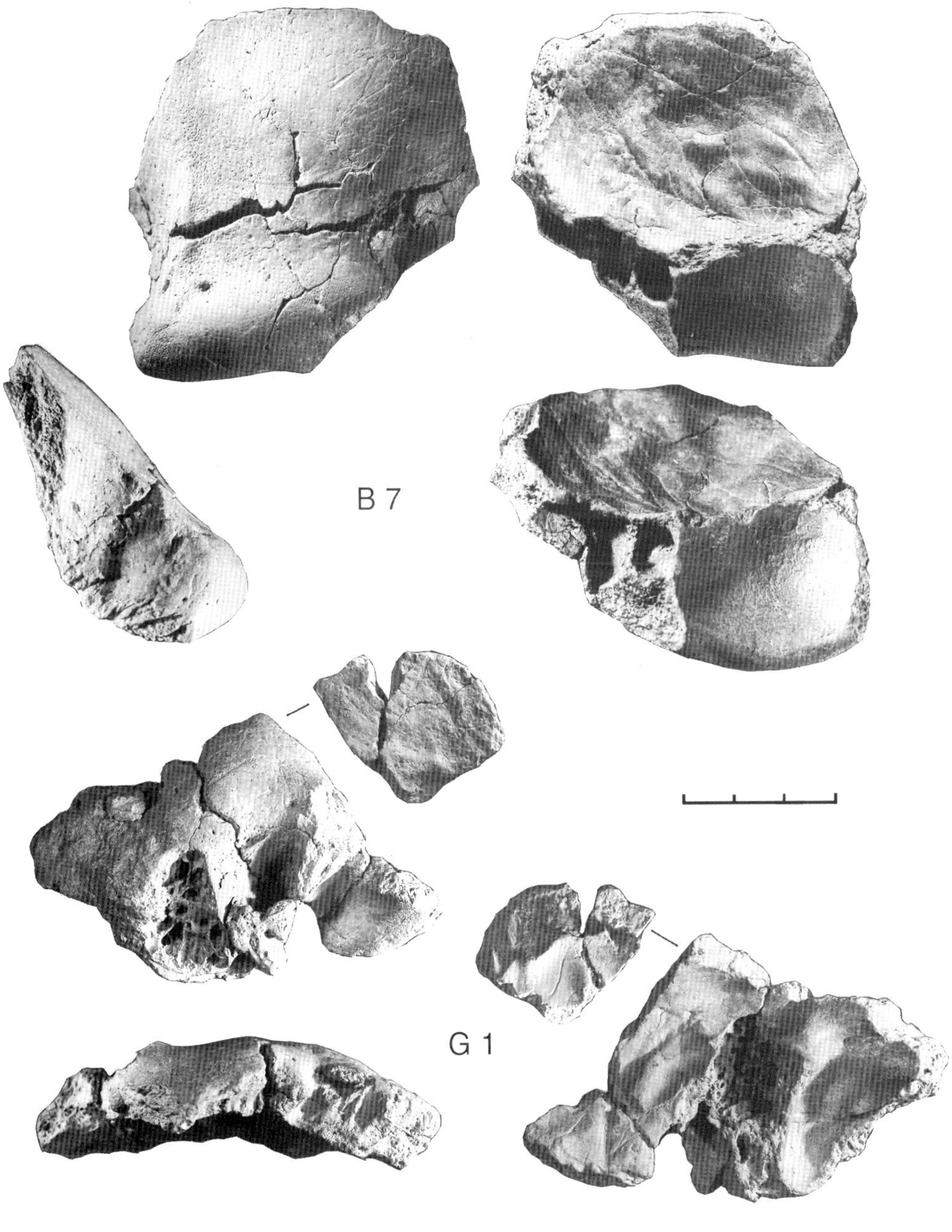

Figure 17 - Bilzingsleben. Remains of human skull.

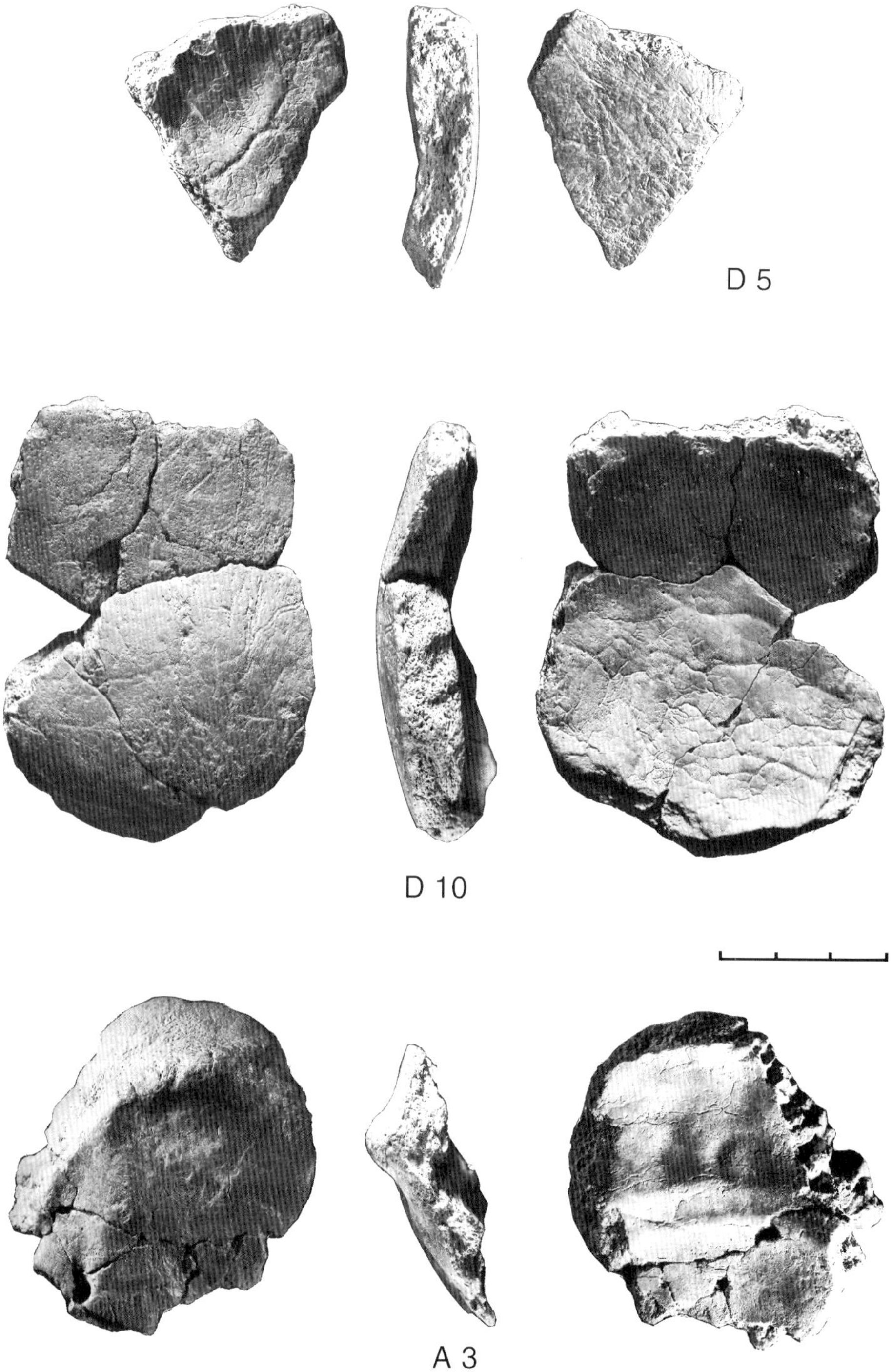

Figure 18 - Bilzingsleben. Remains of human skull.

References

BÖHME, G. (1997): Reste von Amphibien und Reptilien aus der Fundstelle Bilzingsleben und die quartären fossilen Herpetofaunen Thüringens. In: *Bilzingsleben* V, 113-120. Bad Homburg,Leipzig.

CEPEK, G.A. (1986): Quaternary stratigraphy of the German Democratic Republic. *Quaternary Science Revue* 5, 359-364.

DIEBEL, K. & PIETRZENIUK, E. (1980): Pleistozäne Ostracoden vom Fundort des *Homo erectus* bei Bilzingsleben. *Ethnographisch-Archäologische Zeitschrift* 21, 26-35.

EISSMANN, L. & LITT, T. (ed.) (1994): *Das Quartär Mitteldeutschlands*. Altenburg (Altenburger Naturwissenschaftliche Forschungen 7).

ERD. K. (1978): Pollenstratigraphie im Gebiet der skandinavischen Vereisungen. *Schriften für Geologische Wissenschaften* 9, 99-119.

ERD, K. (1997): Pollenanalytische Datierung des Seekalk/Fundhorizonts von Bilzingsleben. In: *Bilzingsleben* V, 107-112. Bad Homburg, Leipzig.

FISCHER, K. (1991): *Sus scrofa* L. (Mammalia: Artiodactyla, Suidae) aus der altpaläolithischen Fundstelle Bilzingsleben in Thüringen. Wildrinderreste (Bovidae, Artiodactyla, Mammalia) aus dem mittelpleistozänen Holstein-Komplex von Bilzingsleben. In: *Bilzingsleben* IV, 131-147. Berlin (Veröffentlichungen des Landesmuseums für Vorgescheschichte Halle, Band 44).

FISCHER, K. (1997): Vögel (Aves) und mittelgroße und kleine Carnivoren (Mammalia) aus der Holsteinwarmzeit von Bilzingsleben. In: *Bilzingsleben* V, 183-188. Bad Homburg, Leipzig.

GUENTHER, E.W. (1991): Die Gebisse der Waldelefanten von Bilzingsleben. In: *Bilzingsleben* IV, 149-174. Berlin (Veröffentlichungen des Landesmuseums für Vorgescheschichte Halle, Band 44).

HEBIG, W. (1983): Die Fischreste von Bilzingsleben. *Ethnographisch-Archäologische Zeitschrift* 24, 558-569.

HEINRICH, W.D. (1994): Zur Paläoökologie und Biostratigraphie der fossilen Kleinsäugetiere aus dem Travertinkomplex von Bilzingsleben. *Ethnographisch-Archäologische Zeitschrift* 34 (1993), 582-590.

HEINRICH, W.D. (1997a): Zur Taphonomie, Paläoökologie und Biostratigraphie fossiler Kleinsäugerfaunen aus dem mittelpleistozänen Travertinkomplex Bilzingsleben II in Thüringen. In: *Bilzingsleben* V, 121-134. Bad Homburg, Leipzig.

HEINRICH, W.-D. (1997b): Über *Trogontherium cuvieri* (Mammalia, Rodentia: Castoridae) aus dem mittelpleistozänen Travertinkomplex Bilzingsleben II in Thüringen. In: *Bilzingsleben* V, 135-182. Bad Homburg, Leipzig.

LITT, T. & TURNER, C. (1993): Arbeitsergebnisse der Subkommission für Europäische Quartärstratigraphie: Die Saalesequenz in der Typusregion. *Eiszeitalter und Gegenwart* 43, 125-128.

MAI, D.H. (1983): Die fossile Pflanzenwelt des interglazialen Travertins von Bilzingsleben (Kreis Artern, Thüringen). In: *Bilzingsleben* II, 45-129. Berlin (Veröffentlichungen des Landesmuseums für Vorgescheschichte Halle, Band 36).

MAI, D.H. (1988): Einige exotische Gehölze in den Interglazialfloren der Mitteleurropäischen Florenregion. *Feddes Repertorium* 99, 419-461.

MAI, D.H. (1989): Die Travertinflora von Bilzingsleben. Weitere Funde und ihre vegetationsgeschichtliche Bedeutung. *Ethnographisch-Archäologische Zeitschrift* 30, 306-310.

MANIA, D. (1973): Paläoökologie, Faunenentwicklung und Stratigraphie des Eiszeitalters im mittleren Elbe-Saalegebiet auf Grund von Molluskengesellschaften. *Geologie,* Beiheft 78/79, 175.

MANIA, D. (1983): Die Molluskenfauna des mittelpleistozänen Travertinkomplexes bei Bilzingsleben und ihre ökologisch-stratigraphische Aussage. In: *Bilzingsleben* II, 131-155. Berlin (Veröffentlichungen des Landesmuseums für Vorgescheschichte Halle, Band 36).

MANIA, D. (1997a): Das Quartär des Saalegebietes und des Harzvorlandes unter besonderer Berücksichtigung der Travertine von Bilzingsleben - Ein Beitrag zur zyklischen Gliederung des eurasischen Quartärs. In: *Bilzingsleben* V, 23-103. Bad Homburg, Leipzig.

MANIA, D. (1997b): Bilzingsleben - Ein kulturgeschichtliches Denkmal der Stammesgeschichte des Menchen. *Praehistoria Thuringica* 1, 30-80.

MANIA, D. (1997c): *125 000 Jahre Klimaentwicklung im Saale-Elbegebiet.* Jena (in press).

MANIA, D. & ALTERMANN, M. (1970): Zur Gliederung des Jung- und Mittelpleistozäns im mittleren Saaletal bei Bad Kösen. *Geologie* 19, 1161-1183.

MANIA, D., ALTERMANN, M. & RAU, D. (1995): *Paläoböden und Stratigraphie des Mittel- und Jungquartärs im mitteldeutschen Trockengebiet.* Sitzung und Exkursion des Arbeitskreises Paläopedologie der Deutschen Bodenkundlichen Gesellschaft 25.-27.5.1995 in Jena.

MANIA, D. & MAI, H.D. (1969): Warmzeitliche Mollusken und Pflanzenreste aus dem Mittelpleistozän des Geiseltales (südlich von Halle). *Geologie* 18, 674-690.

MANIA, D. & MANIA, U. (1988): Deliberate engravings on bone artefacts of *Homo erectus*. *Rock Art Research* 5, 91-107.

MANIA, D., MANIA, U. & VLČEK, E. (1994): Zu den Funden der Hominidenreste aus dem mittelpleistozänen Travertinkomplex von 1987-1993. *Ethnographisch-Archäologische Zeitschrift* 34 (1993), 511-524.

MANIA, D., MANIA, U. & VLČEK, E. (1994): Latest finds of skull remains of *Homo erectus* from Bilzingsleben (Thuringia). *Naturwissenschaften* 81, 123-127.

MANIA, U. (1997): Special aspects of vertebrate fauna from the Middle Pleistocene find horizon of Bilzingsleben. In: *Bilzingsleben* V, 197-200. Bad Homburg, Leipzig.

MUSIL, R. (1991a): Die Bären von Bilzingsleben. In: *Bilzingsleben* IV, 81-102. Berlin (Veröffentlichungen des Landesmuseums für Vorgescheschichte Halle, Band 44).

MUSIL, R. (1991b): Pferde aus Bilzingsleben. In: *Bilzingsleben* IV, 103-130. Berlin (Veröffentlichungen des Landesmuseums für Vorgescheschichte Halle, Band 44).

SCHLOTHEIM, E.F. von (1818): Der Kalktuff als Glied der aufgeschwemmten Gebirgsformation. *Leonhards Mineralogisches Taschenbuch* 12, 315-345.

SCHLOTHEIM, E.F. von (1820): *Die Petrefactenkunde auf dem jetzigen Standpunkt* ... Gotha.

SCHWARCZ, H.P., GRÜN, R.., LATHAM, A.G., MANIA, D. & BRUNNACKER, K. (1988): The Bilzingsleben archaeological site: New dating evidence. *Archaeometry* 30, 5-17.

THIEME, H. (1996): Altpaläolithische Wurfspeere aus Schöningen, Niedersachsen. *Archäologisches Korrespondenzblatt* 26, 377-393.

THIEME, H. (1997a): Altpaläolithische Wurfspeere aus Schöningen, Nordharzvorland. In: *Homo heidelbergensis von Mauer*, 304-312. Heidelberg.

THIEME, H. (1997b): Lower Palaeolithic hunting spears from Germany. *Nature* 385, 807-810.

THIEME, H. & MAIER, R. (1995): *Archäologische Ausgrabungen im Braunkohlentagebau Schöningen, Landkreis Helmstedt.* Hannover.

THIEME, H. & MANIA, D. (1994): Schöningen 12 - ein mittelpleistozänes Interglazialvorkommen im Nordharzvorland mit paläolithischen Funden. *Ethnographisch-Archäologische Zeitschrift* 34 (1993), 610-619.

THIEME, H., MANIA, D., URBAN, B. & KOLFSCHOTEN, T. van (1993): Schöningen (Nordharzvorland). Eine altpaläolithische Fundstelle aus dem mittleren Eiszeitalter. *Archäologisches Korrespondenzblatt* 23, 147-163.

TOEPFER, V. (1983): Ein Oberkieferfragment des Löwen aus dem Travertinkomplex von Bilzingsleben, Kr. Artern. In: *Bilzingsleben* II, 163-173. Berlin (Veröffentlichungen des Landesmuseums für Vorgescheschichte Halle, Band 36).

TURNER, A. (1997): Bears and lions of Bilzingsleben. Palaeoecological implications. In: *Bilzingsleben* V, 189-195. Bad Homburg, Leipzig.

URBAN, B. (1997): Grundzüge der eiszeitlichen Klima- und Vegetationsgeschichte in Mitteleuropa. In: *Homo heidelbergensis von Mauer*, 241-263. Heidelberg.

URBAN, B., LENHARD, P., MANIA, D. & ALBRECHT, B. (1991): Mittelpleistozän im Tagebau Schöningen, Ldkr. Helmstedt. *Zeitschrift der deutschen Geologischen Gesellschaft* 142, 351-372.

VLČEK, E. (1978): A new discovery of *Homo erectus* in Central Europe. *Journal of Human Evolution* 7, 239-251.

VLČEK, E. (1980): Die mittelpleistozänen Hominidenreste von der Steinrinne bei Bilzingsleben. In: *Bilzingsleben* I, 91-130. Berlin (Veröffentlichungen des Landesmuseums für Vorgescheschichte Halle, Band 32).

VLČEK, E. (1991): L'Homme fossile en Europe Central. *L'Anthropologie* 95, 409-472.

VLČEK, E. (1996): Patterns of human evolution. In: J. Svoboda, V. Ložek & E. Vlček, *Hunters between East and West*, 37-74. New York, London.

The Middle Palaeolithic as Development Stage: Evidence from Technology, Subsistence, Settlement Systems, and Hominid Socio-ecology

Nicolas Rolland

Abstract

The term „Middle Palaeolithic", created last century, was never defined and subsequent attempts never reached a consensus. One aim is identifying properties diagnostic of the Middle Palaeolithic as a taxonomic entity by outlining descriptive techno-typological characteristics and their time/space distribution patterns. Another is exploring to what extent Middle Palaeolithic repertoires correlate with other behavioural and adaptive aspects to constitute a developmental stage. Formal characteristics comprise the wide-spread mode 3 (Levallois, disc-core) techniques, and other more episodic or regional ones, such as „Pontinian" and true blade core reduction techniques, and a wide array of bifacial flaking and trimming methods. Middle Palaeolithic assemblages often display increased typological standardization, along with expanding lithic procurement territories, and considerable intra- and inter-regional variability. These express either a mastery of methods for exploiting with parsimonious efficiency preferred lithic resources when circumstances required it, or regionally-clustered isochrestic stylistic trends or traditions. Contrasting interregional repertoires differences, however, do not obscure simultaneous recurrences of basic techno-typological traits which maintain the Middle Palaeolithic's integrity as a broad, polythetic technocomplex, independently of vast geographic distances. Improved geo-chronological resolution and new discoveries have pushed the Middle Palaeolithic time-depth considerably. This expanded span, nevertheless, does not imply technological „stagnation" or change or merely „random walk" kind. Evidence shows assemblage type time patterning, as well as an accumulation of flaking technique innovations, some with progressive properties.

Other behavioural manifestations emerged, besides lithic technology. The subsistence base, in many regions, involved ecological specialization under fluctuating bioclimatic conditions, focusing on exploiting selectively by various techniques a few medium or large gregarious herbivores species, including cold-adapted steppe biome herds, throughout temperate continental Eurasia. Forays or settlement of cold continental and even boreal latitudes beings during Middle Palaeolithic times, evidenced by site geographic locations and prey species. The implied adaptive shifts in hominid land-use strategies and socio-ecology. Shellfish becomes a significant dietary component in some subtropical coastal areas. Seasonal vertical trans-humance in Mediterranean zones can also be traced back to the Middle Palaeolithic. Burials as a method for disposing of the dead becomes documented for the first time, while aesthetic or symbol material expression remains limited in scope and ambiguous.

Hominid settlement systems may have undergone a fundamental re-organization during horizons covering the final Lower Palaeolithic and the transition to Middle Palaeolithic, with the emergence of a home base <u>sensu stricto</u> *residence system by more patterned occupation surfaces and structures, and widespread evidence for regular fire-production, favouring a regular occupation of enclosed (caves, rockshelters) sites. In conclusion, the above properties provide empirical support for retaining use of the term „Middle Palaeolithic" as a broad named taxonomic concept, regarding it as a major formative stage. Many or most of the behavioural parameters observable among recent simple foraging populations were apparently already in place among these ancient hominid societies.*

„...taxonomy should be viewed as one of the most fundamental, and most noble of scientific pursuits - for what can be more basic than the parsing of nature's rich and confusing complexity? ...We must taxonomize to make sense of anything diverse and complex. We therefore classify our activities and professions in various ways, just as we order other organisms by Linnean standards". S.I. Gould (1994): Pride of place. Science without taxonomy is blind. *The Sciences* March/April, 38-39.

„Here, again, we are led to one of the central problems in the writing of history - the problem of periodization...What we should look out for as significant are the differences rather than the similarities, the elements of discontinuity rather than the elements of continuity. ...one of the first tasks ...should be to establish its distinguishing features and its boundaries. ...But it still remains true that unless we keep our eyes alert for what is new and different, we shall all too easily miss the essential - namely, the sense of living in a new period." G. Barraclough (1967): *An introduction to contemporary history,* 12-13. Harmondsworth.

Introduction

We survey evidence on the Middle Palaeolithic, examining to what extent it supports the hypothesis that it represents a developmental stage rather than merely a chronological horizon, though aware that new discoveries may invalidate the empirical generalizations and alter radically the theoretical framework presented here. We illustrate this with material from Western Eurasia, Europe in particular, because it is more abundant and informative but key evidence from Africa, India, Central Asia, Siberia or the Far East will be included. The discussion leaves out fossil hominid remains. The Middle Palaeolithic time-span overlaps with several hominid grades and evolutionary groups (*Homo erectus presapiens, Homo presapiens sapiens* and *neanderthalensis* - Heim,1987: fig. 2).

Major issues

Change and variability pervades the Palaeolithic record, accelerating with time. Disagreement persists as to whether periodization should serve only as a descriptive time unit, or reveal meaningful trends or turning points of cultural history's trajectory.

The classic Palaeolithic subdivisions still used in Eurasia and Africa are inadequate. Their criteria remain too rigid, and narrow, based exclusively on toolmaking repertoires successions. This line of documentation whose usefulness resides in an unbroken record testifying to increased mastery of manufacturing techniques, leaves out other behavioural manifestations. Childe (1944) preferred regarding these Ages of Prehistory as developmental Stages. Toolmaking, from this perspective, constitutes one technological subsystem along with bone, vegetal, or fire-making, whose intrinsic developmental logic is also connected with other anthropic and mesological variables.

Neglect in investigating Middle Palaeolithic emergence (e.g. Ronen, 1982) stems largely from unresolved taxonomic and periodization problems. These involve alternative approaches for selecting definition criteria, and for assessing the Middle Palaeolithic's significance: lumping it within the Early Palaeolithic cycle's continuity (Bordes, 1960; Combier, 1962:77; Müller-Beck, 1987) or stressing major changes (Roe, 1982: 180-181). They bring up fundamental issues and opposite viewpoints, namely regarding the Middle Palaeolithic as part of a qualitatively distinct archaic palaeocultural' stage, persisting until the Upper Palaeolithic and fully modern humans (Binford, 1985; Jelinek, 1982), versus perceiving it as a major turning point (Bosinski, 1982).

Cultural Taxonomy and periodization rely on natural and empirically-based criteria, to avoid circularity. A viable taxonomy should be flexible but consistant, with a plurality of traits, i.e. polythetic. Periodization should highlight novelty, tracing first appearances, instead of continuity. Cultural developments tend to be *homotaxial* rather than hologenetic, with differential change

A. Primary flaking

1. prepared-core (mode 3) = Levallois flakes, blades, points techniques
   = disc-core technique
2. facetting = dihedral, multifacetting
3. planoconvex technique
4. prismatic blade-core technique
5. specialized "Pontinian" or spicchio pebble flaking

B. Secondary trimming

1. bifacial trimming diversification
2. invasive, flat retouch
3. step (Quina) retouch
4. blunting retouch

C. Typology

1. racloirs types morphological diversification
2. Mousterian points *sensu stricto*: (a) pointed tools, (b) elongated weapon points
3. cordiform, triangular, 'Micoquian' handaxes
4. bifacial cutting implement types
5. Blattspitzen
6. endscrapers
7. burins
8. pradniki
9. truncated tools
10. Howiesons Poort large geometric tools
11. Aterian stemmed tools
12. Lupembian lanceolate bipointed implements

D. Assemblages

1. increasing manufacture of exotic lithic materials
2. flake implements standardization
3. importance of resharpened cutting tools/implements
4. specialized assemblages dominated by = racloirs, bifaces, denticulates and notches
5. assemblages dominated by endscrapers and burins
6. marked frequency variations of unretouched non-Levallois flakes or blades

Table 1 - Middle Palaeolithic techno-typological characteristics.

tempos and time lags in different regions, comprising thresholds or intermediate periods bridging stages. The rate of Lower to Middle Palaeolithic transition was more protracted (400 to 320 ky) than Middle to Upper (47 to 30 ky).

The Culture Concept *sensu lato* used here involves the socio-historical transmission of information repertoires whose contents include several 'open' systems and subsystems as heuristic units observable as diachronic series: technology, landuse and settlement, subsistence, social organization, idology/cognition, each interacting directly or indirectly with one another, as well as with ecological, demographic or psychological factors, with different response times.

Technology

Lithic repertoires

I. Taxonomy. Lartet's 'Middle Palaeolithic' lacked archaeological definition, supplied later by de Mortillet's 'Le Moustier' Epoch. Neither describe the Middle Palaeolithic's actual diversity sufficiently but the latter is preferred over 'Mousterian', being taxonomically more comprehensive. Table 1 lists diagnostic techno-typological traits across Eurasia and Africa.

Defining Middle Palaeolithic repertoires properties include diversified flaking and trimming methods to produce more standardized and specialized assemblage, usually dominated by specific and more differentiated types. Many since Commont consider Levallois reduction methods as a diagnostic classification key and as a decisive breakthrough in tool manufacturing skills, e.g. Leroi-Gourhan (1966). Bifacial flaking (Bosinski, 1967) constitutes another recurrent technical theme with a versatile expression. Prismatic blade production was another major innovation. Prepared-core flaking, racloirs and bifacial implement types dominate assemblage contents as dynamic elements with marked frequency variations. Even preexisting types such as denticulates assume a specialized role with variable frequencies.

These Middle Palaeolithic characteristics testify to enhanced mastery of diverse methods for manufacturing cutting tools with more prolonged use spans under varying circumstances, involving when required, reduction strategies (disc-core mass production, planoconvex or step-flaking) devised to exploit preferred i.e. fine-grain lithic resources more parsimoniously. These materials usually include flint but also chalcedony, agate, jasper e.g. in India, indurated shale or silicified sandstone in Southern Africa, chalcedony, jasper, green schist e.g. Eastern Morocco, agate, opal, flint, e.g. Xujiajuo, China (Jia, 1980: 38). Raw material selecting, along with typologically cohesive assemblages, correlate with more extensive procurement ranges and importing exotic materials, trend increasing over time e.g. Geneste (1989), Tavoso (1984). Most stone implements appear to be tools. Evidence for stone weapons remains either scarce, e.g. 'mousterian points' (Cotte St. Brelade, Rheindahlen) or stone ball heaps (Windhoek, Xujiayao) or indirect (slotted javelin tips from Schöningen).

II. Periodization. The 'short' chronological span attributed previously to the Middle Palaeolithic was based on arbitrary and hybrid criteria: an Upper Pleistocene time-stratigraphic boundary, combined with de Mortillet's Epochs, i.e. 'Mousterian', preceded by Lower Palaeolithic 'Acheulian'. Findings from palynology and marine isotopic stages stretched the Upper Pleistocene's duration from 95 to 130 ky, while accumulating 'Pre-Mousterian' discoveries e.g. Rigabe, Abri Vaufrey, Cotte St. Brelade, Ripario Paglicci, Achenheim, Rheindahlen, Hunas, pushed Middle Palaeolithic emergence steadily back to Saalian times, during OIS 9 or 320 ky, with similar trends for the Near East (Tabun, Qatafa, Berekhat Ram), Africa (Cave of Hearths, Lake Ziway sites, Ethiopia) or Siberia (Denisova, Kurtak).

Major Middle Palaeolithic trends, according to periodization criteria stress emergence: mode 3, step-flaking, standardized and diversified implement and assemblage types, imported exotic materials (Callow, 1986; Junkmanns, 1991). These amplify and become widespread, with additional traits, over time. They should also include penecontemporanous 'Epi-Acheulian' occurences where handaxes become numerically less important (Atelier Commont, Orgnac 3, Lazaret, Combe-Genal, La Chaise, Markkleeberg, the Yabrudian and Fauresmith), whose essential structure became 'mousterian'. The term 'Acheulian' is reserved for naming a techno-typological span, straddling the Lower and earlier Middle Palaeolithic, rather than using it for periodization. This applies also to 'levallois', persisting in the transitional (Bohunician) or later Upper Palaeolithic ('Epi-Levallopisian' in Egypt).

Middle Palaeolithic diachronic patterns, including Last Glacial evidence for regional sequences, invalidate notions of a 'stagnating' terminal Lower Palaeolithic horizon. Levallois assumes regional idiosyncrasies while blade-core reduction recurs independently at different times, e.g. Northwest Europe, Levant, Vaal River, Central Asia e.g. Khonako II (Schäfer et al., 1996).

The transitional Lower to Middle or Intermediate horizon (400 - 320 ky) contains innovations such as soft-hammer and lanceolate handaxe manufacture, flake cleavers, 'Proto-Levallois' (Victoria West) in different regions of Eurasia or Africa, an accelerating trend immediately before the Middle Palaeolithic turning point, determined perhaps by long-term technological factors, but also by basic shifts in ancient hominid behaviour.

III. Distribution. Perusing contrasting repertoires differences between regions during Last Glacial e.g. 'Mousterian Complex', 'Keilmessergruppen', 'Crimean Mousterian', 'Levantine Mousterian', may convey the impression that links unifying Middle Palaeolithic occurrences in a single taxonomic entity are tenuous. Closer inspection, however, reveals equally pronounced intra-regional variations, while the repeated presence of basic 'Levallois-Mousterian' trait clusters across Eurasia up to Central Asia, Mongolia (Orkhon) or the Russian Far East (Osinovka), as well as Africa, confirms the reality of a broad unity underlying the Middle Palaeolithic mosaic-patterned *oikoumene*. Dispersed pockets of undiagnostic, unspecialized flake assemblages, e.g. African Rift, Southeast India, could exemplify isolated lagging patterns, within wide homotaxial tendencies.

The Far Eastern boundaries become fuzzy, with attenuated expressions interdigitating perhaps with local idiosyncratic developments, e.g. small bifaces at Zhukoudian locality 15 (Pei, 1939). Intermediate or Middle Palaeolithic traits include flake cleavers (locality 15, Jonggog-ni), step-flaking and racloirs (Dingcun), prepared core flaking, Dadong Cave, S. China (Huang et al., 1995), points, racloirs (Xujiayao, Yiatougou, Juangjiawu, in China) levallois points (locality 15) and/or blades (Leang Burung 2 and Nguom caves, Southeast Asia) (Glover, 1993).

Bone and lignic repertoires

Discerning trends from these industries becomes problematic. Insufficiently developed analytical methods hamper resolution for discriminating bone, ivory or antler artifacts, leaving a grey area and making conclusions tentative. Differential preservation of lignic materials restricts comparisons. These limitations induced negative preconceptions, often reflected by less systematic data retrieval and observation taking.

Organic raw materials, though often possessing less effective shearing properties, substituted or complemented lithic artifacts, frequently as part of composite tool kits (Mania, 1995: fig.7). The Late Lower or Intermediate Palaeolithic suggests the onset of an increased and more varied range of bone and antler modification e.g. Broken Hill, Vértesszölös (Dobosi, 1983), Bilzingsleben (Mania & Vlček, 1987), Casal di Pazzi. Antlers were used as hammers, for prying lithic materials from seams, or for making protective screens (Raj, Roc en Pail). Bone racloirs (Mas des Caves), choppers, chisels replicated stone tools. Points and barbed harpoons are evidenced in Central Africa from 90 ky (Yellen et al., 1995) while bone flakers, chisels and gouges become commonplace in Europe (La Quina). Mammoth bone and tusks were also used for erecting screens (Molodova, Ripiceni Izvor).

Notches, denticulates and other obtuse-angled stone tools appear early in the Palaeolithic record, pointing to a concomitant exploitation of wood, bark and other vegetal matters. Most of the oldest identifiable wooden artifacts, however, coincide with the Intermediate and later: Kalambo Falls, Clacton, Torralba, Bilzingsleben, Schöningen (Thieme, 1997), Bad Canstatt (Wagner, 1990), Lehringen, consisting mostly of thrusting spear remains, although Schöningen also contains javelin specimens with slotted distal extremities, as well as digging tools, while Kalambo Falls and

Florisbad exemplify wooden clubs and throwing sticks. Torralba and Ambrona yielded miscellaneous wooden tool fragments (Harrold, 1983). Direct and indirect evidence points to wood hafting of stone tools in Northwestern Europe (Kärlich, Biache St. Vaast), and in the Levant (Friedman et al., 1994-95). A wooden board remains comes from Gesher Bnot Yakov, while impressions or preserved remains of carrying containers comes from Terra Amata, Arriaga and Abric Romani. Postholes are indicated for Combe-Grenal and Fontmaure.

Fire-making

Opinion remains divided (Perlès, 1987; Schüle, 1992) whether ancient hominids mastered fire-production as early as Late Pliocene in Africa. Early evidence from Europe, e.g. L'Escale, Isernia seems inconclusive but limited fire use is indicated by 700 ky (Stránska Skála). Regular fire-making becomes evidenced and widespread with the Intermediate (Table 3A, Rolland, 1996: table 1). Fire-modified wood use also appears to coincide with Middle Palaeolithic occurrences, e.g. charred logs at Kalambo Falls, fire-hardened spear tip at Lehringen.

Landuse and settlement systems

These lines of evidence or subsystems are interrelated.

Home-ranges extension

Lithic material sourcing and occurrences located in distinct but complementary habitats and containing similar assemblages indicate territories exploited by the same populations. Home ranges become more extensive during Saalian times in Europe, e.g. Schweinskopf, Bouheben, increasing further and more numerous during the Upper Pleistocene (Roebroeks et al., 1988: table 1), reaching 100 to 300 km, in Central and Eastern Europe where colder continental conditions called for more extensive spatial ranging, e.g. Plaidter Hummerich, Zwiezyniec, Kulna, Sólyomkút, Bacho Kiro. They indicate increased mobility with local groups radiating more widely from residence bases, the latter shifting or circulating over more comprehensive year-round ranges. Similar trends appear also in Aquitaine and the Mediterranean (Geneste, 1989) and the Mediterranean e.g. Eastern Morocco (Wengler, 1990).

Steppe habitats and cold latitudes occupation

Ancient hominids dispersing into Eurasia preferred at first tropical or subtropical zones, occupying temperate ones only during warmer interglacials, retreating to southern refugia during ice ages. This nevertheless created opportunities for gradual acquaintance with seasonal variations involving long winters and short daylight spans. Continental influences were already present in North China, e.g. Hihewan basin, adjacent to the Eurasian steppes, or Zhukoudian. Amplified palaeoclimatic cycles since 875 ky (Rea, 1990) stimulated the evolution of faunal communities adapted to subarctic and arctic conditions, the *Mammoth-steppe tundra biome* (Guthrie, 1990) spreading episodically from its Beringida centre in different directions (Bonifay, 1980; Sher, 1992) with wooly mammoth, wooly rhinoceros, reindeer, horse, bison, saiga, arctic fox, lemming. These characteristic steppe-tundra associations appear in Central Europe by OIS 8 and in Southwest France by stage 6.

Bioclimatic changes, combined with hominid home-range expansion, coincide with a progressive adaptation to sharply fluctuating climate and habitat conditions in Northwest and Central Europe during Saalian times (Roebroeks et al., 1992) and in temperate continental Asia. Occurrences during OIS 11 (Schöningen, Bilzingsleben) already document the exploitation of an increasingly diverse range of gregarious herbivore species in meadows or forest steppe under temperate

A. Cold stadial (Isotopic Stage 6) or higher latitudes occupation		
Wannen, Ariendorf I, Schweinskopf, Tönchesberg I, Rheindahlen B3, Cotte St.Brelade (3,6), Beauvais, Achenheim III, Achenheim 'sol' 74, Kurtak (Tazov glacial horizon), Diring Yuriakh		
B. High latitude or alpine habitat Intermediate/Middle Palaeolithic occupations (over 500 m)		
Western and Central Europe	**Zagros**	**Central Asia [1]**
Drachenloch (2450)	Humian (2000)	Pamir sites (>2000)*
Monte Genzana (2100)	Warwasi (ca. 1300)	Sel'ungur (>2000)
Salzofenhöhle (2000)	Kunji (ca. 1170)	Darra-i-Kur (ca. 2000)
Chilchli (1925)	Ghamari (ca. 1170)	Teshik-Tash (>1500)
Ranggiloch (1845)	Ghar Arjeneh (ca. 1170)	Sanghao (ca. 550)
Wildmannlisloch (1628)	Ghar Villa (ca. 930)	
Wildkirchli (1427)	Ghar Huchi (ca. 930)	
Schnurrenloch (1230)	Shanidar (765)	**Southwest China**
Prélétang (1208)	Barda Balka (700)*	Dadong Cave (1630)
Plaints (1120)		
Crvena Stijena (700)		
Cotencher (659)		
Betalov spodmol (537)		
Istállóskö (535)		

* Open-air occurrences
1 Uzbekistan, Kirghizia, Afghanistan, Pakistan Northwest Frontier

Table 2 - Settlement and landuse systems.

conditions. Regular associations of anthropic sites with steppe faunas begin with terminal OIS 9 to initial stage 8 (Bosinski, 1963; Junkmanns, 1991: table 2). The onset of cold wet early stadial conditions and mosaic vegetation in Northwest France provided Middle Palaeolithic occupants with an optimal animal food biomass (Auguste, 1996). The late Riss stadial in Aquitaine shows similar conditions, e.g. Combe-Grenal, La Chaise.

Evidence in continental Asia for early cold latitude occupation includes Kurtak, Ienissei (Derevinako, 1990: 63), and Diring Yuriakh, in the subarctic latitudes of the Aldan (Sakha), probably as a discrete interglacial event, by 300 ky (Waters et al., 1997). Several Middle Palaeolithic occurrences in North China contain cold or steppe-adapted species (przevalskii horse, wooly rhino), also pointing to increasing continentality.

High altitude/alpine settlement and vertical transhumance

More extensive and inclusive home-ranges and radiating settlement strategies resulted in transhumant land uses incorporating high altitude/alpine mountain habitats (Table 2), coinciding with the Upper Pleistocene Middle Palaeolithic in Eurasia (Henry, 1992; Jéquier, 1975; Smith, 1986). A wider habitat range (deserts, forests, mountains, sea coasts) and occupation is also on record throughout Sub-Saharan Africa.

A.	**On-site fire production evidence - Late Mid-Pleistocene intermediate and Middle Palaeolithic occurrences:** Constructed hearths, fire pits, ashbands, ash lenses, burnt blocks, burnt bones, charcoal spreads, diffuse charcoal remains	
	Africa:	Kalambo Falls, Cave of Hearths
	Asia:	Tabun E, Gesher B'not Yakov, 'Umm Qatafa D, Zhukoudian loc.13, loc.1
	Europe:	Cova del Bolomor, Torralba/Ambrona, San Quirce de Pisuerga, Solana del Zamborino, Pech de l'Aze II (6-9), Combe-Grenal (58-60), Fontéchevade, Mas des Caves, Rigabe (X), Orgnac, Terra Amata, L'Observatoire (K), Lazaret, Riparo Paglicci, Ménez-Drégan, Port-Pignot, Roche-Gélétan, Abri Vaufrey (VIII), Biache St.Vaast, Cagny-la-Garenne, Pontnewydd, Beeches Pit, Cotte St.Brelade, Achenheim 'sol' 81, Teting, Maastricht-Belvedere, Hunas, Schöningen, Bilzings-leben, Ehringsdorf (LT), Vértesszölös, Kudaro I(V), Azykh (V)
B.	**Other miscellanous structures**	
	Africa:	Kalambo Falls B/59 floor 5 - dwelling's semi-circular stone infrastructure Arkin - dwelling's semi-circular stone intrastructure
	Europe:	***I. Late Mid-Pleistocene*** Bilzingsleben - oval paved area; 3 simple dwelling floor plans Saint-Colomban - angular block structure Abri Vaufrey (VIII) - food consumption patterned remains Baume-Bonne - oval stone pavements made of introduced silicious pebbles Achenheim 'sol74 - living floor remains ***II. Upper Pleistocene*** Rheindahlen-Westwand - oval pit Königsaue (b) - patterned artifact clusters Ripiceni Izvo - mammoth tusks and bone screen structure Molodova - mammoth tusk and bone screen structure Raj Cave interlocking reindeer antlers for entrance protective screen Roc en Pail - interlocking reindeer antlers for entrance protective screen Combe-Grenal (14) - posthole Pech de l'Aze I(4B) - large limestone blocks structure, (4) - 30cm high dry stone wall patterned artifacts surface distribution Fontmaure - several rounded hut-shaped artifacts surface distribution Cueva Morin (17) - flints and bones patterned clusters stone piles stone wall remain

Table 3 - Site structures.

Cave sites residence

Caves occupation was mostly confined to brief daytime visits during the Lower Palaeolithic e.g. Swartkrans, Cueva Victoria, Kent's Cavern, Belle-Roche, Stránska Skála. Residential occupation, including deep semi-subterranean karstic systems e.g. Pontnewydd, Mas des Caves, Rigabe, La Chaise, Cova del Bolomor and different altitudes, began with Intermediate/early Middle Palaeolithic, probably resulting from several factors, i.e. fire-use, home bases, vertical transhumance, extensive home-ranges.

Habitat structures

These remained rather amorphous or spatially diffuse, e.g. Olorgesailie, Venosa, Soleihac, Fontana Ranuccio (Kind, 1985, Perlès, 1987; Villa, 1988) until Intermediate/Middle Palaeolithic times, when low density e.g. Abri Vaufrey or more elaborate spatial structures appear in the record, suggesting repeated residential occupation, e.g. Bilzingsleben, Terra Amata, Mas des Caves, Orgnac. Fire-making (Table 3A) for heating, producing light or cooking, with evidence ranging from diffuse charcoal, through ash lenses, widespread ashbands and constructed hearths of various types, e.g. Roche-Gélétan, Port Pignot, Pech de l'Aze become a major structural element (Table 3B; Rolland, 1996). Other structures (Table 3B) comprise stone alignments, blocks, postholes, pits, screens.

Subsistence strategies

Ecological specialization

This wide spectrum characteristic of the Middle Palaeolithic in Eurasia and Africa, superimposed over an omnivorous subsistence base, improved the balancing of various essential nutritional requirements, i.e. proteins, fats, carbohydrates. Available documents are weighted in favour of animal food, since systematic recording of plant macrobotanical remains is still exceptional in Palaeolithic research. Single game species focus begins with the Intermediate e.g. red deer (70 % at Zhukoudian loc. 1), rhinos (Bilzingsleben), horse (Schöningen). Selective and organized ungulate exploitation, concentrating on 1-3 prey species within a range of 15-20, in addition to evidence for specialized animal food processing, i.e. butchering, special activity areas and sites, as well as preserved hunting weaponry during Saalian times in Europe puts to rest the notion that pre-modern human meat procurement depended primarily on scavenging (Auguste, 1993). These patterns contrast with the generalized or opportunistic strategies prevailing during the Lower Palaeolithic. Table 4 A illustrates game specialization for Eurasia. Specialized exploitation of juvenile megaherbivores (elephant, rhinos) is indicated at Taubach, Krapina or Tata.

Similar specializing trends, also combined with a game range considerably greater than for Lower Palaeolithic, appear in Subsaharan Africa e.g. bovids (Florisbad), zebras, warthogs (Twin Rivers), primates, rodents (Andalee) - Clark, 1965:843, 1970:138-141).

Alpine species exploitation

These include larger-sized ibex and chamois, found at lower altitudes during Glacials in Europe, as well as higher latitudes representatives, and Siberian mountain goat in Inner Asia during Middle Palaeolithic times (Table 4 B).

Aquatic resources

Scant evidence suggests that fish (Lazaret, Orgnac) or mollusks (Terra Amata) played a marginal dietary role until the Upper Pleistocene (e.g. Abri des Pêcheurs; L'Homme, 1987). Ethnoecolo-

A. Specialized exploitation of one or few species: examples

	Sites	Dominant species
Central Asia:	Teshik-Tasch	Sibirian mountain goat
Southwest Asia:	Tabun B	Persian fallow deer
	Quneitra	auroch, norse, steppe ass
	Douara (early)	gazelle, wild goat
	(later)	camelids, wild goat
	Shanidar D	wild goat
Europe:	***I. Saalian***	
	Atelier Commont	horse, auroch, red deer
	Achenheim (LA moyen)	horse, steppe mammoth
	Biache St. Vaast	auroch
	Montieres	auroch, horse, red deer
	Ehringsdorf	elephas *antiquus,* Merck rhino
	Cova del Bolomor	horse, red deer
	Cotte St. Brelade (3,6)	w.mammoth, w.rhino
	Achenheim 'sol' 74	horse, w.mommoth
	Combe Grenbal 62-57	reindeer
	Abri Suard	reindeer
	II. Upper Pleistocene	
	Krapina	Merck rhino (?)
	Salzgitter-Lebenstedt	reindeer, w.mammoth
	Achenheim (L recent)	horse, w.mammoth
	Mont-Dol	horse, w. mammoth
	Sempigny	horse, w.mammoth, w.rhino
	Molodova I	w.mammoth
	Tata	w.mammoth
	Il'skaia	bison
	Staroselie	steppe ass
	Combe Grenal 31, 28-17, Vergisson IV, Roch de Marsal IV-XI, Régourdou, Esquicho Grapaou, Hauteroche 1, Roc-en Pail 5, Chapelle aux Saints	reindeer
	La Quina C2-C3, Mauran, Le Roc, Les Ourteix, Sandougne, Combe Grenal 15-13, 12-10	bison
	Combe Grenal 50-52	red deer
	Grotte de Bise, Ioton	horse
	Etouvy	bison, horse
	Abri Romani	horse, red deer
	Cova Negra	red deer, horse, ibex
	Grotta di Gosto, Grotta Tadeo, Barma Grande, Grotta del Poggio	red deer
	Mützig	reindeer, horse, w. mammoth
	Asprochaliko 3,5	red deer, fallow deer

B. Alpine species exploitation (ibex, chamois, *Sibirian mountain goat) - Exemples from late Mid-Pleistocene and/or Upper Pleistocene occurrences

Present.	Crni kal, Krapina, Adji koba, Torre Nave, Ramanelli, Monte Circeo, Sant'Agostino, Riparo Fumane, Ccmbe-Grenal, Abir Suard, Le Moustier(lower), Pech de l'Aze II, Chapelle aux Saints, Ranc-Pointu, *Okladnikov cave, *Denisova cave
Dominant:	L'Observatoire, Lazaret, Bau de l'Aubesier, Grotte des Pecheurs
Strongly dominant:	Bacho Kiro, Subalyuk, Baume-Bonne, Les Peyrards, l'Hortus, *Teshik-Tash

Table 4 (page 324) - Subsistence: Herbivores exploitation patterns.

gical observations, however, make it likely that freshwater fish must have been a critical dietary component, despite its numerical unimportance in cold temperate or boreal zones. Shell middens appear in the record from raised beaches or coastal sites only with Last Interglacial at the Sea Harvest and Hoedjies Punt sites, South Africa (Volman, 1978), and at several sites, e.g. Los Aviones, Las Perneras, in Levantine Spain (Montes Bernárdez, 1989) while The Gibraltar sites and coastal Italy also contain shellfish food remains.

Plant foods

The Palaeolithic record remains silent concerning this undoubtedly important range of staples, making trend detection problematic. Carbonized materials and sites locational patterns, however, suggest a shift to specialized geophyte exploitation during Last Glacial Middle Palaeolithic in South Africa (Deacon, 1989: 557). Roasting expanded considerably the dietary range by releasing nutrients, detoxifying or making digestible various plants (Clarke, 1976:450), a development probably coinciding with the Intermediate/early Middle Palaeolithic. Geophagy or clay eating, a widely used practise by higher primates and traditional human societies as emergency food, nutritional supplement or for detoxifying tubers, roots or acorns, especially when combined with roasting (Johns, 1991) could imply that acorn consumption became a regular staple during the Middle Palaeolithic.

Hominid socio-ecology

Figure 1 outlines a hypothesis (Rolland, 1996) contrasting Lower and Intermediate/Middle Palaeolithic social and settlement structures, namely local group locational organization relating to landuse and year-round activities. It is adapted and modified from other models (Clark, 1987; Perlès, 1987) and based on comparative observations of sites contents and spatial patterns (section ***Landuse ans settlement systems***).

Lower Palaeolithic populations inherited from their primate ancestors a system characteristic of ground-living species, stressing stability and habitat familiarity, safety in open landscape settings by aggregating, and by segregating daytime and night-time activities, reconciling this with traits analogous with those of social carnivores (Schaller, 1967: 264), and acquired by coevolution. The Lower Palaeolithic 'core-area' system's long history testifies to its adaptability.

The Intermediate/Middle Palaeolithic *home-base* system (with associated hearths, used artifacts, animal food remains) builds on preceding repertoires but represents a major organizational shift whose manifold repercussions deserve exploring further. Comparative releasing from previous constraints fostered residential intensification and group fission, expressed by more extensive home ranges and more diverse habitat exploitation, by closer articulation of social units with seasonal variations (wet/dry; warm/cold), by transhumance and local group flux. These in turn made migratory dispersals, colonizing new habitats and perhaps, stochastic population growth, more likely.

Regular fire-making played a decisive role as a necessary though not sufficient condition, by prolonging artificially daylight duration (year-round in the tropics, during winter in higher latitudes), roasting plant and/or animal foods, heating and keeping predators at bay during the night. Earlier populations, who probably used fire more incidentally, may have perceived these latent benefits but preferred relying on a pre-existing, time-tested system, being aware as well that regular fire-making entailed new constraints such as fuel gathering and maintaining hearths and fire.

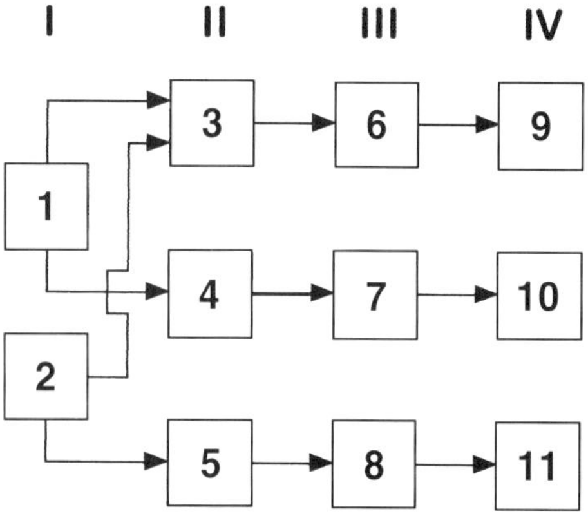

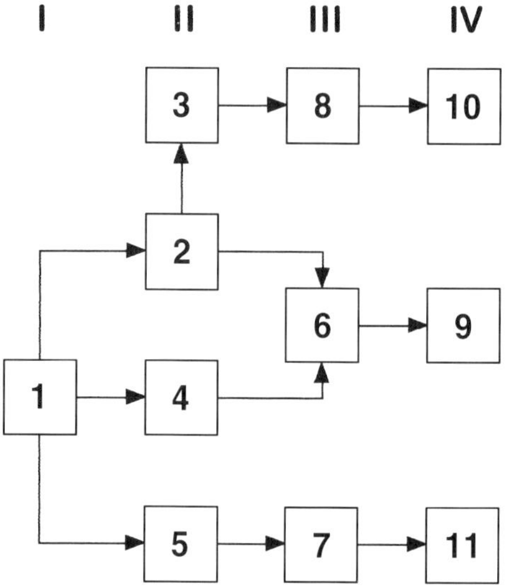

Figure 1 - Ancient hominid socio-ecological organization models.

A. Lower Palaeolithic 'core-area' system *(left)*

I. Initial conditions: *1. Ancient hominids inherit ground-living higher primates' Omnivorous and diurnal socio-ecological patterns requiring strict segregation of daytime and night activities. 2. Acquired meat-eating procurement strategies and coevolutionary interactive patterns of social carnivores. Both 1. and 2. developed at first in Subsaharan Africa and became part of hominid repertoires Colonising Eurasia.*

II. Activities organisation: *3. Various subsistance and domestic activities carried out exclusively during daylight hours. Animal food acquisition during daytime would be adaptatively advantageous, minimising risky confrontational interaction and competition with dangerous carnivores. 4. Darkness hours (and daytime rest periods) is devoted strictly to sleep or rest, coinciding with the time when potential predators or competitors become active and more aggressive, while diurnal hominids become vulnerable. 5.Episodic migrations between 'core-are as' involves movements by entire local groups, optimising safety by numbers.*

III. Locational aspects: *6. Most activities carried out within the comparatively protected mixed woodland/ grassland 'core-areas', containing trees and other convenient landscape features, as well as basic water and food resources. Foraging beyond into open settings would involve rapid forays by a few individuals. Time spent at anthropic or carnivores kill sites would be minimized, carrying back portable edible animal food remains, whenever feasible. 7. Nights or rest times spent in safe places offering natural protection or scope for rudimentary artificial sheltering or protection such as tree tops, branches where wooden platforms or nests could be built, or nests protected by thorny bushes, or blocks concentrations. Bringing back meat or leaving animal food refuse in these locations would be avoided, being too dangerous by attracting scavengers or predators. 8. Migratory moves would take place across less protected, open areas, without prolonged stays.*

IV. Archaeological evidence: *9. Most identifiable sites coincide with major daytime activity loci, with rather unstructured spatially diffuse or palimpsests of on-site deposits of lithic and/or animal bones, and occasional simple structures. Little or no fire-making evidence or hearths, except for possible burnt areas (daytime fires to keep carnivores at bay). Forays beyond 'core-areas' would leave little or low density lithic artifact scatters, » re rarely, kill/butchering activity traces. 10. Night-time sites would leave few if any preservable or recognizable direct evidence. 11. Migratory movements could leave low-density off-site lithic material scatters.*

B. Intermediate/Middle Palaeolithic 'home base' reorganizational shift *(right)*

I. New system: *1. Major reorganizational shift of ancient hominid anthropoid-like socio-ecology into an emergent residential and landuse system more aking to simple recent foraging societies, with more time/ space activities overlap.*

II. Activities organizations: *2. Increased independence from daylight/night-time separation constraints, amalgamating into 'home-bases'. Regular fire-making for food-preparation, heating and protection,* *contributes in enhancing tendencies for carrying out domestic activities simultanously and together/ regardless of daylight hours. 3. Foraging and procurement activities performed by smaller social units, radiating dayly or over a few days, from residential bases, or by longer transhumant journeys. 4. Sleeping or rest is no longer separated strictly in time or place from domestic activities. 5. Movements between home-bases may involve seasonal or scheduling group segmentation or flux, aggregating or dispersing, in addition to to entire group migrations.*

III. Locational aspects: *6. Home-bases where hitherto time-segregated activities overlap. Home-bases locations vary according to increased diversity of habitats exploited. 7. Transit sites where activity ranges, though less comprehensive, are no longer separated. 8. Specialized material or food procurement activities, as well as exploratory sites (using smoke signal communication).*

IV. Archaeological evidence: *A wider documentary range, from high material densities to off-site dispersed residues, becomes available. 9. Home-bases, combining, lithic artifacts, food refuse, hearths/ ashbands, and possible sleeping areas, indicated by empty spaces with claws left by fur-bearing animal pelts, or macrobotanical remains from plant bedding. Occasional structure remains (pits, postholes, animal bone, tusks or antlers screens). 10. Better preserved, recognisable activitiy sites e.g. kill/butchering, stone quarries, occasionally with simple structure or fire-making remains. 11. Transit sites with low density residues/ and off-site artifact scatters.*

Cognitive, ideological and esthetic expression

Resolution relating to these themes remains problematic. Opinion is polarized because analytic verification methods lack stringency, leaving ambiguous results or unsupported claims (Mellars, 1996: ch.12). Dogmatic appeals to theory and negative preconceptions has lead to rejecting unilaterally any evidence which would have been acceptable in an Upper Palaeolithic context. Lines of evidence include modified or contextually significant objects, or patterned structures, and disposal of the dead testimonies.

Our survey points to an increasingly wider range of Middle Palaeolithic material culture which, however, remains modest when compared with the Late Palaeolithic/Mesolithic record. Material expression of collective representations among foraging societies is liable to considerable variability, and may not yield substantial archaeological traces. Concrete actualization of symbolic or esthetic behaviour going back to Middle Palaeolithic times would tend to be sparse and not always recognizable.

Late *erectus* populations corresponding with Intermediate/early Middle Palaeolithic display highly developed brains, implying speech and abstract thought capabilities (Kien, 1991; Ullrich, 1995: 372-373). In societies with comparatively less rich material inventories, expressing ideological concepts may be confined to endowing unmodified objects or patterns of the landscape features e.g. cave stalagmites, or the 'sacred sites' of recent Aboriginal Australia, with symbolic contents. Studying minerals, fossils, or non-utilitarian objects, ranked according to physical properties, scarcity, esthetic appeal or 'appreciation', shows their distinctive beginning with the Middle Palaeolithic (Schäfer, 1996). Use of colouring matter appears during late Middle Palaeolithic. Selective exploitation of dietarily significant ungulate species could equally be linked with symbolic appreciation, a body of important information for group survival transmitted orally over many generations (Minc, 1986), long before being visually represented in cave or home art. Similarly quality lithic materials can assume an esthetic or symbolic dimension among foraging populations in cold latitudes.

Table 5 A,B lists confirmed or probable instances of ideological/esthetic expression for the Middle Palaeolithic. The Berekhat Ram figurine's uniqueness could relate as much to its more perishable but more easily carved material, its intermittent production, as much as to identification

A. Portable Objects		
	Africa:	Howiesons Poort - ochre pencils
	Western Asia:	Berekhat Ram - late Mid-Pleistocene carved basalt anthropomorphic figurine
	Europe:	West Tofts - late Mid-Pleistocene handaxe from exotic flint, trimmed around a fossil invertebrate Vindija - late Mousterian bone 'flute' fragment Tata - ochre stained mammoth ivory rounded piece; imported fossil numulite Chez Pourrez - imported invertebrate fossil Les Merveilles, La Ferrassie, La Quina, Pech de l'Aze I, Gare de Couze, Ccmbe-Capelle - manganese bioxyde and ochre fragments with scraping/polishing traces
B. Structures without evident utilitarian purpose		
	Africa:	El Guettar - piles of hundreds of spheroids stuffed with flints and bones
	Asia:	Djebel Qafezeh (11) - human burial with fallow deer antlers Skhul (V) - human burial holding a wild boar mandible Amud (7) - infant burial with red deer maxilla deposited on pelvis Shanidar - infant burial associated with flower remains
	Europe:	Arcy sur Oure (Grotte de l'Hyene) - 10 spheroid piles mixed with mammoth Maras (3) - 1 m diameter ochre floor with reindeer bones and an ochre-stained hammerstone La Ferrassie - family burial with ochre-stained stone plaquettes Regourdou - brown bear femurs with beveled extremities, arranged crosswise Bruniquel - late Mousterian broken stalagmite and stalagtites arrangement Grotte du Prince - 'Mousterian cairn' (?)

Table 5 - Material expression of ideological or esthetic behaviour.

difficulties, or to investigative biases neglecting to sort through sites with a thick matrix for retrieving similar objects. Expression by woodworking, common in the Middle Palaeolithic, would also leave few traces.

The disposal of the dead theme is equally rife with polarized viewpoints. New investigations (Belfer-Cohen & Hovers, 1992; Ullrich, 1986, 1995) refute the notion that systematic disposal did not occur until the Upper Palaeolithic. Varied mortuary practices begin with the Intermediate and during the Middle Palaeolithic. They include throwing away, intentional deposition, burying selected human bones, deposition or burial of entire corpses, as well as defleshing and disarticulating techniques (Ullrich, 1995). Burying entire individuals remained uncommon until Late Palaeolithic/ Mesolithic, e.g. Ibero-Maurusian ossuaries, increasing slowly from the Upper Pleistocene (Qafzeh, Skhul). Intentional grave goods are scarce (Qafzeh, Skhul, Amud, Shanidar, Ferrassie) before the Upper Palaeolithic.

The onset of mortuary practices indicates a behaviour related directly to death awareness (Ullrich, 1995) which coincides, interestingly enough, with concrete indications for home-base settlements and the Intermediate's onset.

Concluding discussion

The survey's findings summarized on Figure 2 reveal a succession of emergence clusters for several cultural systems and subsystems from Intermediate and throughout the Middle Palaeolithic. New discoveries and dates may alter the order of this tentative time series arrangement. They indicate, nevertheless, that the null hypothesis, i.e. no major differences separating the Middle from the Lower Palaeolithic, must be rejected. It invalidates previous notions that basic transformations took place only with the Upper Palaeolithic (e.g. Zvelebil, 1984).

Interpreting these empirical patterns may result in alternative conclusions: (1) Regarding the accumulating traits and the covariations between systems or subsystems as loosely connected, 'hybrid', incremental series of events into a corpus of repertoires; (2) Exploring whether most of these series and systems/subsystems' contents, interacting directly or indirectly with distinct time-scales and responses, constituted meaningfully organized developments, diagnostic of a developmental stage. This task, beyond the present paper's scope, should be targeted by future investigations.

We propose, as a provisional hypothesis, that the emergence of home-base residence, mortuary practices, long-term home-range expansion trends, exploring and exploiting new habitats and gradual specializing or diversifying of basic food staples, regular occupation of higher latitudes, seasonal flux and vertical transhumance, and more extensive socio-demographic spacing and networks, may have been linked causally. These reorganizational shifts, involving more anchored residence, would also provide additional scope for elaborating on innovations such as fire-use, food-preparation, mortuary practices or ideological elaboration.

Figure 3 relates the standard periodization of Prehistory with its time scale. It produces a trajectory fitting an exponential curve (see also Isaac, 1972: figs. 4,5, or Movius, 1956: fig. 1 for analogous results), which may imply that the accelerating tempo of repertoires differentiation was determined largely by the magnitude of preceding ones. The initially slow developmental rates would thus be the consequence of more restricted antecedents. This also means that innate biological determinisms ceased to play a leading role early in hominid cultural developments (Boas, 1940: 248-249; Hayden, 1993).

The above describes reality schematically, stressing a basic trajectory blurring gradual or punctuated change patterns. Lower Palaeolithic repertoires, however, display considerable internal variability, indicating behavioural versatility rather than specialization (Belfer-Cohen & Goren-Inbar, 1994). Time-related stochastic variations also overshadow directional change (Isaac, 1969: fig. 4, 1977: fig. 33). The Intermediate/Middle Palaeolithic, on the other hand, combines variability and specialization, while its time-span corresponds with accelerating, cumulative change and broadening cultural systems, rather than oscillating shifts focussed on particular systems (contra Mithen, 1996: figure 33) - which would suggest a comparatively grater influence of neuro-psychological factors on culture change. Middle Palaeolithic specialization and broadening trends are not reducible to a phyletic branching model (e.g. Braidwood, 1960: fig. 1). They fit better with a long-term tendency towards increasing integration and connectedness (e.g. Semënov, 1970: fig.1).

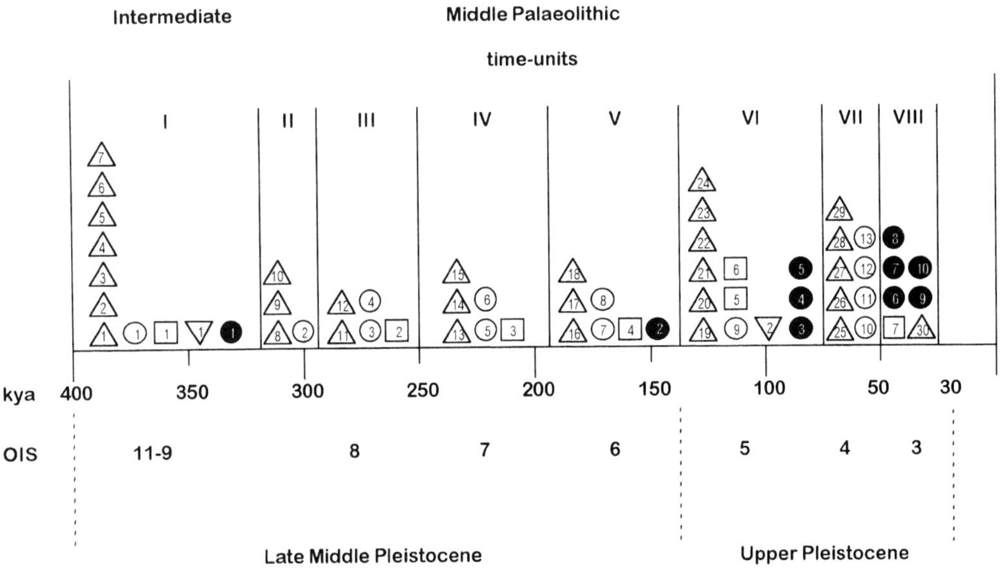

Figure 2 - Summary of Intermediate/Middle Pleistocene emergent system and subsystems.

Systems: *Δ technological; ○ landsuse/settlement; □ subsistence; ∇ socio-ecological; ● cognitive/ideological/esthetic*

Chronological scale and time-span units (N = 8): Pleistocene subdivisions; oxygen isotopic stages; 1000 y intervals = Late Middle Pleistocene (Saalian complex in Europe), OIS 11 to 6, 400-130 ky; Upper Pleistocene (Last Interglacial and Early Glacial 1 and Lower Interpleniglacial), OIS 5 to 3, 130-35 ky.

Δ : *Late Middle Pleistocene*

Time-unit I - OIS 11-9, 400-320 ky: *1. Flake cleavers (in Eurasia); 2. Victoria West or Proto-Levallois (in Africa and peninsula India); 3. Soft-hammer trimming, lanceolate handaxes; 4. Wooden clubs, throwing sticks, digging sticks; 5. Composite wooden javelins; 6. Regular fire-production; 7. Bone, antler artifacts.*

Time-unit II - terminal OIS 9, 320-297 ky: *8. Flake implements and assemblages standardisation; 9. Prepared-core (Levallois and disc-core) flaking; 10. Facetting (dihedral, multifacetting).*

Time-unit III - OIS 8, 297-250 ky: *11. Diversified and exotic lithic material use; 12. Parsimony of preferred lithic materials, and emphasis on resharpening cutting tools.*

Time-unit IV - OIS 7, 250-195 ky: *13. Step-flaking; 14. Planoconvex flaking; 15. Prismatic blade-core flaking.*

Time-unit V - OIS 6, 195-130 ky: *16. Elongated Mousterian points as weapon tips; 17. 'Pontinian' flaking; 18. Wooden boards.*

Upper Pleistocene

Time unit VI - OIS 5e, 130-115 ky; OIS 5d-a, 115-75 ky: *19. Specialized handaxe types differentiation; 20. Buring/endscrapers-dominated assemblages (Levant); 21. Beginning of pronounced inter-regional repertoires differentiation, in addition to preexisting intra-regional facies variations; 22. Stone balls missiles heaps; 23. Bones harpoons (Central Africa); 24. Lupembian bifacial lanceolate points (Subsaharan Africa).*

Time-unit VII - OIS 4, 75-45 ky: *25. Blattspitzen; 26. Pradniki; 27. Increased exotic lithic materials use; 28. Bitumen hafted tools; 29. Bone punces, gouges, retouchers.*

Time-unit VIII - OIS 3, 45-35 ky: *30. Aterian stemmed tools (Magreb,Sahara).*

○ ***: Late Middle Pleistocene***

Time-unit I, OIS 11-9, 400-320 ky: 1. Regular caves residence.

Time-unit II - terminal OIS 9, 320-297 ky: *2. Semi-circular huts stone infrastructures.*

Time-unit III - OIS 8, 297-250 ky: *3. Beginning of home-ranges extension; 4. Regular cold-steppe habitat occupation.*

Time-unit IV - OIS 7, 250-195 ky: *5. Episodic (interglacial) occupation of boreal zones (Siberia); 6. Elaborate hearths construction.*

Time-unit V OIS 6, 195-130 ky: *7. Occupation of mid-latitude habitats during cold stadials; 8. Paved cave floors made of imported river pebbles.*

Upper Pleistocene

Time-unit VI - OIS 5e, 130-115 ky; OIS 5d-a, 11575 ky: *9. Vertical seasonal transhumance landuse.*

Time-unit VII - OIS 4, 75-45 ky: *10. Considerable home-range extension; 11. Pit structures in dwellings; 12. Use of mammoth tusks,bones and reindeer antlers for dwelling protective screens; 13. Seasonal settlement flux.*

□ ***: Late Middle Pleistocene***

Time-unit I - OIS 11-9, 400-320 ky: *1. Exploitation of wider diversity of open landscape gregarious herbivores species.*

Time-unit III - OIS 8, 297-250 ky: *2. Selective exploitation of 1-3 ungulate prey species,*

Time-unit IV - OIS 7, 250-195 ky: *3. Selective hunting/trapping of juvenile megaherbivores.*

Time-unit V - OIS 6, 195-130 ky: *4. Regular exploitation of mammoth/steppe-tundra biomes species in mid-latitude Eurasia.*

Upper Pleistocene

Time-unit VI - OIS 5e, 130-115 ky; OIS 5d-a, 115-75 ky: *5. Earliest shell midden archaeological sites; 6. Regular exploitation of high and mid-altitude alpine ungulates.*

Time-unit VII - OIS 4, 115-75 ky: *7. Specialized geophytes exploitation (Southern Africa).*

∇ ***: Late Middle Pleistocene***

Time-unit I - OIS 11-9, 400-320 ky: *1. Home-base residence reorganizational shift.*

Upper Pleistocene

Time-unit VI - OIS 5e-a, 130-75 ky: *2. Possible population growth resulting from long-term consequences of home-base socio-ecological and organisational shift on landuse and settlement systems.*

● ***: Late Middle Pleistocene***

Time-unit 11-9, 400-320 ky: *1. Earliest deliberate mortuary practices.*

Time-unit V - OIS 6, 195-130 ky: *2. Carved anthropomorphic basalt figurine (Levant).*

Upper Pleistocene

Time-unit VI, 130-115 ky: *3. Burials of entire deceased individuals; 4. Earliest identifiable burial offerings; 5. Spheroid piles stuffed with lithic or bone objects.*

Time-unit VII - OIS 4, 75-45 ky: *6. Use of coloring matters; 7. Importing 'unusual' fossil objects; 8. Importance and use of esthetically-pleasing lithic materials.*

Time-unit VIII - OIS 3, 45-35 ky: *9. Bone 'flute' (Vindija); 10. Stalagmite and stalagtite framents artificial structure (Bruniquel).*

The diagram's succession of Intermediate/Middle Palaeolithic emerging systems and subsystems as punctuated (e.g. technology during OIS 11-9, 5e and 4) or gradually spreading (e.g. landuse/settlement systems), showing accelerating trends (e.g. ideological/esthetic subsystems) during the Upper Pleistocene (Last Glacial, especially) is a tentative time-series ordering or working hypothesis which may be re-arranged because of future discoveries or new datings. These trends undergo considerable amplification and intensification during the Upper Palaeolithic, building on Intermediate/Middle Palaeolithic formative thresholds, in addition to other emerging characteristics. This emphasizes the underlying exponential growth trajectory of Stone Age Prehistory.

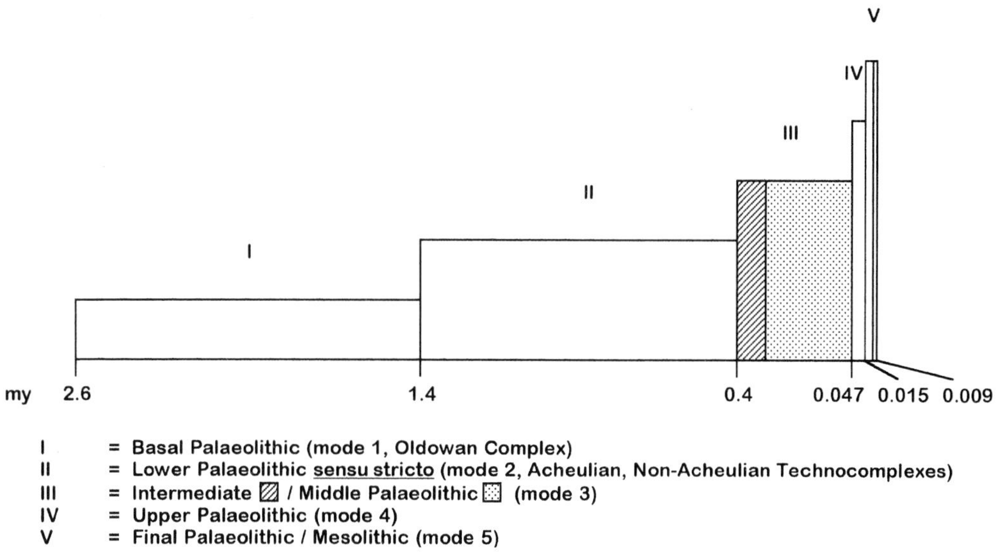

Figure 3 - *Stone Age subdivisions time-spans and diachronic trajectory.*
I - Basal Palaeolithic (mode 1, Oldowan complex);
II - Lower Palaeolithic senso stricto *(mode 2, Acheulian, Non-Acheulian technocomplexes);*
III - Intermediate (left) / Middle Palaeolithic (right) (mode 3);
IV - Upper Palaeolithic (mode 4);
V - Final Palaeolithic/Mesolithic (mode 5).

References

AUGUSTE, P. (1993): Acquisition et exploitation du gibier au paléolithique moyen dans le Nord de la France. Perspectives paléo-écologiques et palethnologiques. *In: Exploitation des animaux sauvages à travers le temps,* 49-62. Juan-Les-Pins.

AUGUSTE, P. (1996): Adaptations biologiques et culturelles des prénéandertaliens et des néandertaliens aux modifications paléoécologiques et fauniques en France spetentrionale. In: M. Otte (éd.), *Nature et culture,* 99-116. Liège.

BELFER-COHEN, A. & GOREN-INBAR, N. (1994): Cognition and communication in the Levantine Lower Palaeolithic. *World Archaeology* 26 (2), 144-157.

BELFER-COHEN, A. & HOVERS, E. (1992): In the eye of the beholder: Mousterian and Natufian burials in the Levant. *Current Anthropology* 33, 463 - 471.

BINFORD, L.R. (1985): Human ancestors: changing views of their behavior. *Journal of Anthropological Archaeology* 4, 292-327.

BOAS, F. (1940): *Race, language and culture.* New York.

BONIFAY, M.-F. (1980): Relations entre les données isotopiques océaniques et l'histoire des grandes faunes europées plio-pleistocènes. *Quaternary Research* 14, 251-262.

BORDES, F. (1960): Evolution in the Palaeolithic cultures. In: S. Tax (ed), *The evolution of man. Mind, culture, and society,* 9-110. Chicago.

BOSINSI, G. (1963): Eine mittelpaläolitische Formengruppe und das Problem ihrer geochronologischen Einordnung. *Eiszeitalter und Gegenwart* 14, 124-140.

BOSINSKI, G. (1967): *Die mittelpaläolitischen Funde im westlichen Mitteleuropa.* Köln, Graz.

BOSINSKI, G. (1982): The transition Lower/Middle Palaeolithic in Northwestern Germany. In: A. Ronen (ed.), *The transition from Lower to Middle Palaeolithic and the origin of modern man,* 165-175. Oxford.

BRAIDWOOD, R. (1960): Levels in prehistory: A model for the consideration of the evidence. In: S.Tax (ed.), *The evolution of man. Mind, culture, and society,* 143-151. Chicago.

CALLOW, P. (1986): The La Cotte industries and the European Lower and Middle Palaeolithic. In: P.Callow & J.M. Cornford (eds.), *La Cotte de St. Brelade 191-1978. Excavations by C.B.M. McBurney,* 377-388. Norwich.

CHILDE, V.G. (1944): Archaeological ages as technological stages. *Journal of the Royal Anthropological Institute* 74. 7-24.

CLARK, J.D. (1965): The later Pleistocene cultures of Africa. *Science* 150, 833-847.

CLARK, J.D. (1970): *The prehistory of Africa.* London.

CLARK, J.D. (1987): Transitions. *Homo erectus* and the Acheulian: the Ethiopian sites of Gadeb and the Middle Awash. *Journal of Human Evolution* 16, 809-826.

CLARKE, D. (1976): Mesolithic Europe: the economic basis. In: G. de Sieveking, I.H. Longworth & K.E.Wilson (eds.), *Problems in economic and social archaeology*, 449-481. London.

COMBIER, J. (1962): Chronologie et systématique du moustérien occidental: données et conceptions nouvelles. In: *Atti del VI Congresso Internazionale delle Scienze Preistoriche e Protostoriche (Roma)* 1, 77-96.

DEACON, H.J. (1989): Late Pleistocene palaeoecology and archaeology in the Southern Cape, South Africa. In: P. Mellars & C. Stringer (eds.), *The human revolution. Behavioural and biological perspectives on the origins of modern humans,* 547-564. Edinburgh.

DEREVIANKO, A.P. (1990): *Paleolithic of North Asia and the problem of ancient migrations.* Novosibirsk.

DOBOSI, V.T. (1983): Die Knochenartefakte von Vértesszölös. *Ethnographisch-Archäologische Zeitschrift* 24, 349-361.

FRIEDMAN, E., GOREN-INBAR, N., ROSENFELD, O., MARDER, O. & BURIAN, F. (1994-45): Hafting during Mousterian times - Further indication. *Journal of the Israel Prehistoric Society* 26, 8-31.

GENESTE, J.-M. (1989): Economie des ressources lithiques dans le Moustérien du Sud-Ouest de la France. In: M. Otte (éd.), *L'Homme de Néandertal.* Vol. 6: *La Subistance,* 75-97. Liège.

GLOVER, I.C. (1993): Tools and culture in Late Paleolithic Southeast Asia. In: G. Burenhult (ed.), *The first humans. Human origins and history to 10,00 BC,* 128-129. San Franciso.

GUTHRIE, R.D. (1990): Late Pleistocene faunal revolution. A new perspective on the extinction debate. In: L.D Agenbroad, J.I. Mead & L.W. Nelson (eds.), *The mammoth site of Hot Springs, South Dakota, Inc.,* Vol. 1,42-53. Hot Springs.

HARROLD, F.G. (1983): *Wooden artifacts from Torralba and Ambrona, Spain.* Paper presented at the XI International Congress of Anthropological and Ethnological Sciences, 9 pp. Vancouver.

HAYDEN, B. (1993): The cultural capacities of Nandertals: A review and re-evaluation. *Journal of Human Evolution* 24, 113-146.

HEIM, J.-L. (1987): Les hommes de Néandertal: un nouvel essai d'interprétation phylogénique. *Comptes Rendus de l'Académie des Sciences Paris* 4 (4), 305-323.

HENRY, D.O. (1992): Transhumance during the late Levantine Mousterian. In: H.L. Dibble & P. Mellars (eds.), *The Middle Palaeolithic: Adaptation, behavior, and variability,* 143-162. Philadelpia.

HUANG, W., XINGIANG, S., HOU, H., MILLER-ANTONIO, S. & SCHEPARTZ, L.A. (1995): Excavations at Panxian Dadong, Guizhou Province, Southern China. *Current Anthropology* 36, 844-846.

ISAAC, G.L. (1969): Studies of early culture in East Africa. *World Archaeology* 1 (1), 1-28.

ISAAC, G.L. (1972): Chronology and the tempo of cultural change during the Pleistocene. In: W.W. Bishop & J.A. Miller (eds.), *Calibration of hominoid evolution,* 381-430. Edinburgh.

ISAAC, G.L. (1977): *Olorgesailie. Archaeological studies of a Middle Pleistocene lake basin in Kenya.* Chicago.

JELINEK, A.J. (1982): The Tabun cave and Palaeolithic man in the Levant. *Science* 216, 1369-1375.

JÉQUIER, J.-P. (1975): *Le moustérien alpin.* Yverdon.

JIA, L. (1980): *Early man in China.* Bejing.

JOHNS, T.(1991): Well-grounded diet. The curious practice of eating clay is rooted in its medicinal value. *The Sciences* September/October, 38-43.

JUNKMANNS, J. (1991): Die Steinartefakte aus Achenheim in der Sammlung Paul Wernert. *Archäologisches Korrespondenzblatt* 21, 1-16.

KIEN, J. (1991): The need for data reduction may have paved the way for the evolution of language ability in hominids. *Journal of Human Evolution* 20, 157-165.

KIND, C.-J. (1985): *Die Verteilung von Steinartefakten in Grabungsflächen. Ein Modell zur Organisation alt-und mittelsteinzeitlicher Siedlungsplätze.* Tübingen.

L'HOMME, G. (1987): L'Abri des Pêcheurs à Casteljau (Ardèche). *Ardèche Archéologie* 4:1.

LEROI-GOURHAN, G. (1966): Le paléolithique moyen. In: A. Leroi-Gourhan, G. Bailloud, J. Chavaillon & A. Laming-Emperaire (eds.), *La préhistoire*, 94-112. Paris.

MANIA, D. & VLČEK, E. (1987): *Homo erectus* from Bilzingsleben (GDR). His culture and environment. *Anthropologie* (Brno) 25, 1-45.

MANIA, U. (1995): The utilisaton of large mammal bones in Bilzingsleben - a special vaiant of Middle Pleistocene man's relationship to his environment. In: H. Ullrich (ed.), *Man and environment in the Palaeolithic,* 239-246. Liège.

MELLARS, P. (1996): *The Neandertal legacy. An archaeological perspective from Western Europe.* Princeton.

MINC, L. (1986): Scarcity and survival: The role of oral tradition in mediating subsistence crisis. *Journal of Anthropological Archaeology* 5, 39-113.

MITHEN, S. (1996): *The prehistory of the mind.* London.

MONTES, BERNÁRDEZ, R. (1989): La grotte „de Los Aviones", Cartagena (Espagne). *Bulletin de la Société Préhistorique Française* 86 (2), 40-44.

MOVIUS, H.L. (1956): The Old Stone Age. In: H.L. Shapiro (ed.), *Man, culture, and society,* 49-93. New York.

MÜLLER-BECK, H. (1987): The ecosystem of the „Middle Paleolithic" (late Lower Paleolithic) in the Upper Danube Region: A stepping-stone to the Upper Paleolithic. In: H.L. Dibble & A. Montet-White (eds.), *Upper Pleistocene prehistory of western Eurasia,* 233-254. Philadelphia.

PEI, W.C. (1939): A preliminary study on a new Palaeolithic station known as locality 15 within the Choukoutien region. *Acta Geologica Sinica* 19, 147-187.

PERLÈS, C. (1987): La naissance du feu. *L'Histoire* 105, 28-33.

REA, D.K. (1990): Aspects of atmospheric circulation: The Late Pleistocene (0-950,000 yr) record of eolian deposition in the Pacific Ocean. *Palaeogeography, Palaeoclimatology, Palaeoecology* 78, 217-227.

ROE, D.A. (1982): The transition from Lower to Middle Palaeolithic, with particular reference to Britain. In: A. Ronen (ed.), *The transition from Lower to Middle Palaeolithic and the origin of modern man,* 177-191. Oxford.

ROEBROEKS, W., KOLEN, J. & RENSINK, E. (1988): Planning depth, anticipation and the organization of Middle Palaeolithic technology: The „archaic natives" meet Eve's descendants. *Helinium* 28 (1), 17-34.

ROEBROEKS, W., CONARD, N.J., & KOLFSCHOTEN, T. van (1992): Dense forests, cold steppes, and the Palaeolithic settlement of Northern Europe. *Current Anthropology* 33, 551-586.

ROLLAND, N. (1996): Later Pleistocene complexity: the Middle Palaeolithic antecedents. In: D.A. Meyer, Dawson, P.C. & Hanna, D.T. (eds.), *Debating complexity,* 193-201. Calgary.

RONEN, A. (ed.) (1982). *The transition from Lower to Middle Palaeolithic and the origin of modern man.* Oxford.

SCHÄFER, J. (1996): Die Wertschätzung außergewöhnlicher Gegenstände (non-utilitarian objects) im Alt-und Mittelpaläolithikum. *Ethnographisch-Archäologische Zeitschrift* 36, 173-190.

SCHÄFER, J., SOSIN, P. & RANOV., V.A. (1996): Neue Untersuchungen zum Lößpalaolithikum am Obi-Mazar, Tadzikistan. *Archäologisches Korrespondenzblatt* 26, 97-109.

SCHALLER, G.B. (1967): *Golden shadows and flying hooves.* New York.

SCHÜLE, W. (1992): Vegetation, megaherbivores, man and climate in the Quaternary and the genesis of closed forests. In: J.G. Goldhammer (ed.), *Tropical forests in transition,* 45-76. Basel.

SEMËNOV, S.A. (1970): The forms and functions of the oldest tools. *Quartär* 21, 1-20.

SHER, A. (1992): Beringian fauna and early Quaternary mammalian dispersal in Eurasia: ecological aspects. *Courier Forschungs-Institut Senckenberg* 153, 125-133.

SMITH, P.E.L. (1986): *Palaeolithic archaeology in Iran.* Philadelphia.

TAVOSO, A. (1984): Réflections sur l'économie des matières premières au moustérien. *Bulletin de la Société Préhistorique Française* 81, 79-82.

THIEME, H. (1997): Lower Palaeolithic hunting spears from Germany. *Nature* 385, 807-810.

ULLRICH, H. (1986): Manipulations on human corpses, mortuary practice and burial rites in Palaeolithic times. *Anthropos* (Brno) 23, 227-236.

ULLRICH, H. (1996): Mortuary practices in the Palaeolithic - reflections of human-environment relations. In: H. Ullrich (ed.), *Man and environment in the Palaeolithic,* 363-378. Liège.

VILLA, P. (1988): A propos d'analyse spatiale et de taphonomie. *Les Nouvelles de l'Archéologie* 31, 55-59.

VOLMAN, T.P. (1978): Early archeological evidence for shellfish collecting. *Science* 201, 911-913.

WATERS, R., FORMAN, S.L. & PIERSON, J.M. (1997): Diring Yuriakh: A Lower Paleolithic site in Central Siberia. *Science* 275, 1281-1284.

WAGNER, E. (1990): Ökonomie und Ökologie in den altpaläolitischen Travertinfundstellen von Bad Canstatt. *Fundberichte aus Baden-Württemberg* 15, 1-15.

WENGLER, L. (1990): Économie des matières premières et territoire dans le moustérien et l'atérien maghrébins. Exemples du Maroc Oriental. *L'Anthropologie* 94, 335-360.

YELLEN, J.E., BROOOKS, A.S., CORNELISSEN, E., MEHLMAN, M.J. & STEWART, K. (1995): A Middle Stone Age worked bone industry from Katanda, Upper Semliki Valley, Zaire. *Science* 268, 553-556.

ZVELEBIL, M. (1984): Clues to recent human evolution from specialized technologies. *Nature* 307, 314-315.

Contrasting Lifestyles and Survival Strategies between the Gravettian and the Epigravettian in Central and South-eastern Europe

Janusz K. Kozlowski

Abstract

The paper evaluates the influence of the Last Glacial Maximum (LGM) on changes in subsistence strategies and lifestyles in central European transept north-south from southern Poland to Greece. The LGM was reflected in a different way in the periglacial zone of Europe and in its Mediterranean zone. In the former the cycle of climatic changes with a number of short oscillations was more conspicuous, whereas in south-eastern Europe such oscillations are hardly noticeable while changes are manifested by progressive aridity of climate rather than by cooling of the climate. Consequently, in the periglacial Europe there is a contrast between pre-Pleniglacial Gravettian and post-LGM Epigravettian which is seen in different hunting strategies, different social organization, ways of occupation and exploitation of territory, in the different systems of lithic raw material procurement, or even in the sphere of spiritual and symbolic culture. In south-eastern Europe the contrast is much weaker, cultural evolution took place within fairly isolated microregional frameworks where models of basic economic spheres and social structures show continuity. The Middle Danube Basin played an important role: its northern part displays changes typical of the periglacial zone in the consequence of which we observe a discontinuity of the cultural evolution due to the migrations of large population groups from about 20 to 18 Kyr B.P. In southern part of the Middle Danube Basin, on the other hand, core residential areas shifted gradually to the south reaching the northern Balkans.

Introduction

In recent years numerous publications have been devoted to the influence of Vistulian Pleniglacial II on the evolution of Palaeolithic societies inhabiting Europe at that time (comp. especially: Gamble & Soffer, 1990; also Mussi & Roebroeks, 1996). Considerable attention has been paid to the influence of Pleniglacial II on the decline of occupation in central Europe during the maximum transgression of the Scandinavian ice-sheet, to migrations necessitated by the shifts of ecological zones, and to the issues of the Late Glacial re-settling of the European Lowland. Relatively less consideration has been given to territorial variability in the response to the impact of ecological changes as reflected in diverse adaptation strategies of the Upper Palaeolithic or to the diachronic dynamics of these processes between 20 to 10 Kyr years B.P. Similarly, the comparison of the pre-Pleniglacial adaptation strategies with the strategies developed immediately after the LGM has been given relatively modest space.

In this paper I shall attempt to discuss the influence of Pleniglacial II and resultant transformations of subsistence strategies and lifestyles in the central European south-north transept, from southern Poland as far as Greece.

The role of this territory is particularly interesting as the palaeogeographical consequences of the maximum transgression of the Scandinavian ice-sheet and mountain glaciers in the Alps and the Carpathians were most dramatic. This zone played a decisive role in the division of post-Pleniglacial Europe into two distinctive cultural provinces: the western – Magdalenian, and the eastern – Epigravettian.

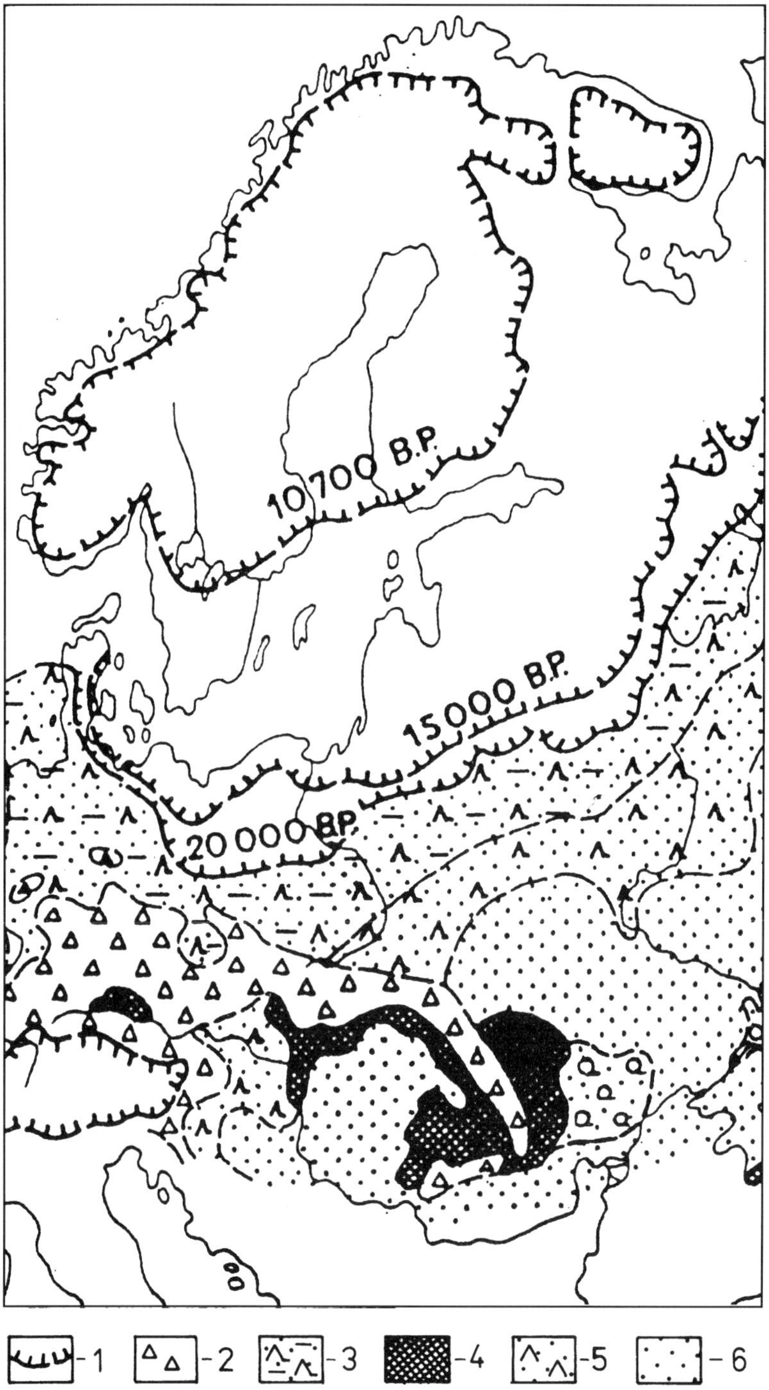

Figure 1 - Map of the Last Glacial Maximum (ca 20-18 Kyr B.P.) in central Europe.
1 - ice-sheet boundary, 2 - alpine vegetation, 3 - forest-tundra, 4 - forests with birch, pine and larch, 5 - steppe forest, 6 - steppe.

Palaeoecological conditions

The alleged uniform rhythm of cooler and warmer climatic oscillations over the whole European continent during the Interpleniglacial, Pleniglacial II and the Late Glacial that has been proposed by palynologists meets today with increasing criticism. This refers particularly to the time-span from 30 to 15 Kyr B.P. On the one hand, the cool episodes before the LGM seem to be of less importance in the light of recent investigations, especially in the conditions of steppe-tundra with its large biomass (Guthrie, 1990). On the other hand, the Scandinavian ice-sheet transgression must have been much more rapid and intensive than it has until recently been believed, especially in view of the fact that southern Scandinavia was free of ice-cover as late as about 22 Kyr B.P. It seems that in the northern part of central Europe a stronger negative impact on the conditions of occupation was exerted by the sedimentation of loess blown from river valleys – on other words the growing aridity of climate – rather than by large decrease in temperature (Kozlowski, 1996). In south-eastern Europe, notably in the Balkans, at the same time, the palaeogeographical conditions display a much bigger stability of humidity and temperature than some palynological estimations would indicate (Carciumaru, 1989; Bottema, 1974 and others).

Consequently, the impact of the maximum ice-sheet transgression in the northern part of central Europe must have been powerful and rapid, but the same time brief. The permafrost zone and the snow-cover spread restricting the territory of the „mammoth-steppe-tundra". The influence of the extreme ice-sheet transgression was less strongly connected with the drop in the annual mean temperatures. The impact of the LGM was undoubtedly weaker in south-eastern Europe where it was manifested in the refuge character of this part of the continent which absorbed yet another „reflux" of population from the territories in the Middle Danube Basin which were being gradually abandoned.

Pre- and post-Pleniglacial development of the northern part of central Europe

The contrast between the pre-LGM Gravettian and the post-LGM Epigravettian is much sharper and more important than the depopulation of the northern part of central Europe during the LGM itself. The progress of investigations in the territory of Poland, Slovakia, Moravia, Austria and Hungary has shown that depopulation of these territories was not synchronous. It took place first in the south, in the middle Danube basin (about 25 to 20 Kyr) and latest in southern Poland (about 20 to 17 Kyr B.P.). I have tried to account for this apparent paradox by connecting the process of abandoning these territories with the periods of extreme intensity of eolian loess sedimentation (Kozlowski, 1996).

Independently of migrations or shifts of core residential areas central Europe during the LGM saw changes encroaching upon a number of spheres of the life of Palaeolithic societies. These changes constitute adaptations necessitated by sudden and rapid deterioration of palaeogeographical conditions. The transformations that occurred between the Gravettian (understood as the Pavlovian together with the Late Gravettian shouldered points horizon) and the Epigravettian affected:

1. Location of settlements: Almost nowhere are Epigravettian sites located in the same place as the sites of the Pavlovian or the shouldered points horizon (Dobosi, 1986; Hromada & Kozlowski, 1995).
2. Type of settlement/camp: Large residential settlements with large dwelling structures (built of timber, stones, bones and skins – Klima, 1983) are replaced by small, short-term camps represented by individual kshemenitsas – without preserved habitation structures or with features such as light tents or shelters (Banesz et al., 1992).
3. Systems of occupation of a territory: In place of large settlements with several dwelling structures existing at the same time, individual, small camps appear which sometimes form clusters or are superimposed on one another as people repeat visits to the same areas.

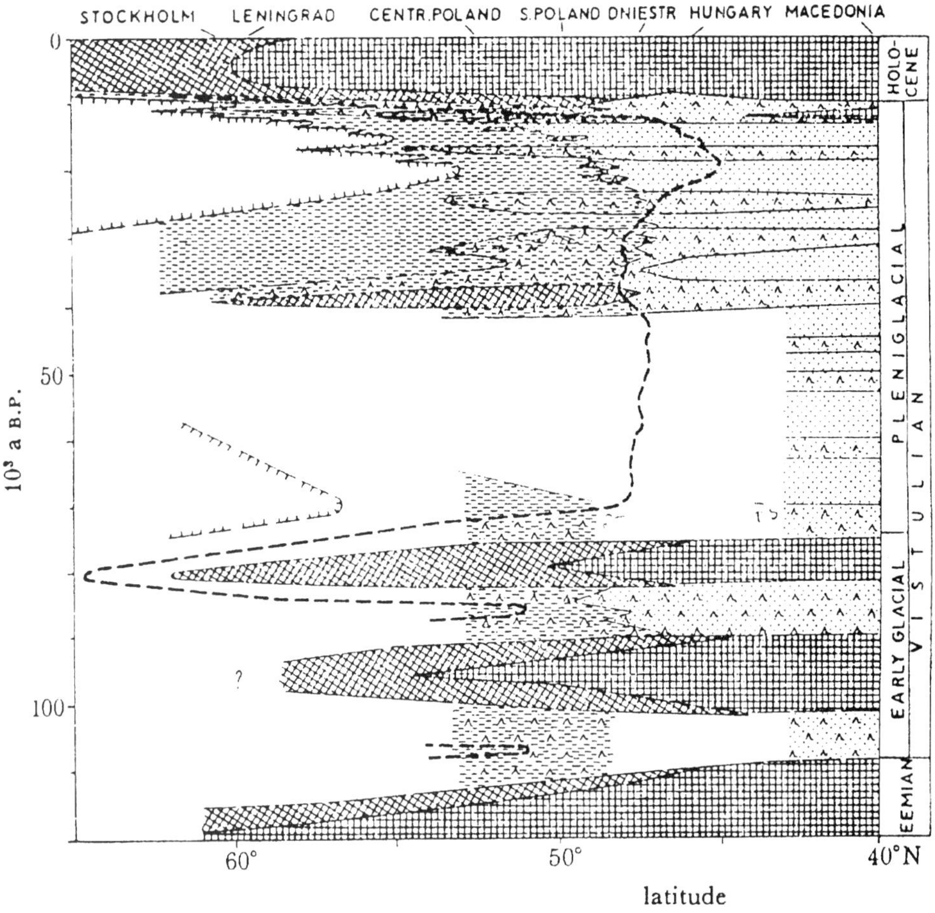

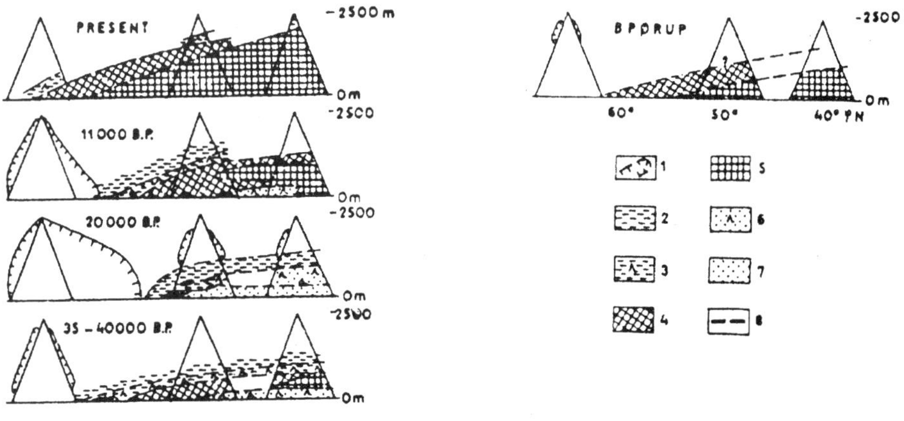

Figure 2 - *Changes in vegetation in North-South European transept.*
1 - ice-sheet, 2 - tundra and polar desert, 3 - forest-tundra, 4 - boreal forest, 5 - decideous forest, 6 - steppe-forest, 7 - steppe, 8 - permafrost limit. According to L.Starkel.

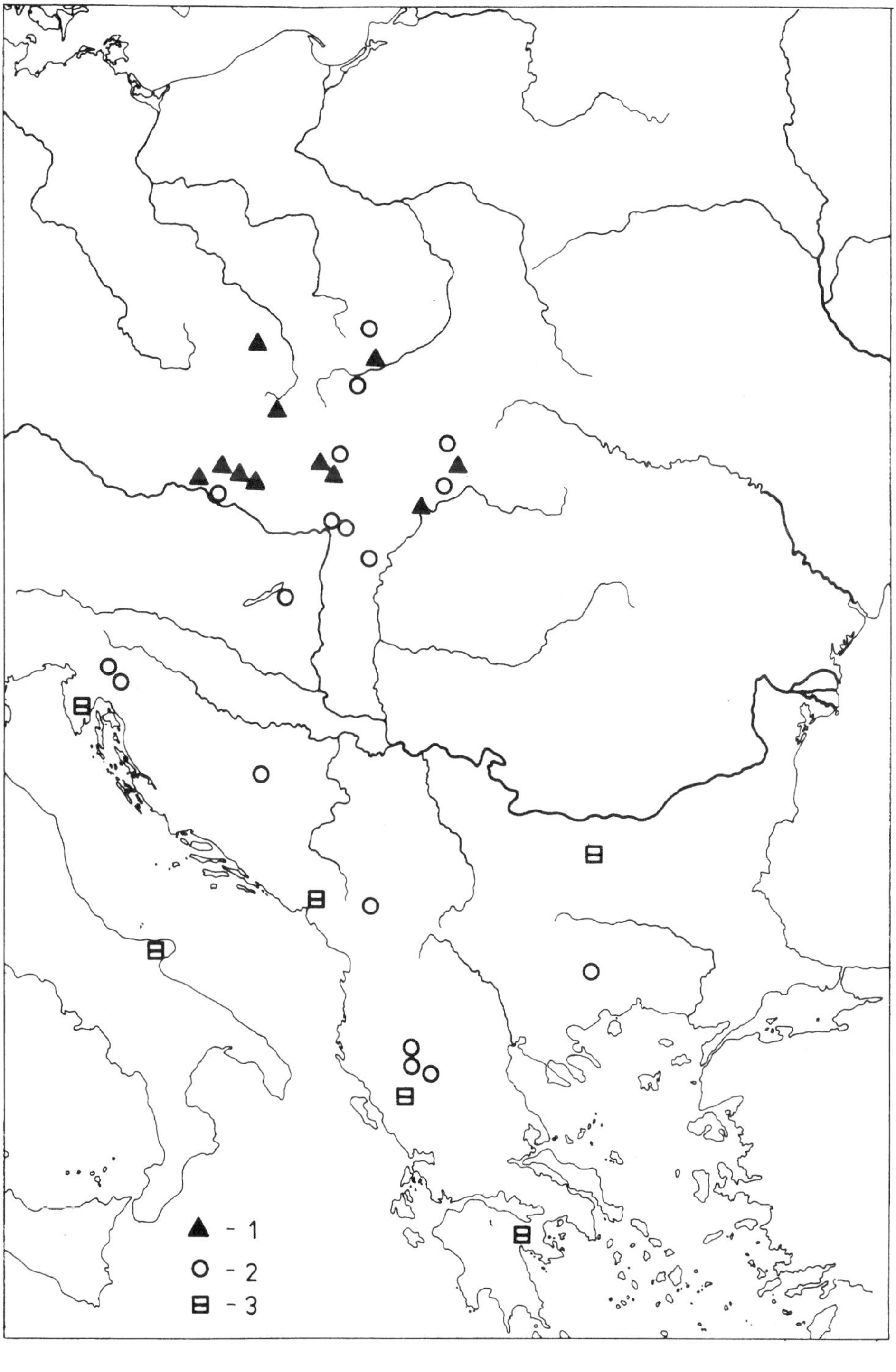

Figure 3 - Most important Gravettian and Epigravettian sites in Central Europe and the Balkans. 1 - Gravettian, 2 - Epigravettian, 3 - stratified Gravettian/Epigravettian sites.

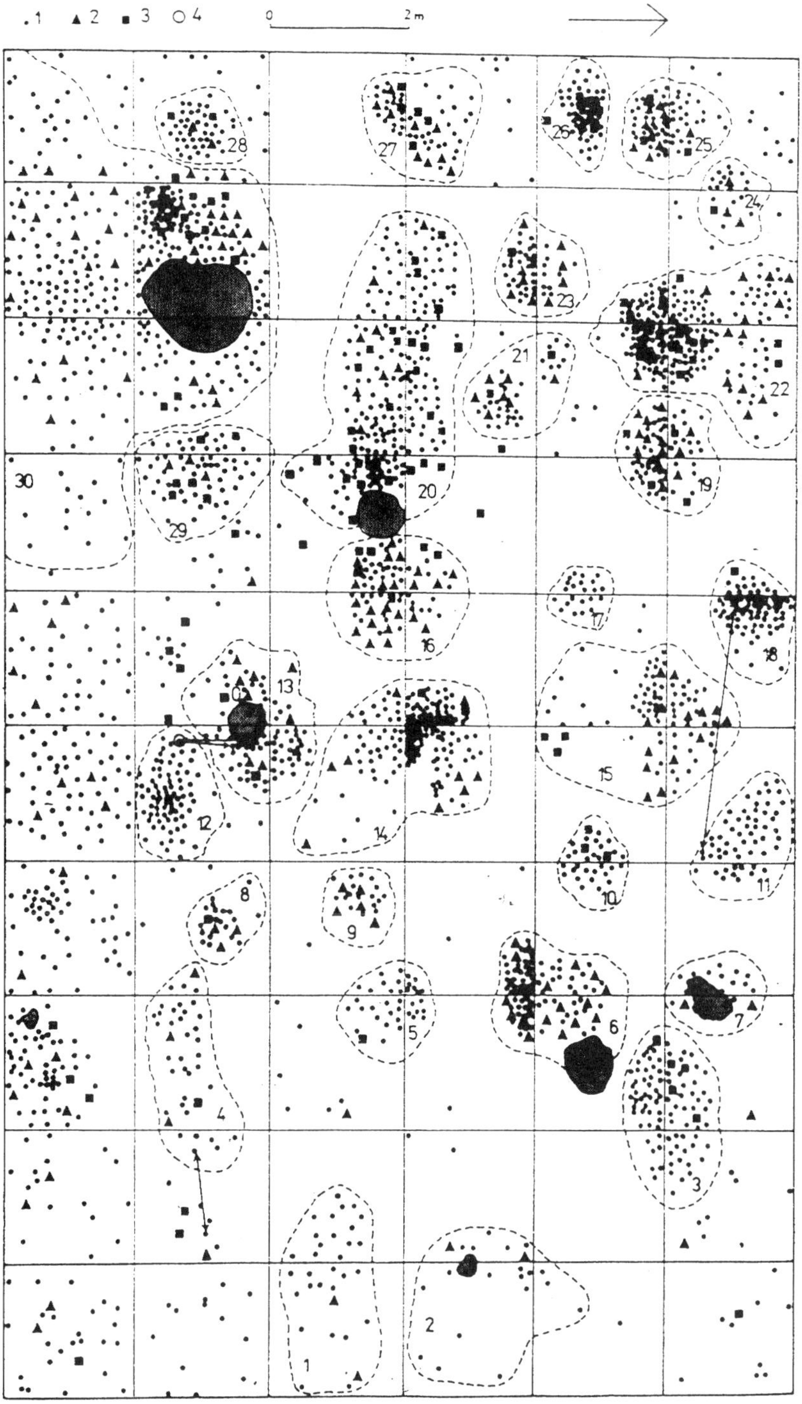

Figure 4 - Horizontal distribution of the ksemenitsas in one sector of the Epigravettian (upper) layer of the site at Kasov (Eastern Slovakia).
1 - flakes and blades, 2 - retouched tools, 3 - cores, 4 - refitted pieces. Hearths are shaded.

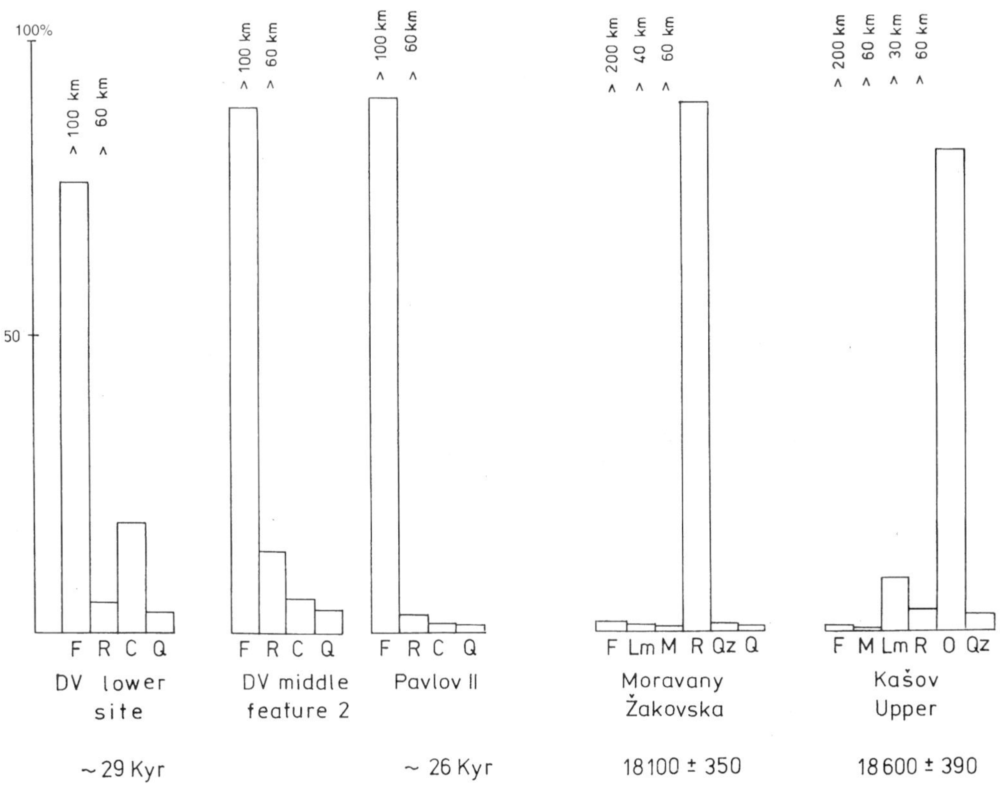

Figure 5 - *Raw material structure in selected Gravettian (DV - Dolní Věstonice, Pavlov) and Epigravettian (Moravany-Zakovska, Kasov, upper layer) sites .*
F - flints, R - radiolarites, C - west-moravian hornstone, Q - quartz, Lm - limnoquartzite, M - menilithic hornstone, Qz - quartzite, 0 - obsidian.

4. Base camps with a diverse spectrum of activities – evidenced by functional differentiation of tools – are replaced by short-term camps with a few tools, often with a limited range of functions.

5. Instead of large habitation structures with several hearths there are individual kshemenitsas and small hearths which point not only to a greater mobility of groups but constitute a proof of transformation of social organization wherein bands or extended families are replaced by nuclear families.

6. Subsistence strategies, too, undergo changes: specialized mammoth hunting – so important in the shouldered points horizon (Soffer, 1993) – is no longer practised. The Epigravettian people hunt a broader range of animals with an increasing emphasis on horse and reindeer (West, 1996).

7. The system of lithic raw materials procurement changed essentially. In the Gravettian a systematic supply of extra- and meso-local raw materials operated on the basis of specialized workshops at extraction points (Kozlowski, 1986). These workshops were regularly visted by groups representing peoples who inhabited regions frequently a considerable distance away from workshops. In the Epigravettian local and mesolocal raw materials gain importance. These raw materials are processed at the base camps. Extra-local raw materials are less common, unsystematically supplied as they were accidentally collected during hunting trips.

8. Social contacts following the meridional axis are predominant in the Gravettian, connected with systematic trips in search of raw materials, whereas during the Epigravettian contacts along a latitudinal

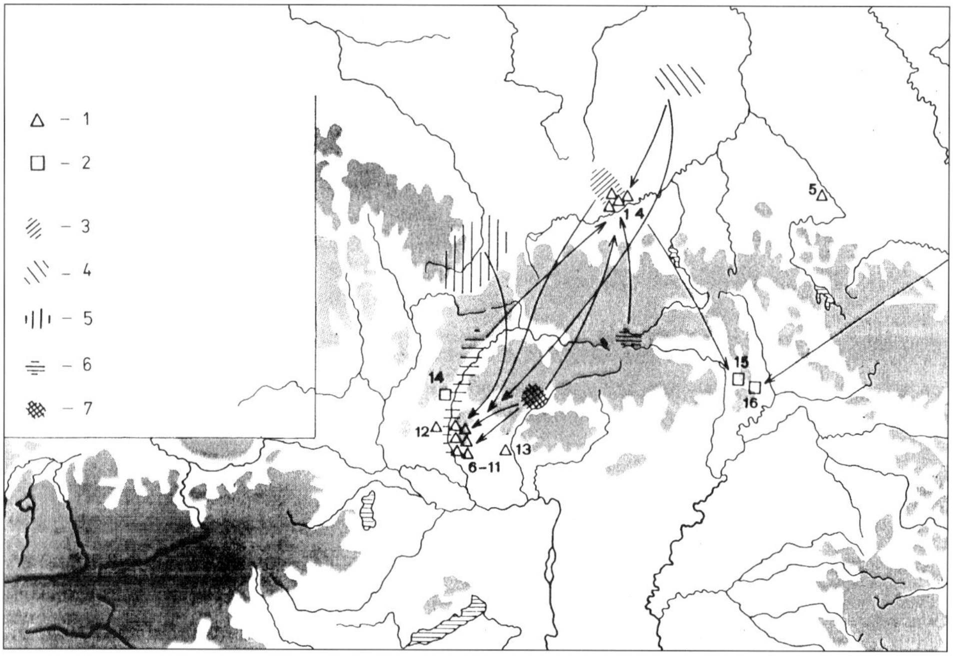

Figure 6 - Transcarpathian links reconstructed on the base of raw material circulation during the Late Gravettian shouldered point horizon (23-21 Kyr B.P.).
1 - Gravettian with shouldered points, 2 - other Late Gravettian groups, 3 - area of deposits of Jurassic flint, 4 - area of outcrops of the „chocolate" flint, 5 - area of outcrops of the erratic „northern" flint, 6 - area of outcrops of radiolarites, 7 - area of outcrops of limnoquartzites.

axis acquire more importance (Kozlowski, 1992). This is, possibly, related to seasonal migrations of, for example, reindeer from low-altitude pastures to mountain regions the Alps and in the Carpathians.

9. Generally, the Epigravettian groups display a greater mobility than did the Gravettian people. For the latter groups is characteristic the „logistic mobility" basing on the co-existence of large residential settlements with short-term specialized activity camps (e.g. stone processing workshops, hunting camps). The Epigravettian occupation employed a model closer to „residential mobility" consisting in frequent shift of base settlements (Binford, 1980). It can, thus, be seen, that the Gravettian socioeconomic strategies involved planning and internal division of work within larger groups. In the Epigravettian this system was replaced by more „opportunistic" strategies used by nuclear families.

The transformations in subsistence strategies, social relations and behaviour that we described are also reflected in the sphere of artistic expression. In place of the realistic, zoo- and anthropomorphic art characteristic for the Gravettian (the Pavlovian style), especially for the shouldered points horizon (the Kostenkian style), symbolic art emerges with geometric motifs and schematic anthropomorphic representations adapting natural forms, such as pebbles (Kozlowski, 1997).

The social, economic, and cultural changes in the northern part of central Europe do not seem to have been the effect of the population exchange. True, during the LGM migrations to the south and to the east can be seen manifested in the continuation of the shouldered points horizon both in the Adriatic basin and in the Russian Plain. But, there is little evidence of re-immigration to

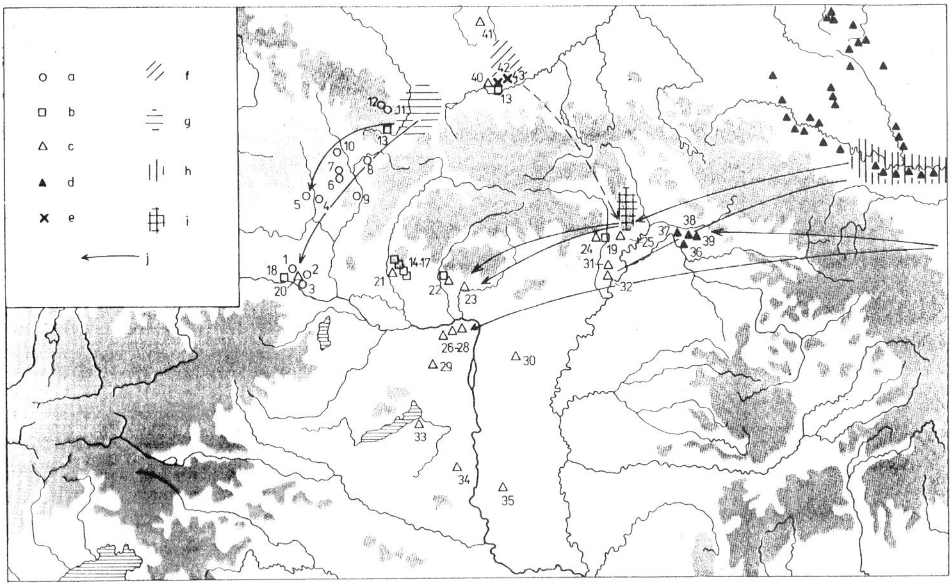

Figure 7 - *Transcarpathian links based on raw material circulation during the Epigravettian (19-15 Kyr B.P.).*
a - Late Aurignacian, b - Gravettian (shouldered point horizon), c - Epigravettian of the middle Danube Basin, d - Molodovian (with Lipa group), e - Epigravettian workshops, f - areas of deposits of Jurassic flint, g - areas of deposits of erratic „northern" flint, h - areas of deposits of Cretaceous flint from Volhynian Plateau and Dniester Basin, i - areas of deposits of carpathian obsidian, j - raw material circulation routes.

central Europe after the LGM. For this reason we are inclined to associate the emergence of the Epigravettian, especially in the Middle Danube Basin with the perseverance of the local substratum and its adaptation to the conditions of Pleniglacial II. It can, therefore, be assumed that the new subsistence strategies and new social structures were indispensable conditions for the survival of some population groups. The process of partial depopulation extended over a different time-span in the various parts of central Europe, but refugia in the middle Danube basin were the source of re-colonisation of Central Europe already in the period from 18 to 17 Kyr B.P.

Pre- and post-Pleniglacial evolution of south-eastern Europe

We have attempted elsewhere to justify a hypothesis that autonomous centers of backed blade industries existed in the Balkans and the Adriatic basin (Kozlowski & Otte, 1997). These centers, obviously, are hardly comparable with the central European Pavlovian in respect of the level of cultural evolution and socio-economic advancement. On the other hand, the north-eastern part of the Balkans maintained contacts with Gravettian centers in the middle Danube basin. This is particularly well evidenced by a complete Gravettian and Epigravettian sequence from the period between 29 to 14 Kyr B.P. excavated in the Temnata cave in northern Bulgaria (Ginter et al., 1992, 1994). At the very beginning of the LGM (about 19 Kyr B.P.) Danubian influences reinforced: thus at the transition of the Gravettian/Epigravettian the typical Gravettian shouldered points were widespread throughout the Adriatic-Balkan zone.

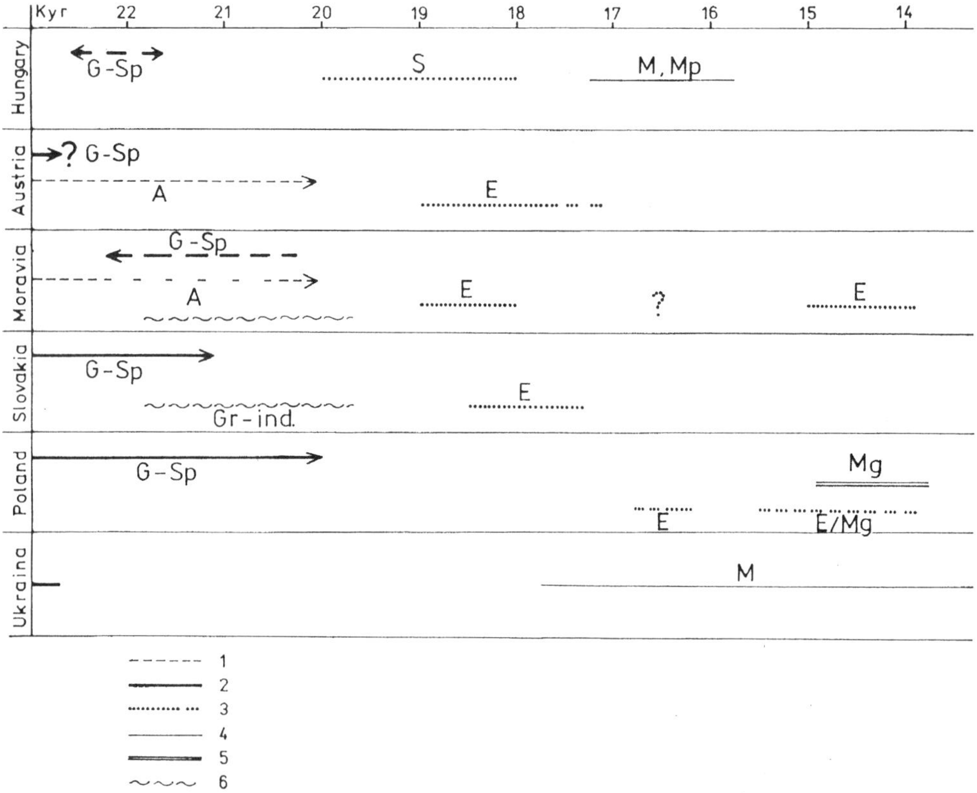

Figure 8 - *Evolution of the various taxanomic units before and after the LGM in Central Europe. 1 - Latest Aurignacian, 2 - Gravettian - shouldered point horizon, 3 - Danubian Epigravettian (E) and the Sagvarian (S), 4 - Molodvian (M, Mp), 5 - Magdalenian, 6 - indetermined Late Gravettian.*

When we begin to look for a contrast between the pre-Pleniglacial Gravettian and the post-Pleniglacial Epigravettian in this zone the conclusion is inevitable that no radical changes either cultural or socio-economic took place in between 20 to 18 Kyr B.P. that would coincide with the changes of climate. In the territories where pre-Pleniglacial Gravettian occupation continues to occur (Epirus – Bailey et al., 1983; northern Bulgaria – Ginter et al., 1994; Puglia, Veneto – Palma di Cesnola, 1993; Broglio, 1994) it was characterized by:

1. A lack of large base camps.
2. Continuation of occupation of the same sites from the period before 20 Kyr and also after 20 Kyr. B.P.
3. The use of large habitation structures with stone foundations, mostly built in caves.
4. Although it is difficult to evaluate how long a given camp continued to function in the conditions of cave sediments with low sedimentation rate, wherever such an estimation is possible – it indicates a number of short-term overlying sojourns in a cave.
5. A greater importance of hunting within a small radius around a cave, although these were animals inhabiting all types of biotopes. As a result the proportions between the various species are fairly equal, reflecting the mosaic of biotopes in the immediate surroundings of a camp.

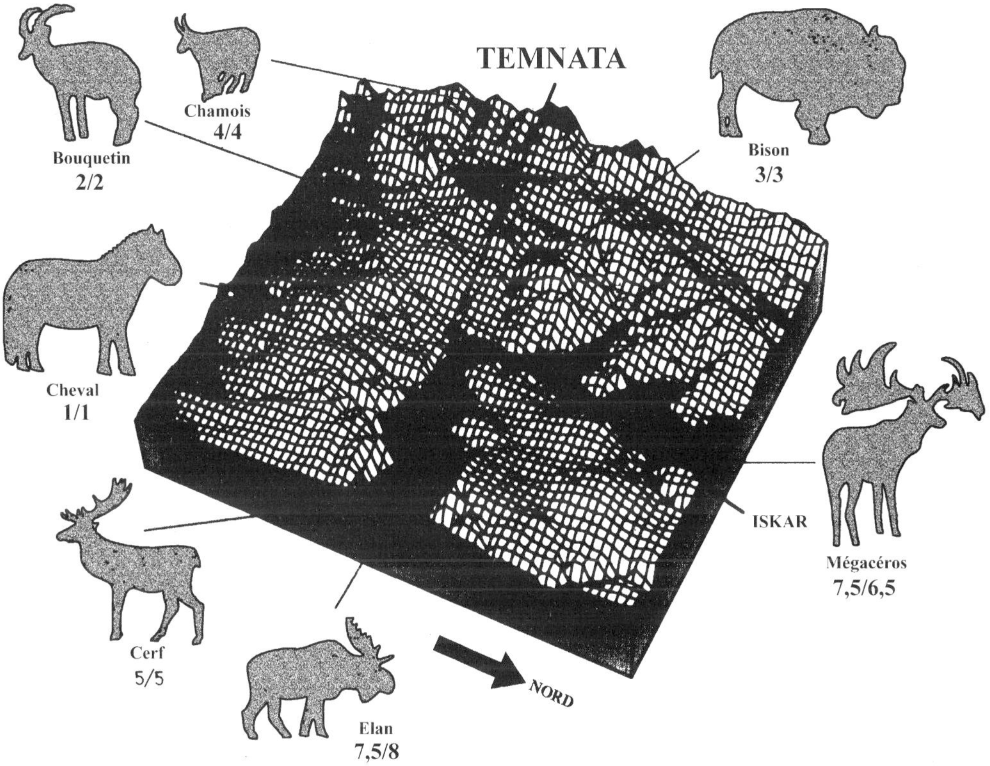

Figure 9 - *Example of the hunting strategies in the Gravettian and Epigravettian layers of the Temnata Cave (Northern Bulgaria).*
Arrows indicate areas of hunting on particular animals in the vicinity of the cave. Numbers indicate the order of frequency of particular species in the Gravettian (first number) and Epigravettian layers (second number) . According to F.Delpech and J.-L.Guadelli.

6. Considerable role of local raw materials, with a relatively small proportion of extralocal raw materials whose frequency changed in each occupation level depending on the direction of seasonal shifts.

7. Manifestations of artistic expression and personal adornements are poor.

The occupation in south-eastern Europe in the pre-Pleniglacial differs from the situation typical of central Europe from 30 to 20 Kyr B.P., and resembles more the type of occupation of northern central Europe after the LGM. Most importantly pre-Pleniglacial south-eastern Europe does not display features of planned „logistic" mobility that is so typical of the „Loess" Gravettian of central Europe.

Noticeable transformations can be seen in the Balkans and the Adriatic basin only after 17 Kyr B.P. when regional groups emerge. These groups adhered to a planned system of exploitation of the various zones of altitude from the littoral planes to high mountains. Such systems based on seasonal migrations of whole groups and complementary exploitation of the various latitudinal zones developed in Epirus (Bailey et al., 1983), Montenegro (Whallon, 1994), Istria, Abruzzo and other territories (Kozlowski, 1992a). The reconstruction of how these regional systems functioned is often hindered as the littoral sites belonging to these settlement systems had been submerged by post-glacial sea transgression and are beyond our reach. We can only conjecture

	Level	Camp type	Raw materials	Seasonality	Activities
GRAVETTIAN	IV	Scattered artefacts hunting camps	61,6; 5,3,4; 1,7,6; 12	?	projectile points
	Vb	small concentrations base camp	61,7; 4,5; 2,3,7 6,1; 8,10,13,14,17	?	projectile points wood, hide
	VI	base camp	62,9; 4,5,6,3; 2,7; 9a,10,11,12,13,14,15,17,20	end of winter spring summer	hide, wood, meat plant cutting and grinding projectile points
	VIIa	base camp (preserved only in fossil gulley)	69,8; 4,5,6,3; 2,7; 9a,16	end of winter spring end of summer	hide, meat, wood projectile points
	VIIb 21200±380	dwelling structure	74,1; 4,5,6,3; 2,7; 8,12a,14,15	end of winter early spring end of summer	hide, wood, meat projectile points
	VIII 28900+1100 29700±1600	two generalized „kshemenitsas" with flat hearths	72,0; 4,5,6; 2,7,3; 10,13,14,17,18,19	?	wood hide
	IXa 28900±1400	two specialized „kshemenitsas"	67,8; 4,5,6; 10,11,14,21	?	
	IXb	cilcular stone dwelling structure and pavement	80,0; 4,5; 6; 10,13,14,20,21	?	
	X	?	82,0; 4,5; 6; 13,14	?	

Figure 10 - *Changes in camp types, raw material procurement systems, seasonality and activities reconstructed on the base of use wear analysis in particular Gravettian layers of the Temnata Cave (Bulgaria).*
Shadowed zone in raw material frequency diagrames express the ratio of local raw materials. The numbers on the left side of the same column are mesolocal raw material varieties; the numbers below circle - extralocal raw materials.

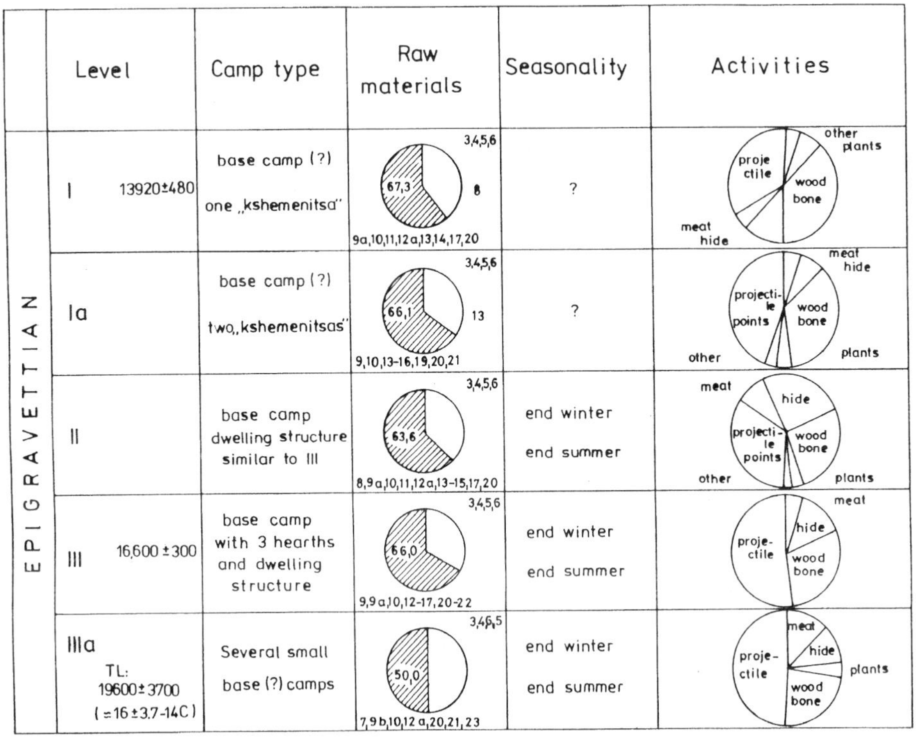

Figure 11 - Changes in the camp types, raw material procurement systems, seasonality and activities based on use-wear analysis in particular Epigravettian layers of the Temnata Cave. Please note that no dramatic changes occur between Gravettian (30-20 Kyr B.P.) and Epigravettian (20-14 Kyr B.P.)

that these systems planned their subsistence strategies and functioned on the basis of more complex system of social organization. It should be added that in the Middle and Late Epigravettian of the Adriatic basin (notably in Puglia) more and more numerous items of mainly realistic zoomorphic art appear.

Conclusions

The evidence we have presented above confirms that the evolution of subsistence strategies and concomitant systems of social organization was not linear. Especially, the hypothesis about the general development from „opportunistic" foraging strategies to planned, logistic strategies in the Upper Palaeolithic has not been confirmed.

The Gravettian/Epigravettian transition shows that in the north more complex forms of social organization, planned specialized hunting strategies evolved into more „opportunistic" strategies in which a nuclear family was the basic social unit. At the same time, there were no precisely demarcated hunting territories in the northern zone, especially in the later period. Shifts of core residential areas in a variety of directions can be seen.

In the southern zone the evolution followed a different course: from more „opportunistic" subsistence strategies and possibly simple social organization to strategies based on planning and complementary exploitation of different landscape zones – but invariably within precisely demarcated territories.

References

BAILEY, G., CARTER, P., GAMBLE, C. & HIGGS, H. (1983): Epirus revisited: seasonality and inter-site variation in the Upper Palaeolithic of north-west Greece. In: G. Bailey (ed.), *Hunter-gatherer economy in prehistory. A European perspective,* 64-78. Cambridge.

BANESZ, L., HROMADA, J., DESBROSSE, R., MARGERAND, I., KOZLOWSKI, J.K., SOBCZYK, K. & PAWLIKOWSKI, M. (1992): Le site de plein-air du Paléolithique supérieur de Kašov en Slovaquie orientale. *Slovenska Archeologia* 40,5-28.

BINFORD, L. (1980): Willow smoke and dog's tails: Hunter gatherer settlement systems and archaeological site formation. *American Antiquity* 45, 4-20.

BOTTEMA, S. (1974): *Late Quaternary vegetation history of northwestern Greece.* Groningen.

BROGLIO, A. (1994): Il paleolitico superiore del Friuli-Venezia Giulia. Considerazioni sur popolamento umano del territorio tra Penisola Italiana e Penisola Balcanica. In: *Atti della XXIX Riunione Scientifica, 37-56.*. Firenze.

CARCIUMARU, M. (1989): Contexte stratigraphique, paléoclimatique et geochronologique des civilisations du Paléolithique supérieur en Roumanie. *L'Anthropologie* 93, 99-122

DOBOSI, V. (1986): Historical sketch of the Late Palaeolithic in Hungary. In: M. Dey & R. FOLEY (eds.), *The Pleistocene Perspective,* vol. I. 2-16. London.

GAMBLE, C. & SOFFER, O. (eds.) (1990): *The World at 18 000 B.P.* London.

GINTER, B., KOZLOWSKI, J.K. & LAVILLE, H. (eds.) (1992, 1994): *Temnata Cave – Excavations in Karlukovo Karst Area, Bulgaria.* Vol. I-II. Kraków.

GINTER, B., KOZLOWSKI, J.K., LAVILLE, H. & SIRAKOV, N. (1994):. Les occupations et les activités humaines dans la séquence gravettienne et épigravettienne. In: J.K. Kozlowski, B. Ginter & H. Laville (eds.),*Temnata Cave – Excavations in Karlukovo Karst Area. Bulgaria,* vol. II. 327-335. Kraków.

GUTHRIE, R. (1990): *Frozen fauna of the mammoth steppe.* Chicago, London.

HROMADA, J. & KOZLOWSKI, J.K. (eds.) (1995): *Complex of Upper Palaeolithic sites near Moravany, Western Slovakia.* Vol. I: *Moravany–Zakovska (Excavations 1991–1992).* Kraków.

KLÍMA, B. (1983): *Dolní Věstonice tábořiště lovců mamůtu.* Praha.

KOZLOWSKI, J.K. (1986.): Changes in raw material economy of the Gravettian technocomplex in northern central Europe. In: *International Conference on Prehistoric Flint Mining and Lithic Raw Material Identification in the Carpathian Basin. Budapest-Sümeg, 20-22 May 1986,* 65-80.

KOZLOWSKI, J.K. (1992): Le paléolithique des Carpates occidentales. *Preistoria Alpina* 28, 113-126.

KOZLOWSKI, J.K. (1992a): Le rythme climatique du Pléistocène supérieur et la présence humaine dans les montagnes. *Preistoria Alpina* 28, 37-47.

KOZLOWSKI, J.K. (1996): The Danubian Gravettian as seen from the northern perspective. In:J. Svoboda (ed.), *Palaeolithic of the Middle Danube Region,* 11-24. Brno.

KOZLOWSKI, J.K. Le deuxième Pléniglaciaire et l'évolution de l'art paléolithique. *L'Anthropologie* 101, 24-35.

KOZLOWSKI, J.K. & OTTE, M. (1997): Transformations culturelles en Méditerranée septentrionale en fin d'époque glaciaire en relations avec les modifications paléoclimatiques. *Colloque Banolas.* (in print).

MUSSI, M. & ROEBROEKS, W. (1996): The big mosaic. *Current Anthropology* 37.

PALMA di CESNOLA, A. (1993): *Il paleolitico superiore in Italia. Introduzione allo studio.* Firenze.

SOFFER, O. (1993): Upper Palaeolithic adaptations in Central and Eastern Europe and man-mammoth interactions. In: O. Soffer & N.D. Praslov (eds.), *From Kostenki to Clovis,* 31-50. New York.

WEST, D. (1996): Horse hunting, processing, and transport in the Middle Danube. In: J. Svoboda (ed.), *Palaeolithic in the Middle Danube Region,* 209-246. Brno.

WHALLON, R. (1994): Lithic industries at Badanj and their significance for the identification of cultural „territories" around the Adriatic and Ionian Seas at the end of the Pleistocene. In: *Palaeolithic of Greece and adjacent areas.* Ioannina. (Preprint).

Dursunlu - A New Lower Pleistocene Faunal and Artifact-bearing Locality in Southern Anatolia

Erksin Güleç, F. Clark Howell & Tim D. White
with assistance of Mustafa Karabiyikodlu, Gerçek Saraç, Engin Ünay,
Antoine Louchart, Cécile Mourer-Chauviré, Garniss H. Curtis & Gary Scott

Abstract

The Dursunlu locality, a disused quarry opened in the course of lignite exploitation over a decade ago, is situated in northern Taurus foothills in the eastern reaches of the Aksehir basin and immediately west of recent Lake Ilgin. It has afforded a diverse vertebrate fauna, particularly of birds and small and large mammals, indicative of the Biharian faunal stage. Some vertebrate fossils were obtained in the course of commercial exploitation, and more recently in intensive survey of the quarry margins and adjacent tip heaps. Cores, drilled to a depth of - 50 meters, reveal a substantial suballuvial succession of limnic sediments and afford a paleomagnetic record of the (late) Matuyama (R) chron. Clastics of foreign origin, including some bearing traces of human workmanship and testifying to human transport, afford the first and by far the oldest evidence thus far of hominid occupation in Anatolia from a well-defined stratigraphic context.

The now-abandoned, partially flooded quarry of Dursunlu is situated nearly directly south and 5.5 km distant from the eponymous village. The village is 17 km north-northwest of Ilgin (1030 m), and Ilgin is 54 km northwest of Konya town in southern Anatolia. In the course of drilling for a well, over a decade ago, exploitable lignite deposits were encountered in the drill cores. Subsequent commercial exploitation opened a large quarry, exposed part of a subsurface Pleistocene sedimentary infill, and afforded some vertebrate fossils. No systematic collections were made during such exploitation, and the recovered sample was small and limited in diversity. In the course of paleoanthropological and paleontological survey activities, in association with the University of Ankara and M.T.A. scientists, the locality was visited in 1993 and in 1994 (and more recently) and a program of intensive and extensive surface collection was undertaken from tip heaps and overburden dumps adjacent to the quarry. These activities greatly enlarged the fossil sample, including its biotic diversity, and also led to the recovery, *in situ* in sediment blocks, of extraneous clastics, including artifacts.

This area of southern Anatolia constitutes the Lycaonian plateau, in the northern foreland reaches of the western Taurus mountains. Earlier studies of this area, and of the Pisidic lakes district, have been those largely of geomorphology by Wenzel (1932, 1935) and also by Chaput (1936) in the course of his various geological itineraries. The area is bordered to the south by the Sultan Dag mountain, of uplifted Paleozoic-Mesozoic rocks, sharply delineated on its northern flank by the nearly linear Sultan Dag fault, an oblique-slip northwest/southeast trending structure active in geologically recent (Miocene-Pliocene) time. The immediate area lying to the north is the greater Aksehir basin, considered a strike/slip wedge graben, down thrown northwards, in which successively occur (eastwards) the Eber, Aksehir and Çavuscu (or Ilgin) lakes (*golu*) (Lahn, 1948). Paleo-Mesozoic highlands border the graben to the north and northeast, as do also Neogene formations to the northeast. In the course of the Quaternary there was extensive development of substantial lakes within this region (Figure 1) (Lüttig & Steffens, 1976).

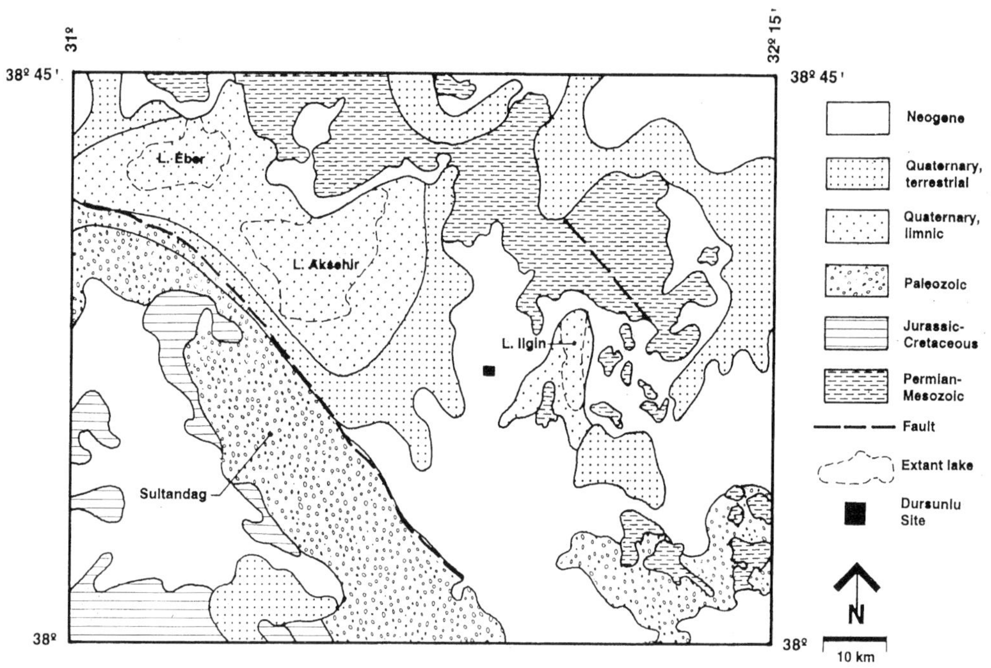

Figure 1 - *Geological aspects of the southern Lycaonian plateau and eastern Pisidic lakes area. After Pamir & Erentoz (1975), Ankara Sheet of the Geological Map of Turkey, and Map 7 (Pleistocene) in Lüttig & Steffens (1976).*

The Ilgin freshwater lake, unlike the others, trends North-South, north of Ilgin town, having an extent of some 50 km², but exhibiting much variation in size, even to extreme desiccation in the mid-19th century. It is fault-bounded, its western boundary having several thermal sources and experiencing substantial seismic activity only some 75-65 years ago. High hills, 1300 - 1700 m, surround the sub-basin to the north and east. Dursunlu town and the eponymous quarry lie directly west of the northern reaches of the Ilgin lake.

The geological setting and upper Cenozoic stratigraphy and sedimentation of these outer intramontane basins of the western Taurus arc are most recently and extensively discussed by Bering (1971). Bering distinguished within the provincial Neogene three successive sedimentary series (Schichten): (a) lower terrestrial-fluviatile series (up to 50-70 m., limited discontinuous distribution), (b) limnic series (extensive distribution, lacustrine limestones, marls, claystones, up to 120-150 m.; often with invertebrates, only rare mammals), and, (c) upper terrestrial-fluviatile series (extensive, but discontinuous and marginal distribution, up to 150-100 m., thinning distally, and having known subsurficial occurrence(s) in boring(s)). In a subsequent formational reclassification of such occurrences (as in Ünay et al, 1997) the following formational designations (successively upwards) replace the previous informal unit appellations: (a) *Asaga Cigil* Fm. (in part, demonstrably early Miocene); (b) bipartite *Argithana* Fm. (lacustrine limestones, and lacustrine claystones); and (c) *Doganhisar* Fm. (alluvial fan-conglomerates, sandstones, mudstones).

In the immediate environs of the Dursunlu area all these formations are known to be represented (Figure 2). Formation (a) is known only in the far north, north of the Karacali Meso-Paleozoic

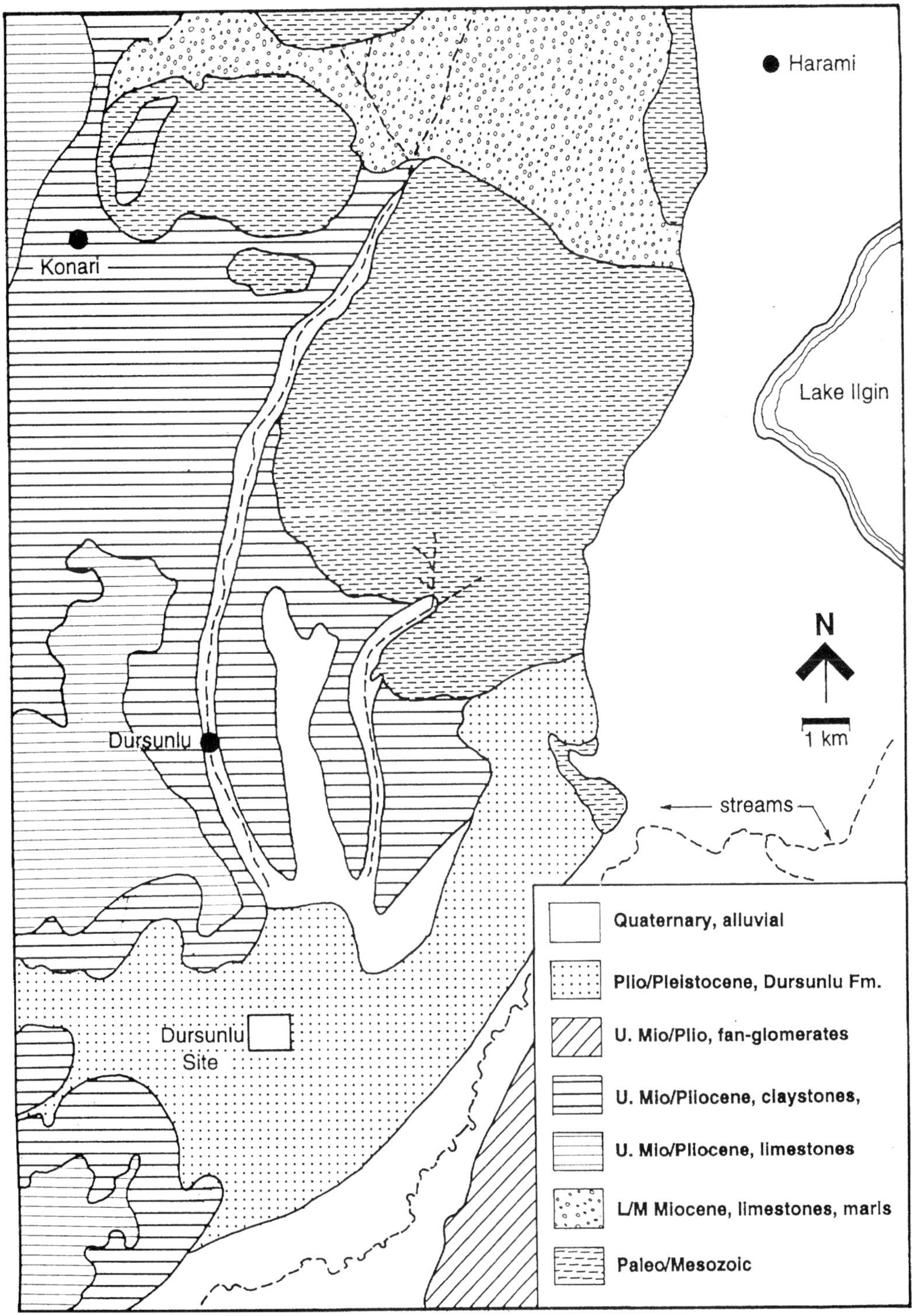

Figure . - Simplified geological map of the environs of the Dursunlu locality.(After Ünay et al., 1997).

upland block. Formation (b) has a more western, marginal distribution of its lower (b1) part (member) and more central distribution of its upper (b2)part. Formation (c) occurs only toward the southern reaches of the sub-basin.

The quarry-exposed deposits are now recognized as formationally distinct, as the *Dursunlu* Formation, comprising (lower) Pleistocene lacustrine limestones, marls and clays with lignite beds. The quarry, now flooded, exposes only some 10 meters of this formation. However, two cores drilled on our behalf by the M.T.A. and Water Resources Department, logged 45 m (Core 2) and 50 m (Core 1) of sedimentary infill. Core 1 is immediately adjacent to the quarry, and Core 2 some distance away. The latter reveals sedimentary infill to some 38 meters underlain by limestone, presumptively of Neogene age and comprising the Argithana ('limnic' series) formation.

Core 1 reveals a quite detailed stratigraphy and has as well afforded a preliminary paleomagnetic record (Figure 3). The basal (+ 5 m) portion comprises fine reddish mud, presumably alluvial and largely or wholly of terrestrial origin. (It might be speculated that it constitutes a (upper) part of the 'upper terrestrial-fluviatile' series of Bering.) The overlying sediments (nearly 45 m) constitute a shallowing upwards lacustrine sequence, incorporating intermittent episodes of fluctuating lake level(s), shallowing and even subaerial exposure(s). The lower part of the sequence (-44 to -23.6, or Z-zone) is greenish fine clay, having laterally continuous tabular sub-beds, infrequently laminated, with surficial root traces and only local bioturbation. The sedimentary context was evidently one of stagnant water and deposition in suspension in a shallow ponded or larger lacustrine setting. This depositional/sedimentary environment recurs subsequently in five instances (zones Y to U) of greatly subordinate thicknesses, 1.4 m (Y), 2.90 m (X), 1.10 m (W), 3.5 m (V), and, finally, 1.5 m (U). The extent to which these manifestations were paleoclimatically induced or, equally likely, were tectonically effected and attendant on subsidence is still to be ascertained.

In the upper (above -15 m) reaches of the succession there are three successive instances of another variant, sedimentary regime. These units constitute clayey silts, sandy silty clay or fine sands, variably bedded, occasionally weakly laminated and bearing dense root marks, and evidence of substantial bioturbation. There may also be associated peat accumulations or weak lignite seams or lenses. Molluscan remains are common. The depositional environment was evidently both shallower and more marginal, suggestive of a marshy and near-littoral setting. These occurrences vary (upwards) from 2.1 m. to 1.3 m and 1.4 m, respectively.

Above the middle reaches of the core there is a bipartite occurrence, between about -19.3 and -17.5 m, the earlier five times (1.2 m vs. 0.25 m) greater than the later, of near-shore or littoral pebbly sands. These are the only such sedimentary units within this succession. However, they are paralleled in Core 2, between -33.0 and -37.8 m, of a similar occurrence in which medium-coarse sands (1.35 m) are succeeded by clayey sands (0.75 m) overlain by fine (sandy) gravels (1.4 m), all overlying the basal limestone and clays (below -38. m) of the lowest reaches of the core. Although analogous in depositional origin, and clearly reflective of emergence and external inflow, there is no assurance of synchroneity, unless the Core 2 succession is much compressed, and/or lacking continuity in deposition.

Successive lignite members occur in both Core 1 and Core 2. In Core 1 a lower lignite (1) occurs between -23.6 to -21.25 m. (= 2.35 m thick) and a thicker bed (2) succeeds it at -12.5 to -9.45 m (= 3.05 m. thick). In Core 2 the respective depths are -27.55 to -27.4 m (= 15 cm) and -22.0 to -21.55 m (= 45 cm) (for 1a, 1b), with a substantial succession of clays and a pebbly sand lens intervening between the beds, and the upper (2), massive accumulation at -19.75 to -16.6 m (= + 3 m), and obviously matching the upper occurrence in Core 1. The lignites are evidently authoctonous accumulations of plant matter in a water-logged situation, presumably a lakeside marginal swamp. In each instance there is sharp inferior and superior contacts with the antecedent

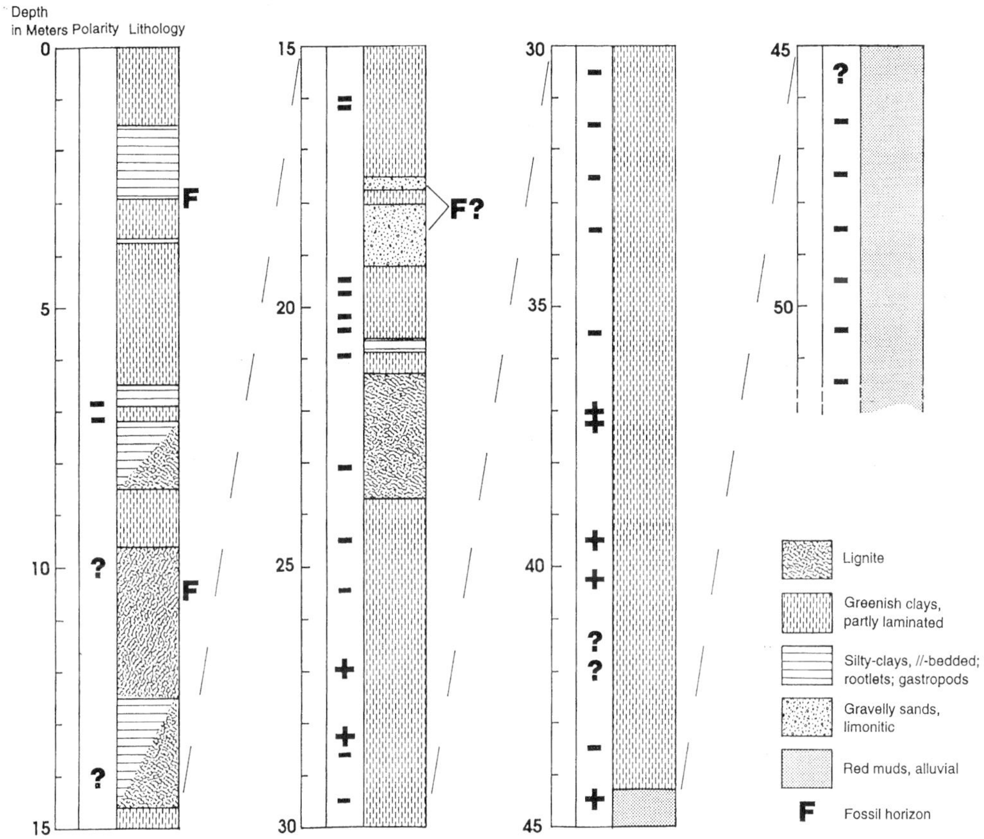

Figure 3 - Simplified log of Dursunlu Core 1, with lithological units and polarity measurements.

and succeedent beds. These units are massive, blocky and comprise wholly organic matter, and are of low calorie, high ash content. Clay partings or lenses are present, but few and thin, indicative of slight intermittent flooding. These occurrences have proved to be rich in pollens, both arboreal and herbaceous (terrestrial and aquatic).

In Core 2 vertebrate bone was sampled within the upper lignite (2) occurrence. From the evidence afforded by the condition and coloration of surface-collected fossils, and from an abortive excavation (1994) with power equipment adjacent to the present quarry, in proximity to Core 1, occurrence of vertebrate fossils in at least two horizons is confirmed. A lower, and principal occurrence is within the upper lignite member, i.e., at ~ -10 to -12 m. Vertebrates are also encountered at the basal part of the massive, bedded silty-clays, at ~ -4 m. These sediments vary from dark gray to yellowish gray into more yellowish. A group of fossils (see below) exhibit a yellowish to rusty brownish coloration, and might thus equate with this horizon. There is, however, also the possibility of derivation from the yellowish, ochreous sands in the lower reaches of the succession. Hence, the precise provenience remains unresolved.

The paleomagnetics of portions of Core 1 have been assessed, after thermal demagnetization, at the Berkeley Geochronology Center. This incomplete sampling and measurement restricts any preliminary interpretation to the level of the chron. The upper nearly 7 meters are un-measured.

At either side of -7 m. are two records of R, representing quarry wall samples, indicating that the Matuyama (R) chron is represented. Thus, whether the B/M boundary is recorded higher in the core is still an open question. Several determinations downward to -14 m. are indeterminate. At -15 m, -16 m and repeatedly (5 samples) between -19 and -21m, and again near -23 m and at -24.5 m and -25.5 m, R determinations are documented. Thus, by extrapolation, we have at least 18 m of fine-grained sediments exhibiting R-inclination.

At - 27 and -28.25 m are two records of N-polarity. The magnetization is weak, and thus it is difficult to ascertain whether it is a question of residual present normal polarity (overprinting) or a past (N) subchron. Downwards, between -28 and -35.5 m are 7 records of apparent R- inclination, spaced approximately a meter apart. Further inferiorly, from -37 to -40.25 m, are two pairs of apparent N-inclination - at around -37 m and either side of -40 m. Another, individual record of possible N-inclination occurs at -44.5 m. (There are two indeterminate assessments at -41.5 and -42.0 m). Between -46.5 and -51.5 m only R-inclination is recorded by six determinations.

It is tempting to view this dominantly R paleomagnetic record as manifesting two probably N-polarity subchrons, the upper representative of the Jaramillo (N) and the lower the Olduvai (N). These have current ages of 1.07 - 0.99 ma. and 1.95 - 1.77 ma., respectively. These purported N-events occur within a protracted succession of fine, partially laminated lacustrine clays, deposited in open water conditions, although not necessarily of very great depth. The approximate distance between events is 8.5 m (8500 cm), and the minimum age span ca. 700 ky. Thus, this particular interval suggests a sedimentation rate of ca. 12 cm. per 1000 years (or 1.2 per 100 years, and 0.12 mm/year, assuming continuous sedimentation). This is a very slow rate of sedimentation, comparable to varve accumulation in some periglacial lakes, and would perhaps also reinforce an interpretation of enhanced water depth. The single N-inclination at -44.5 m is in position potentially favorable to interpretation as the Réunion (N) subchron, currrently assessed as 2.15 - 2.14 ma.

These thoughts on the potential age span of the sedimentary infill are admittedly speculative at this juncture. Further paleomagnetic assessment is evidently worthwhile, and requisite to any finer temporal resolution. The details of sedimentary processes still require elucidation, and the possibility of finer microstratigraphic delineation of the more tabular, laminate limnic deposits merits elucidation. There is every expectation that a useful palynological record may be obtained from the core(s) as a whole, and not only from the lignite members. The invertebrate, and particularly ostracod fauna also calls for in-depth study, in respect to its limnological and paleoecological implications.

The Dursunlu vertebrate fauna is quite diverse, although still quite imperfectly known. Lower vertebrates are uncommon, but include Pisces, Amphibia and Reptilia (Chelonia), all still undetermined. At this juncture the Aves and Mammalia classes are the most informative.

A quite sizeable sample of birds are now known from the Dursunlu locality. These have been studied by Louchart (1997) in association with Mourer-Chauviré. In all there are 13 families represented, only 2 of which are passerines, each with a single species. In all there are 29 genera and 41 species, exclusive of the passerines. The bulk of the taxa, over 80 %, represent extant species, whereas in 7 instances there is distinction at the subspecific or even specific level. Overall there is a predominance of taxa of aquatic habitat(s), indicative of lacustrine, ponded and marshy environments. Among these families there are the following species (**) representations:

Podicipedidae (grebes)	(4)	Anatidae (geese, ducks)	(3, 17)
Phalicrocoracidae (cormorants)	(1)	Rallidae (rails)	(5)
Ardeidae (herons, egrets, etc.)	(5)	Scolopacidae (waders)	(3)
Threskiornithidae (ibis, spoonbills)	(1)	Laridae (gulls)	(2)

DURSUNLU (TURKEY)
MICROMAMMALS

Insectivora
- Insectivora ***gen. et sp. indet.***

Lagomorpha
- ***Oryctolagus cuniculus****
- Leporidae ***gen. et sp. indet.***
- Ochotonidae ***indet.***
- ***Lepus sp.***

Rodentia
- Castoridae
  - ***Trogontherium cuvieri***
  - ***Castor fiber***
- Cricetidae
  - Promethiomyinae
    - ***Elliobus anatolicus****
- Cricetinae
  - ***Cricetulus migratorius****
  - ***Mesocricetus auratus*** *
- Arvicolinae
  - ***Microtus (Allophaiomys) nutiensis***
  - ***Mimomys savini***
  - ***Lagurus arankae n. ssp.*** *

Muridae
- ***Apodemus sp.***
- ***Micromys n. sp.***

Spalacidae
- ***Spalax leucodon****

Sciuridae
- ***Spermophilus sp.***

Dipodidae
- ***Allactaga euphratica*** *
- Dipodidae, sp. indet. -A *

*** open/steppic/xeric habitat(s)**

Table 1 - Micromammal species recovered from the Dursunlu Formation.

In addition, there are represented Accipitridae (eagles, falcons, etc.) (1); Phasianidae (game birds) (1); and Otididae (bustards) (1-2). Among these and others are terrestrial species indicative of open, steppic circumstances. There is some, slight evidence, perhaps indicative of cooler conditions.

A micromammal fauna of some diversity has been obtained by sediment washing. It was studied initially by A. van der Meulen (Utrecht) and subsequently by Engin Ünay. It includes a few insectivores, lagomorphs and more abundant rodents (Table 1). The six families of rodents include Castoridae (extinct *Trogontherium, Castor*), two species of Muridae (*Apodemus* and *Micromys* n. sp.), a single spalacid (*S. leucodon*), a single sciurid (*Spermophilus*), and two dipodids (*Allactaga*, and another indeterminate species). Cricetids are the most diverse, including a promethiomyine (*Elliobus*), cricetine hamsters (*Cricetulus*, and *Mesocricetus*), and at least three arvicolines. These voles represent the rooted *Mimomys savini, Microtus (Allophaiomys) nutiensis* and the lagurine,

ARVICOLINAE
WESTERN ASIA PLIO-PLEISTOCENE

						AEGEAN	ISLANDS
• = extinct	Ubeidiya	Eskisehir	Dursunlu	Emirkaya-2	Yarimburgaz	Kalymnos	Chios: Latomi-1
Arvicolini							
Allophaiomys			*nutiensis* •			*sp. 1, 2* •	
Mimomys		*sp. 1, 2* •	*savini* •	*savini* •			
Arvicola				*cantiana* •	*terrestris*		*praeceptor*
Clethrionomyini							
Clethrionomys				*glareolus*	*glareolus*		
Microtini							
Microtus							
(Terricola)				*sp. indet.*	*sp.*		
(Chionomys)				*nivalis*			*nivalis*
(Tibericola)	*jordanica* •						
(Stenocranius)				*arvalis* *oeconomus*	*cf. rossiae–* *meridionalis* *guentheri*		*arvalis* *guentheri*
Pitimyini							
Pitymys							
Lagurini							
Lagurodon	*arankae* • *praepannonicus* •		*arankae* •				
Lagurus					*transiens*		*lagurus*
Kalymnomys						*major* •	

Table 2 - Arvicolines from earlier Pleistocene localities in western Asia.

Lagurus arankae n. ssp. *Arvicola* is lacking, and this is taken to be biostratigraphically significant rather than the consequence of sampling error.

Among micromammals the rodents are particularly informative, both biostratigraphically and biogeographically. Arvicolines are generally employed in efforts at biozonation. The known arvicoline taxa from seven faunal localities, four from Turkey, two from near-shore Aegean islands, and one from the southern Levant ('Ubeidiya) are set out in Table 2.

The oldest locality is clearly that of Yukari-Sögütönü (Eskisehir), on macrofauna grounds, attributable to MN17, and from which two, unfortunately as yet unidentified *Mimomys* species are recorded (also, the murid *Orientalomys galacticus* has been purportedly recognized in this assemblage). 'Ubeidiya, with a very diverse macro- and microfauna, is commonly attributed now to the lower Biharian (or Tamanian of eastern Europe), and more particularly to the Betfia phase; it yields the distinctive microtine, *M. (Tibericola) jordanica*, and two extinct lagurines, *Lagurodon arankae* and *L. praepannonicus*. Similarly, the assemblage largely of microvertebrates, from Kalymnos Island, is considered of this same biostratigraphic age, and contains two (unidentified) arvicoline species of *Allophaiomys*, and the distinctive lagurine, *Kalymnomys major* n.g. et. n. sp.

Emirkaya-2, a fissure-filling in the northern Taurus foothills (near Seydisehir, south-central Anatolia), affords some 36 mammal species, including abundant *U. deningeri* among 5 carnivores, and a varied micromammal fauna, including rodent taxa. It has been originally attributed to the upper (= late) Biharian (Sen et al, 1991) but is perhaps best referred to the subsequent Toringian (Montuire et al, 1994). Among 7 arvicolines it has some diversity of *Microtus* (3) species, and two arvicolini, *Arvicola cantiana* and *Mimomys savini*. No lagurines are recorded. Thus, on the basis of the arvicolines of Dursunlu - which include *L. arankae*, and the arvicolines *Allophaiomys nutiensis* and *Mimomys savini* (but not *Arvicola*), and (seemingly) lacks microtines - this locality would antedate Emirkaya-2. It thus clearly falls within the Biharian mammal age. The rich and diverse fauna of Yarimburgaz cave (Turkish Thrace) evidently postdates Emirkaya-2 and is broadly mid- to upper Middle Pleistocene in age.

The Biharian age, however, encompasses a long time span, from ~ 1.7 to 0.5 ma., according to authors (Fejfar et al, 1997; Maul, 1990), and constitutes the *Microtus-Mimomys* Superzone. It is considered to extend from the Eburonian into and including the Elsterian (glacial) stages. Its upper boundary is constituted by the FAD of *Arvicola*; its lower boundary is represented by the FAD of *Microtus* (*Allophaiomys*). In greater Europe its *lower* (Early) part constitutes the (conjoint) *M. (M.) savini/M. (Cseria) pusillus* zone (type locality Villàny-5), and its *upper* (Late) part the *M. (M.) savini* zone (type locality Voigtstedt). The approximate boundary between its successive segments is taken to be ~ 0.85 ma, close in age to the profound end-Villafranchian Dispersal Event, of extensive faunal replacement and extinctions. The absence of *M. (C.) pusillus* could suggest an age later, rather than earlier in the Biharian. *M. (A.) nutiensis*, often considered a derivative of *M. pliocaenicus*, is a west Mediterranean species best known in localities in Andalucian Spain (type = Casablanca-3), in southern France (Les Valerots) and in central Italy (Monte Peglia). Its presence suggests a broadly mid-Biharian age, in an approximate sense. Overall the rodent (9 of 13 species) and lagomorph (2 species, at least) fauna strongly suggests an open, steppic habitat; although a more mesic, wooded situation, perhaps riparian, is reflected in the presence of *Trogontherium*.

The condition of large mammal specimens indicates that preservation circumstances were most favorable, and hence the condition of most specimens is indicative of little transport and minimal abrasion. However, the known assemblage - with the exception of a limited series recovered by excavation (in 1994) - often comprises isolated or fragmented elements due to the commercial exploitation by power equipment, and the attendant breakage and/or dissociation of skeletal parts. There is certainly some evidence to suggest that associated elements, and at least partially intact individual(s) existed in the pristine burial situation(s).

The large mammal fauna is relatively diverse (Table 3). Four orders are represented, comprising a total of twelve families. The extent of incompleteness and fragmentation of specimens, coupled with the common lack of diagnostic cranio-dental remains, substantially inhibits taxonomic resolution in several groups (see appendix).

A single proboscidean is relatively poorly represented (App.-1). Fragmentary molar parts, and two complete individual upper and lower molars indicate affinity with *Mammuthus*, and provisional attribution to *M.* cf. *trogontherii*. Probably the available sample is inadequate for resolution at the specific level. *M.* aff. *trogontherii* is documented in the broadly penecontemporaneous fauna of Akhalkalaki (southern Georgia) (Vekua, 1887).

Rhino is very poorly represented (App.-2). For the moment these few remains are attributed to Rhinocerotidae indet., although a partial maxilla fragment and lower teeth (P_4, M) may enable attribution at the species level.

Similarly, there are only scant remains of hippo, all (identified) representing fragments of dental elements (App. -3). Hence attribution is only to Hippopotamidae indet. Suidae are similarly rare, only 5 cranio-dental parts. The (4) teeth indicate attribution to *Sus* cf. *strozzi*, according to J. van der Made (Utrecht) who kindly examined the relevant sample.

Other artiodactyls, both Cervidae and Bovidae, are more numerously represented. However, again the issue of incompleteness and rarity of diagnostic elements (antler, horn cores) hinders species attributions.

Among Cervidae there appear to be at least three size classes/morphs represented (App. -4). There is a single record, a distal tibia, that may represent *Dama* sp. More numerous, including some antler fragments, several teeth, and diverse postcrania, are remains referred here to *Cervus*

DURSUNLU (TURKEY)
LARGE MAMMALS

Proboscidea	
Mammuthus* cf. *trogontherii	Carnivora
	Hyaenidae
Perissodactyla	Hyaenidae **gen. et. sp. indet.**
Rhinocerotidae indet.	
	Canidae
Equidae	***Canis lupus aff. mosbachensis***
Equus caballus mosbachensis	***Canis aff. aureus***
Equus altidens	***Vulpes praecorsac***
Artiodactyla	***Vulpinae indet.***
Hippopotamidae indet.	
	Ursidae
Suidae	***Ursus* sp.**
Sus scrofa	
	Felidae
Cervidae	***Panthera aff. pardus***
Megaloceros sp. nov.	***F. (Lynx) aff. issiodorensis***
Cervus sp. ex. gr. elaphus	***Felis ex. gr. caracal***
aff. Dama sp.	Felinae indet. (small)
Bovidae	Mustelidae
Bos aff. primigenius	***Lutra*** sp. indet.
aff. Gazella sp.	
cf. Gazellospira sp.	

Table 3 - Large mammals recovered from the Dursunlu Formation.

sp. ex. gr. *elaphus*, that is red deer. Most numerous, over two dozen elements, are remains representative of a species of the extinct giant stag, *Megalaceros*. D. Geraads (personal communication), who has also examined some of this latter sample, also considers its specific identification — whether the previously (and long) known *M. verticornis* or an altogether new species - as still unresolved. The latter is suggested by dental features (premolar/molar morphology), but the fragmentary antler parts are unhelpful in this regard. *M. verticornis* is recorded at 'Ubeidiya (Jordan valley) and at Akhalkalaki (south Georgia), but is wholly absent at Dmanisi.

Among Bovidae there are also, minimally, at least three superspecific taxa represented (App. - 5). Most numerous are remains of a bovine, the size of *Bos primigenius*. A (partial) horn, elements of the dentition, and most postcranial parts of both fore- and hindlimb support this attribution. There are no elements demonstrably referrable to *Bison*. A few (only) postcranials are of both a size and morphology to represent *Gazella* sp. indet. A single frontal fragment with basal portion of horn core, of substantial size, exhibits an anterior keel and a counter-clockwise twist. It is thus distinctive and suggestive of goat-antelope, *Gazellospira* (provisionally recorded at 'Ubeidiya, and, purportedly, at Bethlehem) to which it might be compared. It should also be mentioned that the Asian genus *Spiroceros*, attributed to Tragelaphini, has previously been listed from the north Anatolian locality (Merzifon basin) of Kamisli; it was initially considered (by O. Sickenberg) to instead represent *Gazellospira*. And, Vekua (1987) attributes to the Asian genus (?) *Sinoreas*, of uncertain tribal affinity, some fragmentary remains, including a horn core, from Akhalkalaki.

Thus, there is some measure of uncertainty as to whether it is a question of a caprine or tragelaphine, in these instances.

There is also a single example, a distal metacarpal, of size and morphology perhaps most appropriate to *Giraffa*. This genus is already documented at 'Ubeidiya.

Remains of Equidae constitute the most abundant group of larger mammals (App. -6). Most parts of the skeleton are represented, though imperfectly due to their often incomplete state. Two taxa, clearly distinctive in size and morphology, are represented. The less common is typically caballine and readily attributed to *Equus caballus mosbachensis*. There are no cranial parts, only a limited number of cheek teeth, but more numerous postcranial elements. Caballine horses are immigrants into Eurasia from North America. The species is widely represented in western Eurasia in Galerian-stage local faunas. Among the earliest representatives in (central) Asia is the occurrence at Lakhuti-2 (southern Tadjikistan), in a loess-paleosoil complex of post-Jaramillo sc age (~ 0.9 ma).

The second, more abundant form is well represented by dentition and postcrania. The fossils are usually limonite-stained suggesting the possibility of derivation not from the lignite horizon (s), but largely from the underlying gravelly sands. It is thoroughly distinct in morphology, size and lack of skeletal robusticity. This evidently represents *Equus* aff. *altidens*, a form with distinctive hemionine features, as has often been noted. The metapodials are elongate, very slender, having hemionus size and proportions, with distinctive broadening of the distal articulation relative to the supra-articular breadth and that of the mid shaft, and hence producing a rather 'waisted' form. On the distal metapodials the posterior (volar) face is flat, and there is no ridge-like extension there superiorly of the distal trochlear keel. The dentition is markedly hypsodont. In the upper P/M series the enamel is scarcely plicated, the post-fossette is closed, and the protocone lacks or has only a weak heel and a slightly incurved lingual margin. In the lower P/M series the ectoflexids are generally shallow or moderately deep (premolars) to deep and extended into the double knot (molars), and the entoflexids exhibit a shallow to widened and splayed-out V shape.

The type of *Eq. altidens* von Reichenau 1915 derives from early glacial-age deposits of Süßenborn (Thuringia), and occurs within the upper Matuyama (R) chron. It is also present at Voigtstedt. Other referred or attributed occurrences include Sainzelles (Haute-Loire), Abbeville (Somme), Norfolk Forest Bed (Cromerian of West Runton, Trimingham in East Anglia), Venta Micena and Cullar de Baza (ex. - *E. s. granatensis*) (Andalucia), Venosa (Basilicata), Kislang and Varberg (Hungary), Irimesti (Rumania), Kitzkani (Moldavia), and Taman Peninsula (Russia). Some other attributions are difficult to verify, including those of Podumci-I (Dalmatia), Akhalkalaki (*Eq. hipparionoides* of A. Vekua) and possibly Dmanisi; my examination of the Pirro Nord (Apulia) and Selvella (Val de Chiana) samples was not convincing of referral to this taxon. In central Asia it is identified at Lakhuti-1 (southern Tadjikistan) in a loess-paleosoil complex of Jaramillo (N) sc age. *Eq. altidens* is often linked to the ancestry of the subsequent *Equus hydruntinus*, another hemione.

Carnivora are also present in the faunal sample, in some number of elements, but frequently incomplete and lacking more readily diagnostic body parts. These include canids, hyaenid, felids and mustelids, and, possibly an ursid (App.-7). A substantial number of postcranial fragments are referable to Carnivora, but still await (hopefully) finer identification.

Canidae are represented by a few upper teeth and a small number of partial postcrania. Minimally three species are represented, according to size and morphology. These include a small wolf, *C. l.* aff. *mosbachensis* (Mosbach wolf) presumably, and a small jackal, probably

C. aff. *aureus* (golden jackal). Perhaps other elements of such taxa are among still undetermined specimens. A single M1 testifies to the presence of *Vulpes praecorsac* Kormos, a species having also much overall similarity to *Vulpes praeglacialis* Kormos. The former is represented presumably only in eastern central Europe, as at Naghharsányhegy (type), Villány, Beftia-2 (Episcopia), Gundersheim, and Deutsch-Altenburg-2.

Ursidae are exceptionally rarely represented by a single phalanx.

Mustelidae are neither common nor diverse. The only identifiable remains, comprising jaw parts, dentition, and some postcrania, represent an otter of the genus *Lutra.* It remains to be determined if this represents the extant or *L. lutra* only well-documented in the late Pleistocene, or *Lutra simplicidens* (Thenius) recorded at Voigtstedt (Thuringia) and at Hundsheim.

Hyaenidae are surprisingly poorly represented. Several coprolites surely bear witness to some species. The only other few dental elements (C) and several incomplete postcrania are indeterminate as to species affiliation.

Felidae is best represented among Carnivora. There are minimally four taxa represented according to size and available morphology within the sample. Except for distalmost limb elements all parts are largely incomplete specimens. The largest is a leopard-sized species, referable on scant dental and postcranial evidence to *P.* aff. *pardus.* A lynx-sized feline is represented by few specimens; a complete C confirms the presence of *F.* (*Lynx*) *issiodorensis.* A single tooth and a few postcranials are appropriate to caracal [*F.* (*Lynx*) *caracal*], but are inadequate to prove its presence. A tooth (P) and femoral head attest to the presence of an indeterminate small species of *Felis*, wild or desert-cat in size.

Only two major occurrences of vertebrate fossils (mammals) in Anatolia have been reported previously in sedimentary contexts and attributed a Plio-Pleistocene age (Sickenberg, 1975). These are in the Eskisehir (w. Anatolia) and Merzifon (n. Anatolia) basins and both were a focus of our survey activities (1992-1994). In each instance some vertebrates were collected, but in neither instance were these situations paleanthropologically promising or significant due to the sedimentary regime (coarse to medium grade clastics of high energy derivations) and the lack of demonstrable lithic artifacts.

The Eskisehir (or Yukari Sögütönü) vertebrates derive from a series of (now disused) quarries, some 10-20 kms west of Eskisehir town, that expose 5+ m of fluvial sands and light gravels, disconformably above Neogene limnic basin fill, and that are overlain by several meters of lower energy fluvial silts and sands. These occur at 910-850 m (a.s.l.) and constitute part of an extensive basin filling, not well exposed, in the present upper Persuk drainage system. A brief publication (Becker-Platen & Sickenberg, 1968) is devoted to the Eskisehir locality and its vertebrates, a further report by Sickenberg (1972) remains unpublished; also, other MTA collections by Gerçek Saraç from this locality remain unpublished. Saraç and the writer have inventoried all of the Eskisehir vertebrates so that a complete study of the sample is now possible. Less than two dozen vertebrate taxa are identified overall, and something over a dozen at the generic level. The faunal assemblage is broadly of middle Villafranchian attribution (Table 4).

The Merzifon basin was most fully described by Irrlitz (1972) who distinguished there (as elsewhere in this intermontane area) the lower and upper Pontus series (Folge). The latter is extensively expressed in the southern, northern and northeastern reaches of the basin, reaching a thickness of nearly 200 m in the southern sector. The Kamisli vertebrate assemblage derives from a series of interbedded fine-to-medium grained sandstones with silts in the upper third of this succession, in exposures some 16 kms south of Merzifon, and near the eponymous village. The vertebrates from the Kamisli locality have never been published, although there is an unpublished report on the initial collections by Sickenberg (1969); these are presumably still in Hannover (Germany) as they do not seem to have been returned to MTA, Ankara. Again, Saraç and the writer have inventoried both his own previous and our newer collections from Kamisli (all in the

	Yukari–Sögütönü (Eskisehir)	Kamisli (Merzifon basin)
Aves	*Struthio sp. (shells)*	
Insectivora	Soricidae indet.	
Primates	*Macaca sp.*	
Lagomorpha	Leporidae indet.	
Rodentia	*Mimomys sp. 1,2*	*Hystrix sp.*
	Orientalomys galacticus	
Carnivora		
Ursidae		*Ursus. sp. (small)*
Canidae	*Canis etruscus/mosbachensis*	*Canis etruscus/mosbachensis*
	Vulpes sp.	
Hyaenidae		*cf. Pliocrocuta perrieri*
Felidae		*Panthera sp. (leopard–size)*
Proboscidea	*Anancus arvernensis*	Proboscidea indet.
	Mammuthus meridionalis	
	M. ? planifrons	
Perissodactyla		
Rhinocerotidae	*Dicerorhinus sp. (small)*	*Gen et sp. indet.*
Equidae	*Hipparion sp.*	*Hipparion sp.*
	Equus (A.) stenonis	*Equus (A.) stenonis*
Artiodactyla		
Camelidae	*Paracamelus sp.*	
Cervidae	*Eucladoceros sp./sp. nov. ?*	*Eucladoceros sp. ? (large)*
	Cervus elaphus	*Cervus elaphus*
	Cervidae indet.	Cervidae indet.
Bovidae		*cf. Spiroceros sp.*
		Caprini indet.–? *Ovis*
	cf. Leptobos sp.	*cf. Bos/Bison*
	Gazella borbonica	

Table 4 - Mammal species from Lower Pleistocene localities in the Eskisehir and Merzifon basins of Anatolia.

MTA museum). Once the Hannover collection is re-examined this locality too will be more adequately known and can be suitably published. The Kamisli vertebrate assemblage is attributed broadly to an upper Villafranchian/lower Biharian faunal stage. It appears to share little with Dursunlu and is distinctive in respect to its equids and, perhaps, also artiodactyls.

It should be mentioned also that Plio-Pleistocene vertebrate localities are known from three now insular situations off western Anatolia. Plio-Pleistocene vertebrates occur at the Antimachia and Kardamena localities, southern Kos (Istankoy) Island, derived from continental sediments (also with macroflora remains) intercalated between marine deposits. The initial discovery of continental vertebrates was reported by Desio (1931), and their description was first by Airaghi (1928) and subsequently by Charrier & Giglio (1969). Eleven mammal taxa have been recognized, including Carnivora (3), Proboscidea (2), stenonine equid, cervids (2+), and 3 other artiodactyl taxa. The assemblage is of broadly mid-Villafranchian attribution.

Kos, now insular, formed sometimes a continuation of the Bodrum peninsula in Plio-Pleistocene times as also demonstrated by its balanced, continental vertebrate fauna. A younger fauna of (earlier) Biharian affinities, and also of continental aspect, derives from Kalymnos Island (Kuss & Storch, 1978) to the north; it includes up to 3 artiodactyls, bats, an insectivore, and 12 species of 9 rodent genera (6 families represented). A younger but mid-Pleistocene fauna derives from a fissure fill (Latomi-1) (Storch, 1975; Schneider, 1975) on Chios Island, off the Karaburun peninsula, opposite Izmir, from which a diversity of herps and 17 species of 14 genera (8 families) of rodents are known. This assemblage also constitutes a balanced mainland type fauna. The aforesaid occurrences on Kos and Kalymnos certainly would support the likelihood of finding in future vertebrates of Plio-Pleistocene age in western Anatolia.

Clastics of foreign origin, including some bearing evidence of human workmanship, were discovered by T. D. White and others at Dursunlu in 1993. In many instances these were contained within blocks of indurated silty sediments, sometimes with associated gastropods on tip heaps left from the earlier lignite exploitation. Such sediments must therefore derive from peaty-silts now known to occur *below* the several lignites and *above* the lower-lying (fossiliferous) gravelly-sands, both of which apparently lack any trace of such similar clastics/artifacts. (The latter gravelly-sands include quartz elements, but of different texture, condition and quality, and of smaller size and much worn and rounded.) There is for the present no definitely established local source for the artifactual quartz or for some (few) specimens on flint. However, the Paleozoic and Mesozoic rocks of the adjacent uplands would afford such raw materials. The available sample will be described elsewhere by J. D. Clark and E. Güleç. The collection, thus far, numbers some 175 pieces, almost exclusively in quartz, only four of limestone and three flint or chalcedony. It includes small chunks or blocks, with modification, reduction or not; cobbles; flakes, probable flakes, and some modified flakes; flake fragments and many small chips. A subspheroid (limestone), multi-platform core (flint) and flint flake clearly demonstrate the presence of an earlier Lower Paleolithic industry.

Taphonomic examination of recovered faunal elements has still be carried out. One distal metatarsus of a large species of Aves does exhibit a series of largely transverse, bilateral incisures across the upper mid-diaphysis. These can only be interpreted as humanly-induced cut-marks according to their depth and conformation.

The Dursunlu locality, through a set of serendipitous circumstances, affords important new evidence bearing on earlier Pleistocene paleontology and archeology of Anatolia. It affords a diverse vertebrate fauna, particularly of birds and mammals, of demonstrable stratigraphic provenience in a substantial, suballuvial sedimentary succession for which paleomagnetic polarity determinations are feasible. It also affords the first, and so far oldest indication of ancient, Lower Pleistocene hominid occupation in Anatolia in a well-defined stratigraphic context. Hopefully, this will constitute a harbinger of further paleoanthropological localities still awaiting future recognition in these still largely uninvestigated reaches of southern Anatolia.

Ackowledgements

This interim report on the Dursunlu locality has been drafted by the second author on behalf of those involved with its investigation. The joint research effort is part of a paleontological cum paleoanthropological survey project, since 1992, of the M.T.A. (Ankara), Ankara Universitesi and University of California - Berkeley (LHES/MVZ). The splendid assistance and logistical and institutional support of M.T.A. in the course of this work is gratefully acknowledged by us all. The financial support of the Ann and Gordon Getty Foundation enabled this project to go forward, for which we all express our gratitude and deepest appreciation.

References

AIRAGHI, C. (1928): Mammiferi pliocenici dell' isola di Coo (Dodecaneso). *Atti Società Italiani di Scienze Naturali e del Museo Civico di Storia Naturale in Milano,* 67, 125-135.

BECKER-PLATEN, J.-D. & SICKENBERG, O. (1968): Die unterpleistozänen Kiese von Eskisehir (Anatolien) und ihre Säugetierfauna. *Mitteilungen des Geologischen Instituts der Technischen Hochschule in Hannover* 8, 87-20.

BERING, D. (1971): *Lithostratigraphie, tektonische Entwicklung und Seengeschichte der neogenen und quartären intramontanen Becken der Pisidischen Seenregion (Südanatolien).* Hannover (Beiheft 101 des Geologischen Jahrbuches).

CHAPUT, E. (1936): *Voyages d'études géologiques et géomorphogéniques en Turquie.* Istanbul (Mémoires de l'Institut Francais d'Archéologie de Stamboul).

CHARRIER, G. & GIGLIO, A. (1969): Primi risultati di una campagna di rilevamento geologico nell' isola de Coo (Sporadi meridional; Mare Egeo). *Bullettino della Associazione Mineraria Subalpina, Torino* 6, 482-516.

DESIO, A. (1931): *Le isole italiane dell' egeo.* Roma.

FEJFAR, O., HEINRICH, W.-D., PEVZNER, M.A. & VANGENHEIM, E.A. (1997): Late Cenozoic sequences of mammalian sites in Eurasia: an updated correlation. *Palaeogeography, Palaeoclimatology, Palaeoecology* 33, 259-288.

IRRLITZ, W. (1972): *Lithostratigraphie und tektonische Entwicklung des Neogens in Nordostanatolien.* Hannover (Beiheft 120 des Geologischen Jahrbuches).

KUSS, S.E. & STORCH, G. (1978): Eine Säugetierfauna (Mammalia: Artiodactyla, Rodentia) des älteren Pleistozäns von der Insel Kalymnos (Dodekanés, Griechenland). *Neues Jahrbuch für Geologie und Paläontologie, Monatshefte* 978, 206-227.

LAHN, E. (1948): Contribution à l'étude géologique et géomorphologique des lacs de la Turquie. *Maden Tetkik ve Arama Entitüsü Yayinlarindan,* Series B., no. 12, 1-178. Ankara. (Turkish and French).

LOUCHART, A. (1997): *Les oiseaux de Plèistocène inférieur de l'ancienne carrière de Dursunlu (Turquie).* Montpellier (Mémoire de D.E.A. Paléontologie. Université Montpellier II).

LÜTTIG, G. & STEFFENS, P. (1976): *Explanatory notes for the paleogeographic atlas of Turkey from the Oligocene to the Pleistocene. Paleogeographic Atlas of Turkey 1:1 500 000.* Hannover (Bundesanstalt für Geowissenschaften und Rohstoffe).

MAUL, L. (1990): Überblick über die unterpleistozänen Kleinsäugerfaunen Europas. *Quartärpaläontologie* 8, 153-191.

MONTUIRE, S., SEN, S. & MICHAUX, J. (1994): The Middle Pleistocene mammalian fauna from Emirkaya-2, Central Anatolia (Turkey): Systematics and paleoenvironment. *Neues Jahrbuch für Geologie und Paläontologie, Abhandlungen* 193, 107-144.

PAMIR, H.N. & ERENTOZ, C. (eds.) (1975): *Explanatory text of the geological map of Turkey. Ankara sheet. 1:500,000.* Ankara.

SCHNEIDER, B. (1975): Eine mittelpleistozäne Herpetofauna von der Insel Chios, Agais. *Senckenbergiana Biologica* 56, 191-198.

SEN, S., BONIS, L. De, DALFES, N., GERAADS, D., JAEGER, J.-J. & MAZIN, J.M. (1991): Première déscouverte d'une faune de vertébrés pléistocènes dans une fissure karstique en Anatolie centrale. *C. R. Acad. Sci. Paris* (II), 313:127-132.

SICKENBERG, O. (1969): *Die Säugetierfauna von Kamisli im Becken von Merzifon (Vilayet Amasya, Zentralanatolien).* Archiv-Bericht, Bundesanstalt für Bodenforschung, Hannover, Nr. 20713/18. (unpublished).

SICKENBERG, O. (1972): *Die Säugetierfauna der Lok. Eskisehir (Vilayet Eskisehir, West-Zentralanatolien, Türkei).* Archiv-Bericht, Bundesanstalt für Bodenforschung, Hannover, Nr. 219. (unpublished).

SICKENBERG, O. (ed.) (1975): *Die Gliederung des höheren Jungtertiärs und Altquartärs in der Türkei nach Vertebraten und ihre Bedeutung für die internationale Neogen-Stratigraphie.* Hannover (Geologisches Jahrbuch, Reihe JB., Heft 15).

STORCH, G. (1975): Eine mittelpleistozäne Nager-Fauna von der Insel Chios, Agais (Mammalia: Rodentia). *Senckenbergiana Biologica* 56, 165-189.

ÜNAY, E., KARABIYIKODLU, M., KAZANCI, N. & SARAÇ, G. (1997): The Dursunlu open cast mine. In: S.A.G. Leroy & C. Ravazzi (eds.), *Volume of Abstracts, Excursion Guide. INTER-INQUA Colloquium, Ankara (Turkey), March 29 - April 1, 1997. INQUA Commission on Palaeoclimate. Working Group on Milankovitch and Plio-Pleistocene vegetation succession from 2.6 to 0.9 Ma*, 69-74. Ankara.

VEKUA, A. (1987): The Lower Pleistocene mammalian fauna of Akhalkalaki (southern Georgia, USSR). *Palaeontographia Italica* 74, 63-96,

WENZEL, H. (1932): *Sultan Dag und Aksehir Ova. Eine landeskundliche Untersuchung in Inneranatolien.* Kiel (Schriften des Geographischen Instituts der Universität Kiel, Band 1, Heft 1).

WENZEL, H. (1935): *Forschungen in Inneranatolien. I. Aufbau und Formen der Lykaonischen Steppe.* Kiel (Schriften des Geographischen Instituts der Universiät Kiel, Band V, Heft 1).

Appendix

1. The proboscidean remains, as identified, include individual upper (M) and lower (M_1) molars and eight such fragments, and portions of a tusk. There are fourteen postcranials including rib (proximal), metacarpal, incomplete metapodials (2), eight carpal/tarsals, and two phalanges.

2. Rhino is represented by various partial cheek teeth (5), lower premolar and molar, and partial maxilla. the three (identified) postcranials include a carpal, cuboid, and mt-3.

3. Hippo, as identified thus far, is represented only by dental parts, two partial canines and five incomplete cheek teeth.

4. Among identified Cervidae *Megaloceros* is most numerous. Skull parts include seven pieces of antler, none substantial, and seven upper and lower cheek teeth. The 13 postcranials include a mc, astragali (3), tarsals (5 naviculars), partial mt's (4) and a phalanx. A *C. elaphus*-size morph is represented by antler (2) parts, dentition (3), and forelimb (1) and hindlimb (tibia, astragalus, mt, and phalanges) elements. Only a few hindlimb fragments may be attributed to fallow deer.

5. Two dozen specimens are referred to *Bos* aff. *primigenius*. These include horn core (partial), upper (5) and lower (10) cheek teeth, and 14 postcranials of the forelimb (partial radioulna, 4 carpals, 2 mc) and hindlimb (partial tibia, 3 calcanei, mt), and indeterminate mp and a phalanx. The *Gazella*-like postcrania (5) include mc, tibia parts (2), astragalus (1) and partial mp. Undetermined, larger bovid remains are represented by lower teeth (2) and a tarsal (cuboravicular).

6. *Equus* remains are quite numerous. There is only a single associated maxilla/mandible. However, a full set of associated upper and lower cheek teeth of *Eq. altidens* are preserved. Of the cheek teeth uppers (18, 16) and lowers (17, 18) are otherwise about equally common, and at least 5 deciduous teeth are present. Among postcranials there are few axial elements (2), and single coxal and rib parts. All parts of the forelimb are present - humeri (6), ulna (1), radius (2), carpals (1, others unidentified as yet), and mc's (24). The hindlimb includes femur (3), patella (2), tibiae (11), astragali (11), calcanei (2), tarsals (14) and mt's (20). There are 34 phalanges, and 18 undetermined, incomplete mp's.

7. The total number of identified canid parts are nine. These include upper dentition (3), ulna (1), mc (1), femur (1), and calcanei (3). The identified lutrine elements include mandible (1,) upper (3) and lower (2) dentition and humerus (2) and astragalus (1), among a larger, unstudied sample. The rare hyaenid remains are only dental (2) and postcranial (2) elements, the latter a partial ulna and phalanx. The felid remains are most numerous. They include indeterminate mandible fragments, upper (4) and lower (6) dentition, and various postcranials (scapula, radius, 5 femoral parts, 1 tibia, 4 mt's, astragalus and calcaneus). Undetermined (even indeterminate) carnivore parts include mandible fragments (9), teeth fragments (5+), scapular parts (6), mc's (2) of the forelimb, and femur (1), tibia (2), tarsals (1+), mt's (2), astragalus (1+), partial phalanges (13), and many mp parts (27), all mostly representative of smaller species.

The Problems of Interpreting Hominid Subsistence Strategies at Lower Palaeolithic Sites: Miesenheim I - a Case-study from the Central Rhineland of Germany

Elaine Turner

Abstract

Despite intensive archaeozoological and taphonomical research, we still appear to be no closer to defining the role of early humans in the accumulation and modification of faunal assemblages at European Lower Palaeolithic sites, let alone the mode of animal procurement. At many localities dating to this period, site interpretation is often hampered by superimposed sequences of deposition, faunal remains subjected to varying degrees of modification by different agents and a lack of strong evidence of human interaction with the animal bones (cut-marks; percussion marks). The Lower Palaeolithic site at Miesenheim I in the Central Rhineland of Germany was excavated in a series of campaigns at the end of the 1980's and beginning of the 1990's. The site produced an interglacial fauna, dated biostratigraphically to between 400,000 and 450,000 BP, in association with a small assemblage of lithic artefacts. Some of the well-preserved faunal remains had been chewed by carnivores. Several lines of evidence suggest that the find-deposition took place in the back-waters of the Rhine Valley, close to a flood-plain pond. All of the difficulties of site interpretation mentioned above were encountered during the ensuing analysis of the locality and its finds. This paper addresses the issues surrounding the application of body part representation, age-profiles and spatial distribution of bones and stone tools in reconstructing the circumstances of bone accumulation and modification at sites like Miesenheim I, where not only hominids, but also carnivores and the location of the site all played a role in the deposition of the animal remains recovered during excavation.

Introduction

Despite intensive archaeozoological and taphonomical research, investigators still have difficulties in defining the role played by early humans in the accumulation and modification of faunal remains at European Lower Palaeolithic sites (Gaudzinski & Turner, 1996), let alone the way in which they procured animal resources. Although the recent discovery of several wooden spears at Schöningen, Germany (Thieme et al., 1993) does show that early humans were capable of hunting their prey, evidence of the further exploitation of animal remains by humans, such as cut marked bones, hammerstone-induced percussion notches and bone tools, have only been sporadically recorded at Lower Palaeolithic sites. The paucity of evidence is not related to lack of sites or poor bone preservation, as the number of Lower Palaeolithic sites which have been excavated and taphonomically analysed in the last few years shows. The fundamental problem is the sites themselves, with superimposed sequences of deposition and faunal remains subject to varying degrees of modification by different agents. Although many researchers now recognize that human activities were only one of several different factors in the input and output of faunal remains at Palaeolithic sites, much additional work is needed before complex faunal records can be teased apart.

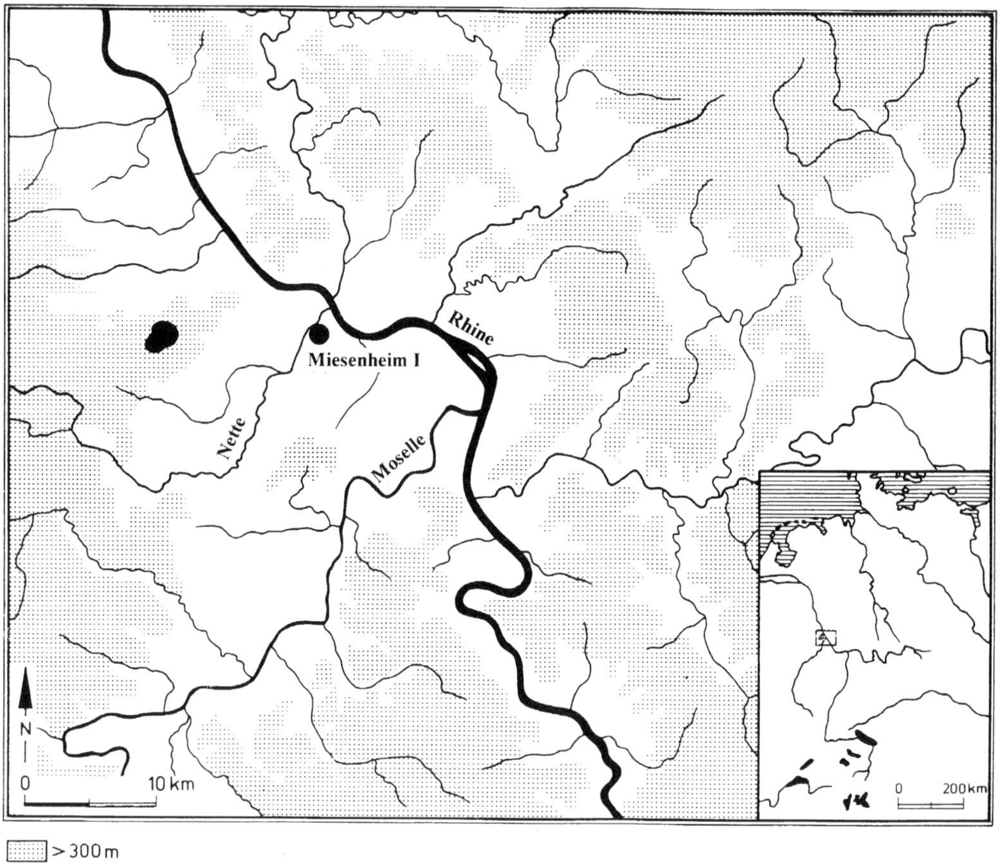

Figure 1 - Location of the Miesenheim I site in the Central Rhineland of Germany.

All of the difficulties of site interpretation mentioned above were encountered during analysis of the Lower Pleistocene site at Miesenheim I in the Central Rhineland of Germany. Excavations at this site site produced an extremely well-preserved interglacial fauna, dated biostratigraphically to between 400,000 and 450,000 BP, in direct association with a small number of lithic artefacts. Several lines of evidence suggest that the site was located in the back-waters of the Rhine Valley, close to a flood-plain pond, during the phase of find-deposition. Traces of carnivore gnawing were observed on some bones, but there was little evidence of human modification of the faunal remains.

In the absence of strong evidence of human modification of bone, researchers resort to body-part representation, age-profiles and the spatial distribution of bones and lithic artefacts to interpret their sites. This paper addresses the issues surrounding the application of these methods in reconstructing the circumstances of bone accumulation and modification at sites like Miesenheim I, where not only hominids, but also carnivores and site location all played a role in the deposition of the animal remains recovered during excavation.

Site background

Miesenheim I is located to the south-west of the Neuwied Basin, a major geomorphological feature of the Central Rhineland of Germany. The site is situated in the valley of the Nette Stream, 60 m above the present-day floor of the valley and just over 2 kilometres to the east of the confluence of the Nette Stream and the Rhine River (Figure 1). Miesenheim I was discovered by a local hobby archaeologist in 1982, after the site had been revealed during the quarrying of pumice deposits. Test excavations were carried out in 1982 and 1983 (Boscheinen et al., 1984) and extensive systematic excavations were undertaken by this author from 1984-86 and again from 1990-91 (Bosinski et al., 1988; Turner, 1995, 1996). At the end of investigations in 1991, five non-contiguous areas of varying size, totalling 436 m^2 of the site, had been excavated.

Stratigraphy, palaeoenvironment and age of the deposits

Geological and sedimentological analyses of sections exposed during excavation show that the site was situated in the flood-plain of the Rhine River during the phase of find deposition. The oldest geological unit exposed is a fluvial deposit up to ten metres thick (Figure 2), consisting of a homogenous fine to medium grained sand facies laid down by the Rhine River and called the Miesenheim sands (Hoselmann & Willems, 1991). The Miesenheim sands represent a phase of fluviatile activity and indicate a direct association between the site and the main channel(s) of the river prior to find deposition. Later, terrestrial conditions prevailed and during this phase, a fossil „Braunerde“ formed on the sands (Müller, n.d.). The soil was truncated by erosion and silty deposits (base of layer G), already showing the effects of waterlogging (gleying), accumulated on top. Reworked sediments were deposited on top of these (upper part layer G) and bones and lithic artefacts were embedded in this unit. The clay-like deposit (layer F) was formed by autochthonous (*in situ*) organic materials and allochthonous (reworked) silty clays accumulating along the banks of an open water or transported here during phases of flooding. An open body of calm, or sluggishly-moving, water was present at the site, and under these extremely wet conditions a sludge developed out of the organic materials and mineral particles in the shallows. Finds continued to become embedded at the site even during this phase. Later, the area around the pond began to silt up and layer F gleyed, forming a hydromorphic soil or Anmoorgley. By this time, find deposition had ceased. Soliflucted sediments stratified above layer F and a loess-palaeosol deposit separate the archaeological level from a sequence of tephra deposits (pumice and basaltic ash). The tephra are covered by reworked sediments and a series of younger loesses with intercalated palaeosols.

The sedimentological interpretation of the site is supported by results of analyses of the molluscan and vertebrate faunas. Bulk sampling produced 73 molluscan taxa, including typical interglacial species (*Cepea nemoralis*) in layer G. Freshwater species dominate, indicating the presence of an open body of water surrounded by marshes during the accumulation of layers G and F. The open water body appears to have been well-oxygenated and rich in vegetation (Roth, n.d.). Abundant algae in samples taken for pollen also indicate open water surrounded by luxuriant vegetation (Bittmann, n.d.).

The remains of several species of smaller and larger mammals (Table 1), birds and a single vertebra from a fish comprise the vertebrate assemblage recovered from layers F and G. The fauna is diversified, with remains of aquatic species such as beaver, which would have lived at or very close to the excavated site, occurring alongside the remains of animals which would have primarily inhabited wooded (roe deer) or open biotopes (horse) along the edges of the flood-plain. Wild boar (*Sus scrofa*) and dormice (*Eliomys quercinus* and *Muscardinus avellanarius*) indicate warm climatic conditions. Although diversified, the faunal assemblage is representative of the type of fossil vertebrate community which would have been living in and around the Rhine valley during an interglacial phase.

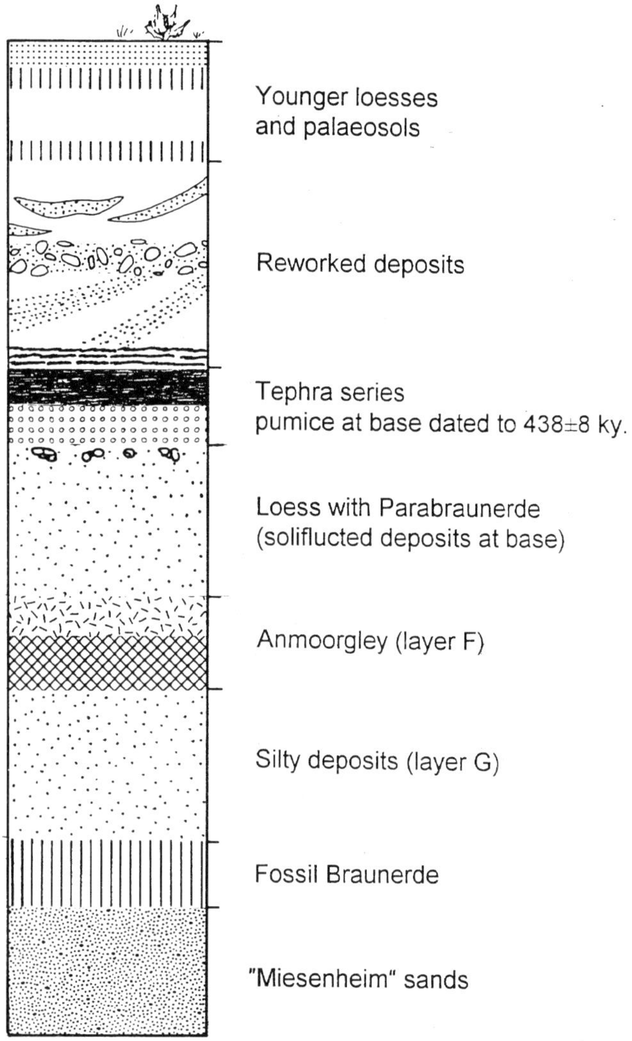

Figure 2 - Lithic artefacts from Miesenheim I. A - fine-grained quartzite flake with basal retouch (2:3); B - quartzite burin de siret 2:3; C1 - (above) quartz core 2:3 and flake (below) 1:1; C2 - quartz core with cortex remains 2:3.

Arvicolid molars from Miesenheim are assigned to the most primitive form of the genus, *Arvicola terrestris cantiana* and remains of *Mimomys* are absent. Kolfschoten (1990, n.d.) has correlated the Miesenheim fauna with the „*Arvicola terrestris cantiana-Sorex (Drepanosorex)* sp. Concurrent-range-subzone". This subzone is correlated in part with the younger part of the Cromer Complex. The interglacial represented at Miesenheim I thus belongs to the youngest warm phase of the Cromer Complex - called the Cromer IV Interglacial - thought to date to between 400,000 and 450,000 BP. This corresponds well with an $^{40}Ar/^{39}Ar$ date of 438±8 ky for a tephra from a neighbouring site (Kärlich) which is mineralogically and chemically identical to the one stratified above the archaeological level at Miesenheim I (Lippolt et al., 1986).

Smaller mammals	
Neomys cf. *newtoni*	*Eliomys quercinus*
Sorex minutus	*Muscardinus avellanarius*
Sorex runtonensis	*Muscardinus* sp.
Sorex (Drepanosorex) savini	*Sicista* sp.
Talpa europaea	*Cricetus major*
Talpa minor	*Allocricetus bursae*
Desmana sp.	*Lemmus lemmus*
Lepus sp.	*Pliomys episcopalis*
Trogontherium cuvieri	*Clethrionomys glareolus*
Castor fiber	*Arvicola terrestris cantiana*
Sciurus sp.	*Microtus (Terricola) arvalidens*
	Microtus (T.) gregaloides
	Microtus arvalis/M. agrestis
	Microtus oeconomus
	Microtus gregalis
	Apodemus sylvaticus
	A. aff. *maastrichtiensis*
Larger mammals	
Canis lupus mosbachensis	Elephantidae gen. et sp. indet.
Ursus cf. *deningeri*	*Equus* sp.
Ursus (Plionarctos) stehlini	*Stephanorhinus* sp.
Mustela (P) putorius	*Sus scrofa*
Mustela cf. *eversmanni*	*Cervus elaphus*
Meles sp.	*Capreolus capreolus*
Felis cf. *sylvestris*	*Alces* sp.
	Bos/Bison sp.

Table 1 - The mammalian fauna from Miesenheim I.

Characteristics of the find assemblage

The faunal remains and the lithic artefacts form a single horizon of finds stratified in the upper part of layer G and in layer F to a depth of between 50 and 70 cms. About 1,700 three-dimensionally recorded finds were thinly distributed across the site. Loss of finds was due, in part, to two main phases of post-burial disturbance. Firstly, a phase of erosion which had truncated the find-level in places, and secondly, the loss of tectonically displaced sedimentary sequences during pumice quarrying. Intact sediments and their finds were preserved in area 3 and in the western part of area 1, but even here find density was often less than ten finds per square metre. Paradoxically area 2, although partially eroded, produced the densest concentration of faunal and lithic materials at the site (Figure 5).

Close spatial clusters of anatomically connected bone elements and refits between lithic debitage and cores, show that both categories of material in the find assemblage were deposited *in situ* in the fine-grained sediments at the site. The vertical distribution of the finds, evidence of deposition of some faunal remains during different seasons of the year, the age-discreteness of seasonally-birthing taxa and superimposed incidences of deposition indicate that the assemblage accumulated over a period of time, in other words is attritional. Short conjoins between bone fragments which had been fractured in the dry or fossil state show that the find-assemblage remained relatively undisturbed after burial, up to the time of the younger erosion and recent quarrying described above.

Raw material	Number	Per cent of total number artefacts (113) including unstratified finds
Quartz	105	92.8
Quartzite	5	4.6
Siliceous slate	2	1.8
River flint	1	0.8
Totals	**113**	**100.0**

Table 2 - Numbers and percentages of lithic artefacts at Miesenheim I.

The lithic assemblage and human modification of bone

The presence of hominids at the site is testified by a small assemblage of 113 lithic artefacts (Table 2; Figure 3). Locally-occurring raw materials - quartz, quartzites and lydite - were utilized. A single flake from a small nodule of relict River Terrace flint may have been reworked into the deposits. Quartz was the favoured raw material. All the artefacts, except the flint one, are fresh in appearance. The bulk of the assemblage consists of debitage (flakes, flakelets and angular debris), along with a few cores and a small, elongated pebble which bears traces of chipping on one end probably consistent with its use as a hammerstone. Two large flakes are retouched.

The presence of cortex on quartz cores and on pieces of quartz debitage, refits between quartz flakes and cores and cortex on some quartzite and siliceous slate artefacts, indicates that these materials had been worked at the site. Lithic artefacts were distributed in both layers G and F, but the majority of the quartzes were located in the upper part of layer G. Five fist-sized quartz cobbles which had been used as cores could be distinguished, and their distribution indicates activity zones in areas 2 and 3 on what must have been temporary, stable landsurfaces in the flood-plain (Figure 5). Two artefacts showed traces of having been burnt, although charred bones were not observed and the remains of hearths were not recovered.

Fine, linear incisions were observed on two fragments of bone from layer G and one fragment from layer F, and these finds were submitted for examination on a scanning electron microscope in an attempt to determine whether they had been produced by stone artefacts during butchering activities (Cook n.d.). The marks on all three specimens are probably of natural origin, produced by contact with coarse fraction clasts in the sediment. Although bones fractured in the green or fresh bone state were present at the site, many of these had traces of carnivore gnawing and impact notches produced by carnivore teeth. Part of an impact notch preserved on a fragment of a long bone from either a horse or a bovid, was the only notch similar to those produced by humans when opening bone shafts with hammerstones to obtain marrow.

The larger vertebrate assemblage and species representation

A total of 1,148 bone, teeth and antler fragments could be identified. The remains represent 13 different vertebrates, whereby the elephant was represented by teeth fragments which could not be definitely identified to either species or genus. As the number of elements for each species was not particularly high, more exact numbers of individuals could be assessed by taking into account not only body-side but also including, for example, wear stages and morphology on teeth, counting unfused juvenile bones separately from adult ones and using differences in bone size of elements from the same body-side. Counts of specimens and numbers of individuals for each of the species identified at Miesenheim I are listed in the following table (Table 3).

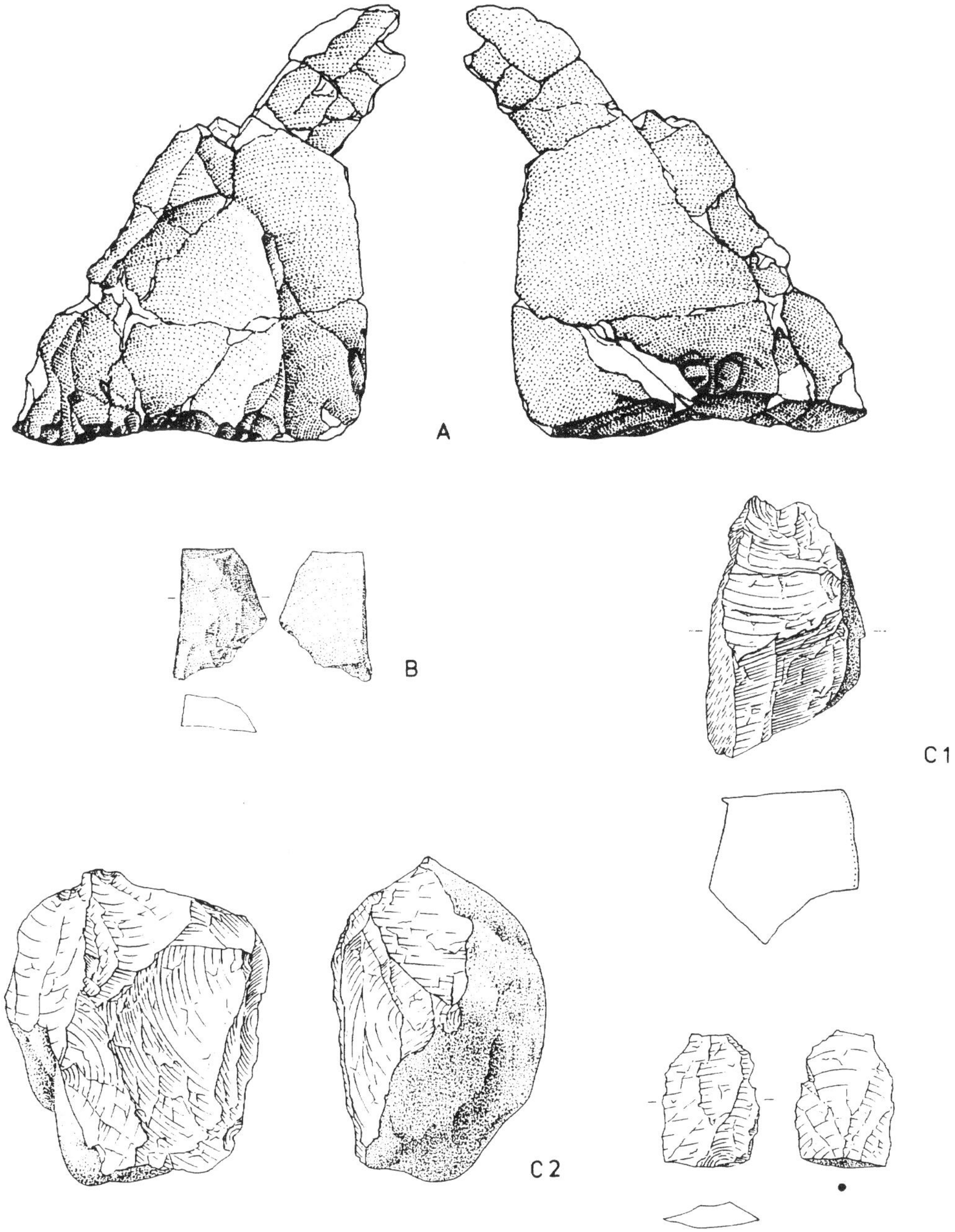

Figure 3 - Schematical section through the deposits revealed during excavation at Miesenheim I.

Typical of this fauna are relatively low numbers of identifiable fragments combined with high numbers of individuals per species. This is particularly obvious in the counts for rhinoceros, horse, roe deer and red deer. Horse, roe and red deer and the bovids have the highest NISP's and, along with rhinoceros, form the dominant group of species at the site based on numbers of

Species	Total NISP	NISP	NI
Badger	8	8	1
Wild cat	1	1	1
Wild boar	5	5	1
Elk	2	2	1
Mosbach wolf	11	10	2
Deninger's bear	12	11	2
Stehlin's bear	34	32	2
Elephant	2	2	2
Rhinoceros	36	32	3
Bovid	304	212	3
Horse	110	84	4
Red deer	317*	168	7
Roe deer	306*	188	9
Totals	**1,148**	**739**	**38**

*Table 3 - Characteristics of the larger vertebrate fauna from Miesenheim I, listed in increasing order of number of individuals. Total NISP - number of fragments before refitting; NISP - number of fragments after refitting; NI - number of individuals per species. *Shed antlers and antler fragments are included in the total NISP counts, but are omitted in the NISP counts for roe and red deers.*

individuals. Expressed as a percentage of the total number of 38 individuals, roe deer (23.6 %) dominates the fauna, followed by red deer (18.4 %), horse (10.5 %) and rhinoceros and bovids (both species 7.8 %). Species occurrence in contexts such as the one found at Miesenheim I, could be due to a number of reasons, including the animals subsceptibility to drowning during times of flood or its vulnerability to predation. The basic factor in such considerations is the animal's relative frequency in the local biotope and habitat utilization. Seasonal evidence from the site, combined with site location may give some information about this factor.

Miesenheim I was located in the Rhine flood-plain close to a flood-plain pond during the period of find-deposition. Water-dependancy can have a strong effect on carcass distribution, and the more water-dependant a species, the more likely it is to concentrate and die in areas near to water. Horses are a good example of a species dependant on water and this factor could probably explain the presence of several individuals at the site of a species usually associated with open grassland.

The presence of avian species (the crake *Porzana porzana,* the corncrake *Crex crex* and the song thrush *Turdus philomelos*) which are only summer visitors to this region today, a single red deer antler frontlet [carried generally by this cervid from September to March or May (Schmid, 1972)] and the estimated age at death for some red deer and roe deer individuals based on tooth eruption and wear stages gives a rough estimate of seasonality, and the deposition of some faunal remains, from summer to early spring.

This is the period of the rut in red deers, a time when the core-units, in which these deers generally live, change because of the mating season. In wooded conditions, a typical red deer core unit consists of a hind and her young. In September or October dominant males become territorial and attempt to maintain a harem of several females, thus temporarily increasing group-size. Red deer do frequent flood-plains, seeking out watering places and grooming themselves by taking mud-baths. Mud-wallowing is favoured particularly by rutting stags. Roe deers are at their most gregarious at the end of the rut in September to October, when mixed groups of males, females and young, sometimes numbering up to 20 animals can form. Roe deer also utilise flood plains as part of their natural habitat, visiting marshes, feeding on sedges and seeking refuge in shrubs along river banks. Miesenheim I can therefore be seen as a watering-place which would have been attractive for a number of species of animals and especially those that are water-dependant. In the case of the deers, which would have utilised the flood-plain as part of their natural habitat anyway, seasonal changes in group-size could have temporarily led to larger numbers of individuals being present. The information derived from seasonality does not neccesarily imply that several individuals of a species died together as a group, it simply indicates that during certain seasons of the year the relative frequency of some animals in the local biotope surrounding the site would have increased. The attraction of the site as a watering-hole, combined with a seasonal increase in numbers of individuals of some species utilising part of their natural habitat, would have produced greater numbers of potential prey for both carnivores and humans and greater numbers of potential victims during natural catastrophes such as flooding.

Carnivores and carnivore attrition

The presence of non-hominid carnivores at Miesenheim I is attested not only by their skeletal remains in the faunal assemblage, but also by a low number (8 %) of gnawing marks on carnivore remains and those of other mammals at the site. The high occurrence of bone with intact (= non weathered) surfaces meant that observation and identification of gnawing traces was relatively easy, and, more importantly, that the percentage of gnawing traces in the excavated assemblage is also very close to that originally present in the buried assemblage.

Taking into account that bone accumulation at Miesenheim I was attritional, then it may be that carnivores played a more definite role only during some of the phases of deposition. Although the number of gnawed elements is generally low, gnawed bones expressed as a percentage of the total minimum number of elements for a species, show differential patterns between the species. Only a single element from rhinoceros had been gnawed (5.2 % of the total MNE). Higher percentages were recorded for the Mosbach wolf and Deninger's bear, but these were probably biased due to the small total amount of elements involved (wolf: 1 out of ten elements = 10 %; bear: 3 out of nine elements = 33 %). Gnawing marks were observed on 14 % of the roe deer remains and 17.2 % of horse remains. Red deer bones appear to have been gnawed the most (26.7 %).

Of the carnivores probably responsible for gnawing bone at Miesenheim, patterns of damage appear to have been mostly caused by wolf. Wolves do not mark up compact bone as heavily as hyenas do, and on the whole, the faunal remains from Miesenheim show the less destructive pattern of carnivore modification produced by wolves. The canid at Miesenheim I was a smaller form of the recent wolf *Canis lupus*, and its ecology and habits were probably much the same as the grey wolf. The lower carnassials of the Mosbach wolf are comparable in size to those of the living dhole *Cuon alpinus,* which weighs 17 kilos and is about 90 cm long from head to rump (Macdonald, 1985). The Mosbach wolf, although small in stature, would probably have been able to kill ungulates the size of red deer, as hunting success depends, as with modern carnivores, such as wolf, dhole and wild dog, more on pack than individual size. One point that could play a role here is the obvious difference in jaw sizes between the Mosbach and the grey wolf, and the implications of this when considering the amount of destruction caused by bone gnawing. The percentage of carnivore attrition on the bones of small-medium-sized ungulates, such as roe and red deer, may be connected to the gnawing activities of a canid which due to its size, could crack open the bones of these animals more easily than those of larger ungulates such as horse.

Species	Juvenile	Adult	Old	Total NI
Badger	0	1*	0*	1
Wild cat	0	1*	0*	1
Wild boar	0	1*	0*	1
Elk	1	0	0	1
Mosbach wolf	1	1*	0*	2
Deninger's bear	1	1*	0*	2
Stehlin's bear	0	2	0	2
Elephant	1	1	0	2
Rhinoceros	1	2*	0*	3
Bovid	1	1	1	3
Horse	1	1	0	4 (2 horses could not be aged)
Red deer	3	4*	0*	7
Roe deer	3	5	1	9
Sum	**13**	**10**	**2**	**38**
Per cent total NI	**34.2**	**55.2**	**5.2**	

Table 4 - *Age representation of larger mammals in the fauna from Miesenheim I. The species are listed in increasing order of numbers of individuals. Asterixes in the adult and old age columns indicate that the numbers of individuals are probably provisional, as these species were aged using post-cranial epiphyseal fusion.*

Age representation

The ages of some animals were assessed using post-cranial epiphyseal fusion (Habermehl, 1961; Schmid, 1972) (Table 4). Approximate ages using teeth were applied to the elephant and Stehlin's bear, but detailed ages based on tooth eruption and wear could only be applied to the roe deers and the bovids. Permanent teeth from adult roe deer and bovids were classed into the „old" group, when the tooth was more than half worn away (Stiner, 1990).

Just over half (55.2 %) of all of the ageable individuals are prime adults, and this pattern of dominance is especially clear amongst the ungulates such as rhinoceros, horse, red deer, roe deer and bovids (Table 4). Only two adults could be classed as old (5.2 %). However, as most post-cranial bones fuse early in life, and after this point are not so useful in separating adults from old adults, exact definitions of adult and old individuals could only be applied where permanent teeth were present. The low proportion of old adults is probably biased due to this factor, and it is possible that some of the individuals placed in the adult group because their bones were fused, actually belong in the old group. Juveniles form, with 34.2 %, just over a third of the total number of individuals. Slightly higher counts of adults among the red and roe deers could also be linked to death of these animals during times of seasonal increases in numbers of adult individuals when harems are formed (red deer) or when mixed groups form after the rut (roe deer).

Assuming for the moment that despite some biasing of the adult and old categories, the age representation shown in Table 4 does give a relatively good picture of the demographies of animals whose remains were deposited at the site. This age-structure does not compare with either „L-shaped" age profiles produced during natural catastrophies, or with „U-shaped" age-profiles commonly produced by the predatory activities of non-human carnivores. Prime-adult dominated age profiles in populations of large ungulates result from selective mortality over a long period of time or less lengthy non-selective mortality affecting a declining population (Haynes, 1991). The presence of several juveniles in the assemblage is inconsistent with a declining population, so that selective mortality over a period of time would appear to explain the dominance of adults at Miesenheim I. Prime-age mortality is usually attributed to human predation using ambush techniques, as carnivores do not consistently kill prime-adult prey (Stiner, 1994). However, a seasonal abundance of primarily adult, small-medium sized ungulates, combined with the presence of a small canid which preferably hunted these ungulates, could possibly account for the dominance of adults amongst roe and red deers over a period of time.

Skeletal part representation

All body parts including teeth, fragmentary crania and mandibles, post-cranial bones and antler were recovered during excavation at the site. Long bones were differentially preserved, including intact bones, fragments of bone shafts, distal and proximal articular ends and bone shafts with either proximal or distal end attached. Even after intensive refitting, many bones were incomplete. These specimens typically ended in dry fractures, so that it was very difficult to attempt to assess the amount of destruction the bone assemblage had undergone. For example, Binford (1981) developed a method to gauge the effective destruction of bone assemblages by plotting in a diagram numbers of proximal versus distal ends of humeri, this being a part of the skeleton preferentially chewed by carnivores, for example. If the humeri from Miesenheim I are plotted in this manner, the assemblage is, with a total of only two proximal ends, well within the „zone of destruction". However, the „destruction" factor at Miesenheim I, simply reflects the number of incomplete bones in the assemblage. It gives no indication of whether the proximal ends are missing due to carnivore chewing, removal by humans or simply due to the proficiency of the researcher who was undertaking refitting and the amount of time allowed within the project for refitting.

In Table 5, the minimum number of elements (MNE) of each species are divided into four main body parts, - head, axial, limbs, feet - in order to present some general information about skeletal representation, while reducing biasing caused by incomplete bones. A recurring pattern of skeletal part representation, regardless of animal body-size, is present in the Miesenheim I assemblage. Axial elements are rare, and represent only 9.0 %, limb bones 17.7 % and head elements, mainly teeth, 23.2 % of the total MNE. The assemblage is dominated by foot bones, which represent almost half (49.8 %) of the total number of elements.

The different percentages of the four body parts in the assemblage could be due to a several factors. Canids and hyenids preferentially carry meat-rich limbs away from a kill-site, leaving the bulkier, less meaty axial skeleton at the kill-site. The presence of isolated limb bones bearing traces of carnivore gnawing and several carnivore-chewed, articulated foot and limb bones units, indicates that some bones in the assemblage were accumulated at the site by carnivores which had been feeding at carcasses nearby. On the other hand, small bones, such as carpals, smaller tarsals and phalanges are uncommon in assemblages of bone accumulated by hyenas (they are usually eaten and dissolved in the gut) and wolves (often eaten and passed as scat away from the consumption site) and these elements are well-represented in the assemblage from Miesenheim I. In fact, the very small carpal bones from roe deers were so common, that counts of roe deer bodies were established on duplicates of one of the carpal bones of this species.

Blumenschine (1986 a,b) has shown that extant non-human carnivores consume edible parts of ungulate carcasses in a particular sequence which deviates only slightly in relation to consumer size and age and consumer type. Late access by hominids to partly eaten carcasses would be

Species	Head	Axial	Limbs	Feet	Total MNE
Wild cat	1	0	0	0	1
Elephant	2	0	0	0	2
Elk	2	0	0	0	2
Wild pig	1	1	0	3	5
Badger	1	0	0	7	7
Deninger's bear	2	1	0	6	9
Mosbach wolf	3	0	3	4	10
Rhinoceros	2	1	4	13	20
Stehlin's bear	22	0	0	0	22
Horse	6	4	10	38	58
Bovid	7	24	12	37	80
Red deer	25	6	22	48	101
Roe deer	35	4	25	74	138
					429
Sum	**100**	**39**	**76**	**214**	
Per cent total MNE	**23.2**	**9.0**	**17.7**	**49.8**	

Table 5 - Skeletal part representation of the larger mammals at Miesenheim I listed in decreasing order of the total MNE.

reflected in the bones which humans accumulated, namely those eaten later on in the sequence by the initial consumers, such as head and limb parts. According to this, the skeletal remains at Miesenheim I in which foot and head elements dominate and axial elements are rare or absent, would be consistent with humans taking remains from carcasses defleshed by other carnivores and leaving them behind after consuming the remains of meat. However, a low frequency of ribs, vertebrae and pelves can also be characteristic of bone assemblages discarded by humans and subsequently ravaged by carnivores (Marean et al., 1992).

The differential skeletal part representation could also have resulted from the effects of hydrodynamic sorting and transport by water after the assemblage had been deposited at the site by other agents. Transport experiments first conducted by Voorhies (1969) showed that skeletal elements could be placed into three „transport groups" which move according to the velocity of the current and the weight and/or density of the bone itself. In low-velocity conditions, such as those existing during the phase of find-deposition at Miesenheim I, the first bones to move are axial elements such as the ribs, vertebrae, sacrum and sternum (group 1) and transitional to these elements, scapulas, phalanges and ulnas (group 1 and 2). Behrensmeyer (1975) suggests that in addition the calcaneum, podials, patella and sesamoids could be included in group 1 for animals comparable to horse in size and smaller. Thus, the low representation of axial elements could be due to sorting and transportation by water, indicating that the bone assemblage at Miesenheim I had been lightly winnowed. However, the presence of phalanges, calcanei, and sesamoids, elements which also have the potential to disperse early or are transitional to group 2, suggests that either winnowing was restricted to certain elements or to certain areas of the site, or that other factors led to the differential skeletal part representation observed at Miesenheim I.

The spatial distribution of partial carcasses and stone artefacts

One interesting aspect of the bone assemblage from Miesenheim I was the presence of a single, almost intact carcass of a bovid and partial carcasses of roe deer, bovid and horse. Their distribution is depicted in Figure 4. Bovid A is a young animal which died between 6 months and 1½ years of age. Its remains are stratified mainly in layer F and it is represented by mandible and tooth fragments, a few limb and foot bones and some cervical vertebrae. The vertebrae and four other bones have been gnawed by carnivores. Its remains overlap those of horse D in layer G. Horse D is also a juvenile animal which died before it attained 3-3½ years of age. It is represented by limb and foot bones and two vertebrae. The elements show intensive carnivore modification. Bovid B is a senile animal, stratified towards the top of layer F. Its carcass is represented by many body-parts including numerous rib fragments, sternal bones and vertebral fragments, parts of the head, and limb and foot bones spread over several metres in area 2. Only a few elements have been gnawn by carnivores. Roe deer D is a robust, adult male deer. Traces of carnivore gnawing were observed on one element from this partial carcass and a carnivore tooth notch may be present on the radius.

Natural deaths, such as death by drowning in floodwaters of the Rhine river, could have resulted in the deposition of complete carcasses at the site. We would not expect repeated catastrophic deaths on a scale comparable to those recorded recently for herd-oriented ungulates in African rivers (Behrensmeyer, 1975), as the species present at Miesenheim I do not herd together in such great numbers. However, single incidences of death by drowning of isolated individuals over a period of time, could have led to the presence of the four carcasses described above. Death by miring or bogging down in alluvial muds could have likewise led to the inclusion of carcasses in the deposits at this site.

Haynes (1988) describes characteristics of carcasses at natural death sites in more detail. Carcasses are usually located close to water sources, on pond edges or lakes. Bone distribution is usually in the form of tight circular scatters, characterized by whole ribs and vertebrae, some legs are partly missing and the skull is usually present. Skeletal elements can be in anatomical order and, in cases where a glut of carcasses are produced, tend to show minimal evidence of carnivore gnawing. Although similar in some aspects to carcasses from natural death sites, such as articulated sequences of elements and tight scattering of bone, the remains at Miesenheim I are not fully comparable with these descriptions.

Partial carcasses, such as those recorded at Miesenheim, are more suggestive of kill or non-predation scavenged sites. Haynes (1980, 1982, 1985) gives detailed descriptions of his observations on recent wolf kill and scavenging episodes of ungulates over 300 kg body-weight in North America in optimal or normal ecosystems. Carcasses at kill-sites are characterized by the presence of lower legs, vertebral units, scapulae with vertebral borders damaged, broken rib and long bone fragments, most of the cranium and mandibles or separate portions of toothrows. Assuming that conditions during deposition at Miesenheim I were optimal, body-part representation of the partial carcasses still differs in some aspects to that recorded in carcasses at kill-sites. The lack of axial elements, heads and some limb bones in the carcass of the young bovid, adult roe deer and young horse are not consistent with patterns of skeletal representation found in carcasses at kill sites.

One thing is however clear, the articulated units of limbs and feet and the tight clusters of limb bones from a single animal indicate that these carcasses were deposited intact at the site and that those parts missing, mainly axial elements, occasionally the head and some portions of the legs, were removed *from* the carcass. Carnivores do not generally remove axial elements from intact carcasses, even though they have been observed carrying off articulated segments of limbs. Axial elements could have been removed by water sorting and transport, but this factor does not account

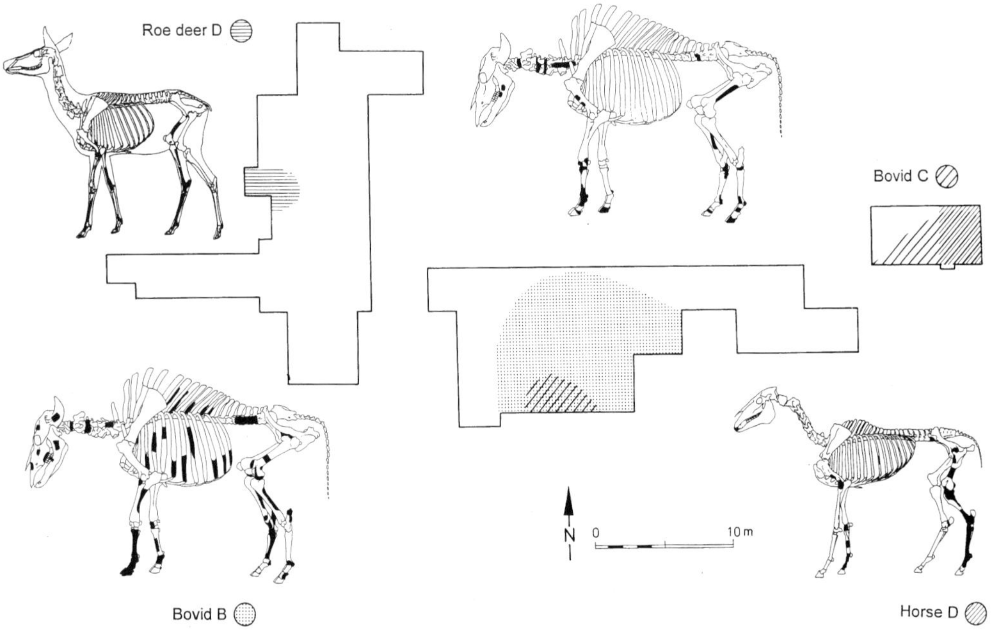

Figure 4 - *Distribution of intact and partial carcasses in the excavated areas at Miesenheim I.*

for the presence of some post-cranial bones which also move in low-velocity conditions. Did humans remove axial and other elements from the carcasses? According to the consumption sequence axial elements are the ones eaten early on by carnivores at carnivore kills. If humans were removing axial parts along with other portions of the carcass, this would indicate first or early access to carcasses by humans. First access by humans to an animal carcass could indicate human hunting activities, a theory which might be supported by the dominance of adults in the age-structure of the mammalian assemblage. Turner (1989) argues that early humans may have practiced organized scavenging in order to gain high-utility body-parts, rather than being content with collecting low-uitility ones. Organized scavenging would be neccessary to provide access to carcasses, presumably killed by non-human carnivores, before the original carcass occupiers had removed too much meat, or to provide first access to carcasses of animals that had died due to non-predation causes and prior to either natural decomposition processes taking place or non-human carnivores arriving at the carcass. Access by humans to intact carcasses obtained either by hunting, organized scavenging at natural death sites or swift intervention at carnivore kills, would produce assemblages of animal bones probably indistinguishable from each other.

Does this theory of early access to the carcass hold up when the spatial distribution of lithic artefacts and partial carcasses is considered? In the vertical projection, the largest and most informative group of lithic artefacts, the quartzes, are predominantly stratified in layer G. Horse remains were the ones predominantly deposited in this layer. If there was a close association between hominids and horse remains, then this should be apparent in the skeletal representation of this species when compared with that of the other animals present at the site. This is not the case, and with the exception of bovid B, the remaining mammals have a skeletal representation not greatly dissimilar to that of the horses. In the horizontal projection, activity zones depicted by concentrations of lithic artefacts are located in the westernmost part of area 3 next to the partial carcasses of horse D and bovid A, and in area 2 where the bone/lithic artefact association is somewhat different (Figure 4).

Bovid B is the only carcass in area 2 and this carcass is the one which has produced a large amount of axial elements including all those usually transported by water in the first instance, such as ribs, vertebrae and sternum bones. In contrast to the other carcasses, its remains appears to represent an relatively undisturbed bone assemblage, but could the differential preservation of this carcass be simply due to localized conditions of deposition? Early humans would not have stood around in shallow water to produce their tools, and concentrations of their artefacts probably indicate areas of dry, stable land surfaces upon which they could move around. The carcass of bovid B is microstratigraphically younger than the main concentration of lithic artefacts in area 2 and its remains appear to represent an incident of deposition not connected in any way with the presence of humans at the site. At the time of deposition of bovid B's carcass, a stable, dry, flood-plain surface may still have existed in area 2. There was, however, no geological evidence for a shore-line in the sequence of deposits exposed at the site, but it is likely that in a flood plain location the boundaries between dry, temporary stable surfaces and wetter areas were subject to constant fluctuation not reflected in a change of sediment. Artefact distribution and an undisturbed carcass does indicate that area 2 was dry throughout the phase of deposition, water not reaching and modifying the bovid carcass, but modifying the carcasses of other animals laid around in the shallows.

Conclusion

Even though excavation, multidisciplinary analyses and a detailed analysis of all categories of finds, did show that three main factors, the location of the site in the flood-plain of a river, the presence of carnivores and the presence of humans, probably all played a part in shaping the bone assemblage, many aspects of site interpretation remained inconclusive.

In an article about the taphonomy and zooarchaeology of the Locus 2 Mousterian faunal assemblage from La Quina, Chase and others (1994) write, „It is true that, for each hypothesis tested, the assemblage does not meet *all* predictions. This may be because of sampling error or because we have not found the correct hypothesis“ ... „This implies that the reason the Locus 2 assemblage fits none of the patterns expected is because each agent has superimposed its own pattern on those left by earlier processes“ (Chase et al., 1994: 301). The same factors apply to the faunal assemblage from Miesenheim I, where not only did attempts to interpret the faunal remains using the results of age-profiles and skeletal part representation fail to meet all the predictions but could also, under the known circumstances of deposition, be interpreted in several different ways. Thus, the lack of axial elements could be interpreted as light winnowing or, in the case of the partial carcasses, removal of axial and other elements by humans because they reached the carcass first. In the case of age-representation, a dominance of adult individuals could be interpreted as selective hunting by humans, for the roe and red deers the dominance of adults is probably related to other factors such as the seasonal increases in numbers of adult individuals living and dying at the site.

Despite this, a number of inferences can be drawn from the evidence produced during analysis. The presence of gnawed bones and articulating units of gnawed limb and foot bones from various species, show that carnivores were not only modifying the bone assemblage, but probably accumulating and dispersing parts of it around the site. The low total percentage of gnawed bones and the presence of small bone elements does suggest that carnivores did not originally accumulate all of the assemblage. Similarly, the partial carcasses did not resemble carnivore kills, but may have been overprinted by winnowing.

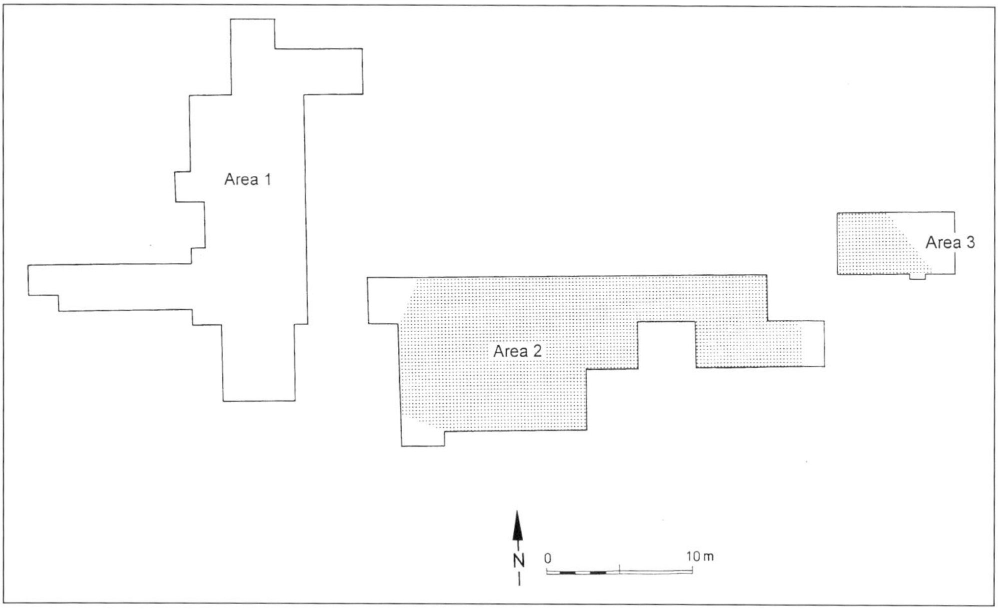

Figure 5 - *Distribution of chipped quartzes and conjoining cores (= cobbles 1-5) and debitage in areas 1, 2, 3 and 4 at Miesenheim I.*

The absence of axial elements may have been caused by light winnowing, a theory that was supported to some extent by attempts to establish areas of drier land where winnowing could not have taken place, by comparing the distribution of lithic artefacts and intact and partial carcasses. The mammal remains at Miesenheim I did not accumulate as part of mass deaths due to drowning as the species present at the site do not herd together. It is possible that the partial carcasses represent the remains of single instances of natural deaths over a long period of time.

The presence of humans is indicated by a small assemblage of lithic artefacts, some of which were produced at the site. The low number of artefacts and the lack of any form of structure or hearths suggests that humans visited Miesenheim I for a very short period of time. The foot and head dominated skeletal representation could be interpreted as humans scavenging from carnivore kills. On the other hand, the absence of axial elements in the partial carcasses could indicate first access of humans to carcasses, in turn implying hunting or organized scavenging.

Is there any hope of interpreting the role of humans in faunal assemblages from Lower Palaeolithic sites? Bone tools recovered from several sites in Italy (Segre & Ascenzi, 1984; Pitti & Radmilli, 1989; Anzidei et al., 1995) and at Bilzingsleben in Germany (Mania, 1986) are unequivocal evidence of the human exploitation of animal bone. Cut marks and the systematic fracturing of bone to obtain marrow have also been recorded at a number of Lower Palaeolithic sites in Europe such as Cagny l'Epinette (F) (Tuffreau et al., 1995), Boxgrove (UK) (Roberts et al., 1995) and Isernia La Pineta (I) (Anconetani, 1996). However, none of this evidence gives clear indications about the way in which the animals were procured by humans, if at all. One exception is the reindeer-dominated fauna from the Lower Palaeolithic site of Caune d'Arago (F) (Moigne & Barsky, n.d.). The dominance of reindeer (75 %) of the faunal remains, could be interpreted as the result of selective hunting activities by man. In contrast to the faunal record, wooden spears recently uncovered at the site of Schöningen (G) (Thieme et al., 1993) provide indisputable evidence of early hunting activities during a warm phase dating to between the Holstein Complex and the Saalian Cold Stage.

References

ANCONETANI, P. (1996): Lo studio arcezoologico del sito di Isernia La Pineta. In: C. Peretto (ed.), *I reperti paleontologici del giacimento paleolitico di Isernia La Pineta. L'uomo e L'ambiente*, 87-186. Isernia.

ANZIDEI, A., VILLA, P. & CERILLI, E. (1995): La Polledrara di Cecanibbio (Roma). Date preliminari sull' analisi tafonomica dei reperti faunistici. *Preistoria e Protoistoria in Etruria. Atti del secondo incontro di studi.* Vol. 2. *Ricerche e scavi*, 27-35.

BEHRENSMEYER, A.K. (1975): The taphonomy and paleoecology of Plio-Pleistocene vertebrate assemblages East of Lake Rudolf, Kenya. *Bulletin of the Museum of Comparative Zoology* 164 , 473-573.

BINFORD, L.R. (1981): *Bones. Ancient men and modern myths*. New York.

BLUMENSCHINE, R.J. (1986a): *Early hominid scavenging opportunities*. B.A.R. International Series 283.

BLUMENSCHINE, R.J. (1986b): Carcass consumption sequences and the archaeological distinction of scavenging and hunting. *Journal of Human Evolution* 15, 639-659.

BOSCHEINEN, J., BOSINSKI, G., BRUNNACKER, K., KOCH, U., KOLFSCHOTEN, TH. van, TURNER, E. & URBAN, B. (1984): Ein altpaläolithischer Fundplatz bei Miesenheim, Kreis Mayen-Koblenz/Neuwieder Becken. *Archäologisches Korrespondenzblatt* 14, 1-16.

BOSINSKI, G., KOLFSCHOTEN , TH. van & TURNER, E. (1988): Miesenheim I. Die Zeit des *Homo erectus. Andernacher Beiträge* 2. Andernach.

BITTMANN, F. (n.d.): Botanische Untersuchungen in Miesenheim I. In: E. Turner (ed.), *Miesenheim I. Excavations at a Lower Palaeolithic site in the Central Rhineland of Germany.*

CHASE, P., ARMAND, D., DEBÉNATH, A., DIBBLE, H. & JELINIK, A.J. (1994): Taphonomy and zooarchaeology of a Mousterian faunal assemblage from La Quina, Charente, France. *Journal of Field Archaeology* 21, 289-305.

COOK, J. (n.d.): Marked bone fragments from Miesenheim I. In E. Turner (ed.), *Miesenheim I. Excavations at a Lower Palaeolithic site in the Central Rhineland of Germany.*

GAUDZINSKI, S. & TURNER, E. (1996): The role of early humans in the accumulation of European Lower and Middle Palaeolithic bone assemblages. *Current Anthropology* 37, 153.156.

HABERMEHL, K.-H. (1961): *Die Altersbestimmung bei Haustieren, Pelztieren und jagdbaren Wild.* Berlin, Hannover.

HAYNES, G. (1980): Prey bones and predators: Potential ecologic information from anaylsis of bone sites. *Ossa* 7, 75-97.

HAYNES, G. (1982): Utilization and skeletal disturbances of North American prey carcasses. *Arctic* 35, 266-281.

HAYNES, G. (1985): On watering holes, mineral licks, death, and predation. In: D. Meltzer & J.I. Mead (eds.) *Environments and extinctions in late Glacial North America*. 53-71. Orono.

HAYNES, G. (1988): Mass deaths and serial predation: Comparative taphonomic studies of modern large mammal death sites. *Journal of Archaeological Science* 15, 219-235.

HAYNES, G. (1991): *Mammoths, mastodonts and elephants*. Cambridge.

HOSELMANN, C. & WILLEMS, N. (1991): Die Miesenheimer Sande (Mittelrhein): Eine petrographisch-, sedimentologisch- und gammaspektromische Untersuchungen. In: W. Boenigk & A. Aktas (eds.), *Festschrift zur Vollendung des 70. Lebensjahres von Karl Brunnacker. Geologisches Institut der Universität zu Köln, Sonderveröffentlichung* 82, 139-161. Köln.

KOLFSCHOTEN, TH. van (1990): The evolution of the mammal fauna in The Netherlands and the Middle Rhine area (Western Germany) during the Late Middle Pleistocene. *Mededelingen Rijks Geologische Dienst* 37, 3.

KOLFSCHOTEN, TH. van (n.d.): Mittelpleistozäne Kleinsäugerfauna von Miesenheim I. In: E. Turner (ed.), *Miesenheim I. Excavations at a Lower Palaeolithic site in the Central Rhineland of Germany.*

LIPPOLT, H.J., FUHRMANN, U. & HRADETZKY, H. (1986): $^{40}Ar/^{39}Ar$ age determinations on sanidines of the Eifel Volcanic Field (FRG): Constraints on age and duration of a Middle Pleistocene cold period. *Chemical Geology (Isotope Geosciences Section)* 59, 187-204.

MACDONALD, D. (ed.) (1985): *All the world's animals: Carnivores.* New York, Toronto.

MANIA, D. (1986): Die Geweihartefakte des *Homo erectus* von Bilzingsleben. In: D. Mania & T. Weber (eds.), *Bilzingsleben III*, 233-256. Berlin.

MAREAN, C.W., SPENCER, C.M., BLUMENSCHINE, R.G. & CAPALDO, S.D. (1992): Captive hyena bone choice and destruction, the Schlepp-Effect and Olduvai archaeo-faunas. *Journal of Archaeological Science* 19, 101-21.

MOIGNE, A.-M. & BARSKY, D.R. (n.d.): Large mammal assemblages from Lower Palaeolithic sites in France: La Caune de l'Arago, Terra Amata, Orgnac 3 and Cagny l'Epinette. In: S. Gaudzinski & E. Turner (eds.), *The role of early humans in the accumulation of European Lower and Middle Palaeolithic bone assemblages.*

MÜLLER, D. (n.d.): Geologische Untersuchungen in Miesenheim I. In: E. Turner (ed.), *Miesenheim I. Excavations at a Lower Palaeolithic site in the Central Rhineland of Germany.*

PITTI, C. & RADMILLI, A.M. (1984): La stazione del Paleolitico inferiore a Castel di Guido presso Roma. *Atti della XXIV Riunione Scientifica Istituto Italiano di Preistoria e Protostoria* 73-87.

ROBERTS, M.B., GAMBLE, C.S. & BRIDGLAND, D.R. (1995): The earliest occupation of Europe: the British Isles. In: W. Roebroeks & Th. van Kolfschoten (eds.), *The Earliest occupation of Europe,* 165-191. Leiden.

ROTH, G. (n.d.): Die Mollusken der mittelpleistozänen Fundstelle Miesenheim I. In: E. Turner (ed.), *Miesenheim I. Excavations at a Lower Palaeolithic site in the Central Rhineland of Germany.*

SCHMID, E. (1972): *Atlas of animal bones*. Amsterdam, London, New York.

SEGRE, A. & ASCENZI, A. (1984): Fontana Ranuccio: Italy's earliest Middle Pleistocene hominid site. *Current Anthropology* 25, 230-233.

STINER, M.C. (1990): The use of mortality patterns in archaeological studies of hominid predatory adaptations. *Journal of Anthropological Archaeology* 9, 305-51.

STINER, M.C. (1994): *Honor among thieves. A zooarchaeological study of Neanderthal ecology.* Princeton.

THIEME, H., MANIA, D., URBAN, B. & KOLFSCHOTEN, TH. van. (1993): Schöningen (Nordharzland). Eine altpaläolithische Fundstelle aus dem mittleren Eiszeitalter. *Archäologisches Korrespondenzblatt* 23, 147-163.

TUFFREAU, A., ANTOINE, P., CHASE, P., DIBBLE, H.L., ELLWOOD, B.B., KOLFSCHOTEN, TH. van, LAMOTTE, A., LAURENT, M., McPHERRON, S.P., MOIGNE, A.-M. & MUNAUT, A.V. (1995): Le gisement acheuléen de cagny-L'Epinette (Somme). *Bulletin de la Société Préhistorique Française* 92, 169-199.

TURNER, A. (1989): Sample selection, schlepp effects and scavenging: The implications of partial recovery for interpretations of the terrestrial mammal assemblage from Klasies River Mouth. *Journal of Archaeological Science* 16, 1-11.

TURNER, E. (1995) Miesenheim I. In: G. Bosinski, M. Street & M. Baales (eds.), *The Palaeolithic and Mesolithic of the Rhineland 15. Inqua Congress, Quaternary field trips in Central Europe,* vol. 2, 891-894. München.

TURNER, E. (1996): Miesenheim I: Lower Palaeolithic artefacts in early Middle Pleistocene deposits. In: K.W. Beinhauer, R. Kraatz & G.A. Wagner (eds.), *Homo erectus heidelbergensis from Mauer. Kolloquium I.* 29-34. Sigmaringen.

VOORHIES, M. (1969): *Taphonomy and population dynamics of an early Pliocene vertebrate fauna, Knox County, Nebraska.* Contributions in Geology, Special Paper No. 1. Laramie.

Lower Palaeolithic Throwing Spears and Other Wooden Implements From Schöningen, Germany

Hartmut Thieme

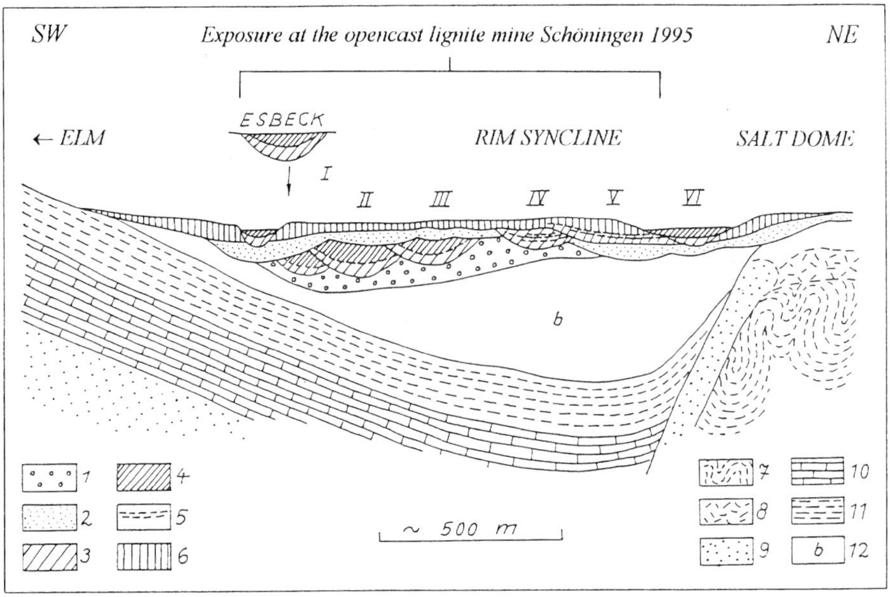

Figure 1 - *The brown-coal mine Schöningen.*
Section through the cyclical Quaternary sedimentary sequence, in the depression along the Staßfurt-Helmstedter Saltdome (Thieme & Mania, 1993). The distance between Cycle VI and the Saltdome is reduced in this figure; the actual distance is about 2 km.
1 - Elsterian glacial deposits, 2 - Saalian glacial deposits, 3 - Lacustrine deposits, 4 - Limnic telmatic sequences, 5 - Soil complexes, 6 - Loess deposits, 7 - Evaporites, 8 - Gypsum caprock, 9 - Buntsandstein, 10 - Triassic limestone (Muschelkalk), 11 - Triassic deposits (Keuper), 12 - Tertiary deposits.

Abstract

Our knowledge of the organic component of Lower and Middle Palaeolithic technologies is extremely limited, and this applies especially to wooden tools. This is a result of the unfavourable conditions prevailing in most sedimentary contexts after burial. In Europe, only two well preserved examples from this period are known; the lance tip from Clacton-on-Sea, Essex, UK (Oakley et al., 1977), discovered in 1911, and the lance from Lehringen, Lower Saxony, Germany, excavated in 1948 (Thieme & Veil, 1985), both made of yew (Taxus). These two examples have been dated to the Middle Pleistocene Holsteinian warm stage and the Late Pleistocene Eemian warm stage, respectively. This paucity of material highlights the importance of two Lower Palaeolithic sites, excavated since 1992, in the brown coal mine at Schöningen, near Helmstedt. Within these sites (Schöningen 12 and Schöningen 13 II–4) numerous exceptionally well preserved wooden implements have been found. More than half a dozen of wooden spears – the oldest complete

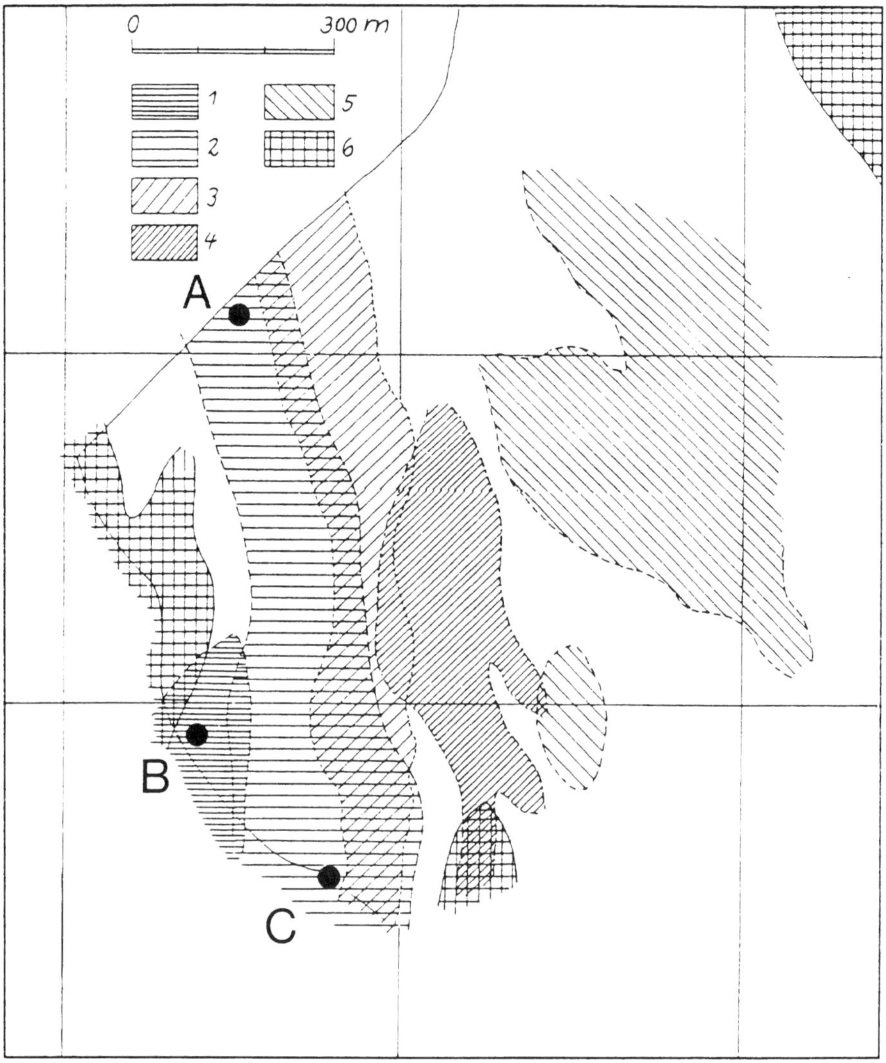

Figure 2 - *Course of the Middle and Late Pleistocene channels in the southern part of the Schöningen open cast mine.*
1 - Schöningen I, 2 - Schöningen II, 3 - Schöningen III, 4 - Schöningen IV, 5 - Schöningen V, 6 - Schöningen VI. The Elsterian glacial deposits lie beneath all the channels (see Figure 1); the channel of Schöningen III is covered by the glacial series of the Saalian glaciation sensu stricto (Drenthe). Channels Schöningen VI contain Holocene deposits.
Location of Lower Palaeolithic sites mentioned in the text: A: Schöningen 12 (1992) with two archaeological find horizons. B: Schöningen 13 I (1994). C: Schöningen 13 II–4 (since autumn 1994), the "spear site".

hunting weapons so far discovered – has been excavated since 1995 in association with stone artefacts and butchered remains of at least 15 horses. These finds provide new insights into the development and culture of early hominids about 400,000 years ago.

Figure 3 - Composite schematic stratigraphical sequence through the Schöningen deposits, which cover the period from the Middle Pleistocene to the Holocene. 1 - Denudation horizon, 2 - Gravelly sands, 3 - Sands, 4 - Lacustrine deposits, 5 - Limnic organogenic sediments, 6 - Peat, 7 - Soils (Lessivé, Pseudogley), 8 - Humic zones, 9 - Fine grained alluvial deposits, 10 - Travertine, 11 - Periglacial structures, 12 - Loess, 13 - Ground moraines, 14 - Laminated clay deposits. Black dots: archaeological find horizons.
The spears are from level 4 within the Schöningen II sequence and date from the end of the Reinsdorf Interglacial. Lg: Late glacial. Plg: Pleniglacial. Eg: Early glacial. Igl: Interglacial. 1–5: Upward shallowing sequences in the Reinsdorf Interglacial.

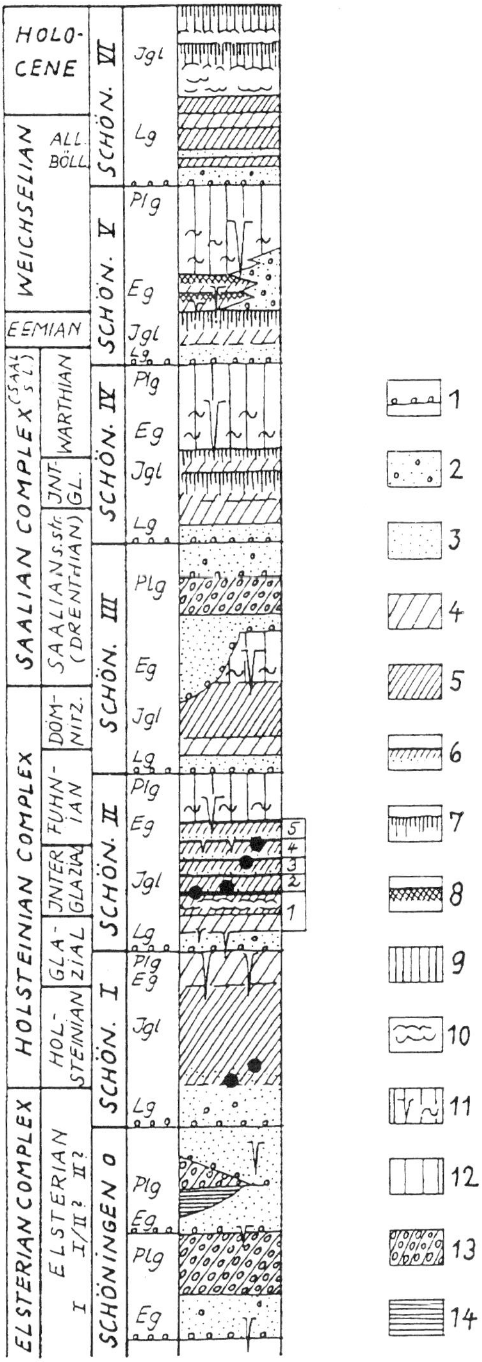

Location and stratigraphy

Since 1983 the Institut für Denkmalpflege, Hannover, has led long-term archaeological excavations under the direction of the author in the area of the Schöningen brown-coal mine. An area in excess of 350,000 m^2 has been excavated in the c. 6 km^2 of the mine complex (Thieme et al., 1987, 1992). During the course of the mining operation and excavation of Holocene sites from the Neolithic to Iron Age (Thieme & Maier, 1995: 108–187), the Pleistocene exposures were constantly monitored and analysed by geological and environmental specialists (Urban et al., 1988, 1991 a, b).

The Schöningen brown-coal opencast mine is situated in the northwestern part of Central Europe, about 100 km east of Hannover, in the northern foreland of the Harz mountains, at the southeastern edge of the Triassic limestone ridge called the Elm (Thieme et al., 1987; Thieme & Maier, 1995). This area belongs to the northern region of the 70 km long sub-hercynic basin between Helmstedt and Staßfurt.

The oldest Pleistocene deposits exposed in the mine are up to now the sediments of the Elster Glaciation (Figure 1). Above these sediments a series of six major erosional channels have been documented since 1992 in the southern part of the Schöningen opencast mine (Thieme & Mania,

1993; Mania, 1995 a). The channels and their associated sediments represent a series of interglacial/glacial cycles that have been named Schöningen I–VI (Figures 1, 2) and suggest an age range from the Holsteinian to the Holocene. Channels I–III, which contain limnic sediments, date to a period between the Elster and Saale glacial *sensu stricto* (Fig. 1).

The oldest interglacial sediments (Schöningen I) in the southern part of the pit (Figure 2) probably date to the Holsteinian, documented in the northern part of the Schöningen mining area (Figure 1: Esbeck) (Urban et al., 1991 b). The Schöningen II channel (Figure 1) is filled by sediments of the Reinsdorf Interglacial (Thieme et al., 1993; Thieme & Mania, 1993) and the ensuing Fuhne cold stage (Figure 3). The depositional sequence contains five levels of organic muds and peats (1–5). Level 1 represents both the early and interglacial maxima of the Reinsdorf. The upper levels represent cool temperate phases and exhibit frost structures between Levels 4 and 5 (Figure 3). The Reinsdorf Interglacial is a new biostratigraphical unit between the Elster and Saale *sensu stricto* (Urban, 1993, 1995 a, b): Palynological analysis indicates that its vegetational history differs from both the preceding Holsteinian and the following Schöningen interglacials (Schöningen III), which is correlated to the Dömnitz Interglacial (Urban, 1993, 1995 a) (Figure 3). The mollusc fauna of Level 1 of the Reinsdorf Interglacial is a thermophilous fauna rich in species with Mediterranean and SE-European elements (*Helicigona banatica* fauna), indicating temperatures 2 to 3 degrees warmer than the present day (Thieme & Mania, 1993; Mania, 1995 b). The Schöningen IV channel is younger than the Saale glacial *sensu stricto* (Drenthe) and consists of an extensive double soil complex (Figure 1, 3). The infill of channel V is correlated to the Eemian Interglacial, whilst the sixth channel infill is of Holocene age. Work by D. Mania has established a correlation between the Schöningen sequence and the terrace-travertine series at Bilzingsleben (Thuringia, Germany) (Mania, 1993, 1995 a, b).

The Lower Palaeolithic sites at Schöningen

Since 1992, several Lower Palaeolithic sites have been discovered in Middle Pleistocene interglacial sediments, from 8–15 m below the present ground surface (Thieme & Maier, 1995).

Schöningen 13 I

The oldest evidence of human occupation in Schöningen, a lakeshore site (Figure 2, B), dates to the earliest part of the Holsteinian complex (Schöningen I) (Figure 3). This site was discovered in alluvial sands with occasional gravels and partially excavated (120 m^2) in the course of three months in 1994 (Thieme, 1995 a). It comprises flint tools, flakes and numerous burnt flints together with faunal remains of steppe elephant (*Mammuthus trogontherii*), bovids, horse and red deer (Schöningen 13 I–1).

On the surface of an overlying organic mud the rests of a *Bison* skull and several tracks of large mammals were discovered (Schöningen 13 I–2) (Figure 3).

Schöningen 12

The Lower Palaeolithic site Schöningen 12 (find layer 1) was excavated in 1992 in the course of three months with more than 150 m^2 (Figure 2, A) (Thieme et al., 1993; Thieme & Mania, 1993; Thieme, 1995 b). Lakeshore deposits with gyttja sands from the Reinsdorf Interglacial (Schöningen II), Level 1 (Figure 3) contained numerous flint artefacts and three worked branches (Figure 4) of the common silver fir *Abies alba* (Schoch, 1995). The wooden tools (length 170–320 mm, max. width 36–42 mm) have a diagonal groove cut into one end; the grooves vary in length between 33 and 47 mm in the front and 10 to 30 mm at the back. The groove breadth varies between 4 and 9 mm at the distal end. A forth worked branch (length 113 mm, width 28 mm),

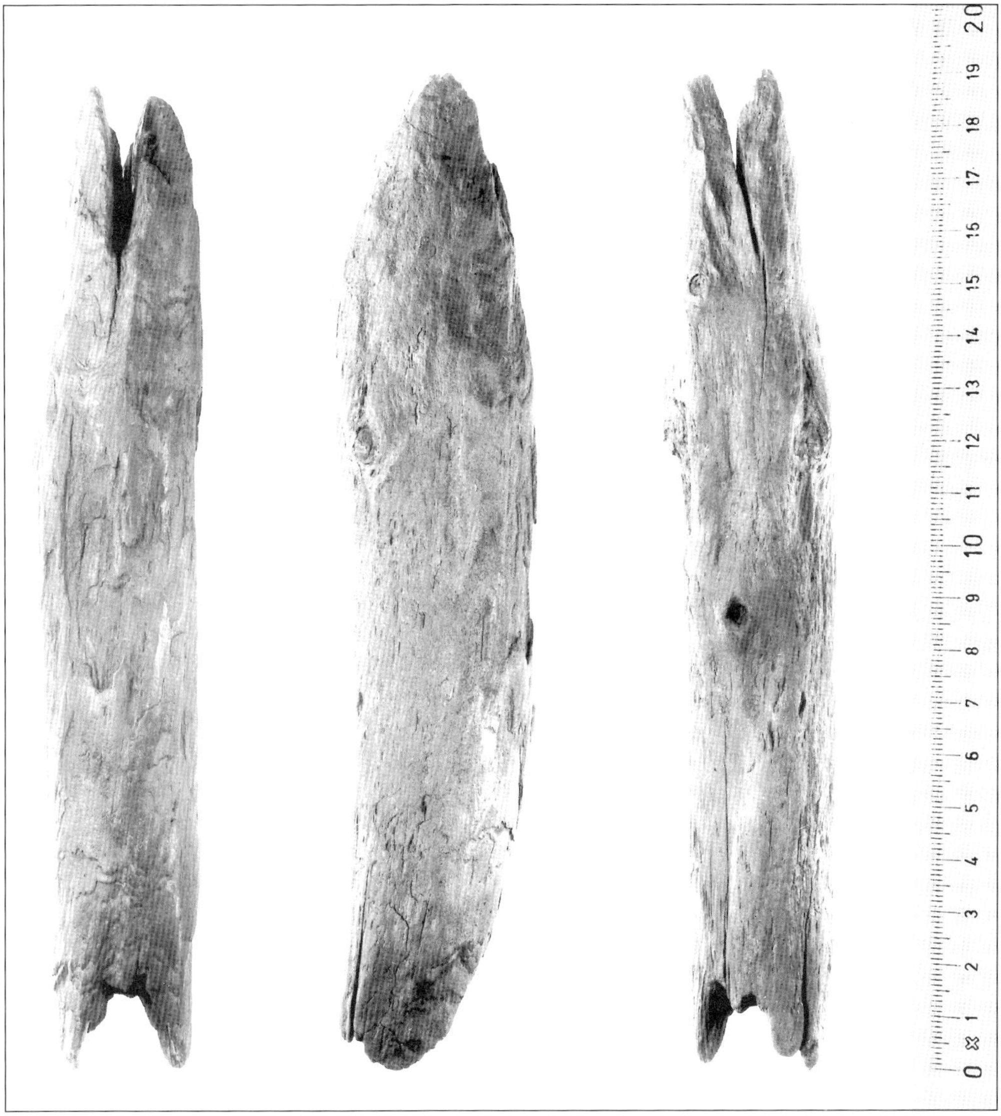

Figure 4 - Schöningen 12 (find layer 1). Lower Palaeolithic wooden implement: a broken branch of the common silver fir (Abies alba) with a diagonal groove cut into the distal end. Scale in cm.

discovered 1996, has grooves cut into both ends. It is postulated that the grooves were for holding flint tools or flakes. If this supposition is correct, these implements represent the oldest composite tools in the world.

Associated with these finds were more than one thousand large mammal bones, some of which exhibit cut marks from butchery. The mammal fauna (van Kolfschoten, 1993, 1995) includes *Ursus* sp., straight tusked elephant *Palaeoloxodon antiquus*, rhinoceros *Stephanorhinus kirchbergensis*, *Equus mosbachensis, Sus scrofa, Cervus elaphus, Capreolus capreolus* and *Bos/ Bison*. There are also numerous small mammals, including the water vole *Arvicola terrestris*

Figure 5 - *Situation of the site Schöningen 13 II–4 in May 1995 (view from southwest), with the archaeological horizon on top of a high sediment socle, containing the complete sedimentary sequence (6–8 m) of the Reinsdorf Interglacial (without Level 5). In the background the power-station Offleben.*

cantiana and the beaver-like *Trogontherium cuvieri*, together with the remains of birds, fish and reptiles. Analysis of the *Arvicola* molars from Schöningen II, Level 1 (the Reinsdorf Interglacial) suggests a correlation with the Bilzingsleben *Homo erectus* site (van Kolfschoten, 1993; Heinrich, 1993).

During excavations of find layer 1 a second archaeological horizon was discovered 2 m to 3 m higher up in peat sediments, with flint artifacts and butchered remains of large mammals in 30 m^2 (Schöningen 12, find layer 2) (Figure 3).

Schöningen 13 II–4

Further archaeological material has been recovered since autumn 1994 from Level 4 of the Schöningen II (Reinsdorf Interglacial) channel (Figures 1, 3) (Thieme, 1995 c). The finds of this site (Schöningen 13 II–4; Figure 2, C) lie in an organic mud which underlies a peat horizon. Preliminary analyses by D. Mania of the molluscan fauna and the pollen spectra by B. Urban,

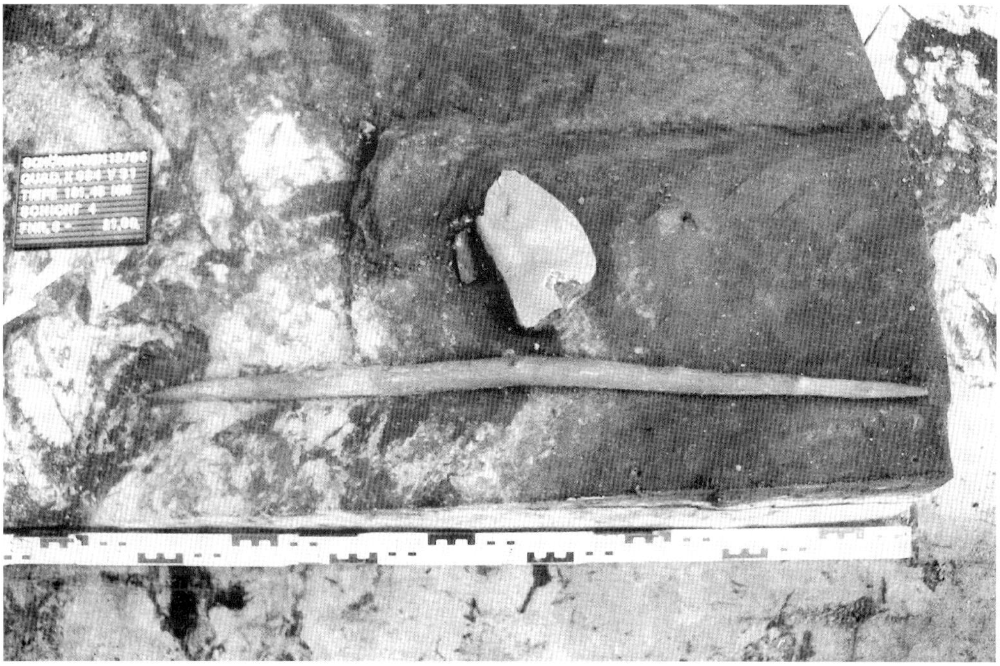

Figure 6 - *Schöningen 13 II–4. View of a Lower Palaeolithic throwing stick (?) in the field in October 1994. Close to the stick is a larger bone fragment with on it's left a flintscraper. Scale in cm.*

Implement	Length	Tip length	Maximum diameter	Point of maximum diameter*	Midpoint thickness	Thickness at 1/3 length*	Thickness at 2/3 length*
Spear I	2.25 m	>600 mm	47 mm	700 mm	36 mm	47 mm	35 mm
Spear II	2.30 m	>600 mm	37 mm	800 mm	35 mm	37 mm	34 mm
Spear II	1.82 m	>250 mm	29 mm	300 mm	24 mm	27 mm	23 mm
Throwing stick?	0.78 m	140 mm 120 mm	30 mm	260 mm	27 mm	30 mm	23 mm

Table 1 - *Wooden implements from Schöningen 13 II-4 (Channel II, Level 4). These data are preliminary: The wood has been slightly deformed due to sediment pressure and the influence of permafrost. *Measurements taken from the distal end of the spear.*

suggests a boreal, cool-temperate climate; the vegetation is a mix of meadows and forest steppes with an abundance of pine, spruce and birch woodland.

Until summer 1997, the excavation (Figure 5) yielded much more than 15,000 well preserved faunal remains, mainly of more than 15 horses (*Equus mosbachensis*), from an area of 2,000 m^2, with densities of up to one hundred pieces per metre square. Many of the bones display traces of butchery, in the form of cut marks and fracturing. The stone artefacts were all made on flint: The assemblage includes some points, carefully retouched scrapers and hundreds of spalls. In the

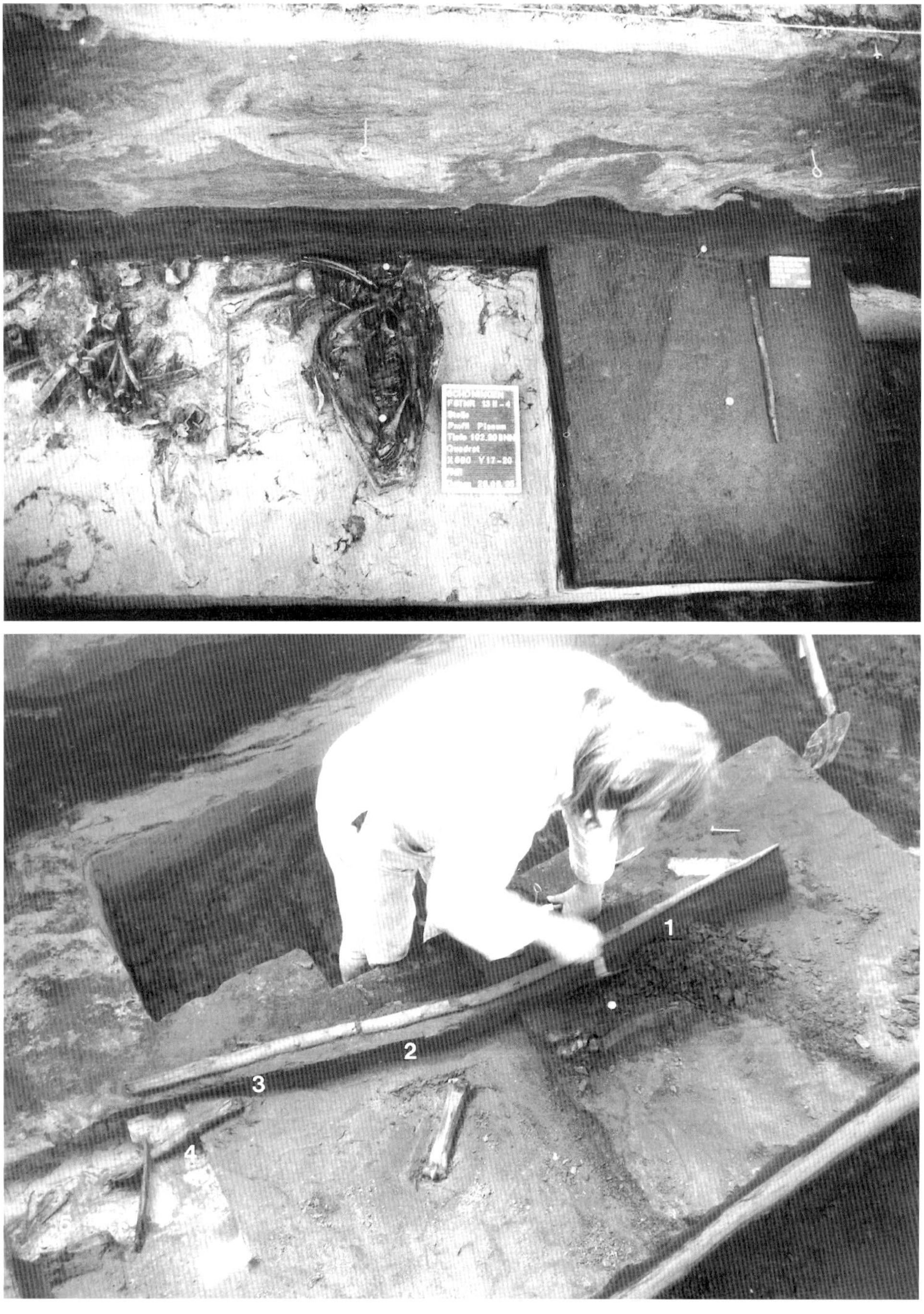

Figure 7 - *Schöningen 13 II–4. (a) Lower Palaeolithic find level end of August 1995 with a dense concentration of bones and a horse skull (left); on the right the base of spear I. Length of the wood: about 60 cm. Spear I during excavation (b). Post-sedimentary processes led to fragmentation of the spear in five parts: 1 (base, see Fig. 7 a) to 5 (tip). Bones are also visible, with one long bone on top of fragment 4.*

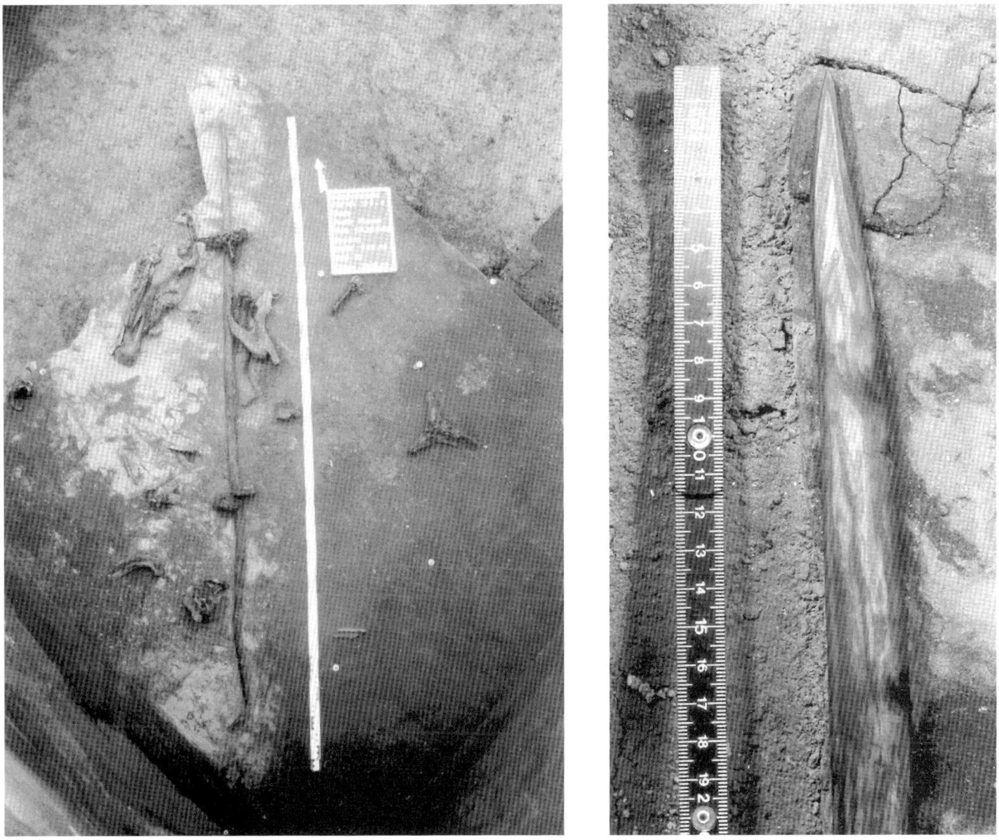

Figure 8 - *Schöningen 13 II–4 (a). View of spear II, 2.30 m long, between butchered horse remains, with a horse skull to the left of the spear. The tip of the spear points upwards. Detail of the well worked tip of spear (b)II. Scale in cm.*

southwestern part of the excavation several fire-places were discovered; the chalky sediments in this areas had reddened under the influence of the heat of the fire and display many fossilised drying cracks.

The wooden tools from Level 4 (Thieme, 1995 c, 1996, 1997) comprise an implement of indeterminate function (probably a throwing stick), with both ends sharpened to a point (Figure 6 and Table 1) and three exceptionally well preserved spears (Figures 7–9 and Table 1), found in autumn 1995. They are all made from spruce *Picea* sp. and were found in conjunction with abundant faunal remains (Figure 7 a). The wood selected, exhibits a dense concentration of growth rings, implying slow growing conditions in a cool environment. The spears are made from individual trees, which were felled, de-barked and de-branched; the tip/distal ends are worked from the hardest part of the wood at the base of the tree. All three spears, although of different lengths (Table 1), were manufactured to the same pattern with the maximum thickness and weight situated at the front end of the spear. The tails are long and taper towards the proximal end (Figure 9): In all respects they resemble modern javelins and were made as projectile weapons rather than thrusting spears or lances. In 1996 and in January 1997 two other spears were discovered. Spear IV is uncomplete until now, spear V is more than 2 m long.

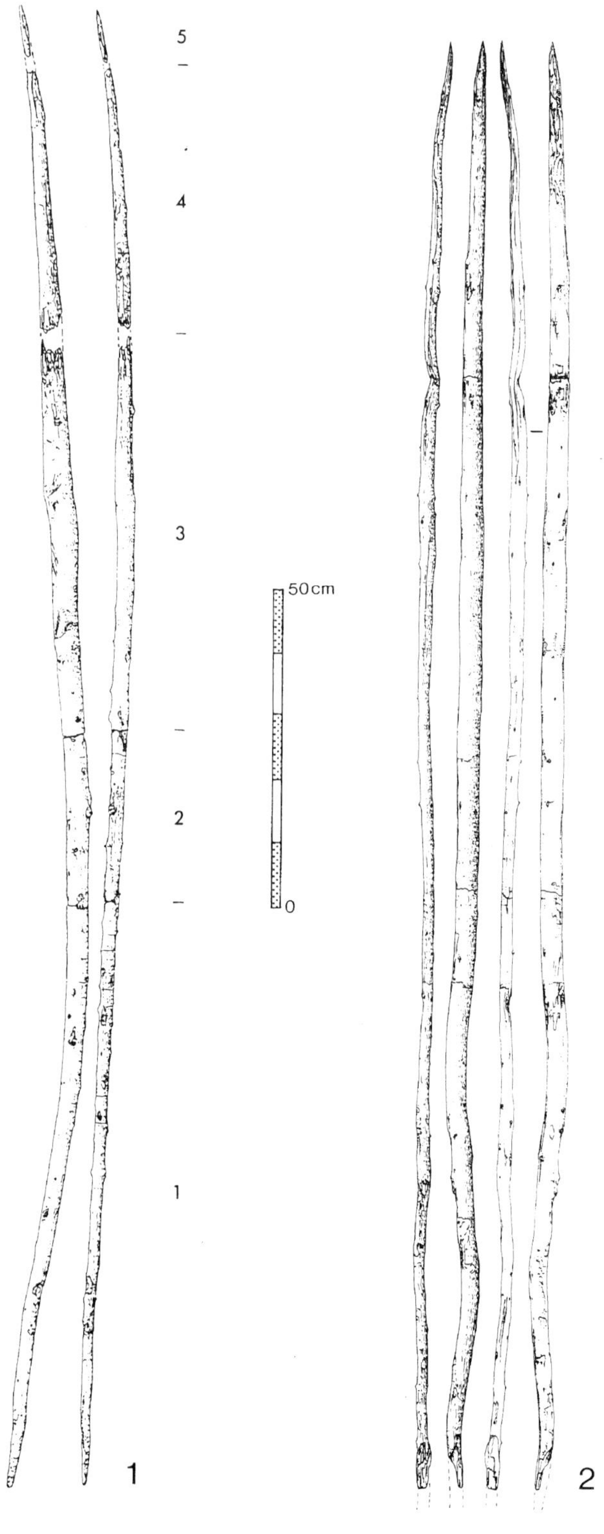

Figure 9 - *Schöningen 13 II–4. 1 - Spear I (length 2.25 m) with its fragments 1–5. 2 - Spear II (length 2.30 m).*

Figure 10 - Schöningen 13 II–4. Situation of spear VI in May 1997, view to northwest.

Just ten days after the symposium in Weimar, spear VI came to light, with a length of more than 2.5 m (Figure 10). And in the beginning of June 1997 spear VII with uncomplete base was discovered, also more than 2 m long.

Conclusions

These finds constitute the world's oldest wooden throwing spears and the oldest complete hunting weapons in one so far discovered to have been used by humans. As these spears are sandwiched between deposits of the Elsterian and the Saalian glaciations and situated within a well studied sedimentary sequence, their age can be estimated as probably 400 kyr BP. Studies of the litho-, pedo- and biostratigraphy of the pit have shown, that the findbearing sediments of the new discovered Reinsdorf Interglacial date from a temperate period that is older than oxygen isotope stage (OIS) 7, and are at least of OIS 9 age. Correlations of the Schöningen sequence to other areas (Mania, 1993, 1995 b) suggest for the time being that the sediments of the Reinsdorf Interglacial were deposited during the forth last interglacial (Figure 3) and are dating the spears probably at the end of OIS 11. Judging from the mammal fauna the Reinsdorf assemblage has to be younger than the Lower Palaeolithic site of Boxgrove, West Sussex, UK (Roberts et al., 1994; Roebroeks & van Kolfschoten, 1995).

Hitherto, the oldest complete known "spear" was one discovered in Eemian deposits at Lehringen, Lower Saxony, Germany in 1948 (Thieme & Veil, 1985); with an age of about 125 kyr BP (OIS 5e), this thrusting spear was recovered from between the ribs of a straight tusked elephant, and was made from yew (*Taxus*). The Schöningen spears are probably three full interglacial/glacial cycles older (Figures 1, 3). The throwing spears bring fresh light to the debate on the development of hunting capacities and on subsistence strategies of Middle Pleistocene hominids by showing that well-balanced sophisticated hunting weapons were very common from an early period of the Middle Pleistocene onwards. Accordingly, meat from hunting may have provided a larger dietary contribution than many workers have been prepared to acknowledge (Gamble, 1987; Nitecki, 1987).

The Schöningen wooden tools and hunting weapons dramatically illustrate how little is known about the "organic" component of early hominid material culture.

Acknowledgments

We are grateful to the Braunschweigische Kohlen-Bergwerke AG (BKB), Helmstedt, for the long term technical and organizational support of the archaeological excavations in the Schöningen mine, especially for their efforts to enable us to dig a 3,000 m^2 large area at Schöningen 13 II–4 (Figure 5). Thanks are due to W. H. Schoch (Adliswil, Switzerland) for the wood identifications, D. Mania (University of Jena) for the drawings of Figures 1–3, B. Kaletsch (Landesamt für Denkmalpflege Hessen, Marburg) for the drawing of Figure 9, P. Pfarr and C. S. Fuchs (Institut für Denkmalpflege, Hannover) for the photographs (Figures 4, 6–8, 10) and K. Cornelius (BKB, Helmstedt) for the photograph of Figure 5.

References

GAMBLE, C. (1987): Man the shoveler. In: O. Soffer (ed.), *The Pleistocene Old World. Regional perspectives,* 81–98. New York.

HEINRICH, W.-D. (1993): Zur Paläoökologie und Biostratigraphie der fossilen Kleinsäugetiere aus dem Travertinkomplex von Bilzingsleben. *Ethnographisch-Archäologische Zeitschrift* 34, 582–590.

KOLFSCHOTEN, T. van (1993): Die Vertebraten des Interglazials von Schöningen 12B. *Ethnographisch-Archäologische Zeitschrift* 34, 623–628.

KOLFSCHOTEN, T. van (1995): Faunenreste des altpaläolithischen Fundplatzes Schöningen 12 (Reinsdorf-Interglazial). In: H. Thieme & R. Maier, *Archäologische Ausgrabungen im Braunkohlentagebau Schöningen, Landkreis Helmsted,* 85–94. Hannover.

MANIA, D. (1993): Die Terrassen-Travertin-Sequenz von Bilzingsleben. Ein Beitrag zur Stratigraphie des Mittel- und Jungpleistozäns im Elbe-Saale-Gebiet. *Ethnographisch-Archäologische Zeitschrift* 34, 554–575.

MANIA, D. (1995 a): Die geologischen Verhältnisse im Gebiet von Schöningen. In: H. Thieme & R. Maier, *Archäologische Ausgrabungen im Braunkohlentagebau Schöningen, Landkreis Helmstedt,* 33–43. Hannover.

MANIA, D. (1995 b): The earliest occupation of Europe: The Elbe-Saale region (Germany). In: W. Roebroeks & T. van Kolfschoten (eds.), *The earliest occupation of Europe,* 85–101. Leiden.

NITECKI, M. H. (1987): The idea of human hunting. In. M. H. Nitecki & D. V. Nitecki (eds.), *The evolution of human hunting,* 1–9. New York/London.

OAKLEY, K.P., ANDREWS, P., KEELEY, L.H. & CLARK, J.D. (1977): A reappraisal of the Clacton spearpoint. *Proceedings of the Prehistoric Society* 43, 13–30.

ROBERTS, M. B., STRINGER, C. B. & PARFITT, S. A. (1994): A hominid tibia from Middle Pleistocene sediments at Boxgrove, UK. *Nature* 369, 311–313.

ROEBROEKS, W. & van KOLFSCHOTEN, T. (1995): The earliest occupation of Europe: A reappraisal of artefactual and chronological evidence. In: W. Roebroeks & T. van Kolfschoten (eds), *The earliest occupation of Europe,* 297–315. Leiden.

SCHOCH, W. H. (1995): Hölzer aus der Fundschicht 1 des altpaläolithischen Fundplatzes Schöningen 12 (Reinsdorf-Interglazial). In: H. Thieme & R. Maier, *Archäologische Ausgrabungen im Braunkohlentagebau Schöningen, Landkreis Helmstedt,* 73–84. Hannover.

THIEME, H. (1995 a): Der altpaläolithische Fundplatz Schöningen 13 I (Holstein-Interglazial). In: H. Thieme & R. Maier, *Archäologische Ausgrabungen im Braunkohlentagebau Schöningen, Landkreis Helmstedt,* 57-61. Hannover.

THIEME, H. (1995 b): Die altpaläolithischen Fundschichten Schöningen 12 (Reinsdorf-Interglazial). In: H. Thieme & R. Maier, *Archäologische Ausgrabungen im Braunkohlentagebau Schöningen, Landkreis Helmstedt*, 62–72. Hannover.

THIEME, H. (1995 c): Ein altpaläolithischer Lagerplatz aus der Zeit des Urmenschen von Schöningen 13 II (Reinsdorf-Interglazial). In: H. Thieme & R. Maier, *Archäologische Ausgrabungen im Braunkohlentagebau Schöningen, Landkreis Helmstedt*, 95–106. Hannover.

THIEME, H. (1996): Altpaläolithische Wurfspeere aus Schöningen, Niedersachsen. Ein Vorbericht. *Archäologisches Korrespondenzblatt* 26, 377–393.

THIEME, H. (1997): Lower Palaeolithic hunting spears from Germany. *Nature* 385, 807–810.

THIEME, H. & MAIER, R. (1995): *Archäologische Ausgrabungen im Braunkohlentagebau Schöningen, Landkreis Helmstedt*. Hannover.

THIEME, H., MAIER, R. & URBAN, B. (1987): Archäologische Schwerpunktuntersuchungen im Helmstedter Braunkohlerevier (ASHB). Zum Stand der Arbeiten 1983–1986. *Archäologisches Korrespondenzblatt* 17, 445–462.

THIEME, H., MAIER, R. & URBAN, B. (1992): Neue Erkennntnisse zum urgeschichtlichen Siedlungsgeschehen. *Archäologie in Deutschland* 8, 2, 26–30.

THIEME, H. & MANIA, D. (1993): „Schöningen 12" – ein mittelpleistozänes Interglazialvorkommen im Nordharzvorland mit paläolithischen Funden. *Ethnographisch-Archäologische Zeitschrift* 34, 610–619.

THIEME, H., MANIA, D., URBAN, B. & van KOLFSCHOTEN, T. (1993): Schöningen (Nordharzvorland). Eine altpaläolithische Fundstelle aus dem mittleren Eiszeitalter. *Archäologisches Korrespondenzblatt* 23, 147–163.

THIEME, H. & VEIL, S. (1985): Neue Untersuchungen zum eemzeitlichen Elefanten-Jagdplatz Lehringen, Ldkr. Verden. *Die Kunde* N.F. 36, 11–58.

URBAN, B. (1993): Mittelpleistozäne Interglaziale im Tagebau Schöningen. *Ethnographisch-Archäologische Zeitschrift* 34, 620–622.

URBAN, B. (1995 a): Vegetations- und Klimaentwicklung des Quartärs im Tagebau Schöningen. In: H. Thieme & R. Maier, *Archäologische Ausgrabungen im Braunkohlentagebau Schöningen, Landkreis Helmstedt*, 44–56. Hannover.

URBAN, B. (1995 b): Palynological evidence of younger Middle Pleistocene Interglacials (Holsteinian, Reinsdorf and Schöningen) in the Schöningen open cast lignite mine (eastern Lower Saxony, Germany). *Meded. Rijks Geol. Dienst* 52, 175-186.

URBAN, B, ELSNER, H., HÖLZER, A., MANIA, D. & ALBRECHT, B. (1991 a): Eine eem- und frühweichselzeitliche Abfolge im Tagebau Schöningen, Landkreis Helmstedt. *Eiszeitalter und Gegenwart* 41, 85–99.

URBAN, B., LENHARD, R., MANIA, D. & ALBRECHT, B. (1991 b): Mittelpleistozän im Tagebau Schöningen, Ldkr. Helmstedt. *Zeitschrift der deutschen geologischen Gesellschaft* 142, 351–372.

URBAN, B, THIEME, H. & ELSNER, H. (1988): Biostratigraphische, quartärgeologische und urgeschichtliche Befunde aus dem Tagebau „Schöningen", Ldkr. Helmstedt. *Zeitschrift der deutschen geologischen Gesellschaft* 139, 123–154.

Hominid Ranging Patterns and Dietary Strategies

Clive Gamble & James Steele

Abstract

Lithic transport data are used to estimate home range areas of Palaeolithic hominids at Caune de l'Arago and at Grotte Vaufrey. Modern comparisons demonstrate that these home ranges were of a magnitude consistent with use by local groups of a few families, and whose dietary focus was on meat. This is consistent with faunal evidence from Lower and Middle Palaeolithic Europe, and also with stable isotope evidence from Neanderthal bone collagen. Lithic transport data, technology, and faunal evidence all point to a shift to more intensive, specialized procurement of large, social herbivores after about 300, 000 BP.

Introduction.

Hominid ranging patterns and dietary strategies are of central concern to the study of human evolution yet they have proved difficult to quantify. Among archaeologists, the study of hominid ranging behaviour has been dominated in recent years by discussion of Isaac's 'central place foraging' model of site formation in the East African Early Palaeolithic (Isaac, 1978, 1980). This model was important for focusing attention on the potential of raw material transfers to provide a measure of the distances involved. The result of such studies in Africa and Europe have now received a major review (Féblot-Augustins, 1997).

Our aim here is to supplement Isaac's seminal approach by broadening it to consider ranging behaviour among other taxa and to use raw materials as a means to develop our understanding of early hominid landscape use. Our starting point is the rich literature which exists on ranging behaviour in extant mammals and which focuses primarily not on the identification and explanation of focal activity points (refuges, sleeping sites, locations of scarce resources) analogous to archaological 'sites', but rather on the ecological determinants of the total areal extent of the 'home range'.

In this context, the 'home range' denotes the area normally traversed by an individual animal or group of animals during activities associated with feeding, resting, reproduction, and shelter-seeking: infrequent exploratory movements outside this area are not considered part of this normal ranging area (Burt, 1943: cf Harestad & Bunnell 1979). In this paper, we shall summarize some aspects of the evolutionary ecology of hominid foraging strategies (relating to encephalization and to diet) for which estimates of home range area would have a direct relevance. We shall then introduce simple mathematical techniques of home range estimation which can be applied to raw material source data, and use them to derive home range area estimates for two sites from Lower and Middle Palaeolithic Europe. Finally, we shall draw some conclusions from this exercise concerning possible shifts in the dietary strategies of hominids in Lower and Middle Palaeolithic Europe.

Encephalization, dietary strategy and home range size

Hominid evolution is characterized by the trend to increased brain size, both in absolute terms and relative to the size of the body (Table 1). Since the brain accounts for 2 per cent of our body weight yet 20 per cent of our energy intake (Aiello & Wheeler, 1995) it is expected that

Sample	Temporal Range (Kyr BP)	Encephalization Quotient
Early PLeistocene	1,800-1,200	3.46
Late Early to early Middle Pleistocene	1,150-600	3.40
Middle Middle Pleistocene	550-400	3.82
Late Middle Pleistocene	300-200	4.26
Early Late Pleistocene	150-100	4.73
Skhul-Qafzeh	90	5.29
Late archaic *Homo sapiens*	75-36	4.78
Early Upper Palaeolithic	35-21	5.35
Late Upper Palaeolithic	21-10	5.41

Table 1 - The global pattern of evolution of brain-body size ratios in Pleistocene hominids. Data from Ruff et. al. (1997, Table 1).

encephalization will result in changes to dietary behaviour and subsistence organisation. Moreover, if we are to understand the reasons why encephalization had local selective advantages for Pleistocene hominids, then we have to understand both the costs of such a strategy, and the possible benefits.

With regard to metabolic costs, Aiello & Wheeler (1995) have argued that primates with normal basal metabolic rates cannot expand their level of energetic investment in one metabolically expensive tissue without compensating reduction in some other such tissue. They argue that an expansion in the relative size of the human brain has been obtained by corresponding reduction in the volume of gut tissue, and that this reduction in gut surface area and digestive capacity must have entailed a corresponding shift to a higher quality diet with less bulk and more rapid assimilation (namely, a shift to greater dependence on animal protein). The reduction in gut size that accompanied encephalization can be seen in contrasting the reconstructed shape of the thorax in *Australopithecus afarensis* (conical, impying a larger, chimpanzee-like gut compartment), and in *Homo erectus* (barrel-shaped, implying a human-like reduced gut compartment) (Figure 1).

Dietary shifts of the kind predicted by Wheeler & Aiello would have entailed corresponding shifts in the scale of the home range area. Comparative studies show clearly that among primates, higher-quality diets tend to be associated with greater home range areas: Thus frugivores have greater home ranges than folivores, reflecting the more dispersed, patchy and seasonal distribution of ripe fruits compared with leaves (Clutton-Brock & Harvey, 1977; Harvey & Clutton-Brock, 1981). In wider comparisons among living mammals, it is also clear that carnivores have far greater home range areas for their body size than herbivores and omnivores, due to the lesser productivity and biomass density of prey at the higher trophic level (Lindstedt et al., 1986; Figure 2). So, following Aiello & Wheeler's predictions, we should expect to see indications of a shift to greater dependence on animal protein not just in the faunal remains from Palaeolithic sites, but also in the total areal extent of the home range. Furthermore, carnivores in the less productive high latitudes have larger home ranges for their body size than carnivores living at lower latitudes (Figure 2). Looked at from a slightly different perspective, carnivores inhabiting large ranges for their group biomass tend not to be territorial: with greater range areas, range defence is not practicable (Grant et al., 1992; Figure 3).

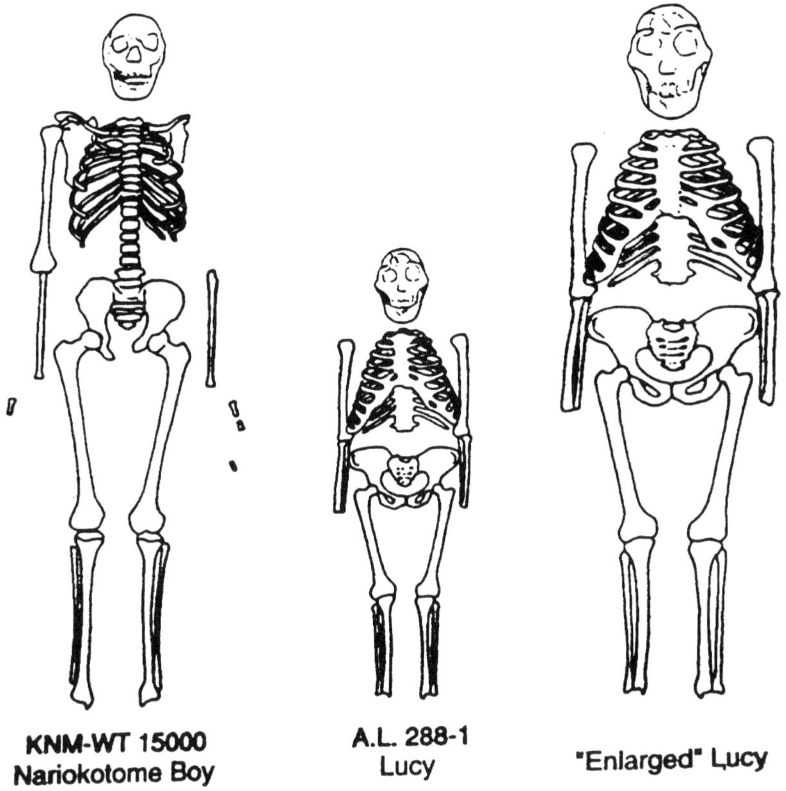

Figure 1 - *Reconstructed morphology of the thorax of Australopithecus afarensis and Homo erectus suggests that Homo erectus had a much reduced gut compartment, consistent with a greater dependence on animal protein. After Ruff (1993), Figure 2.*

With regards to adaptive benefits of encephalization, Byrne (1995) has recently reviewed the competing hypotheses for differences in brain size and relative neocortex size among extant non-human primates. He argues that while the 'Machiavellian intelligence' hypothesis (Byrne & Whiten, 1988), that primate cognitive abilities are selected for primarily as adaptations to competition within large social groups, may work for monkeys, it appears less effective as an explanation of the enhanced cognitive abilities of the great apes. For the great apes, he suggests, a better explanation may be in terms of adaptation to environmental challenges, including complex tool-assisted manipulative processing of food items, and the capacity to exploit complex diets and large home ranges by making economical route choices in foraging (as seen in chimpanzees and orangutans). By implication, this argument that hominoid brain size and cognitive ability is an adaptation to environmental challenges should also be extended to the hominines. Again, we would therefore expect to see an expansion of the home range as a correlate of the increased cognitive abilities of the more encephalized later hominids associated with the Palaeolithic record.

Home range area estimation, then, has relevance both for testing models of the physiological costs of hominid encephalization (since these imply dietary shifts to a strategy necessitating more extensive ranging patterns), and for testing models of the selective advantages of encephalization which relate it to resource competition by superior environmental knowledge (Byrne, 1995).

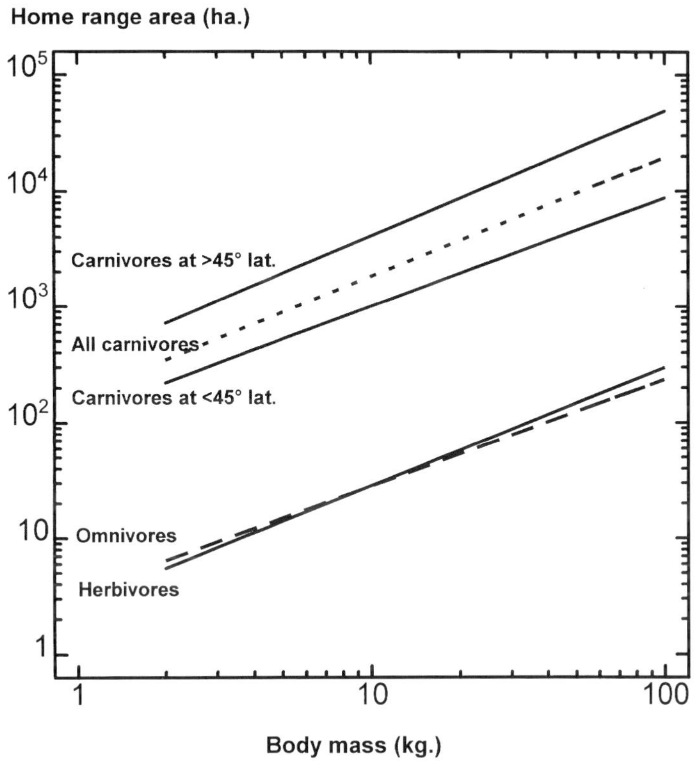

Figure 2 - Regression lines fitted to data on home range area and body mass for modern carnivores, herbivores and omnivores. Regression equations from Lindstedt et al. (1986).

Using raw material source data to estimate home range area

The principal method used to derive home range area estimates in the studies of primate and wider mammalian dietary groups is called the 'minimum convex polygon' method (cf Jennrich & Turner, 1969). This method takes all available point observations of an animal's location, and connects the outermost points to circumscribe the area, subject to the rule that the polygon thus formed has no inner angles greater than 180°.

Of course, other techniques exist for estimating home ranges of animals from tracking data, including those which attempt to delimit zones of most intensive use of the range - for example, by defining the core area containing an arbitrary proportion (such as 90 % or 75 %) of the observations (e.g. Anderson, 1982; Schoener, 1981; Worton, 1989). However, since the studies of comparative home range size in primates and other mammals cited above have generally used the largest estimates of home range size, usually as estimated by the Minimum Convex Polygon method, we shall adopt that method ourselves.

For our examples in this study, we shall use data derived from two sites in France (Figure 4), the Lower Palaeolithic site of Caune de l'Arago (Lebel, 1992; Wilson, 1988), and the early Middle Palaeolithic site of Grotte Vaufrey (Geneste, 1988a). The relevant levels from Caune de l'Arago (levels D, E, F and G) are correlated with OIS 12, the Anglian-Elster-Oka glaciation of northern Europe dated to 427,000-478,000 BP. Grotte Vaufrey has Levallois-based assemblages in Levels VIII through II, across the approximate date range ~142,000--~74,000 BP (Blackwell & Schwarcz

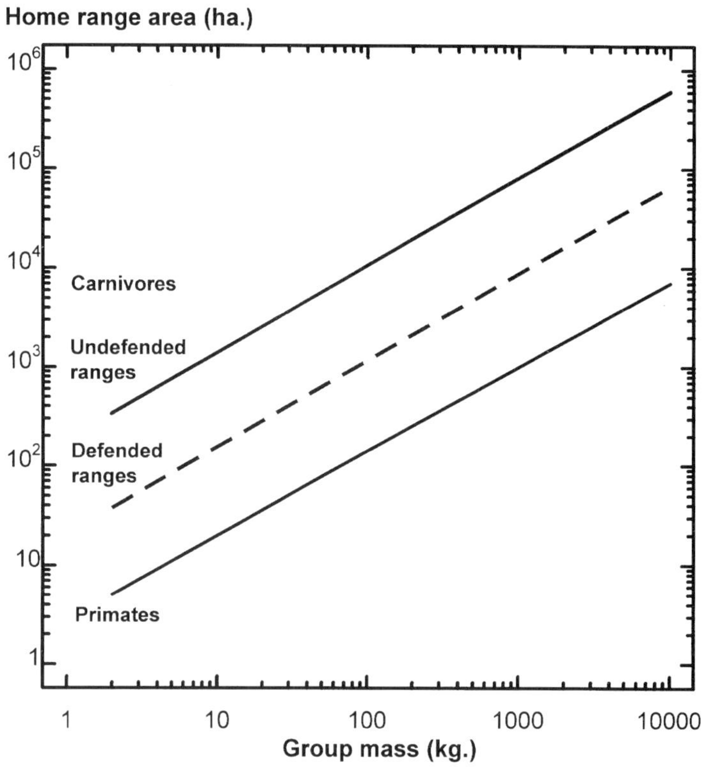

Figure 3 - Regression lines fitted to data on home range area and group mass for modern carnivores with defended and undefended ranges, and for primates. Regression equations from Grant et al. (1992).

1988): We shall focus on the earliest and richest artifact level in that range, level VIII.

The data which we shall use as point observations of hominid movements across the home ranges which included these two sites are raw material source data. That is, for a given site and level we have one point observation of activity - the site itself - and additional point observations of activity at the nearest outcrop or suitable secondary deposit of each of the rock types used to make the artefacts and waste found on the site. For example, at Arago lithic materials were brought in from at least eight locations: This gives us nine independent point observations of hominid location in the home range which included that site (Figure 5). In Level VIII of Grotte Vaufrey, there are lithic materials from at least seven different locations, giving us at least eight independent point locations of hominid activity within the home range that included that site (Figure 6).

Using the Minimum Convex Polygon method to fit boundaries to the home ranges defined by these observations, we can derive estimates of the home range areas of the hominids who visited these two Palaeolithic sites. For Arago, this method produces a home range area estimate of 97,600 hectares (Figure 7); for Level VIII of La Grotte Vaufrey, the estimate is 202,500 hectares (Figure 8). When these are compared to the relevant value from the curve for carnivores with undefended home ranges (Figure 9) we find that the expected group biomass for a carnivore inhabiting such ranges would be 1258 kg. and 1892 kg., respectively.

These estimates locate the ranges of these hominids squarely in the order of magnitude expected. If we assume that the average hominid social group consisted of an age-sex mixture in the

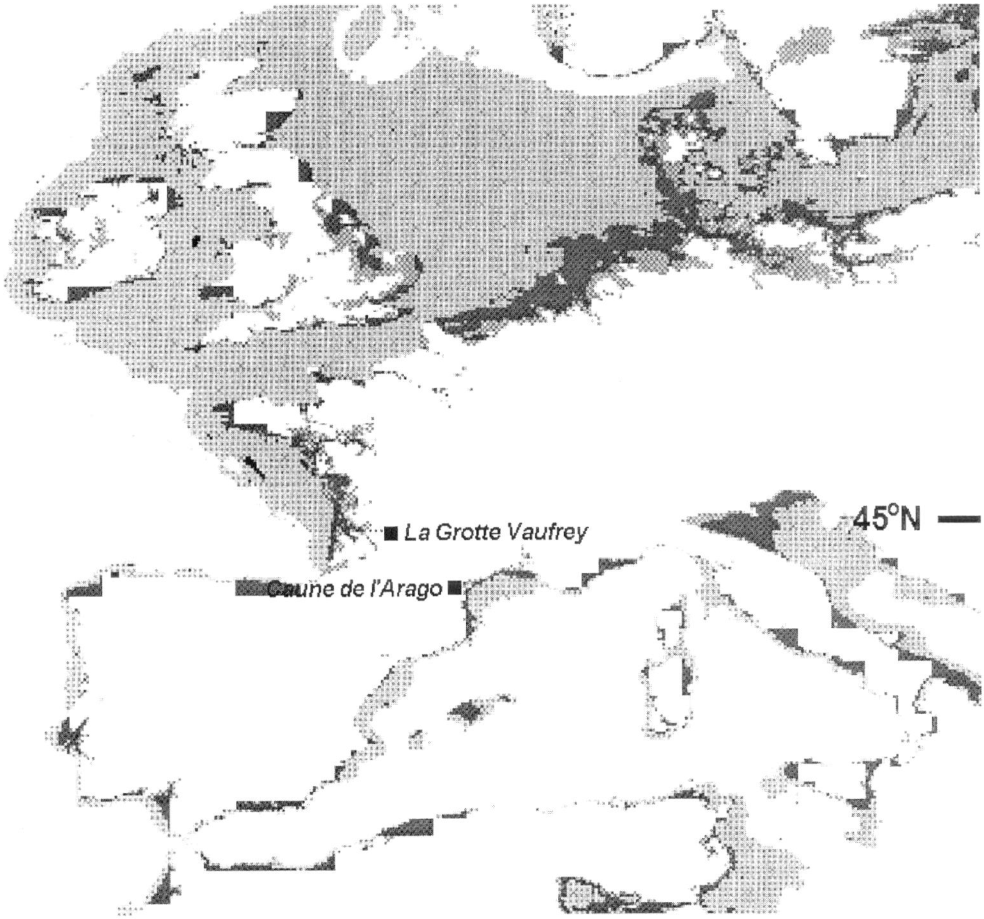

Figure 4 - Locations of Caune de l'Arago and Grotte Vaufrey, the two seites mentioned in the text.

proportions one adult male: one adult female: two children, with body weights as estimated for archaic *Homo sapiens* of 60 kgs for males, 52 kgs for females and children weighing half the mean adult weight (McHenry, 1996, 1988), then we can translate these hominid group biomass estimates into estimated group sizes of 30 (for Caune de l'Arago) and 69 (for Grotte Vaufrey). Because of the compounded errors and uncertainties in such estimations - not least in predicting the value of the independent variable from the estimated value of the dependent variable, and in using a late Holocene set of reference standards - we should perhaps conclude no more than that this is in the right sort of ball-park figure for hominid group sizes as conventionally estimated and suggesting that our comparative methodology may have some value for further refinement (Steele, 1996).

Other methods which are sensitive to intensity of use of the range can also be applied to these data, by repeating the point observations of activity at each raw material source location proportionally to the quantity of material recovered. To take the example of Level VIII of La Grotte Vaufrey, application of the adaptive kernel method (Worton, 1989) with a cut-off at 90% of

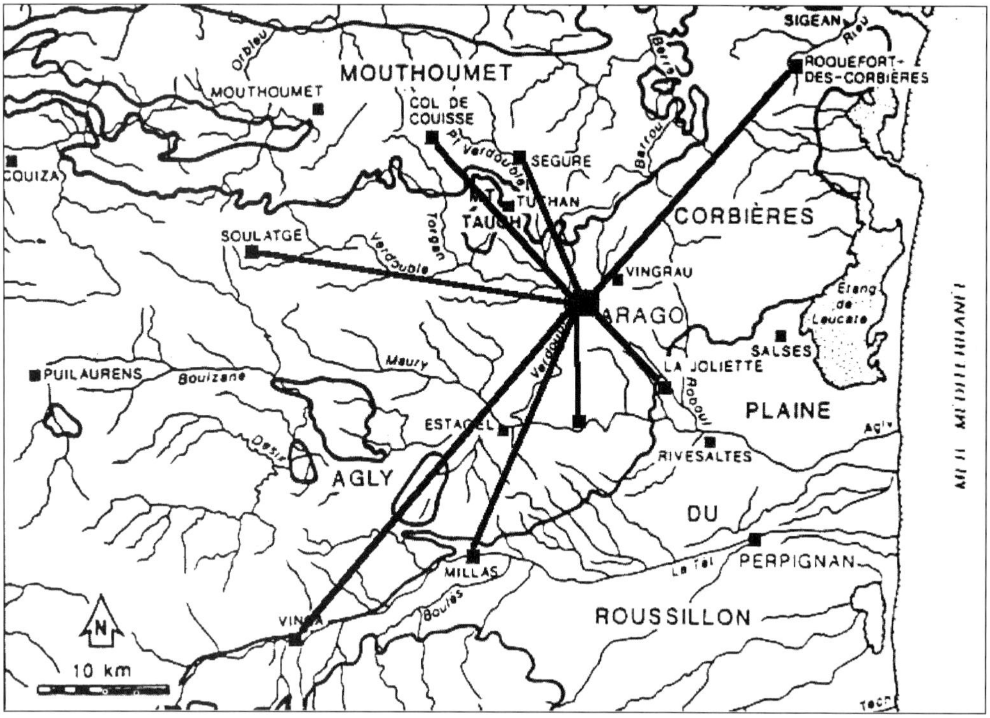

Figure 5 - Raw material sources for lithic finds from Caune de l'Arago, Levels D,E,F and G. Base map from Lebel (1992), Figure 4.

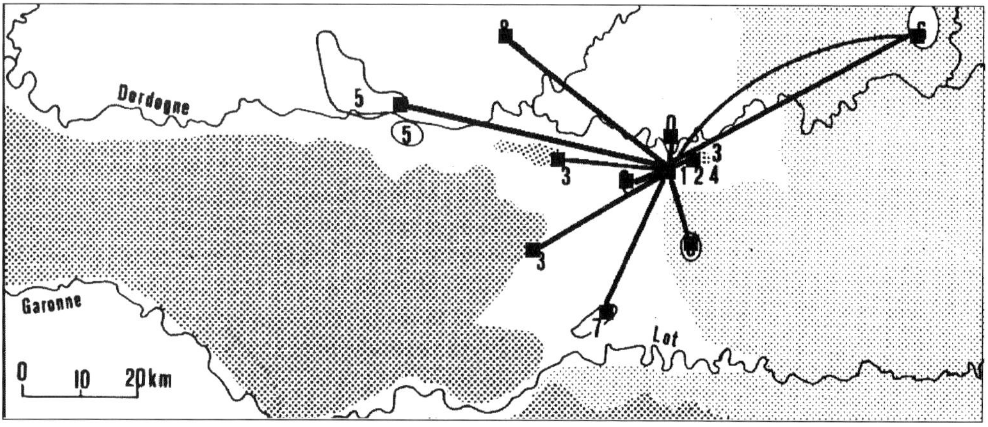

Figure 6 - Raw material sources for lithic finds from Grotte Vaufrey, Level 8. Base map from Geneste (1988a), Figure 2.

such observations substantially reduces the range estimate to an area focused on the site itself, with a radius of about 5 kilometres (Figure 10). This reflects the fact that the vast majority of material found on the site comes from the closest raw material sources. Indeed, the finding that most material comes from distances of 5 kilometres or less has already been reported by Geneste

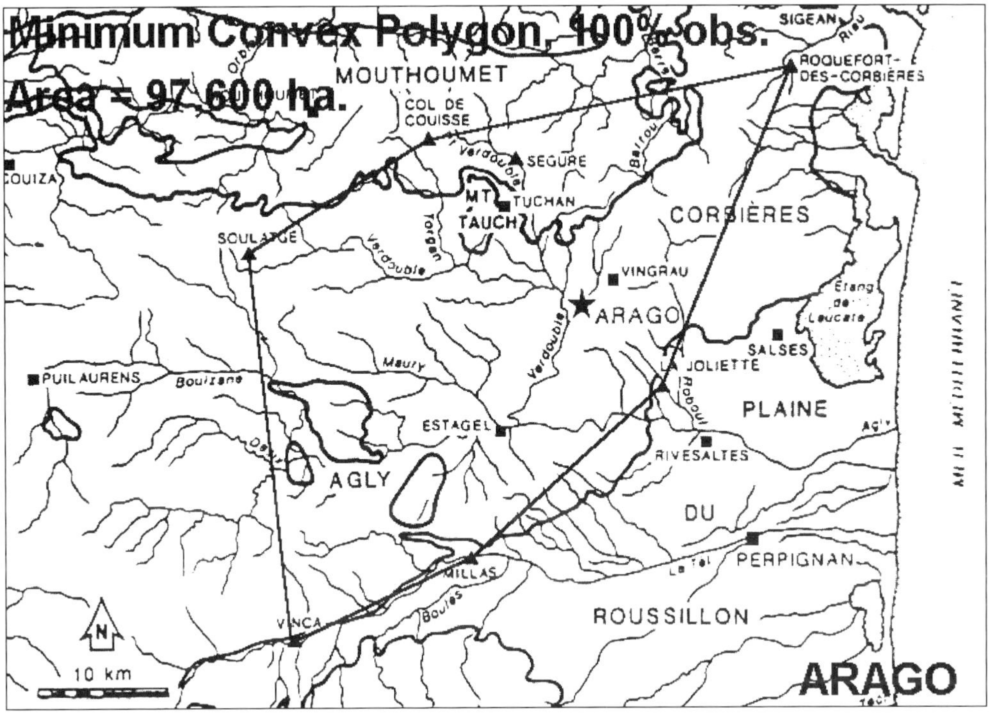

Figure 7 - *The Minimum Convex Polygon fitted to the Arago data, giving a home range estimate of 97,600 hectares.*

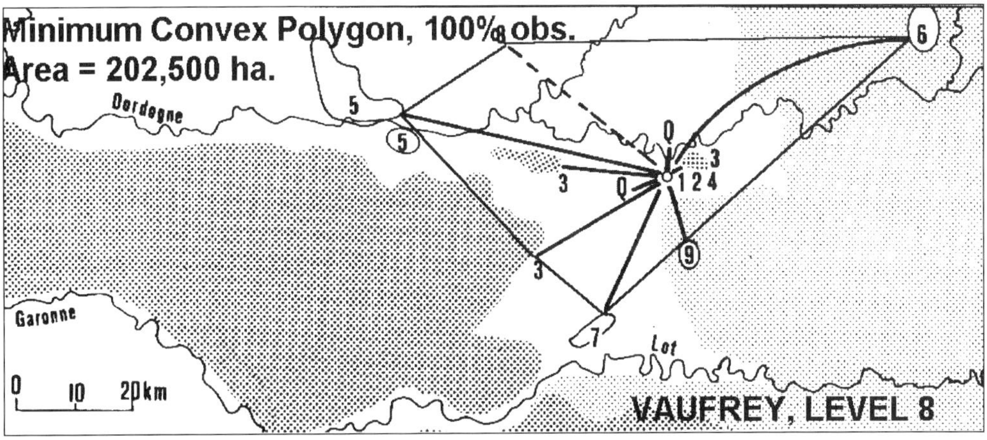

Figure 8 - *The Minimum Convex Polygon fitted to the Grotte Vaufrey data, giving a home range estimate of 202,500 hectares.*

in a detailed study of raw material transfers in a number of Middle Palaeolithic sites in south west France (Geneste, 1990, 1989, 1988a, 1988b), and he has also noted that material from the more distant sources is most likely to consist of retouched tools without the accompanying débitage. A similar pattern can also be seen at the much earlier site of Arago, where 80 % of all lithic material

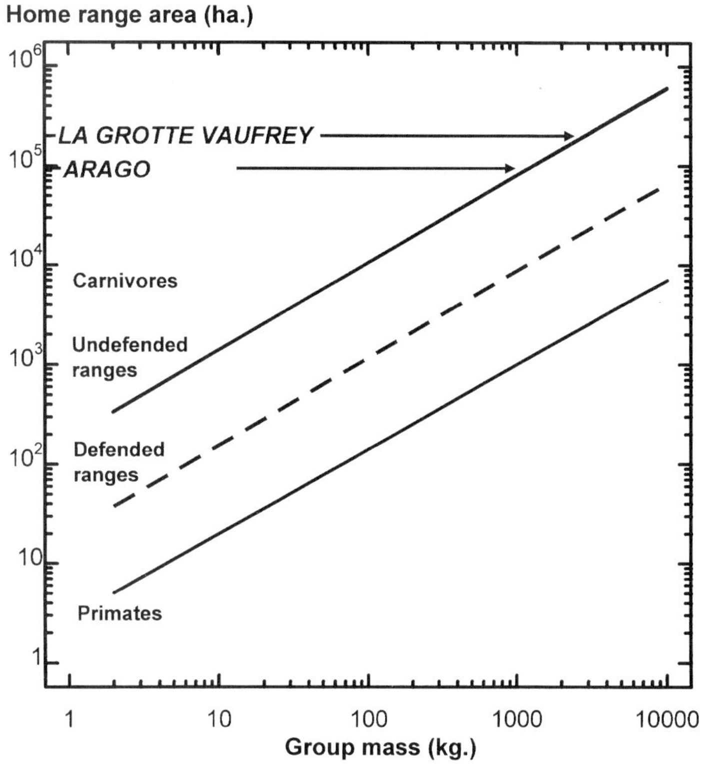

Figure 9 - The home range areas of Arago and Grotte Vaufrey, estimated using the Minimum Convex Polygon method, in relation to modern data from species at potentially comparable trophic levels.

came from within 5 kilometres of the cave (Lebel, 1992: 63). But while defining these local foci of intense procurement activity may tell us about the economics of radial mobility from the site itself, seen as a central place, it devalues the information about wider (and not site-centred) ranging patterns contained in the rarer representation of artefacts from the more distant sources. Presumably there were other places within the range, closer to these other raw material sources, where these artefacts were preferentially manufactured and where the waste was more frequently discarded.

Discussion.

Taken in isolation, these examples simply point to the potential for future studies of hominid ranging patterns requiring precise home range area estimates. To illustrate the direction such studies might take we present here an analysis based on Féblot-Augustins' (1997) survey of raw material transfers from Palaeolithic of Europe which draws among others on the work of Geneste (1988b), Roebroeks et al. (1988), Turq (1988), and Floss (1994).

The compilation in Table 2 shows some suggestive patterns. It confirms, in broad terms, that the colonisation of northern latitudes necessitated a carnivore-scale ranging pattern. This is indicated by the doubling of greatest ranging distances between the Lower Palaeolithic samples from Africa and Europe. This difference does not of course imply that early *Homo* in Africa was not exploiting

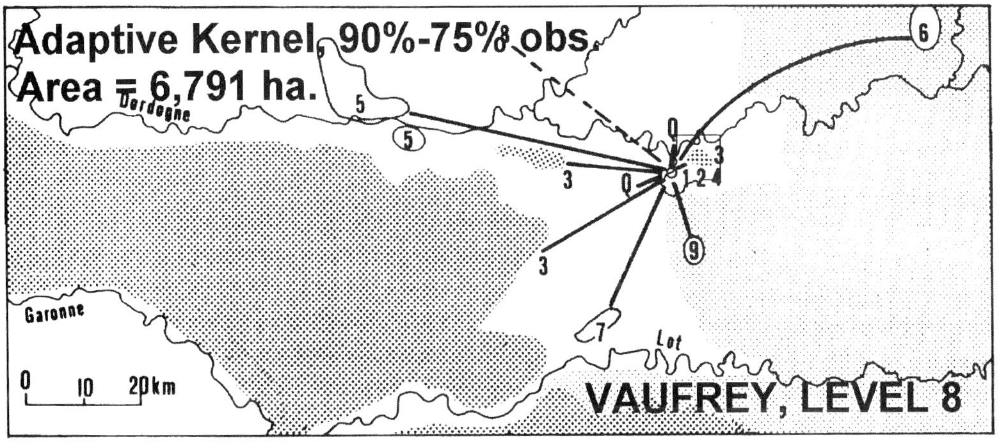

Figure 10 - The Adaptive Kernel method of home range area estimation, with a cut-off at anywhere between 90 % and 75 % of all observations, gives an estimate of 6,791 hectares.

animal protein as predicted by the expensive tissue hypothesis. The maximum transfer distances still imply home ranges substantially greater than those of comparable large-bodied primates. What we could be observing in Table 2 is the effect of latitude on the ecology of variation in Lower Palaeolithic foraging behaviour. Such variation in home range area along a similar latitudinal transect is also seen both in living carnivore species (e.g. Gompper & Gittleman, 1991; Linstedt et al., 1986), and (for both range size and residential mobility) among modern hunters and gatherers (Kelly, 1995).

In addition to such traces of variation in home range area in space, we have suggestive evidence of evolutionary trends in the extent of ranges within geographical regions over time. We have already seen that the home range estimated for Grotte Vaufrey was more than twice the area of that estimated for the much older site of Caune de l'Arago. Some of this contrast may be due to uncertainties in the sourcing of distant raw materials found at Grotte Vaufrey. But another possible implication is that the two home ranges were being used at different levels of intensity and in alternative ways. When we calculate the same average for the maximum distance to raw material sources across a larger sample of sites/layers we find indications that ranging areas became yet more extensive in the Late Middle Palaeolithic, and then again in the Early and Late Upper Palaeolithic (Table 2).

We might predict that any shift to more intensive but stable exploitation of selected resources would have to be counterbalanced either by their more active management or by an expansion of the area harvested, to give the resource available within the home range sufficient time to replenish itself. Could these apparent changes in Palaeolithic ranging patterns correlate with such shifts in dietary focus towards more intensive exploitation of single species, and/or of prime age animals? Or are these signs of mobility over greater distances the result of increased social contacts, largely independent of any further significant dietary shifts (Gamble, 1993)?

A number of lines of independent evidence may be relevant here. Stable isotope analysis of preserved collagen and bone apatite may enable reconstruction of the diet of Middle and Upper Palaeolithic hominids in Europe, and there is evidence already that Neanderthal bone collagen shows an isotopic signature that clusters with that of the carnivores (Fizet et al., 1995; Bocherens, this volume). It would also be interesting to see if there is faunal evidence for successive phases

	Africa	Europe
Lower Palaeolithic (1 myr - 300 kyr BP)	**14.6 km** n = 52	**28.2 km** n = 33
Early Middle Palaeolithic (300 - 128 kyr BP)		**37.8 km** n = 24
Late Middle Palaeolithic (128 - 40 kyr BP)		**57.6 km** n = 110
Early Upper Palaeolithic (40 - 20 kyr BP)		**114.7 km** n = 198
Late Upper Palaeolithic (20 - 10 kyr BP)		**125.01 km** n = 112

Table 2 - Average greatest distances for the transfer of lithic raw material at different periods. N = number of sites/layers in the sample. Data from Féblot-Augustins (1997).

of intensification of hunting practices during the Lower and Middle Palaeolithic. Certainly animals were killed during the earliest periods of persistent human occupation of northern Europe, as the recent find of spears from Schöningen demonstrates (Thieme, 1997): This site with its many horse remains will provide very important insight into hunting preferences. But hunting during the period 500,000-300,000 BP generally appears to have been directed at single animals (as at Neumark Nord - Mania et al., 1990) or to have consisted of discrete episodes where animals of different age, sex and condition were apparently stampeded, as at La Cotte (Scott, 1986). Stone tools are found with single megafauna (rhino at Boxgrove - Pitts & Roberts, 1997; mammoth, elephant at Notachirico - Belli et al., 1991) or with smaller herd species but where the human involvement remains ambiguous, as at Miesenheim (Turner, 1989).

By contrast, the Levallois technologies which appear with increasing frequency after 300,000 BP, of which Grotte Vaufrey is an example, also appear to be transported technologies associated with greater planning in use of lithic resources within the range (Geneste, 1989; White & Pettitt, 1995). This greater range size and change in the organisation and transport of aspects of technology seems to correlate with widespread evidence for the repeated hunting of prime age animals after 300,000 years ago (Gaudzinski, 1995, 1996; Hoffecker et al., 1989; Stiner, 1994). A selection of such sites is shown in Table 3, which points to the importance from the Early Middle Palaeolithic of large herd species (bison, reindeer, red deer). It is not clear whether such a shift should be attributed to an increased dietary focus on meat per se, or to a change in procurement strategy associated with changing social systems (and perhaps increases in group size and group mass). Certainly the data in Table 3 suggest that the subsequent, dramatic increases in greatest transfer distance in the Upper Palaeolithic had more to do with social relations and exchange than with fundamental dietary imperatives (Stringer & Gamble, 1993), since there is no evidence of any comparably large corresponding shift to a more 'small-gut, large-brain' physiology in modern humans as compared with their Neanderthal predecessors - despite recent estimates suggesting a substantially greater muscularity and body mass for the Neanderthals (Ruff et al., 1997, cf. Table 1).

Undoubtedly there will prove to be undetected errors in the assumptions we have made in this exercise, and undoubtedly there is scope for far more careful analysis of the variability in patterns

Site	OIS Age	No of IDs	No of carnivore species/ bones	No of herbivore species/ bones	Species	Percent of identified bones	MNI	Reference
Coudoulous I, Level 4	OIS 6-5				Bison	98	94	Jaubert (1994)
La Borde	OIS 6-5	440	1/10	4/430	Aurochs	93		Jaubert et al. (1990)
Il'skaya	OIS 5d-4	1446	5/35	7/1411	Bison	82	51	Hoffecker et al. (1991)
Érd	OIS 4	14862	9/13699	12/1163	Bear	89	c. 480	Gabori-Csank (1968: 62)
Wallertheim E	OIS 5c	>190	2/?	2/?	Bovid	c. 85		Conard (1995)
Wallertheim F	OIS 5c	>467	2/?	4/?	Horse	c. 78		Conard (1995)
Wallertheim B1	OIS 5d	1131	3/3	5/1128	Bison	76	52	Gaudzinski (1992: Table 4)
Grotte Vaufrey VIII	OIS 6	1009	6/109*	5/894	Red deer	74		Delpech (1988)
Salzgitter-Lebenstedt	OIS 5d-4		1/?	5/?	Reindeer	72	c. 80	Kleinschmidt (1953: 180)
Wallertheim AII	OIS 5d	2436	3/86	7/2350	Bison	67		Gaudzinski (1996:Table 2)
Abri des Canalettes 3	OIS 5a	1222	7/38	9/1184	Red deer	53	19	Patou-Mathis (1993: Table 40)
Biache St Vaast II base	OIS 7	435	2/92	8/343	Bovid	48	8	Auguste (1992: Table 10)
Abri des Canalettes 2	OIS 5a	382	2/6	9/376	Red deer	47	11	Patou-Mathis (1993: Table 31)

* Does not include bear.

Table 3 - Specialized faunas 300,000 - 60, 000 BP.

of home range use both across ecological zones and through time. Nonetheless, we propose that this initial exercise shows that the archaeological record is capable of yielding the right data for testing general conceptual models of hominid behaviour and its evolution.

Acknowledgements

We are grateful for the insightful comments of our colleagues in Weimar on the version of this paper that was presented at the meeting. We thank Dr. Herbert Ullrich for his kind invitation to participate in the Symposium.

References

AIELLO, L. & WHEELER, P. (1995): The expensive-tissue hypothesis: The brain and the digestive system in human and primate evolution. *Current Anthropology* 36, 199-221.

ANDERSON, D.J. (1982): The home range: A new non-parametric estimation technique. *Ecology* 63, 103-112.

AUGUSTE, P. (1992): Etude archéozoologique des grands mammiferes du site pleistocène moyen de Biache-Saint Vaast (Pas-de-Calais, France). *L'Anthropologie* 96, 49-70.

BELLI, G., BELLUOMINI, G., CASSOLI, P.F., CECCHI, S., CUARZI, M., DELITALA, L., FORNACIARI, G., MALLEGNI, F., PIPERNO, M., SEGRE, A.G. & SEGRE-NALDINI, E. (1991): Découverte d'un femur humain acheuléen a Notachirco (Venosa, Basilicate). *L'Anthropologie* 95, 47-88.

BLACKWELL, B. & SCHWARTZ, H.-P. (1988): Datation des spéleothems de la grotte Vaufrey par la famille d'Uranium. In: J.-P. Rigaud (ed.), *La Grotte Vaufrey à Cenac et Saint-Julien (Dordogne). Paléoenvironnements, chronologie et activités humaines*, 365-380. Paris (Mémoires de la Société Préhistorique Française vol. 19).

BURT, W.H. (1943): Territoriality and home range concepts as applied to mammals. *Journal of Mammalogy* 24, 346-352.

BYRNE, R.W. (1995): *The thinking ape*. Oxford.

BYRNE, R. W. & WHITEN, A. (eds) (1988): *Machiavellian intelligence: Social expertise and the evolution of intellect in monkeys, apes and humans*. Oxford.

CLUTTON-BROCK, T.H. & HARVEY, P.H. (1977): Primate ecology and social organization. *Journal of Zoology (London)* 183, 1-39.

CONARD, N.J. (1995): The Middle Palaeolithic faunal assemblages from the 1991-1994 excavation in Wallertheim, Rheinhessen. Paper presented at the Neuwied conference on „*The role of early humans in the accumulation of European Lower and Middle Palaeolithic bone assemblages*".

DELPECH, F. (1988): Les grands mammifères de la grotte Vaufrey à l'exception des Ursides. In: J.-P. Rigaud (ed.), *La Grotte Vaufrey à Cenac et Saint-Julien (Dordogne). Paléoenvironnements, chronologie et activités humaines*, 213-290. Paris (Mémoires de la Société Préhistorique Française vol. 19).

FÉBLOT-AUGUSTINS, J. (1997): *La circulation des matières premières au paléolithique*. Liège (E.R.A.U.L. 75).

FIZET, M., MARIOTTI, A., BOCHERENS, H., LANGE-BADRÉ, B., VANDERMEERSCH, B., BOREL, J.P. & BELLON, G. (1995): Effect of diet, physiology and climate on carbon and nitrogen stable isotopes of collagen in a Late Pleistocene anthropic ecosystem: Marillac, Charente, France. *Journal of Archaeological Science* 22, 67-79.

FLOSS, H. (1994): *Rohmaterialversorgung im Paläolithikum des Mittelrheingebietes*. Bonn.

GABORI-CSANK, V. (1968): *La station du paléolithique moyen d'Erd, Hongrie*. Budapest.

GAMBLE, C.S. (1993): Exchange, foraging and local hominid networks. In: C. Scarre & F. Healy (eds.), *Trade and exchange in prehistoric Europe*, 35-44. Oxford.

GAUDZINSKI, S. (1992): Wisentjäger in Wallertheim: Zur Taphonomie einer mittelpaläolithischen Freilandfundstelle in Rheinhessen. *Jahrbuch des Römisch-Germanischen Zentralmuseums Mainz* 39, 245-423.

GAUDZINSKI, S. (1995): Wallertheim revisited: A reanalysis of the fauna from the Middle Paleolithic site of the Wallertheim (Rheinhessen/Germany). *Journal of Archaeological Science* 22, 51-66.

GAUDZINSKI, S. (1996): On bovid assemblages and their consequences for the knowledge of subsistence patterns in the Middle Palaeolithic. *Proceedings of the Prehistoric Society* 62, 19-39.

GENESTE, J.-M. (1988a): Les industries de la Grotte Vaufrey: technologie du débitage, économie et circulation de la matière premiére lithique. In: J.-P. Rigaud (ed.), *La Grotte Vaufrey à Cenac et Saint-Julien (Dordogne). Paléoenvironnements, chronologie et activités humaines*, 441-518. Paris. (Mémoires de la Société Préhistorique Française vol. 19).

GENESTE, J.-M. (1988b): Systemes d'approvisionnement en matières premières au paléolithique moyen et au paléolithique supérieur en Aquitaine. In: M. Otte (ed.), *L'Homme de Néandertal*, vol. 8, 61-70. Liège.

GENESTE, J.-M. (1989): Economie des ressources lithiques dans le mousterien du sud-ouest de la France. In: M. Otte (ed.), *L'Homme de Néandertal*, vol. 6, 75-97. Liège.

GENESTE, J. (1990): Development des systems de production lithique au cours du paleolithique moyen en aquitaine septentrionale. *Memoires du Musée de Prehistoire d'Il-de-France* 3, 203-213.

GENESTE, J., TEXIER, J. & RIGAUD, J. (1989): Les plus anciens vestiges de la presence humaine en aquitaine. In: *Les premiers europeens*, 11-26. Paris.

GOMPPER, M.E. & GITTLEMAN, J.L. (1991) Home range scaling: Intraspecific and comparative trends. *Oecologia* 87, 343-348.

GRANT, J.W.A., CHAPMAN, C.A. & RICHARDSON, K.S. (1992): Defended versus undefended home range size of carnivores, ungulates and primates. *Behavioural Ecology and Sociobiology* 31, 149-161.

HARESTAD, A.S. & BUNNELL, F.L. (1979): Home range and body weight - a re-evaluation. *Ecology* 60, 389-402.

HARVEY, P.H. & CLUTTON-BROCK, T.H. (1981): Primate home range size and metabolic needs. *Behavioral Ecology and Sociobiology* 8, 151-155.

HOFFECKER, J. F., BARYSHNIKOV, G.F. & PATAPOVA, O.R. (1989): Mousterian bison hunters of the northern Caucasus: analysis of faunal remains from Il'Skaya 1. *CRP* 6, 69-72.

HOFFECKER, J. F., BARYSHNIKOV, G.F. & PATAPOVA, O.R. (1991): Verebrate remains from the Mousterian site of Il'skava 1 (Northern Caucasus, U.S.S.R): New analysis and interpretation. *Journal of Archaeological Science* 18, 113-147.

ISAAC, G. (1978): The food sharing behaviour of proto-human hominids. *Scientific American* 238, 90-108.

ISAAC, G. (1980): Stone age visiting cards: Approaches to the study of early land use patterns. In: N.Hammond, G.Isaac & I.Hodder (eds.), *Pattern of the past: Studies in honour of David Clarke*, 131-55. Cambridge.

JAUBERT, J. (1994): Environmental background of Middle Palaeolithic occupation and Middle Palaeolithic settlement systems: Basin de la Garonne: Region Midi Pyrenées. Paper presented at the *„ESF Workshop on the European Middle Palaeolithic, Arras"*.

JAUBERT, J., LORBLANCHET, M., LAVILLE, H., SLOTT-MOLLER, R., TURQ, A. & BRUGAL, J. (1990): *Les chasseurs d'aurochs de La Borde*. Paris (Documents d'Archeologie Francaise, vol. 27).

JENRICH, R.I. & TURNER, F.B. (1969): Measurement of non-circular home range. *Journal of Theoretical Biology* 22, 227-237.

KELLY, R. (1995): *The foraging spectrum: Diversity in hunter-gatherer lifeways*. Washington, London.

KLEINSCHMIDT, A. (1953): Die zoologischen Funde der Grabung Salzgitter-Lebenstedt, 1952. *Eiszeitalter und Gegenwart* 3, 166-188.

LEBEL, S. (1992): Mobilite des hominides et systemes d'exploitation des ressources lithiques au paléolithique ancien: La Caune de l'Arago (France). *Canadian Journal of Archaeology* 16, 48-69.

LINDSTEDT, S.L., MILLER, B.J. & BUSKIRK, S.W. (1986): Home range, time, and body size in mammals. *Ecology* 67, 413-418.

MANIA, D., THOMAE, M., LITT, T. & WEBER, T. (1990): *Neumark-Gröbern. Beiträge zur Jagd des mittelpaläolithischen Menschen*. Berlin (Veröffentlichungen des Landesmuseums für Vorgeschichte Halle, Bd. 43).

MCHENRY, H. M. (1988): New estimates of body weight in early hominids and their significance to encephalization and megadontia in „robust" australopithecines. In: F.E. Grine (ed.), *The evolutionary history of the robust australopithecines*, 133-48. New York.

MCHENRY, H. (1996): Sexual dimorphism in fossil hominids and its socioecological implications. In: J. Steele & S. Shennan (eds.), *The archaeology of human ancestry*, 91-109. London.

PATOU-MATHIS, M. (1993): Etude taphonomique et palethnographique de la faune de l'abri des Canalettes. In: L. Meignen (ed.), *L'Abri des Camalettes. Un habitat mousterien sur les grands Camsses (Nomt, Aveynon)*, 199-237. Paris.

PITTS, M. & ROBERTS, M. (1997): *Fairweather Eden*. London.

ROEBROEKS, W., KOLEN, J. & RENSINK, E. (1988): Planning depth, anticipation and the organization of Middle Palaeolithic technology: the „archaic natives" meet Eve's descendants. *Helinium* 28, 17-34.

RUFF, C.B. (1993): Climatic adaptation and hominid evolution: the thermoregulatory imperative. *Evolutionary Anthropology* 2, 53-60.

RUFF, C.B., TRINKAUS, E. & HOLLIDAY, T.W. (1997): Body mass and encephalization in Pleistocene *Homo*. *Nature* 387, 173-176.

SCHOENER, T.W. (1981): An empirically based estimate of home range. *Theoretical Population Biology* 20, 281-325.

SCOTT, K. (1986): The large mammal fauna. In: P. Callow & J.M. Cornford (eds.), *La Cotte de St. Brelade 1961-1978: excavations by C.B.M. McBurney*, 109-39. Norwich.

STEELE, J. (1996): On predicting hominid group sizes. In: J. Steele & S. Shennan (eds.), *The archaeology of human ancestry*, 230-252. London.

STINER, M. C. (1994): *Honor among thieves: A zooarchaeological study of Neandertal ecology*. New Jersey.

STRINGER, C. & GAMBLE, C. (1993): *In search of the Neanderthals: Solving the puzzle of human origins*. London.

THIEME, H. (1997): Lower Palaeolithic hunting spears from Germany. *Nature* 385, 807-810.

TURNER, E. (1989): Miesenheim I: A lower Palaeolithic site in the Middle Rheineland (Neuwied Basin), FRG. *Ethnographische-Archäologische Zeitschrift* 30, 521-31.

TURQ, A. (1988): L'approvisionnement en matières premières lithiques du magdalénien du Quercy et du Haut-Agenais: étude préliminaire. *Colloque de Chancelade10-15 Octobre 1988*, 301-308.

WHITE, M. J. & PETTITT, P. (1995): Technology of early Palaeolithic western Europe: Innovation, variability and a unified framework. *Lithics* 16, 27-40.

WILSON, L. (1988): Petrography of the Lower Palaeolithic tool assemblage of the caune de l'Arago (France). *World Archaeology* 19, 376-387.

WORTON, B.J. (1989): Kernel methods for estimating the utilization distribution in home-range studies. *Ecology* 70, 164-168.

Reconstruction of Neandertal Diet Using Bone Collagen Carbon and Nitrogen Stable Isotopes

Hervé Bocherens

Abstract

In a given individual, natural abundances in ^{13}C and ^{15}N of bone collagen are linked to those of its diet. This isotopic information can be linked to dietary parameters such as the plants at the basis of the food web and the position in the trophic web. In the case of opportunistic species, such as hominids, this approach can provide valuable information on the relative proportions of food originating from plant and animal organic matter, as well as on the relative proportions of terrestrial and marine foodstuff. However, this palaeodietary approach is valid only if the original isotopic abundances of fossil collagen are preserved despite the diagenetic evolution of fossil bones. The verification of collagen isotopic integrity is done by controling the biochemical purity of the organic matter extracted from fossil bones, and by checking that the isotopic differences observed in modern mammals are indeed measured in the fossil samples. Sites with a good isotopic preservation of collagen older than 100,000 yr BP have been recognized in temperate and arctic environments. Thus, some Neandertal specimens fall within the range of use of this technique. A review of the isotopic features of the Upper Pleistocene mammal fauna will be presented as well as examples and perspectives of applications to the knowledge of the diet of Neandertals.

Introduction

The transition from the Middle to the Upper Paleolithic is an important moment in human evolution in Europe which takes place between 40,000 and 30,000 BP. This change is linked to the replacement of Neandertal populations by modern human populations (Hublin, 1983; Hayden, 1993; Stringer & Gamble, 1993; Hublin et al., 1996; Otte, 1996). Changes in dietary adaptations have often been proposed as an important factor in this transition. For instance, it has been suggested that Neandertals were less effective predators than Upper Paleolithic humans (Binford, 1968). However, increasing evidence of specialized hunting for large bovids (e.g. Farizy & David, 1989; Gaudzinski, 1996) makes Neandertal's behaviour closer to that of modern humans than previously thought, at least on this aspect of its subsistence strategy (David & Enloe, 1993). It is thus crucial to assess quantitatively the dietary behaviour of Neandertals.

Traditional approaches used for the determination of the diet of ancient humans deal mainly with bone remnants, plant fragments and artefacts recovered from archaeological excavations. However, the archaeological record is usually biased against perishable items, such as plant material. As a consequence, bone is over-represented. Moreover, the remnants recovered during an excavation are not representative of the consumed species during the whole year, since most of the sites are only seasonally occupied by hunter-gatherers (Tauber, 1981). An additional complication arises from the fact that in most cave sites bones have been accumulated by other carnivores as well as by humans, and this taphonomic factor may confuse the reconstruction of human food-getting behaviour (Webb, 1989). For all these reasons, the dietary remnants are never preserved in archaeological sites in proportions reflecting exactly their proportions in the diet of ancient humans. An additional source of palaeodietary information is provided by some morphological characters of the human skeletal remnants themselves, such as dental microwear and skeletal pathologies also provide dietary information (Puech et al., 1980; Lalueza Fox & Pérez-Pérez, 1993; Lalueza et al., 1993; Skinner, 1996), but this information is qualitative at best (Ambrose, in press).

The chemical composition of bones and teeth do record some information about the dietary input of an individual, since the skeletal tissues are formed from this material. This chemical record is also partly dependent on the physiological and environmental conditions experienced by the individual. Thus, the chemical composition of fossil bones and teeth might provide valuable information about ancient ways of life, but only if the relationships between diet and the chemical composition of skeletal tissues are well understood on modern material, and provided that diagenetic alteration did not interfere with the biogenic record.

This paper will focus on the isotopic composition of carbon and nitrogen in collagen, a protein that can survive fossilisation for thousands of years in some specimens, and about the information that it can provide on Neandertal's diet. Emphasis will be put on the conditions in which this approach can be used safely.

Diet reconstruction with stable isotopes of collagen

Carbon and nitrogen stable isotopes

Both carbon and nitrogen present two stable isotopes, with slightly different atomic weights, carbon-12 (^{12}C) and carbon-13 (^{13}C), and nitrogen-14 (^{14}N) and nitrogen-15 (^{15}N) respectively. In each case, the heavier isotope (^{13}C and ^{15}N) is much less abundant in nature than the lighter one (^{12}C and ^{14}N). Relative to the whole corresponding element on earth, ^{13}C ranges from 0.99 to 1.15 % whereas ^{15}N ranges from 0.348 to 0.395 % (Éliot et al., 1993). The isotopes of a given chemical element present similar chemical characteristics, but the small differences in the weight of their respective cores lead to small differences in behaviour during the chemical reactions in which they are involved. When a chemical reaction takes place in a mixture of molecules containing one or the other isotope of a given chemical element, the reaction will be biased in favor of one isotopic type relative to the other, leading to an isotopic fractionation. The consequence is that the product of the chemical reaction will present a slightly different amount of a given isotope when compared to the starting product. Mass spectrometry allows the measurement of isotopic enrichments or depletions of very small amplitude, thanks to a relative measurement relative to internationally defined standards. The following isotopic notation „delta" (δ) is used to express the measured isotopic amounts: $\delta EX = (R_{sample}/R_{standard} - 1) \times 1000$ (‰), where X stands for C or N, E stands for 13 or 15 respectively, and R stands for the isotopic ratios $^{13}C/^{12}C$ and $^{15}N/^{14}N$ respectively. The standard is a marine carbonate (PDB = PeeDee Formation Belemnite) for carbon (Craig, 1953) and atmospheric N_2 (AIR) for nitrogen (Mariotti, 1983). A positive value of δ corresponds to the case of a sample enriched in the heavy isotope relative to the standard, whereas in the case of a negative value of δ the sample is depleted in the heavy isotope relative to the standard.

Carbon isotopes in food webs

The $\delta^{13}C$ values of the whole body of an animal are slightly higher (+ 0 to 1 ‰) than those of its diet (DeNiro & Epstein, 1978). In a given ecosystem, the $\delta^{13}C$ values of the animals are thus linked to those of the plants at the basis of the trophic webs to which they belong. In temperate and arctic ecosystems, terrestrial plants present an average $\delta^{13}C$ value of -27.1 ± 2.0 ‰ (O'Leary, 1981), with variations due to changes in the isotopic composition of their source of inorganic carbon (atmospheric CO_2) and to some environmental conditions (Deines, 1980; Tieszen, 1991). The atmospheric CO_2 presents an average $\delta^{13}C$ value of -7.8 ‰, with temporal and spatial variations. This value decreases since the middle of the nineteenth century due to the addition of CO_2 from industrial fossil fuel combustion and deforestation, and this CO_2 has a $\delta^{13}C$ value close to -25 ‰. The $\delta^{13}C$ value of atmospheric CO_2 thus decreased from around -6.5 ‰ in 1850 to -6.7 ‰ in 1956, and to -7.8 ‰ in 1989 (Marino & MacElroy, 1991). These isotopic shifts are recorded by the plants and along the whole food web, and thus have to be taken into account when comparing isotopic values obtained from modern and ancient specimens. Under a closed canopy forest the CO_2 generated by the respiration and the decay of organic matter with $\delta^{13}C$ values around -25 ‰ leads to a local depletion of the atmospheric CO_2 $\delta^{13}C$ values. Consequently the underbush plants are significantly ^{13}C-depleted relative to plants at the top of the canopy or plants in an open environment, with $\delta^{13}C$ values ranging from -28 to -35 ‰ (Van der Merwe & Medina, 1991). Marine plants use mostly dissolved bicarbonate with $\delta^{13}C$ values close to 0 ‰ and present $\delta^{13}C$ values around -20 ‰ (Deines, 1980). An addtional factor of variation for terrestrial plants $\delta^{13}C$ values is their photosynthetic pathway. Nearly all native plants in temperate and arctic environments use the so-called „C3" photosynthetic pathway, and present $\delta^{13}C$ values around -27 ‰ as described above. On the other hand, tropical grasses, including maize, sugar cane and sorgho, as well as some plants adapted to salty environments use the „C4" photosynthetic pathway and their $\delta^{13}C$ values average -13.1 $\pm$ 1.2 ‰ (O'Leary, 1981). A few plants (mostly succulent plants such as cactuses) exhibit a third photosynthetic pathway, the so-called CAM plants (CAM means Crassulacean Acid Metabolism). The $\delta^{13}C$ values of these CAM plants cover the whole range of values of C3- and C4 plants, since they use alternatively both photosynthetic pathways depending on the environmental conditions (O'Leary, 1981). However, high latitude ecosystems are completely dominated by C3-plants and thus C4 and CAM plants will be negligeable in the forthcoming discussions.

Nitrogen isotopes in food webs

Contrarily to carbon, the $\delta^{15}N$ values of animals are significantly higher than those of their diet (DeNiro & Epstein, 1981). Thus $\delta^{15}N$ values increase at each trophic level, in every ecosystem (Minagawa & Wada, 1984; Schoeninger & DeNiro, 1984). Controlled diet experiments have found an increase of the $\delta^{15}N$ values between diet and animal of 3.3 ‰ (DeNiro & Epstein, 1981). The measured difference between the $\delta^{15}N$ values of collagen from herbivorous and carnivorous mammals in the same ecosystem range from 3 ‰ (Schwarcz, 1991) to 5.7 ‰ (Ambrose & DeNiro, 1986). The $\delta^{15}N$ values most frequently measured in temperate and arctic environments are 2 to 7 ‰ for terrestrial herbivores, 7 to 12 ‰ for terrestrial carnivores, and 12 to 20 ‰ for marine vertebrates. However, the influence of local ecological conditions on the $\delta^{15}N$ values precludes the determination of exact ranges of variations according to an animal trophic level on a global scale. Environmental parameters such as aridity and soil acidity lead to important local variations of the $\delta^{15}N$ values of animals and humans. The collagen $\delta^{15}N$ values of a given herbivorous species are significantly higher for individuals living in an arid area relative to the individuals inhabiting a more humid region, independantly to the $\delta^{15}N$ values of the local plants (Ambrose & DeNiro, 1986; Heaton et al., 1986; Sealy et al., 1987). This effect is probably due to a water and/ or dietary stress, leading to recycling of nitrogen within the herbivore body, and thus producing a ^{15}N-enrichment, which is more pronounced with increasing stress intensity (Ambrose, 1991). Such an effect can interfer with the determination of the trophic level, but it may also provide additional information concerning the environmental conditions. Moreover, local conditions can lead to significant variations in plant $\delta^{15}N$ values, and thus a shift at every trophic level of the

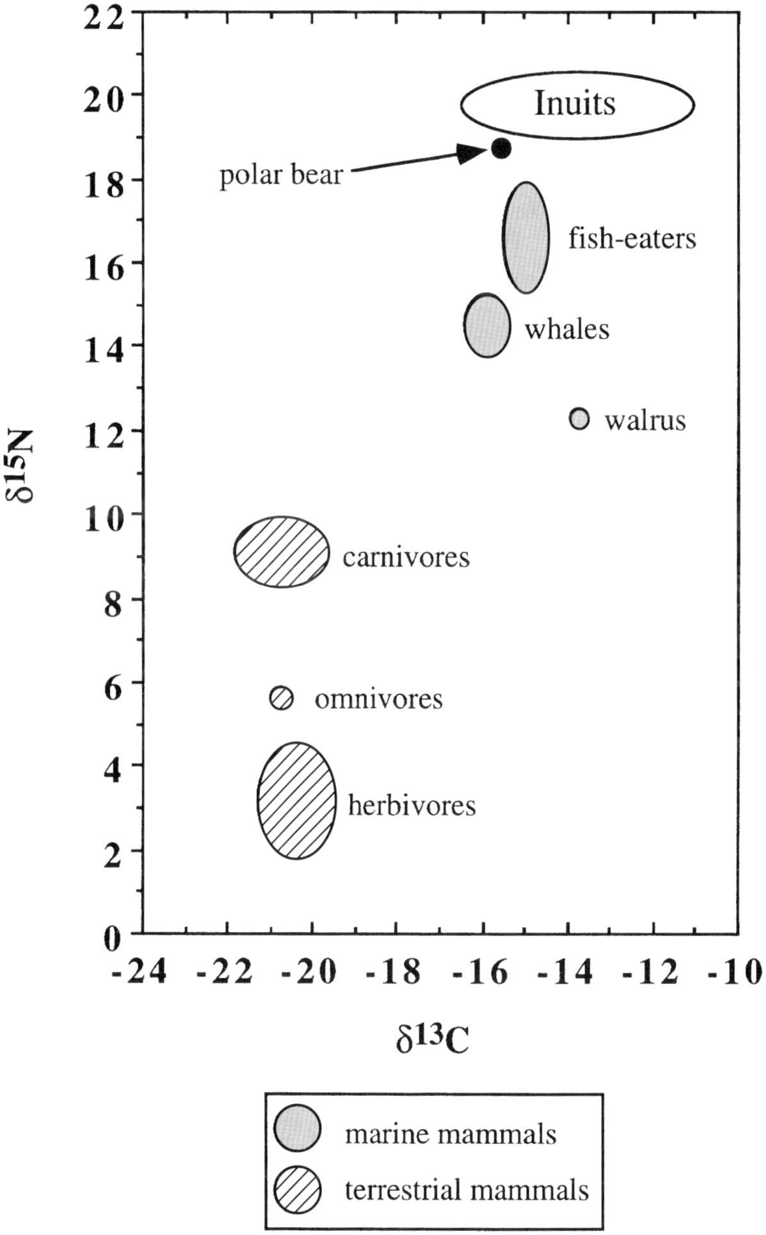

Figure 1 - Discrimination of trophic categories in arctic environment using $\delta^{13}C$ and $\delta^{15}N$ values of modern mammal collagen. Ellipses stand for average ± one standard deviation for each trophic group (calculated using values from Chisholm et al., 1983; Schoeninger & DeNiro, 1984; Nelson et al., 1986a; Bocherens, 1992; Bocherens et al., 1996).

corresponding trophic web. For instance, the plants of the Dourdan forest (south of Paris, France) exhibit very negative $\delta^{15}N$ values, probably because of the soil acidity, and these very low plant $\delta^{15}N$ values are reflected in the $\delta^{15}N$ values of the forest roe deer which show collagen $\delta^{15}N$ values as low as -3 ‰ (Rodière et al., 1996).

Stable isotopes in collagen

Within a given organism, the different biochemical fractions exhibit significant differences in their isotopic composition. For instance, carbohydrates show roughly the same isotopic composition than the average diet, whereas lipids are ^{13}C-depleted of around 4 ‰ and proteins are ^{13}C-enriched of around 2 ‰ relative to the average diet (DeNiro & Epstein, 1978). A systematical enrichment of around 5 ‰ has been measured in collagen relative to the average body carbon isotopic composition in large mammals (Vogel, 1978; Van der Merwe & Vogel, 1983). However, this is true only when dietary protein and non-protein come from an isotopically homogenous pool, since collagen is derived mainly from dietary proteins rather than from the whole diet, as demonstrated by controled feeding experiments on rodents (Ambrose & Norr, 1993; Tieszen & Fagre, 1993). The C3-plant dominated terrestrial ecosystems fall into this category, but the situation is different if the bulk of the proteins come from an isotopically distinct source than the bulk of the carbohydrates (e.g. proteins from marine food versus carbohydrates from terrestrial C3-plants).

Collagen presents the same nitrogen isotopic composition than the whole body. Thus, combined measurements of bone collagen $\delta^{13}C$ and $\delta^{15}N$ values can be used to determine the position of a given animal or human within its trophic web (marine or terrestrial). Bone collagen $\delta^{13}C$ values point out to the plants at the beginning of the food web whereas $\delta^{15}N$ values indicate the trophic level of the studied specimen. An example is given for the modern arctic area (Figure 1). Animals belonging to terrestrial and marine food webs are clearly distinguished by the $\delta^{13}C$ values of their bone collagen ($\delta^{13}C$ around -20 ‰ and -15 ‰ for terrestrial and marine mammals respectively) and they present increasing $\delta^{15}N$ values according to their upper trophic level. For instance, shellfish-eater walrus present lower $\delta^{15}N$ values than fish-eaters, which have a higher trophic level. It is noteworthy that polar bear and Inuits present collagen $\delta^{13}C$ and $\delta^{15}N$ values indicating a trophic position as top predator in a marine food web, although for both of them, the skeleton characteristics would indicate terrestrial mammals.

In tissues synthesized at different periods of an individual life, different $\delta^{13}C$ and $\delta^{15}N$ values can be recorded if the individual experienced diet changes in the mean time. Such a change occurs systematically in mammals at weaning, when the young animal shifts its diet from milk (enriched in $\delta^{15}N$ by around 3 ‰ relative to its mother's diet) to its adult diet, isotopically similar to its mother's diet (Fogel et al., 1989). This change in $\delta^{15}N$ values can be followed in the tissues formed at different ages in one individual of a given mammal species. This phenomenon leads to a ^{15}N-enrichment in dentine collagen relative to bone collagen in mammal species in which teeth start their growth before weaning is completed and stop their growth afterwards (Bocherens et al., 1994a; Bocherens & Mariotti, 1997). Indeed, dentine partly records a suckling isotopic signal which is not eliminated when the tooth growth is completed, whereas bones turn-over their constituents all life long, and thus reflect in their isotopic abundances an average value of their diet during the most recent period of their life. In the case of teeth with a prolonged growth period, such as in horses, the dentine of an adult individual will also be synthesized after weaning and thus no isotopic difference will be recorded between bone and dentine $\delta^{15}N$ values (Bocherens et al., 1994a; Fizet et al., 1995). As it will be exposed later, this difference between bone and dentine collagen $\delta^{15}N$ values in some mammals can be used to check the preservation of prystine isotopic signals in collagen from ancient samples.

Collagen preservation in Upper Pleistocene skeletal remains

The fact that collagen fragments can be recovered from fossil bones of Upper Pleistocene age has been recognized some decades ago. Collagen is present in large amounts in fresh bone and dentine (around 20 % dry weight) and this protein is highly insoluble in its native state, making it a good candidate for preservation for at least thousands of years. Palaeobiochemical investigations on bones from different sites in the 1960s demonstrated the preservation of collagen with a characteristic amino acid pattern (e.g. Ho, 1965; Wyckoff, 1972). Such ancient collagen is nowadays routinely used as a reliable support for radiocarbon dating (e.g. Sellstedt et al., 1966; Longin, 1971; Stafford et al., 1987; Hedges & Law, 1989). However, ancient bones present very variable states of preservation, and it is necessary to verify the quality of the residual collagen before interpreting isotopic data obtained from this material in terms of palaeodiets.

Bone diagenesis and collagen preservation

The only commun point between bones and teeth excavated from Upper Pleistocene sites is the fact that they escaped the general law of recycling in the sedimentary cycle taking place at the boundary between biosphere, hydrosphere, atmosphere and lithosphere on earth. Any condition limiting the destructive effects of weathering (such as fast burial in sediment or deposit in cave), microbial activity (such as anaerobic conditions or dryness) or chemical corrosion (such as neutral or high pH sediments) and any condition increasing the stability of bone mineral (such as permineralizing fluids) will contribute to the preservation of skeletal material as fossils (e.g. Weigelt, 1927; Behrensmeyer & Hill, 1980; Martill, 1991; Patou-Mathis, 1994). However, not all such processes are equally favorable to the preservation of collagen in fossil bones and teeth. As far as collagen preservation is concerned, two processes are especially important: collagen degradation and organic contaminants. Although native collagen is highly insoluble, hydrolysis cuts the long chains of amino acids into smaller peptides, which are much more soluble and can be easily leached out of the fossil (Hare, 1980). Hydrolysis of collagen is accelerated by microbial activity (Piepenbrink, 1984), and this activity can modify the isotopic composition of the organic residue (Turban-Just & Schramm, in press). A side effect of this microbial activity is contamination by organic compounds, which can be added to extraneous organic contamination originating from the surrounding sediment (Bocherens et al., in press b).

Ancient collagen extraction methods

Extraction techniques had to be defined to deal with these problems and to purify the remaining collagen from ancient bones. Besides removing the mineral fraction of fossil bone and dentine as well as possible exogenous minerals, such as clay or calcite, these techniques need to remove organic contaminants as well, but without removing the residual collagen fragments. It must be also simple enough to be used on a large number of samples in a reasonable amount of time and it must not contaminate the sample with exogenous carbon or nitrogen during the extraction process. A preliminary step is to measure the amount of nitrogen in whole bone, which can be performed with an elemental analyzer on around 5 mg of bone, in order to screen out the samples devoid of collagen (Bocherens et al., in press b). Indeed, samples containing no nitrogen will contain no collagen and can be discarded. The most common techniques of collagen extraction aim at collagen purification by the way of differential solubilisation of the different fractions of fossil bone in different chemical solutions. Figure 2 presents the two major ways of treating fossil bone in order to remove these phases, either by hydrochloric acid (HCl) at various concentrations, followed by soaking in a basic solution in order to eliminate humic acids, and finally gelatinization (Longin, 1971; Bocherens et al., 1991b), or by soaking in EDTA, a chelating agent for calcium (Tuross et al., 1988; Bocherens et al., 1995b). The first technique is faster and does not bring any organic contamination, but it may destroy a large proportion of the collagen if it is fragilized (Iacumin et al., 1996). Variants of this technique have been proposed to deal with samples containing fragilised collagen (Ambrose, 1990). The second technique is gentler (neutral pH), but takes

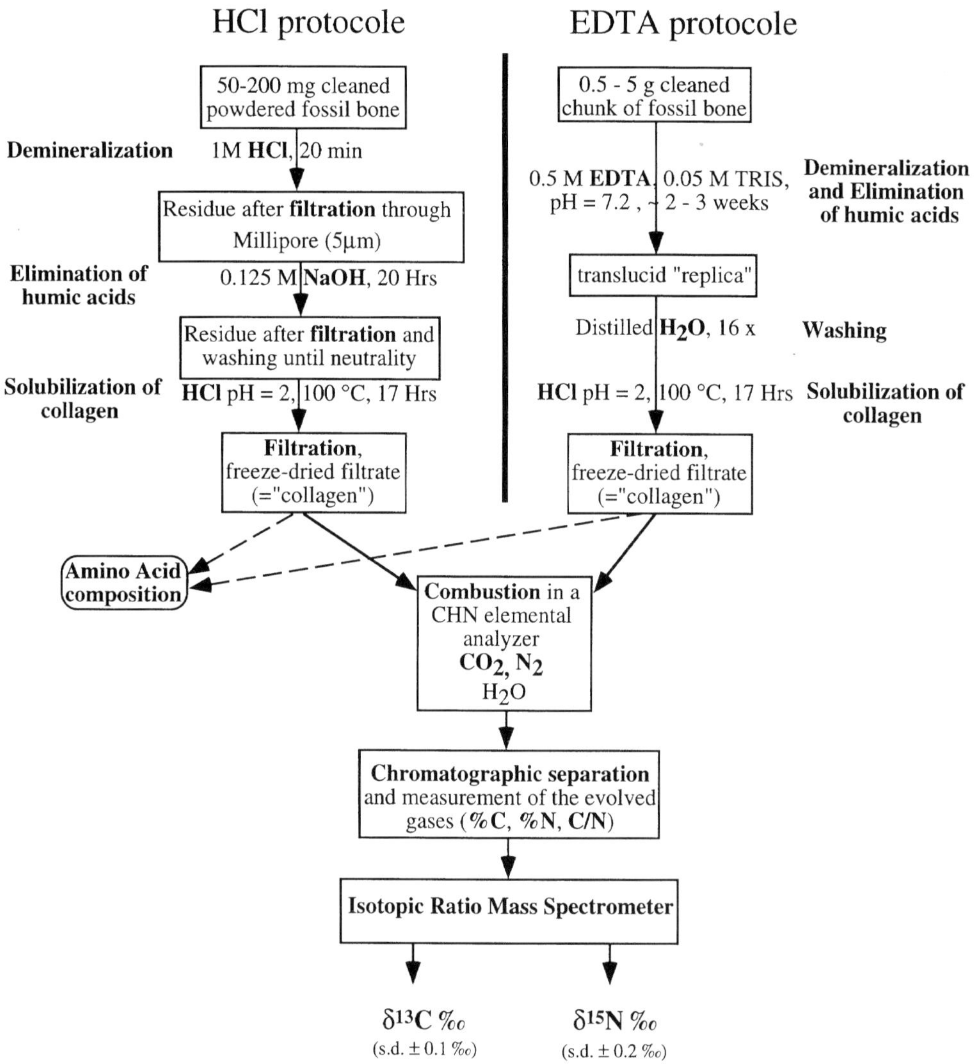

Figure 2 - Schematic presentation of the two major techniques used to extract collagen from fossil bone and dentine.

much longer and requires extensive washing to prevent addition of extraneous carbon during extraction. Using the most developed analytical techniques available today, 1 mg of collagen is enough for the measurement of carbon and nitrogen isotopic abundances, which typically corresponds to an amount of fossil ranging from 50 mg to 500 mg depending on the quality of collagen preservation. Checking the collagen biochemical integrity can be performed by checking that the amino acid composition of the extracted „collagen" corresponds to that of fresh collagen, since collagen exhibits a highly specific amino acid composition. If the amino acid composition is changed, alteration of the isotopic composition is likely, since the isotopic abundances of individual amino acids are very variable within collagen (Hare et al., 1991), and contamination by exogenous products is also possible in such cases. Routinely, such a check is performed through the carbon and nitrogen amounts of the „collagen", and through the C/N ratio. This C/N ratio is very stable in collagen (2.9 to 3.6) and this kind of measurement has been proposed as a marker of biochemical (and thus isotopic) alteration (DeNiro, 1985). It is however better to also check the actual C and N amounts in the extracted collagen, in order to verify that no contamination by mineral fractions occurred (Ambrose, 1990). Indeed, some minerals can trap small amounts of nitrogen containing contaminants with isotopic abundances very different from those of the sample (Bocherens et al., 1997).

Using these extraction techniques, preserved collagen has been extracted for stable isotopic investigations from several Upper Pleistocene sites from Eurasia and North America (Bocherens et al., 1991a, 1991b, 1994a, 1994b, 1995a, 1995b, 1996, in press a; Koch, 1991; Fizet et al., 1995; Matheus, 1995; Heaton, 1996; Ambrose, in press). From these studies, it seems that frozen soils, cave deposits and loess are favorable to collagen preservation for dozens of thousands of years, especially under cold conditions. Incidentally an attempt of collagen extraction from a Neandertal specimen from Syria failed, due to unfavorable burial conditions (Ambrose, in press). Starting in the late 1970s, stable isotopes in collagen extracted from archaeological bones less than 10,000 years old have been used first to decipher dietary problems in America, especially because it allowed the quantification of the amount of maize (a C4-plant) consumed by ancient people living in C3-plant environments (e.g. Vogel & Van der Merwe, 1977; Burleigh & Brothwell, 1978; Van der Merwe & Vogel, 1978). An other kind of isotopic distinction used in archaeology is the difference between terrestrial and marine food sources, for instance in Europe (Tauber, 1981; Clutton-Brock & Noe-Nygaard, 1990; Lidén & Nelson, 1994), North America (Chisholm et al., 1982, 1983; Nelson et al., 1986b), Australia (Hobson & Collier, 1984) and Southern Africa (Sealy & Van der Merwe, 1985). Applications of isotopic biogeochemistry have been a powerful tool in Holocene archaeological research, but the recovery of collagen from much older bones widens the potential field of this new approach. However, this will be possible only if this old collagen has retained its biogenic signature.

Is collagen recovered from Upper Pleistocene bones and teeth suitable for isotopic palaeodietary reconstructions?

When collagen can be extracted from some Upper Pleistocene bones using the techniques described above, it remains to determine whether this collagen has kept its original biogenic carbon and nitrogen isotopic compositions. In fresh bone, the amount of collagen is around 21 % dry weight, which means that 1 gramme of bone yields around 210 mg collagen (210 $mg.g^{-1}$). In Upper Pleistocene bones, yields are sometimes close to the value in fresh bone. Spectacular examples are the bones preserved in arctic frozen grounds (Bocherens et al., 1995a, 1996). However, the yields are usually lower than in fresh bones. It appeared nevertheless that the amino acid composition did not change significantly according to the amount recovered, even for yields as low as 1 $mg.g^{-1}$, indicating that even very low amounts of preserved collagen can be biochemically pure (Bocherens et al., 1991a; Fizet et al., 1995).

Independently from the biochemical preservation of collagen, the preservation of its biogenic isotopic abundances has to be checked. It can be done by comparing the carbon and nitrogen isotopic compositions measured on different bones of a given species which yielded very different quantities of collagen in the same site. Such a comparison performed on horse and reindeer from Marillac (40,000-45,000 BP, Charente, France) showed that the observed isotopic variations are not due to diagenetic alteration of the preserved collagen (Fizet et al., 1995). A similar test can be performed when several species are present in the site by comparing the deviation of collagen

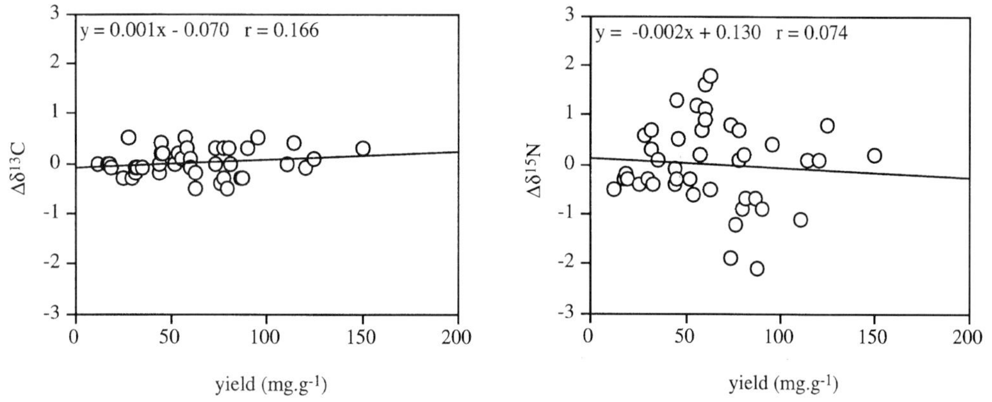

Figure 3 - *Deviation of carbon ($\Delta\delta^{13}C$) and nitrogen ($\Delta\delta^{15}N$) isotopic compositions relative to the average value for a given species (horse, bovine, rhinoceros, cave bear, hyena) and for a given tissue (bone and dentine). Values are from Bocherens et al. (in press a).*

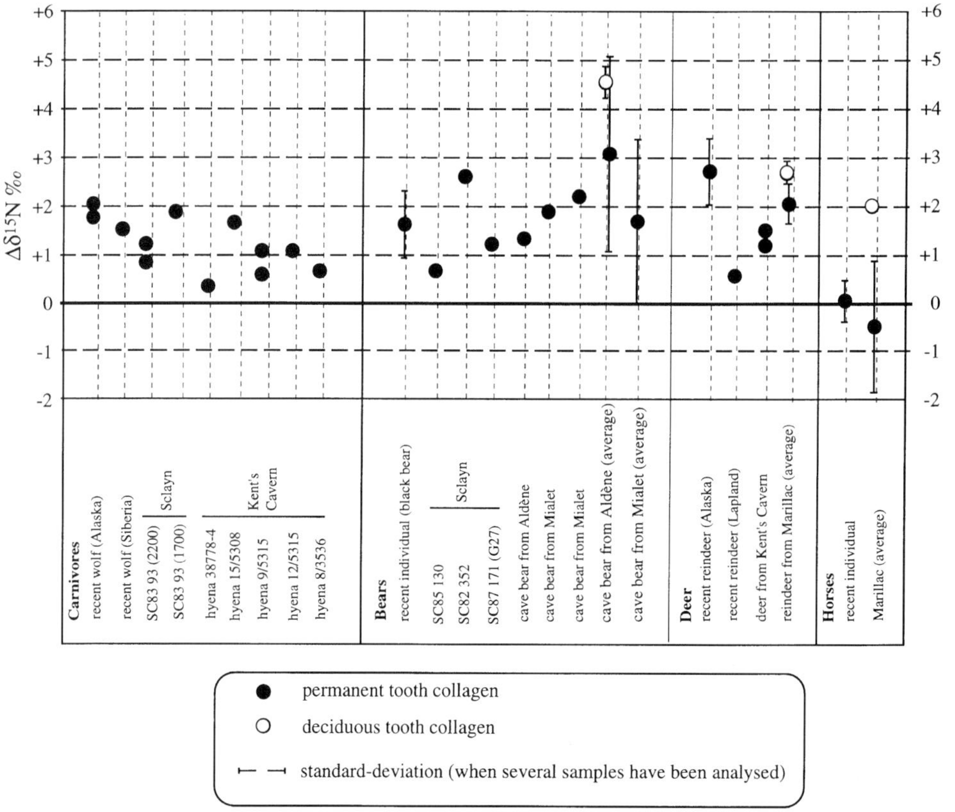

Figure 4 - *Differences between the nitrogen isotopic compositions of bone and dentine collagen of Kent's Cavern and Sclayn 1A hyenas, Southern France and Sclayn 1A cave bears, Marillac reindeer and Marillac horses (data from Bocherens et al., 1994a, 1995b, in press a).*

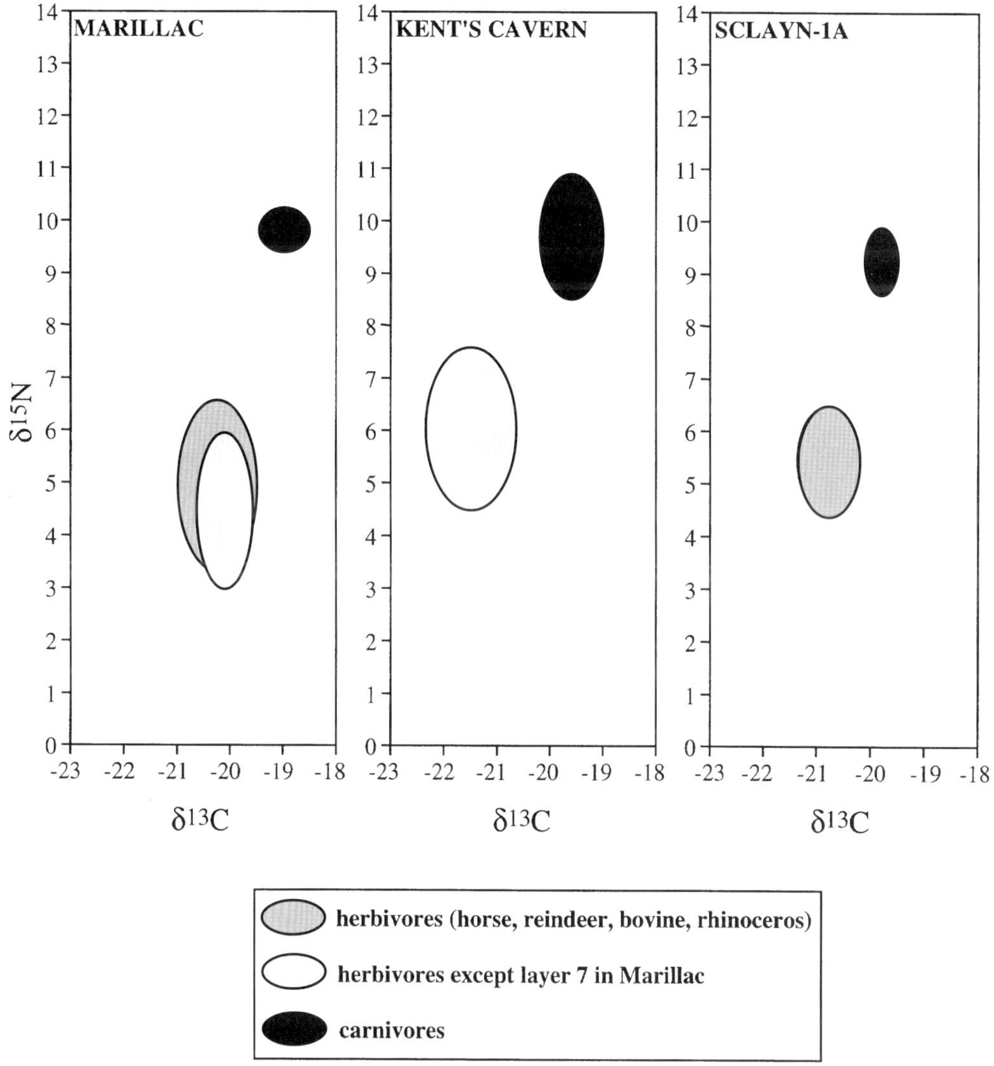

Figure 5 - Comparison of the carbon and nitrogen isotopic composition of bone collagen in herbivorous and carnivorous taxa in Marillac (40,000 to 45,000 BP, data from Fizet et al., 1995), in Kent's Cavern (35,000 to 60,000 BP, data from Bocherens et al., 1995b) and in Sclayn 1A (40,000 BP, data from Bocherens et al., in press a). Ellipses stand for average ñ one standard deviation for each trophic group. In the case of Marillac, herbivores have been considered with and without individuals from layer #7, where the climate was colder and presumably drier (Fizet et al., 1995).

carbon and isotopic compositions relative to the average value for each species. Figure 3 shows an example for mammals from layer 1A in Sclayn (40,000 BP, Belgium: Bocherens et al., in press a). In this case, there is no tendency in deviation with decreased collagen yield for neither carbon nor nitrogen. This indicates that collagen is isotopically equivalent independantly of the preserved amount in bone or dentine.

A predictable isotopic difference that can be used to check the isotopic preservation of collagen is the higher^{15}N values in dentine relative to bone in individuals of deer and carnivores due to suckling in young mammals and definite tooth growth whereas bone continuously turn-overs, as explained previously. Such a difference in δ^{15}N values between dentine and bone has been observed in reindeer from Marillac, in hyenas from Kent's Cavern and Sclayn, and in cave bears from Southern France and from Sclayn (figure 4). The differences between dentine and bone δ^{15}N values of the studied fossil individuals are comparable to those observed in modern equivalent specimens. When deciduous teeth have been analyzed, they present even higher δ^{15}N values than permanent teeth, reflecting the fact that they are entirely formed during the suckling period. Such an isotopic difference was not observed in horses from Marillac, due to the continuous growth of teeth in this species, as in modern horses.

Finally, the preservation of biogenic isotopic compositions can also be checked if the isotopic differences observed between modern herbivores and carnivores are actually found in ancient specimens of still extant species or of species closely related to extant ones, for which the diets are well known, such as reindeer, bison, horse, woolly rhinoceros, wolf or hyena. Examples of such a checking approach are presented for the sites of Marillac, Kent's Cavern and Sclayn (Figure 5). The specimens from layer 7 in Marillac have been treated separately due to the increased δ^{15}N values of herbivores in this layer due to colder and probably drier conditions during its deposit (Fizet et al., 1995). This emphasizes the fact that such comparisons should be done as much as possible within individual stratigraphical layers, in order to avoid the mixture of herbivorous populations leaving under different climatic conditions which may interfere with the trophic layer isotopic differences. Such an interference is probably to be blamed for the poorer isotopic discrimination between herbivores and carnivores in Kent's Cavern (Figure 5), due to insufficient stratigraphical resolution of the studied specimens. Indeed, collagen from three different bones from Kent's Cavern used in this study have been radiocarbon dated and the ages range from 40,600 to 27,640 BP (Hedges et al., 1995).

Thus, there are different ways to assess the quality of preservation of the isotopic signature of collagen and, depending on the samples available for analysis, as many independant ways of checking this preservation should be used before trying to interpret the measured isotopic variations in terms of palaeoecology. In the present state of research, every time collagen corresponding to the established criteria could be extracted from Upper Pleistocene bones, even in very small quantities and from very old bones, the isotopic composition appeared to reflect that of the living individual.

Isotopic ecology of the Upper Pleistocene fauna

The Upper Pleistocene sites studied so far and relevant for the study of Neandertal diet are located in Alaska (Bocherens et al., 1994b, 1995a), Siberia (Bocherens et al., 1996) and western Europe (Bocherens et al., 1991a, 1991b, 1994a, 1995a, 1995b, in press a). Although the geographical distribution of these sites is wide, they correspond to a well defined palaeoecological unit refered as the „mammoth steppe" (Guthrie, 1982, 1990). This area covering middle and southern Europe, northern Asia, and Alaska was during the upper Pleistocene cold phases boarded to the northwest and to the east by icecaps, and was characterized by a rather homogenous faunal assemblage dominated by *Mammuthus, Bison* and *Equus* (Guthrie, 1982). The species diversity of this fauna and the gigantism of its elements reminds more of the African savannah than of today's arctic toundra. Neandertals were on the other hand restricted to the western part of this palaeoecological unit, but the conclusions that could be drawn from the isotopic study of the mammoth steppe as a whole can be relevant for the more geographically restricted Neandertals, at least during the cold climatic phases.

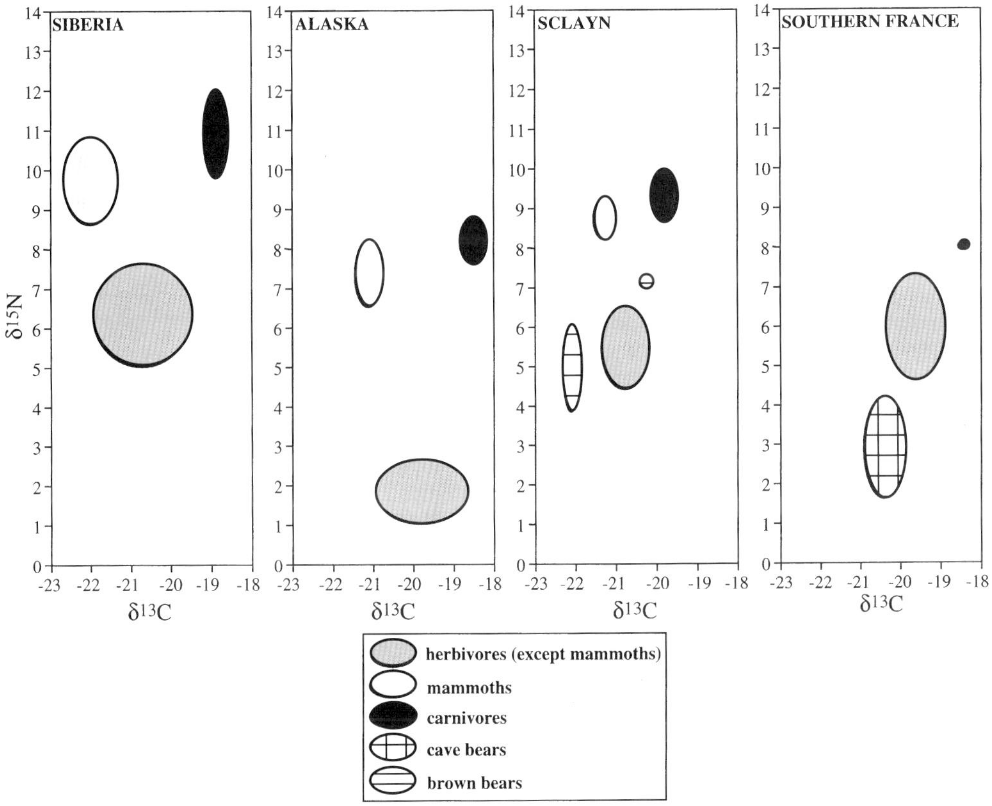

Figure 6 - *Comparison of the carbon and nitrogen isotopic composition of bone collagen in herbivorous species except mammoths, cave bears, brown bears and carnivores in Upper Pleistocene Siberia (data from Bocherens et al., 1996), Alaska (data from Bocherens et al., 1994b, 1995a), Southern France (data from Bocherens et al., 1994a) and Sclayn 1A (data from Bocherens et al., in press a). Ellipses stand for average ± one standard deviation for each trophic group.*

Isotopes in the mammoth steppe fauna

The comparison of the isotopic abundances of different species of co-occurring herbivore species in different upper Pleistocene sites emphasized some consistent patterns of variations that can be tentatively linked to dietary or ecological characteristics of these species. A spectacular example is given by the woolly mammoths relative to the other herbivores in Siberia, in Alaska and in Sclayn (Figure 6). It clearly appears that mammoths exhibit in all cases unexpectedly high $\delta^{15}N$ values when compared to coeval other herbivorous species. Indeed, mammoths $\delta^{15}N$ values are close to those of carnivores. However, their $\delta^{13}C$ values are more comparable to those of other herbivores. A similar pattern occurred in Eastern Europe (Ambrose, in press), showing that this pattern is widespread in the mammoth steppe fauna. In the current state of research, this difference has been considered as a possible consequence of the nature of the diet of mammoth, which consists of large amounts of low nitrogen plant material, probably leading these animals to recycle a large proportion of their dietary nitrogen, thus increasing their collagen $\delta^{15}N$ values (Bocherens et al., 1996).

An other isotopic difference observed between different species of herbivores is the fact that reindeer exhibit less negative $\delta^{13}C$ values than horses and possibly other herbivorous species (Fizet et al., 1995). In this case, the difference has been tentatively linked to the large amount of lichens in reindeer diet, since lichen are known to present less negative $\delta^{13}C$ values than other plants (Fizet et al., 1995). Other isotopic differences between upper Pleistocene herbivorous species are suspected but additional data are necessary to check that they are not artefacts due to the still small number of analyses performed until now. It is hoped that further research in this direction will allow to distinguish the different prey species consumed by carnivorous species with stable isotopes.

As far as omnivorous species are concerned, brown bears exhibited $\delta^{15}N$ values intermediate between those of herbivores and carnivores in Sclayn 1A (Bocherens et al., in press a) or as high as those of coeval carnivores in Kent's Cavern (Bocherens et al., 1995a). On the contrary, cave bears yielded very low $\delta^{15}N$ values in all western European sites studied so far, indicating a vegetarian diet (Bocherens et al., 1994a, in press a). A similar conclusion had already been reached using morphological characters (Kurtén, 1976). These clear differences exemplify the potentials of the isotopic approach in reconstructing the actual diet of omnivorous species, with a large range of potential dietary resources.

A case study of Upper Pleistocene omnivorous mammal diet: Alaskan bears

Bears present an interesting group to exemplify diet reconstruction using collagen isotopic composition, since like hominids they can eat a wide range of vegetal and animal products, and they present the advantage to have left great numbers of fossil remnants which can be sampled quite easily. An interesting case deals with an extinct bear species from North America, the short-faced bear (*Arctodus simus*), which exact diet was difficult to determine using traditional approaches. Indeed, this 1.80 meters high at the shoulder and 800 kg species has a feline like skull, which made Kurtén (1967) conclude that this species had a feline way of life, as an active predator. However, a more recent study suggested a vegetarian diet, based primarily on a comparison of short-faced bear's anatomical characters with those of the extant spectacled bear *Tremarctos ornatus*, a relative of the short-faced bear, and a peaceful vegetarian (Emslie & Czaplewski, 1985). Two independant isotopic investigations have been conducted on this species (Bocherens et al., 1995a; Matheus, 1995). The first one aimed at comparing collagen isotopic composition of short-faced bear with those of coeval herbivores (reindeer, moose) and carnivores (lion) (Bocherens et al., 1995a). The obtained results clearly showed similar isotopic values in short-faced bears and Upper Pleistocene Alaskan lions, which indicated a carnivorous diet for short-faced bears (Bocherens et al., 1995a). The second isotopic study compared collagen isotopic composition of fossil brown bears with those of short-faced bears from Alaska and Yukon, and interpreted those values by comparison with those of modern arctic mammals (Matheus, 1995). Results of both studies have been combined on Figure 7 and compared to collagen isotopic composition of modern and fossil herbivores and carnivores, where modern collagen isotopic values have been corrected for the effect of CO_2 release from industrial fuel combustion in modern atmosphere. It is clear that the variability of short-faced bear collagen isotopic composition is much less than for brown bear. Short-faced bear isotopic values are close to those of carnivores, whereas brown bear isotopic values are very variable and range from those of herbivores to those of marine predators. For brown bears, this reflects different diets for different individuals, including almost fully vegetarians, terrestrial carnivores and fish eaters (Matheus, 1995). Similar studies can be envisaged for fossil humans such as Neandertals when a sufficient number of specimens can be analysed. The next section will describe the first and only successful isotopic study of the paleodiet of Neandertals so far.

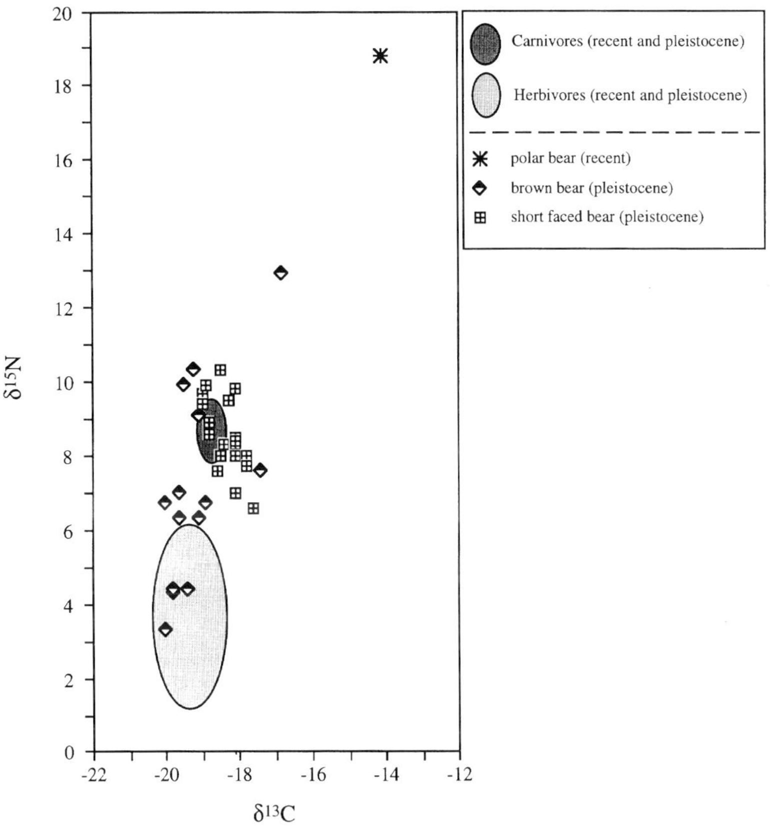

Figure 7 - Paleodiet reconstruction of Upper Pleistocene brown bears and short-faced bears in Alaska. Ellipses stand for average ± one standard-deviation of herbivores and carnivores. In the case of modern individuals, $\delta^{13}C$ values have been corrected from the fuel combustion effect on atmospheric CO_2 (+ 1.5 ‰) (data are from Bocherens et al., 1994a, 1995a; Matheus, 1995).

Isotopic determination of Neandertal's diet in Marillac

In a similar way than for Upper Pleistocene bears, the reconstruction of Neandertal diet was performed by comparing the $\delta^{13}C$ and $\delta^{15}N$ values of human samples with those measured on coeval herbivores and carnivores. In Marillac (Charente, France), it appeared clearly that for both analysed Neandertal specimens, the $\delta^{15}N$ values were close to those of the carnivorous taxa from the same site (Figure 8; Bocherens et al., 1991a; Fizet et al., 1995). Does this result mean that Neandertals had a diet similar to those of wolves and hyenas ? The conclusion is probably not as straightforward. Indeed, a purely carnivorous diet leads to severe metabolic disruption in humans (Delluc et al., 1995). One factor to be accounted for is the higher amount of protein in meat than in plant foodstuff, and the fact that collagen is synthesized mostly on the proteic fraction of the diet (Parkington, 1991; Ambrose & Norr, 1993). For instance, with a diet containing 60 % meat, more than 70 % of the body proteins are of animal origin (calculation based on data from Delluc et al., 1995, for hunter-gatherers). Moreover, seasonal dietary shifts can provoke fast collagen turn-over in case of abundant meat in the diet at particular times of the year (Parkington, 1991; Ambrose & Norr, 1993; Tieszen & Fagre, 1993). Besides these complicating factors, the isotopic

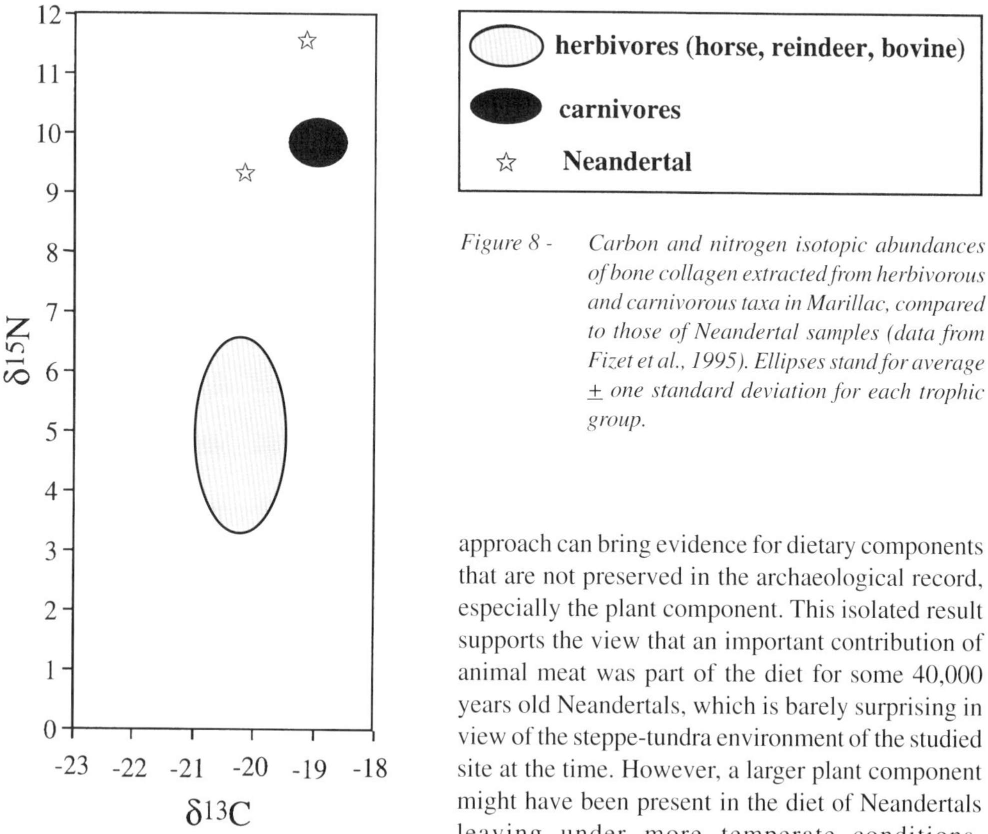

Figure 8 - *Carbon and nitrogen isotopic abundances of bone collagen extracted from herbivorous and carnivorous taxa in Marillac, compared to those of Neandertal samples (data from Fizet et al., 1995). Ellipses stand for average ± one standard deviation for each trophic group.*

approach can bring evidence for dietary components that are not preserved in the archaeological record, especially the plant component. This isolated result supports the view that an important contribution of animal meat was part of the diet for some 40,000 years old Neandertals, which is barely surprising in view of the steppe-tundra environment of the studied site at the time. However, a larger plant component might have been present in the diet of Neandertals leaving under more temperate conditions. Preliminary results from an isotopic investigation in progress on a 80,000 years old Neandertal specimen from a temperate paleoenvironment in Sclayn (Bonjean et al., 1994) suggests that the meat component was also high in this individual (Bocherens et al., in prep.).

Conclusions

The isotopic approach is only at a beginning stage as far as its applications to Upper Pleistocene Europe and Neandertals are concerned. Preliminary investigations demonstrate that the picture „herbivores versus carnivores" is probably more complicated than expected and that further studies will provide valuable information about the structure of food webs among the mammoth steppe mammal fauna. Moreover, several tests have been defined to check whether the measured isotopic values reflect collagen biogenic records, and usable samples as old as 130,000 years B.P. have been found. However, age is not the only parameter and some much younger specimens did not preserve their collagen properly. Approaches using the isotopic signals in the mineral phase of ancient bones and teeth, presumably more stable than the organic phase, may provide complementary data.

One recommendation is to use this approach within a palaeoecological framework, by analysing herbivorous and carnivorous animals from the same sites than the humans, for two reasons: first, to check that biogenic isotopic compositions are indeed preserved, and second, to replace the ancient humans within their trophic web. Isolated isotopic values obtained on humans would have little paleodietary relevance in the absence of isotopic data on the coeval fauna.

Although the methodology presented here does destroy some material, the amounts necessary are already quite low, and near future improvements in the technology will certainly allow to decrease the material necessary to minimum amounts on the milligramme order of magnitude. Especially interesting would be to multiply the isotopic investigations on Upper Paleolithic Modern Humans, in order to compare their dietary strategies with those of Middle Paleolithic Neandertals. The results of these investigations would allow to test some of the proposed hypotheses for the replacement of Neandertals by Modern Humans in Europe at the transition between Middle and Upper Paleolithic.

References

AMBROSE, S. H. (1990): Preparation and characterization of bone and tooth collagen for isotopic analysis. *Journal of Archaeological Science* 17, 431-451.

AMBROSE, S. H. (1991): Effects of diet, climate and physiology on nitrogen isotope abundances in terrestrial foodwebs. *Journal of Archaeological Science* 18, 293-317.

AMBROSE, S. H. (in press): Prospects for stable isotopic analysis of Later Pleistocene hominid diets in West Asia and Europe. In: T. Akasawa, K. Aoki and O. Bar-Yosef (eds.), *Neanderthals and humans in West Asia*. New York.

AMBROSE, S. H. & DeNIRO, M. J. (1986): The isotopic ecology of East African mammals. *Oecologia* 69, 395-406.

AMBROSE, S. H. & NORR, L. (1993): Isotopic composition of dietary protein and energy versus bone collagen and apatite: purified diet growth experiments. In: J. B. Lambert & G. Grupe (eds.), *Prehistoric human bone - Archaeology at the molecular level*, 1-37. Berlin.

BEHRENSMEYER, A. K. & HILL, A. P. (eds.) (1980): *Fossils in the making. Vertebrate taphonomy and paleoecology.* Chicago, London.

BINFORD, L.R. (1968): Early upper Pleistocene adaptations in the Levant. *American Anthropologist* 70, 292-327.

BOCHERENS, H., 1992: Biogéochimie isotopique (^{13}C, ^{15}N, ^{18}O) et paléontologie des vertébrés : Applications à l'étude des réseaux trophiques révolus et des paléoenvironnements. Thèse de Doctorat de l'Université Paris 6, *Mémoire des Sciences de la Terre*, 92-6.

BOCHERENS, H. & MARIOTTI, A. (1997): Comments on: Diet, physiology and ecology of fossil mammals as inferred from stable carbon and nitrogen isotope biochemistry: Implications for Pleistocene bears by Bocherens et al. - Reply. *Palaeogeography, Palaeoclimatology, Palaeoecology* 128, 362-364.

BOCHERENS, H., FIZET, M., MARIOTTI, A., BILLIOUT, D., BELLON, G., BOREL, J. P. & SIMONE, S. (1991a): Biogéochimie isotopique (^{13}C, ^{15}N, ^{18}O) et paléoécologie des ours pléistocènes de la grotte d'Aldène. *Bulletin du Musée d'Anthropologie et de Préhistoire, Monaco* 34, 29-49.

BOCHERENS, H., FIZET, M., MARIOTTI, A., LANGE-BADRÉ, B., VANDERMEERSCH, B., BOREL, J.-P. & BELLON, G. (1991b): Isotopic biogeochemistry (^{13}C, ^{15}N) of fossil vertebrate collagen: implications for the study of fossil food web including Neandertal Man. *Journal of Human Evolution* 20, 481-492.

BOCHERENS, H., FIZET, M. & MARIOTTI, A. (1994a): Diet, physiology and ecology of fossil mammals as inferred from stable carbon and nitrogen isotope biogeochemistry: Implications for Pleistocene bears. *Palaeogeography, Palaeoclimatology, Palaeoecology* 107, 213-225.

BOCHERENS, H., FIZET, M., MARIOTTI, A., GANGLOFF, R. A. & BURNS, J. A. (1994b): Contribution of isotopic biogeochemistry (^{13}C, ^{15}N, ^{18}O) to the paleoecology of mammoths (*Mammuthus primigenius*). *Historical Biology* 7, 187-202.

BOCHERENS, H., EMSLIE, S. D., BILLIOU, D. & MARIOTTI, A. (1995a): Stable isotopes (^{13}C, ^{15}N) and paleodiet of the giant short-faced bear (*Arctodus simus*). *Comptes Rendus de l'Académie des Sciences, Paris* 320, 779-784.

BOCHERENS, H., FOGEL, M. L., TUROSS, N. ZEDER, M. (1995b): Trophic structure and climatic information from isotopic signatures in a Pleistocene cave fauna of Southern England. *Journal of Archaeological Science* 22, 327-340.

BOCHERENS, H., PACAUD, G., LAZAREV, P. & MARIOTTI, A. (1996): Stable isotope abundances (^{13}C, ^{15}N) in collagen and soft tissues from Pleistocene mammals from Yakutia. Implications for the paleobiology of the mammoth steppe. *Palaeogeography, Palaeoclimatology, Palaeoecology* 126, 31-44.

BOCHERENS, H., GRUPE, G., MARIOTTI, A. & TURBAN-JUST, S. (1997): Molecular and isotopic preservation of mesolithic human finds from the Ofnet Cave (Bavaria, Germany). *Anthropologischer Anzeiger* 55, 121-129.

BOCHERENS, H., BILLIOU, D., PATOU-MATHIS, M., BONJEAN, D., OTTTE, M. & MARIOTTI, A. (in press a): Isotopic biogeochemistry (^{13}C, ^{15}N) of fossil mammal collagen from Scladina cave (Sclayn, Belgium). *Quaternary Research* .

BOCHERENS, H., TRESSET, A., WIEDEMANN, F., GILIGNY, F., LAFAGE, F., LANCHON, Y. & MARIOTTI, A. (in press b): Bone diagenetic evolution in two French Neolithic sites. *Bulletin de la Société Géologique de France* .

BONJEAN, D., OTTE, M. & TOUSSAINT, M. (1994): L'homme de Sclayn. *Archeologia* 299, 26-30.

BURLEIGH, R. & BROTHWELL, D. (1978): Studies on Amerindian dogs. 1. Carbon isotopes in relation to maize in the diet of domestic dogs from early Peru and Ecuador. *Journal of Archaeological Science* 5, 355-362.

CHISHOLM, B. S., NELSON, D. E. & SCHWARCZ, H. P. (1982): Stable-carbon isotope ratios as a measure of marine versus terrestrial protein in ancient diets. *Science* 216, 1131-1132.

CHISHOLM, B. S., NELSON, D. E. & SCHWARCZ, H. P. (1983): Marine and terrestrial protein in Prehistoric diets on the British Columbia Coast. *Current Anthropology* 24, 396-398.

CLUTTON-BROCK, J. & NOE-NYGAARD, N. (1990): New osteological and C-isotope evidence on Mesolithic dogs: Companions to hunters and fishers at Star Carr, Seamer Carr and Kongemose. *Journal of Archaeological Science* 17, 643-653.

CRAIG, H. (1953): The geochemistry of stable carbon isotopes. *Geochimica et Cosmochimica Acta* 3, 53-92.

DAVID, F. & ENLOE, J. G. (1993): L'exploitation des animaux sauvages de la fin du Paléolithique Moyen au Magdalénien. In: J. Desse & F. Audouin-Rouzeau (eds), *Exploitation des animaux sauvages à travers le temps*, 29-47. Juan-les-Pins.

DEINES, P. (1980): The isotopic composition of reduced organic carbon. In: P. Fritz & J. Ch. Fontes (eds.), *Handbook of environmental isotope geochemistry*. Vol.1, *The terrestrial environment*, A. 329-406. Elsevier.

DELLUC, G., DELLUC, B. & ROQUES, M. (1995): *La nutrition préhistorique*. Périgueux.

DeNIRO, M. J. (1985): Postmortem preservation and alteration of in vivo bone collagen isotope ratios in relation to palaeodietary reconstruction. *Nature* 317, 806-809.

DeNIRO, M. J. & EPSTEIN, S. (1978): Influence of diet on the distribution of carbon isotopes in animals. *Geochimica et Cosmochimica Acta* 42, 495-506.

DeNIRO, M. J. & EPSTEIN, S. (1981): Influence of diet on the distribution of nitrogen isotopes in animals. *Geochimica et Cosmochimica Acta* 45, 341-351.

ÉLIOT, É., LÉTOLLE, R. & ROTH, É. (1993): Analyses isotopiques. Applications. *Les Techniques de l'Ingénieur* P3740, 1-22.

EMSLIE, S. D. & CZAPLEWSKI, N. J. (1985): A new record of giant short-faced bear, *Arctodus simus*, from western North America with a re-evaluation of its paleobiology. *Natural History Museum Los Angeles County, Contribution to Sciences* 371, 1-12.

FARIZY, C. & DAVID, F. (1989): Chasse et alimentation carnée au paléolithique moyen, l'apport des gisements de plein air. In: L. Freeman & M. Patou (eds.), *L'homme de Neandertal*, vol.6, 59-62. Liège .

FIZET, M., MARIOTTI, A., BOCHERENS, H., LANGE-BADRÉ, B., VANDERMEERSCH, B., BOREL, J. P. & BELLON, G. (1995): Effect of diet, physiology and climate on carbon and nitrogen isotopes of collagen in a late Pleistocene anthropic paleoecosystem (France, Charente, Marillac). *Journal of Archaeological Science* 22, 67-79.

FOGEL, M. L., TUROSS, N. & OWSLEY, D. W. (1989): Nitrogen isotope tracers of human lactation in modern and archeological populations. *Annual Report of the Director of the Geophysical Laboratory, Carnegie Institution of Washington 1988-1989*, 111-117.

GAUDZINSKI, S. (1996): On bovid assemblages and their consequences for the knowledge of subsistence patterns in the Middle Palaeolithic. *Proceedings of the Prehistoric Society* 62, 19-39.

GUTHRIE, R. D. (1982): Mammals of the mammoth steppe as paleoenvironmental indicators. In: D.M. Hopkins, J. V. Matthews Jr., C. E. Schweger & S. B. Young (eds.), *Paleoecology of Beringia*, 307-326. Academic Press.

GUTHRIE, R. D. (1990): *Frozen fauna of the Mammoth Steppe*. Chicago.

HARE, P. E. (1980): Organic geochemistry of bone and its relation to the survival of bone in the natural environment. In: A. K. Behrensmeyer & A. P. Hill (eds.), *Fossils in the making. Vertebrate taphonomy and paleoecology*, 208-219. Chicago and London.

HARE, P. E., FOGEL, M. L., STAFFORD, T. W. Jr., MITCHELL, A. D. & HOERING, T. C (1991): The isotopic composition of carbon and nitrogen in individual amino acids isolated from modern and fossil proteins. *Journal of Archaeological Science* 18, 277-292.

HAYDEN, B. (1993): The cultural capacities of Neandertals: a review and re-evaluation. *Journal of Human Evolution* 24, 113-146.

HEATON, T. H. (1996): Interpretation of $\delta^{13}C$ values from vertebrate remains of the Alexander Archipelago, Southeast Alaska. *Current Research in the Pleistocene* 12, 95-97.

HEATON, T. H. E., VOGEL, J. C., CHEVALLERIE, G. v. l. & COLLETT, G. (1986): Climatic influence on the isotopic composition of bone nitrogen. *Nature* 322, 822-824.

HEDGES, R. E. M. & LAW, I. A. (1989): The radiocarbon dating of bone. *Applied Geochemistry* 4, 249-253.

HEDGES, R. E. M., LEE-THORP, J. A. & TUROSS, N. C.(1995): Is tooth enamel carbonate a suitable material for radiocarbon dating? *Radiocarbon* 37, 285-290.

HO, T.-Y. (1965): The amino acid composition of bone and tooth proteins in late Pleistocene mammals. *Proceedings of the National Academy of Sciences* 54: 26-31.

HOBSON, K. A. & COLLIER, S. (1984): Marine and terrestrial protein in Australian aboriginal diets. *Current Anthropology* 25, 238-240.

HUBLIN, J.-J. (1983): Les origines de l'homme de type moderne en Europe. *Pour la Science* 64, 62-71.

HUBLIN, J.-J., SPOOR, F., BRAUN, M., ZONNEVELD, F. & CONDEMI, S. (1996): A late Neandertal associated with Upper Paleolithic artefacts. *Nature* 381, 224-226.

IACUMIN, P., BOCHERENS, H., MARIOTTI, A. & LONGINELLI, A. (1996): An isotopic palaeoenvironmental study of human skeletal remains from the Nile Valley. *Palaeogeography, Palaeoclimatoly, Palaeoecology* 126, 15-30.

KOCH, P. L. (1991): The isotopic ecology of Pleistocene proboscidians. *Journal of Vertebrate Paleontology, Abstracts* 11(3), 40A.

KURTÉN, B. (1967): Pleistocene bears of North America. 2. Genus *Arctodus*, short-faced bears. *Acta Zoologica Fennica* 117, 1-60.

KURTÉN, B. (1976): *The cave bear story, life and death of a vanished animal.* New York.

LALUEZA FOX, C. & PÉREZ-PÉREZ A. (1993): The diet of the Neandertal child Gibraltar 2 (Devil's Tower) through the study of the vestibular striation pattern. *Journal of Human Evolution* 24, 29-41.

LALUEZA, C., PÉREZ-PÉREZ A. & TURBON, D. (1993): Microscopic study of the Banyoles mandible (Girona, Spain): diet, cultural activity and toothpick use. *Journal of Human Evolution* 24, 281-300.

LIDÉN, K. & NELSON, E. D. (1994): Stable carbon isotopes as dietary indicator, in the Baltic area. *Fornvännen* 89, 99-120.

LONGIN, R. (1971): New method of collagen extraction for radiocarbon dating. *Nature* 230, 241-242.

MARINO, B. D. & MacELROY, M. B. (1991): Isotopic composition of atmospheric CO_2 inferred from carbon in C4 plant cellulose. *Nature* 349, 127-131.

MARIOTTI, A. (1983): Atmospheric nitrogen is a reliable standard for natural ^{15}N abundance measurements. *Nature* 303, 685-687.

MARTILL, D. M. (1991): Bones as stones: The contribution of vertebrae remains to the lithological record. In: S. K. Donovan (ed.), *The processes of fossilization,* 270-292. London.

MATHEUS, P. E. (1995): Diet and co-ecology of Pleistocene short-faced bears and brown bears in Eastern Beringia. *Quaternary Research* 44, 447-453.

MINAGAWA, M. & WADA, E. (1984): Stepwise enrichment of ^{15}N along food chains: Further evidence and the relation between ^{15}N and animal age. *Geochimica et Cosmochimica Acta* 48, 1135-1140.

NELSON, B. K., DeNIRO, M. J., SCHOENINGER, M. J., De PAOLO, D. J. & HARE, P. E. (1986a): Effects of diagenesis on strontium, carbon, nitrogen and oxygen concentration on isotopic composition of bone. *Geochimica et Cosmochimica Acta* 50, 1941-1949.

NELSON, D. E., CHISHOLM, B., LOVELL, N., HOBSON, K. & SCHWARCZ, H. P. (1986b): Paleodiet determinations by stable carbon isotope analysis. In: J. S. Olin and M. J. Blackman (eds.), *Proceedings of the 24th International Archaeometry Symposium,* 49-54. Smithsonian Institution Press.

O'LEARY, M. H. (1981): Carbon isotope fractionation in plants. *Phytochemistry* 20, 553-567.

OTTE, M. (1996): *Le paléolithique inférieur et moyen en Europe.* Paris.

PARKINGTON, J. (1991): Approaches to dietary reconstruction in the Western Cape: Are you what you have eaten ? *Journal of Archaeological Science* 18, 331-342.

PATOU-MATHIS, M. (ed.) (1994): *6ᵉ Table Ronde Taphonomie/Bone Modification.* Artefacts 9, Editions du Centre d'Études et de documentation archéologiques, Triegnes, Belgium.

PIEPENBRINK, H. (1984): Beispiele biogener Dekompositionserscheinungen an Knochen unter längerer Liegezeit. *Anthropologischer Anzeiger* 42, 241-251.

PUECH, P.-F., PRONE, A. & KRAATZ, R. (1980): Microscopie de l'usure dentaire chez l'homme fossile: Bol alimentaire et environnement. *Comptes rendus de l'Académie des Sciences, Paris* 290, 1413-1416.

RODIÈRE, É., BOCHERENS, H., ANGIBAULT, J.-M. & MARIOTTI, A. (1996): Particularités isotopiques chez le chevreuil (*Capreolus capreolus* L.): Implications pour les reconstitutions paléoenvironnementales. *Comptes rendus de l'Académie des Sciences, Paris* 323, 179-185.

SCHOENINGER, M. J. & DeNIRO, M. J. (1984): Nitrogen and carbon isotopic composition of bone collagen from marine and terrestrial animals. *Geochimica et Cosmochimica Acta* 48, 625-639.

SCHWARCZ, H. P. (1991): Some theoretical aspects of Isotope paleodiet studies. *Journal of Archaeological Science* 18, 261-275.

SEALY, J. C. & Van der MERVE, N. J. (1985): Isotope assessment of Holocene human diets in the southwestern Cape, South Africa. *Nature* 315, 138-140.

SEALY, J. C, Van der MERWE, N. J., LEE-THORP, J. A. & LANHAM, J. L. (1987): Nitrogen isotopic ecology in southern Africa: Implications for environmental and dietary tracing. *Geochimica et Cosmochimica Acta* 51, 2707-2717.

SELLSTEDT, H., ENGSTRAND, L. & GEJVALL, N.-G. (1966): New application of radiocarbon dating to collagen residue in bones. *Nature* 212, 572-574.

SKINNER, M. (1996): Developmental stress in immature hominines from Late Pleistocene Eurasia: Evidence from enamel hypoplasia. *Journal of Archaeological Science* 23, 833-852.

STAFFORD, T. W. Jr., JULL, A. J. T., BRENDEL, K., DUHAMEL, R. C. & DONAHUE, D. (1987): Study of bone radiocarbon dating accuracy at the University of Arizona NSF accelerator facility for radioisotope analysis. *Radiocarbon* 29, 24-44.

STRINGER, C. & GAMBLE, C. (1993): *In search of the Neandertal.* New York.

TAUBER, H. (1981): ^{13}C evidence for dietary habits of prehistoric man in Denmark. *Nature* 292, 332-333.

TIESZEN, L. L. (1991): Natural variations in the carbon isotope values of plants: implications for archaeology, ecology, and paleoecology. *Journal of Archaeological Science* 18, 227-248.

TIESZEN, L. L. & FAGRE, T. (1993): Effect of diet quality on the isotopic composition of respiratory CO_2 bone collagen, bioapatite and soft tissues. In: J.B. Lamber & G. Grupe (eds.), *Prehistoric human bone - Archaeology at the molecular level.,* 121-155. Berlin.

TURBAN-JUST, S. & SCHRAMM, S. (in press): Stable carbon and nitrogen isotope ratios of individual amino acids give new insights into bone collagen degradation. *Bulletin de la Société géologique de France.*

TUROSS, N., FOGEL, M. L. F. & HARE, P. E. (1988): Variability in the preservation of the isotopic composition of collagen from fossil bone. *Geochimica et Cosmochimica Acta* 52, 929-935.

Van der MERWE, N. J. & MEDINA, E. (1991): The canopy effect, carbon isotope ratios and foodwebs in Amazonia. *Journal of Archaeological Science* 18, 249-259.

Van der MERWE, N. J. & VOGEL, J. C. (1978): ^{13}C content of human collagen as a measure of prehistoric diet in woodland North America. *Nature* 276, 815-816.

Van der MERWE, N. J. & VOGEL, J. C. (1983): Recent carbon isotope resaerch and its implications for African archeology. *African Archeology Review* 1, 33-56.

VOGEL, J. C. (1978): Isotopic assessment of the dietary habits of ungulates. *South African Journal of Science* 74, 298-301.

VOGEL, J. C. & Van der MERWE, N. J. (1977): Isotopic evidence for early maize cultivation in New York State. *American Antiquity* 42, 238-242.

WEBB, R.E. (1989): A reassessment of the faunal evidence for Neandertal diet bases on some western European collections. In: L. Freeman & M. Patou (eds.), *L'Homme de Neandertal.* Vol. 6, *La subsistance,* 155-178. Liège.

WEIGELT, J. (1927) : *Rezente Wirbeltierleichen und ihre paläobiologische Bedeutung.* (Translation by J. Schaefer, *Recent vertebrate carcasses and their paleobiological implications.* Chicago 1989).

WYCKOFF, R. W. G. (1972): *The biochemistry of animal fossils.* Bristol.

Gathering Fruits as a Way of Food Supply During the Palaeolithic Period

Klaus-Dieter Jäger & Dieter Klaus Schäfer

Abstract

Founded by ethnographic observations a stage of hunters and food gatherers in the history of mankind has been postulated characterizing the appropriative economy of the Palaeolithic period. Hunting activities of Palaeolithic man are proved by the archaeological evidence of animal bones and their osteological examination. On the contrary archaeological evidence of food gathering is scarcely available. However, particular observations are provable since the Middle Palaeolithic period. This may be examplified by the observation and palaeobotanical examination of charred putamina of the corne (Cornus mas L.) from the well-known travertine site of Weimar-Ehringsdorf in Thuringia.

In the present time we observe a very considerable part of vegetable food in the totality of human nutrition. This is true not only in the case of agricultural populations, but also in the case of non-agricultural ones. There the situation may be examplified by two populations of Australian aborigines. Informations on the components of their nutrition have been published by the British ethnographer Rose (1966, p. 125). According to his informations the inland population of the Dieri consumed ca. 18 % vegetable foods, whereas the insular inhabitants of Groote Eylandt used 24 %. They preferred the fruits of a tree called Burrawang (*Cycas media*), whereas in Queensland the nuts of another tree, called Bunya-Bunya (*Araucaria bidwillii* HOOK.) were highly esteemed. Comparable reports, even with reference to other components of vegetable food have been published from other continents, p. ex. the subarctic Canada (Kuehnlein & Turner, 1991). Gathering as field of action has been a task preferably of women according to all available informations (Obermaier, 1926; Hahn, 1929; Rose, 1966, p. 107 and 1969, p. 24; Owen, 1996).

After all'such observations support the ethnographic conception of an appopriative economy based by hunting and gathering in populations which did not dispose of agriculture and cattle-breeding (Treide, 1964 b). This conception includes the presumption that this appropriative economy has originated a long time before agriculture and cattle-breeding. Moreover, it should have been dominated the entire economy during Palaeolithic and Mesolithic times worldwide (Dittmer, 1954; Treide, 1964 a).

Evidence for Palaeolithic hunting has been discussed already by palaeontological and archaeological investigations, p. ex. related to the skeletal remains from Thuringian interglacial travertine sites (Soergel, 1922: Taubach near Weimar; Guenther, 1975, 1977: Weimar-Ehringsdorf). These sites prove a distinct preference of young elephant individuals: More than 60 % of them are less than 40 years old, whereas the natural durability of life ranges to and exceeds 96 years. Already Soergel (1922) concluded that the distribution of ages observed in the elephants population of Taubach may be caused by Palaeolithic hunting by means of pitfalls. Especially younger individuals dispased of less experience as well as ability to defend. Thus, frequently they could be victimized by such traps.

At present we dispose of some better defined evidence of Palaeolithic hunting from the sites (p. ex. Schöningen in Lower Saxony: Thieme, 1996, 1997) whereas conclusions based on the composition of the Thuringian travertine faunas may be discussed yet. Thus, already Kahlke (1957: 104-105) has noted that the bones claimed by Soergel's statistics were „indeed only a selection of accidently well preserved exposition objects in great demand". Another important aspect of discussion has been pointed out by Roebroeks (1988: 145): „One of the major problems encountered in the interpretation of these - and other - travertine sites is of course that we are dealing with faunal assemblages formed in sedimentary environments which favour the preservation of faunal elements 'produced' by a wide range of accumulations agents: natural death, various non-human predators, hominids, and geological processes. The 'sites' excavated in such sedimentary environments are the products of a complex series of depositional events of which hominid activities form only a part, but one that is given most of the credit for the statistics encountered ..."

Moreover, further aspects to be considered in the interpretation of such statistics have been reflected by Müller-Beck (1982: 336) and by Haynes (1993: 213, 216). Therefore, the final conclusion of Guenther (1989: 420) concerning the faunal assemblages from the interglacial travertines in the Thuringian Ilm valley is rather reserved. „Near Weimar-Ehringsdorf and Taubach evidence of [ancient] man is proved, and, thus, one may assume that at least a proportion of bones and teeth may derive from the bag of human hunting."

However, if hunted or scavenged one has to consider a remark of Hahn (1925) pointing at the fact that human beings are not adapted generally to dominating in nutrition position by meat. Such nutrition threatens man with scorbute. This means that at any time a biological necessity has required the completion of hunting or scavenging by gathering vegetable foods. Moreover, the generalization of recent observations of ethnographers, p.ex. as mentioned related to Australian aborigines, to Palaeolithic conditions elsewhere - and on that account to Europe, too - is supported by strong circumstantial evidences.

The substantial evidence of this conception of an appropriative economy during the European Palaeolithic requires on the base of archaeological finds:

- not only the prove of contemporaneous hunting or scavenging by means of examinations of bone materials and archaeological analysis,
- but also prove of contemporaneous gathering by botanical examinations.

Thus, their interdisciplinary treatment is required.

These demands meets one of the Middle Palaeolithic sites in Central Germany. It is the well-known open site in the interglacial travertine of Weimar-Ehringsdorf (Thuringia), where also skeletal remains of early men as well as artefacts have been found .

The geological age, the chronological position of the Ehringsdorf interglacial is still disputable (a.o. Jäger & Heinrich, 1979; Schäfer, 1991; Jäger, 1992). Nevertheless, several important facts are beyond all dispute:

- A lot of palaeontological objects is of interglacial age (comprising Steiner, 1981), irrespectively of the definition of its position in the course of Pleistocene.
- There are archaeological traces of a contemporaneous presence of man (Behm-Blancke, 1960; Steiner, 1981; Schäfer, 1993).
- The archaeological dating is Middle Palaeolithic (Schäfer, 1993, Fig. 59).

Special interest require in this context botanical macroremains. Plant imprints and incrustations (Vent, 1955, 1957, 1958, 1974) are preserved in the travertine and typical finds in many Central European travertine sites. Less frequent is in their sequences the preservation of carbonized plant remains. Such preservation may be examplified by pieces of charcoal from the evidently Eemian travertine site of Tata in northwestern Hungary (Stieber, 1964; Járai-Komlodi et al., 1964). Charcoal remains have been found in the travertine site of Weimar-Ehringsdorf, too (Jacob, 1961, 1984). Worth mentioning is the evidence of *Thuja occidentalis* SCHLECHTENDAHL. However, one has to add to this charred remains of diaspores, i.e. seeds (putamina) of the cornel (*Cornus mas* L.).

The evidence of this species in a travertine sequence is not unique. Thus, p.ex. another sample from Thuringia has been published by Mai (1983, p. 93-94) from Bilzingsleben, but there is only some probability for proving the imprint of a putamen. However, further sites yielding the cornel are cited by Mai (ibidem). All these sites prove the contemporaneous presence of the shrub in Central Europe. Nevertheless, this does not include the utilization or gathering of cornel by man.

This is the peculiarity of the finds from the quarry of Weimar-Ehringsdorf. Their preservation is different from all the leaf or putamen imprints from other sites. The carbonized one means that there fire must have been. The use of fire by man at this site is repeatedly testified by burn marks and layers (Steiner, 1981, p. 158-165). We owe the discover of these finds Dr. Schäfer from Innsbruck.

A first publication of the found charred putamina has been presented by Schäfer & Jäger (1984).

The determination is based and secured:

- by shape and size of the approximately cylindrical objects recalling carbonized grains of cereals, but without any exterior embryo (as opposed to grains of cereals),
- by the superficial sculpture with alternating longitudinal bulges and furrows;
- by the spongy porous anatomical structure visible in the cross-section of each specimen.

These are exactly the features characterizing recent putamina of the cornel (*Cornus mas* L.). However, the anatomical structure is more precisely visible in the comparable recent specimens, whereas it has collapsed by carbonization in the fossil ones (Figure 1). These are smaller than the recent samples, i.e. the length of the fossil objects is between 7 and 11 mm, whereas comparable seeds of recent cornel shrubs range according to Brouwer & Stählin (1973: 143) from 11 to 16 mm. The diameter is in both the cases about 5 mm.

The deviation of length and the corresponding values of diameters are explicable by actualistic oberservations of Hopf (1957, p. 17-20) by means of carbonized grains of cereals from prehistoric and recent times (wheat: *Triticum monococcum* L.; *Tr. dicoccum* Schübl.; barley: *Hordeum vulgare* L.). The typical effect of carbonization was in these cases characterized by shortening and disproportion of grains. Considering this fact one may state a rather good comparability of the prehistoric and recent values.

Moreover, the size values permit a clear discrimination of three European species of the shrub genus *Cornus* (*C. mas* L.; *C. sanguinea* L., *C. suecica* L.). According to these criteria the putamina of *Cornus mas* L. can be identified. Within the fruits they are embraced by a juicy pulp like a cherry-stone by the soft pulp of the cherry.

Approximately 20 scattered finds of such putamina have been distributed among a very restricted partial area of the travertine site. Charred remains of other plant species are not represented among them. Thus, we dispose of a pure find complex of *Cornus mas* L. without any admixture, but in carbonized state. The influence of fire excludes any other reason of accumulation, p.ex. by birds (Hegi, 1906-1931, V, 2, p. 552).

The later utilization of the cornel is proved since the classical antiquity: A comprehensive review of antique written tradition has been compiled by Schmidt (1901). Remarkable is in this context the significance of this fruit for winter storage as mentioned by several antique authors (Columella, De re rustica VII, 10.3; Ovid, Metamorphoses, VIII, 665). Thus, there is a tradition line of use from the Palaeolithic period to the classical antiquity. The more medieval evidences by palaeoethnobotanical investigations in the Czechlands, in France, and in Hungary for the 8th to the 17th century are compiled by Opravil (1976a), Hajnalová (1989) and Kroll (1996, p. 31), who mention also other finds from Neolithic times in Greece and Slovakia and from the Iron Age in the Czechlands and in Slovenija.

Figure 1 - *Seeds (putamina) of the cornel (Cornus mas L.).*
The characterizing morphological sculpture and anatomical structure of recent specimen from the environs of Dresden (a) provide the features for the determination of charred fossil objects from the upper travertine at Weimar-Ehringsdorf (b).

The more, the tradition line visible in these fossil finds may be continued until present times in regions, where the cornel shrub not only grows but also meets the thermal supply necessary for an annual ripening of fruits.

However, the finds from Weimar-Ehringsdorf prove that this tradition line has started at least during the Middle Palaeolithic period respectively during the Middle Pleistocene, most probably in the time slice of oxygene stage 7.

The share of vegetable foods in the bill of fare of any recent or previous population depends first of all on their regional and temporal disponibility. It is modified by cultural aspects and customs as well as by taste and nutritive quality (Karg & Haas, 1996, p. 429). These facts influence the archaeological probability of gathering activities. It depends on preservable remains of gathered vegetable foods.

In Central Europe plant macroremains from later periods of prehistory and history frequently are placed at disposal due to good prerequisites of preservation by damp soil conditions including also parts of fruits. These chances have been used to the investigation of the economical base of Neolithic and post-Neolithic agrarian societies and populations. The respective examinations have been opened by the detection of lake-shore settlements (so-called pile-dwellings or in German Language „Pfahlbauten") by Keller (1854) and the subsequent treatment of plant remains rescued here by Heer (1865).

Elsewhere the examination of plant macroremains from archaeological sites as of agrarian populations as of non-agrarian ones depends on carbonization. With reference to diaspores, i.e. fruits and seeds, this means that under dry soil conditions only the remains of species are preservable which consist of hard (sclerotic) material or embrace hard components like the putamen within the fruit of cornel.

The evidence of gathering activities in non-agrarian populations and societies is rendered more difficult by this situation. Therefore, observations as referred in the preceding report dispose of scarcity value. In the case of Weimar-Ehringsdorf they meet the palaeontological evidence of hunting activities. This means, here we dispose of findings for both the fields of action in the economical system of hunter-gatherers. Furthermore the period for which this system is proved is not younger than the Middle Palaeolithic one. Last not least the conception of this system is not an unfounded transfer of ethnographic experiences to archaeological interpretations but an object of good archaeological evidence.

References

BEHM-BLANCKE, G. (1960): *Altsteinzeitliche Rastplätze im Travertingebiet von Taubach, Weimar, Ehringsdorf.* Weimar.

BROUWER, W. & STÄHLIN, A. (1973): *Handbuch der Samenkunde für Landwirtschaft, Gartenbau und Forstwirtschaft.* Frankfurt/Main. 2. Aufl.

DITTMER, K. (1954): *Allgemeine Völkerkunde - Formen und Entwicklung der Kultur.* Braunschweig.

GUENTHER, E.W. (1975): Die Backenzähne der Elefanten von Ehringsdorf bei Weimar. In: H.-D. Kahlke (Hrsg.), *Das Pleistozän von Weimar-Ehringsdorf*, Bd. 2, 399-452. Berlin (Abhandlungen des Zentralen Geologischen Institutes 23).

GUENTHER, E.W. (1977): Die Backenzähne der Elefanten von Taubach bei Weimar. In: H.-D. Kahlke (Hrsg.), *Das Pleistozän von Taubach bei Weimar*, 265-304. Berlin (Quartärpaläonthologie 2).

GUENTHER, E.W. (1989): Die Gebisse der Waldelefanten von Bilzingsleben. *Ethnographisch-Archäologische Zeitschrift* 30, 416-450.

HAHN, E. (1925): Dreistufentheorie. In: M. Ebert (Hrsg.), *Reallexikon der Vorgeschichte*, Bd. 2, 460-462. Berlin.

HAHN, I. (1929): Wirtschaft § 12. In: M. Ebert (Hrsg.), *Reallexikon der Vorgeschichte*, Bd. 14, 336-338. Berlin.

HAJNALOVÁ, E. (1989): *Súčasné poznatky z archeobotaniky na Slovensku.* Nitra (Acta Interdisciplinaria Archaeologica 6).

HAYNES, G. (1993): *Mammoths, mastodonts and elephants.* Cambridge.

HEER, O. (1865): Die Pflanzen der Pfahlbauten. In: *Neujahrsblatt der Naturforschenden Gesellschaft auf das Jahr 1866,* 1-54. Zürich.

HEGI, G. (1906-1931): *Illustrierte Flora von Mitteleuropa mit besonderer Berücksichtigung von Deutschland, Österreich und der Schweiz.* München.

HOPF, M. (1957): Botanik und Vorgeschichte. *Jahrbuch des Römisch-Germanischen Zentralmuseums Mainz* 4, 1-22.

JACOB, H. (1961): Informationen über Versuche pollenanalytischer Untersuchungen und Holzkohlebestimmungen der Interglazialablagerungen in der Umgebung von Weimar. *Geologie* 10, 546-549.

JACOB, H. (1984): Neue botanische Befunde von Interglazial-Ablagerungen ausThüringen. *Alt-Thüringen* 20, 7-14.

JÄGER, K.-D. (1992): Weimar-Ehringsdorf. In: L. Eissmann & Th. Litt (Hrsg.), *The Saalian sequence in the type region (Central Germany),* 36-40. Halle/Saale.

JÄGER, K.-D. & HEINRICH, W.-D. (1979): Aktuelle Aspekte und Probleme bei der quartärstratigraphischen Einordnung der mittelpaläolithischen Travertinstation von Ehringsdorf bei Weimar. *Ausgrabungen und Funde* 24, 261-26.

JÁRAI-KOMLODI, M., SKOFLEK, I. & STIEBER, J. (1964): Die stratigraphische Wertung der botanischen Beobachtungen. In: L. Vértes (Hrsg.), *Tata - Eine mittelpaläolithische Travertin-Siedlung in Ungarn,* 83-86. Budapest.

KAHLKE, H.-D. (1957): Eine Statistik paläolithischer Jagdbeute-Reste aus Brandschichten von Weimar-Ehringsdorf. *Ausgrabungen und Funde* 2, 104-105.

KARG, S. & HAAS, J. N. (1996): Indizien für den Gebrauch von mitteleuropäischen Eicheln als prähistorische Nahrungsressource. In: I. Campen, J. Hahn & M. Uerpmann (Hrsg.), *Spuren der Jagd - Die Jagd nach Spuren,* 429-435. Tübingen (Festschrift für Hansjürgen Müller-Beck; Tübinger Monographien zu Urgeschichte 11).

KELLER, F. (1854): Die keltischen Pfahlbauten in den Schweizerseen. *Mitteilungen der Antiquarischen Gesellschaft in Zürich,* 9, 67-100. Zürich.

KROLL, H. (1996): Literature on archaeological remains of cultivated plants (1995/96). *Vegetation History and Archaeobotany* 6, 25-67.

KUHNLEIN, H.V. & TURNER, N.J. (1991): *Traditional plant foods of Canadian indigenous peoples nutrition, botany and use.* Paris.

MAI, D.H. (1983): Die fossile Pflanzenwelt des interglazialen Travertins von Bilzingsleben (Kreis Artern, Thüringen). In: D.H. Mai, D. Mania, T. Nötzold, V. Toepfer, E. Vlček & W.-D. Heinrich, *Bilzingsleben II,* 45-129. Berlin (Veröffentlichung des Landesmuseums für Vorgeschichte in Halle, Bd. 36).

MÜLLER-BECK, H. (1982): Late Pleistocene man in northern Eurasia and the mammoth-steppe biome. In: *Palaeoecology of Beringia,* 329-352. Academic Press.

OBERMAIER, H. (1926): Jagd: A. Paläolithikum. In: M. Ebert (Hrsg.), *Reallexikon der Vorgeschichte,* Bd. 6, 134-142. Berlin.

OPRAVIL, E. (1976 a): Die Kornelkirsche (*Cornus mas* L.) aus archäologischen Funden des ČSSR-Gebietes. *Folia Quarternaria* 47, 25-27. (Kraków).

OPRAVIL, E. (1976 b): *Cornus* L. s. l. v. Československérn kvartéru. *Časopis Slezského Muzea* 25, 97-106. (Opava).

OWEN, L.R. (1996): Geschlechterrollen und die Rekonstruktion von Werkzeuggebrauch im europäischen Jungpaläoliuthikum. In: I. Campen, J. Hahn & M. Uerpmann (Hrsg.), *Spuren der jagd - Die Jagd nach Spuren,* 25-33. Tübingen (Festschrift für Hansjürgen Müller-Beck;Tübinger Monographien zur Urgeschichte 11).

ROEBROEKS, W. (1988): *From find scatters to early hominid behaviour; A study of Middle Palaeolithic riverside settlements at Maastricht-Belvédere (The Neatherlands).* Leiden (Analecta Praehistorica Leidensia 21).

ROSE, F, (1966): *Ureinwohner, Känguruhs und Düsenclipper - Fünfundzwanzig Jahre unter Australiern.* Leipzig.

ROSE, F. (1969): *Die Ureinwohner Australiens - Gesellschaft und Kunst.* Leipzig.

SCHÄFER, D. (1991): Weimar-Ehringsdorf: Diskussionsstand zur geochronologischen und archäologischen Einordnung sowie aktuelle Aufschlußsituation. *Quartär* 41/42, 19-43.

SCHÄFER, D. (1993): Grundzüge der technologischen Entwicklung und Klassifikation vor-jungpaläolithischer Steinartefakte in Mitteleuropa. *Berichte der Römisch-Germanischen Kommission* 74, 49-193.

SCHÄFER; D. & JÄGER, K.-D. (1984): Verkohlte Steinkerne der Kornelkirsche (*Cornus mas* L.) aus dem Paläolithikum des oberen Travertins von Weimar-Ehringsdorf. *Alt-Thüringen* 20, 15-22.

SCHMIDT, M.C.P. (1901): Cornus. In: Pauly & Wissowa (Hrsg.), *Real-Encyclopädie der Classischen Altertumswissenschaft,* Neue Bearbeitung. Col. 1633-1635. Stuttgart.

SOERGEL, W. (1922): *Die Jagd der Vorzeit.* Jena.

STEINER, W. (1981): *Der Travertin von Ehringsdorf und seine Fossilien.* Lutherstadt Wittenberg. 2. Aufl.

STIEBER, J. (1964): Die anthrakotomische Untersuchung der Holzkohlen. In: L. Vértes (Hrsg.), *Tata - Eine mittelpaläolithische Travertin-Siedlung in Ungarn,* 79-82. Budapest.

THIEME, H. (1996): Altpaläolithische Wurfspeere aus Schöningen, Niedersachsen. - Ein Vorbericht. *Archäologisches Korrespondenzblatt* 26, 377-393.

THIEME, H. (1997): The world's oldest spears. *Nature* 385, 787-788, 807-810.

TREIDE, D. (1964 a): Die Entwicklung der Kultur. In: W. Liedtke, E. Lips, W. Stegner & D. Treide (Hrsg.), *Völkerkunde für jedermann,* 44-53. Gotha, Leipzig.

TREIDE, D. (1964b): Gruppen ohne Bodenbau und Viehzucht. In: W. Liedtke, E. Lips, W. Stegner & D. Treide (Hrsg.), *Völkerkunde für jedermann,* 54-67. Gotha, Leipzig.

VENT, W. (1955): Über die Flora des Riß-Würm-Interglazials in Mitteldeutschland unter besonderer Berücksichtigung der Ilmtaltravertine von Weimar-Ehringsdorf. In: *Wissenschaftliche Zeitschrift der Friedrich-Schiller-Universität Jena, Mathematisch-Naturwissenschaftliche Reihe* 4 (1954/55), 467-485.

VENT, W. (1957): Die pflanzliche Umwelt der prähistorischen Menschen im Ilmtal bei Weimar-Ehringsdorf-Taubach zur Unstrut-Warmzeit. *Thüringer Heimat* 2 (1), 8-15. Weimar.

VENT, W. (1958): Die Pflanzenwelt der Ilmtaltravertine von Weimar-Ehringsdorf zur Unstrut-Warmzeit. *Alt-Thürngen* 3 (1957/58), 16-28.

VENT, W. (1974): Die Flora der Ilmtaltravertine von Weimar-Ehringsdorf. In: H.-D. Kahlke (Hrsg.), *Das Pleistozän von Weimar-Ehringsdorf,* Bd. 1, 259-321. Berlin (Abh. des Zentralen Geologischen Institutes 21).

Comments on the Osteological Criteria for Hunting and Scavenging in the Middle Palaeolithic

Bodil Bratlund

Abstract

The subsistence economy of the Middle Palaeolithic has during the last decade been subject to vigorous discussion. One of the recurrent themes has been the relative importance of scavenging as opposed to active hunting of the large mammals.

Some of the most tangible arguments are based on the interpretation of specific patterns in the archaeological faunal assemblages. Too often, however, the interpretations are based on assumptions originating from recent actualistic observations, which have not been considered in relation to the full range of taphonomic problems specific for Stone Age assemblages. The significance of the patterns could preferably be critically reviewed against later archaeological faunal assemblages, where the subsistence economy is reasonably well known.

Besides, several of these arguments were essentially well-known to the European research already the beginning of the century, only to be „re-discovered" in the wake of the mainly Anglo-American New Archaeology. Most recently, researchers working on Stone Age faunal assemblages tend to be increasingly reserved concerning specific arguments allegedly allowing positive identification of either hunting or scavenging.

In this paper some of the most common arguments for hunting or scavenging will be reviewed. New results from an archaeozoological analysis of the faunal assemblage from the Taubach travertine near Weimar, Germany, are presented, which strongly suggests planned and repetitive megafauna hunting by Neandertal man.

Introduction

Discussions of the cultural capacities of Neandertals has for years been linked - directly or indirectly - to the prevailing suppositions about their subsistence economy. Bones of possible prey animals found in the same archaological context as artefacts and other cultural remains have been used as a starting point for often very far-reaching interpretations of what Neandertals did for a living, and - compared to evidence from modern humans - how efficiently they seemed to manage the tasks.

It is needless here to repeat the portrayals of cavemen with crude hunting tools (and limited social skills) en vogue from the late 19th century, - and in popular form persisting well beyond the middle of this century. More interesting is, how after a fairly long period after World War II where a more positive perception of the abilities of prehistoric man was prevalent, the last two decades have seen a trend to dehumanize Neandertals. The emphasis on what modern humans was seen to be able to do, and the Neandertals not, did of course not leave out interpretations of basic subsistence practices. A fascinating fact is, however, that the reversal was not so much brought about by striking new discoveries, but was based on re-interpretations of long known finds.

The first serious challenge to the view of primitive, but universal big-game hunting in the Middle Palaeolithic was posed by Binford, whose work in the early eighties questioned the evidence for active hunting and proposed scavenging as an important (perhaps even more important) way of obtaining animal protein (Binford, 1981, 1984, 1988). It is Binfords undisputed merit to question the up till then tacitly assumed equation between animal bones found in an archaeological context and cultural waste, and consequently demand (or rather provoke) a more thorough taphonomic analysis of Middle Palaeolithic sites, if, however, we to a certain extent see a recycling of arguments used earlier in the discussion of Lower Palaeolithic subsistence. This provocation to actually present the hard evidence for hunting has indirectly provided us with a wave of actualistic studies, most important on scavenging by carnivores in the wild, traces of gnawing or tool use and the treatment of carcasses by recent traditional hunters.

In a recent paper Brian Hayden has presented a review and re-evaluation of the arguments presented by various authors for the more dehumanized portrayals of Neandertals (Hayden, 1993).

He is concerned mainly with evidence relevant to those organizational and cognitive issues, which have been considered to show significant differences between the Middle and Upper Palaeolithic, and he does succesfully demonstrate the existence of an abundance of reasonable indications for sophisticated cultural capacities of Neandertal man.

However, it is only briefly mentioned that most researchers [with reference to the works of Mellars (1989) and Chase (1989)] „ought to agree" on the point that Neandertal man was capable of efficient hunting, although an extensive, and very interesting, discussion of the differences between Middle and Upper Palaeolithic hunting, especially respecting indications of the organization level, is given.

I would like to approach the first issue here, as a glance through the most recent publications shows that the importance of hunting in Middle Palaeolithic subsistence is by far not a settled question. Also it is apparent that the general picture of the subsistence economy is still highly influenced, and probably too much so, by material from the caves and rockshelters of Southwestern Europe.

Hunting *versus* scavenging or hunting and scavenging?

A very important point here is to specify the qualities of the hunting skill. If we take „hunting", simply as the ability to intentionally kill a medium-sized animal and to do so on a regular basis, there would probably be a general agreement on Neandertal hunting skills. The differences of opinion arise when we try to include prey animals from the so-called megafauna or try to estimate the relative importance of hunting versus scavenging.

The analysis of the bone remains from archaeological sites have in the latter decades become increasingly sophisticated concerning the analysis of the find context and the taphonomic qualities of the assemblages. Following the evaluation of all possible non-human influencing factors and a sorting out of the relevant fractions of the bone assemblages we do, however, still have to face the problem of interpreting whatever remains to the archaeologist as cultural data.

For Palaeolithic sites this should basically mean a discussion of the evidence for hunting (including different ways of catching the prey) and also a consideration of the possible contribution of human scavenging to the bone assemblage.

The easiest way out is simply to avoid the problem and proceed as if hunting is already proven by the presence of the bone remains. In most European reports on excavated material, even very recent ones, the reader will search in wain for any discussion telling precisely which arguments lead the researcher to assume that active hunting alone was relevant for the interpretation. But for a better description of the local stratigraphy and taphonomic processes in general this may be regarded a relapse to, or perhaps more correct, a continuation of the „common sense" tradition regarding the determined bones from a site to be direct evidence of the hunters activities.

A few studies do, however, take the problem serious. Bone assemblages from French cave sites comprising the remains of middle sized or large ungulates have been analysed by Chase, who mainly on the basis of the limited number of represented species and the carcass exploitaion, could argue for active and seasonally organized hunting (Chase, 1990). The faunal remains from Grotte Vaufrey, which only in 1988 had been interpreted by Binford as indicative of scavenging (Binford, 1988), was only a few years later submitted to a taphonomic by Grayson & Delpeche (1994), who concluded that Binfords claims could not be supported.

It may be more generally accepted now than just a few years ago, that hunting of middle-sized ungulates most probably was an important element of the Neandertal economy. Many authors do, however, still appear to feel uncomfortable at the notion of Neandertals actively hunting very large and dangerous prey, like e.g. rhinos or bears (Stringer & Gamble, 1993). Recent analysis of material from open-air sites like Biache-Saint-Vaast or Taubach, which are rich in cutmarked and processed bone remains from rhinos and brown bears, should, however, contribute to disperse such reservations (Auguste, 1992; Bratlund, in print).

The opinion pendulum has thus within few years swung from one extreme: Neandertal man as *the* mega-fauna hunter, to another: a scavenger hardly capable of active hunting. It is apparently now wavering in an opportinistic middle position, where hunting of medium-sized animals, and sometimes very large prey, is accepted, but scavenging on the other hand not excluded.

The called for „correct approach" to an assemblage should for the moment apparently be to try to evaluate not whether hunting *or* scavenging, but how much of either, can be found at a particular site or at the sites in a given region.

Unfortunately, I have not as yet been able to find any reports, where claims of Neandertal scavenging could be convincingly substantiated. The so far best candidates are found in the work of Mary Stiner who lately has presented several publications, which have been based on taphonomic analyses of faunal material from cave sites on the Italian coast. She has there argued for the importance of scavenging in the Middle Palaeolithic economy. However intelligent the analytical approach in this case may be, the basic archaeological data does in my opinion appear to be insufficient and the impressive models consequently become impaired. Before proceeding I would like to give a few short comments on these reservations.

The material which Mary Stiner analyses come from cave sites excavated 50 to 30 years ago. We may take the Grotta Guattari case as an example as the information given on the basic find context here is fairly detailed (Stiner, 1991). According to Stiner, the site was first investigated by Blanc, who excavated by geological levels, and later by Cardini, who applied thin arbitrary units. Later „Segre was able to unify the provenience data of Blanc and Cardini into seven major geological levels". A short description af these units, 1-7 from the top, is given, but later in the presentation only the names G0 to G7 are used. We must assume that the units G4-G5, supplying the hominid-associated material are the same as the geological levels 4 and 5, and the units G0, G1 and G2, containing hyena-den material corresponds to the cave floor and the two first geological layers. In this case, I find it somewhat surprising, that the layer G3 can be described as „a sterile layer 10-75 cm thick that unequivocally separates levels G4 an G5 from the overlying depositional sequence" (Stiner, 1991: 105) as the geological stratum 4 (= G4 ?) is given as „partially cemented fill with fossil and modern root casts" (Stiner, 1991: Tab. 1, p. 105) ?

The total number of fragments still found in the museum collections and ascribed to the two units G4-G5 are 102 (Stiner, 1994, Tab. 4.7) hereof are only 26 bone fragments the rest being teeth. Stiner does, however, mention that: „the faunal collections are biased in favour of bones with anatomical landmarks. Most of the unidentifiable fraction including shaft fragments, was throwm away at the time of the excavation" (Stiner, 1991, note 4, p. 106). Also Stiner found „ so many discrepancies between Cardini´s list of species by stratigraphic level (in Blanc and Segre 1953) and my own observations that I am convinced that the list should be disregarded (Stiner, 1991, note 2, p.105).

The 26 bone fragments from G4-5 are seen as hominid-asssociated, as they have a lower percentage of gnawed fragments than the one found in the hyena-den material from the layers above, and most (about two thirds of the total) of the stone artefacts from the cave come from these units. Due to a dominance of head bone fragments the 26 fragments are, by analogy to head-dominated assemblages from hyena den assemblages considered to show that a similar procurement technique, namely scavenging, was used by the humans.

In the comments on the article surprisingly few of the reviewers express uneasiness concerning the basic data. D´Errico (p.127) observes, that the rich bone assemblages from the upper layers may be due to the exceptional preservation conditions (the upper layers are rich in limestone blocks and concretions) and not as Stiner argues preservation is good throughout the section. Both Conard (p.125) and Donahue (p.128) are reserved concerning the use of such small samples.

I believe that the material from Grotta Guattari is inadequate for the following reasons: The stratigraphical units had to be re-constructed, and consequently the original context of individual pieces may not be precisely known, the material has been sorted during excavation and even that which could be found in the museum does not match earlier reports, the high percentage of teeth indicates either vigorous excavation sorting or insufficient preservation conditions. Last not least, the presence of artefacts in the same general geological unit does not show us who was the bone accumulation agent.

I think that the possibility should be considered that the small assemblage from Guattari G4-G5 might be from scavenging, but by hyenas, or perhaps more likely, represents the leftovers of an assemblage so worn down as to mimick the structure of the heavily damaged hyena materials.

Criteria for economic interpretations

Some of the criteria which lately have been presented, whether speculative or more firmly based on recent actualistic observations, as diagnostic of either hunting or scavenging, may appear seducively clear in theory, but they still fail to convince experienced archaeozoologists. One of the main obstacles is the often observed fact that the same kind of pattern in a bone sample can come into being in different ways, and not only by the procedure encountered in a recent actualistc study or experiment. I would like to present some comments on the most frequently encountered arguments, well aware that many things mentioned are for the experienced colleagues simple truisms.

Leaving out the circumstantial site evidence, which from case to case may differ, only a few precise criteria could be gathered from the literature. I think the most tangible indicators to look for in a bone material are: marks of weapon impacts, cut marks and/or fragmentation patterns, dominance of fragments from the head and feet, „unusual“ species patterns (dominance of one or two species) and „unusual“ age profiles (dominance of presumably fit animals). These indicators should be relevant, not only to a Middle Palaeolithic material.

Weapon impact

Hunting lesions in bone containing embedded fragments of stone points would off course be the easiest way to demonstrate active hunting. From the European Middle Palaeolithic so far no secure weapon impacts in bone are known, but recently Milo have reported „the probable tip of a stone point embedded in a vertebra of *Pelerovis antiquus*“ from Klasies River Mouth Cave 1, an early Late Pleistocene site inhabited by near modern *Homo sapiens* (Milo, 1996).

Use-wear studies on assemblages from the Levantine Mousterian by Shea did, however, show impact fractures and wear from hafting on pointed artefacts, i.e. the traces were not restricted to types like the Levallois point, which one spontanously would suppose to be weapon armatures (Shea, 1993).

Wooden spears like the wellknown example from Lehringen may have been in more universal use, thus reducing the possibilities of finding diagnostic impacts with stone fragments.

In later times, however, such hunting fractures are still rare. Even in the excellently preserved Mesolithic material from the Danish Islands no more than one or two specimens are likely to turn up in a bone assemblage (Noe-Nygaard, 1974), but only a small part of this material has so far been scrutinized under magnification and more specimens may lie hidden in the museums.

From the Late Glacial reindeer kill site of Stellmoor in Northern Germany two arrowhead impacts containing flint fragments have long been known. Recent cut mark studies, however, have produced merely as a by-product a fair sample of hunting lesions. Of the 27 flint-containing hunting lesions now known from the Ahrensburgian assemblage of Stellmoor only 8 could have killed or disabled the reindeer immediately, while the rest represented more superficial wounds. Just for comparison did the excavated approx. 12,000 bones and antlers represent an MNI of at least 650 dead reindeer. Also, most of the impacts were found in vertebrae and ribs. The Stellmoor case seems to indicate that a fairly high proportion of the fragile bones surrounding the most vulnerable parts of the ungulate body would greatly improve the prospects of finding hunting lesions in a Stone Age material (Bratlund, 1990).

Also, the number of excavated Middle Palaeolithic sites with large bone assemblages embedded in soft, waterlogged sediments are still limited. We may suppose that the lack of hunting lesions in Middle Palaeolithic bones is more a question of prevailing conditions of preservation and site type than of prehistoric subsistence activities.

Cut mark and fragmentation patterns

In a recent report on the bones from Shield Trap Cave which originated from animals naturally trapped in the cave (which can be described as a natural pit fall) Oliver has given an excellent description of how cut marks, fragmentation and polish of bones may come into existence without the help of man (Oliver, 1989). Concerning cut marks and fragmentation we should not only be able to give an anatomical or mechanical explanation of the type of mark, but (hopefully) should also be able to demonstrate from a series of similarly cut marked bones that it was part of a planned procedure, and thus a man made pattern.

In my experience defleshing marks on the diaphysis of long bones tend to be superficial and rare (or absent) even on marrow-fractured bones, and, generally speaking, the deeper cutmarks tend to be situated very near the articular ends. Mostly this was apparently due to difficulties cutting the short ligaments going across the joints from bone to bone. In some cases a few bones in a sample will show an atypical profusion of marks, which should be considered in terms of dried, frozen or insufficiently cooked joints rather than filleting of fresh carcasses.

Leaving a cut mark in a bone during butchery of fresh carcasses can be considered more an accident than a necessity, as it may damage the cutting edge on your tool. Even in excellently preserved kill site assemblages from soft, waterlogged sediments of a considerably younger dating than the Middle Palaeolithic we see that not all butchered bones got cut marks. In the Ahrensburgian sample from Stellmoor e.g. only 120 out of 300 distal ends of marrow fractured humeri from reindeer had recognisable cut marks, i.e. only about a third of these bones which had been through the whole processing procedure show marks.

For a few other skeletal elements (like atlas or the last thoracic vertebra and first four lumbar ones) cut marks were even more common as the partially butchered carcasses (which still carried more or less intact limbs), had been subjected to some selective filleting or dismemberment. For the majority of bones, however, cut mark frequencies were between 10 and 25 %. It is my impression from ongoing studies that Mesolithic material dominated by remains of fully exploited animals from single kills may have slightly higher occurrences of marks, than the reindeer mass kill assemblages.

Depending on the composition of the bone assemblage in question, cut mark frequencies of 25-30 % may be the highest we can reasonably expect from a kill and processing site, even before decomposition of the bone surface has been taken into account (Bratlund, 1996)

Given a lesser quality of preservation the cut marks will disappear with the bone surface, and it must be expected that the deeper cuts near the articulations successively will become over-represented, thus leaving the false impression of a preoccupation with joints of already defleshed carcasses.

Another point is, that the size of the studied sample should be considered. The different cut mark frequencies for various species from the relatively large Taubach assemblage (Bratlund, in print) could suggest, that a minimum of 100-150 bones with a very good surface preservation and of each one species are needed, before we can say with confidence that a lack of cut marks in a given bone sample is a significant observation.

This is a criteria which many Stone Age assemblages do not meet and we cannot in such cases exclude man as a bone collector. On the other hand, gnawed fragments as in the hyena den cases reported by Stiner do also come in frequencies (in Buca della Iena 35 % for the ungulate remains from all levels and even less, 21 %, for the carnivores) (Stiner, 1994).

In short, many Stone Age bone assemblages are small and damaged and will never provide sufficient information for any determination of whether hominids, hyenas or both were the responsible bone collectors.

Dominance of head and feet

A dominance of bone fragments from the head and distal part of the legs has been proposed as an indication of scavenging. This argument was initially brought forward by Binford, and is based on the observation that scavenging means secondhand exploitaion of an animal carcass. Logically, the starting point for the hominids would then be incomplete carcasses, and depending on the intensity of the preceding carnivore activity their share would become more or less biased towards the marginal parts (Binford, 1984). Blumenschines extensive fieldwork and detailed analysis of recent African carnivore kills supports the idea so far that heads will mostly be the last useful leftovers of a kill. Bones of the lower part of the legs may as well be found relatively late in the consumption sequences (Blumenschine, 1986).

Despite the context problems which may exist for the Italian cave finds considered by Stiner (Stiner, 1994) to represent scavenging (and which severely limits the value of a far-reaching economic model) these very small bone accumulations mimicks what we could be looking for, if - on a purely theoretical basis - we were to set out on the search for traces of scavenging. In numbers the bones should be few and, by analogy, the representation of skeletal elements similar to that of wild carnivore scavengers, i.e. many head fragments.

I doubt that the head-transport-analogy proposed by Stiner is sound. From the unfortunately few reports on scavenging by humans I have read so far, I do not remember any instances of transports of heads, what, however, there could be explained by meaty joints being available, and thus a choice still possible. If a carcass had been exploited so intensively that the head was the only nutritionally interesting bit available, we must suppose the kill to be already several hours, or days, old. In a temperate or moderately warm climate the soft tissues of the head can be expected to start decomposing quickly (and before a similar sized piece of muscle) and thus host a variety of micro-organisms, including very dangerous bacteria (like *Clostridium*). Even if eating such a head would still be a treat for a carnivore with the stomach of a canid or a hyena, for a hominid it could quite likely become his last experiment ever.

	Bjørnsholm		Tybrind Vig	
Red deer				
Head	11	17.74 %	12	13.04 %
Feet	28	45.16 %	27	29.35 %
Ribs & vertebrae	15	24.19 %	14	15.22 %
Other bones	8	12.90 %	39	42.39 %
Total	62	99.99 %	92	100.00 %
Roe deer				
Head	31	25.62 %	7	13.21 %
Feet	48	39.67 %	14	26.42 %
Ribs & vertebrae	12	9.92 %	3	5.66 %
Other bones	30	24.99 %	29	54.72 %
Total	121	100.02 %	53	100.01 %

Table 1 - Late Mesolithic, Denmark. Coastal settlements, Ertebølle culture.
Bjørnsholm, Northjutland.
Shellmidden: Activity/habitation area(s).
Approx. 90 % of all fragments smaller than 5 cms.
Approx. 16 % identified to species.
Tybrind Vig, Funen.
Submerged coastal site, waste accumulation in marine deposits.
Approx. 80 % of all fragments larger than 5 cms (or intact bones).
Approx. 75-80 % identified to species.

A predominance of head and feet fragments is moreover a very dangerous argument to use in an archaeological context. For bone studies there is unfortunately a very big source of error working on the material, who needs to see a minimum of recognisable features on each given bone fragment in order to determine the type and species. In samples of fragmented or even chemically altered bones the head and the lower part of the legs are more likely to supply identifiable fragments than other parts of the skeleton. Experienced researchers will actually on the basis of a list of vertebrate species and the skeletal elements identified from an assemblage be able to give a fair estimate of the general state of preservation.

It is not accidental that zoological systematics are built up around an evaluation of skeletal elements of the head and the ends of the extremities. These are the immediate points of contact between the animal and the environment, and provide the more evident hints of what the animal did for a living and how it moved about. Also these parts of the skeleton provide us with relatively sturdy pieces, notably teeth, massive crania bits and metapodials. You may e.g. be reasonably sure to determine whether a two-centimeter long piece of metapodial belonged to either roe deer or sheep or goat, but a bit of vertebrae or rib of the same size will hardly bring you any closer than „small ungulate". The latter being of a spongeous bone it may in fact have dissolved and disappeared long before the metapodial fragment, if buried in a suboptimal environment. These things are banalities to most excavators but we must take the consequences also of banalities in the interpretation of our bone lists.

The point here is, that the Neandertals should not be subjected to tests dismissing their hunting activities which even Mesolithic (or later) hunters can not pass adequately. To illustrate this I have summarized the differences between two Late Mesolithic samples in Table 1. Both are from coastal settlements of the Danish Ertebølle culture (Andersen, 1985, 1993,1995) where we feel confident in describe the economy as based on active hunting, fishing and gathering, and the only indications of scavenging are related to the occasional exploitation of large, and most probably stranded, whales.

The Bjørnsholm sample consists of animal bone fragments from one of Denmarks largest shellmiddens situated in Northern Jutland. In such a shellmidden bones are embedded in and between shells along with ceramics, stone and antler tools as well as manufacturing waste, hearths, ash layers, etc., the whole structure representing a complex succession (vertically and horizontally) of more or less well preserved activity areas and living floors. Due to the calciferous environment chemical erosion of bones and antlers is usually neglectable, except in top layers just under the surface soil. Unfortunately we do not have a preserved waste discard zone in front of the Bjørnsholm shellmidden for direct comparison, but a sample of bones from this part of the now submerged coastal site Tybrind Vig by Funen may do instead. The site waste here ended up in soft waterlogged sediments in the beach area, and was preserved as part of an submerged site. Due to stream erosion of the sediment the site was discovered and recently excavated under water. Generally the preservation of the bones is excellent. A fraction has during the sediment erosion have been subject to some surface abrasion minimizing the possibilities for cutmark studies, but hardly ever to a degree that would affect species determination. Due to careful excavation techniques and sampling for fish bones small bone fragments and antler chips are plentiful from both sites. In both cases the tables represent only large samples of excavated material, not the entire bone assemblage.

There, however, the similarities stop. The Tybring Vig sample can be described as the normal dump waste - prey animals processsed, waste, including bones, disposed of and forgotten. In Bjørnsholm the situation is more complex as we are dealing with remains of living floors. In short, I interprete the bones from Bjørnsholm to be the remains after the butchering, marrowfracturing, cooking etc. had been going on near the hearths, but here also a cleaning-up, however superficial, had taken place and cumbersome bones carried away.

The small size of the fragments in general, about 90 % were less than five centimers long, and a tremendous bone loss (when we look at the MNI values for the larger prey species) indicate an extraction of material. The presence of some larger, „normal" bone fragments can be explained by the crevices and hollows found between large shellheaps, but the large majority of fragments will fit in or in between the large *Ostrea*-shells which constituted the „sediment" of the living floor (Bratlund, 1991).

The relevance for Middle Palaeolithic studies of this excursion into the Mesolithic may not be so obvious at first sight. When we look at the representation of body parts in the two samples, however, the samples may become more interesting. In table 1 the results for red deer and roe deer, representing a medium sized and a small cervid resp. are listed.

In the Bjørnsholm living floors head and feet fragments dominate, together constituting approx. 65 % of the identified fragments, whereas in the Tybring Vig dump the situation is different with head and feet only constituting approx. 40 %.

I resist the conclusion that Ertebølle people at Tybrind Vig where great hunters, while at the same time in Bjørnsholm scavenging should be important.

The Late Mesolithic samples should so far be relevant for comparison as they represent material from unspezialized hunting. You may, however, produce similar tables using samples from later time periods, from e.g.. from an Iron Age or Middle Age context, provided the animal skeletons have been submitted to a suffient amount of destructive agents before being excavated.

Moreover the shellmidden living floors are the result of a rapid accumulation of shells making the identification of stratigraphical units from relatively short periods of time easier than in open cave floors and the shell material provides a better chemical bone environment, than what can be expected from most cave or rockshelter sediments.

Considering megafauna I guess that most readers can imagine how many identifiable fragments of the body skeleton you will have a chance of getting, if an aurochs or other very large animal is chopped to pieces and, especially, if subsequently cumbersome pieces are removed from the living floor and the rest submitted to weathering and some chemical destruction.

Thus, due to the greater resistance of teeth and compact bone from head and feet and the easier recognition of the species features in these pieces one ought to be reluctant to accept far-reaching models built on scanty bone assemblages. We should here be able to conclude that the present shape (and size) of a material may have been influenced by so many not readily recognisable and equally possible variables as to be unsuited to answer all our questions.

Dominance of one or two species, and dominance of a particular age class

I have chosen to deal with these two criteria under the same headline as both basically refers to a discrepancy between the species representation and/or the mortality pattern observed in the excavated death assemblage and the pattern(s) one would suppose to see in a natural (including carnivore influenced) assemblage. As most archaeological faunal assemblages are small (total MNI´s rarely going beyond 100 and often being less than 10-20 individuals), these criteria are usually employed for the same bone remains of one species. Perhaps we should refer to our material as a sample more often, as what we have on the table is only the excavated fraction of the discovered faunal assemblage, which again is only the taphonomically influenced remains of a kind of sample of the once living fauna. Using the noun „sample“ could make it easier sometimes to apply the adjectives „biased“ or „inadequate“.

Concerning mortality patterns the differences between the basic two types, catastrophic (where all age classes should be represented like expected from a living population wiped out in a single moment) and attritional (where natural deaths have been accumulated, giving a dominance of the „most-dying“ ages, the very young and very old) have been outlined and their limitations discussed in several textbooks, and there is no need to present a bibliography on the subject here. When we limit ourselves to surveying large ungulate species, mortality patterns dominated by prime age adults and subadults do indeed appear „unnatural“. In the cases where we can exclude an admixture from a natural catastrophe into our archaeological assemblage (like e.g. the deposition of decomposed cadavers from drowning upstream onto a river terrace occupation floor) a dominance of individuals not belonging to the physically most vulnerable age groups appears to be a valid criteria for active hunting by man. The use of distance weapons means that Neandertals (and earlier hominids before them) can shift the tactical emphasis from out-tiring to more out-manouvering of the prey. Even when the hunting activities contribution to the sample can be assumed to have been „attritional“, the end sum will make the hunter appear as a „catastrophic agent“.

The most important obstacle for a correct evaluation of the evidence is here - and this concerns age class overrepresentation as well - that we need to be very informed about what can be considered normal resp. unusual patterns, in short the etology of the prey species concerned. The prehistoric hunter was an expert on this subject, where most of us are bleeding amateurs, except perhaps on a few species, and concerning species of the extinct faunas we have no better choice than to apply more or less helpless analogies from their relatives now surviving in other climates and/or on other continents.

In the Mammoth Steppe the wild fauna can be supposed to be numerically dominated by herbivores living in migrating herds or bands for most of the yearly cycle. Under such conditions a few hours or days of hunting suffice to produce a prey sample dominated by individuals from one or perhaps two species.

The emergence of hunting sites with the beginning of OIS7 (and it may be added, provided the datings are correct, their becoming abundant after the Last Interglacial) may thus not be so surprising, as they are defined as such by the criteria of single species dominance.

Under woodland conditions where more animals tend to be non-migratory and groupings seasonally limited, it may be difficult to accumulate bones for a clear species or age class dominance to emerge before a site is given up. Species dominance may occur at a particular site, and we are then obliged to consider the more sophisticated possibility, namely of repeated, targetted hunting expeditions more closely in this case.

Monospecific assemblages may thus to a large extent be a reflection of the behaviour of the best available prey animals. Single species dominance is quite acceptable for a hunting site, and may be considered much more likely for a hunted than for a possibly partly scavenged or mixed economy assemblage, but it can not be an exclusive argument for the existence of hunting activities. If, just for comparison, we really employed single species dominance as a paramount criteria for identifying hunting activities the interpretation of many Stone Age faunal assemblages from the time after the Late Glacial would indeed provide problems. We do not need to subject the Neandertal assemblages to test criteria or model constructions which tried on later material under better context control would be readily dismissed.

Conclusion

In the short rewiev above it was already in the beginning stated that the possibilities for a positive demonstration of scavenging activitites are limited.

The small bone assemblages theoretically likely to come from scavenging would hardly be given attention if incorporated into a larger hunted assemblage, and, if found alone, the most basal methodological reservations against small numbers of finds from open deposits should be enough to check our ambitions.The discussion of the different criteria in use for the identification of either hunting or scavenging shows, that we are much better equipped to accumulate arguments for hunting from an archaeological assemblage.

References

ANDERSEN, S.H. (1985): Tybring Vig. A Preliminary report on a submerged Ertebølle settlement on the west coast of Fyn. *Journal of Danish Archaeology* 4, 52-67.

ANDERSEN, S.H. (1993): Bjørnsholm. A stratified køkkenmødding on the Central Limfjord, North Jutland. *Journal of Danish Archaeology* 10, 59-96.

ANDERSEN, S.H. (1995): Coastal adaption and marine exploitation in Late Mesolithic Denmark - with special emphasis on the Limfjord region. In: A. Fischer (ed.), *Man & sea in the Mesolithic. Coastal settlement above and below present sea level.* 41-66. Oxford.

AUGUSTE, P. (1992): Étude archéozoologique des grands mammifères du site pléistocène moyen de Biache-Saint-Vaast (Pas-de-Calais, France): Apports biostratigraphiques et palethnographiques. *L´Anthropologie* 96, 49-70.

BINFORD, L.R. (1981): *Bones. Ancient men and modern myths.* New York.

BINFORD, L.R. (1984): *Faunal remains from Klasies River Mouth.* Orlando.

BINFORD, L.R. (1988): Étude taphonomique des restes fauniques de la Grotte Vaufrey. In: J.-P. Rigaud (ed), La Grotte Vaufrey à Cénac et Saint-Julien (Dordogne): Paléoenvironnements, chronologie et activités humaines. *Mémoires de la Société Préhistorique Francaise* 19, 213-290.

BLUMENSCHINE, R. J. (1986): Carcass consumption sequences and the archaeological distinction of scavenging and hunting. *Journal of Human Evolution* 15, 639-659.

BRATLUND, B.E. (1990): Rentierjagd im Spätglazial. Eine Untersuchung der Jagdfrakturen an Rentierknochen von Meiendorf und Stellmoor, Kreis Stormarn. *Offa* 47, 7-47.

BRATLUND, B.E. (1991): The bone remains of mammals and birds from the Bjørnsholm-Shellmound. A preliminary report. *Journal of Danish Archaeology* 10, 97-104.

BRATLUND, B.E. (1996): Hunting strategies in the late Glacial of northern Europe: A survey of the faunal evidence. *Journal of World Prehistory* 10 (1), 1-48.

BRATLUND, B.E. (in print): Anthropogenic factors on the thanatocoenosi of the last Interglacial travertines at Taubach, Germany. In: E. Turner & S. Gaudzinski, S. (eds.), *Die Rolle des Menschen in Faunenakkumulationen des Alt- und Mittelpaläolithikum Europas.* Mainz.

CHASE, P. (1989): How different was Middle Palaeolithic subsistence? In: P. Mellars & C. Stringer (eds.), *The human revolution,* 321-337. Edinburgh.

CHASE, P. (1990): Scavenging and hunting in the Middle Palaeolithic. In: P. Mellars (ed.), *The emergence of modern humans*, 225-232. Ithaca.

GRAYSON, D.K. & DELPECH, F. (1994): The evidence for Middle Palaeolithic scavenging from couche VIII, Grotte Vaufrey (Dordogne, France). *Journal of Archaeological Science* 21, 359-375.

HAYDEN, B. (1993): The cultural capacities of Neandertals: A review and re-evaluation. *Journal of Human Evolution* 24, 113-146.

MELLARS, P. (1989): Major issues in the emergence of modern humans. *Current Anthropology* 30, 349-385.

MILO, R.G. (1996): Human-animal interactions in southern African prehistory: A microscopic study of bone damage signatures. *AAC 9514010 ProQuest - Dissertation Abstracts.* 2pp.

NOE-NYGAARD, N. (1974): Mesolithic hunting in Denmark illustrated by bone injuries caused by human weapons. *Journal of Archaeological Science* 1, 217-248.

OLIVER, J.S. (1989): Analoques and site context: Bone damages from Shield Trap Cave (24CB91), Carbon County, Montana, U.S.A. In: R. Bonnichsen & M. Sorg (eds.), *Bone modification*, 73-98. Orono.

SHEA, J.J. (1993): Lithic use-wear evidence for hunting by Neandertals and early modern humans from the Levantine Mousterian. In: G.L. Peterkin, H.M. Bricker & P. Mellars (eds.), *Hunting and animal exploitation in the later Palaeolithic and Mesolithic of Eurasia,* 189-198. Archaeological Papers of the American Anthropological Association Number 4.

STINER M., WHITE, T. & TOTH, N. (1991): The cultural significance of Grotta Guattari reconsidered. *Current Anthropology* 32, 103-138.

STINER, M. (1994): *Honor among thieves. A zooarchaeological study of Neandertal ecology.* Princeton.

STRINGER, C. & GAMBLE, C. (1993): *In search of the Neandertals. Solving the puzzle of human origins.* London.

Jagd- und RohstofföKonomie im Mittelpaläolithikum Mährens

Karel Valoch

Abstract: Hunting and Raw Materials Economy in the Middle Palaeolithic in Moravia

There are three cave sites in Moravia which we can base our thinking on. The value of the evidence obtained from them, however, varies. Maška made excavations at the Šipka Cave between 1879 and 1883. In 1880 he found a fragment of the lower jawbone of a Neanderthal child, containing 6 teeth. The stone industry and abundant animal remains there were not published in detail until 1965. In 1906 a student, Kubasek, found the nearly complete mandible of a Neanderthal man, together with animal bones, in the Švédův stůl (Swede's Table) Cave. Systematic research in the cave was undertaken by Klíma between 1953 and 1955, and published as a monograph in 1962. The Kůlna Cave was excavated between 1961 and 1976; the upper jawbone with four teeth, a parietal bone and three milk-teeth of Neanderthals were discovered there. A monograph was published in 1988.

Musil has determined the fauna for all three sites, and also evaluated them in terms of ecological demands and chronological position. Švédův stůl Cave is a typical hyena den, and the animal remains found there together with human remains were dragged inside by hyenas. The negligible quantity of stone implements testifies to the entirely episodic, albeit recurring, presence of humans in the cave. Hunting activity was probably not significant. Šipka Cave was, by contrast, a bear cave. The bones of cave bears completely prevail among the osteological materials collected there, and represent individuals of all age categories, with juveniles predominant, however. The number of stone industries there gives evidence of the recurrent presence of humans in the cave, and ungulate bones indicate their hunting activities. Kůlna Cave was a true Neanderthal settlement. Remains of bears and carnivores are not abundant, and traces of biting and chewing appear in smaller numbers. Most of the cervine, bovine and equine bones were, however, chopped up by humans, and the not very numerous mammoth remains (mostly molars and tusks) were carried in.

The strategy of obtaining mineral supplies was diverse. In Šipka Cave local chert was predominant, and was obtained from within a distance of several hundred meters. There are some pieces of flint from the glacifluvial gravels found within a radius of ten kilometres, and occasional radiolarite stones, coming perhaps from greater distances. In Švédův stůl Cave there is probable evidence of the use of local limestone. Quartzite cobbles may also have been collected from within a radius of ten kilometres, while cretaceous spongolite came from an area about 30 kilometres distant. The origin of higher quality chert or flint (?) is not known. The composition of raw materials in Kůlna Cave is quite diverse, though spongolite is predominant in the Micoquien, and spongolite, quartzite and quartz have a similar prominence in the preceding Taubachien. Other kinds of rocks, even if represented only by a few pieces, testify to contact with places distant by more than 100 kilometres to the north, 60 to 100 kilometres to the northeast, around 60 kilometres to the west, and more than 30 kilometres to the south.

In Mähren (Tschechische Republik) befinden sich drei Höhlenfundplätze des Mittelpaläolithikums, die sowohl menschliche Skelettreste als auch reiche archäologische und paläontologische Funde erbracht haben. Sie wurden aber zu verschiedenen Zeiten gegraben und sind dementsprechend unterschiedlich dokumentiert.

Als erste wurde die Šipka-Höhle (am Berge Kotouc bei Štramberk in Nordostmähren) von K.J. Maška in den Jahren 1879-1883 gegraben. Im Jahre 1880 wurde ein Unterkieferfragment mit 6

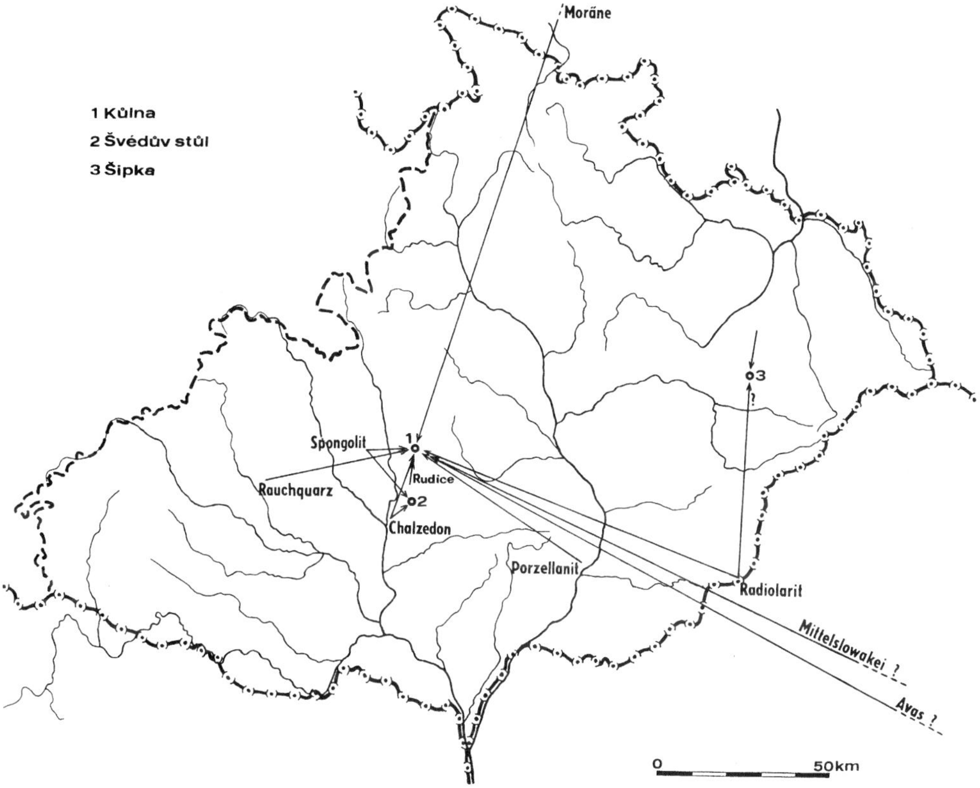

Abb. 1 - *Lage der drei behandelten Höhlen in Mähren.*
1 - Kůlna, 2 - Švédův stůl, 3 - Šipka. Die Linien mit Pfeilen deuten die vermutliche Herkunft einzelner Rohstoffe an. Für die Kůlna wurde das gesamte Mittelpaläolithikum zusammengefaßt. Porzellanit ist nur für das Taubachien (Schicht 11) spezifisch, die Fernverbindungen nach dem SO gehören dem Micoquien Schicht 7a an.

Zähnen eines Neandertalerkindes gefunden, welches durch die Publikationen von Maška (1882, 1886) und Schaaffhausen (1880) in der damaligen Auseinandersetzung über das Wesen der Neandertaler eine Rolle mitgespielt hat. Maška unterschied in der Höhle vier Schichten mit paläolithischen Artefakten: I - Magdalénien (wohl Federmesser, Valoch 1957), II - Solutréen (Gravettien, Valoch 1957), III und IV - Moustérien. Von den insgesamt fast 600 mittelpaläolithischen Artefakten ist nur ein Teil von Maška so bezeichnet, daß man sie zu einer der beiden Schichten mit Sicherheit zuweisen kann (vergl. Valoch 1965). Zeitlich dürften keine großen Unterschiede zwischen ihnen sein, da das gesamte Mittelpaläolithikum aus wechselnden Lagen dunkler Lehme an der Basis der Höhlenablagerungen stammt, wo (wie aus Maškas Notizen hervorgeht, Valoch 1965) keine getrennten Fundhorizonte zu erkennen waren. Ihr Alter wurde, unter Berücksichtigung sowohl der Fauna als auch der Typologie, früher (Musil 1965; Valoch 1965) als Würm 1/2 bezeichnet, heute würde ich eher an frühglaziale Schwankungen (analog dem PK II der Lößprofile) denken.

Die Höhle Švédův stůl (Schwedentischgrotte) befindet sich im südlichen Teil des Mährischen Karstes, etwa 15 km NO von Brünn. In den Jahren 1886 und 1887 grub M. Kríz als erster in dieser Höhle; er fand jedoch nur sehr zahlreiche Tierknochen, aber keine Spuren menschlicher Anwesenheit. Dies gelang erst 1905 einem Studenten, K. Kubasek, der zusammen mit Tierresten auch einen menschlichen Unterkiefer geborgen hatte (sog. Ochoser Unterkiefer). Diesen veröffentlichte Rzehak (1905, 1906) und bezeichnete ihn als neandertaloid. Bei einer späteren Nachgrabung (1964) fand J. Vañura zwei Schädelbruchstücke und einen Zahn. Mittelpaläolithische Artefakte wurden erst während einer gründlichen Erforschung der Höhle durch B. Klíma in den Jahren 1953-1955 an den Tag gebracht (Klíma 1962). Ähnlich wie in der Šipka-Höhle wurden auch hier Artefakte aus zwei voneinander getrennten Lagen in dunklen Lehmen geborgen. Aus der oberen stammen etwa 33, aus der unteren 7 Stück. Auf Grund von pedologischen, paläontologischen und archäologischen Untersuchungen wurde der untere Teil der dunklen Lehme dem letzten Interglazial, der obere Teil dem Mittelwürm (Würm 1/2) zugewiesen (Pelíšek 1965; Musil 1962; Klíma 1962). Auch in diesem Falle neige ich heute dazu, die obere Lage anstatt mittelwürmzeitlich eher als frühwürmzeitlich zu betrachten.

In der Kůlna-Höhle haben Grabungen von M. Kríz und J. Knies in der Zeit von 1881 bis 1913 zahlreiche mittelpaläolithische Artefakte geborgen. Eine genaue stratigraphische Gliederung der bis 15 m mächtigen, vorwiegend jungpleistozänen Ablagerungen ermöglichte erst eine ausgedehnte Grabung von 1961 bis 1976. Aus 11 sedimentologisch unterschiedlichen Schichten stammen Spuren menschlicher Anwesenheit während des Mittelpaläolithikums, sechs davon erbrachten größere Artefaktkollektionen. Dem reichsten, stratigraphisch zweitjüngsten Micoquieninventar (Schicht 7a) gehörten auch Skelettreste des Neandertalers an: ein Oberkieferfragment mit 4 Zähnen, der rechte Parietalknochen und 3 Milchzähne. Die Fundschichten umfassen einen Zeitraum vom Ende der vorletzten Vereisung bis zum Ende des ersten Würm-Pleniglazials (Valoch 1988).

Jagdaktivitäten

Die drei behandelten Höhlen repräsentieren drei völlig unterschiedliche Verhaltensmuster der Neandertaler. Die hochgewölbte und breitgeöffnete Kůlna-Höhle stellt einen typischen Rastplatz für wiederholte kürzere oder längere Aufenthalte dar. Dem entsprechen auch die Tierreste in den einzelnen Schichten. „Fleischtiere" (Equiden, Cerviden, Capriden, Boviden, Mammut, Nashorn) dominieren in allen Tiergemeinschaften. In einigen kommen manche Arten sehr häufig vor, z.B. in den beiden jüngeren Schichten (6a, 7a) Ren, Mammut und Wollnashorn, in den mittleren Schichten (7c, 7alpha) Pferd und Mammut, in den tieferen (8a, 9b) sind nur Pferde am häufigsten und in Schicht 11 zusammen Pferd und Elch. Höhlenbären sind in allen Schichten vertreten, an Raubtieren (oder „Pelztieren?") gibt es Wölfe, Füchse, Hyänen und in den jüngeren sowie mittleren Schichten auch den Vielfraß. Raubkatzen sind nur durch Einzelfunde dreimal nachgewiesen. Diese Gemeinschaften weisen darauf hin, daß sie vornehmlich als Jagdbeute auf die Fundstelle gelangt sind, wo sich die Raubtiere - meist wohl während menschlicher Abwesenheit - an ihr miternährt haben. Die Höhlenbären lebten zu jener Zeit in den der Kůlna benachbarten Slouper Höhlen sowie in den meisten kleineren Höhlen des Öden Tales, an dessen Beginn die Kůlna liegt (Musil in Valoch et al. 1969; Valoch 1989).

Die Lage der Höhle am Nordrande des Karstgebietes ermöglichte den Menschen, verschiedene Ökosysteme zu nutzen. Neben den engen Karsttälern und den dazwischen liegenden Hochflächen war es auch das sanfte Hügelland des angrenzenden Gebietes, deren Biotope gewiß zu jeder Zeit ihre Eigenart bewahrt haben. Auf den Hochflächen und im Hügelland konnte man sowohl Mammuten als auch Boviden und Cerviden begegnen. Es ist jedoch auffallend, daß die Anwesenheit des Mammuts in der Höhle vorwiegend durch Zähne (Molaren und Stoßzähne) belegt ist, wogegen Skelettreste ziemlich selten sind. Das würde bedeuten, daß man in die Höhle nur Fleisch mit wenigen Knochen und Zähne (als Trophäen?) mitgebracht hat. Dabei ist es gleichgültig, ob

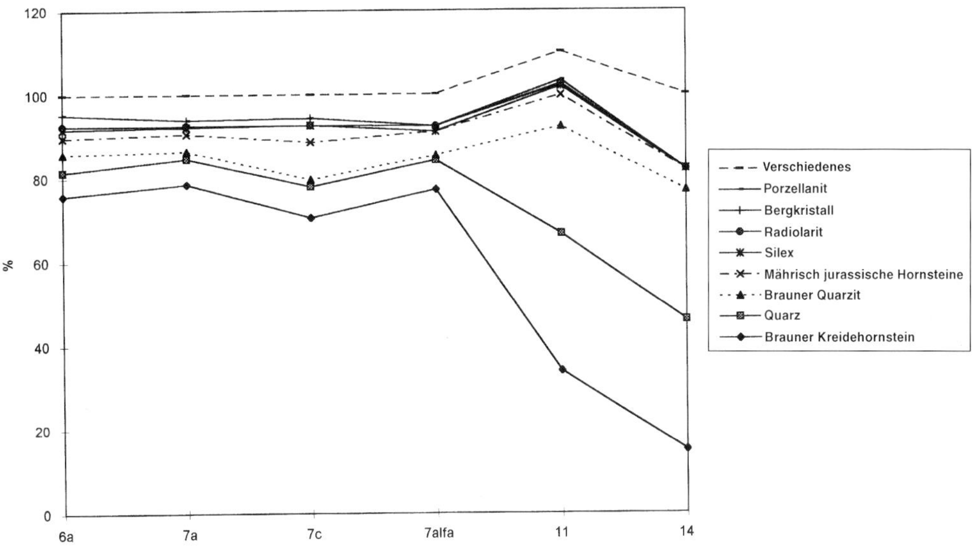

Abb. 2 - Anteile verschiedener Rohstoffe in den mittelpaläolithischen Industrien der Kůlna-Höhle.

man die Neandertaler als richtige Mammutjäger oder nur als Aas- und Knochensammler betrachtet, wie dies in letzter Zeit für die Menschen des Gravettien postuliert wurde (Soffer 1993). Eine archäozoologische Studie der gesamten Tierreste aus dem Mittelpaläolithikum der Kůlna, von der man nähere Angaben über das Verhalten der Menschen erwarten kann, wird gegenwärtig von M. Patou-Mathis und P. August (IPH Paris) durchgeführt. Man kann jedoch vermuten, daß die Knochenakkumulation in der Höhle vorwiegend auf menschliche Tätigkeit zurückgeführt werden kann und daß Raubtiere an ihr nur in geringem Maße beteiligt gewesen waren.

Grundsätzlich unterschiedlich war die Nutzung der Šipka-Höhle. Die geborgenen Knochen stammen überwiegend (etwa 80 %) vom Höhlenbär, wobei es sich um Individuen aller Altersklassen (in Schicht 4 gab es besonders viele juvenile Tiere) handelt. Die Šipka war also eine Bärenhöhle, wo die Bären die Hibernisation verbrachten und wo auch die Jungen zur Welt kamen. Die Morphologie der Höhle mit drei schmalen Eingängen und einem größeren Mittelraum war wohl für sie günstig. Die Reste der vermutlichen menschlichen Jagdbeute stammen in Schicht III vom Hirsch, Pferd, Esel, Wollnashorn, Mammut und von Boviden, in Schicht IV fehlt das Mammut, es kommen aber dazu Reh, Wildschwein, Biber, Moschusochse und Gemse. In beiden Schichten kamen außerdem zahlreiche Knochen von Raubtieren (Hyäne, Raubkatzen u.a.) vor (Musil 1965).

Die Menschen besiedelten die Höhle abwechselnd mit Höhlenbären. Daraus kann man schließen, daß es wohl oft wiederholte kurzfristige Begehungen im Sommer und Herbst waren. Außerdem wurde die Höhle ziemlich häufig von Raubtieren aufgesucht. Abgesehen von der natürlichen Akkumulation der Höhlenbärenkadaver ist anzunehmen, daß Raubtiere auch zur Entstehung der dortigen Thanatozönose wesentlich beigetragen haben.

Die Höhle Švédův stůl ist ein Hyänenhorst, was zahlreiche von diesem Tier benagte Knochen bezeugen. Der Höhlenbär war jedoch unter den Knochenresten am meisten vertreten. Die gesamte Tiergemeinschaft ist artenreich (Musil 1962). Die Höhle diente wiederholt den Menschen als kurzfristiger Unterschlupf; die menschlichen Skelettreste dürften jedoch von Hyänen hineinge-

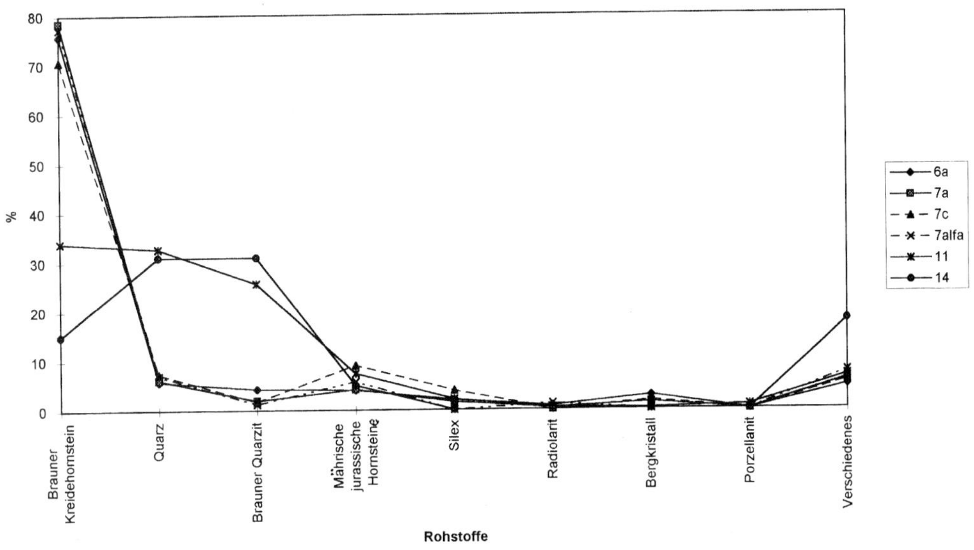

Abb. 3 - Änderungen im Rohstoffspektrum mittelpaläolithischer Industrien der Kůlna-Höhle.

schleppt worden sein. Über Jagdaktivitäten läßt sich ohne genaue Untersuchungen der Knochen nichts aussagen, da man menschliche Jagdbeute von jener der Hyänen nicht unterscheiden kann.

Rohstofförkonomie

Für die Frage nach der Rohstofförkonomie sind besonders die genauen, durch die Analyse der Inventare aus der Kůlna-Höhle erzielten Ergebnisse aussagefähig.

Kůlna-Höhle

Aus der ältesten Schicht 14 standen 100 Artefakte zur Verfügung. Davon entfielen auf Quarz und braunen (sog. Drahaner) Quarzit je 31%, auf Grauwackengerölle 17% und auf gelbbraunen Kreidehornstein (Spongolit) 15%. Ferner waren noch grauer Jurahornstein mit 4%, schwarzer Jurahornstein und unbestimmter Silizit mit je 1% vertreten. Es handelt sich durchweg um Gesteine, die entwerder in unmittelbarer Nähe der Höhle (Quarz, Quarzit, Grauwacke) oder höchstens in einem Umkreis von etwa 10 km gesammelt werden konnten.

Das Taubachien des Schichtkomplexes 11 enthält 1713 Artefakte. Auch hier bildeten in nächster Umgebung gesammelte Gesteine die absolute Mehrzahl (Quarz 34,09%, Spongolit 33,22%, Quarzit 15,35%). An Quarzen konnte L. Slezák drei Arten unterscheiden: am häufigsten sind Gangquarzstücke aus dem nördlich liegenden Kulm vertreten, an zweiter Stelle Quarz, oft in Form von Geröllen aus klastischen Ablagerungen des Unteren Devon, die im Bereich des Karstes oder in seiner Nähe vorkommen. Selten gibt es Quarzgerölle aus mesozoischen Ablagerungen (sog. Ruditzer Schichten) von Rudice. Es ist sonderbar, daß die Menschen dieses zum Spalten wenig geeignete Material aus verschiedenen Quellen geholt haben.

Außerdem gibt es ein buntes Spektrum verschiedener, nur mit wenigen Stücken vertretener Rohstoffe, die zum Teil aus beträchtlicher Entfernung stammen. Aus dem Umkreis von etwa 10 km konnten noch grauer (2,98%) und schwarzer (2,51%) Jurahornstein, Grauwacke (1,17%), Hornstein Typus Býcí skála (1,05%) und gegebenenfalls auch, falls es sich um kleine Stücke handelt, Bergkristall (0,18%) gesammelt werden.

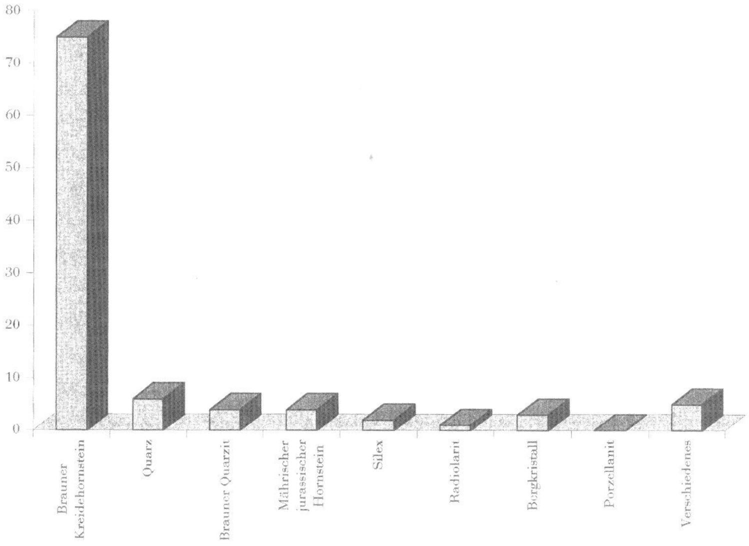

Abb. 4 - Kůlna-Höhle. Rohstoffspektrum im Micoquien der Schicht 6a.

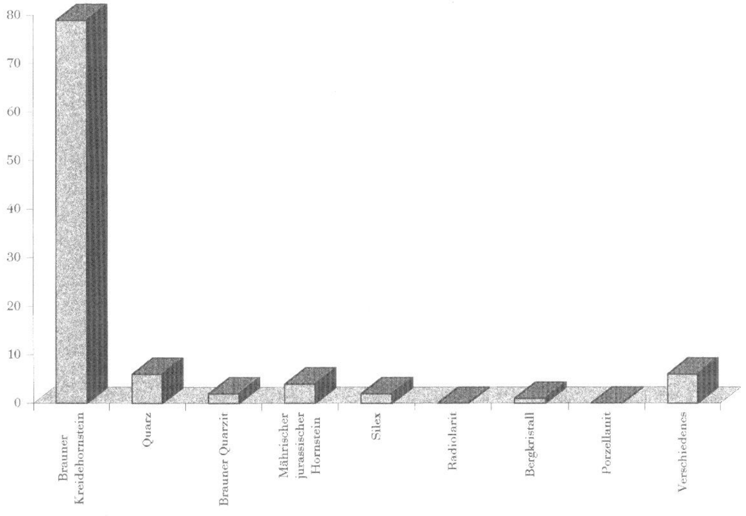

Abb. 5 - Kůlna-Höhle. Rohstoffspektrum im Micoquien der Schicht 7a.

Für die Erkenntnis der Wanderwege und Beziehungen paläolithischer Jäger sind von besonderer Wichtigkeit jedoch solche benutzten Rohstoffe, die vielleicht nur in sehr wenigen Stücken oder sogar vereinzelt auftreten, aber aus größerer Entfernung geholt wurden. Im Taubachien ist die Vielfalt solcher Artefakte auffallend. Darunter gibt es sowohl solche, die mit ziemlicher Sicherheit bestimmt werden konnten, als auch jene, deren Herkunft noch unsicher ist. Zur ersten Gruppe gehören graue Hornsteine vom Typus Kromauer Wald (Krumlovský les), die verläßlich nur nach schwarzen Kortexresten bestimmt werden können (0,70%). Ihre primären Fundstätten

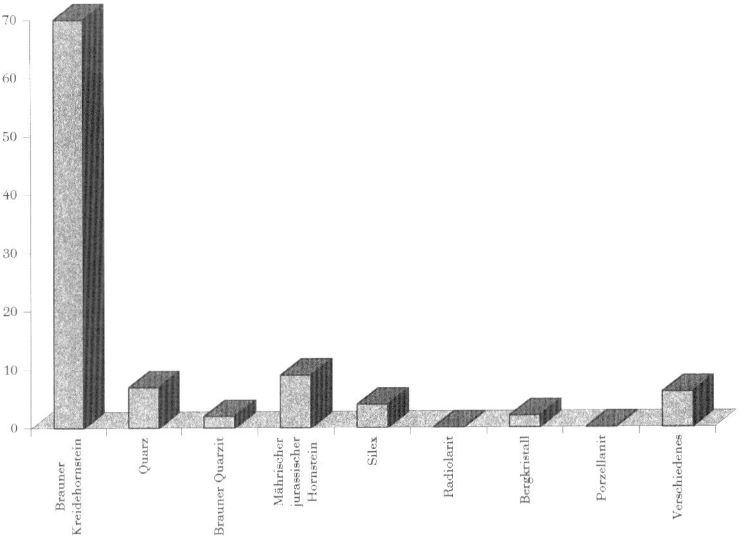

Abb. 6 - Kůlna-Höhle. Rohstoffspektrum im Micoquien der Schicht 7c.

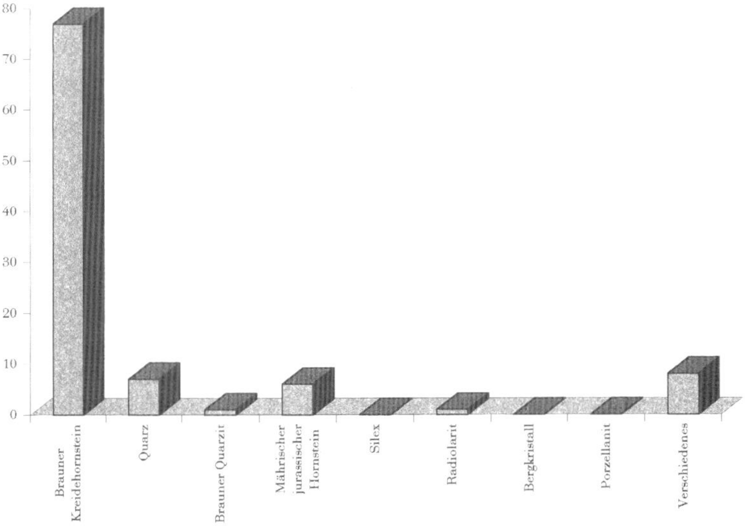

Abb. 7 - Kůlna-Höhle. Rohstoffspektrum im Micoqiuien der Schicht 7alpha.

sind miozäne Ablagerungen Südmährens, sekundär sind sie dann in pleistozänen Flußschottern eingebettet. Am nächsten konnten sie südlich vom Mährischen Karst im Raum von Brünn gesammelt werden, also etwa 30 km entfernt. Die Herkunft des Radiolarits (0,70%) wird meist im Raum des Vlára-Passes in den Weißen Karpaten, etwa 100 km NO, gesucht. Aus derselben Richtung, aber aus kleinerer Entfernung von etwa 60 km, dürfte Porzellanit (0,82%) in zwei Farbvarietäten - bräunlichgelb und ziegelrot - stammen. Es ist das nächste bekannte Vorkommen, alle anderen liegen viel weiter. Dann folgen zwei Silexarten, braun (0,70%) und durchscheinend

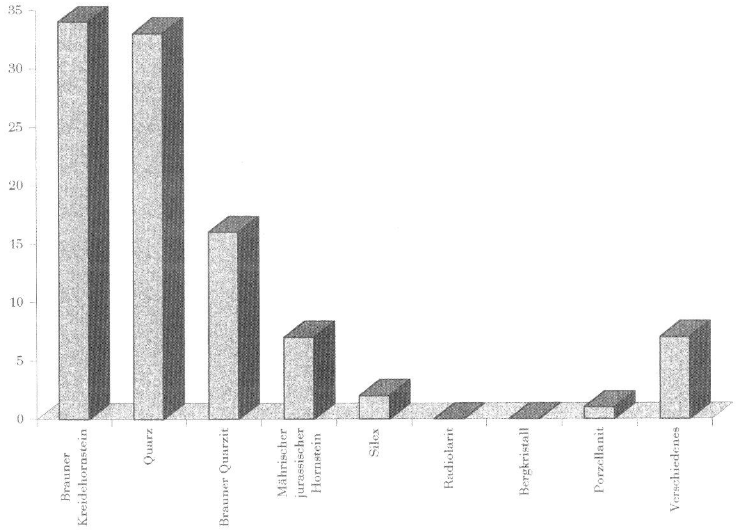

Abb. 8 - Kůlna-Höhle. Rohstoffspektrum im Taubachien der Schicht 11.

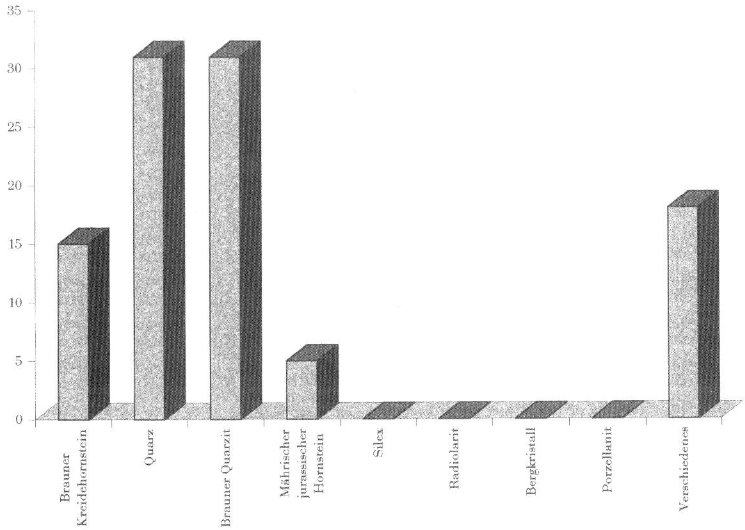

Abb. 9 - Kůlna-Höhle. Rohstoffspektrum im älteren Mittelpaläolithikum der Schicht 14.

grau (1,40%); zwei Stück davon wurden von A. Prichystal als aus den Moränen stammender Feuerstein bestimmt. Sie wurden also aus einer Entfernung von etwa 150-200 km geholt. Es ist der früheste Nachweis der Nutzung des Moränenfeuersteins im mährischen Paläolithikum. Schließlich gibt es noch mehrere Stücke, die noch nicht bestimmt wurden und die vorläufig als „verschiedene Hornsteine" (1,58%), „verschiedene Quarzite" (0,17%) und „verschiedene Silizite" (1,80%) bezeichnet worden sind.

Im gesamten Micoquien ändert sich das Rohstoffspektrum beachtlich. Dominierend erscheint der kreidezeitliche Spongolit, dessen primäre Lager etwa 10 km westlich bei Boritov zugänglich gewesen waren. Untersucht wurden drei Inventare (7c + 7 alpha - 335 Stück, 7a - 2463 Stück, 6a - 400 Stück), und in allen ist Spongolit mit mehr als 70% vertreten. An zweiter Stelle steht Quarz, dessen Anteil von unten nach oben allmählich sinkt (7,16%, 6,05%, 5,75%). Im Inventar 7c + 7 alpha folgt schwarzer Jurahornstein (6,57%), der in den beiden übrigen höchstens 1,5% erreicht, wogegen in 6a an dritter Stelle der braune Quarzit steht (4,25%), der sonst mit weniger als mit 2% vertreten ist.

Alle anderen Rohstoffe übersteigen nicht den Anteil von 2%. Vorhanden sind Radiolarit, Bergkristall und auch Moränenfeuerstein, im Vergleich mit dem Taubachien fehlt der Porzellanit. Durch neue Untersuchungen von A. Prichystal konnten einige überraschende Bestimmungen vorgenommen werden. Zwei kernförmig bearbeitete Stücke, die früher vorläufig als westmährischer Opal bezeichnet wurden, scheinen eher ein Opal/Chalzedon mittelslowakischer (?) Herkunft zu sein; dazu gehört vermutlich auch ein buntfarbiger Biface. Weitere Artefakte, die früher als „braune Hornsteine" bezeichnet wurden, stellen eher eine opal/chalzedonische limonitische Verwitterungsmasse dar, die sehr ähnlich den Rohstoffen vom Avasberg in Miskolc (NO-Ungarn) ist. Ein Faustkeil ist aus einem Andesitgeröll hergestellt, welches ebenfalls auf eine östliche Herkunft hinweist, da in der Nähe des Karstes (etwa 20 km SW von der Höhle) nur kleine Andesitadern bekannt sind, und größere Gesteinsmassive es entweder in NO-Mähren oder in der Mittelslowakei gibt. Ähnlichen Ursprungs könnte auch ein nicht näher bestimmter aphanitischer saurer Vulkanit sein. Ein größerer grünlichgrauer Quarzitabschlag scheint auch nicht aus Mähren zu sein. Ob er aus den mannigfachen böhmischen Quarzitlagern oder von anderswo stammt, bleibt ungewiß. Interessant ist auch die Feststellung mehrerer Artefakte aus einer Chalzedonart, die bisher nur aus Geröllen am NO-Rande von Brünn (Brno-Líšen) bekannt war und deren Qualität eher als mittelmäßig bezeichnet werden kann. Aus dem Raum von Trebíc, etwa 60 km westlich von der Kůlna, holte man größere Kristalle von Bergkristall und Rauchquarz, aus denen in 7a je ein Biface hergestellt wurde. Natürlich gibt es auch mehrere unbestimmte Hornsteine, Quarzite und andere Gesteinsarten. Es ist bezeichnend, daß es von den exotischen Rohstoffen, aus denen z.B. die beiden letzlich erwähnten Faustkeile hergestellt sind, keine Abschläge oder Abfallstücke gibt, so daß man annehmen kann, daß sie als Fertigprodukte in die Höhle gebracht worden sind.

Die Schichten 7c und 6a brachten fast keine auffälligen Rohstoffe. In 6a gibt es zweifellos Moränenfeuerstein und in 7c wiederum einen Fäustel aus braunem Quarzit, der nicht mährischer Provenienz zu sein scheint.

Die Menschen des Micoquien nutzten intensiv die Hornsteinquellen bei Boritov, verwendeten auch den lokalen Quarz sowie einen aus relativer Nähe stammenden Chalzedon und hatten vermutlich erstaunlich weite Kontakte nach dem Osten (bis 300 km, wenn man die Möglichkeit der Rohstoffe aus der Mittelslowakei und vom Avasberg in Betracht zieht). Auch bei ihnen kann man den Kontakt an die schlesischen Moränen sowie die Kenntnis der Radiolaritlager belegen. Das Gesamtspektrum ist jenem des Taubachien ähnlich; neu treten Kontakte in westlicher Richtung (Opal, Rauchquarz) hinzu. Es ist aber auffallend, daß sich durch die bisher bestimmten Rohstoffe keine Beziehungen nach Böhmen (vielleicht zwei Quarzitstücke?) oder nach dem Süden außerhalb Mährens andeuten. Ebenso überraschend ist die Abwesenheit von jurassischen Siliziten aus dem Raum von Krakau, obwohl es dort eine ziemlich ausgebreitete Micoquien-Besiedlung gegeben hat.

In beiden Kulturstufen, im Taubachien und im Micoquien, wurden überwiegend örtliche und aus dem Umkreis von 10-20 km stammende Rohstoffe verarbeitet. Es ist aber überraschend, daß es ziemlich viele, nur mit wenigen Stücken, gegebenenfalls allein durch Faustkeile oder Abschläge vertretene Rohstoffarten gibt, die aus größerer Entfernung stammen. Wurden sie auf Streifzügen zufällig aufgehoben und mitgenommen oder als Tauschgut von anderen Menschengruppen erworben? Läßt sich auf Grund der Rohstoffe die Richtung, von wo die Menschen auf den untersuchten Rastplatz gekommen sind, erschließen oder begrenzen sie das Territorium, welches von jener Gruppe genutzt wurde? Hatten Geräte oder auch nur Abschläge aus besonderem Material einen außergewöhnlichen, nicht utilitären Wert für die Menschen (z.B. Miniatur-Blattspitzen aus farbigem Porzellanit im Taubachien oder Bifaces aus Andesit, Bergkristall, Rauchquarz und Chalzedon im Micoquien)? Viele interessante Fragen könnten genaue Kenntnisse der Herkunft von Rohstoffarten beleuchten. Jedenfalls zeugen die Rohstoffspektren, wie Féblot-Augustins (1993) in einem Vergleich von Aquitanien und Mitteleuropa dargelegt hat, von einer nicht geahnten Mobilität der Neandertaler.

Šipka-Höhle

Das gesamte Mittelpaläolithikum der Kůlna-Höhle ist, trotz Vorherrschens lokaler Rohstoffe, durch eine ziemlich große Mannigfaltigkeit der Gesteinsarten gekennzeichnet. Im Gegensatz dazu steht das Inventar aus der Šipka-Höhle (etwa 500 Stück) als beinahe homogen da. Dort wurde ein grauer bis dunkel graugrünlicher Sandstein, dessen beste Varietäten einem Hornstein ähneln, verarbeitet, der auf dem gegenüber der Höhle befindlichen Hügel, kaum 1 km entfernt, an den Tag tritt. Er bildet etwa 95% der Artefakte. Ferner ist mit mehreren Stücken Quarz, weniger ein patinierter Silizit, der vermutlich aus nahen glazifluvialen Schottern stammt, und auch Radiolarit vertreten. Die Rohstoffstrategie dortiger Neandertaler war also recht einfach, da nur der Radiolarit auf eine Entfernung von 50 km hinweist.

Höhle Švédův stůl

Das bescheidene Inventar der Höhle Švédův stůl ist dagegen bunter. Aus der unteren Schicht gibt es neben fünf ziemlich fraglichen Kalksteinartefakten nur zwei braune Quarzite, aus der oberen werden 33 Artefakte erwähnt (Klíma 1962). Die verwendeten Gesteinsarten wurden neu bestimmt und mit A. Prichystal konsultiert. Das Rohstoffspektrum ist, in Hinsicht auf die geringe Stückanzahl, auffallend verschiedenartig. Es überwiegen Quarz (8 Stück) und brauner Quarzit (6), dann folgen jurassische Hornsteine (5), kreidezeitlicher Spongolit (4), Grauwacke (3) und zu je 1 Stück schwarzgrauer Hornstein Typus Olomucany, schwarzer Lydit, dunkelblaugrauer Quarzit und Kreidemergel. Die drei restlichen als „grauer Hornstein" bezeichneten Artefakte (Klíma 1962) sind in der Anthropos-Ausstellung installiert und konnten nicht genau angesehen werden. Höchstwahrscheinlich handelt es sich ebenfalls um jurassische Hornsteine.

Trotz der Mannigfaltigkeit der Rohstoffarten befanden sich ihre Quellen nicht allzuweit von der Höhle entfernt. Der braune Drahaner Quarzit sowie der Spongolit konnten auf den Karsthochflächen gesammelt werden, die jurassischen Hornsteine entweder in den hochgelegenen Schottern bei Brünn oder in Ablagerungen bei Rudice, wo auch der Hornstein Typus Olomucany vorkommt. Quarz, Grauwacke, dunkelgrauer Quarzit und vielleicht auch Lydit konnten in den polymikten Konglomeraten des Kulm, welche im Ríckatal, wo sich die Höhle befindet, wenige Kilometer südlich und nördlich an den Tag treten, gefunden werden. Alle diese Gesteine konnte man in einem Umkreis von etwa 15 km gewinnen. Nur der Kreidemergel dürfte aus einer Entfernung von etwa 30 km südlich stammen.

Die relativ große Buntheit der Rohstoffe dieser kleinen Kollektion erweckt den Eindruck, daß die Menschen während ihrer wohl wiederholten Begehungen der Höhle zufällig, ohne eine gezielte Auswahl, von ihren Streifzügen in naher Umgebung gefundene Gesteine mitgebracht und bearbeitet haben.

Zusammenfassend läßt sich ein markanter Unterschied im Verhalten verschiedener Gruppen der Neandertaler beobachten. An langfristig benutzten Rastplätzen (Kůlna) wurden sowohl eine ausgiebige Rohstoffquelle systematisch genutz als auch kleine Mengen verschiedener Gesteine von weither geholt. Andererseits konnte man sich mit einem zwar weniger geeigneten, jedoch in unmittelbarer Nähe des Rastplatzes in rauhen Mengen zur Verfügung stehendem Gestein begnügen (Šipka). Man könnte sich fragen, ob die Träger verschiedener „Kulturen" (Micoquien in der Kůlna, gezähntes Moustérien in der Šipka) auch ein unterschiedliches Verhalten in der Rohstoffbesorgung aufgewiesen haben. Die zwar wiederholten, jedoch kurzen Aufenthalte in der Höhle Švédův stůl sind dagegen durch *ad hoc* in naher Umgebung gesammelte Rohstoffe gekennzeichnet.

Literatur

FÉBLOT-AUGUSTINS, J. (1993): Mobility strategies in the Late Middle Palaeolithic of Central Europe and Western Europe: Elements of stability and variability. *Journal of Anthropological Archaeology* 12, 211-265.

JELÍNEK, J. (1962): Der Unterkiefer von Ochoz. Ein Beitrag zu seiner phylogenetischen Stellung. In: *Die Erforschung der Höhle Švédův stůl 1953-1956,* 261-284. Brno (Anthropos 13, N.S.5).

JELÍNEK, J. (1965): Der Kiefer aus der Šipka-Höhle. *Anthropos* 17, N.S.9, 135-179.

JELÍNEK, J. (1988): Anthropologische Funde aus der Kůlna- Höhle. In: *Die Erforschung der Kůlna-Höhle 1961-1976,* 261-284. Brno (Anthropos 24, N.S. 16).

KLÍMA, B. (1962): Die archäologische Erforschung der Höhle Švédův stůl. In: *Die Erforschung der Höhle Švédův stůl 1953-1955,* 9-96. Brno (Anthropos 13, N.S.5).

MAŠKA, K.J. (1882): Über den diluvialen Menschen in Štramberg. *Mitteilungen der Anthropologischen Gessellschaft in Wien* 12, 32-38.

MAŠKA, K.J. (1886): Über den Fund eines menschlichen Unterkiefers in der Šipka-Höhle. *Verhandlungen der Berliner Anthropologischen Gesellschaft* 1886, 341-344.

MUSIL, R. (1962): Die Höhle Švédův stůl, ein typischer Höhlenhyänenhorst. In: *Die Erforschugn der Höhle Švédův stůl 1953-1955,* 97-260. Brno (Anthropos 13, N.S.5).

MUSIL, R. (1965): Wertung der früheren paläontologischen Funde aus der Šipka- Höhle. *Anthropos* 17, N.S.9, 127-134.

PELÍŠEK, J. (1965): Quartärsedimente der Höhle „Švédův stůl" im südlichen Teil des Mährischen Karstes. In: *Die Erforschung der Höhle Švédův stůl 1953-1955*, 285-297. Brno (Anthropos 13, N.S.5).

RZEHAK, A. (1905): Über den Homo primigenius Wilseri im mährischen Diluvium. *Verandlungen der k.k. Geologischen Reichsanstalt* 1905, 329-331.

RZEHAK, A. (1906): Der Unterkiefer von Ochos. Ein Beitrag zur Kenntnis des altdiluvialen Menschen. *Verhandlungen des Naturforschenden Vereins Brünn* 44, 91-114.

SCHAAFFHAUSEN, A. (1880): Ueber die in der Šipka-Höhle bei Stramberg in Mähren gemachten Funde und insbesondere über das daselbst gefundene Bruchstück eines menschlichen Unterkiefers. *Sitzungs-Berichte der Niederrheinischen Gesellschaft in Bonn* 1880, 260-264.

SOFFER, O. (1993): Upper Palaeolithic adaptations in central and Eastern Europe and man-mammoth interactions. In: O.Soffer & N.D.Praslov (eds.), *From Kostenki to Clovis,* 31-49.

VALOCH, K. (1957): Jeskyne Šipka a Certova díra u Štramberku. I. Mladší paleolit. *Časopis Moravského Musea,* sc.soc., 42, 5-24.

VALOCH, K. (1965): Die Höhle Šipka und Certova díra bei Štramberk. *Anthropos* 17, N.S.9, 5-125.

VALOCH, K. (1988): Die Erforschung der Kůlna-Höhle 1961-1976. *Anthropos* 24, N.S.16, 7-200.

VALOCH, K. (1989): Osídlenii a klimatické zmeny v poslední dobe ledové na Morave. *Časopis Moravského Musea,* sc.soc., 74, 7-34.

VALOCH, K., PELÍŠEK, J., MUSIL, R., KOVANDA, J. & OPRAVIL, E. (1969): Die Erforschung der Kůlna-Höhle bei Sloup im Mährischen Karst. *Quartär* 20, 1-45.

Behaviour and Survival Strategy in Moravian Early Gravettians. Mammoth Hunters or Scavengers?

Jan Jelínek

Abstract

The critical study of the Mid-Pleistocene hominid behaviour has questioned the hunting of large mammals even for Upper Palaeolithic times.

When considering the possibility of scavenging activities we find three theoretical possibilities:

1) Gravettians were mammoth hunters and not scavengers. When knowing the generally opportunistic character of human behaviour it seems improbable that the Gravettian hunters would be an exception avoiding any scavenging opportunity. There is no sound reason not to profit from a chance-find of a fresh cadaver with good meat.

2) Hunters and scavengers. The distance from the found fresh cadaver from the living site was important for practical exploitation of such a find. The contribution for the population was limited if not near to living site.

3) To consider scavenging as the principal way of food supply would not be reasonable considering the frequent selection for young individuals when analysing mammoth bone accumulations. Further, scavenging opportunity in periglacial zone was less frequent than in tropical savannah environment.

Several points speak against natural origin of large bone accumulations and support human activity as their origin:

1) Strategical position of the Gravettian sites nearby the flood plains reflects the economical interest of those people.

2) Strategically situated long lasting sites allowed the advantageous killing of mammoths in near distance of camping sites thus avoiding long, strenuous and problematic transport.

3) The age composition of the bones in the accumulation, where the younger animals often prevail, supports the selection of hunted prey.

4) If all accumulated mammoth bones should be considered as collected in the landscape than the remaining bones of other animals are never sufficient as food remains of corresponding living site.

5) Two kinds of bone accumulations were recognised: (a) as the part of the living floor; in this case the long bones are often with broken epiphyses or split for marrow; (b) nearby the living site; in this case the accumulations are of varying size, sometimes counting tenths, nearly a hundred and even more individuals.

6) On several living sites we find bone charcoal in fireplaces. This means that fresh bones containing fat were used as fuel.

7) Bones used for constructions/habitations and large anvil bones with holes were older dry bones. It would be peculiar to use fresh bones for such a task.

8) Long characteristic ivory spearheads (up tu 40 cm long) are frequent. If no other large animals were regularly hunted, what would be their use? They were appropriate for mammoth hunting.

All these arguments point to diversified use of mammoth bones in periglacial Moravia both fresh bones and older collected bones as well. Like in other activities also in mammoth bone accumulation the Gravettian people were opportunistic. These accumulations are not natural but man made of various origin and purpose.

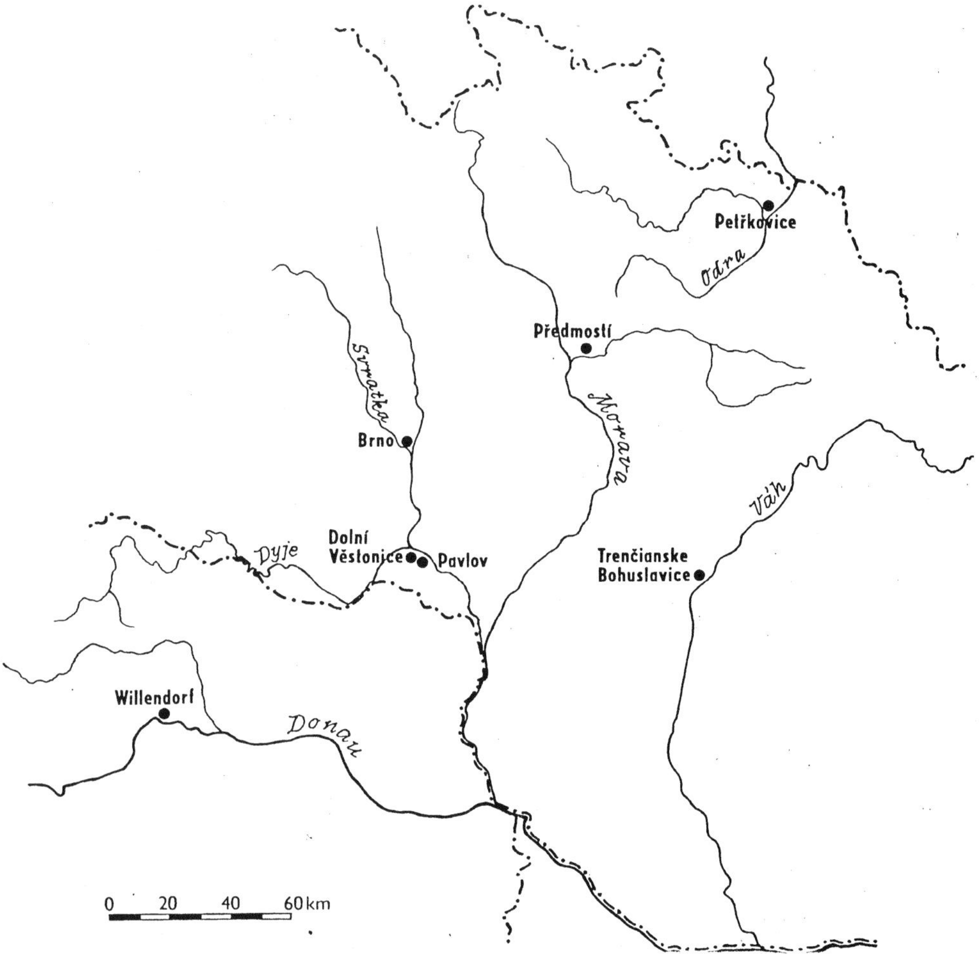

*Figure 1 - The principal Gravettian localities in Moravia.
The most important are Dolní Věstonice, Pavlov Milovice and Předmostí. Smaller are Petřkovice and Trenčianské Bohuslavice (in Slovakia). In Brno, only an isolated Gravettian burial Brno 2 was discovered, not a living site.*

Introduction

The critical study of the Mid-Pleistocene hominid behaviour has questioned the hunting of large mammals not only in the Middle Pleistocene (Binford, 1981, 1987; Klein, 1987) but also in the Upper Palaeolithic period (Vereschagin, 1979, 1981). The hunting capacity of the Middle and Upper Palaeolithic populations was examined focusing mainly on the origin of mammoth bone accumulations in Moravian, Ukrainian and Russian Gravettian sites. Consequently several important studies on the proboscidean behaviour and proboscidean fossil bone accumulation were published, some studying the North American situation and some the Central and East European sites (Haynes, 1991; Vereschagin, 1971, 1977, 1979, 1981; Vereschagin & Ukrainceva, 1985; Saunders, 1980).

Whereas the original archaeological studies of the Moravian sites considered the accumulations of mammoth bones as food remains (Absolon, 1945; Maška, 1886, 1889; Klíma, 1963, 1969,

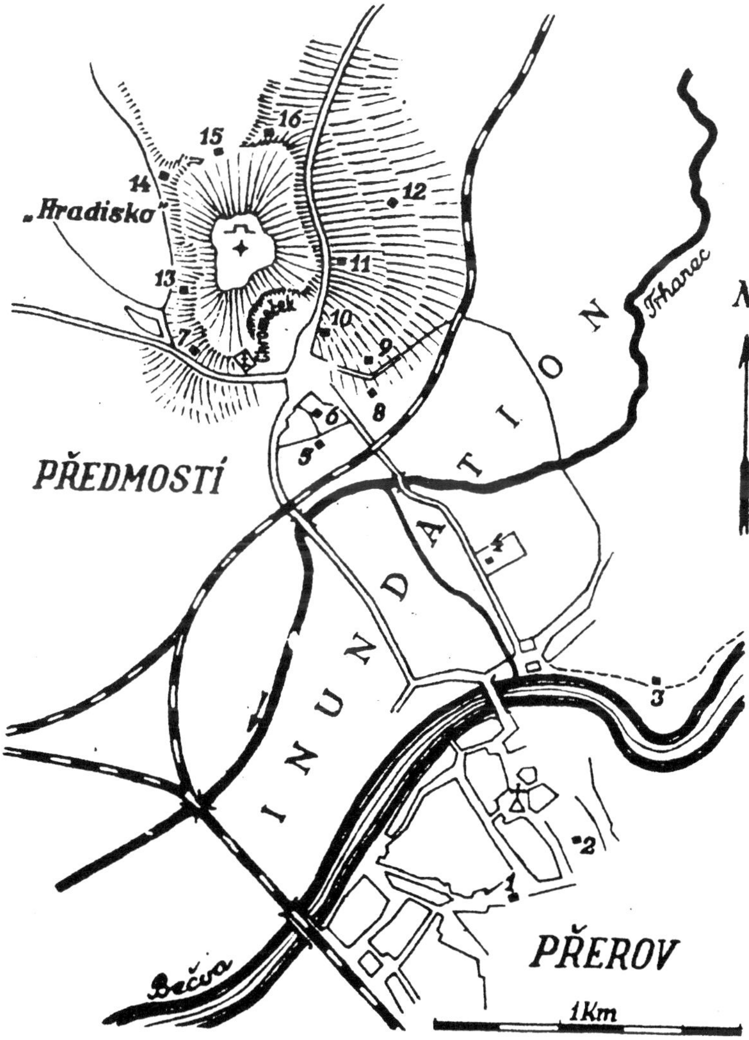

Figure 2 - The disposition of the Gravettian locality Předmostí (after Absolon & Klíma, 1977).

1983; Musil, 1955, 1959a,b) some more recent contributions (Soffer, 1985, 1993; Vereschagin, 1977, 1979, 1981) based mainly on the Russian sites mentioned also gathering activities or natural factors participating in the origin of these bone accumulations. As in most other studies, also here the either/or simplifying approach did not resolve the problem.

In the present study the strategical, taphonomical, palaeontological and behavioural aspects are combined to get better approach. As for the hunting and scavenging activities of the Gravettians following three possibilities are considered:

1) The Moravian Gravettians were mammoth hunters, not scavengers.
2) They were both, hunters and scavengers.
3) They were mammoth scavengers, not mammoth hunters.

As for the origin of the bone accumulation following features were considered:

1) Character and strategical position of the living sites.
2) The character and composition of bone accumulations.
3) The economical base of the Gravettians.
4) The need and use of bones.

I do not consider here the demographic composition of the mammoth remains in the bone accumulations as we know, that the natural mortality (see e.g. Berelekh compared with Mezin, Dobranichevka and Radomyshl) and human hunting can leave the same demographic pattern behind (Soffer, 1985, p.307).

The size, number and position of living sites

Because the intensive prehistoric research in Moravia has started fairly early (the Předmostí site was excavated by J. Wankel already in 1880-1982) we have today a relatively good knowledge of the Gravettian settlement of Moravia. The principal living sites are (Figure 1):

Předmostí

Situated in central Moravia near the Bečva river on place geographically called Moravian Gate (Figure 2). This is the dividing water-shed between the Bečva river flowing south and Odra river flowing north, where in Southern Poland rich raw flint material was found in glacial and fluvial gravels. The visually central point of the Předmostí site was a conspicuous rock which gave the local name to this spot, Skalka (the Rock). Another important feature of this site is a mineral spring. For the Palaeolithic man it was no doubt very atractive feature.

Petřkovice

Northwards, in Northern Moravia is another Gravettian living site Petřkovice near the confluence of the Opava and Odra rivers. This site, studied by Klíma (1955, 1969) is known through the find of several Gravettian habitations and through the find of female figurines cut in the haematite stone.

Plavatisko

Near the Mladeč village in Central Moravia, well known through its caves with famous early Aurignacian anthropological finds, is a Gravettian site situated on a flat promontory elevated above the confluence of the Hradčná creek with the Morava river overlooking the inundation terrain of this principal Moravian river. Unfortunately no bones were conserved here and only surface finds of Gravettian stone tools were collected here.

Věstonice, Pavlov, Milovice

All these sites are large and rich South Moravian sites at the foot of Palava hills near the confluence of the Svratka and Dyje rivers in a position controling the nearby river valleys.

Settlement strategy

Only unimportant finds are known from the Moravian Karst region, e.g. Kůlna cave (Valoch, 1988), which was settled mainly by the Neanderthals.

Evident preference for the environment near the flat, broad and periodically inundated river valleys, combined with the great numbers of mammoth bones at those Gravettian sites where the bones were conserved, signals the economical reasons for this settlement strategy.

The other fact supporting this specific strategy is that the isolated small sites of short stay of small groups of people are nearly absent. The Moravian terrain is well surveyed for prehistory so that the map presenting our contemporary knowledge of the Gravettian sites - even when not

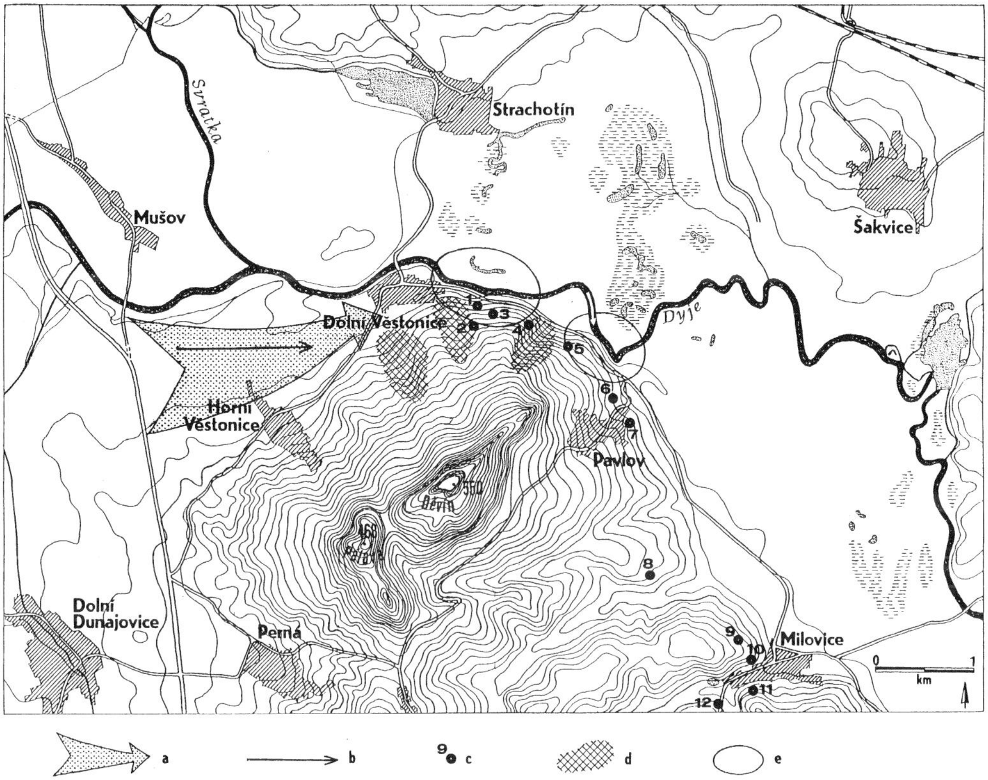

Figure 3 - *The disposition of Gravettian localities at the foot of the Palava Hills.*
Note, that the sites situated on the elevated terrain follow the flood-plain. The best hunting terrain is near the sites of Dolní Věstonice in the direction against the wind which comes here mainly from the west. The critical killing sites are in gullies where the river approached the high loess banks.
Explanations: a) roaming animals direction, b) principal wind direction, c) the most important Gravettian sites, d) three principal loess gullies, e) two critical places with high river banks.

complete - reflects the real settlement situation, namely the striking concentration of the Gravettian population into few large living sites. If we try to object that those sites can be sites of temporary agglomeration of some smaller groups then we lack the normal living sites of such smaller groups and of course all the characters of the living floors contradict such interpretation.

Still anothor supporting factor is the existence of specialized hunting with other Upper Palaeolithic populations as was horse (Solutrée), bison (Amvrosievka) or reindeer (Magdalenian sites) hunting.

Unfortunately not all sites conserve the osteological material and the largest and richest Předmostí site was excavated too early to supply modern documentation as we need it now for our specific study. We have only the excavation diary of Maška, some preliminary publications of Wankel, Maška, Kříž, Absolon and Klíma and the large monograph published by Absolon & Klíma as late as 1977. So the best knowledge comes from Dolní Věstonice I, II, Pavlov I, II and Milovice, all situated in one chain of sites at the foot of the Palava hills (Figure 3). Here we can follow the detailed topographic situation illustrating the specific position of those sites and their advantage for the hunting economy of large pachyderms.

Figure 4 - The reconstruction of the foot of the Palava Hills in Upper Palaeolithic times (after B.Klíma). Note the gullies of the slopes. They give the possibility of easier and safer mammoth killing when the mammoth tried to traverse them.

The soft loess sediments on north-west slopes of these hills were furrowed in their lower parts by errosion and the river approached twice these slopes undercutting high banks in them. Even when knowing that the position of the riverbed has changed frequently in the past, the geomorphology and the sediments supply a good evidence on the past situation (Klíma, 1995). Western winds prevail here and the animals when roaming in the lowland on the right side of the river, if advancing along the river in SW direction they reached these narrow places. When advancing further they had only the possibility to mount the eroded wet and unsafe slope gullies (Figure 4). Seen with the eyes of a hunter such moment was advantageous for hunters attack. It gave the possibility of safer higher position for the hunters when attacking the animals during their crossing of a slippery and uneven ground. The defence of those pachyderms was very difficult if not impossible in such a position and they were highly vulnerable. Another important advantage of this place was that it was situated from the camp site against the wind so that the animals could not scent the people and even not see the camp. The third important advantage was that those killing sites were near the camping ground. This limited the uneasy transport of the quarry. The crossing of the river and of the swampy valley was no doubt very difficult. With the heavy load it would be practically impossible. For this reason mammoth hunting far from the camp was less profitable and on the other side of the river even problematic.

If we turn to the settlement character, the largest sites like Dolní Věstonice I, Pavlov I or Předmostí with well preserved animal bones are very rich in exceptional artistic objects. Some of them, e.g. the fine front bands made of very thin ivory plates were too fragile to be used in everyday life. More probably they were for special occasions, ceremonies. Sometimes, like with ceramic figurines, we have proofs that they were made on spot and the quantity of discovered fragments demonstrates the intensive long lasting production, therefore the long stay. Also the large ash places and remains of diversified activities support this view. We can find a very rich spectrum of stone, bone and ivory tools and objects. Especially important for our considerations are finds of long ivory points, the spearheads. Some were 40 cm long and they were certainly not for hunting of small game. They were effective even for hunting of such animals like mammoths. This is especially clear if we consider the composition of the animal remains found in the cultural layer as part of the living floor or of the bone accumulations.

All this supports the mammoth hunting economy. But we know, that human behaviour is characteristically opportunistic. It would be naive to suppose, that they principally refused to use a good fresh carcass when found in useful condition. Anyhow, the biomass living in periglacial environment is different from the biomass living e.g. in tropical savannah environment. The animals are less numerous, in their number and density as well. Larger numbers we find only with some animals, especially in certain periods of the year. Scavenging is less profitable, especially for people. There are no vultures and kites, the best signal-giving animal scavangers, omnipresent in tropical environment. True, dead animals, especially in winter times can be consumed longer time in low temperatures but this advantage does not outweight lower numbers and less numerous kinds of animals. There is nothing which would support the full scavenging economy of local Gravettians. Some help brings us the analysis of bone accumulations found at the living sites or nearby.

Character of the bone accumulations

The best to study are Dolní Věstonice group of sites (Dolní Věstonice I, II; Pavlov I, II; Milovice) and Předmostí.

We find two kinds of bone accumulations: those being the part of the living floor and those situated nearby the settlement. The accumulations found as a part of the living floor are usually smaller than the corresponding accumulations situated nearby and out of the living floor. The mammoth bones from the living floor are frequently split or with broken-off epiphyses. Such damage is much less frequent in the accumulations out of this living floor. A good example of such a situation is Dolní Věstonive I. On the living floor prevail the mammoth bones but the share of the remains of other animals, however rare, is here bigger than in nearby accumulation, out of living floor. It is interesting to compare the situation in Dolní Věstonice I and Dolní Věstonice II sites.

Dolní Věstonice I living site excavated in 1924-1938 by Absolon and in 1948-1973 by Klíma is much bigger and the cultural layer is much more rich in every comparable feature: in stone-, bone- and ivory tools, in artistic objects, in habitations, workshops, ash places and primitive ceramic-kilns, mammoth bones, shortly in all human activities. The size and rich cultural layer was conditioned by long lasting supply of food.

The Dolní Věstonice I living floor bone accumulation

The bone accumulation found and excavated by Absolon in 1926 is a part of the living floor and covered 52 square metres (Figure 5). 28 mammoth tusks and fragments were found here and only one horse tibia, ulna and scapula fragment, further one wolverine humerus, bones of a paw of a lion (in anatomical position) and also some wolf bones, always on the margin of the accumulation: four mandibles, few vertebra and two leg bones. All other bones (some 200 pieces) belong to mammoth, dismembered, not in anatomical position (Figure 5). The biomass of mammoth remains is evident and if these bones would be considered as old bones collected by the Gravettians in the landscape there would be no other culinaristic remains sufficient for survival of local inhabitants. It is interesting that the wolf remains (a scavenger) are always following the rim of the accumulation.

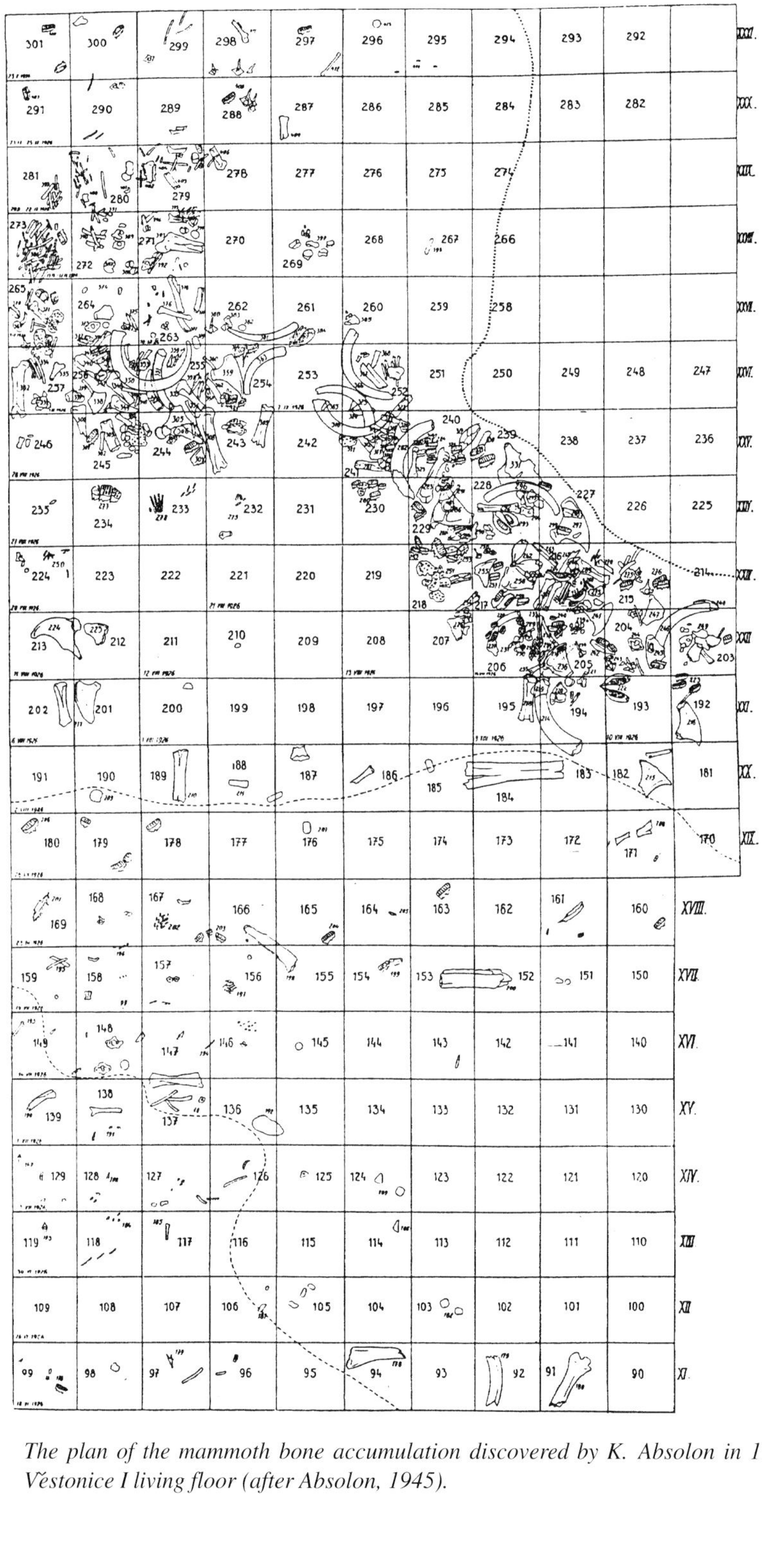

Figure 5 - The plan of the mammoth bone accumulation discovered by K. Absolon in 1926 at Dolní Věstonice I living floor (after Absolon, 1945).

This does not seem to be a hazard. The fact, that the lion is represented only by paw bones supports the possibility that they represent the remains of a fell. It is also important to stress that between the bones were fairly numerous tools and even stone tool workshop, further several ash and fire remains. The red-burned loess under the accumulation proves that people were here before all those accumulation activities which therefore cannot be of natural origin but the result of human activities.

The composition of accumulated mammoth remains is irregular and illustrates certain man-made selection. This is especially the case with mammoth tusks. Some roots of those tusks are undamaged, and it is clear that they were loose and easy to take out from the alveols, therefore they were collected as old items in the landscape. Most other roots are heavily damaged. In fresh cadaver is the root of the tusk firmly rooted and it is nearly imposible to take it undamaged out of the alveolus. It follows that ivory was not rare in those times but nevertheless, it was also collected in the landscape.

Předmostí living floor bone accumulations

Let us now look for the comparision on the situation in the largest Moravian Gravettian site, Předmostí. Unfortunately the excavations of this site started fairly early, in 1880-1882 (Wankel, 1884, 1886) and than continued irregularly by different excavators (Maška, 1884, 1894, 1896; Kříž 1894, 1896; Absolon 1924-1930) till 1930. Some valuable observations we can find in diaries of J. Knies who often visited and as a keen palaeontologist followed the excavations. The research of Klíma after the second World War verified only some stratigraphical questions even when some archaeological and palaeontological material was also obtained (Klíma, 1973, 1977). From Maškas diary it is clear that the mammoth and other bones found as a part of the living floor were accumulated by man. In 1893 Maška has found in one place four complete tusks and few days later other 13 tusks evidently accumulated by man, the same as other four tusks and three skulls without tusks. Similar like in the bone accumulation of the Dolní Věstonice living floor, also here has Maška found in 1890 several lion and wolverine paws, bones in anatomical position, probably remains of the fells (Absolon & Klíma, 1977). All archaeologists have found here more numerous animal species represented as the part of the living floor, than Absolon and Klíma have found at Dolní Věstonice I living floor. Certainly also here were the mammoth remains the most numerous but wolf, wolverine, ice fox, are not exceptional. Rare are reindeer or lion but the frequence of animals changes in different places. In the living floor excavated during 1926 and 1927 the remains of at least of 18 mammoths are reported, belonging to individuals of different ages, some bones well preserved but majority no doubt intentionally split. Two fully broken mammoth skulls (for brain or as the result of their pneumatisation?), 26 tusks, 89 molars, 16 shoulder-blades, 14 radii and two ulnae, 18 shin- and thigh-bones found only in fragments. Between the bones as the part of the living floor were always found stone tools, workshops, ivory and bone objects and tools, fire places, haematite pieces etc. Absolon explicitly stated that there was no difference in the living floor bone accumulations in Předmostí and Dolní Věstonice. Evidently the situation was very similar, only the percentage of other animal species was different. This seems to reflect the environmental and/or climatic conditions. This is supported when comparing with the Central Russian sites (see further on).

Dolní Věstonice I bone accumulation nearby the living site

In 1969 Klíma started to excavate a large bone accumulation situated in a wet furrow of the terrain next to the living site (Figure 6). This position has brought two questions. 1) Is not this accumulation the result of natural death of stressed pachyderms, found and exploited only

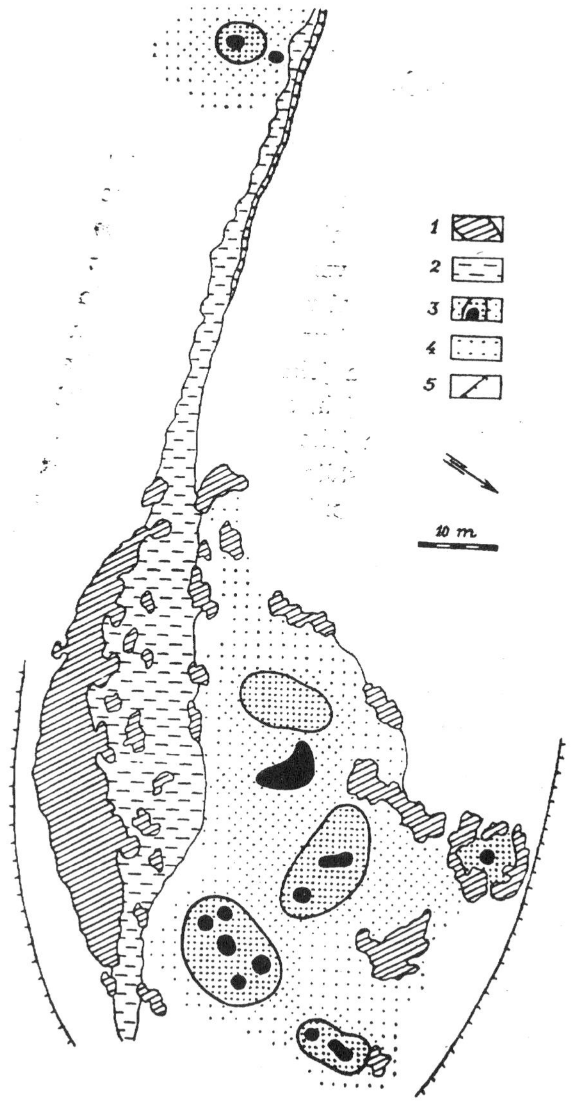

Figure 6 - The disposition of the living site and associated mammoth bone accumulation in Dolní Věstonice I. Explanations: 1) bones, 2) water in the loess furrow, 3) fireplaces, 4) the living floor.

afterwards by the Gravettians? 2) Isn't it a killing site with the Gravettians settled next to it afterwards?

Both these possibilities can be easily refused. First, under the bones were found traces of fire, sometimes the loess was burnt to red colour, and also the radiolarite stone tool workshop was discovered there (Klíma, 1969). Therefore the settlement was the primary feature here and the accumulation of bones followed later. Second, all bones were dismembered, not a single one was in anatomical contact with other bone.

According to detailed stratigraphy Klíma rightly recognized that this accumulation was not the result of a single event but of gradual activities. Nearly exclusively only mammoth remains are represented (Figure 7). There were counted 6335 mammoth bones, one half of a reindeer mandible, one half of a wolf mandible, a skull, mandible, phalanx and few other bones of horse and some bone fragments of a hare. Klíma supposed, that the remains represented nearly 100 mammoth

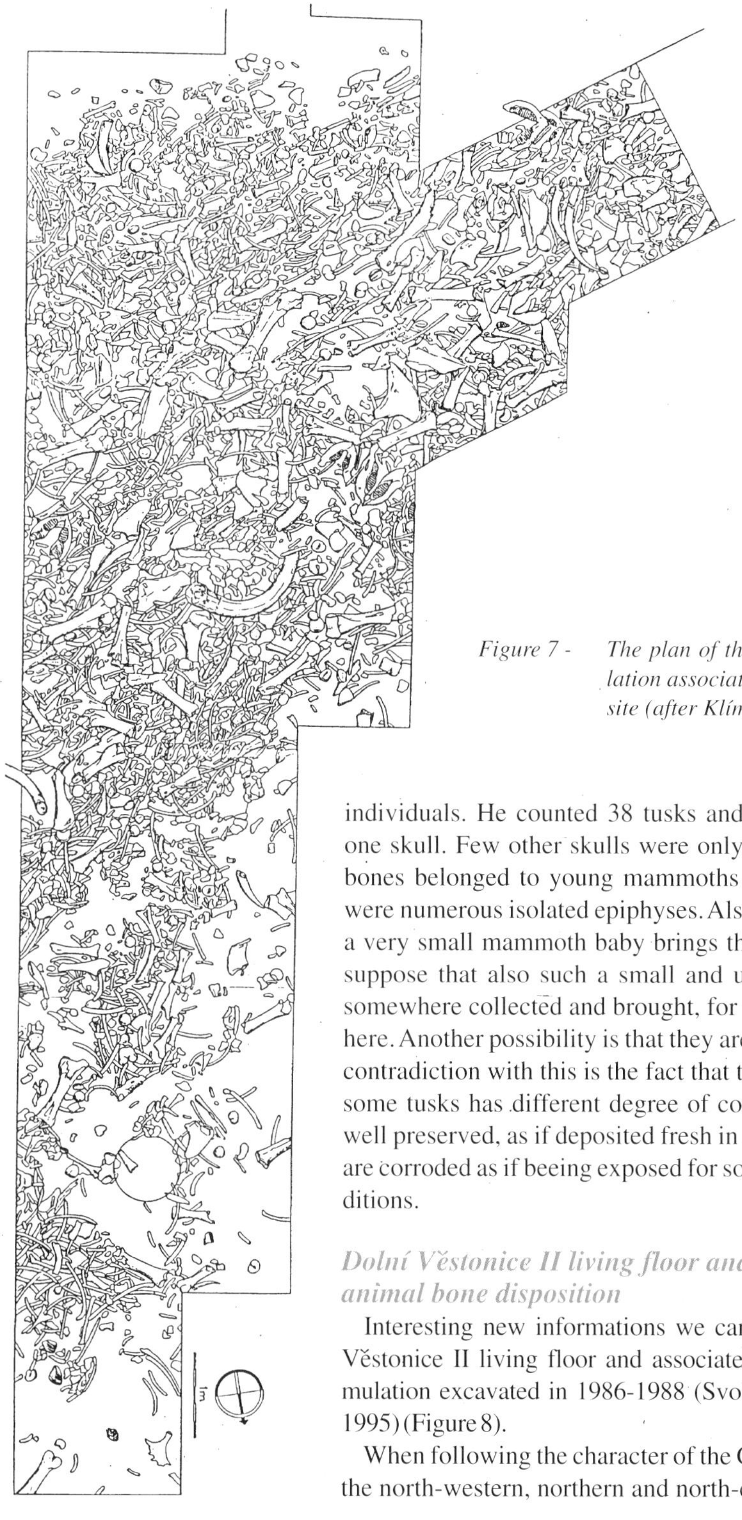

Figure 7 - The plan of the mammoth bone accumulation associated to the Dolní Věstonice I site (after Klíma, 1983).

individuals. He counted 38 tusks and 48 mandibles but only one skull. Few other skulls were only in poor remains. Many bones belonged to young mammoths and between the bones were numerous isolated epiphyses. Also isolated carpal bone of a very small mammoth baby brings the question if we should suppose that also such a small and unimportant bones were somewhere collected and brought, for any use, to be deposited here. Another possibility is that they are culinaristic remains. In contradiction with this is the fact that the surface, especially in some tusks has different degree of corrosion. Some tusks are well preserved, as if deposited fresh in the accumulation. Some are corroded as if beeing exposed for some time to climatic conditions.

Dolní Věstonice II living floor and animal bone disposition

Interesting new informations we can obtain from the Dolní Věstonice II living floor and associated mammoth bone accumulation excavated in 1986-1988 (Svoboda, 1989, 1991; Klíma, 1995) (Figure 8).

When following the character of the Gravettian occupation of the north-western, northern and north-eastern slopes of Pavlov

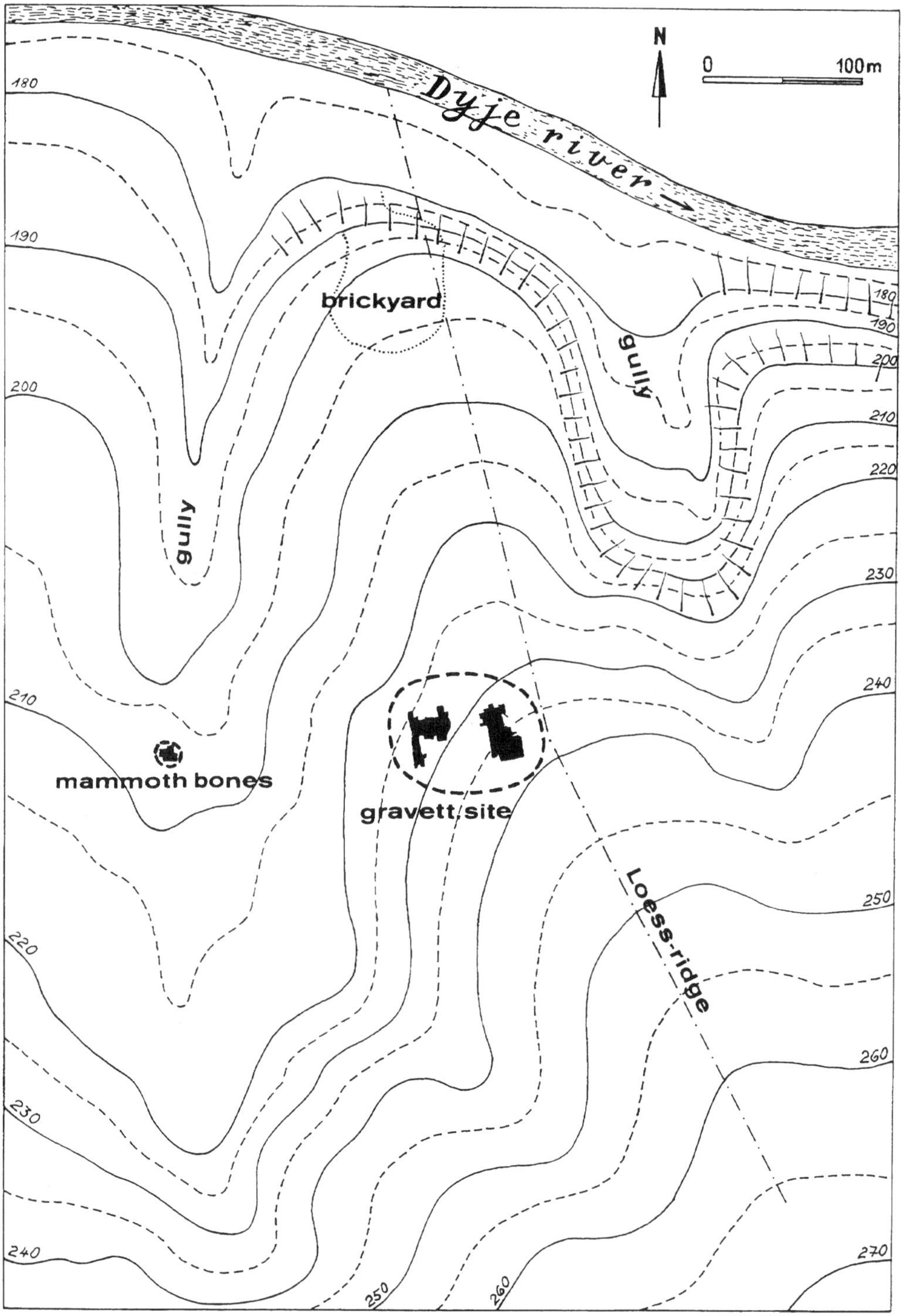

Figure 8 - The disposition of the living floor and the associated mammoth bone accumulation in Dolní Věstonice II (adapted after Klíma, 1995).

Hills and when considering the Dolní Věstonice I site as occupaying the central, north facing position, than the Dolní Věstonice II site is situated westernmost, facing north-west. The place is in an excellent strategical position overlooking to the west the whole broad inundation terrain. Most winds are comming from west and therefore, the animal roaming on this side of the river could not smell the hunters waiting eastwards. When following the stream the mammoths more and more approached the loess slopes with the Dolní Věstonice II position.

The principal features of the site and its dating was similar like in Dolní Věstonice I, that is the living floor contained some habitation remains, fireplaces, broken bones and stone tools and nearby this site was a small mammoth bone accumulation situated in a wet erosional furrow of the terrain.

But this similarity holds only for principal features. After detailed study we find more differences. On the living floor we find only few decorative pieces (possibly corporal decorations) not a single art object comparable to those found at Dolní Věstonice I or Předmostí. The Dolní Věstonice II is smaller than both supersites of Dolní Věstonice I or Předmostí but even when taking this fact into consideration we are surprised by the limited number of animal bones scattered here. The participating palaeontologist has counted alltogether 28,000 bones and bone fragments (Seitl, 1995). Nearly 18,000 pieces are mammoth remains. The high number should not surprise us as it covers all fragments found there. The personal approach of the palaeontologists and archaeologists, the absence of the minimal numbers of individuals, of the numbers of complete bones, of damaged bones and of fragments makes the objective evaluation difficult. But if we observe the excavated living floor of Dolní Věstonice II we see how poor it appears when compared with Dolní Věstonice I. It is clear that this place was used for a shorter period of time by a less numerous group.

The other than mammoth bones belong to reindeer, wolf, horse some hare and ice fox. Interesting is the find of three halves of the ice fox mandibles found in one spot, all deprived of their canines. This brings again the possibility, that other profit than meat (canines, fell, possibly some non-rational reasons) could be responsible for hunting these scavengers. We should remember that in the female burial Dolní Věstonice 3 found in 1948 (Jelínek, 1954) in the hand of this individual were 6 ice fox teeth. Not drilled, not a neckless or a pendant but plain teeth. It was certainly a special reason that the burried woman had them in her hand.

Isolated bones of brown bear, lion, hyena, wolverine and beaver are also exceptionally found dispersed on the living floor.

Climatic difference or the difference in the period of the year, ecological situation, sample size and chance factor, all can be responsible for the differences found during such comparison of meagre finds.

Mammoth bone accumulation nearby the living floor Dolní Věstonice II

Nearly some 200 m west from the site an isolated mammoth bone accumulation was found. As with the living floor also this accumulation is poor. The dating shows that both, the living site and the accumulation are simultaneous. The bones are situated in two crossing zones and they belong most probably to two individuals as the distribution of the vertebrae, ribs and other bones suggests. Several long bones, flat bones (hip-bones, shoulderblades), vertebrae, ribs, one skull, one tusk and two mandibles were found. It seems that mammoths were originally complete here (Figure 9). If so, than they do not represent accumulated bones, collected somewhere and brought here. They were dismembered on spot. Some few (6) stone tools and six flakes were dispersed between the bones and further five atypical coarse pieces represent here a special group of heavy duty tools. What was their use here when the bones are complete, not split. Were they used for dismembering and meat filetting? What other use they could serve?

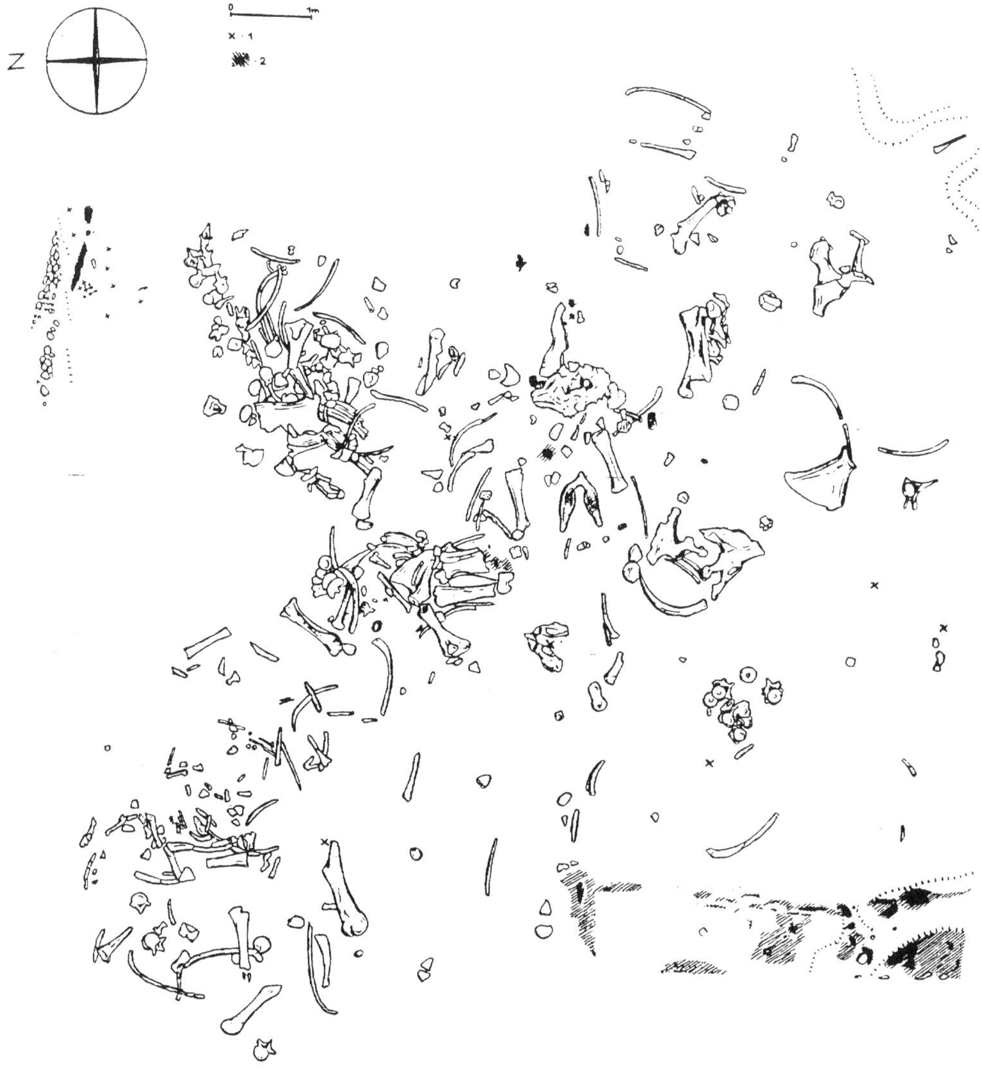

Figure 9 - The plan of the mammoth bone accumulation associated with the Dolní Věstonice II site (after Svoboda, 1991a). Explanations: 1) stone tools, 2) charcoal. .

In the erosional furrow together with these bones were found also the shells of water snails illustrating the original environmental condition of that place.

The striking difference between this accumulation and the living floor is in complete and not split bones found here and split and otherwise worked bones, fragments, numerous splinters and flakes found at the living floor of Dolní Věstonice II (see Svoboda, 1991, p.48).

Milovice living site bones

This site is situated north-east from Dolní Věstonice I and Pavlov I and Oliva (1988, 1989) has discovered here the largest mammoth bone accumulations after Dolní Věstonice I and Předmostí sites. More than 90% of bones belong to mammoth. The site is situated on the mild eastern slope

Figure 10 - The view on the part of the mammoth bone accumulation at Milovice discovered by M. Oliva.

of a low hill, some 2 km south from the river valley. The other known sites Dolní Věstonice I, II and Pavlov are situated nearer to the inundation terrain.

The living site is dated 25,220 ± 280 B.P. The charcoal found here demonstrate the existence of high grown trees, fir, pine and ash. Bones of horse, reindeer and isolated bones of lion and wolf are the part of the cultural layer but mammoth bones are far the most numerous. A circle of bones represents a construction remains: 14 shoulderblades, 13 hip-bones, 7 mandibles, 4 fragmented skulls and 24 long bones. One hip bone has an artificially prepared hollow, a pit, probably to

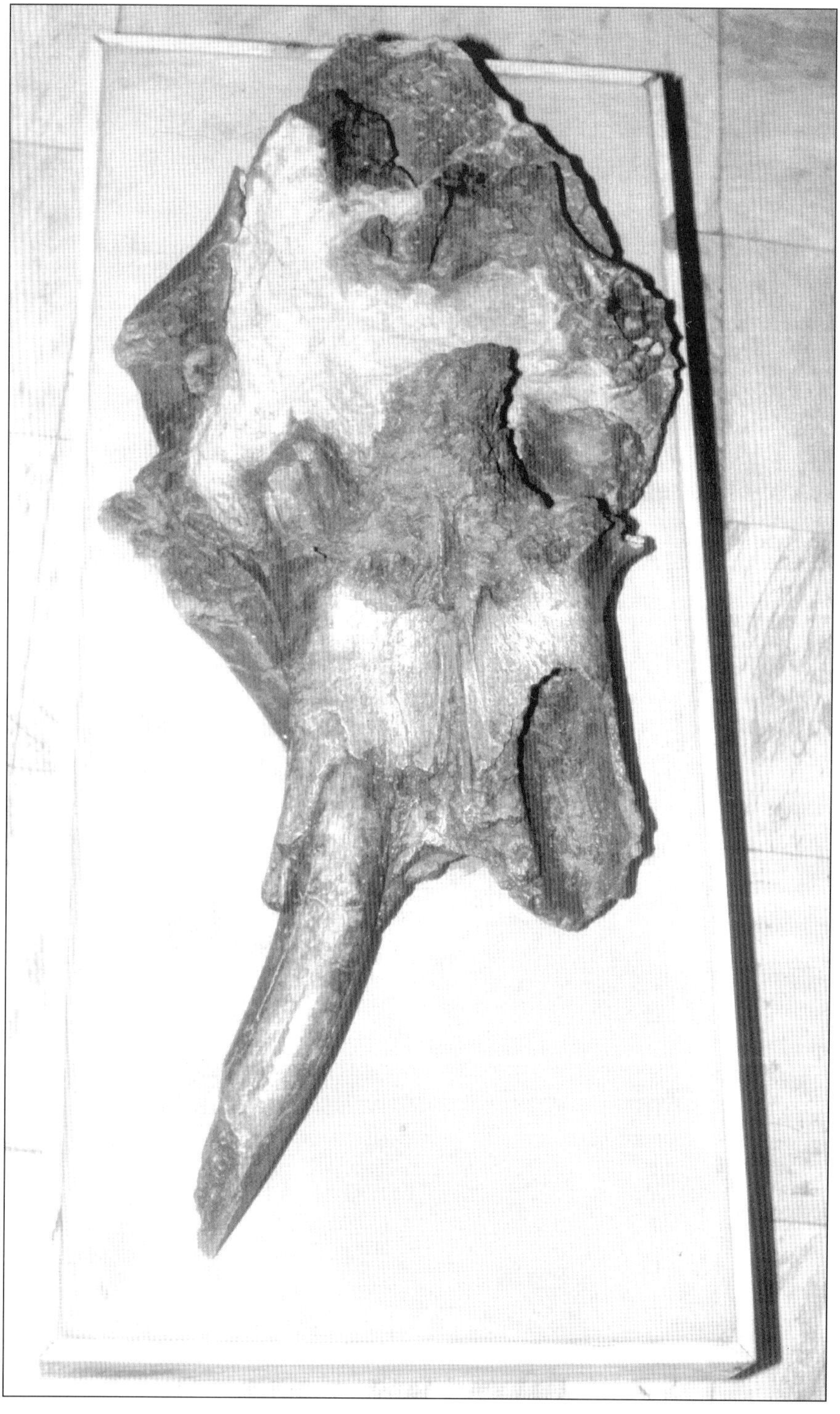

Figure 11 - The mammoth skull found at Milovice. One tusk was missing and the other hewn off.

anchor stick or pole in the construction of the habitation. An interesting find are the group of vertebrae and the tail vertebrae of a small mammoth baby, all in anatomical position. These could not be fully decomposed remains collected somewhere in the landscape. The fireplace was situated in front of the entry of this dwelling. Not workshops with stone flakes or cores were found, only small chips demonstrate the final stone working. It is also surprising that no bone tools, so typical for Gravettian sites, were present here.

Bone accumulation

In the western direction there is a large mammoth bone accumulation (Figure 10). Not only long bones but also hip bones, shoulderblades and fragmented skulls were found here. One skull has one empty alveol deprived of the tusk and in the second alveol is only a stump of the ivory as the tusk was hewn off (Figure 11). This means that one tusk was taken out but the second one was still solid in the alveol. The skull was collected somewhere and manipulated, being still not fully decomposed. Further nearby was found a large flaked piece of ivory, a large reindeer antler with broken-off tine, the bones of lions paw (in anatomical position) and a tusk. These and other bones accumulated here do not represent a living floor but probably some concentrated material prepared for - and/or resulting from working activities. Rough heavy duty tools made of local limestone, marlstone and hornstone support this idea.

Another mammoth bone accumulation in Milovice

Another large mammoth bone accumulation was discovered downslope. It also has not the character of a living floor. The bones are situated on a slightly gleyed base of the loess. Only rare stone tools are found between the bones but also some coarse heavy duty tools and large pebbles.There are traces of ash and some traces of sorting can be observed. Majority of bones belong to young individuals and more than 90 % of bones are mammoth bones.

Pavlov I bone accumulation

Before considering the bones found at the living floor of this important site, it is interesting to mention the find of a small zemlyanka, an isolated find in Moravian Gravettian (Figure 12). The small dimensions and 80 cm depth, the circular shape, its content (bones), small fireplace and characteristic entry are similar like with the Ukrainian finds where such type of zemlyanka is fairly frequent.

Only a small part of the site was excavated and studied palaeontologically by Musil (1959b). As in the other Gravettian living sites (Předmostí, Dolní Věstonice I, II and Milovice) also here the animal species are more numerous than in the bone accumulations out of the living floor. Mammoth, even when far prevailing in the body mass is only the third in the number of individuals: Hare 21 %, wolf 16 %, mammoth 9 %, ice fox 8 %, reindeer 4 %, wolverine 3 %, horse 3 %, other 5 %. Alltogether mammoth, ice fox, wolf and hare count 75 % of all animal bones found.

Wolf, even when less frequent than in Předmostí is numerous in his remains. 13 tibiae and 17 astragals were found. Of the ice fox seven left radii, 17 femora (not a single complete), 43 tibiae (only one complete), 17 astragali, 32 calcaneuses and also numerous individual metacarpal bones and phalanges were found. Wolverine was represented only with three arm bones. The selection of young mammoths is proved by majority of long bones without or with unfused epiphyses and by majority of subadult molars. The age distribution of mammoths was following: 12 % - 1-2 years, 12 % - 2-6 years, 50 % - 6-9 years, 25 % - 15-20 years (the oldest).

Many mammoth bones were flaked probably to produce bone tools. Horse long bones were often split for marrow.

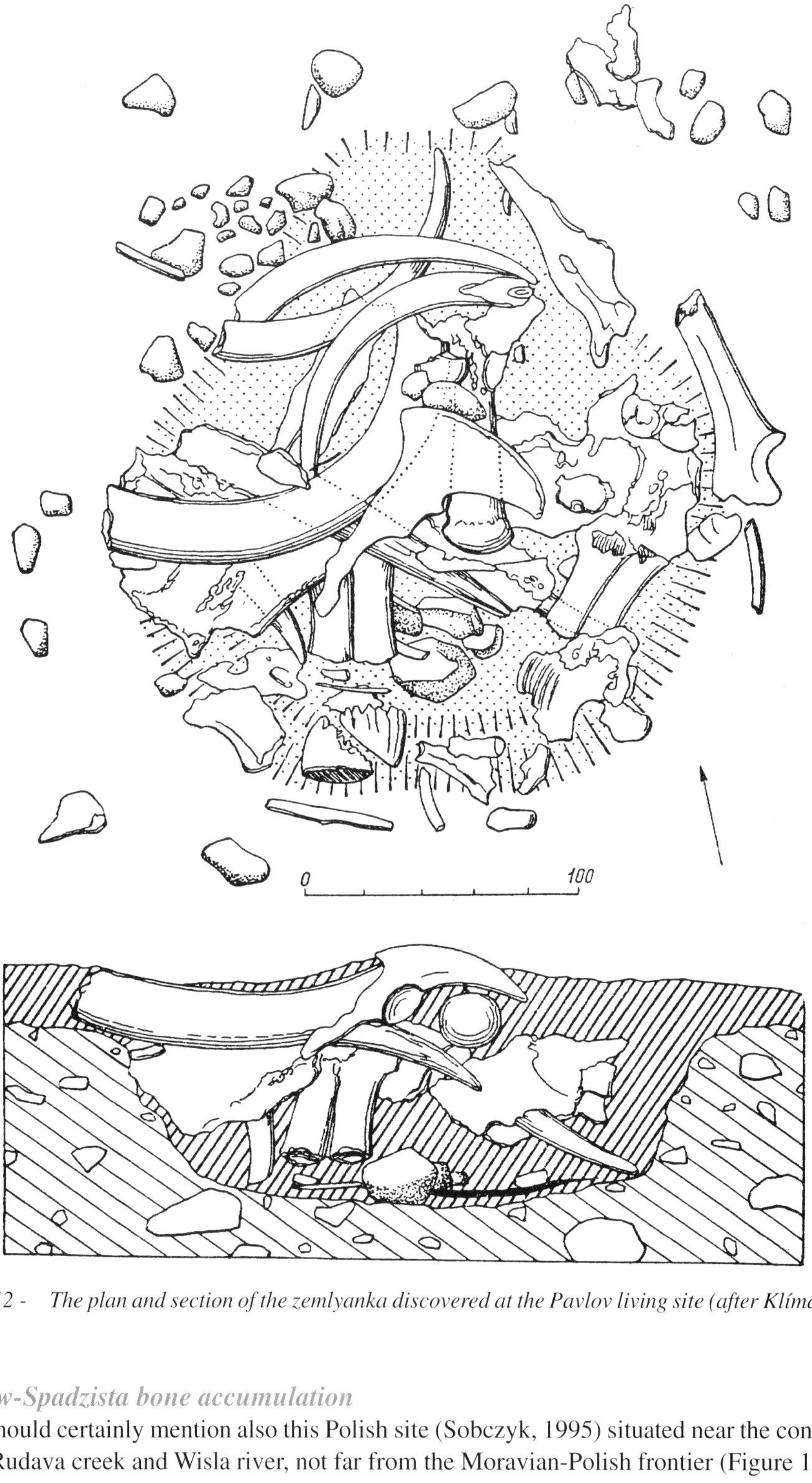

Figure 12 - The plan and section of the zemlyanka discovered at the Pavlov living site (after Klíma, 1977).

Krakow-Spadzista bone accumulation

We should certainly mention also this Polish site (Sobczyk, 1995) situated near the confluence of the Rudava creek and Wisla river, not far from the Moravian-Polish frontier (Figure 13).

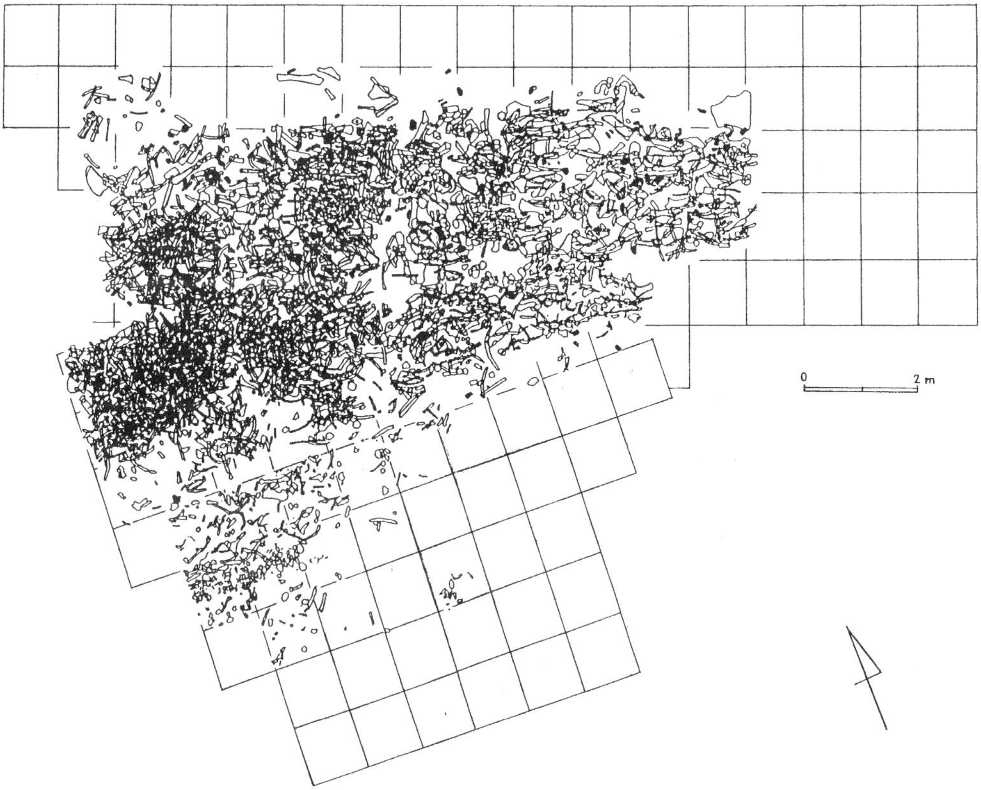

Figure 13 - Mammoth bone accumulation found at Spadzista Gravettian site in Cracowie (after Sobczyk, 1995). Note the amount and the accumulation of isolated bones..

Sobczyk counted in local bone accumulation 71 mammoth individuals representing 99.9 % of all animal bones. It was a typical accumulation of individual mammoth bones, none in anatomical position, comparable to the Moravian accumulations situated nearby the living sites. Between the bones some stone tools were irregularly found and also some traces of fire. Most of the bones belong to young animals: 26 individuals (37 %) - 2-8 years, 21 individuals (30 %) - 14-18 years, 23 remaining animals (33 %). These data demonstrate an intentional selection and exludes the possibility of exclusive collection of bones in the surroundings.

Sobczyk rightly considers this accumulation as a result of long term cumulative human activities. All features (age selection, strong prevalence of one species, individual, not anatomically situated bones and the strategical position of the locality) support the idea that two activities are responsible for the origin of this accumulation, namely hunting and collecting.

Analysis of hunting and scavenging possibilities

When considering the economical background of the Moravian Gravettian sites we follow the first possibility, that the Gravettians were hunters, not scavengers:

Several features often support the hunting hypothesis especially the strategical position of most sites so advantageous for hunters of big game living in inundation terrain of rivers in broader valleys and nearest surroundings. Also the size of such sites and rich social life represented by

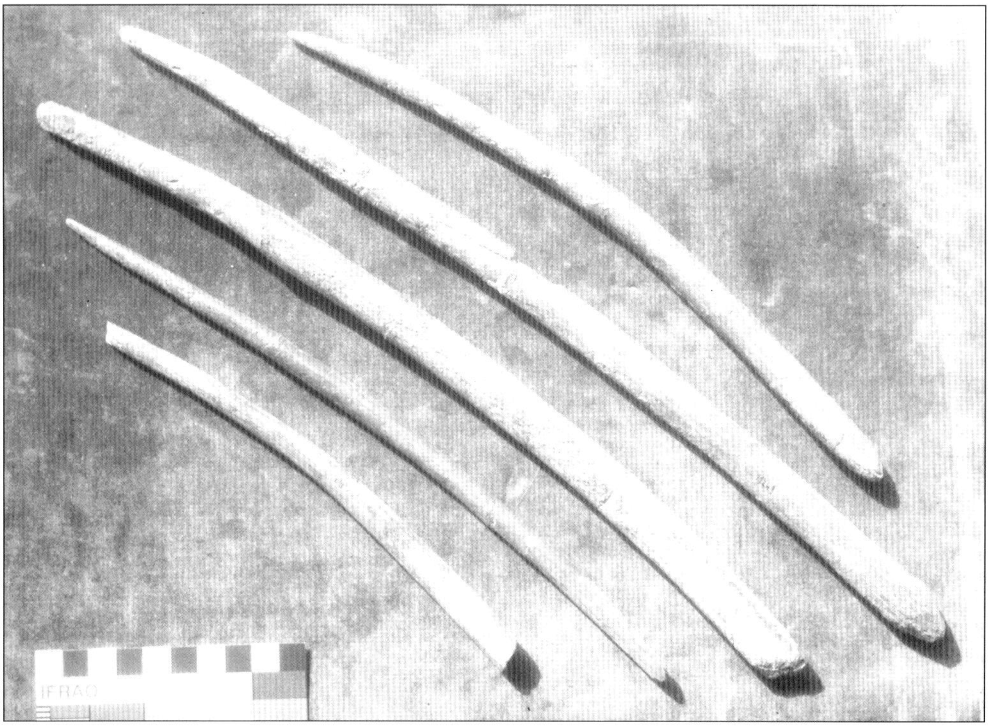

Figure 14 - Selection of typical Gravettian ivory spear-points, appropriate for hunting of mammoths.

rich archaeological finds demonstrate rich and long lasting activities at those sites which required corresponding supply of nutrition. When eliminating the mammoth bones we do not find other sufficient nutrition remains for the local group of hunters. It is also most probable, that the finds of paw bones of wolf, wolverine or lion found in anatomical position represent a pelt remains. These could result from a hunt and the case of a lion illustrates the hunting capacity at least of some of those hunters. Finally the characteristic large mammoth ivory spearheads were certainly good hunting weapons even for killing large pachyderms (Figure 14). If we do not accept this use, we have no other reasonable use for such a strong and frequent hunting tool at Moravian Gravettian localities.

If we consider the other idea, namely that of scavenging we find many characters contradicting this idea and not one supporting it:

First the periglacial environment does not give sufficient natural deaths plus remains of predators catch to cover the economy of local people by scavenging only. Most natural deaths appear at the end of the winter period or at early spring but those carcasses are very lean and bring the danger of protein poisoning if not counterbalanced with fet. Split bones, possibly for marrow, we find as part of the living floor but not as parts of the largest bone accumulation nearby the site. If Gravettians should scavenge after the predators activities, than the gnawing traces after the predators should be find on some bones. No such traces were found. If the palaeontological analysis of bone accumulations demonstrate the economical specialisation than such accumulations contradict the scavenging possibility just by that specialisation. Single or strongly prevalent species preference in bone accumulation does not reflect different kind of carcasses avaiable for scavenging.

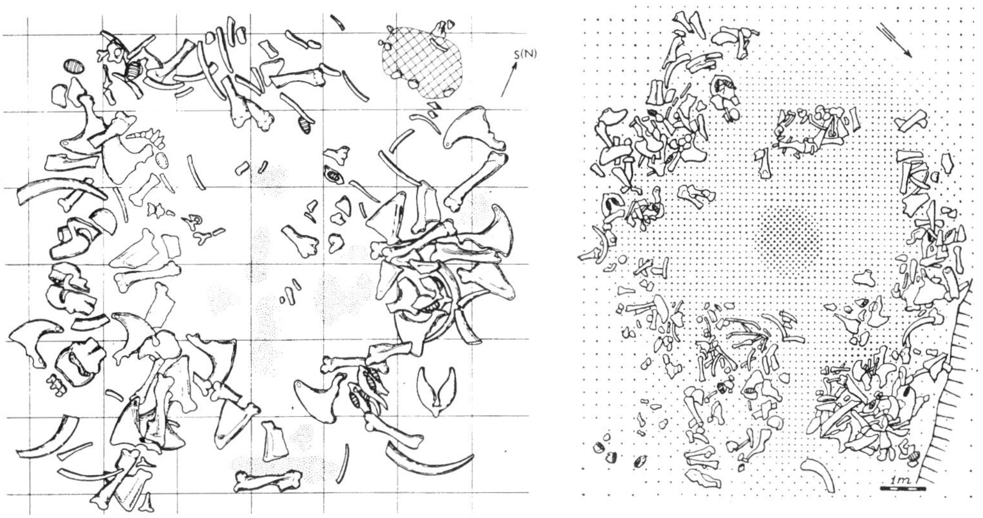

Figure 15 - Milovice. a) Plan of the remains of the mammoth bone habitation (after Oliva, 1989). b) Plan of the habitation excavated at Dolní Věstonice I by Absolon in 1928 and recognized by Klíma in 1963 (after Klíma, 1963).

Taking into consideration the opportunistic behaviour of Palaeolithic hunters we cannot a priori exclude the exceptional scavenging possibility if there were a good opportunity for such a quarry, but this can be only an exception not the regular way how to survive.

Conclusions

Considering the character of bone accumulations:

If the Gravettians are for several reasons considered as possible mammoth hunters than especially larger part of those mammoth bone accumulations found as a part of the living floor should be considered as culinaristic remains. If the mammoth meat would not be their important food than we have no other sufficient food remains. The rare bones of fur bearing animals were most probably result of opportunistic hunting for fur. Also the canines were often used for decorative reasons.

The presence and splitting and flaking of mammoth bones on the living floor support their culinaristic explanation and their use for making bone tools, for habitation constructions and as anvils for wooden constructions (see Valoch, 1987). The use of bones for habitation constructions was different fom such a use in Ukrainian or Russian sites. In Moravia so far only in Milovice (Oliva, 1989) and Dolní Věstonice I (Klíma, 1995) the use of big bones forming the circular circumference of a habitation was recognized (Figures 15a,b). This use was much more modest than in the well known Ukrainian and Russian sites with their complex constructions. It is also illustrative that we find only one case of characteristic zemljanka structure, that one found in Pavlov I (Klíma, 1977) (Figure 12), whereas such structures with mammoth bones are quite frequent in Ukrainian and Russian sites.

The nearly absent large pits with bones and frequent large mammoth bone accumulations at Moravian sites and contradictory situation at Ukrainian and Russian sites correspond most probably to different climatic and environmental conditions.

The fur bearer remains found as a part of the living floor or as a part of the nearby accumulations need not be and most probably are not the culinaristic remains. I mentioned already other reasons supporting their other origin.

The differential weathering of some bones and especially of the tusks supports the collecting idea as for their origin. Also the undamaged roots of some tusks can result only from old and disarticulated finds, collected somewhere in the surroundings.They cannmot come from a fresh carcass.

Contradictory, the find of isolated small bones of mammoth babies or finds of blocks of bones in their anatomical position (however rare) cannot represent collected bones. They speak against the gathering hypothesis.

When considering the origin of those accumulations situated nearby the living sites we find as explanation their multiple character:

A reserve of collected bones and tusks in the surroundings.

A refuse of some culinaristic remains.

In some cases it can be the remains of the butchering site.

Such an accumulation was a reserve as fuel or as any other raw material used in everyday activities of Gravettian people.

All this demonstrate the variable character of a cultural feature, that is of the mammoth bone accumulations, according to local conditions, needs and traditions, climatic and environmental conditions and other. They cannot be explained saying they are nothing more than or they are either/or. In the future they should be analysed as any other archaeological feature, in a detailed and complex way. This could not be the goal of this contribution. I tried only to stress the variable and important character of this feature illustrating the mode of life and survival strategy of Gravettian people.

References

ABSOLON, K. (1945): *Výzkum diluviální stanice lovců mamutů v Dolních Věstonicích na Pavlovských kopcích na Moravě. Pračovní zpráva za třetí rok 1926.* (Die Erforschung der diluvialen Mammutjäger-Station von Unter-Wisternitz in den Pollauer Bergen in Mähren. Arbeitsbericht über das dritte Grabungsjahr 1926). Brno. 252 S.

ABSOLON, K. & KLÍMA, B. (1977): *Předmostí. Ein Mammutjägerplatz in Mähren.* Praha (Fontes Archeol. Moraviae. Vol. 8).

BINFORD, L.R. (1981): *Bones: Ancient men and modern myths.* New York.

BINFORD, L.R. (1987): Were there elephant hunters at Torralba? In: M.H. Nitecki & D.V. Nitecki (eds.), *The evolution of human hunting*, 47-105. New York.

HAYNES, G. (1991): *Mammoths, mastodonts and elephants: Biology, behavior, and the fossil record.* Cambridge.

JELÍNEK, J. (1954): Nález fossilního člověka Dolní Věstonice 3. The Fossil Man of Dolní Věstonice 3. *Anthropozoikum* 3 (1953), 37-92.

KLEIN, R.G. (1987): Problems and prospects in understanding how early people exploited animals. In: M.H. Nitecki & D.V. Nitecki (eds.), *The evolution of human hunting*, 11-45. New York.

KLÍMA, B. (1955): Výsledky archeologického výzkumu na tábořište lovců mamutů v Ostravě Petřkovicích v roce 1952-1953. *Časopis Slezského Muzea v Opave* 4, 1-35

KLÍMA, B. (1963): *Dolní Věstonice Výsledky výzkumu tábořište lovců mamutů v letech 1947-1952.* Praha.

KLÍMA, B. (1969): *Die große Anhäufung von Mammutknochen in Dolní Věstonice.* Praha (Přírodovědné práce ústavu ČSAV v Brně, 3/6).

KLÍMA; B. (1973): Archeologický výzkum paleolitické stanice v Předmostí u Přerova v r. 1971. *Památky archeologické* 1964, 1-23.

KLÍMA, B. (1977): Malaja poluzemljanka na paleoliticheskoj stojanke Pavlov v Tchechoslovakiji. In: *Problemy paleolita vostochnoj i centralnoj Jevropy,* 144-148. Leningrad.

KLÍMA, B. (1983): *Dolní Věstonice. Tábořište lovců mamutů.* Praha (Památníky naší minulosti, vol.12).

KLÍMA, B. (1995): *Dolní Věstonice II. Ein Mammutjagerplatz und seine Bestattungen*. Liège (ERAUL, vol.73).

KRÍZ, M. (1894): Die Lösslager in Předmostí bei Prerau. *Mittheilungen der Anthropologischen Gesellschaft in Wien*, 24, 40-50.

KRÍZ, M. (1896): Mé výzkumné práce v Předmostí a jejich hlavní výsledky. *Časopis vlastivedného spolku muzea olomuckého* 13, 1-9, 51-61, 87-102.

MAŠKA, K.J. (1884): Neue Lössfunde bei Předmost in Mähren. *Korrespondenzblatt der Deutschen Gesellschaft für Anthropologie, Ethnologie und Urgeschichte*, 15.

MAŠKA, K.J. (1886): *Der diluviale Mensch in Mähren*. Neutitschein.

MAŠKA, K.J. (1889): Über die Gleichzeitigkeit des Mammuts mit dem diluvialen Menschen in Mähren. *Sitzungsberichte der Anthropologischen Gessellschaft in Wien* 19, 82-87.

MAŠKA, K.J. (1894): Ausgrabungen in Předmost. *Mittheilungen der k.k. Central-Commission fur Kunst- und historische Denkmale* 20, 129.

MAŠKA, K.J. (1896): Postup výzkumných prací v Předmostí. *Časopis vlasteněckého muzejního spolku olomuckého* 45, 77-80.

MUSIL, R.(1955): Osteologický materiál z paleolitického sídlište v Pavlově. *Práce brněnské základny Československé Akademie Věd* 27, 279-319.

MUSIL, R. (1959a): Bemerkungen zum paläontologischen Material aus Dolní Věstonice. *Anthropozoikum* 8, 73-82.

MUSIL, R. (1959b): Das osteologische Material aus der paläolithischen Siedlungsstätte in Pavlov. *Anthropozoikum* 8, 83-106.

OLIVA, M. (1988): A Gravettian site with mammoth-bone dwelling in Milovice (Southern Moravia). *Anthropologie* (Brno) 26, 105-112.

OLIVA, M. (1989): Excavations in the Palaeolithic site of Milovice (Southern Moravia) in the year 1988. *Anthropologie* (Brno) 27, 265-271.

SAUNDERS, J. (1980): A model for man-mammoth relationship. *Canadian Journal of Anthropology* 1, 87-98.

SEITL, L. (1995): Tierknochen. In: B. Klíma, *Dolní Věstonice II. Ein Mammutjagerplatz und seine Bestattungen*, 32-33. Liège (ERAUL, vol 73).

SOBCZYK, K. (1995): *Osadnictwo wschodniograweckie w dolinie Wisly pod Krakowem*. Kraków.

SOFFER, O. (1985): *The Upper Paleolithic of the Central Russian Plain*. Orlando.

SOFFER, O. (1993): Upper Paleolithic adaptations in Central and Eastern Europe and man-mammoth interactions. In: O. Soffer & N.D.Praslov (eds.), *From Kostenki to Clovis*, 31-49. New York, London.

SVOBODA, J. (1989): Výzkum mamutí skládky u Dolních Věstonic (okr. Břeclav). *Přehled Výzkumů 1986*, 7-8. Brno.

SVOBODA, J. (ed.) (1991a): *Dolní Věstonice II, Western Slope*. Liège (E.R.A.U.L. 54)

SVOBODA, J. (1991b): Beendigung der Untersuchungen der Anhäufung von Mammutknochen bei Dolní Věstonice. *Přehled Výzkumů 1988*, 9-10. Brno.

VALOCH, K. (1987): Contribution to the architecture of Upper Paleolithic dwellings. *Anthropologie* (Brno) 25, 115-116.

VALOCH, K. (1988): *Die Erforschung der Kůlna Höhle 1961-1976*. Brno (Anthropos, vol.24).

VERESCHAGIN, N.K. (1971): Okhota pervobytnovo cheloveka i vymiraniye Pleistotsennovih mlyekopitayuschih v SSSR. In: N.K.Vereschagin (ed.), *Materialy po faunam Antropogena SSSR*, 123-193. Leningrad (Trudy Zoologicheskogo Instituta, vol. 49).

VERESCHAGIN, N.K. (1977): Berelyohskoye kladbische mamontov. In: A.O. Skarlato (ed.), *Mamontovaya fauna Russkoi ravniny*, 5-50. Leningrad (Trudy Zoologicheskogo Instituta, vol. 72).

VERESCHAGIN, N.K. (1979): *Pochemu vymerli mamonty*. Leningrad.

VERESCHAGIN, N.K. (1981): *Zapiski paleontologa*. Leningrad.

VERESCHAGIN, N.K. & UKRAINCEVA, V.V. (1985): Proizchozhdenie i stratigrafia berelekhskovo kladbishtscha mamontov. *Trudy Instituta Zoologii A.N. SSSR* 131, 194-113.

WANKEL, J. (1884): První lidské stopy na Moravě. *Časopis vlasteněckého muzejního spolku olomuckého* 1, 2-7, 41-49, 89-100, 137-147.

WANKEL, J. (1886): Die Mammuthlagerstätte bei Předmost in Mähren. *Correspondezblatt der Deutschen Gesellschaft fur Anthropologie* 17, 149-153.

Survival Strategies in the Eastern Gravettian

Sandra L. Pichler
In memoriam Wilhelm Schüle †

Abstract

A large number of Upper Palaeolithic sites have been excavated during the last century, yielding a wealth of information on stone and bone inventories and site features as well as on lithic raw materials, artefact typologies, and local units („cultures") within individual technocomplexes. Serious attempts to come to comprehensive analyses of the available data have not been made until very recently, however. Such investigations aim at realizing interacting parameters and reconstructing diachronic processes at the level of cultural ecology. Synthetic approaches enable us to draw conclusions on the way of life and on living conditions of Palaeolithic communities which are not to be gained by isolated studies of individual sources.

Based on a comparative evaluation of Central and Eastern European sites assigned to the Eastern Gravettian technocomplex, three issues of central importance for estimating subsistence practices of Upper Palaeolithic groups are approached:

1. What practices of resource exploitation ensured the survival of human groups in ice-age Europe?

2. Are similar subsistence strategies observable in Central and Eastern Europe?

3. Were specific patterns of resource exploitation upheld under environmental stress, or does climatic change necessarily induce technological change?

The analysis indicates that there are distinct differences in resource use between Eastern Gravettian groups in Central and Eastern Europe both regionally and diachronically. Whereas local resources are steadily exploited in Central Europe, a greater degree of specialization is observable in Eastern European sites. Diachronically, opposing developments become manifest in the two study areas. While Eastern Gravettian inventories are gradually substituted by Magdalenian artefacts in Central Europe, there is an apparent intensification of resource use in the east. These differences are attributed mainly to the impact of the climatic maximum.

Introduction

In the Upper Palaeolithic of Europe, the Eastern Gravettian[1] represents a unique phenomenon. Both it's geographical and chronological range are unparalleled. A large number of sites with rich and diverse inventories have been excavated since the end of the last century. Singular finds have aroused attention far beyond the narrow limits of the archaeological sciences. Our knowledge of the technocomplex in general appears quite satisfactory.

Yet, on second sight, it becomes apparent that this knowledge is based primarily on studies of a few exceptional sites like Dolní Věstonice (Czech Republic) or Mezin (Ukraine). There (and elsewhere), archaeological investigations have mostly focused on traditional aspects, like artefact typologies, lithic raw materials, or on the portable art, especially the so-called „Venus"-figurines (Šovkopljas, 1965; Pidopličko, 1969; Delporte, 1979; Kozlowski & Kozlowski, 1979; Otte, 1981; Klíma, 1983; Gvozdover, 1985; Gamble, 1986; Svoboda, 1991, Iakovleva, 1995).

Comprehensive studies of the available data beyond these classical fields of interest have been attempted only recently (Soffer, 1985; Pichler, 1996). Such syntheses aim at including all the available data and gaining an understanding of the processes these reflect at the level of cultural ecology , i.e. of how human communities interact both with their natural environment and with other human groups (Jochim, 1981; Campbell, 1983; Butzer, 1987). Regardless of regional or chronological aspects, such analyses give us insights into the living conditions and the way of life of past populations that are not to be gained by isolated studies of individual sources of data.

For the Eastern Gravettian, the prevailing practice of intra-site analyses leaves us with a wealth of valuable information on individual classes of artefacts and on site features, but we know little of inter-site relations, nor of the processes underlying the differences in archaeological inventories both regionally and diachronically. The problem of the survival strategies of Eastern Gravettian groups reflects both the abundance of single-site data and the lack of specific knowledge of a more general character, especially on a meta-regional level.

Subsistence and survival - The role of plants and animals

The role of plant use in the Palaeolithic, especially during glacial times, is still highly controversial. Due to their physiological needs, humans cannot exist without a certain amount of vegetable food for longer periods of time. This essential requirement as well as accounts of the amount and type of vegetables consumed by traditional communities in arctic and boreal environments and of the methods employed in securing such commodities make it clear that vegetables contributed to the diet of human beings at all times (Eidlitz, 1969; Dennell, 1979; Parkington, 1991; Delluc et al., 1995; Bocherens, 1997). There is little discussion on this point, even though in archaeological contexts direct evidence is at best scarce prior to the Mesolithic.

The situation changes, however, if one considers objects made from materials of vegetable origin. Even though a wide spectrum of such artefacts is known from the ethnographic literature (and often enough from the same sources as the dietary information), many archaeologists are still unwilling to see analogies. On the contrary, there are not few researchers who insist that conclusions „ex silencio“ are quite impermissible in that specific context. One may argue that long-term survival is not possible without such basic equipment as tent poles, ropes, baskets, nets, containers etc. made of wood, bark, fibres, and other organic materials, sceptics insist on material proof for all the things needed in everyday life (cf. Semenov, 1970; Tyldesley & Bahn, 1983; Gamble, 1986; Owen, 1992; Vencl, 1994; Pichler, 1996).

Lately, however, extraordinary finds helped to fill in some of the blanks in the mosaic knowledge we have of the everyday life of Palaeolithic communities. The discovery of Lower Palaeolithic wooden insert tools and throwing spears in Germany, probably dating to approximately 400 kyr, fueled discussions both on organic inventories and subsistence practices of pre-modern hominids (Gamble, 1987; Thieme, 1997). Recent findings by an American-Czech research team who discovered several types of weaver's knots impressed in fragments of burned clay from the Pavlovian (i.e., Eastern Gravettian) settlement of Pavlov I (Czech Republic) practically revolutionized our knowledge on the art of cordmaking (and weaving?; Hoffmann, 1964; Adovasio et al., 1996; Pringle, 1997). In the light of this new information, a bone artefact from the site of Molodova 5, which the excavators interpreted as a weaver's shuttle, suddenly seems much more substantial (Cernyš, 1961; Klein, 1973). Even if they are quite singular (for the present at least), these finds challenge long-cherished beliefs and prejudices and may help to change established patterns of thinking concerning the abilities and technological competence of Pleistocene hominids.

All in all, however, the kind and the extent of plant use by human groups in ice-age Europe can only be speculated on. As direct evidence for exploitation of floral resources is lacking for the vast majority of Eastern Gravettian sites, we shall concentrate our consideration of the survival strategies on the one source of data that is nearly ubiquitously present: on the archaeozoological assemblages.

Existing models of resource exploitation

Inspired by discoveries of large accumulations of animal bones, especially of the whooly mammoth, at Předmostí (Czech Republic), Gontsy (Ukraine), and other sites excavated around the turn of the century, Eastern Gravettian groups were soon addressed as highly specialized big game or „mammoth hunters“ (Vereshchagin, 1967; Polikarpovič, 1968; Klein, 1969; Pidopličko, 1969; Kozlowski et al., 1974; Absolon & Klíma, 1977; Klíma, 1983; Soffer, 1985). This simplistic concept began to change after the 1960s, reflecting the growing importance of scientific analyses in the archaeological sciences. Improved excavation techniques, especially the introduction of fine-mesh sieving, as well as the closer attention paid to small animal species in archaeozoological analyses and the more precise identification of cutmarks and traces of natural vs. anthropogenic origin on bones from archaeological contexts contributed to a more differentiated picture of the subsistence practices of human groups in the late Pleistocene (Binford, 1981; Weniger, 1982; Davis, 1987; Davis & Reeves, 1990; Herrmann, 1994; Newell & Constandse-Westermann, 1996; Pichler, 1997).

The heroic image of tall, brave men daring to hunt mammoths with close-range weapons, though clearly rooted in nineteenth-century ideals, has nevertheless survived for a surprisingly long time. Only lately have new discoveries suggesting the use of nets and „boomerangs“ in animal hunting shed more light on the complexity of hunting, survival strategies, and social organization of Upper Palaeolithic communites (Evers & Valde-Nowak, 1994; Adovasio et al., 1996; Pichler, 1996; Rozua, 1996; Alt et al., 1997; Pringle, 1997).

Specific information on the exploitation of faunal resources by Eastern Gravettian groups is nevertheless scarce beyond a site-centered level. Basically, there are two competing models of survival strategies in the Eastern Gravettian in general. One distinguishes between central and eastern European groups who specialize in different animals (Otte, 1985), the other describes the existence of north, central, and east-central European groups exploiting their respective environments more or less generally (Kozloswki, 1986).

Both models are consistent, yet they convey completely contradictory views of the mode of resource exploitation. In central Europe, for example, Eastern Gravettian groups either were specialists, or they were generalists - they cannot have been both at the same time. Either scheme implies fundamentally different strategies for survival, as well as opposing ways of reacting to environmental (or social/demographic) stress. Furthermore, the two models are more or less static, because they are based on a geographical differentiation only. In view of the technocomplex's „life-span“ of about 18 millenia, of the degree of climatic (and environmental) change within that era, and considering all we know of the flexible subsistence strategies of modern hunter-gatherers (cf. Lee & DeVore, 1968; Winterhalder & Smith, 1981; Milton, 1985; Politis, 1996), the absence of diachronic adaptive processes would seem rather unlikely. In an attempt to solve that problem, a number of Eastern Gravettian sites from central and eastern Europe are selected for a comparative analysis of the exploitation of faunal resources.

Testing the differential exploitation of faunal resources - The data base

To test the hypothesis of possible geographic and diachronic differences in the survival strategies of Eastern Gravettian groups, the archaeozoological assemblages of fourteen central European and eighteen eastern European sites centering on Moravia and Ukraine as study areas are submitted to a comparative analysis (Fig. 1). The majority of the sites were excavated prior to the 1950s. For twenty-seven of the thirty-two sites, thermoluminescence or radiocarbon dates are available, few of which were obtained by radiocarbon accelerator dating (cf. Pichler, 1996). All of this results in the usual drawbacks: Most sites were excavated by inefficient field methods, in some cases records

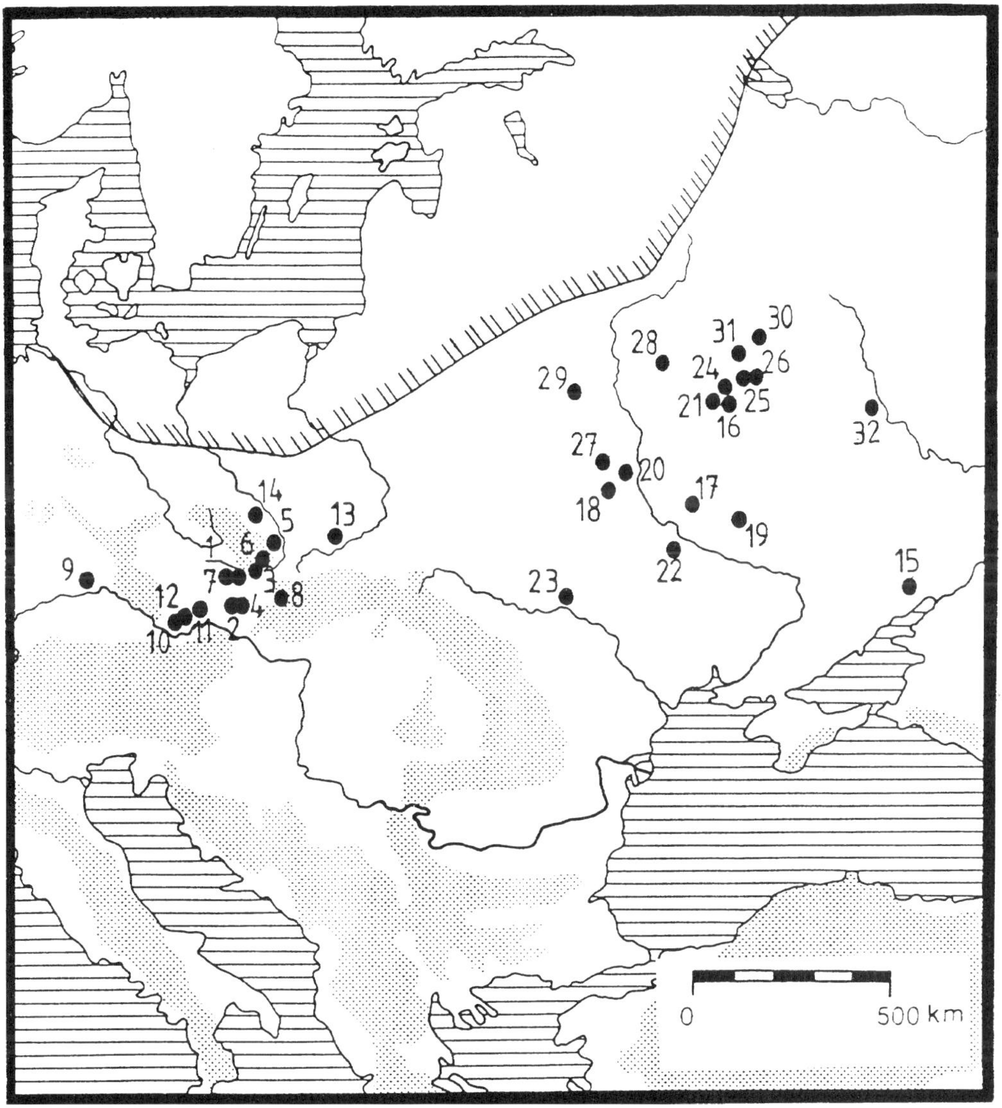

Figure 1 - Eastern Gravettian sites included in the present study. 1: Brno; 2: Dolní Věstonice; 3: Kůlna cave/Sloup; 4: Pavlov; 5: Petřkovice; 6: Předmostí; 7: Stránska Skála; 8: Moravany; 9: Weinberghöhlen/Mauern; 10: Aggsbach; 11: Ruppersthal; 12: Willendorf; 13: Kraków; 14: Wójcice; 15: Amvrosievka; 16: Čulatovo; 17: Dobraničevka; 18: Fastov; 19: Goncy; 20: Kiev; 21: Mezin; 22: Mežiriči; 23: Molodova; 24: Novgorod-Severskij; 25: Pogon; 26: Puškari; 27: Radomyšl'; 28: Berdyž; 29: Jurovići; 30: Chotylevo; 31: Eliseeviči; 32: Kostenki ; hatched: maximum extent of the ice sheet (after Frenzel et al., 1992).

are fragmentary or missing, inventories may neither be complete nor even representative of the material originally preserved at the sites, and radiocarbon dates assigned to specific cultural layers may vary as much as four, in one case even eleven, millenia. That, however, simply represents the list of problems always encountered in comparative analyses.

Herbivores:		**Carnivores:**	
Rangifer tarandus	[Rangifer]	*Canis lupus*	[Lupus]
Alces alces	[Alces]	*Alopex lagopus*	[Alopex]
Cervus elephus	[Cervus]	*Alopex corsac*	[Corsac]
Megaceros sp.	[Megaceros]	*Vulpes vulpes*	[Vulpes]
Bison priscus	[Bison]	*Ursus arctos*	[Arctos]
Bos primigenius	[Bos]	*Ursus spelaeus*	[Ursus]
Capra ibex	[Ibex]	*Mustela* sp.	[Mustela]
Rupicapra rupicapra	[Rupicapra]	*Gulo gulo*	[Gulo]
Ovibos moschatus	[Ovibos]	*Felis silvestris*	[Felis]
Saiga tatarica	[Saiga]	*Felis lynx*	[Lynx]
Equus sp.	[Equus]	*Panthera spelaea*	[Leo]
Sus scrofa	[Sus]		
Coelodonta antiquitatis	[Coelodonta]	**Birds:**	
Mammuthus primigenius	[Mammuthus]	*Anas* sp.	[Anas]
		Cygnus sp.	[Cygnus]
Small Mammals:		*Anser* sp.	[Anser]
Lepus sp.	[Lepus]	*Larus* sp.	[Larus]
Marmota bobak	[Marmota]	*Lagopus* sp.	[Lagopus]
		Lyrurus tetrix	[Lyrurus]
Fish:		*Bubo bubo*	[Bubo]
Pisces[1]	[Pisces]	*Corvus corax*	[Corvus]

[1] - Because they are seldom listed in Palaeolithic archaeozoological assemblages, all fish are counted as a single „species".

Table 1 - Animals regularly exploited in the Upper Palaeolithic.
Term in [ ] corresponds to designation of species in figures.

In order to counterbalance the bias contained in the data, the sites are assigned to three time intervals (A, B, and C) of approximately six millenia each. Delimitation of the intervals follows Gričuk's (1969) classification of Upper Pleistocene vegetation according to changes in temperatures and precipitation. Interval A denotes the Denekamp/Dunayevo interstadial, interval B comprises the era of the rapid growth and maximum extent of the Eurasian ice shield, and interval C represents the gradual amelioration of glacial conditions starting with the Lascaux/Somino-Ula interstadial (Zubakov, 1974; Leroi-Gourhan, 1980; Gerasimov & Veličko, 1982; Mangerud, 1991; Velichko, 1992).

Analyses focus on the economically important fauna which enabled Upper Palaeolithic communities to survive under glacial conditions. Relying both on archaeozoological data and accounts in the ethnographic literature, thirty-six species of animals regularly exploited by human groups are used to estimate patterns of resource exploitation in central and eastern Europe (Table 1). Divided into groups of large and medium-sized herbivores, carnivores, and small prey species (mammals, birds, fish), these species are investigated in view of regional and diachronic differences between the study areas.

Regional analysis - Do subsistence bases differ in central and eastern Europe?

A comparison of the exploited faunal species indicates that there are differences in the spectrum of animals used between central and eastern Europe. Figure 2 shows the group of the large and

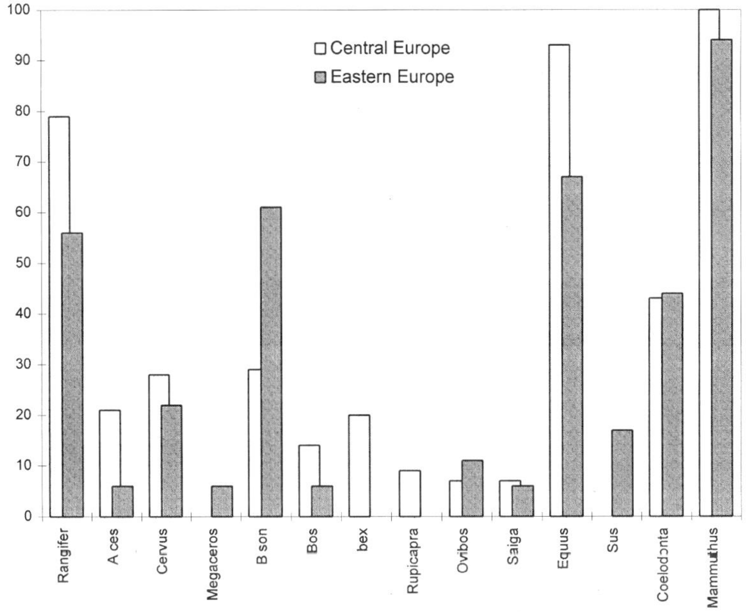

Figure 2 - Regional analysis: Comparison of large and medium-sized herbivores exploited in central and eastern Europe. Percentage values relate to number of sites in specific study area: central Europe = 14, eastern Europe = 18; designation of species see Table 1.

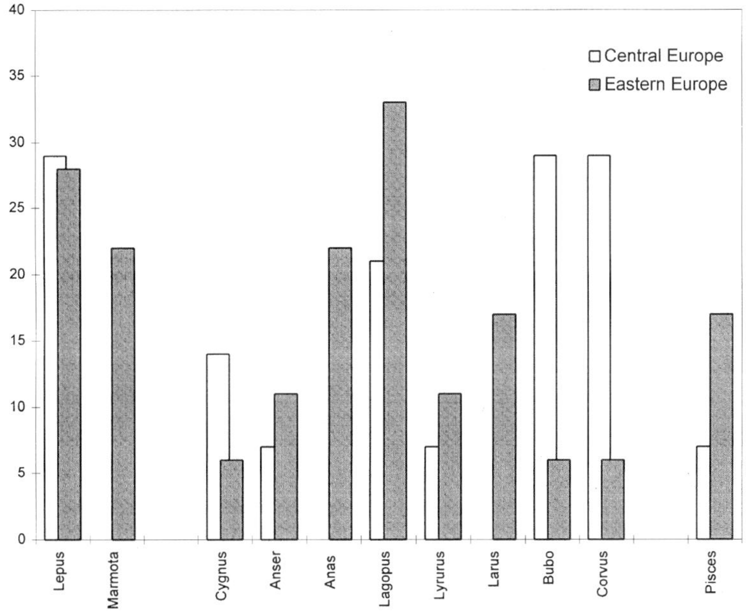

Figure 3 - Regional analysis: Comparison of small prey species exploited in central and eastern Europe. Percentage values relate to number of sites in specific study area: central Europe = 14, eastern Europe = 18; designation of species see Table 1.

medium-sized herbivorous animals, which are primarily hunted for their meat. From among that group, the herd animals are most important. In both central and eastern Europe, reindeer, horse, and mammoth are most frequently encountered in archaeozoological assemblages, as is the bison in eastern Europe. In the east, seasonally migratory species occur in similar numbers at all sites. These are also most frequently encountered in central Europe. Yet the regular occurence of relatively stationary species, like aurochs and ibex, may indicate a stronger orientation towards locally available resources as compared to the east.

Small prey species, which are killed for their meat and fur as well as for other uses, are contrasted in Figure 3. As one would expect, hare and ptarmigan stand out in numbers from the remaining species. For other animals, there seems to be a pronounced preference either in eastern or in central Europe. All in all, small prey species appear to have been used more regularly in the east than in the west. However, the animals in this group were most obviously subject to differential selection during sedimentation, preservation, recovery, and analysis of archaeozoological inventories (Pichler, 1996). The data base for this group is, therefore, frequently quite unsatisfactory.

With the exception of the felines, carnivores are more frequently recorded in eastern Europe (Fig. 4). Especially wolf and arctic fox, which are present in more than 70 % of the sites, were certainly hunted systematically for their furs. In central Europe, these two species also occur frequently, but there species relations are generally more balanced than in the east.

We can conclude that both in central and eastern Europe, large, herbivorous, migratory mammals represent the base of subsistence of Eastern Gravettian communites. For central European sites, a more systematic exploitation of species which are locally available year-round seems to emerge. In contrast, there are indications for a higher degree of specialization to specific animals in eastern European sites.

To return to the initial question: There actually do seem to be differences in the spectrum of animals used in central and eastern Europe, indicating that the subsistence base of Eastern Gravettian groups was not uniform throughout the technocomplex. This basic statement has interesting and far-reaching implications with regard to the survival strategies of the specific communities. More interesting still is the problem of how these groups managed to cope with the environmental stress imposed by the ice-age climate, especially with the progressive deterioration of interstadial conditions after cira 24,000 BP. A diachronic analysis of the archaeozoological assemblages should provide some clues as to how the groups reacted to the changes in their specific enviroments.

Diachronic analysis - Adaptive processes to the changing climate

The comparative analysis of the archaeozoological material by time-intervals reveals that there is, indeed, a substantial chronological component to the exploitation of faunal resources in the study areas. In central Europe, the number of species of large and medium-sized herbivores recorded from the sites declines steadily from interval A to interval C (Fig. 5). Whereas twelve species are listed for interval A, only seven are observed for interval B. In interval C, a mere three species of large herbivores remain.

In the group of the carnivorous animals, this development is even more pronounced (Fig. 6). Of the ten species of carnivores present in interval A, only four are left in interval B. For interval C, carnivore remains are no longer recorded from the selected sites in central Europe. Small prey species are not analyzed, because in the central European sites considered in this study they are recorded for interval A only.

In eastern Europe, there is a completely different development. There, the number of large and medium-sized herbivores exploited increases from eight species in interval A to eleven species in interval C (Fig. 7). The same development is visible in the group of the small prey species (Fig.

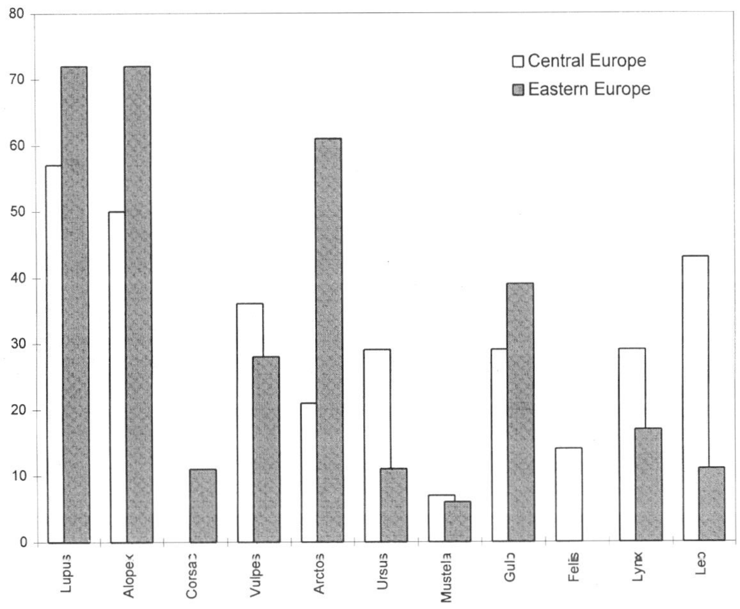

Figure 4 - *Regional analysis: Comparison of carnivores exploited in central and eastern Europe. Percentage values relate to number of sites in specific study area: central Europe = 14, eastern Europe = 18; designation of species see Table 1.*

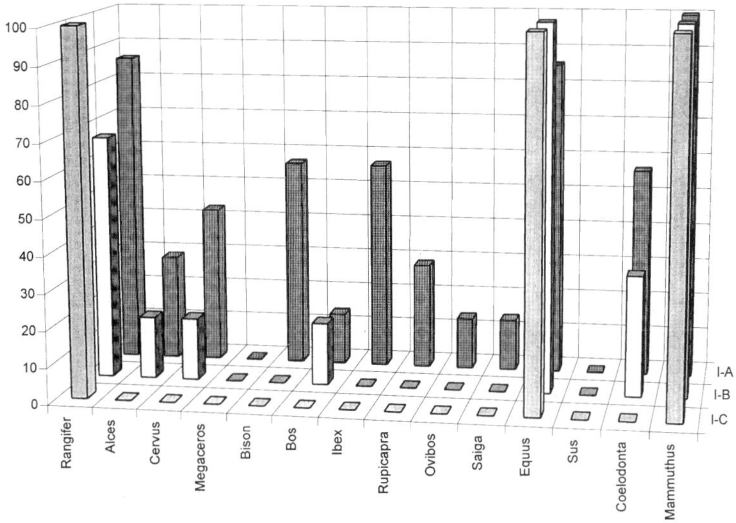

Figure 5 - *Diachronic analysis: Large and medium-sized herbivores exploited in central Europe during time intervals A to C (I-A, I-B, I-C; see text):*
Percentage values relate to number of sites considered in specific interval: I-A = 7, I-B = 6, I-C = 1; designation of species see Table 1.

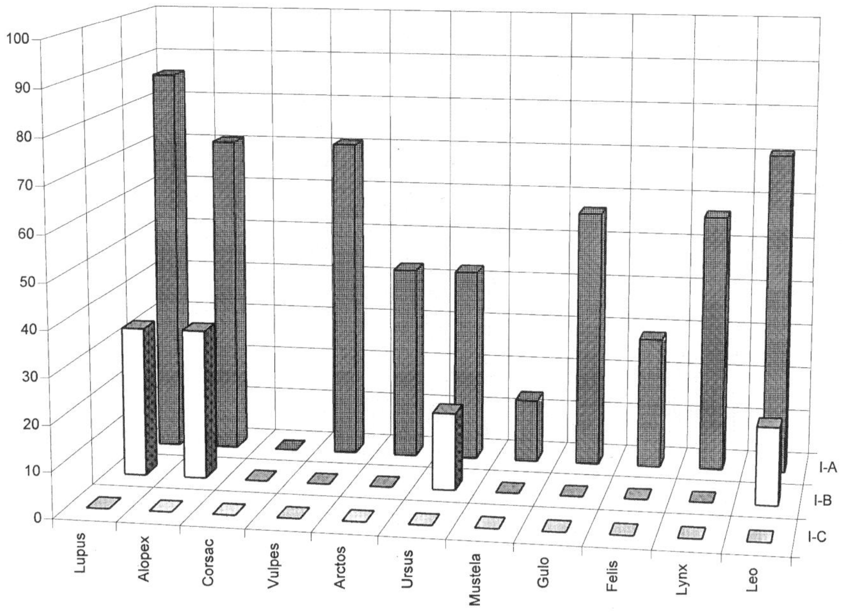

Figure 6 - *Diachronic analysis: Carnivores exploited in central Europe during time intervals A to C (I-A, I-B, I-C; see text). Percentage values relate to number of sites considered in specific interval: I-A = 7, I-B = 6, I-C = 1; designation of species see Table 1.*

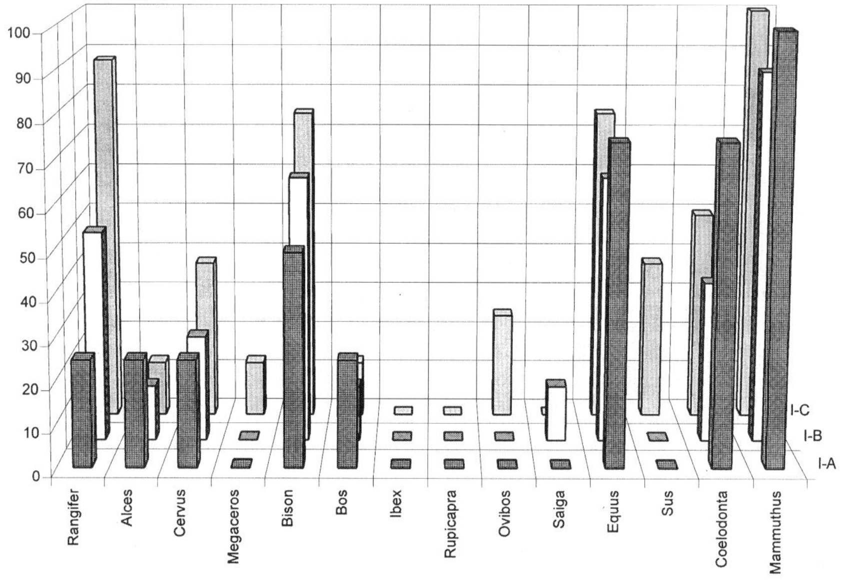

Figure 7 - *Diachronic analysis: Large and medium-sized herbivores exploited in eastern Europe during time intervals A to C (I-A, I-B, I-C; see text). Percentage values relate to number of sites considered in specific interval: I-A = 4, I-B = 8, I-C = 8; designation of species see Table 1.*

8). Furthermore, parallel to the growing number of species used, the number of sites for which small animal remains are recorded also increases continually from interval A to interval C. The carnivorous animals reflects this process as well (Fig. 9). The number of carnivores exploited in eastern Europe doubles from five species in interval A to ten species in interval C.

In summary, the results of the diachronic analysis are that the study areas pass through opposing developments in the course of the Eastern Gravettian occupation. In central Europe, animal species appear to have been exploited more or less evenly during interval A. In interval B, the exploitation of faunal resources seems to be narrowed down to a few, i.e. the most „profitable" species already. As a consequence, the further reduction in the number of species used which is observable in interval C is consistent with the emerging development, and it is, therefore, probably not due to missing values or to the overall poverty of the data base. Furthermore, the three species of animals still exploited in interval C are the species which the regional analysis showed to be the mainstay of Eastern Gravettian groups in central Europe: reindeer, horse, and mammoth.

In eastern Europe, on the other hand, the steady increase in the number of species exploited from interval A to interval C indicates a growing importance of locally available resources. Of the thirty-six species included in the analysis, nineteen are recorded in interval A, twenty-five in interval B, and as many as thirty-three in interval C. Moreover, the species that are „missing" from eastern European assemblages, i.e. ibex, chamois, and wild cat, are excluded from the ensemble of possible prey species for ecological reasons. So we may conclude that at the end of the Pleistocene, Eastern Gravettain groups in eastern Europe exploited all the faunal resources available to them.

Strategies for survival in the Eastern Gravettian

The comparative analysis of the patterns of resource exploitation of Eastern Gravettian groups shows that considerable differences exist between the central and eastern European study areas. Large, herbivorous, seasonally aggregating mammals form the basis of subsistence economies of Eastern Gravettian (as well as other Upper Palaeolithic) communities everywhere. A regional analysis shows that, beyond that basic fact, human groups in central and eastern Europe exploited different spectra of animals. In central Europe, the local component appears to have been more important, whereas a greater degree of specialization seems to have occured in the east.

A diachronic analysis of the archaeozoological assemblages reveals differences of a more fundamental character. In the course of the investigated time-span, opposing developments in the exploitation of faunal resources by Eastern Gravettian communities emerge in central and eastern Europe. In central Europe, resources seem to have been exploited in a generalistic way during the Denekamp/Dunayevo interstadial. In that strategy for survival, species which were locally available throughout the year appear to have played an important role. After the deterioration of the „mild" interstadial climate, findings indicate that Moravia was used in a far more limited manner, and that human groups may have been present seasonally only. This restricted use of the study area becomes most evident in sites post-dating ca. 23,000 BP. During and directly after the height of the last glaciation, Moravia might have been exploited for the specialized hunting of a limited number of species only. This finding needs to be corroborated by more detailed investigations, however.

The decline in the exploitation of the central European study area can be attributed to the climate, i.e. to the sinking mean annual temperatures and reduced precipitation during the glacial maximum. The Eurasian ice-shield advancing from the north as well as the growth of the ice-caps of the Alps and the Carpathian mountains reduced productive areas and led to a decisive deterioration of the living conditions of „producers" and „consumers" alike.

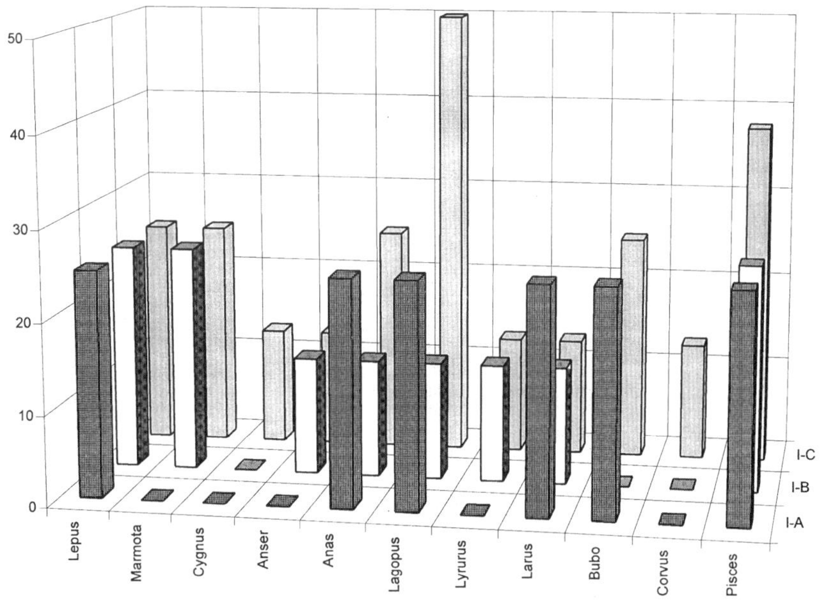

Figure 8 - Diachronic analysis: Small prey species exploited in eastern Europe during time intervals A to C (I-A, I-B, I-C; see text).
Percentage values relate to number of sites considered in specific interval: I-A = 4, I-B = 8, I-C = 8; designation of species see Table 1.

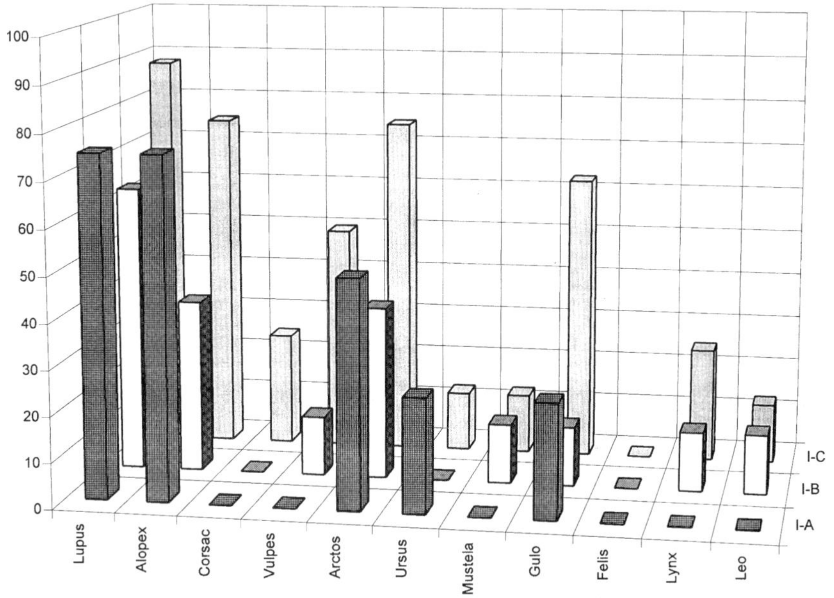

Figure 9 - Diachronic analysis: Carnivores exploited in eastern Europe during time intervals A to C (I-A, I-B, I-C; see text).
Percentage values relate to number of sites considered in specific interval: I-A = 4, I-B = 8, I-C = 8; designation of species see Table 1.

In eastern Europe, a contrary trend becomes manifest as the number of exploited species increases steadily. From a limited number of animals used at the onset of the Eastern Gravettian occupation, the survival strategy of eastern European groups changed to include an ever-growing range of prey species. By the end of the Pleistocene, Eastern Gravettain communities exploited practically all the faunal resources of their environment. Contrary to formerly held opinions on the scarcity or even absence of corresponding sites, this development is traceable even through the time of the glacial maximum. Furthermore, there are indications of a growing complexity in the seasonal and regional exploitation of game species. This process corresponds well to the increasing complexity of site features observable in such late Pleistocene Ukrainian sites as Mezhirich or Mezin.

In its basic lines, that development was certainly made possible by the geography of the practically barrierless plains of eastern Europe. On these plains, migrating herds (in the wake of the recolonization of the northern latitudes by a more thermophilous vegetation) are not impeded by impassable mountain ranges, and animal species are able to evade climatic (environmental) stress by shifting their ranges to the south unhindered. The expansion of their base of subsistence allowed Eastern Gravettian communities in eastern Europe to compensate both for considerable oscillations of the climate and for dramatic changes in environmental productivity. The development which becomes manifest in eastern Europe makes it obvious that Eastern Gravettian groups developed flexible and successful strategies for survival in a glacial environment.

In the present study, the environmental changes prior to, during, and after the glacial maximum emerge as being the determining factor in the opposing developments that become manifest in central and eastern Europe. In view of the palaeoclimatic background, one could expect the climatic component to be of major importance in the exploitation of faunal resources in the study areas. However, both the extent of the adaptive processes and the flexibility of the reaction of Eastern Gravettian groups, especially those in eastern Europe, exceed what one would expect from considering our knowledge of the technocomplex in general. Eastern Europe is supposed to have served as a „refugium" to human groups during the glacial maximum (Klein, 1973; Soffer, 1987). It might well be worthwile to make a comparison of the processes leading to a growing complexity in the ecological and cultural adaptation of Eastern Gravettain groups in eastern Europe, and the changes Upper Palaeolithic groups underwent in the Mediterranean area, the other major „refugium" in ice-age Europe.

Acknowledgement

I am indebted to Dr. Herbert Ullrich for his invitation to contribute to the present volume.

Note

[1] In the following, the use of the term „Gravettian" follows Kozlowski's (1986) broad, comprehensive definition of the technocomplex. According to this definition, the Eastern Gravettian (including the Epigravettian) comprises „the industries with backed points, blades, and bladelets that fall in the middle and late phases of the Upper Paleolithic and that occur in the territory extending from the Rhine to the Urals, between the southern limit of the ice sheet to the north and the Alps, the Balkans, and the Black Sea to the south. Chronologically, they fall at the end of the Interpleniglacial, the Upper Pleniglacial, and the beginning of the Late Glacial [i.e. ca. 14,500 BP, S.LP.]." (Kozlowski 1986,131).

References

ABSOLON, K. & KLÍMA, B. (1977): *Předmostí, ein Mammutjägerplatz in Mähren*. Praha (Fontes Archaeologiae Moraviae 8).

ADOVASIO, J.M., SOFFER, O. & KLÍMA, B. (1996): Upper Palaeolithic fibre technology: Interlaced woven finds from Pavlov I, Czech Republic, *c.* 26,000 years ago. *Antiquity* 269, 123-131.

ALT, K.W., PICHLER, S., VACH, W., KLÍMA, B., VLČEK, E. & SEDLMEIER, J. (1997): Twenty-five thousand year-old triple burial from Dolní Věstonice: An ice age family? *American Journal of Physical Anthropology* 102, 123-131.

BINFORD, L.R. (1981): *Bones. Ancient men and modern myths.* New York.

BOCHERENS, H. (1997): Isotopic biochemistry as a marker of Neandertal diet. *Anthropologischer Anzeiger* 55, 101-120.

BUTZER, K.W. (1987): *Archaeology as human ecology.* Cambridge.

CAMPBELL, B. (1983): *Human ecology.* London.

CERNYŠ, A.P. (1961): *Paleolitična stojanka Molodove V.* Kiev.

DAVIS, L.B. & REEVES, B.O.K. (1990): *Hunters of the recent past.* London (One World Archaeology 15).

DAVIS, S.J.M. (1987): *The archaeology of animals.* London.

DELLUC, G., DELLUC, B. & ROQUES, .M (1995): *La nutrition préhistorique.* Périgueux.

DELPORTE, H. (1979): *L'image de la femme dans l'art préhistorique.* Paris.

DENNELL, R.W. (1979): Prehistoric diet and nutrition: some food for thought. *World Archaeology* 11, 121-135.

EIDLITZ, K. (1969): *Food and emergency food in the circumpolar erea.* Uppsala (Studia Ethnographica Upsaliensia 32).

EVERS, D. & VALDE-NOWAK, P. (1994): Wurfversuche mit dem jungpaläolithischen Wurfgerät aus der Oblazowa-Höhle in den polnischen Karpaten. *Archäologisches Korrespondenzblatt* 24, 137-144.

FRENZEL, B., PÉCSI, M. & VELICHKO, A.A. (eds.) (1992): *Atlas of paleoclimates and paleoenvironments of the northern hemisphere.* Stuttgart.

GAMBLE, C. (1986): *The Palaeolithic settlement of Europe.* Cambridge.

GAMBLE, C. (1987): Man the shoveler: Alternative models for Middle Pleistocene colonization and occupation in northern latitudes. In: O. Soffer (ed), *The Pleistocene Old World*, 81-98. New York, London.

GERASIMOV, I.P. & VELIČKO, A.A. (1982): *Paleogeografija Evropy na poslednie sto tysjac let.* Moskva.

GRIČUK, V.P. (1969): Gljacial'nye flory i ich klassifikacija. In: I.P. Gerasimov (ed), *Poslednij lednikovyj pokrov na severo-zapade evropejskoj časti SSSR*, 57-70. Moskva.

GVOZDOVER, M.D. (1985): Ornament na podelkach Kostenkovskoj kul'tury. *Sovetskaya Archeologiya* 1, 9-22.

HERRMANN, B. (1994): *Archäometrie. Naturwissenschaftliche Analyse von Sachüberresten.* Berlin.

HOFFMANN, M. (1964): *The warp-weighted loom.* Oslo.

IAKOVLEVA, L. (1995): Les représentations féminines du paléolithique supérieur de Mézine (Ukraine). *L'Anthropologie* 99, 273-285.

JOCHIM, M.A. (1981): *Strategies for survival.* New York.

KLEIN, R.G. (1969): *Man and culture in the late Pleistocene.* San Francisco.

KLEIN, R.G. (1973): *Ice-Age hunters of the Ukraine.* Chicago.

KLÍMA, B. (1983): *Dolní Věstonice. Tábořište lovců mamutů.* Praha (Památníky naší minulosti, vol.l2).

KOZLOWSKI, J.K. (1986): The Gravettian in central and eastern Europe. In: F. Wendorf & A.E. Close (eds): *Advances in world archaeology,* vol. 5, 131-200. New York.

KOZLOWSKI, J.K. & KOZLOWSKI, S.K. (1979): *Upper Palaeolithic and Mesolithic Europe. Taxonomy and palaeohistory.* Kraków (Prace Komisji Archeologicznej 18).

KOZLOWSKI, J.K., Van VLIET, B., SACHSE-KOZLOWSKA, E., KUBIAK, H. & ZAKREZEWSKA, G. (1974): *Upper Paleolithic site with dwellings of mammoth bones - Cracow, Spadzista Street B.* Kraków (Folia Quaternaria 44).

LEE, R.B. & DeVORE, I. (1968): *Man the hunter.* Chicago.

LEROI-GOURHAN, A. (1980): Les interstades du Wurm superieur. In: J. Chaline (ed), *Problemes de stratigraphie quaternaire en France et dans les pays limitrophes*, 192-194. Dijon.

MANGERUD, J. (1991): The last interglacial/glacial cycle in northern Europe. In: L.C.K. Shane & E.J. Cushing (eds): *Quaternary landscapes*, 38-75. London.

MILTON, K. (1985): Ecological foundations for subsistence strategies among the Mbuti pygmies. *Human Ecology* 13, 71-78.

NEWELL, R.R. & CONSTANDSE-WESTERMANN, T.S. (1996): The use of ethnographic analyses for researching late Palaeolithic settlement systems, settlement patterns and land use in the Northwest European Plain. *World Archaeology* 27, 372-388.

OTTE, M. (1981): *Le gravettien en Europe Centrale*. Brugge (Dissertationes Archaeologicae Gandenses 20).

OTTE, M. (1985): Le gravettien en Europe. *L'Anthropologie* 89, 479-503.

OWEN, L.R. (1992): Der Gebrauch von organischen Materialien bei den Indianern und Inuit und seine Bedeutung für die Urgeschichte. *Ethnographisch-Archäologische Zeitschrift* 33, 25-34.

PARKINGTON, J. (1991): Approaches to dietary reconstruction in the Western Cape: Are you what you have eaten? *Journal of Archaeological Science* 18, 331-342.

PICHLER, S. (1996): *Paläoökologie des Östlichen Gravettien. Quellen und Ansätze ökologischer Rekonstruktionen der jungpleistozänen Umwelt.* Bonn (Universitätsforschungen zur Prähistorischen Archäologie 35).

PICHLER, S. (1997): Zähne und Kiefer in der Archäozoologie - Relevanz und Methoden. In: K.W. Alt & J.C. Türp (Hrsg.), *Die Evolution der Zähne: Phylogenie, Ontogenie, Variation,* 731-779. Berlin.

PIDOPLIČKO, I.G. (1969): *Pozdnepaleolitičeskie žilišča iz kostej mamonta na Ukraine.* Kiev.

POLIKARPOVIČ, K.M. (1968): *Paleolit verchnego podneprov'ja.* Minsk.

POLITIS, G.G. (1996): Moving to produce: Nukak mobility and settlement patterns in Amazonia. *World Archaeology* 27, 492-511.

PRINGLE, H. (1997): Ice age communitites may be earliest known net hunters. *Science* 277, 1203-1204.

ROZUA, Sch. (1996): Kop'emetalka i luk doistoričeskich ochotnikov. Technika i sravnitel'naja demografija. *Rossiya Archaeologiya* 2, 5-21.

SEMENOV, S.A. (1970): *Prehistoric technology.* Bath.

SOFFER, O. (1985): *The Upper Paleolithic of the Central Russian Plain.* Orlando.

SOFFER, O. (1987): Upper Paleolithic connubia, refugia, and the archaeological record from Eastern Europe. In: O. Soffer (ed), *The Pleistocene Old World*, 333-348. New York.

ŠOVKOPLJAS, I.G. (1965): *Mezinskaja stojanka.* Kiev.

SVOBODA, J. (1991): *Dolní Věstonice II - Western Slope.* Liège (E.R.A.U.L. 54).

THIEME, H. (1997): Lower Palaeolithic hunting spears from Germany. *Nature* 385, 807-810.

TYLDESLEY, J.A. & BAHN, P.G. (1983): Use of plants in the European Palaeolithic: A review of the evidence. *Quaternary Scientific Revue* 2, 53-81.

VELICHKO, A.A. (1992): Correlation of the Late Pleistocene events within glaciated areas of the northern hemisphere. In: B. Frenzel, M. Pécsi & A.A. Velichko (eds), *Atlas of paleoclimates and paleoenvironments of the northern hemisphere,* 101-105. Stuttgart.

VENCL, S. (1994): Nádoby z organických hmot v archeologických nálezech. *Archeologické Rozhledy* 46, 527-532.

VERESHCHAGIN, N.K. (1967): Primitive hunters and Pleistocene extinction in the Soviet Union. In: P.S. Martin & H.E. Wright (eds), *Pleistocene extinctions: The search for a cause,* 365-398. New Haven.

WENIGER, G. (1982): *Wildbeuter und ihre Umwelt.* Tübingen (Archaeologica Venatoria 5).

WINTERHALDER, B. & SMITH, E.A. (1981): *Hunter-gatherer foraging strategies.* Chicago.

ZUBAKOV, V.A. (1974): *Geochronologija SSSR. 3: Novejšij etap.* Leningrad.

Dietary Shifts and the Origins of the European Upper Palaeolithic

Susan Cachel

Abstract

Five lines of evidence support the widespread existence among non-ruminant mammals of physiological constraints to protein consumption, unaccompanied by other food types. High latitude modern humans with traditional diets must maintain a critical ratio of dietary fat to protein. An analysis of fifteen New World Arctic groups reveals that the abundance and predictability of dietary fat affects eight social variables. Assuming that the physiology of extinct Pleistocene humans resembles that of anatomically modern humans, if geochemical reconstruction of diet from human teeth and bone demonstrates very high protein consumption in Palaeolithic Europeans, it is possible to recognize limits to behavior that were determined by the necessity of maintaining a critical ratio of dietary fat to protein.

Subtle dietary shifts may have affected Pleistocene hominid biology and social behavior, with selection pressure exercised principally through population growth. Evidence from modern humans and a number of non-human primate species indicates that female fertility responds quickly to and is enhanced by increasing dietary fat. An initial increase in dietary fat at the end of the European Middle Palaeolithic would increase population size, with very little lag-time. Later in the Upper Palaeolithic, dietary breadth expanded. After human population grew in response to an increase in dietary fat, seasonally available plant species could finally be regularly incorporated into human diet, because larger social groups with a greater division of labor could engage in some exploitation of fine-grained resources. Any dietary carbohydrates contributed by plant species would lessen the impact of dietary fat on human behavior by making the immediate consumption of fat less crucial. A relaxation of selection pressure for maintaining skeletal robusticity could result in high protein diets without intolerable calciuria. The interacting consequences of population increase eventually result in a change identified as the Upper Palaeolithic. The following archaeological signatures of the Upper Palaeolithic could be generated: larger, more permanent human aggregations; a more permanent human presence in regions where humans were previously only ephemeral occupants; greater evidence of ritual and ceremony; ranking; and the occurrence of regional stone tool industries and variants. In the eastern Canadian Arctic, the transition between the Dorset Culture and Thule Culture offers a recent analogue to the European Middle Palaeolithic/Upper Palaeolithic transition.

There are four general implications that follow from the critical fat/protein argument. 1. If similar physiology is assumed, different human populations may have independently acquired traits of anatomical and behavioral modernity. 2. The geographic dispersion and permanent settlement of habitats associated with anatomical modernity may have been initiated simply by population growth and concomitant social complexity, and need not be thought to reside in novel intellectual or linguistic abilities. 3. Traits of modernity need not be acquired as a suite: a disjunction of traits may occur. 4. In tropical or subtropical areas, the transition to anatomical and behavioral modernity may be gradual. In temperate areas, the transition is more abrupt because dietary carbohydrates are seasonally restricted. Greater fat availability consequently has a more profound impact on human biology and behavior.

Introduction

The European Middle Palaeolithic/Upper Palaeolithic transition has been the focus of intensive research and analysis for the last decade, and the rate of research appears to be accelerating (e.g., White, 1982; Smith & Spencer, 1984; Trinkaus, 1986; Gowlett, 1987; Stringer & Andrews, 1988; Mellars, 1989, 1996; Mellars & Stringer, 1989; Trinkaus, 1989; Aiello, 1993; Conkey, 1993; Frayer et al., 1993; Howells, 1993; Stringer, 1993; Stringer & Gamble, 1993; Trinkaus & Shipman, 1993). This focus illustrates a general anthropological trend for investigating the origins of anatomical modernity. The advent of mitochondrial DNA analysis also contributed to this interest, because it initially appeared to support a history of a population of anatomically modern humans expanding from Africa and completely replacing earlier human populations in the Old World (Cann et al., 1987). The mitochondrial DNA evidence has since been challenged (Maddison, 1991; Maddison et al., 1992; Goldman & Barton, 1992; Templeton, 1993).

Although it is clear from archaeological evidence that no necessary connection exists between anatomically modern morphology and the advent of modern behavior (Klein, 1989; Frayer et al., 1993: 20-21), in Europe novel morphology and behavior appear more firmly linked and relatively abrupt in origin. Researchers therefore continue to use the European fossil and archaeological records to support a catastrophic transition to anatomical modernity and Upper Palaeolithic industries. I will argue here that a relatively subtle shift in subsistence may explain the Middle Palaeolithic/Upper Palaeolithic transition in Europe.

Protein and fat

Five lines of evidence demonstrate that the daily amount of protein ingested by humans must be limited during seasonal episodes of resource stress, when fat or carbohydrates are scarce. These are the following: ethnographic reports (summarized by Speth, 1991, and Speth & Spielmann, 1983); experiments with energy metabolism in humans and other non-ruminant mammals (Newsholme, 1978; Blaxter, 1989); clinical treatment for severe protein malnutrition (Whitehead, 1992); ethological evidence documenting reduced protein intake and preferential fat consumption in that mammalian carnivores that often consume only the skin and fat of prey species, and discard the rest of the carcass entirely (Stirling & Guravich, 1988; Pond, 1989; Court, 1996); and palaeontological data on the convergent evolution of bone-crushing adaptations in mammalian carnivores has been linked to the necessity of obtaining fat from bone marrow (Martin & Martin 1988). These lines of evidence support the widespread existence in non-ruminant mammals of physiological constraints to protein consumption, unaccompanied by other food types. This effect should be especially marked in high latitudes, where carbohydrates are seasonally unavailable, and where traditional human diets contained virtually no carbohydrates (Cachel in press a, b).

Selection pressures in high latitudes

A number of selection pressures operate in high latitudes which have a profound impact on human diet. A higher caloric intake supports human existence in cold climates. Basal metabolic rate rises upon sustained exposure to cold temperatures. *Ad libitum* food intake in humans and experimental animals increases upon exposure to cold temperatures. There is also evidence from domesticated animals and wild birds that food digestibility is reduced by exposure to cold temperatures. Carbohydrates are rare or impossible to obtain, so human diet is essentially a combination of protein and fat. The effects of maximal protein ingestion are countered with fat. The complete story of human subsistence in high latitudes lies in the ease of obtaining, storing, and transferring lean meat and fat. Protein may fluctuate, but a certain critical ratio of fat to protein must be present. It is the amount of available dietary fat which ultimately determines the success of aboriginal high latitude adaptation. Constraints on human existence are ultimately imposed by physiology. There is an upper limit to the rate at which the liver metabolizes proteins and synthesizes urea, and to the body's tolerance of the simultaneous water loss caused by these physiological activities. Fuel is absolutely necessary, but fuel may be constrained by lack of timber, and other vegetable fuel may be scarce. Fat or grease-laden bone may be the only available fuel, and if these resources are burned, they cannot be eaten.

It is clear that fat and oil are critical resources in high latitudes. Yet research summarized by Pond and Pereira (Pond, 1978; Pereira & Pond, 1995) unequivocally shows that fat from terrestrial mammals and birds is quantitatively limited. Fat reserves in each species are patchily distributed in time and space, because they do not accumulate until just prior to mating, giving birth, hibernating, or migrating. Fish have lean meat, unless the species is anadromous, and fish are caught while returning upstream to spawn. Sea mammals consistently maintain a much higher percentage of total body weight as subcutaneous fat. Thus sea mammals, unlike land mammals, birds, or most fish species, are abundant and predictable sources of fat and oil.

Behavioral variables affected by the abundance and predictability of fat

The increasing availability and predictability of dietary fat affects social organization in high latitudes, as demonstrated by fifteen New World Arctic hunter-gatherer groups (Cachel in press a, b). If more quantitative dietary analyses existed for modern high latitude peoples foraging for aboriginal foods, and especially if the amount of dietary fat were quantitatively determined and sources of this fat were distinguished, then one would be better able to assess the degree of linkage between the quantity of dietary fat and oil and specific elements of social organization in high latitude peoples.

The groups examined demonstrate that a number of behavioral variables are affected by the abundance and predictability of fat. Sedentism, the size and permanence of human aggregations, and such social variables as the permanence and complexity of leadership, ranking, the elaboration of marriage, residence, and descent rules, the importance of partnerships and other alliance mechanisms, the relative importance of biological or fictive kin, and xenophobia are all fundamentally affected by dietary fat in modern high latitude groups with traditional diets. Relatively detailed reconstruction of social organization may be possible for extinct high latitude human groups, if the critical fat/protein model is utilized and applied to the archaeological record.

Late Pleistocene application of the critical fat/protein model

It might be possible to apply the critical fat/protein model to the Late Pleistocene. However, some researchers explicitly reject the idea that behavioral data generated from modern human groups can be applied to archaic *Homo sapiens* (Klein, 1973: 123; Binford, 1982: 26-27; Klein, 1989: 263-264, 309-311). Behavioral data generated from modern Arctic groups also suffer the burden of nineteenth and early twentieth century theorizing about „the Chancelade race" and the idea of Upper Palaeolithic technology surviving among the Inuit, who were considered probable direct descendants of certain Upper Palaeolithic groups (e.g., Sollas 1924: 589-599). I intend no such arguments; my behavioral comparisons to modern humans are analogies, and not intended to demonstrate either genetic relatedness or the survivorship of Upper Palaeolithic „types" among living humans. With these explicit cautions, I nevertheless believe that archaeological application of the critical fat/protein model may be possible in the Late Pleistocene. One would need objective dietary information, and one would need to assume physiological similarity to modern humans. The widespread correspondence in protein metabolism among living non-ruminant mammals indicates that assuming homologous physiology in Late Pleistocene humans is not far-fetched.

Because sites are not in high latitudes, timber and other vegetable fuel was probably generally available. Another feature of Late Pleistocene European climate is that sunlight was not seasonally restricted by high latitude. This contrasts with the high Arctic, and is another factor ameliorating selection pressure on Palaeolithic humans exercised through diet. Reconstruction of local Pleistocene vegetation using modern analogues is difficult, because novel cloud-cover, regional temperature, and local wind regimes existed. Yet, unlike the case in circumpolar regions, plants almost certainly could provide hominids with dietary carbohydrates, as well as fuel. Palaeolithic hominid dietary reconstruction would therefore need to be based on the chemistry of prehistoric human teeth or bones in the context of food webs from extinct ecosystems.

Collagen geochemistry of animal and human remains has been used by Bocherens and his colleagues to reconstruct an extinct food-web and infer diet in Neanderthal specimens (Bocherens et al., 1991; Dorozynski & Anderson, 1991). A highly carnivorous diet is indicated for these Neanderthal individuals. Assuming physiological similarity to modern humans and other non-ruminant mammals, one can infer that the critical fat/protein ratio exercised an influence on humans with such diets, and that fat required to maintain this ratio was a crucial resource. The fat imperative might influence Neanderthal behavior in a manner analogous to that seen in modern humans with high-protein diets. Given that contemporary human and animal bones can be investigated to document ancient food-webs, a specific geochemical test for fat consumption is not needed by palaeoanthropologists. If the chemistry of prehistoric human bone and teeth documents a great deal of protein consumption, low reliance on carbohydrates, and no indications of marine resources in the diet, then adequate terrestrial fat must be an essential dietary component for prehistoric humans. The seasonality of sites, deliberate selection of fat-rich prey animals and body parts, the degree of bone breakage, and the incidence of scavenging will demonstrate the intensity of selection pressure exercised through fat.

No reliance on marine resources or anadromous fish has ever been documented for Mousterian humans. All dietary fat was supplied from land mammals or birds. Consequently, if one assumes that fat exercised constraints on social organization similar to those seen in the modern New World Arctic groups examined, the closest modern analogue would be found in groups exploiting terrestrial land mammals and not trading for sea mammal fat. For Mousterian humans, dietary fat would be limited, but conditions would be ameliorated by easy local access to timber and other vegetable fuels, and the existence of dietary carbohydrates. Once unequivocal evidence of prehistoric diet is given by the geochemical analysis of teeth and bones, relatively detailed inferences about social organization may be tested through a variety of means (Cachel in press b).

Neanderthal humans and other archaic peoples may have had an additional physiological problem - the need to counteract high protein intake in order to limit the loss of skeletal calcium. In modern humans, a very high-protein diet results in calciuria, probably as a result of a high phosphorus intake with little corresponding intake of calcium (Draper, 1977; Klepinger, 1992). If urinary calcium output increases, the rate of bone resorption accelerates, and there is a decline in bone mass. The thick cortical bone of Neanderthals, generated by extremes of activity rarely encountered in modern people, is based on the ability to mobilize sufficient amounts of serum calcium and to transfer it to living bone, which is being constantly remodeled. Yet serum levels of calcium cannot drop below a critical level without muscle tetany and seizures occurring. As a result, it seems reasonable that Neanderthal humans may have experienced more physiological stress from very high-protein diets than modern humans.

In Europe, the major change in human subsistence occurs during the late Upper Palaeolithic. Three intergrading stages characterize subsistence during the European Middle and Upper Palaeolithic (Table 1). These changes have a fundamental effect on human population size. Popu-

Middle Palaeolithic	
Diet	***Physiological variables***
Land mammals hunted for protein and fat.	High activity-levels mandate skeletal robusticity; calciuria from high protein intake; serum calcium level cannot fall with neurological consequences.
Transitional Period	
Diet	***Physiological variables***
Dietary fat and protein augmented: female fertility affected; population increase and division of labor make gathering plant foods cost effective; hunting by drives and surrounds can occur.	Relaxation of selection pressure for skeletal robusticity; calciuria from high protein intake no longer major physiological constraint.
Upper Palaeolithic	
Diet	***Physiological variables***
Land mammals and birds hunted; fishing, shellfish and plant foods gathered; nuts and seeds gathered and ground.	High protein intake counteracted by other fat sources and carbohydrates; critical fat/protein ratio can decline; tolerance for loss of bone mass.

Table 1 - Three subsistence stages and the Middle/Upper Palaeolithic transition.

lation size then affects hunting techniques, general foraging behaviors, and individual activity-levels. Dietary abundance affects human population size, perhaps by affecting female fertility through nutritional changes reducing the age at menarche and intervals between births. Increasing body fat causes the onset of puberty, because adipose tissue secretes leptin, which has been experimentally shown to trigger female reproductive function (Chehab et al., 1997). This explains the very short lag-time between dietary change which increases the percentage of body fat and population growth - the onset of female puberty is directly affected. Reduction in the age of menarche occurs in modern humans following nutritional changes accompanying industrialization; these changes involve an increase in dietary protein and fat (Frisancho, 1993). Similar dietary controls on reproduction occur in the yellow baboon (*Papio cynocephalus*), a non-human primate species often studied in laboratories engaged in research on human reproductive physiology. Other, free-ranging primate species which are provisioned also show population increase. This makes inferences about the physiology of Palaeolithic humans less problematic. Human stature increases in the European Upper Palaeolithic. This may represent either generally better nutritional status, or a reduction of negative selection pressures in comparison to the Middle Palaeolithic. This is how the secular trend in size is explained in modern industrialized populations (Frisancho, 1993: 415-417). Because even high socioeconomic classes are affected by this trend under industrialization, general dietary change is implicated, and not merely an increase in calories or improvements in public health (Harrison et al., 1988: 390-395). Upper Palaeolithic Europeans are taller than their Mesolithic successors, and more robust (Frayer, 1984). Whatever the social function of the Upper Palaeolithic Venus figurines may have been, they sometimes show anatomical collapse of the knee joints, and are thus anatomically accurate renditions of human obesity. They therefore attest to the beginning of dietary bounty, at least for some individuals.

Anatomical	Archaeological
Skeletal gracilization	Population expansion
Decline in traumatic injuries	More material and symbolic culture
Increase in stature	Long-distance trade
Dental changes	More inter-group differences
Decline in cranial capacity	More dietary diversity
	Sedentism

Table 2 - Variables explained by the critical fat/protein model.

Relative dietary abundance and an increase in the abundance and predictability of dietary fat just before the advent of the Upper Palaeolithic would ameliorate extremes of individual activity, because constant hunting to maintain critical fat ratios and highly laborious food preparation (breaking of long bones to extract marrow and pounding of epiphyses to boil out bone grease) would immediately be reduced. A decrease in hunting frequency, and circulating mobility through a succession of recurrently used seasonal camps in an annual cycle is inferred for anatomically modern humans utilizing the Mousterian industry in the Levant (Lieberman & Shea, 1994: 316-319). Long-distance movement to scout out novel hunting opportunities, or to establish or maintain essential hunting contacts or partnerships with other, remote groups might be lessened if the physiologically based fat imperative were reduced. Nomadism would decrease. Trinkaus's analysis of paleopathology and paleodemography support this behavioral reconstruction (Trinkaus, 1995: 138-139). Skeletal robusticity maintained through high activity levels (Ruff et al., 1993) would decline.

Relative dietary abundance would also indirectly affect the structure and composition of social groups. Palaeolithic humans would diffuse through a geographic area from local nuclei of population growth. Once human population in these nuclei reached a certain size, division of labor within social groups or between groups within the local nucleus could further reduce individual activity-levels and ameliorate individual selection pressures. Larger human populations would also ensure an increase in knowledge and social complexity, because the loss of information occurring as genetic drift removes knowledgeable individuals in small populations would no longer take place, and there might be a greater degree of generation overlap. Individual selection pressures would further ameliorate, because solutions to tasks and technological innovations and refinements would not periodically need to be re-discovered or re-invented.

Using the critical fat/protein model in a more general fashion, the European Middle Palaeolithic/Upper Palaeolithic transition can be analyzed in terms of relative dietary abundance, including a greater availability of fat. Shifting selection pressures initiated by subtle dietary changes may account for the concatenation of anatomical and archaeological traits associated with the Upper Palaeolithic (Table 2). I will now discuss the cascading effects of these selection pressures.

Some researchers have denied that competent or even purposeful hunting existed in the Middle Palaeolithic (Binford, 1982,1984; Klein 1973, 1989). Other researchers detect no significant increase in hunting prowess, foresight, or planning with the advent of the Upper Palaeolithic (S. Binford, 1968; Chase & Dibble, 1987; Chase, 1989; Straus, 1992; Enloe, 1993).

It is possible that the Upper Palaeolithic saw the introduction of new dietary carbohydrate resources, such as nuts and grains. Plant oils extracted from these items would be valuable dietary items. Additional dietary carbohydrates would further reduce the fat imperative. Furthermore, new cooking techniques (e.g., stone boiling) which appear in the Upper Palaeolithic

Thule culture	European Upper Palaeolithic
Population increase	Population increase
Rapid geographical dispersal	Rapid geographical dispersal
Permanent human presence	Permanent human presence
Increased sedentism	Increased sedentism
Larger sites	Larger sites
Denser archaeological record	Denser archaeological record
Technological innovation	Technological innovation
Abrupt disappearance of earlier cultural substratum	Abrupt disappearance of earlier cultural substratum
Displacement? extinction?	Displacement? extinction?
Absorption ? of earlier peoples	Absorption ? of earlier peoples
**Decrease in art	**Increase in art

Table 3 - Comparisons of Thule and Upper Palaeolithic traits

would save fuel. Innovations in shelter and hearth fabrication would lessen critical fat or carbohydrate constraints. The presence of numerous hearths, severely burned bone, bone charcoal, and large storage pits at some Upper Palaeolithic sites in the Central Russian plain testify to the density of human occupation, as well as food caching or hoarding, the use of bone as fuel, and refined heating technology (Soffer, 1985: 375-386).

More dietary diversity characterizes the Upper Palaeolithic. New, non-ungulate meat resources (birds, fish, mollusks) and new land mammal resources (suids, felids, Alpine ungulates - Freeman, 1973; Straus, 1987) account for greater amounts of lean meat. Intensive exploitation of a single prey species can also be detected in the Upper Palaeolithic (Binford, 1968; Freeman, 1973, Straus, 1985, 1987, 1992 - but see Chase, 1989). Anadromous fish, group driving of prey, and intensive exploitation of a single species may account for greater amounts of fat. A comprehensive examination of temporal change in subsistence within a single region (northern Spain), reveals that anadromous salmonids and other fish species occur in archaeological sites from the late Upper Palaeolithic of Cantabria (Straus, 1985); evidence of salmon-fishing is inferred from the presence of harpoons and fish vertebrae (Straus, 1987: 176). In addition, new carbohydrate resources, such as nuts and grains, may have been more habitual components of Upper Palaeolithic diet, because tools showing use-wear indicating grinding of vegetable materials become regular components of Upper Palaeolithic sites (Straus, 1992; de Beaune, 1993). Nevertheless, it is clear that the major break between Middle and Upper Palaeolithic subsistence appears late in the Upper Palaeolithic (Straus, 1985, 1987, 1992; Enloe, 1993).

Mousterian points in the Levant apparently functioned as spear-points (Shea, 1989), but Upper Palaeolithic lithic, bone, or antler projectile points, which appear to have been designed for long-distance throwing, may have augmented general hunting efficiency, and contributed to an increase in dietary meat and fat.

The subtle shift in diet itself accounts for changing selection pressure on individuals, and ultimately causes profound changes in behavior and social organization. Because the major division in hominid subsistence appears to occur not between the Middle and Upper Palaeolithic, but between the early and late Upper Palaeolithic (Straus, 1985, 1987, 1992; Enloe, 1993), this also indicates that Palaeolithic dietary shifts were probably subtle.

A recent analogue in the High Arctic

There is a recent analogue in the High Arctic. The transition between the Dorset Culture and Thule Culture people in the eastern Canadian Arctic offers a recent analogue to the European Middle Palaeolithic/Upper Palaeolithic transition. In fact, some researchers have already noted parallels between the Upper Palaeolithic of the Central Russian plain and the Thule Culture, or the culture of the modern Inuit (Soffer, 1985 :259, 416-417; Krupnik, 1993b: 7-9). But the advent of the Thule Culture also parallels the advent of the European Upper Palaeolithic.

Zooarcheological analysis and the appearance of whaling-related artifacts demonstrate that Thule Culture people, who are the immediate ancestors of the historical Inuit, specialized in active, deliberate, and selective open ocean hunting of whales and other sea mammals (Savelle & McCartney, 1991). People of the preceding Dorset Culture were apparently oriented toward exploitation of interior habitats; it is not even certain whether Dorset Culture people had kayaks or other boating technology (Arima,1994: 194). Beginning in 900AD, the Thule Culture abruptly spread across Arctic America: the diaspora from the Bering Straits and northern Alaska eastward across Arctic Canada and into Labrador and Greenland occurred in perhaps as little as two or three generations or 100 years by radiocarbon dating (McGhee, 1978). The earlier Dorset Culture substratum concomitantly disappeared. Population growth, rapid human geographic expansion and dispersal, a permanent human presence in Arctic America, increased sedentism, larger sites, a denser archaeological record, and artifact innovation and elaboration are all associated with the Thule culture (McGhee, 1978; Maxwell, 1985; Dumond, 1987; Krupnik 1993a: 228-229). These events parallel those inferred for the Upper Palaeolithic. Archaeological signatures for and inferences of behavior in the European Upper Palaeolithic nearly duplicate this list (e.g., White, 198; Chase & Dibble, 1987; Mellars, 1989; Klein, 1989).

Ironically, art (ornamented implements, personal adornment) appears to diminish in Arctic North America among the later Thule people, which may vitiate its use in arguments about intellectual advancements in the Upper Palaeolithic (Marshack, 1972a, 1972b, 1985, 1989; Conkey, 1993).

If human fossil evidence were scant or absent in Europe, the parallel to the Dorset/Thule transition would be even more extraordinary (Table 3). If one did not know *a priori* of morphological differences between Neanderthals and their successors, one would more easily recognize how similar the archaeological indicators of the modern Arctic transition are to those indicating the European Middle Palaeolithic/Upper Palaeolithic transition. Even the abrupt disappearance of the preceding Dorset Culture substratum in the North American Arctic as the Thule Culture spread resembles the seemingly abrupt disappearance of the Mousterian in Europe. The modern Arctic example also includes debate about whether the earlier Dorset inhabitants of Canada were displaced, exterminated, or genetically absorbed by later Thule people (McGhee, 1978; Dumond, 1987: 145; Krupnik 1993a: 228). If one did not know that both Dorset and Thule people were anatomically modern, one might easily posit some overwhelming intellectual or cultural advantage to Thule people, rather than merely a different hunting strategy.

Selection pressure and Neanderthal morphology

But we do know of morphological differences between Middle and Upper Palaeolithic peoples. Another possible application of the critical fat/protein model is that it may help to explain a change in hominid morphology by shifting selection pressures operating through diet, rather than explaining morphological change exclusively by extinction and replacement. Subtle dietary shifts may have had a cascading effect on hominid biology and social behavior, principally exercised through population growth. Just as population expansion caused by an increase in dietary fat can explain the Dorset/Thule culture transition in the High Arctic, similar dietary shifts may explain the Middle/Upper Palaeolithic transition. Eastern Europe, Central Russia, or the Levant would be logical places to test this suggestion. If the archaeological record were dense enough and dated finely enough in these regions, one might expect to find evidence of denser populations and population expansion in sites occupied by anatomically modern humans. Hominid specimens which show a mixture of anatomically modern and archaic traits (Corruccini, 1992) might represent individuals in populations experiencing the cascading effect of changes initiated by dietary shifts. Foraging strategy and mobility differences in the Levantine Mousterian between anatomically modern humans and Neanderthals have already been noted (Lieberman & Shea, 1994).

Conclusions

That food resources affect group size and population density is not a novel finding in evolutionary ecology, because it continually surfaces in reviews of non-human primate and carnivore sociality (e.g., Wrangham et al., 1993). In addition, the existence of a positive correlation between human population size and social complexity has long been a fixture in comparative studies of human cultural change (Carneiro, 1967).

Traits associated with anatomical and behavioral modernity are loosely linked together, and explained by increase in human population size, which I think is principally generated in Europe and perhaps the Levant by relatively subtle dietary shifts. There are four implications that follow from this:

1. Assuming similar physiology, different human populations may have independently acquired traits of anatomical or behavioral modernity. There need not be a single source area for anatomical or behavioral modernity.

2. The geographic dispersion and permanent settlement of habitats associated with modernity may have been initiated by population growth and accompanying social complexity. This dispersal need not reflect new intellectual or linguistic abilities. This agrees with the observed lag between the advent of anatomical modernity and major changes recorded in the archaeological record.

3. Traits of modernity need not be acquired as a suite. This agrees with the data from palaeontology and archaeology. The mosaic fashion in which traits of anatomical modernity are acquired, and the lag between anatomical and behavioral modernity, are expected if these changes are ultimately caused by long-term, population-level responses to relatively subtle dietary changes, rather than being a quick, radical transformation of anatomy and behavior in a small founding population.

4. In temperate areas, the transition to anatomical and behavioral modernity is more abrupt. The population expansion triggered by dietary changes is more marked, because dietary carbohydrates are seasonally restricted here. Subtle changes in subsistence have a more profound impact on human biology and behavior.

References

AIELLO, L.C. (1993): The fossil evidence for modern human origins in Africa: A revised view. *American Anthropologist* 95, 73-96.

ALTMANN, J., SCHOELLER, D., ALTMANN, S.A., MURUTHI, P. & SAPOLSKY, R.M. (1993): Body size and fatness of free-living baboons reflect food availability and activity levels. *American Journal of Primatology* 30, 149-161.

ALTMANN, S. (1991): Diets of yearling female primates (*Papio cynocephalus*) predict lifetime fitness. *Proceedings of the National Academy of Sciences (U.S.A.)* 88, 420-423.

ARIMA, E. (1994): Caribou and Iglulik Inuit kayaks. *Arctic* 47, 193-195.

BEAUNE, S.A. de (1993): Nonflint stone tools of the Early Upper Paleolithic. In: H. Knecht, A. Pike-Tay & R. White, *Before Lascaux. The complex record of the Early Upper Paleolithic,* 163-191. Boca Raton.

BINFORD, L.R. (1982): The archaeology of place. *Journal of Anthropological Archaeology* 1, 5-31.

BINFORD, L.R. (1984): *Faunal remains from Klasies River Mouth.* New York.

BINFORD, S.R. (1968): Early Upper Pleistocene adaptations in the Levant. *American Anthropologist* 70, 707-717.

BLAXTER, K. (1989): *Energy metabolism in animals and man*. Cambridge.

BOCHERENS, H., FIZET, M., MAROTTI, R., OLIVE, C. BELLON, G. & BILLIOU, D. (1991): Isotopic geochemistry (^{13}C, ^{15}N) of fossil vertebrate collagen: Application to the study of a past food web including Neandertal man. *Journal of Human Evolution* 20, 481-492.

CACHEL, S. (in press a): Dietary shifts and the European Upper Palaeolithic transition. *Current Anthropology* 38.

CACHEL, S. (in press b): Subsistence factors among Arctic peoples and the reconstruction of social organization from evidence of prehistoric human diet. In: P. Rowley-Conwy (ed.), *Animal bones, human societies*. Oxford.

CANN, R.L., STONEKING, M. & WILSON, A.C. (1987): Mitochondrial DNA and human evolution. *Nature* 325, 31-36.

CARNEIRO, R.L. (1967): On the relationship between size of population and complexity of social organization. *Southwestern Journal of Anthropology* 23, 234-243.

CHASE, P.G. (1989): How different was Middle Palaeolithic subsistence? A zooarchaeological perspective on the Middle to Upper Palaeolithic transition. In: P. Mellars & C. Stringer (eds.), *The human revolution*, 321-337. Edinburgh.

CHASE, P.G. & DIBBLE, H.L. (1987): Middle Paleolithic symbolism: A review of current evidence and interpretations. *Journal of Anthropological Archaeology* 6, 263-296.

CHEHAB, F.F. , MOUNZIH, K., LU, R. & LIM, M.E. (1997): Early onset of reproductive function in normal female mice treated with leptin. *Science* 275, 88-90.

CONKEY, M.W. (1993): Humans as materialists and symbolists: Image making in the Upper Paleolithic. In: D.T. Rasmussen (ed.), *The origin and evolution of humans and humanness,* 95-118. Boston.

CORRUCCINI, R.S. (1992): Metrical reconsideration of the Skhul IV and IX and Border Cave 1 crania in the context of modern human origins. *American Journal of Physical Anthropology* 87, 433-445.

COURT, G.S. (1996): The seal's own skin game. *Natural History* 105(8), 36-41.

DOROZYNSKI, A. & ANDERSEN, A. (1991): Collagen: A new probe into prehistoric diet. *Science* 254, 520-521.

DRAPER, H.H. (1977): The aboriginal Eskimo diet in modern perspective. *American Anthropologist* 79, 309-316.

DUMOND, D.E. (1987): *The Eskimos and Aleuts*. New York.

ENLOE, J.G. (1993): Subsistence organization in the Early Upper Paleolithic: Reindeer hunters of the Abri du Flageolet, Couche V. In: H. Knecht, A. Pike-Tay & R. White (eds.), *Before Lascaux. The complex record of the Early Upper Paleolithic*, 101-115. Boca Raton.

FRAYER, D.W. (1984): Biological and cultural change in the European Late Pleistocene and Early Holocene. In: F.H. Smith & F. Spencer (eds.), *The origins of modern humans*, 211-250. New York.

FRAYER, D.W., WOLPOFF, M.H., THORN, A.G., SMITH, F.H. & POPE, G.G. (1993.): Theories of modern human origins: The paleontological test. *American Anthropologist* 95, 14-50.

FREEMAN, L.G. (1973): The significance of mammalian faunas from Paleolithic occupations in Cantabrian Spain. *American Antiquity* 38, 3-44.

FRISANCHO, A.R. (1993): *Human adaptation and accommodation*. Ann Arbor.

GOLDMAN, N. & BARTON, N.H. (1992): Genetics and geography. *Nature* 357, 440-441.

GOWLETT, J.A.J. (1987): The coming of modern man. *Antiquity* 61, 210-219.

HARRISON, G.A., TANNER, J.M., PILBEAM, D.R. & BAKER, P.T. (1988): *Human biology*. Oxford. 3rd ed.

HOWELLS, W.W. (1993): *Getting here. The story of human evolution*. Washington.

KLEIN, R.G. (1973): *Ice-Age hunters of the Ukraine*. Chicago.

KLEIN, R.G. (1989): *The human career. Human biological and cultural origins*. Chicago.

KLEPINGER, L.L. (1992): Innovative approaches to the study of past human health and subsistence strategies. In: S.R. Saunders & M.A. Katzenberg (eds.), *Skeletal biology of past peoples: Research methods,* 121-130. New York.

KRUPNIK, I. (1993a): *Arctic adaptations. Native whalers and reindeer herders of northern Eurasia*. Hanover (New England).

KRUPNIK, I. (1993b): Prehistoric Eskimo whaling in the Arctic: Slaughter of calves or fortuitous ecology? *Arctic Anthropology* 30, 1-12.

LIEBERMAN, D.E. & SHEA, J.J. (1994): Behavioral differences between archaic and modern humans in the Levantine Mousterian. *American Anthropologist* 96, 300-332.

MADDISON, D.R. (1991): African origin of human mitochondrial DNA reexamined. *Systematic Zoology* 40, 355-363.

MADDISON, D.R., RUVOLO, M. & SWOFFORD, D.L. (1992): Geographic origins of human mitochondrial DNA: Phylogenetic evidence from control region sequences. *Systematic Biology* 41, 111-124.

MARSHACK, A. (1972a): *The roots of civilization*. New York.

MARSHACK, A. (1972b): Cognitive aspects of Upper Paleolithic engraving. *Current Anthropology* 13, 445-477.

MARSHACK, A. (1985): *Hierarchical evolution of the human capacity: The Paleolithic evidence.* New York (Fifty-Fourth Annual James Arthur Lecture on the Evolution of the Human Brain. American Museum of Natural History).

MARSHACK, A. (1989): Evolution of the human capacity. The symbolic evidence. *Yearbook of Physical Anthropology* 32, 1-34.

MARTIN, L.D. & MARTIN, J.B. (1993): Why are there bone-crushing carnivores. *Journal of Vertebrate Paleontology* 13, 48-49 (Abstracts).

MAXWELL, M.S. (1985): *Prehistory of the eastern Arctic.* Orlando.

McGHEE, R. (1978): *Canadian Arctic prehistory.* Scarborough, Ontario.

MELLARS, P. (1989): Major issues in the emergence of modern humans. *Current Anthropology* 30, 349-385.

MELLARS, P. (1996): *The Neanderthal legacy: An archaeological perspective from Western Europe.* Princeton.

MELLARS, P. & STRINGER, C. (eds.) (1989): *The human revolution.* Princeton.

NEWSHOLME, E.A. (1978): Substrate cycles: their metabolic, energetic and thermic consequences in man. *Biochemical Society Symposia* 43, 183-205.

PEREIRA, M.E. & POND, C.M. (1995): Organization of white adipose tissue in Lemuridae. *American Journal of Primatology* 35, 1-13.

POND, C.M. (1978): Morphological aspects and the ecological and mechanical consequences of fat deposition in wild vertebrates. *Annual Review of Ecology and Systematics* 9, 519-570.

POND, C.M. (1989): Bearing up in the Arctic. *New Scientist* 121, 40-46.

RUFF, C.B., TRINKAUS, E., WALKER, A. & LARSEN, C,S, (1993): Postcranial robusticity in *Homo*. I. Temporal trends and mechanical interpretation. *American Journal of Physical Anthropology* 91, 21-53.

SAVELLE, J.M. & McCARTNEY, A.P. (1991): Thule Eskimo subsistence and bowhead whale procurement. In: M.C. Stiner (ed.), *Human predators and prey mortality*, 201-216. Boulder.

SHEA, J.J.(1989): A functional study of the lithic industries associated with hominid fossils in the Kebara and Qafzeh Caves, Israel. In: P. Mellars & C. Stringer (eds.), *The human revolution*, 611-625. Edinburgh.

SOFFER, O. (1985): *The Upper Paleolithic of the Central Russian plain.* San Diego.

SOLLAS, W.J. (1924): *Ancient hunters and their modern representatives.* London. 3rd edition.

SMITH, F.H. & SPENCER, F. (eds.) (1984): *The origins of modern humans.* New York.

SPETH, J.D. (1991): Protein selection and avoidance strategies of contemporary and ancestral foragers: Unresolved issues. *Philosophical Transactions of the Royal Society of London* B 334, 265-270.

SPETH, J.D. & SPIELMANN, K.A. (1983): Energy source, protein metabolism, and hunter-gatherer subsistence strategies. *Journal of Anthropological Archaeology* 2, 1-31.

STIRLING, I. & GURAVICH, D. (1988): *Polar bears.* Ann Arbor.

STRAUS, L.G. (1985): Stone Age prehistory of Northern Spain. *Science* 230, 501-507.

STRAUS, L.G. (1987): Upper Paleolithic ibex hunting in Southwest Europe. *Journal of Archaeological Science* 14, 163-178.

STRAUS, L.G. (1992): *Iberia before the Iberians. The Stone Age prehistory of Cantabrian Spain.* Albuquerque.

STRINGER, C.B. (1993): New views on modern human origins. In: D.T. Rasmussen (ed.), *The origin and evolution of humans and humanness*, 75-94. Boston.

STRINGER, C.B. & ANDREWS, P. (1988): Genetic and fossil evidence for the origin of modern humans. *Science* 239, 1263-1268.

STRINGER, C.B. & GAMBLE, C. (1993): *In search of the Neanderthals. Solving the puzzle of human origins.* New York.

TEMPLETON, A.R. (1993): The „Eve" hypothesis: A genetic critique and reanalysis. *American Anthropologist* 95, 51-72.

TRINKAUS, E. (1986): The Neandertals and modern human origins. *Annual Review of Anthropology* 15, 193-218.

TRINKAUS, E. (ed.) (1989): *The emergence of modern humans. Biocultural adaptations in the later Pleistocene.* Cambridge.

TRINKAUS, E. (1995): Neanderthal mortality patterns. *Journal of Archaeological Science* 22, 121-142.

TRINKAUS, E. & SHIPMAN, P. (1993): *The Neanderthals. Changing the image of mankind.* New York.

WHITE, R. (1982): Rethinking the Middle/Upper Paleolithic transition. *Current Anthropology* 23, 169-192.

WHITEHEAD, R.G. (1992): Kwashiorkor in Uganda. In: E.M. Widdowson & J.C.Mathers (eds.), *The contribution of nutrition to human and animal health*, 303-313. Cambridge.

WRANGHAM, R.W., GITTLEMAN, J.L. & CHAPMAN, C.A. (1993): Constraints on group size in primates and carnivores: Population density and day-range as assays of exploitation competition. *Behavioral Ecology and Sociobiology* 32, 199-209.

Ideology-dependent Subsistence in the Aceramic of Cyprus

Avraham Ronen

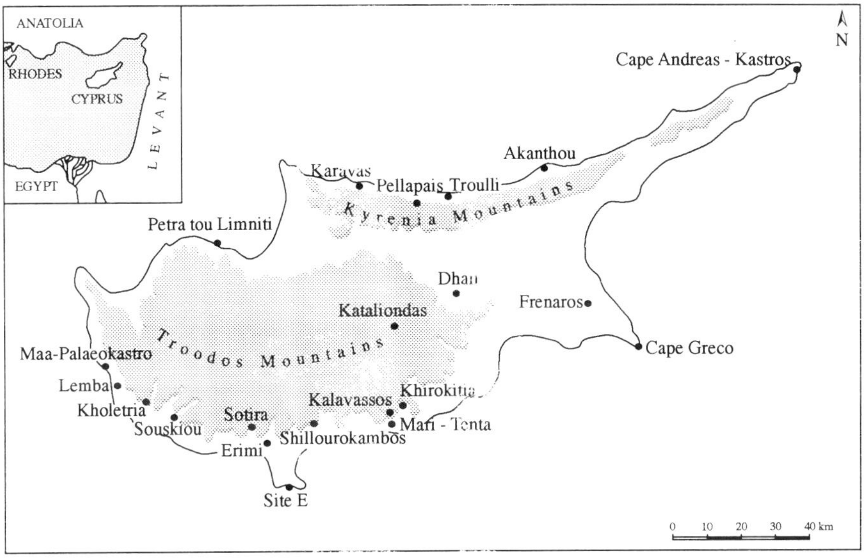

Figure 1 -Major Neolithic sites in Cyprus.

Abstract

The aceramic Neolithic culture of Cyprus appeared on the island around 7,500 years BC. For decades the aceramic culture puzzles researchers by its highly original traits which affect all aspects of life: the house, artifacts, subsistence and art. These affinities were treated exclusively in conjunction with the question of place of origin of the settlers. The possible meaning of these traits, and the behavioural pattern reflected by them, were hardly ever discussed. This paper will focus on the aceramic food animals. Being an oceanic island never connected to the mainland, all edible plants and animals had to be transported from the mainland. Since climatically and environmentally Cyprus is similar to the mainland; theoretically, then, all the food animals of the mainland could have been transported. Hence, the species which were actually chosen for consumption inevitably reflect a deliberate choice. That choice consisted of the wild Dama dama mesopotamica as the major resource, forming 50 % of the faunal remains, completed by domesticated porc (20-30 %) and caprovines (also 20-30 %). Other food animals of the time, like gazelle, ass, cattle or hare, were not included in the aceramic diet. This dietary composition is an extreme deviation from Levantine norms. First, because once animal domestication has been adopted, wild game became an additional resource, never the main resource as in Cyprus. Second, the absence of cattle and of any artistic representation of cattle is specially significant considering the central role of Bos in Levantine Neolithic ritual. It is therefore argued here that the absence of Bos in the aceramic reflects a cultural induced negative attitude toward this animal. Likewise, other odd-looking aceramic behaviours such as preference for deer, circular houses, peculiar human figurines, unnecessary defence walls and the absence of hunting/fighting gear, all became meaningful and neatly woven into a coherent aceramic ideology.

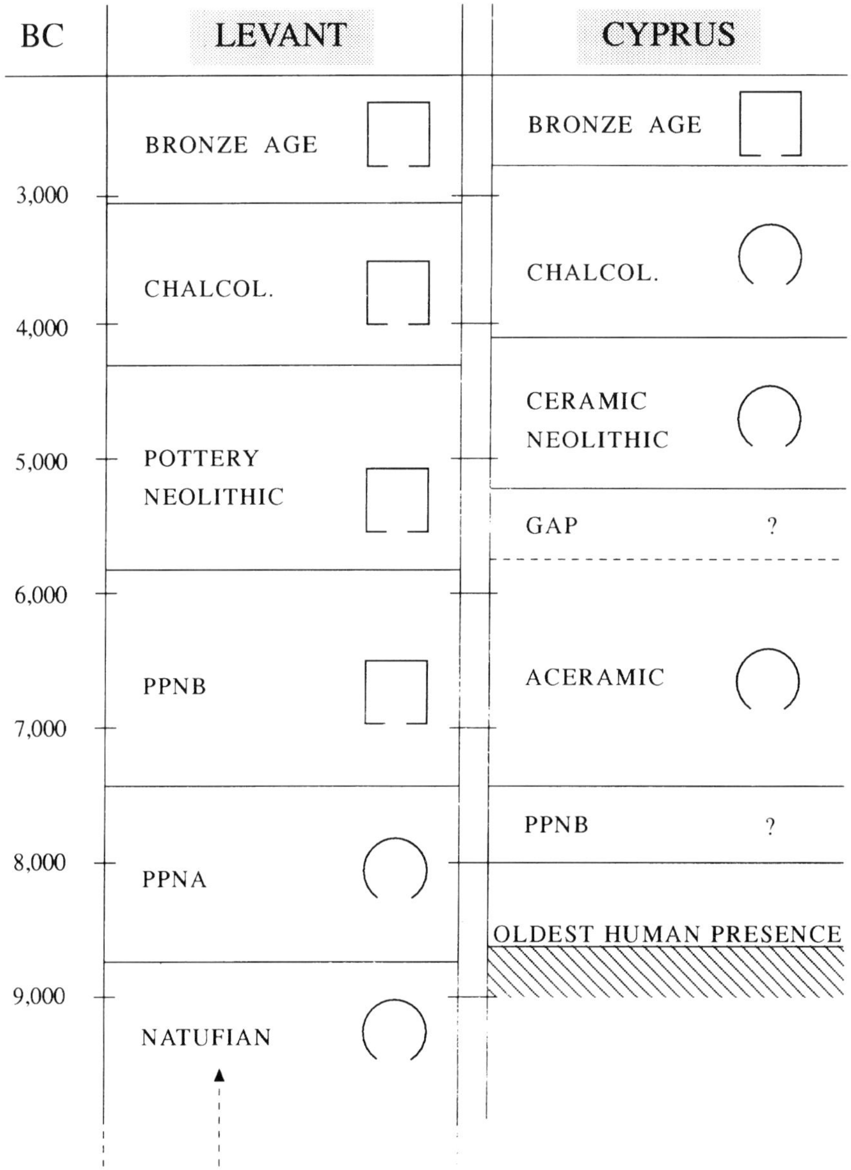

Figure 2 - The dominant house form in the Levant and Cyprus.

The Aceramic Neolithic culture of Cyprus which began ca. 7,500 BC and lasted for some two millennia (Figure 1) had puzzled researchers for decades by its originality, which affected virtually all aspects of life. When examined in a Near Eastern context, all these original traits seem to be antithesis to contemporary mainland norms (Ronen, 1995):

- In Cyprus there are exclusively round houses, long after the round house was almost completely replaced by rectangular structures on the mainland (Figure 2) save for desert areas, where the round house persisted into the third millennium BC (e.g. Bet Arieh, 1986). Some Cypriot circular houses have, in addition, very thick walls relative to their diameter (Le Brun, 1984) (Figure 3).

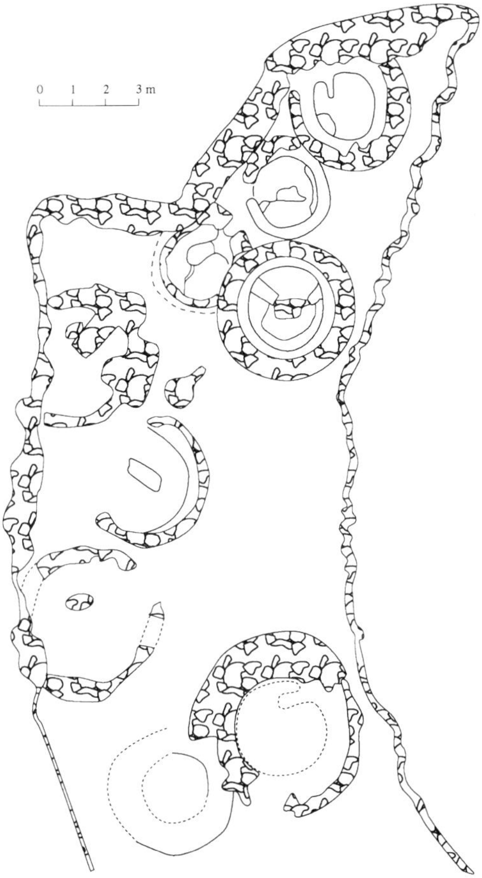

Figure 3 - Schematic rendering of house complexes in Level II of Khirokitia. Mudbricks are indicated by dots. Note the thickness of some structures (after Le Brun, 1984).

- Aceramic stone tools are of simple manufacture and lack the sophisticated knapping techniques of the mainland, e.g. elongated blades or pressure flaking (Le Brun, 1989). Also, Aceramic toolkit is devoid of arms or hunting gear, whereas arrowheads constitute the most typical element of the mainland Neolithic. The Aceramic knapper apparently preferred poor quality raw material (Guilaine et al., 1995a,b; Guilaine, pers. comm.).
- The female sex is not indicated on Aceramic anthropomorphic figurines (Le Brun, 1994), in contrast with the emphasized or exaggerated female sex of the „Mother Goddess" on contemporary mainland (Cauvin, 1994).
- Some Aceramic Cypriot settlements are surrounded by massive walls (Le Brun, 1992; Todd, 1987), a phenomenon unknown on the mainland except for the single, and far older, case of Jericho (Kenyon, 1981).

Hence, the first explanation ever offered of Aceramic behaviour as a meaningful whole concluded that it represents a sectarian ideology (Ronen, 1995). In the present paper I will attempt to show that the curious subsistence pattern of the Aceramic Cypriots fits well into the model of ideological opposition.

Cyprus is an oceanic island which has never been connected to the mainland (Sondaar, 1977: 677). The pre-colonization, Pleistocene fauna of Cyprus consisted of dwarf elephants, pygmy hippopotamus, two murid mice and a soricid shrew (Davis, in Le Brun, 1984: 148). The elephant and hippo became extinct at their earliest encounter with humans in the 10th millennium, preceding the Aceramic (Simmons, 1988, 1991).

None of the edible resources of the mainland, vegetal or animal, had existed in Cyprus, and hence all had to be transported by the settlers. Climatically and environmentally Cyprus is similar to the mainland (Le Houeron, 1992) (Figure 4); theoretically, then, all the food items consumed on the mainland could have been transported to Cyprus. This was indeed the case with vegetal foods (Todd, 1989: 8; Le Brun 1996), but not with animals. Of all the food animals consumed on the mainland around the time of crossing, only four species were allowed into the Aceramic diet: a wild deer and the three domesticates goat, sheep and pig. As none of these animals is likely to have crossed by swimming (Davis 1989), all must have been transported to the island. This four-species choice constitutes a unique composition in the Eastern Mediterranean indicating, in all likelihood, some self-imposed restrictions. The Aceramic choice is outstanding on two grounds: first, by the presence of deer and the absence of gazelle and cattle (Figure 5); secondly, by the considerable reliance on a wild animal - the deer - while posessing domesticated animals.

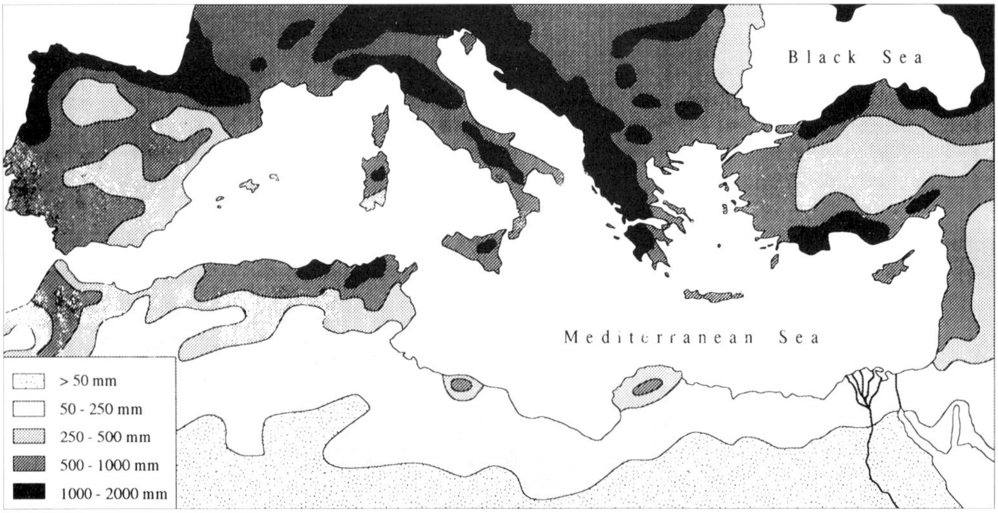

Figure 4 - Circum-Mediterranean precipitation (after Le Houeron, 1992).

The deer

The very choice of the Levantine fallow deer, *Dama dama mesopotamica,* contradicts the common, gazelle-dependent food habits on contemporaneous mainland. Two questions arise: Why was the deer preferred over gazelle, and why was deer so highly valued in the Aceramic meat intake?

During the latest part of the Pleistocene and until domestication, gazelle very largely dominated the food animals in the Levant with deer trending far behind (Davis, 1982) (Figure 6). Still present in the faunal assemblages of the Kebaran and early Natufian, deer has almost disappeared from Late Natufian onward. In the PPNA occupation of Hatula, for example, a single bone of a fallow deer was found among thousands of remains dominated by gazelle (Davis, 1985). Around the onset of the Cypriot Aceramic culture in the 8th millenium BC, bone assemblages in the Mediterranean climatic zone of the Levant contained no more than a few percent of fallow deer while none is found in the desert area (see Figure 5). Yet deer, not gazelle was chosen by the Aceramic. An economical advantage is not clear: both deer and gazelle can adapt, when required, to browsing or grazing (Hemmer, 1990: 169; Baharav, 1974) and both can not be domesticated (Davis, 1983). Deer breeds once a year (Hemmer, 1990: 169), gazelle breeds once or twice a year depending on the availability of water (Baharav, pers. comm.). The preference for deer can not be easily accounted for by economical or environmental considerations.

Not only was deer preferred, but it also had acquired a very important place in the Aceramic meat intake, ranging (in terms of bone counts) between 50 % and 20 % (Todd, 1989: 8; Le Brun et al., 1987: 301; Croft, 1991; Simmons, 1994). This situation is unusual because normally, when the first food animals goat and sheep became domesticated, the domasticates formed the principal meat resource and the importance of wild game abruptly declined. In Jericho, for example (Clutton-Brock, 1979), a ca. 60 % dominance of gazelle in the pre-domestication phase changed after domestication to a 50 % dominance of caprines (Figure 7). Yet in spite of having domesticated pig and caprovines in the Aceramic (Davis, 1989; Croft, 1991: 73), the wild fallow deer continued to form the dominant, or at least a considerable, part of the meat resource.

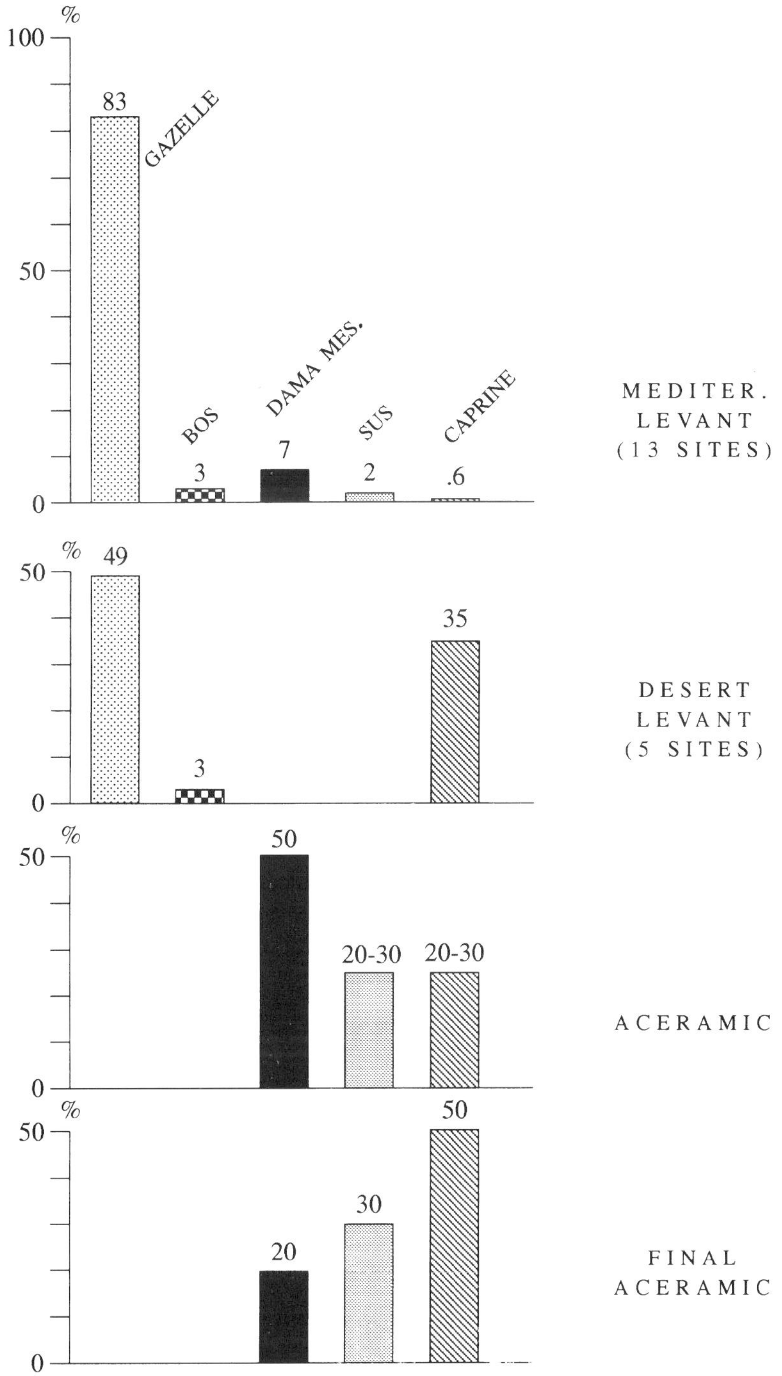

Figure 5 - *Relative frequency of early Holocene food animals in the Mediterranean zone of the Levant (weighted average of 13 sites), the desert zone (weighted average of 5 sites) and in Aceramic Cyprus. (Levantine data after Banning & Byrd, 1989: 176. Cypriot data from Croft, 1991 and Simmons, 1994).*

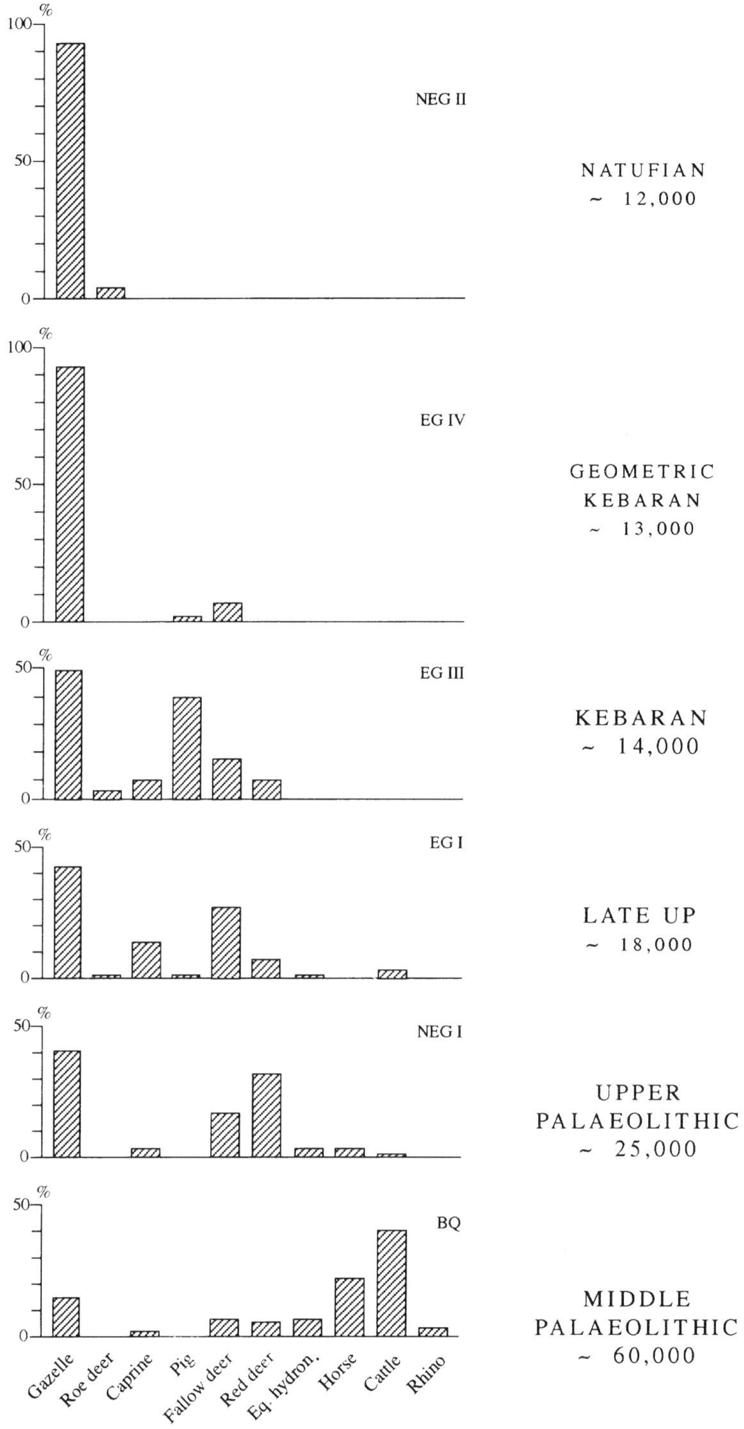

Figure 6 - *Composition of food animals through time in the Upper Pleistocene of Israel (from Davis, 1982).*

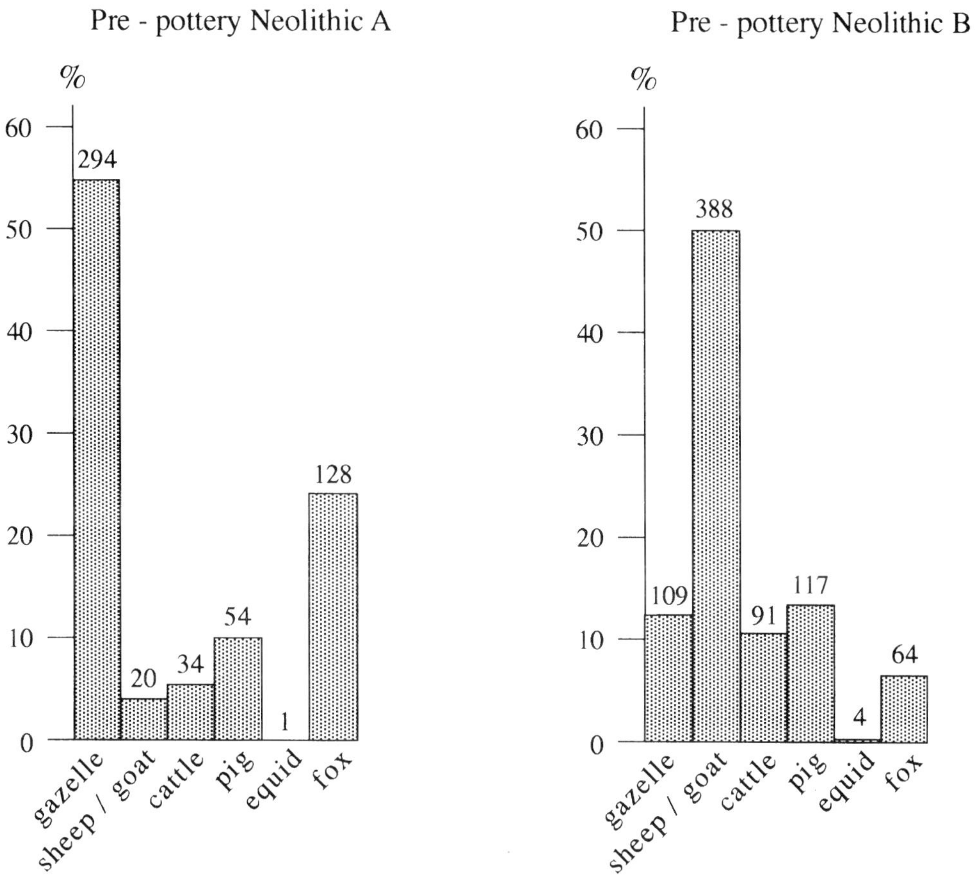

Figure 7 - Jericho food animals before and after domestication (from Clutton-Brock, 1979).

The importance of deer in Cypriot economy far exceeded the chronological boundaries of the Aceramic. It lasted more than four millennia, from ca. 7,500 to 3,000 BC, through the Aceramic Neolithic, Ceramic Neolithic and Chalcolithic phases. This persistent trait was termed by Croft (1991: 63) a"deer-oriented economy", and it raises the question how was this low-breeding animal so successfully exploited for millennia without becoming extinct. This could be atteined, in all likelihood, by some special man-deer relations, possibly a sort of „controlled management". Some management may, indeed, be indicated in layer B of Khirokitia, where many more females than male deer were culled (Davis, 1994). The assumed management could not, however, reach the stage of domestication because, as far as it is known, deer has never been domesticated. Furthermore, the ratio of adult male bones generally found among Aceramic deer, 37 % - 53 %, conforms to random culling (Croft, 1991: 75). Thus, the special man-deer relations emerging for Aceramic Cyprus allowed a controlled exploitation of the deer without atteining the stage of domestication.

A question all too often ignored is how the deer was captured, given the absence of hunting equipment in Aceramic toolkit (Le Brun et al., 1987: 289). The possible use of wooden points or bone points in the place of arrowheads, raised by some researchers, has no support in the Aceramic material culture. In my opinion, deer was caught as part of the „management" mentioned above.

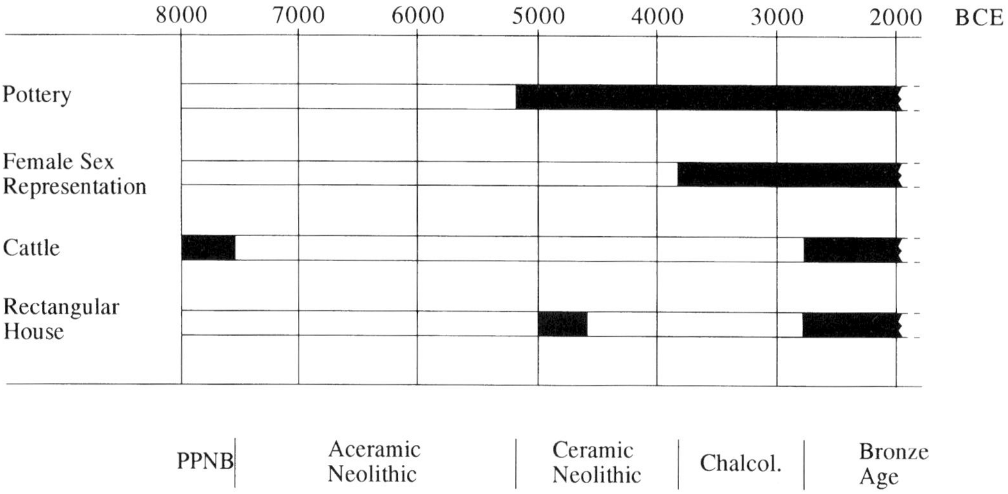

Figure 8 - Principal traits of the Cypriot sect and their duration (modified from Ronen, 1995).

I have suggested earlier (Ronen, 1995) that taming and hand-feeding might have been the key element in the Cypriot man-deer relationship, similarly to the free-roaming yet hand-fed pigs in New Guinea (Sillitoe, 1979 and pers. comm.). Thus while Aceramic people admittedly hunted a wild species, in reality it may have been a somewhat ceremonial hunt. And, like the case of the New Guinea pigs (Sillitoe, 1979), the meat of the „hunted" Cypriot deer might have had a significance beyond simple food. There has been in Cyprus not just a „deer-oriented economy", but a „deer-oriented ideology".

It is in this venue that we have to understand the „odd" preference for deer over the far more common gazelle in the land of origin. Deer is easily tameable (Hemmer, 1990: 169) while gazelle can not be tamed (Dayan, pers. comm.). Gazelle could not be caught without hunting gear, and it simply did not fit the ideologically-determined Aceramic strategy of meat procurement which combined hunting and no violence.

Bos

Cattle are completely absent in the Aceramic of Cyprus, contrary to the mainland (see Figs. 5 and 7) or to other Mediterranean islands, for example Crete (Evans, 1971). Formerly it had seemed as if cattle were not brought to Cyprus before the Bronze Age (Ronen, 1995). That picture has changed by the recent discoveries at Shillourokambos, where a layer of PPNB came to light for the first time in Cyprus (ca. 8,200-7,500 BC), overlain by a classical Aceramic habitation starting around 7,500 BC (Guilaine et al., 1995a,b). The PPNB horizon contains ca. 10 % of cattle bones as against virtually none in the overlying Aceramic (Guilaine, in letteris 12.7.97) [1]. Cattle were brought to the island, then, and survived there for a half millenium prior to the onset of the Aceramic. Therefore, their exclusion from Aceramic diet was most probably voluntary.

Cattle were not present, nor of course consumed, in Cyprus for a remarkably long period of over four millennia, throughout the Aceramic, Ceramic Neolithic and the Chalcolithic (Croft, 1991), parallel to the above mentioned „deer-dependent ideology" (Figure 8). Cattle were finally re-introduced to the island in the Bronze Age (Swiny, 1989).

Not only was the presence of cattle banned in the Aceramic, but any depiction of it in either sculpture or painting was banned as well. The absence of cattle imagery is especially significant

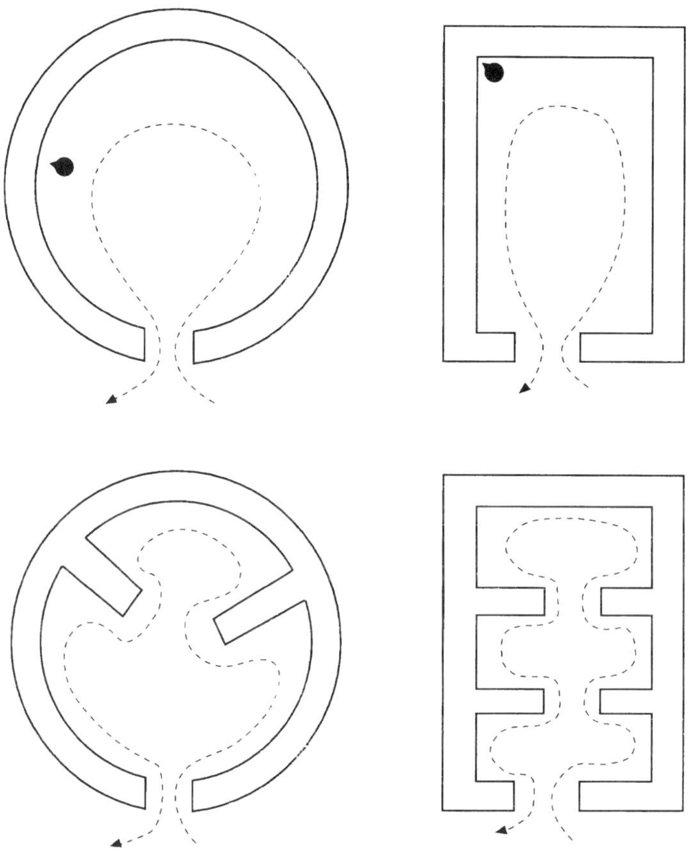

Figure 9 - *Circular and rectangular houses (upper row) have a similar 'circulation pattern' (Banning & Byrd, 1989) but only the latter offers one's own corner (Ronen, 1995). The interior of the Aceramic circular house (lower row) is similar to that of the multi-chamber rectangular house.*

considering the widespread cult of bull in contemporary Levantine and Anatolian Neolithic ritual. From ca. 8,000 BC cattle have acquired an outstanding symbolic and cultic significance in the Levant and in Anatolia (Cauvin, 1994): in Hatula (Lechevallier & Ronen, 1994: 326); in Mureybet (Cauvin, 1977); in Hallam Cemi Tepesi (Rosenberg & Davis, 1992); and at Kefar HaHoresh (Goring-Morris et al., 1995). Viewed in a regional context, the most significant fact is that a negative attitude toward the bull is reflected by the Aceramic precisely when the creature was so widely worshipped on the mainland. The absence of *Bos* in the Aceramic must, therefore, be considered a cultural-ideological phenomenon.

Discussion

It is clear by now that contrary to former claims (e.g. Held, 1989), environmental conditions can not be held responsible for Aceramic characteristics. Furthermore, it is difficult to concieve of an environment that could induce humans to renounce their weaponry. I suggest that along with the absence of weapons, the entire range of Aceramic behaviour was culturally determined as well - the house form, the encircling walls and the subsistence pattern. Aceramic Cypriots resided in an old-fashion circular house and maintained the old-fashion life of hunters.

Yet, Aceramic Cypriots were not Natufians. Most of their houses had inner partitions which made them similar to the multi-room rectangular house (Figure 9). The wild animal they hunted seems to have been tamed, so the hunt, too, has to be viewed as ceremonial and symbolical. A symbolic behaviour emerges as the strongest trait of the Aceramic. Only in this way can we understand, I surmis, the mighty walls sorrounding some Aceramic settlements. With the absence of arrowheads and any other fighting gear, these walls could not have functioned for defence in the normal sense of the word. Furthermore, Le Brun (1992) has convincingly argued that the domestic animals could not have been kept inside the walled settlement of Khirokitia because the only entrance through the wall had steps and was only 80 cm wide, clearly unfit for the passage of herds. With these valued assets outside, what did the wall protect inside? In addition, there is no sign for war, fire, destruction or any act of violence in Cyprus throughout the Aceramic period (Held, 1990). All considered, it does not seem that the Aceramic walls had fulfilled any function which we commonly associate with defence. Rather, the function of these walls has to be sought in the realm of ideology, as suggested earlier (Ronen, 1995: 195). The most straightforward explanation would be that these walls demarkated the boundary between a dangerous 'outside' and a safe 'inside' (Rapoport, 1979: 10; Hodder, 1987). The thick walls of some residential units, mentioned above, may have had a similar task on the individual, or family, level. The walls of Jericho, by the way, have to be explained in a similar way (Ronen, 1995: 195 contra Bar-Yosef, 1986).

Conclusions

We may now better define the 'enigmatic' behaviour of the Aceramic. Its most outstanding character is an ideological and symbolic world which systematically opposed contemporary mainland norms. We observe here a separatist attitude, an opposition. Considered in its entirety, the Aceramic ideological system resembles sectarian behaviour, the main characters of which are simplicity, conservatism and isolation (Hostetler, 1963).

Sects radically condemn the existing social order (Talmon, 1962: 145). They develop their own ideology (Hostetler, 1963: 35) and, seeking away from the influence of 'others', they isolate themselves using mental as well as physical boundaries. The insularity of Cyprus gave Aceramic people the ultimate isolation, making them the ideal folk society „ having communication with no outsider" (Redfield, 1947: 296). We certainly owe it to the laboratory-like conditions offered by this perfect isolation that the sect is so clearly discernible through its archaeological remains.

Sectarian cleavages occur when the socio-political system collapses (Talmon, 1962: 138). Regardless if the Aceramic sect had originated on the mainland or in Cyprus, its inception should be sought within the immense environmental and social changes which accompanied the agricultural revolution. There was the endless and terrifying post-Pleistocene rise of sea level, continuously submerging land and settlements. In the socio-economic sphere territory became valued, communities grew bigger, exchange intensified and economic elites rose to power. Religious practice developed and gave rise to religious bureaucracies. Conflict became more frequent than before (Watkins, 1992). It may now be suggested that some groups may have resisted these developments. They seem to have opposed the new situation by rejecting its main features and by thriving back to origins (Chery, 1959: 386), as did the Aceramic Cypriots.

Note

[1] We are very grateful to Professor Jean Guilaine for allowing to use unpublished information from his excavation. Shillourokambos fauna is studied by J.-D. Vigne and I. Carrere, whom we also thank.

References

BAHAREV, D. (1974): Notes on the population structure and biomass of the Mountain Gazelle, *Gazella gazella gazella. Israel Journal of Zoology* 23, 39-44.

BANNING, E.B. & BYRD, B.F. (1989): Alternative approach for exploring Levantine Neolithic architecture. *Paléorient* 15, 154-160.

BAR-YOSEF, O. (1986): The walls of Jericho: An alternative interpretation. *Current Anthropology* 27, 157-162.

BEIT ARIEH (1986): Two cultures in southern Sinai in the third millennium B.C. *Bulletin of the American Schools of Oriental Research* 263, 27-54.

CAUVIN, J. (1977): Le Moyen Euphrate au VIII millenaire d'apres Mureybet et Sheikh Hassan. In: J.C. Margueron (ed.), *Le Moyen Euphrate, zone de contacts et d'echanges. Colloque 1977*, 21-34. Strasbourg.

CAUVIN, J. (1994): *Naissance des divinites, Naissance de l'agriculture.* Paris.

CHERY, H.-C. (1959): *L'offensive des sectes.* Paris.

CLUTTON-BROCK, J., 1979 The mammalian remains of the Jericho Tell. *Proceedings of the Prehistoric Society* 45, 135-158.

CROFT, P. (1991): Man and beast in Chalcolithic Cyprus. *Bulletin of the American Schools of Oriental Research* 282/283, 63-79.

DAVIS, S.J.M. (1982): Climatic change and the advent of domestication: The succession of ruminant artiodactyls in the late Pleistocene-Holocene in the Israel region. *Paléorient* 8, 5-15.

DAVIS, S.J.M. (1983): The age profiles of gazelles predated by ancient man in Israel: Possible evidence for a shift from seasonality to sedentism in the Natufian. *Paléorient* 9, 55-62.

DAVIS, S.J.M. (1985): The faunal remains. In: M. Lechevallier & A. Ronen (eds), *Le site Natoufien-Khiamien de Hatoula pres de Latroun, Israel.* Jerusalem.

DAVIS, S.J.M. (1989): Some more animal remains from the aceramic Neolithic of Cyprus. In: A. Le Brun (ed.), *Fouilles recentes a Khirokitia (Chypre) 1983-1986*, 189-221. Paris.

DAVIS, S.J.M. (1994): Even more bones from Khirokitia. In: A. Le Brun (ed.), *Fouilles recentes a Khirokitia (Chypre),1988-1991*, 305-333. Paris.

EVANS, J.D. (1971): Neolithic Knossos. The growth of a settlement. *Proceedings of the Prehistoric Society* 37, 95-117.

GORING-MORRIS, A.N., GOREN, Y., HOROWITZ, L.K., HERSHKOVITZ, I., LIEBERMAN, R., SARAL, J. & BAR-YOSE, D. (1995): The 1992 season of excavations at the pre-pottery Neolithic B settlement of Kfar Hahoresh. *Mitekufat Haeven* 26, 74-121.

GUILAINE, J., COULAROU, J., BRIOIS, F. & CARRERE, I. (1995a): Le site néolithique de Shillourokambos (Chypre). *Bulletin Correspondance Hellenique* 119, 737-741.

GUILAINE, J., COULAROU, J., BRIOIS, F. & CARRERE, I. (1995b): L'etablissement neéolithique de Shillourokambos (Parekklisha, Cyprus). Premiers Resultats. *Report of the Department of Antiquities 1995, Cyprus,* 11-32.

HELD, S.O. (1989): Colonization cycles on Cyprus 1: The biogeographic and paleontological foundations of early prehistoric settlemenmt. *Report of the Department of Antiquities 1989, Cyprus*, 7-28.

HEMMER, H. (1990): *Domestication, the decline of environmental appreciation.* Cambridge.

HODDER, I. (1987): Contextual archaeology: an interpretation of Catal Huyuk and a discussion of of the origins of agriculture. *Bulletin of the Institute of Archaeology* 24, 43-56.

HOSTETLER, J.A. (1963): *Amish society.* Baltimore.

KENYON, K.M. (1981): *Excavations at Jericho, III.* Jerusalem (Bristish School of Archaeology in Jerusalem).

Le BRUN, A. (ed.) (1984): *Fouilles recentes a Khirokitia (Chypre), 1977-1981.* Paris.

Le BRUN, A. (1989): Le néolithique de Chypre et sa relation avec le PPNB du Levant. *Paléorient* 15, 161-167.

Le BRUN, A. (1992): Monter, descendre: les acces au village néolithique préceramaique de Khirokitia (Chypre). In: *Studies in Honour of Vassos Karageorghis,* 21-26. Nicosia.

Le BRUN, A. (1994): Les figurines anthropomorphes du néolithique préceramique Chipriote. In: F. Vandenabeele & R. Laffineur (eds.), *Cypriote stone sculpture. Proceedings of the Second International Conference on Cypriote Studies,* 15-21. Brussels, Liège.

Le BRUN, A. (1996): L'économie de Chypre au néolithique. In: *The development of the Cypriot economy from the prehistoric period to the present day,* 1-16. Nicosia.

Le BRUN, A., CLUZAN, S., DAVIS, S.J.M., HANSEN, J. & RENAULT-MISKOVSKI, J. (1987): Le néolithique préceramique de Chipre. *L'Anthropologie* 91, 283-316.

LECHEVALLIER, M. & RONEN, A. (eds) (1994): *Le gisement de Hatoula en Judee occidentale, Israel.* Paris (Mémoires et Traveaux Centre Recherche Francais de Jerusalem No. 8. Association Paleorient).

Le HOUERON, H.N. (1992): Vegetation and land-use in the Mediterranean Basin by the year 2050: A prospective study. In: L. Jeftic, J.D. Milliman & G. Sestini (eds), *Climatic change and the Mediterranean,* 175-232. London.

RAPOPORT, A. (1979): Cultural origins of architecture. In: J.C. Snyder & A.J. Catanese (eds.), *Introduction to Architecture,* 2-20. New York.

REDFIELD, R. (1947): The folk society. *American Journal of Sociology* 52, 293-308.

RONEN, A. (1995): Core, periphery and ideology in Aceramic Cyprus. *Quartär* 45/46, 177-206.

ROSENBERG, M. & DAVIS, M.K. (1992): Hallam Cemi Tepesi, an early Aceramic Neolithic site in eastern Anatolia. *Anatolica* 18, 1-18.

SILLITOE, P. (1979): *Give and take. Exchange in Wola society*. New York.

SIMMONS, A. H. (1988): Akrotiri-Aetoktemnos (Site E). An early prehistoric occupation in Cyprus. *Report of the Department of Antiquities1988, Cyprus*, 15-23.

SIMMONS, A.H. (1991): Humans, island colonization and Pleistocene extinctions im the Mediterranean: The view from Akrotiri-Aetokremnos, Cyprus. *Antiquity* 65, 857-869.

SIMMONS, A.H. (1994): Early Neolithic settlement in western Cyprus: Preliminary report on the 1992-1993 test excavations at Kholetria-Ortos. *Bulletin of the American Schools of Oriental Research* 295, 1-14.

SONDAAR, P.Y. (1977): Insularity and its effect on mammal evolution. In: M.N. Hecht, P.C. Goody & B.M.Hecht (eds.), *Major patterns in vertebrate evolution,* 671-707. New York.

SWINY, S. (1989): From round house to duplex: A re-assessment of prehistoric Cypriot Bronze Age society. In: E. Peltenburg (ed), *Early society in Cyprus,* 14-31. Edinburgh.

TALMON, Y. (1962): Pursuit of the millenium: The relation between religion and social change. *Archives Européenne de Sociologie* 3, 125-148.

TODD, I.A. (1987): Vasilikos Vallay project 6: Excavations at Kalavasos-Tenda I. *Studies in Mediterranean Archaeology* 71/6.

TODD, I.A. (1989): Early prehistoric society: A view from the Vasilikos valley. In: E. Peltenburg (ed), *Early society iin Cyprus,* 2-13. Edinburgh.

WATKINS, T. (1992): The beginning of the Neolithic: Searching for meaning in material culture change. *Paléorient* 18, 63-75.

Planning Capacity and Symbolism as Survival Strategies

Fiorenzo Facchini

Abstract

Both planning and symbolism are essential and typical aspects of culture. They have had and have a great importance in man's relationships with his environment. They assume an essential not marginal adaptive meaning as regards survival even if they are not limited to an adaptive dimension.

To these basic attitudes corresponded in the various phases of mankind very diversified types of behaviour due to different strategies (tools, organization of the territory, social needs, etc.). These strategies have been possible since man was able to plan and enrich his behaviour with symbolic contents by realising symbolic systems of communication that promoted his social life and enabled him to face the impact with the external environment better.

There are different manifestations of symbolism: functional, social and spiritual.

In the prehistory the same knowledge we have at our disposal for the present man is not available. But different elements (cerebral development, technology, economy, social life, etc.) suggest that prehistoric man - starting from Homo habilis *phase - had a set of symbolic system at his disposal. The same can be said for the typical way of symbolic communication represented by language. If we use a global approach in considering morphological aspects (development of the cerebral areas of the language, enlargement of the upper laryngeal space) and the behavioural aspects (technologies and their transmission, social organization) it is legitimate to think that symbolic language has existed since the most ancient age of the humanity.*

We can state: 1) Planning capacity characterized man's technique expressed in the tools making from the Homo habilis *phase; 2) Symbolism is not something added at certain point of man's morphological development in the course of the evolution. On the contrary it is an essential expression of human psychism that always accompanied man whatever its expression may have been. 3) Planning and ability to symbolism represented real survival strategies: They were essential for the population spread and adaptation, but their function is not limited to adaptation to the environment.*

Planning and symbolism, that is culture, have generated increasingly advanced abilities in control of and adaptation to the environment, in complex social organization and in lifestyle.

Culture as an adaptive strategy represents the substantial mediation between man and his environment and is the main reason for the success of the human species in competition with the other living beings and the physical environment. Culture is peculiar to man and there is no organ, with adaptive meaning, to compare with.

Culture envelops every man's behaviour, its contents neither depend on nor are linked to a particular function. Montagu (1967) defines culture as „man's adaptive dimension“. We propose to define culture as “man's ecological niche” (Facchini, 1988, 1992).

As noted by Ullrich (1995a), ”cultural adaptations enabled humans directly and actively to react to changes in natural environmental conditions. The results of cultural adaptations had to be passed on from one living social group to their descendants and these adaptations had to be renewed actively by the descendents“. By speaking of human species we rightly talk about biocultural adaptation, though the concept of culture is not completely exhausted by its relationship with the environment, since there are expressions of culture that transcend the environment itself.

Two elements define culture as man's peculiar activity:

a) Planning technology, as the ability to work out a strategy for using flint as well as making tools, building shelters and organising the territory for living and for food management. Planning technology includes the capacity of behaviour innovation.

b) Symbolism, especially when it is connected with language and spiritual expressions (art, religion, ethics, etc.).

What is planning ?

What marks human activity, in the technological field, is not only how products are made, but also the meaning they take on in hominids life. In flint tools made by man we can see functionality and a specific intention out of the way they were thought of, represented and realised, not only for an immediate, but also for later purposes and therefore it is for this reason that tools are kept by man (Facchini, 1995).

Planning capacity leads to a variety of types, better methods and progression in man's technique. It is easy to see in it a conceptual activity that gives a meaning to the artefacts and brings it into the man's imagination world. The tools produced by prehistoric man are not accessories to be used and then thrown away; but they are systematically produced by introducing further innovations as they are necessary and essential for his survival, as it was stressed by Kitahara Frish (1984).

Consequently their manufactures are neither occasional nor marginal, as we can instead observe in non-human primates or in *Australopithecus*.

Furthermore planning ability concerns not only the lithic industry, but also its spread all over the territory as well as territory organization in relation to its resources particularly of food and raw material for making tools. Whenever we are in presence of planned technology man documents himself. What is fundamental is to recognise the characteristic of project and the meaning of his activity (Facchini, 1995).

What is symbolism ?

Symbolism represents the deep core of human psychism using sign as symbol. Symbol, in fact, goes beyond sign.

As noticed by Ries (1993) „sign and symbol are two essential points of human imagination, a symbol is a sign that refers to an invisible reality with which man is put into communication".

Symbolic activity is revealed by attributing a meaning to a sign or a sound that is not immediately connected with the content of the gesture or the sound. A symbolic capacity is founded on mind cognitive ability and is also revealed indirectly by technique realizations. Besides answering to a project, what is obtained, by means of technique, is also a sign or a recall of a previous use. Every tool can assume its own meaning in man's imaginary world. I propose to speak of *functional symbolism.*

Moreover by means of a symbol it is possible to attribute a meaning to a single sign or sound, to establish and increase relationship. It particularly happens in language and in other kinds of social communication, as well as when the answers to biological needs are enriched by other meanings (for example, cloths having not only protective function but also a social one). Then we can speak of *social symbolism.*

There is another type of symbolism which is not connected with biological needs and it is represented by art, religion and ethics. This is *spiritual symbolism.*

However it is not easy to establish the time when cognitive-symbolic capacity was reached.

According to Noble & Davidson (1991) symbolic activity was not reached before *Homo sapiens*. It is expressed by painting, engravings, etc. and can be found only after 30,000 years ago.

In our opinion functional and social symbolism is to be recognized in early humans.

Planning and symbolism in human adaptation

Both planning and symbolism are interconnected as two faces of the same medal. They have had and still have great importance in man's relations with his environment and take on an essential adapted meaning, never marginal, as regards survival.

The activities of prehistoric man that show planning and symbolic capacity and their connection are the following ones:

- organization of territory (shelters, camps) and economy (hunting and gathering organization);
- technology (making tools, lighting fire);
- language and other symbolic communications;
- ethic and aesthetic activity.

Organisation of the territory and economy

As far as social life is concerned the relations with territory and food strategies, assemblages of bones and tools in the sites occupied by Olduwan and Lower Palaeolithic man point out evidences of a complex organization of hunting, preparing and sharing food and tools made by *Homo erectus* and *Homo habilis* settled in savanna environment (f.e. Olduvai, Melka Kunturé) and by *Homo erectus* in temperate forest environment (f.e. Torralba, Ambrona, Atapuerca, Chou-kou-tien). In relation with territory's temporary resources, seasonal movements were practised by *Homo erectus*. He knew the precise location of animals and how to organize home camps or hunting camps. Moreover climatic changes forced him to look for new territories and to adopt new strategies by using what the environment offered him. It is interesting to point out his flexibility in the use of raw material for making tools for its availability. The use of bones of large mammals for making tools, in consequence of scarcity of lithic material, is documented in some deposits (Bilzinsgleben, Castel di Guido).

In this connection I would remember the researches carried out on Olduvai and Melka Kunturé and a recent analysis of both surface and excavated records in the Okote member of the Koobi Fora Formation dated to about 1.6 million years ago. According to Rogers this analysis provides clues for discerning early *Homo erectus* land use and planning capabilities. A general model of hominid land use includes the anticipation of a future need for stone and flexible lithic reduction strategies contingent upon the sizes of stone available (Rogers, 1997).

A variation in subsistence behaviour among hunters and gatherers as a set of food management strategies has been suggested (Gamble, 1986; Binford, 1980). Particularly Binford supposes there were two settlement systems for hunters and gatherers: a forager system, when camps are established and parties set out on a daily basis to encounter food, particularly plant food (residential camps), and a settlement system with camps and other location (catches, field station where the game is monitored, field camps) which is temporary location centre for a task group operating at some distance from the residential camp (Binford, cit. by Gamble, 1986).

According to Otte (1997), "bone remains found at Mousterian sites show both the diversity of prey, the adaptive possibilities presented by hunting techniques and the possible specialisation of Mousterian predation ... Conversely a wide range of prey shows opportunistic behaviour, that is to say, intimately adapted to the behaviour of each animal".

The organization of the hunt, especially for large mammals, required group coordination, besides knwoledge of animal movements.

All these strategies are in relation with the resources of the environment, but show a cognitive and predictive capacity.

As remarked by Clark (1996): "The great success of early Hominids was ability for social organization, for hunting, foraging and the protection of the group."

Technology

The preservation and the lighting of fire are to be considered a particular sign of cognitive capacity and humans psychism. In the mind of prehistoric man fire had a complex significance as a technological goal that increased his possibility of control and management of environment and it was also a way to communicate and reinforce the bounds of the group.

As far as lithic and bone industry is concerned, planning activity is shown by the innovation and progression of a useful technique in order to save raw material and to reach a specialization of tool which is so important to adaptation and survival. As noted by Gowlett (1995) "although the design content varies, both major Oldowan and Acheulean sites preserve evidence of guiding plans".

In the evolution course the development of tool technology is more evident in transitional periods.

According to Mellars (1989), Byers (1994), Kozlowski (1997) and other scholars, the increasing complexity of tools production in the Middle-Upper Palaeolithic transition and in Upper Palaeolithic suggests that the artisans attached to these tool forms a symbolic meaning either for a specific economic function or for some social or economic context in which the tools were used. But I think that also the technology of Middle and Lower Palaeolithic man shows a kind of symbolic meaning. In the Lower Acheulean culture the tool-kits, with various types, are elements of a structural complex of the cultural system (Fiedler, 1996). „Antler, bones, ivory and wood were taken for different types of tools. This shows that *Homo erectus* had well definid traditions with fixed symbols in the common mind of social groups" (Fiedler, 1995).

The variety of tools correspond to the visual images of artisan in the goal and the existence of tool-making rejects in anthropic sites or in the lithic workshops shows a no-correspondence to the artisan representational contents. According to Wynn (1991), „artisan proceeds until failure, revises his constellation, begins again, and so on. The dynamic interplay between goal and constellation takes place in action ... In contingency planning ... reversibility is a characteristic of thinking ... where failures are anticipated and alternative procedures prepared ahead of time ... modern tool behaviour incorporates both" (cit. by Byers, 1994).

If we consider the bifaces, we can find the concepts of linearity and symmetry and in the Levallois technique, besides tool-planning, also the meaning of intentional behaviour is evident.

Language

Technological man, besides making tools, is also the inventor of symbol. Both anatomical and archaeological evidences can suggest indirectly human language ability. We then agree with the authors who admit relations between tool making, social behaviour, language use and neuronal development. Only a global approach (anatomical, technological, social, etc.) can help to identify the presence of language in fossil man, particularly in early *Homo*.

As far as phonetic capacity is concerned, the descent of the larynx and the expansion of naso-pharynx can be documented in some specimens of *Homo erectus* dating back to a million and a half years ago (Laitman et al., 1979; Laitman, 1985). According to Falk (1983) the region of Broca in *Homo habilis* is very similar to that of modern humans, while no difference has been observed between *Australopithecus* and ape. In *Homo habilis* endocranium Tobias (1983, 1987,

1991, 1996) points out not only a certain development of Broca's area, but also the prints of Wernicke's area, connected with the understanding of spoken or written language and especially with the ideational aspects of language (Eccles, 1989). According to Tobias the development of Wernicke's area would be exclusively owned by humans. An enlargement of the brain size and neuronal changes reflecting a selection for the production of language begins with *Homo habilis* (Deacon, 1992; Schepartz, 1993).

If to all these anatomical evidences, we add the documents of the cultural activity (tool technology, organization of territory, etc.) that show functional symbolism, we can suppose that articulated and symbolic language did characterize early humans, besides *Homo sapiens*. A possible correlation between the development of Broca's area and the ability of making tools is to be pointed out both in *Homo erectus* and in *Homo habilis-rudolfensis*; but, according to Gibson (1991), only Acheulean bifaces made by *Homo erectus* show a cognitive capacity of a higher degree than Ape's.

On the basis of a global approach combining anatomical and cultural evidences we can identify the possibility of human language (Facchini, 1995, 1996). We agree with Calvin (1993) who suggests a unitary hypothesis of language, tool making, intelligence and planning ability in correlation with the development of the brain.

In our opinion human phonemes, simple or together with mimic gesture, can reveal symbolic capacity. Starting from mimic-phonetic language with symbolic meaning prehistoric man developed to a more complex language. Transmission of culture was favoured and enriched by language.

Ethic and aesthetic activity

Ethics is a man peculiar attitude. It is connected with his cognitive ability and freedom. The variety of chances is documented by the variety of his cultural products. In planning activities we can suppose that the alternatives were evaluated.

The ethical attitude is expressed by behaviours that go beyond the ones strictly connected with his survival and are considered as manifestations of spiritual symbolism.

As far as artistic activity is concerned, recent evidences seem to date it earlier than Upper Palaeolithic. According to Mania (1997), in some Bilzinsgleben artefacts (Lower Palaeolithic) deliberately-engraved sequences of lines are documented and clearly show symbolic activity. In some Norfolk Acheulean befaces, a fossil bivalve is preserved in the centre of them (Oakley, 1981). We might recall the ancient use of red-ochre or the non-utilitarian objects from the early Middle Palaeolithic (Schäfer, 1996).

A particular meaning, probably regarding fertility and survival, is to be recognised in the Upper Palaeolithic symbolic representation, such as anthropomorphic figures (female sculptures) and pictured caves. They were of great importance for social life and propitiatory cults. Besides, more ancient evidences of art have recently been reported in Australia (150,000 years ago) and a little figure of a female has been discovered in Israel by Goren Inbar (1986) in the site of Berekhat Ram dating to 250,000 years ago.

Mortuary practices are documented since *Homo erectus* phase about 500,000-300,000 years ago (Ullrich, 1995b) and burial practices are evident in *Homo sapiens* of Middle and Upper Palaeolithic. According to Ullrich (1995b) „burial deposition of the entire intact corpse of the deceased, wich was seldom practised in Middle and Upper Palaeolithic times and coexisted with other mortuary practices, reflects a new way of thinking by humans and a new form of confrontation with human life and death during the daily life of palaeolithic man".

These behaviour generally are not tied to biological needs; they reveal a mentality able to face both life and death.

Phase	Technique	Territory	Economy	Language and social life
Homo sapiens (Middle and Upper Palaeolithic)	Blade production Levallois technique	Collective exploitation Family and hunting camps	Colletive hunting and gathering	Complex language Symbolic system of communication Burial practices
Homo erectus (Lower Palaeolithic)	Core tools Bifaces (tools symmetrically worked) Levallois technique Production of fire	Domestic and collective exploitation Family and seasonal camps Family and hunting camps	Collective hunting and gathering	Language with functional and symbolic meaning Cooperation for the hunting Family ties and division of work Mortuary practices Bands
Homo habilis (Olduwan)	Pebble Culture Variety of tools	Domestic exploitation Family camps	Domestic hunting and gathering Scavenging	Mimic-phonetic language with functional and symbolic meaning Cooperation for the butchering Family ties and division of work

Table 1 - Planning and symbolic activity in the evolution of the genus Homo.

Conclusions

Increases in culture were realised by means of the innovations and accumulations of knowledge, experiences and technique. Particularly the increasing co-operative and social relationships as well as specialization of lithic industry in the course of the evolution were possible and favoured by planning and symbolism (see Table 1).

Culture products show a constant progression to higher levels of technique and social organization, particularly out of the innovations introduced. However, in the course of man's evolution, even if planning and symbolic activity maintained the same performance, we can notice some periodic discontinuities, as to say, some overnight sudden improvements. This suggests to visualize culture evolution by a graphic representation in which the development of culture is represented by straight lines segments which follow one another, shifted in time but maintaining the same slope. In this representation the slope expresses the same aware and creative basic aptitude (planning and symbolic capacity), while the variations of distance from the abscissa axis show the innovations or discontinuities which can be observed during cultural evolution. (Figure 1).

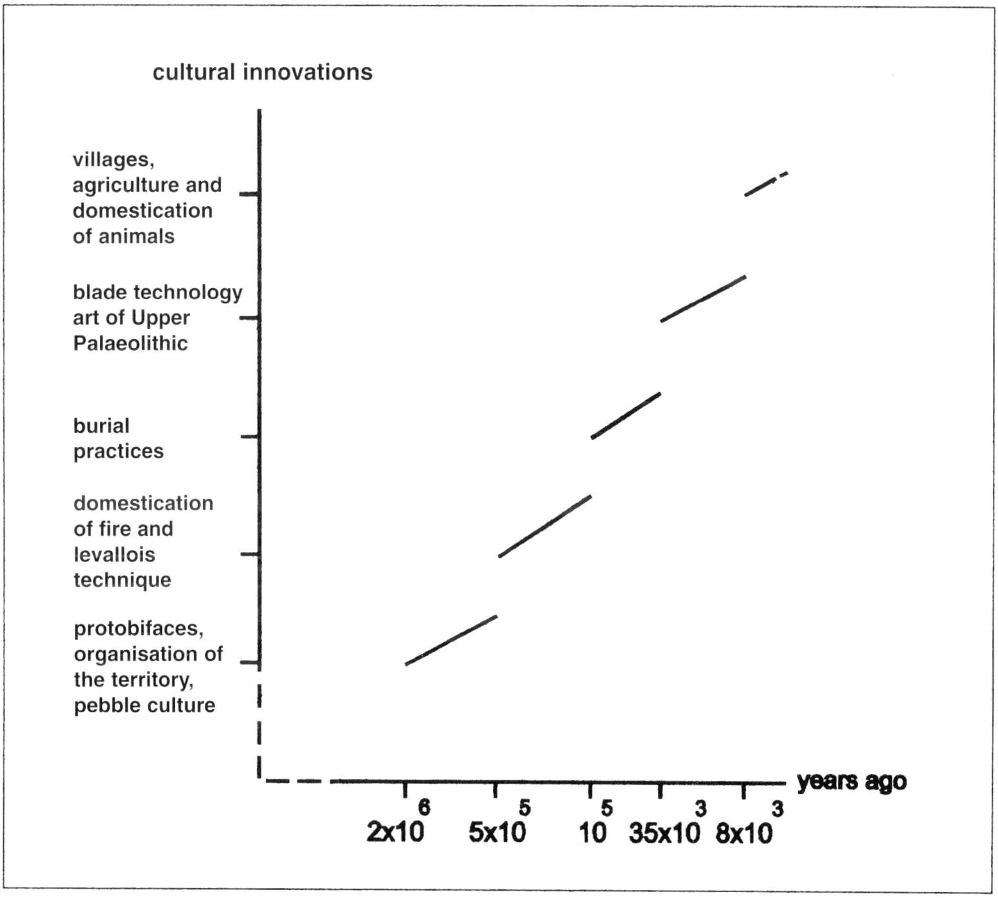

Figure 1 - *Graphic representation hypothesis of the culture evolution. The years are indicated in logarithmic scale, the slope of segments of straights is considered constant. The quantification of the cultural innovations is made assigning different points on the basis of the different importance of the innovations.*

As an hypothesis, we could identify several moments of discontinuity: with the Levallois technique and the domestication of fire (500,000 years ago), with the development of burial practices (approx. 90,000 years ago), with the lithic blade technologies and art of Upper Palaeolithic (35,000 years ago), with the agriculture, domestication of animals and the early villages (8,000 years ago). And perhaps a new discontinuity is taken in place in the modern era (industrial and postindustrial).

In conclusion we can state:

- Planning characterized man's technique as it was expressed by technology, territory organization and food management since *Homo habilis-rudolfensis* phase; technology and language reveal symbolic man. *Homo* is *technologicus* and *loquens*, as *homo symbolicus*.

- Symbolism is not something added to man's morphological development at a certain point of the course of evolution, on the contrary it is an essential expression of human psychism, expressed also in the technology, and in every time accompanied man, whatever his expression was.

- Planning and symbolism ability represented a real survival strategy: they were essential for the adaptation, even if their function was not limited to the environment adaptation.

- Out of planning and symbolic activity not only survival but also humans' spread and their demographic growth was possible.

Planning and symbolism, as culture, increasingly generated advanced abilities to control the environment favouring the adaptation to it, leading man to a more complex social organization and to a better lifestyle and improving his spreading all over the world.

References

BINFORD, L.R. (1980): Willow smoke and dogs tails: Hunter-gatherer settlement systems and archaeological site formation. *American Antiquity* 45, 4-20.

BYERS, A.M. (1994): Symboling and the Middle-Upper Palaeolithic transition. *Current Anthropology* 35, 369-400.

CALVIN, W.H. (1993): The unitary hypothesis, a common neural circuitry for novel manipulations, language, plan-ahead and throwing? In: K. Gibson & T. Ingold (eds.), *Tools, language and cognition in human evolution,* 230-250. Cambridge.

DEACON, T. W. (1992): Primate brains and senses: Human brain. In: S. Jones, R. Martin & D. Pilbeam (eds.), *Human evolution,* Cambridge.

CLARK, J.D. (1996): Lithic industries, language and social behaviour of the first human forms. In: F. Facchini (ed.), *The first humans and their cultural manifestations,* 85-86. Forli (XIII International Congress of Prehistoric and Protohistoric Sciences, Forlì (Italia), 8-14 Sept. 1996).

ECCLES, J. (1989): *Evolution of the brain: Creation of the self.* London, New York.

FACCHINI, F. (1988): Culture et spéciation dans la phylogénèse humaine. *C.R. Academie des Sciences, Paris* 307, s. II, 1573-1576.

FACCHINI, F. (1992): *Premesse per una paleoantropologia culturale.* Milano.

FACCHINI, F. (1995a): About the origins, development and meaning of human language. In: M. Di Bacco, E. Pacciani & S. Borgognini Tarli (eds.), *Statistical tools in human biology*, 203-214. Singapore, New Yersey, London, Hong Kong.

FACCHINI, F. (1995b): Quand la technique instrumentale peut-elle-etre définie activité humaine chez les Hominidés? In: M. Otte (éd.), *Nature et culture*, 945-952. Liège (ERAUL n. 68).

FACCHINI, F. (1996): Structures anatomiques et correlations culturelles dans le developpement du langage humain. In: F. Facchini (ed.), *The first humans and their cultural manifestations,* 125-133. Forli (XIII International Congress of Prehistoric and Protohistoric Sciences, Forlì (Italia), 8-14 sept. 1996).

FALK, D. (1983): Cerebral cortices of East African early hominids. *Science,* 222, 1072-1074.

FIEDLER, L. (1995): Die kulturelle Interpretation der Artefakte aus der Zeit des *Homo erectus*. In H. Ullrich (ed.), *Man and environment in the Palaeolithic,* 231-238. Liège (ERAUL n. 62).

FIEDLER, L. (1997): Conception of lower Acheulean tools. A comparison of three sites of the early handaxe culture. In: H.Ullrich (ed.), *Lifestyles and survival strategies in Pliocene and Pleistocene hominids.* Abstracts, 40-41. Weimar (International Interdisciplinary Symposium, May 4-10, 1997).

GAMBLE, C. (1986): *The Palaeolithic settlement of Europe.* Cambridge.

GIBSON, K. (1991): Tools, language and intelligence: Evolutionary implications. *Man* 26, 255-264.

GIBSON, K. & INGOLD, T. (1991): *Tools, language and cognition in human evolution.* Cambridge.

GOREN-INBAR, N. (1986): A figurine from the Acheulean site of Berekhat Ram. *Mitakufat Haeven* 19, 7-12. (cit. by Schepartz).

GOWLETT, J.A.J. (1995): Psychological words within and without: Human environment relations in early parts of the Palaeolithic. In: H. Ullrich (ed.), *Man and environment in the Palaeolithic*, 29-42. Liège (ERAUL, n. 62).

KITAHARA FRISCH, J. (1984): Ethologie animale et image del'homme. *Nouvelle Revue Theologique* 106, 235-250.

KOZLOWSKI, J. (1997): Contrasting lifestyles and survival strategies between the Gravettian and the Epigravettian in Central and South-East Europe. In: H.Ullrich (ed.), *Lifestyles and survival strategies in Pliocene and Pleistocene hominids.* Abstracts, 55-56. Weimar (International Interdisciplinary Symposium, May 4-10, 1997).

LAITMAN, J.T. (1985): Evolution of the hominid upper respiratory tract. In: P. Tobias (ed.), *Hominid evolution. Past, Present and future*, 281-286. New York.

LAITMAN, J.T., HEIMBUCH, R.C. & CRELIN, E.S., (1979): The basicranium of fossil hominids as an indicator of their upper respiratory systems. *American Journal of Physical Anthropology* 51, 14-34.

MANIA, D. (1997): Geology, palaeontology, archaeology and palaeoecology of the *Homo erectus* site Bilzinsgleben. In: H.Ullrich (ed.), *Lifestyles and survival strategies in Pliocene and Pleistocene hominids*. Abstracts, 124-130. Weimar (International Interdisciplinary Symposium, May 4-10, 1997).

MELLARS, P. (1989): Technological changes at the Middle-Upper Palaeolithic transition: Economic, social and cognitive perspectives. In: P. Mellars & C. Stringer (eds.), *The human revolution*, 321-337. Edinburgh.

MONTAGU, M.F.A. (1967): *The human revolution*. New York.

NOBLE, W. & DAVIDSON, I. (1991): The evolutionary emergence of modern human behaviour: Language and its archaeology. *Man* 26, 223-254.

OAKLEY, K.P. (1981): Emergence of higher thought 3,0-0,2 My BP. In: *The emergence of man*, 205-211. London.

OTTE, M. (1997): Modes of life in the Palaeolithic: Not survival but well-being. In: H.Ullrich (ed.), *Lifestyles and survival strategies in Pliocene and Pleistocene hominids*. Abstracts, 64-67. Weimar (International Interdisciplinary Symposium, May 4-10, 1997).

RIES, J. (1993): *Le religioni*. Milano.

ROGERS, M. J. (1997): Early *Homo erectus* land use and future planning: inferences from lithic discard patterns across a 1.6 million-years-old paleolandscape at East Turkana, Kenya. Abstract. *Journal of Human Evolution* 32, A 17-18.

SCHÄFER, J. (1996): Die Wertschätzung außergewöhnlicher Gegenstände (non-utilitarian objects) im Alt- und Mittelpaläolitithicum. *Ethnographisch-Archäologische Zeitschrift* 36, 173-190.

SCHEPARTZ, L.A. (1993): Language and modern human origins. *Yearbook of Physical Anthropology* 26, 91-126.

TOBIAS, P. (1983): Recent advances in the evolution of the Hominids with especial reference to brain and speech. In: C. Chagas (ed.), *Recent advances in the evolution of Primates*, 85-140. Rom (Pont. Acad. Scientiarum, Città del Vaticano).

TOBIAS, P. (1987): The brain of *Homo habilis*: A new level of organisation in cerebral evolution. *Journal of Human Evolution* 16, 741-761.

TOBIAS, P. (1991): The emergence of spoken language in hominid evolution. In: D. Clark (ed..), *Cultural beginnings. Approach to undertasting early hominid life-ways in the African savanna*, 67-68. Bonn (Römisch-Germanisches Zentralmuseum, Monographien Band 19).

TOBIAS, P. (1996): The evolution of the brain, language and cognition. In: F. Facchini (ed.), *The first humans and their cultural manifestations*, 87-94. Forli (XIII International Congress of Prehistoric and Protohistoric Sciences, Forlì (Italia), 8-14 sept. 1996)

ULLRICH, H. (1995a): Introduction to "Man and Environment in the Palaeolithic". In: H. Ullrich (ed.), *Man and environment in the Palaeolithic*, 1-8. Liège (ERAUL n. 62).

ULLRICH, H. (1995b): Mortuary practices in the Palaeolithic - Reflections of human-environment relations. In: H. Ullrich (ed.), *Man and environment in the Palaeolithic*, 363-378. Liège (ERAUL n. 62).

WYNN, T. (1991): Tools grammar and the archaeology of cognition. *Cambridge Archaeological Journal* 1, 191-206.

Female Proto-symbolic Strategies: A Testable Model for Earliest Symbolic Behaviour

Camilla Power

Abstract

In the later Middle Pleistocene, energetic costs to females of producing larger-brained offspring rose rapidly. This paper models strategies available to archaic Homo sapiens females for extracting increased investment from males. Once ovulation had been concealed in the human lineage, menstruation remained as the most prominent signal in the human cycle. Menstrual bleeding is a 'good' indicator of impending - though not immediate - fertility. It should function together with concealed ovulation to attract relatively 'faithful' males who are prepared to enter longer consortships. But while concealed ovulation withholds information about which females are fertile at any time, the salience of the menstrual signal undermines this effect, marking out imminently fertile females from pregnant or lactating ones. It is suggested that coalitions of late archaic Homo sapiens females resisted male discrimination between fertilisable and non-fertilisable females by cosmetically manipulating menstrual signals. Collective, deceptive and amplified use of red pigments as body paint confused information available to males about female reproductive status, and effectively formed a preadaptation to ritual. The model predicts an explosion of symbolic behaviour, motored by a runaway process of sexual selection, evidenced in the archaeological record by cosmetic artefacts such as haematite crayons.

Introduction

The evolution of the uniquely human facility for symbolic communication, in the forms of language, art and ritual, has probably been the subject of more 'just so' stories than any other sphere of human activity. Tooby & DeVore (1987) discuss the criteria qualifying a model as scientific and therefore helpful in advancing the state of our knowledge. A scientific model should be framed within a coherent and well-tried body of theory; it should enable us to relate fields of data which were previously unrelated; above all, it must generate predictions which are testable in the light of appropriate empirical data.

A compelling account of the emergence of symbolism remains the great challenge facing modern Darwinism (cf. Pinker, 1994; Maynard Smith & Szathmáry, 1995). But what would constitute appropriate test data? So much that is symbolic does not preserve in the archaeological record - words, songs, dances, love-magic. In fact, by definition, the symbolic realm cannot be preserved. In the words of Chase (1994):

„Symbolic culture requires the invention of a whole new *kind* of things, things that have no existence in the 'real' world but exist entirely in the symbolic realm. Examples are concepts such as good and evil, mythical inventions such as gods and underworlds, and social constructs such as promises and football games."

It is too easy when working at the level of the hard evidence of palaeoanthropology and archaeology, the bones and stones, to take symbolism for granted - background to the onward and upward march of hominid progress. Frequently it is assumed that hominids who were capable of making sophisticated tools and weapons, with the skills necessary to hunt large and dangerous animals, *must* have been communicating through symbols. Against this, Chase offers a salutary reminder that there is no reason to suppose symbolism was ever essential to survival in Pleistocene environments (Chase, 1994).

Costliness of symbolic behaviour

Modern 'selfish gene' Darwinism has made its most crucial contribution to evolutionary biology through its focus on costs. From this standpoint, the salient feature of symbolic behaviour such as ritual and art is its high demands on time and energy. Clearly, if we do not pose questions about the relative costs and benefits of symbolic behaviour, we cannot begin to evaluate selection pressures that gave rise to symbolism. Humans regularly invest much time and energy in the production by ritual action of what might be termed 'collective fantasies' or 'deceptions' (Knight et al., 1995). This contrasts starkly with the imaginative life of non-human primates. The great apes, in particular, may display capacities to infer mental states from behavioural cues - mindreading (Byrne, 1995: 144). But they put no time or energy into replication of one another's dreams, fantasies or illusions. While primates may individually fantasize or imagine events, no other individual will be motivated to share such imaginings.

Hunter-gatherer ritual activity has not been subject to rigorous time allocation studies comparable to optimal foraging analyses of subsistence behaviours (see e.g. Hawkes et al., 1992; Hill & Kaplan, 1988). Yet from detailed fieldwork such as that on Ju/'hoan trance activity (Katz, 1982), we can infer that people invest heavily in ritual. In the typical Kalahari case, this may occur all night long, normally continuing into the next night, more than once a month, with both women and men participating in highly energetic clapping and singing or dancing. They subject themselves to considerable stress in entering altered states of consciousness - having spent many years in developing the skills required. Around this activity they weave elaborate ideologies of experiencing death, changing into animals, and roaming across the desert to visit friends and relatives.

Hames (1992: 233) points to the conceptual problems of applying time allocation studies to ritual, political, social and 'non-economic' activities. Taking the common cross-cultural activity of ritual curing, in cases where a curer receives pay, this can reasonably be classed as somatic effort. But in numerous cases, a curer receives nothing „even though he may chant to the spirits all day and night until exhausted and he may have consumed a considerable quantity of hallucinogenic drugs that took additional hours to gather from the forest and to process." (Hames, 1992: 233). Perhaps, Hames suggests, the curer gains prestige, but what currency or proxy measure can be used to translate that prestige into reproductive fitness?

In the example of ritual curing, the expense falls on two sides. While the curer expends time and energy for no apparent gain, the patient, and other attendants, spend similar levels of time and energy engaging in the curer's fictions - the chants to the spirits, addressing an unverifiable 'other' world. From the hard-headed perspective of a chimpanzee this would be sheer time-wasting. In this light, our ancestors appear not as the 'thinking ape' or the 'tool-making ape', but the 'gullible ape'. To account for the evolutionary emergence of such costly engagement with the symbolic realm, we need convincing answers to the following questions. What selection pressures promoted an interest in sharing and propagating conspecifics' illusions? Why did it benefit individuals to share in the unverifiable fantasies of others, rather than develop resistance to what Dawkins (1993) would describe as 'parasitic' memes? Can we model a systematic process in which hominids were driven to expend increasing time and energy on things that did not exist -'deceptions' entertained by groups of individuals?

Why deception matters: Signal evolution theory

The stress here on 'deception' is deliberate. Besides expending high levels of energy in magico-religious rituals constructing our fictitious 'gods', we humans also regularly deploy strings of deceptions in communicating through speech. By this I do not mean primarily lies told for selfish advantage, but the metaphors which persistently characterise human speech. Metaphors are literal untruths. Creative use of metaphor baffles all attempts to program artificial intelligence machines to conduct normal human conversation with its quota of jokes, allusions, slang and catchphrases. In speaking, we communicate via signals that are fictions, blatantly untrue with respect to perceptible reality, to listeners who are willing to probe for the intended meaning behind these fictions (Knight, in press). The human facility for symbolism appears, by this account, to be a facility for manipulating - and sharing - deceptions.

Human language is anomalous in the animal world as a volitionally controlled, conventional - and therefore, cost-cutting - signalling system. It is easy to see the benefits to be gained from adopting conventional signals in that all members of the communicating group are able to cut their energy costs. But 'selfish-gene' methodological individualism reveals the problems that beset the evolution of conventional signals. If signals are cheap, where there are conflicts of interest, some individuals will be inclined to manipulate conventional signals dishonestly. This forces 'honest' signallers to prove their reliability by increasing the costliness of their signals. The requirement that signals be 'hard to fake' undermines any advantages of conventional signalling. Indeed, Zahavi (1993) argues that it is because of this dynamic that there are no conventional signalling systems in the animal world.

The problem, as Knight (in press) puts it, „is that conventional signals depend on trust, whereas those animals intelligent enough to use such signals will also be clever enough to competitively exploit that trust". Despite evident capacities for manipulating symbols (Savage-Rumbaugh & Rumbaugh, 1993), chimpanzees have extremely limited use for conventional signals in the wild (Boesch, 1991). The extent of the great apes' cognitive development comparative to monkeys (Byrne, 1995) contrasts markedly with the lack of parallel development in communicative ability (Ulbaek, in press). Even among the higher primates, vocalisations remain under limbic, not cortical control. Because it is so difficult for monkeys and apes to vocalise dispassionately, their vocal signals are intrinsically more reliable, the cues they provide being emotionally governed and therefore 'hard to fake' (Burling, 1993). The specific context in which primates do exercise their capacities for *cortical* control over vocal signals occurs in cases of tactical deception as, for instance, in the use of a leopard alarm call by a male vervet to distract and deter a potential interloper from joining the troop (Cheney & Seyfarth, 1990: 213-214). Chimpanzees do not appear to falsify vocal signals in this way; it seems they lack even the limited repertoire of coded calls that vervets possess. However, they can successfully silence their vocalisations, as when patrolling near a neighbouring band's territory, deliberately 'shushing' each other (Goodall, 1986: 490-491). The paradoxical implication to be drawn from these examples is that speech as a system of vocal signalling under cortical control has a trajectory in the evolutionary past involving tactical deception; yet, it has emerged as a communication system dependent on listeners who assume honesty (Knight, in press).

These issues need emphasis because there has been a tendency among archaeologists, palaeoanthropologists and linguists to suppose that syntactical speech brings unconditional benefits. According to such 'word-magic' scenarios (cf. Cavalli-Sforza et al., 1988; Mellars, 1991; Bickerton, 1990, and see Pitts & Roberts, 1997), once language had arisen by standard processes of natural selection (Pinker, 1994; Pinker and Bloom, 1990), all the benefits of symbolic culture were conferred on speech-possessing humans. Language, in this view, is the prime mover leading to the development of ritual, religion, art etc. Yet, theoretical linguists (e.g. De Saussure, Chomsky, see Pinker, 1994) too frequently proceed from an 'assumption of communism' (Bourdieu, 1992) - supposing that people are in egalitarian relation to each other and naturally co-operate to transmit and exchange 'good' information. As Nettle (in press) puts it: „they have assumed precisely that which is most important and problematic: the social structure which makes language possible."

The societies of our closest relatives, chimpanzees - and by inference of our ancestors - are riven by hierarchical differences of rank, and rife with competition for status, sex and resources. All the evidence of ape language experiments points to the discrepancy between apparent symbolic cognitive *capacities* of the animals and consistent absence of symbolic *performance* in natural conditions.

Undoubtedly, language as a system of communication arose as an adaptation (Pinker, 1994), but only in conditions where the benefits of using a system of displaced reference - involving transmission of second-hand and uncorroborated information - were not outweighed by the costs. These costs are the dangers of deception - reliance on uncorroborated information *from* others - and also the dangers entailed in entrusting valuable information *to* others. 'Tactical deception' theory (Byrne & Whiten, 1988) does not allow us to assume the requisite levels of trust as a default.

Pinker (1994) has evaded consideration of social contexts by addressing language evolution at the level of mechanisms underlying speech acquisition and production, rather than at the level of selection pressures. Dunbar's 'vocal grooming and gossip' hypothesis (Dunbar, 1996; Aiello & Dunbar, 1993) specifies concrete selection pressures driving the evolution of preadaptations to speech in the context of increasingly complex social groups. As groups got larger, hominids no longer had enough time each day to service their alliances by standard primate means of one to one grooming. Vocal communication, eventually resulting in social information exchange, or 'gossip', took on the function of grooming as a more efficient means of social bonding. In this model, the selective force driving the evolution of larger brain sizes for hominids living in larger social groups is the requirement for 'Machiavellian intelligence' in dealing successfully with increasing numbers of social relationships. This implies that abilities for mindreading, manipulation and tactical deception co-evolve with linguistic abilities. Dunbar has not clarified the question of how in these circumstances of selection for 'super' Machiavellian intelligence sufficient trust can be generated between members of gossiping communities for second-hand uncorroborated information about third-parties to be accepted as reliable. Where such trust does not exist, recipients of gossip would have to run around double-checking their sources, driving up the cost of acquiring reliable social information, and undermining the efficiency of 'gossip' as a mechanism for bonding social groups.

The trouble with the argument that vocal grooming reduces costs of servicing alliances and thus alleviates social time budgets is that primate grooming works as a means of cementing alliances precisely thanks to its costliness. Grooming signals to the groomee in 'hard to fake' terms 'I am spending my valuable time with you, and only you, because you are the one that matters to me'. If vocal signals enable an individual to 'groom' three times as many allies at once, its worth as a signal of commitment to each of those allies is reduced to a third. For the gossip hypothesis to work, we need some independent mechanism providing a costly signal of commitment for establishing coalitionary membership, driving up the costs to any individual who breaks 'faith' with the coalition. Power (in press a) suggests that it is ritual that takes on the 'grooming' function of bonding large-scale coalitions through intense shared physical and emotional experience involving energy-expensive signals like song, dance and cosmetic display (cf. Dunbar, 1996: 146).

Maynard Smith & Szathmáry (1995: 272-273) focus on ritual as a mechanism of crucial importance in sustaining large-scale human systems of reciprocal altruism. Knight (1996; Knight et al., 1995) has used the framework of signal evolution theory to argue that ritual necessarily co-evolves with speech. Human rituals demarcate the boundaries of social groups in all or nothing terms; any individual is either inside or outside a ritual coalition. Ritual - and only ritual - generates and replicates in the participants' heads a shared in-group self-representation (cf. Durkheim, 1915: 230-231). Such ritually structured communities contrast starkly with the fluid and fickle alliances we know from primate politics (de Waal, 1982). It is costly to join human ritual coalitions, sometimes at the price of traumatic, permanent physical or psychological scarring that makes it impossible to leave one such community and join another. Between members of a ritual community who have signalled their commitment at high cost - individuals who demonstrably share the same 'gods' - the necessary trust will exist for habitual use of the low-cost, conventional signals of speech. On this basis, Knight argues that an ingroup/outgroup evolutionary dynamic drove the emergence of 'ritual' signals for group-on-group relations, speech for within-group 'conspiracy'. This turns on its head the idea that language is the prime mover in establishing symbolic culture. On the contrary, it is the ritual structuring of coalitions that provides the matrix for language.

The following model of selection pressures driving the emergence of costly ritual behaviour and the genesis of the symbolic domain is developed through a hypothetico-deductive method. Following Smith (1991: 8), I aim „to employ theory to deduce the conditions under which particular social or behavioral forms will emerge or persist. This deduction must then be framed in testable (operational) terms, and compared with relevant observations".

The model should delimit or constrain what we expect to see as evidence of ritual activity in the archaeological record, indicating correlatives of that evidence for ritual. If the initial conditions for ritual as a high-cost signalling system can be specified with sufficient precision, it may offer predictions testable against the ethnographic database of recently extant ritual. Such a hope would have seemed vain over a decade ago, when an essentially human foraging adaptation was assumed to cover the past two million years. Now, with the stress on the 'human revolution' (Mellars & Stringer, 1989; Stringer & Gamble, 1993), involving the recent emergence of modern human anatomy and a 'modern' human lifestyle, it looks feasible to use ethnography of magico-religious ritual as a test database. It should be stressed that such testing is not a matter of appeal to particularist ethnographic precedents, projected back onto a patchy Pleistocene record. The model is theory-driven, constraining our expectation of the way symbolic systems emerge and are transmitted. If a narrative of symbolic cultural origins fails to offer parsimonious, predictive accounts of symbolic systems it hardly amounts to a theory at all.

Why female strategies matter: Costs of encephalization

The basic premise of the model is the expectation of conflicting reproductive strategies between the sexes. Crudely put, for any sexually reproducing species, males and females get their genes into the next generation by different means. For mammals and especially primates, this involves lengthy periods of gestation and lactation requiring investment of considerable resources for females, while males are not necessarily committed to more than the energy needed to access and impregnate mates (Trivers, 1972). There will be differential trade-offs between the sexes over investment of energy in current offspring (parental effort) as against energy expended for producing future offspring (mating effort). In the case of human evolution, these trade-offs are likely to be especially critical because of the extraordinary energetic costs imposed on hominid mothers by encephalization (Foley & Lee, 1991; Martin, 1990, 1996). The model is succinctly stated: symbolism arose as a signalling system used by female coalitions in strategic response to the reproductive stress experienced as a result of rapid encephalization. It can be said at the outset, males as relatives of female coalition members would be included, whereas males who were potential or actual mates would be excluded from these coalitions. These outgroup males were the targets of the female coalitionary signals, which were designed to 'exploit male muscle power' (cf. Knight et al., 1995; Krebs & Dawkins, 1978) - that is, to motivate males to produce high energy foods for consumption by members of the coalition.

The activity of female coalitions has not loomed large in general theoretical discussions on human behavioural evolution. In their seminal paper on strategic modeling, Tooby & DeVore (1987) omit any reference to the possibility of female coalitions or female bonding as a contributory factor to specific human adaptations. More recently, in addressing the study of behaviour from the fossil and archaeological record, Brooks (1996: 137) lists male-male and male-female relationships as social features which might be inferred from fossil morphology. What of female-female relationships? Primate behavioural ecology assumes as a default that female coalitions form the core of primate societies (Dunbar, 1988: 138-139, 262). Any focus on major evolutionary change in social and sexual strategies forces us to adopt something of a female bias. According to standard socioecological models, changing ecological variables drive changes in mammalian mating systems via changes in *female* strategies rather than male ones (Crook & Gartlan, 1966; Clutton-Brock & Harvey, 1977). Ecological change impacts most directly on female behaviour. While male primates tend to prioritize the search for fertile females over the search for food, mothers distribute themselves independently of males, in accordance with ecologial variables and predation risks, prioritizing resources to feed their young (Wrangham, 1980). Within constraints set by phylogenetic history, the resulting mating system depends on whether females forage in isolation or in groups, and on the nature of interfemale resource competition (van Hoof & van Schaik, 1992). Whether such factors as paternity confidence are high or low depends ultimately on a mating system whose essential features are female-defined.

Large-scale shifts of climate in the late Pliocene appear significant in the emergence of genus *Homo* (de Menocal, 1995). The consequent environmental changes may be responsible for selection for larger group sizes, and hence for larger brain sizes (Aiello & Dunbar, 1993). Encephalization required a high quality diet, permitting reduction in relative gut size and offsetting the increased metabolic costs of relatively larger brains (Aiello & Wheeler, 1995). A high-quality diet in turn requires larger foraging areas (Clutton-Brock & Harvey, 1977) with increased overall energy expenditure (Leonard & Robertson, 1992), likely factors selecting for more efficient bipedality. The major point here is that *all* these changes impact most critically on maternal energetics; it is females who necessarily altered strategies to cope, with males responding to female behavioural change. The corollary of Dunbar's 'vocal grooming' model is that because female hominids experienced most acute stress on social time budgets, they would have come under most pressure to develop more efficient means of servicing their alliances (see Dunbar, 1996: 148ff.). Hence, vocal grooming and 'gossip' served initially to bond female coalitions.

The traditional assumption on phylogenetic grounds that the ancestral human form of social organization was patrilocal (Foley & Lee, 1989; Wrangham, 1987) may be fallacious (see e.g. Dunbar, 1996: 148; Knight et al., 1995: 105; Hawkes, 1996: 298-9). The argument that because our closest relatives, chimpanzees and bonobos, are male philopatric species, therefore so were our hominid ancestors fails to take account of the unique effects of encephalization on female reproductive strategies. Evidence supportive of a matrilocal model emerges from molecular genetics. Among small aboriginal human populations such as East African pygmy groups and Khoisan, the distribution of mtDNA comparative to nucleic DNA shows a pattern of variation consistent with female philopatry (Vigilant et al., 1989; Melnick & Hoelzer, 1993). This pattern differs notably from the genetic structure consistent with female dispersal found in human groups with a long history of property-holding. Any evolutionary account of menopause based on the fitness benefits of 'grandmothering' (mothers improving the reproductive success of children, see Mayer, 1982; Hill & Hurtado, 1991; Rogers, 1993) also relies on a model of daughters staying near mothers once they reach reproductive age (Hawkes, 1996: 298; Peccei, 1995: 432). Hawkes et al. (1997, in press) argue that because it is female foragers who provide the steady stream of nutrients for sharing with offspring, it is the mother's mother who would be the best candidate for improving her daughter's fertility by assisting with provisioning of offspring close to weaning. Such a strategy would bring fitness benefits to females who lived a longer post-reproductive life, i.e. selection for menopause. Hawkes et al. (in press) call for revision of the assumption that nuclear families are fundamental economic units among modern human foragers, let alone ancestral hominids. In conditions where males were not able to supply meat reliably, for instance prior to the development of sophisticated projectile weapons, these grandmother strategies would have been especially important.

Sexual signals and behavioural change

Critical to the reproductive success of females as they came under selection pressure for larger-brained offspring was extracting energy from new sources. The first major increase in brain size occurs with the appearance of early *Homo* over 2 million years ago, culminating in *Homo ergaster*. These costs could have been offset by shifts to high quality diet, allowing reduction of gut size (Aiello & Wheeler, 1995); increases in female body size (McHenry, 1996; Aiello, 1996); and changes in life history variables such as increased longevity, promoting grandmothering (Aiello, in press), and secondary altriciality, slowing down maturation rates of the larger-brained offspring (Foley & Lee, 1991: 70). Investment by males may have been intermittent rather than systematic, and directed as mating effort towards cycling females, rather than pregnant/lactating females (Power & Aiello, 1997, Aiello, in press). A period of over a million years, from the Lower to early Middle Pleistocene is characterised by stasis in relative brain size (Kappelman, 1996; Ruff et al., 1997). The accelerated encephalization rates of the late Middle Pleistocene brought increased reproductive costs, particularly to mothers in early stages of lactation. These steeply increasing costs of reproduction are likely to have driven major social and sexual behavioural changes (see Foley & Lee, 1996: 63-64; Power & Aiello, 1997). Above all, those females who secured increased levels of investment provided by males would have enhanced their fitness. It should be stressed that a female need *not* be concerned which particular male provides energy for herself or her offspring. If a male is providing energy on a basis of persistent mating effort, i.e. to improve his prospects of sexual opportunities, then he too is *less* concerned about paternity issues than if he were targetting investment to specific offspring in the form of parental effort. Male investment in this context operates on a basis of probability of paternity rather than certainty. Evolutionary ecologists such as Hawkes (1991) and Kaplan & Hill (1985) have suggested that mating effort rather than parental effort models characterise male foraging patterns among Hadza and Ache hunter-gatherers.

For females, sexual signals are the primary mechanisms for eliciting behavioural changes in males. It is assumed here that by the stage of archaic grade *Homo sapiens*, overt signs of ovulation had already been lost and the modern pattern of 'loss of oestrus/continuous receptivity' established. These are effective mechanisms for forcing male attentiveness and promoting longer consortships (Alexander & Noonan, 1979) which would have assisted early *Homo erectus/ergaster* females in gaining supplies of energy-rich food and meeting the costs of the earlier phase of brain expansion (c.2 mya). However, concealment of ovulation can never guarantee more than mating effort, since once a female is pregnant/lactating the fact she is *not* ovulating is not concealed. The most prominent remaining signal in the modern human female cycle is menstruation. In a natural fertility population, with most females of reproductive age pregnant or lactating, menstruation occurs relatively rarely and is a good indicator of *imminent* fertility. Pleistocene males who were attentive to recently menstruating females in an effort to improve their mating prospects should therefore enhance their fitness. No male could afford to ignore this signal.

This implies that an archaic female who was menstruating could advertise her signal to males in the vicinity to promote mating effort. Menstrual blood, therefore, would have a material value translatable into energy in the form of male provisioning.

Female coalitionary strategies: Protoritual

For any pregnant/lactating female, a menstrual female is a potential threat capable of diverting male energy and investment away from non-menstruants. One response to this problem, as archaic *Homo sapiens* females experienced increasing reproductive stress, would be to adopt a reciprocal altruistic coalitionary strategy of manipulating menstrual signals. Each female coalition needed to prevent any male from sequestering the imminently fertile female, by surrounding her and restricting sexual access. Given the *economic* value of the signal, rather than hide the menstruant's condition, we would predict the opposite. Whenever a coalition member menstruated, the whole coalition joined in advertising this valuable signal as widely as possible to recruit available male energy to the coalition. The strategy succeeded as long as any mating effort generated by the menstruant's signal flowed into the whole coalition, benefiting both non-menstruants and menstruants. Non-menstruant coalition members would confuse matters by borrowing the menstruant's signal or mimicking it with other blood or blood substitutes. This strategy would effectively prevent males from discriminating in favour of cycling females and undermine attempts by would-be dominant males to monopolise imminently fertile females.

This strategy of cosmetic manipulation of menstrual signals is termed 'sham menstruation'. A dynamic of competition between female coalitions would drive an evolutionary arms race of increasingly elaborate sham menstrual advertising as coalitions aimed to recruit available male energy. This should result in ritualistic amplification of displays, involving use of red pigment to spread the menstrual signal further, with multimedia effects of movement, body paint, song and dance.

Predictions from the 'sham menstruation' model: Archaeological evidence

The main prediction derived from the 'sham menstruation' model of an evolutionary arms race between female coalitions competing for male energy is that the earliest evidence of ritual traditions in the archaeological record will take the form of a cosmetics industry focused on red pigment. Factors of male sexual selection for cosmetically decorated females should also motor an explosive spread of such traditions, while such processes of sexual selection could reinforce speciation of anatomically modern humans (Power, in press b).

Power & Watts (1996) argue for a two-tier process of the evolution of ritual, fundamentally determined by degree of reproductive stress on females. During earlier stages of the brain expansion of archaic *Homo sapiens*, they posit *context-dependent* sham menstruation displays, triggered by the incidence of menstruation in local populations. Female coalitions used these as opportunity arose to attract and retain male support, securing long-term bonds with mates. This strategy implies less planning depth in obtaining materials for cosmetic usage, with correspondingly greater reliance on biodegradable matter, and only occasional traces of utilised ochre. As late archaic to early anatomically modern females endured acute reproductive stress - roughly the period 160,000-130,000BP, coincident with the Penultimate Glacial maximum (Jouzel et al., 1993) - they posit the emergence of a habitual strategy of cosmetic ritual underpinning the sexual division of labour. Greater regularity, planning and organisation of performances would lead us to expect abundant and regular use of ochre. The onset of this earliest symbolic tradition would institutionalise an economic division of labour and forms of social cooperation both between the sexes and between kin groups. Therefore, it should permit relaxation of selection pressures for robusticity, especially in females, and reduce stress levels experienced by juveniles (see Kappelman, 1996: 272; Brooks,

1996: 146-148), possibly permitting earlier weaning. It should promote investment in campsites with females and offspring able to stay 'home' while male hunters depart on logistic hunts.

The archaeological record suggests that archaic grade *Homo sapiens* was the first hominid to use iron oxides. There are fewer than a dozen instances of ochre usage predating the Upper Pleistocene and the majority of these occur in the Late Acheulian and early Middle Stone Age (MSA) in sub-Saharan Africa - within the past 200,000 years (Knight et al., 1995). Beginning with the African MSA2a (possibly from as early as 140,000 BP), there is evidence of regular and widespread ochre use, while in MSA2b cave and rockshelter assemblages (from c.110,000 BP), ochre has become ubiquitous showing an order of magnitude increase across the MSA2a/2b boundary. The scale of ochre use in southern Africa from the MSA2 onwards is unparalleled elsewhere until the Eurasian Upper Palaeolithic. Strong reds occur earliest and massively predominate; black pigments are used very rarely, despite their availability (Power & Watts, 1996). Functional uses for ochre (e.g. in treatment of hides) were then, and remain today, subordinate to ritual and symbolic uses (Watts, nd; Power & Watts, 1996). The use of red ochre in Africa contrasts markedly with the picture in Europe before the Upper Palaeolithic. The amount of pigment found in French Middle Palaeolithic sites is considerably less, and black manganese predominates (Couraud, 1991). It seems that most Neanderthal pigment use occurred during the late Mousterian (60-34,000 BP). Only during the Chatelperronean, when Neanderthals are believed to have been in contact with modern humans, is there significant increase in amounts of red ochre.

Among Khoisan hunter-gatherer groups, application of ochre or haematite, besides vegetable dyes, regularly occurred in ritual contexts and especially menarcheal observances (see Power & Watts, 1996, for review). A /Xam female initiate, on emergence from seclusion, would present the women of the band with lumps of haematite for decorating their faces and cloaks (Lloyd n.d., VI-I 3969 rev 3973) and also adorn the young men in preparation for hunting. In Ju/'hoansi (!Kung) oral narratives, the metaphor for impending ritual action was the sound of women pounding red ochre in camp (Biesele, 1993: 163, 196). Haematite, which produces a rich red colour on grinding, was particularly prized - redness and shininess being the qualities especially valued by both the !Kung and the /Xam (Lewis-Williams & Biesele, 1978). Fischer (1913, cited in Rudner, 1982: 218) notes of Khoisan in Namibia that they painted their faces with red iron oxides at 'the time of menstruation'. In default of the desired haematite, women used other substances such as soot mixed with fat as cosmetics. Wilhelm (1954: 118) writes that !Kung women when menstruating would smear their inner thighs with fat of a large antelope, and paint soot around their eyes. A menstruating !Kung woman would also cut a tonsure in the hair of her youngest child and paint that with fat and soot – as if advertising her imminent fertility after lactational amenorrhea. During menstrual seclusion, a G/wi or G//ana initiate would have her joints cut and a mixture containing her mother's blood rubbed into the cuts – the blood being drawn from her mother's hips (Valiente-Noailles, 1993: 96). For Khoisan initiates generally seclusion is a time of rigorous taboo. Even visual contact between the girl and men is strictly prohibited, and her conduct is carefully monitored by a group of close kinswomen.

Wrong sex, wrong species': The first 'gods'

The 'sham menstruation' strategy involves deceptive signals in that some females who are not imminently fertile pretend to be. Unlike primate tactical deception which is always individualistic and egocentric (Byrne & Whiten, 1988), the deception in this case is sociocentric, being maintained by a collective. As such, it represents a vital step towards sustaining an *imaginary construct* and sharing that construct with others - that is, establishing symbolism on a habitual basis.

Figure 1 - *Representation of Khoisan initiate as 'wrong sex, wrong species', Willcox's Rock Shelter, Drakensberg, Natal (Solomon 1992: 314).*

So long as such deceptive displays are staged only because a local female is menstruating, these signals, however amplified, are not displaced but refer to here-and-now contexts: basically, they form loud advertisements of the presence of an imminently fertile female. But it is easy to see how a female coalition would be pushed into signalling of imagined constructs which are impossibilities, corresponding to no perceptible reality - collective deceptions. Males who are attracted by cosmetic displays advertising imminently fertile females will be reluctant to leave the vicinity; they will instead be inclined to mate-guard. Some males may even be non-cooperative and attempt abduction of menstrual females. These circumstances would force the female coalition (with male kin support) to step up resistance by loudly signalling 'No access' to outgroup males. Knight et al. (1995: 84) argue that the way female coalitions would construct such a 'No' signal would be by reversing the normal parameter settings of the species mate recognition system. Where animals in courtship ritual normally signal 'right species/right sex/right time', systematic reversal by a defiant female coalition would yield 'wrong species' - we are animals, not humans; 'wrong sex' - we are males, not females; and 'wrong time' - we are not fertile right now, but *soon* we will be. This is the predicted signature constructing the potency of the ritual domain and the inviolability, or 'taboo' state, of menstrual or body-painted females. Transmission of such signals counter to perceptible reality will involve energetically expensive, repetitive, iconographic pantomime - high-cost ritual signals sustaining fictitious 'gods'.

The signature of ritual potency: Tests in ethnography

The model predicts periodic female inviolability as the focus of early human ritual traditions, signalled by real blood or cosmetic body paint, inverted sexual attributes, and therianthropic metamorphoses. It expects taboos on access to 'menstruating' ritually potent females, these corresponding to taboos on sex for outgroup males prior to hunting. Because lunar cyclicity provides the most appropriate clock for synchronising ritual action by 'menstrual' female coalitions, the model predicts lunar scheduling of the ritual cycle, with associated beliefs linking lunar periodicity to hunting and menstruation. Light after sundown in the period leading up to full moon may be vital for successful hunting. This dictates an underlying pattern of abstinence from sex or hunting during menstruation around dark moon, prior to big-game hunts in the waxing-to-full moon period when light is optimal.

Generally in the worldwide ethnographic record, ritual potency is predicted to display a characteristic 'time-resistant' signature, revealing its ancestry in menstrual inviolability. Power should be switched 'on' by one set of signals, 'off' by another:

ON	OFF
Loud signals	Weak signals
Waxing moon	Waning moon
Seclusion	Availability
'Other world'	'This world'
Night	Day
Wet	Dry
Bleeding/raw	Cooking/cooked
Hunger/being eaten	Feasting
Flesh taboo	Flesh available
Production	Consumption
Kinship	Affinity
Gender inversion	Heterosexual polarity
Animality	Humanity

Knight (1991: 374ff.) reviews general ethnography on menstrual taboos and associated prohibitions on hunting and cooking. Knight et al. (1995: 92) examine Hadza and Khoisan beliefs concerning hunting, sex, menstrual observances and lunar periodicity. Knight (1997) analyses the menstrual/lunar periodic motifs prevalent in Amerindian mythology, following Lévi-Strauss.

Power & Watts (1997) focus on recurrent gender-inversion associated with African hunter-gatherer initiation rites. From a Darwinian viewpoint, it seems *prima facie* counter-intuitive that a newly marriageable girl would signal that she is 'male' and 'animal'. Despite this improbability, the model forces this prediction. How, then, does this match the ethnographic data? Metaphorically and in ritual practice, Hadza and Khoisan menstruants become hunters - that is, symbolically 'male'. The Hadza say 'She has shot her first zebra' (Woodburn, pers. comm. 1993); the Ju/'hoansi say 'She has shot an eland' (Lewis-Williams, 1981: 51). Yet despite being cast as 'the hunter', upon emergence the Ju/'hoan and /Xam menstruant must *play the part* of the antelope, lowering her eyes so the animal will not see the hunters' approach (Bleek & Lloyd 1911; Lewis-Williams, 1981). A !Xo menarcheal girl shoots a ritual arrow at a gemsbok shield suspended at the back of the seclusion hut to bring luck to the weapons; on emergence, her face is painted with a gemsbok mask (Heinz, 1966). Solomon's (1992) analysis of gender in Khoisan rock art focuses on the recurring motif of ambiguously female figures with menstrual flows - ambiguous because of their therianthropic features and/or male attributes such as a penis or hunting equipment (Figure 1).

Turning to the Hadza, *epeme* is the most sacred ritual, held each dark moon when women are supposed to be menstruating (Woodburn, pers. comm. 1993; Bleek, 1930: 700). The ritual is believed to be vital for success in forthcoming hunts. The name *epeme* also refers to portions of fatty meat exempted from bride-service obligations and reserved for secret consumption by men in the name of 'God'. A Hadza narrative (Woodburn, 1964) portrays the original owner of the sacred *epeme* meat as an old woman, Mambedaka, who dressed as a man, hunted male zebra, and wore the penis of her kills, using this to have sex with her 'wives'. Male hunters had to surrender their meat (the *epeme*) to Mambedaka for distribution among these women. Men were excluded from any share until the old woman's violent overthrow. Mambedaka playing 'God' with her zebra's penis matches precisely the *'wrong sex, wrong species'* construct central to the origins model (for similar myths of 'primitive matriarchy' worldwide see Knight, 1991).

Probably the best known and most widespread aspect of Khoisan menarcheal ritual is the 'Eland Bull Dance', performed during the final days and nights of menstrual seclusion and on the emergence of the 'new maiden' (e.g. Lewis-Williams, 1981; Valiente-Noailles, 1993). Men are sometimes totally excluded; alternatively, one or two older men in affinal relation to the menstruant may play the bull. In an elaborate 'wrong species' pantomime, the women mimic the courtship behaviour of eland cows, dancing around the menarcheal girl, exposing their buttocks to the 'Bull'. The Ju/'hoansi explicitly identify the menarcheal girl as this 'Bull Eland' (Power & Watts 1997). In conformity with predictions, the Khoisan consider such displaced 'animal' sex - in effect, 'anti-sex' - *good for hunting luck*. 'Human' or heterosexual marital sex, here as throughout the world, undermines hunting luck.

Why not a model of male ritual alliances?

The model presented here offers a theoretically coherent account of earliest symbolic behaviour. It yields predictions which are testable not only against the archaeological record, but also against ethnographic data on ritual and myth. Some of these predictions may be directly falsifiable (e.g. ritual power should *never* be enhanced by marital sex; menstruation should *never* be the ideal time for cooking meat). Otherwise, the model's robusticity is demonstrated if it can parsimoniously account for a range of ethnographic phenomena. Features which may be derived from the initial symbolic configuration - that is, ritual identity with the 'wrong sex, wrong species' while blood flows - include totemic practices and associated exogamy rules (Knight, 1991: 88ff.); menstrual taboos and associated hunting prohibitions, as discussed; ritual celibacy and ritual transvestism or so-called 'third gender'; and specific forms of traumatic initiation such as male and female circumcision.

The model challenges much received wisdom in arguing that female strategies drove the emergence of symbolism. Why not propose a model of male ritual coalitions? After all, the ethnographic record shows that most sacred ritual is performed by men, much of it excluding women in general and menstruating women in particular. The model does not preclude the evolution of ritualistic displays by coalitions of male hominids. On Darwinian grounds, we can expect sexually selected male display of alliances among archaic *Homo* species, driven by male-male competition for access to or monopoly of fertile females. But there are several problems with understanding this activity as the root of modern human symbolism.

Firstly, there is the question of the energetics of encephalization. To the extent males were putting their energy towards high-cost signalling of ritual alliances, they would not have been channelling high-energy foods to females burdened by encephalized offspring. The function of the ritual alliances would more likely be to attract cycling, non-pregnant females. Hence, this behaviour is more probably characteristic of less encephalized grades of *Homo* rather than our ancestors who were selected for maximum encephalization. In the model presented, the high-cost ritual signals of an ingroup are designed to exploit the labour power of an outgroup. A situation reversing the dynamic which might be inferred for the late Middle Pleistocene is illustrated in Highland New Guinea, where we find extremely costly ritual display of male coalitions. In these horticultural communities, women are generally the main producers of high-energy foods - pigs - which may be expropriated by men for purposes of formalising alliances through ritual feasting. In these types of 'big man' systems, male-male competition is intensified not reduced, and reproductive differentials between men are increased. In such conditions the area over which sufficient trust exists to sustain a linguistic community is likely to be very localised rather than widespread. These conditions would militate against the emergence of wider forms of social cooperation.

Secondly, the forms of signalling used by such male coalitions raise an even more fundamental objection. How do we explain the emergence of a symbolic domain of collective *deceptions*? We would expect male ritual coalitions to advertise their sexually selected qualities in 'hard to fake' terms, that is in ways that correspond to perceptible reality, visibly demonstrating size of coalition, strength, weaponry and so on. Singing and dancing might be media for such signalling, but there is no reason to suppose these songs or dances would access another world counter to perceptible reality. Why would males be interested in advertising themselves as 'female' or 'animal'? By contrast, a female coalition model straightforwardly yields deceptive sexual signals and habitual construction of 'counter-reality'.

Thirdly, a male ritual model fails to account for what we actually see in the ethnographic record. Why should ritual power celebrate celibacy? Surely, a male ritual model would predict the opposite, maximum access to females. Why should there be any taboos on menstruating women? Surely, male ritual coalitions would be especially anxious to mate-guard imminently fertile females. Why should men imitate female biological signals in their most sacred and esoteric rituals? A male-driven model offers no account for phenomena such as male 'menstruation', subincision and circumcision. By contrast, once the 'gods' are established as monstrous, gender-ambivalent, bleeding therianthropes through female coalitionary strategies, it is logical to expect that male coalitionary counter-strategies would adopt the same language of the 'gods' (see e.g. Knight, 1991: 449ff. on the Aboriginal cult of the Rainbow Snake as a male ingroup conspiracy against women).

Conclusion

What was the original function of ritual? How did it arise? What specific representations did it sustain? The focus here has been on the ritual function of cosmetics. Symbolism arose as a response to increasing levels of reproductive stress experienced by females during the rapid phase of encephalization associated with archaic *Homo sapiens.* Once reliable fertility signals had been phased out, menstrual bleeding was left as the only cue offering males positive information on which females were imminently fertile. Because pronounced menstrual bleeding was valuable for extracting mating effort from males, even non-cycling females 'cheated' by joining in with menstruating relatives whenever blood was flowing, painting up with red pigments to signal 'imminent fertility'. Dance and associated body-painting of this kind long antedates the production of representational imagery on inanimate surfaces. At the point where 'collective deception' was established, such use of movement and pigments constituted symbolism.

The signal of menstruation, appropriated from an individual *by a collective,* communicated for the first time a 'symbolic construct'. The earliest representation of 'divinity' was an amplified and deceptive signal indicating *'wrong sex; wrong species; wrong time'*, constructing taboos on access to 'menstrual' females. Symbolic culture at origin comprised the set of such deceptive signals, designed to exploit male labour power. The archaeological record of ochre use in southern Africa, interpreted in the light of Khoisan and Hadza ethnographic data, matches the predictions of this model.

Acknowledgements

The author would like to thank Herbert Ullrich for the invitation to take part in the Weimar Symposium on 'Lifestyles and survival strategies in Pliocene and Pleistocene hominids'. Leslie Aiello and Chris Knight have contributed to discussion of many ideas in the paper, while Ian Watts has helped especially in reviewing his data on southern African MSA pigment and Khoisan cosmetics.

References

AIELLO, L.C. (1996): Hominine preadaptations for language and cognition. In: P. Mellars & K. Gibson (eds.), *Modelling the early human mind,* 89-99. Cambridge.

AIELLO, L.C. (in press): The expensive tissue hypothesis and the evolution of the human adaptive niche: A study in comparative anatomy. In: J. Bailey (ed.), *Science in archaeology.* London.

AIELLO, L.C. & DUNBAR, R. (1993): Neocortex size, group size, and the evolution of language. *Current Anthropology* 34, 184-93.

AIELLO , L.C. & WHEELER, P. (1995): The expensive tissue hypothesis: The brain and the digestive system in human and primate evolution. *Current Anthropology* 36, 199-221.

ALEXANDER, R.D. & NOONAN, K.M. (1979): Concealment of ovulation, parental care, and human social evolution. In: N. Chagnon & W. Irons (eds), *Evolutionary biology and human social behavior,* 436-453. North Scituate, MA.

BICKERTON, D. (1990): *Language and species.* Chicago.

BIESELE, M. (1993): *Women like meat. The folklore and foraging ideology of the Kalahari Ju/'hoan.* Bloomington.

BLEEK, D. (1930): *Unpublished Hadza notebooks.* African Studies Library, University of Cape Town.

BLEEK, W.H.I. & LLOYD, L.C. (1911): *Specimens of Bushman folklore.* London.

BOESCH, C. (1991): Symbolic communication in wild chimpanzees. *Human Evolution* 6, 81-90.

BOURDIEU, P. (1992): *Language and symbolic power.* Cambridge.

BROOKS, A.S. (1996): Behavior and human evolution. In: W. F. Meikle, F.C. Howell & N.G. Jablonski (eds), *Contemporary issues in human evolution,* 135-166. San Francisco.

BURLING, R. (1993): Primate calls, human language, and nonverbal communication. *Current Anthropology* 34, 25-53.

BYRNE, R. (1995): *The thinking ape. Evolutionary origins of intelligence.* Oxford.

BYRNE, R. & WHITEN, A. (eds) (1988): *Machiavellian intelligence. Social expertise and the evolution of intellect in monkeys, apes, and humans.* Oxford.

CAVALLI-SFORZA, L.L., PIAZZA, A., MENOZZI, P. & MOUNTAIN, J. (1988): Reconstruction of human evolution: Bringing together genetic, archaeological, and linguistic data. *Proceedings of the National Academy of Sciences (USA)* 85, 6002-6006.

CHASE, P.G. (1994): On symbols and the Palaeolithic. *Current Anthropology* 35, 627-629.

CHENEY, D.L. & SEYFARTH, R.M. (1990): *How monkeys see the world.* Chicago.

CLUTTON-BROCK, T.H. & HARVEY, P.H. (1977): Primate ecology and social organization. *Journal of Zoology* 183, 1-39.

COURAUD, C. (1991): Les pigments des grottes D'Arcy-sur-Cure (Yonne). *Gallia Préhistoire* 33, 17-52.

CROOK, J.H. & GARTLAN, J.S. (1966): Evolution of primate societies. *Nature* 210, 1200-1203.

DAWKINS, R. (1993): Viruses of the mind. In: B. Dahlbom (ed:), *Dennett and his critics,* 13-27. Oxford.

DUNBAR, R.I.M. (1988): *Primate social systems.* London.

DUNBAR, R.I.M. (1996): *Grooming, gossip and the evolution of language.* London.

DURKHEIM, E. (1915) [1912]: *The elementary forms of the religious life. A study in religious sociology.* London.

FISCHER, E. (1913): *Die Rehobother Bastarde.* Jena.

FOLEY, R.A. & LEE, P.C. (1989): Finite social space, evolutionary pathways and reconstructing hominid behaviour. *Science* 243, 901-906.

FOLEY, R.A. & LEE, P.C. (1991): Ecology and energetics of encephalization in hominid evolution. *Philosophical Transactions of the Royal Society of London* 334, 223-232.

FOLEY, R.A. & LEE, P.C. (1996): Finite social space and the evolution of human social behaviour. In: J. Steele & S. Shennan (eds), *The archaeology of human ancestry. Power, sex and tradition,* 47-66. London.

GOODALL, J. (1986): *The cchimpanzees of Gombe. Patterns of behavior.* Cambridge, MA.

HAMES, R. (1992): Time allocation. In: E.A. Smith & B. Winterhalder (eds.), *Evolutionary ecology and human behaviour,* 203-235. New York.

HAWKES, K. (1991): Showing off. Tests of another hypothesis about men's foraging goals. *Ethology and Sociobiology* 12, 29-54.

HAWKES, K. (1996): Foraging differences between men and women. Behavioural ecology of the sexual division of labour. In: J. Steele & S. Shennan (eds), *The archaeology of human ancestry. Power, sex and tradition,* 283-305. London.

HAWKES, K., O'CONNELL, J.F. & BLURTON-JONES, N.G. (1992): Hunting income patterns among the Hadza: Big game, common goods, foraging goals and the evolution of the human diet. *Philosophical Transactions of the Royal Society of London.* 334, 243-251.

HAWKES, K., O'CONNELL, J.F. & BLURTON-JONES, N.G. (in press): Hadza women's time allocation, offspring provisioning, and the evolution of post-menopausal lifespans. *Current Anthropology.*

HAWKES, K., O'CONNELL, J.F. & ROGERS, L. (1997): The behavioral ecology of modern hunter-gatherers, and human evolution. *Trends in Ecology and Evolution* 12, 29-32.

HEINZ, H. J. (1966): *The social organization of the !kõ Bushmen.* Unpublished M. A. Thesis, University of South Africa.

HILL, K. & HURTADO, A.M. (1991): The evolution of premature reproductive senescence and menopause in human females: An evaluation of the "grandmother hypothesis". *Human Nature* 2, 313-350.

HILL, K. & KAPLAN, H. (1988): Tradeoffs in male and female reproductive strategies among the Ache, I & II. In: L.L. Betzig, M. Borgerhoff Mulder & P. Turke (eds), *Human reproductive behaviour,* 227-305. Cambridge.

HOOF, J. van & SCHAIK, C. van (1992): Cooperation in competition: The ecology of primate bonds. In: A.H. Harcourt & F.B. de Waal (eds), *Coalitions and alliances in humans and other animals,* 357-389. Oxford.

JOUZEL, J., BARKOV, N.I., BARNOLA, J.I., BENDER, M., CHAPPELLAZ, J., GENTHON, C., KOLLYAKOV, V.M., LIPENKOV, V., LORIUS, C., PETIT, J.R., RAYNAUD, D., RAISBECK, G., RITZ, C., SOWERS, T., STIEVENARD, M., YIOU, F. & YIOU, P (1993): Extending the Vostok ice-core record of palaeoclimate to the penultimate glacial period. *Nature* 364, 407-412.

KAPLAN, H. & HILL, K. (1985): Hunting ability and reproductive success among male Ache foragers. *Current Anthropology* 26, 131-133.

KAPPELMAN, J. (1996): The evolution of body mass and relative brain size in fossil hominids. *Journal of Human Evolution* 30, 243-276.

KATZ, R. (1982): *Boiling energy. Community healing among the Kalahari !Kung.* Cambridge, MA.

KNIGHT, C.D. (1991): *Blood relations. Menstruation and the origins of culture.* New Haven.

KNIGHT, C.D. (1996): Darwinism and collective representations. In: J. Steele & S. Shennan (eds), *The archaeology of human ancestry. Power, sex and tradition,* 331-346. London.

KNIGHT, C.D. (1997): The wives of the sun and moon. *Journal of the Royal Anthropological Institute (N.S.)* 3, 133-153.

KNIGHT, C.D. (in press): Ritual/speech coevolution: A solution to the problem of deception. In: J.R. Hurford, M. Studdert-Kennedy & C.D. Knight (eds), *Approaches to the evolution of language: Social and cognitive bases.* Cambridge.

KNIGHT, C.D., POWER, C. & WATTS, I. (1995): The human symbolic revolution: A Darwinian account. *Cambridge Archaeological Journal* 5, 75-114.

KREBS, J.R. & DAWKINS, R. (1978): Animal signals: Information or manipulation? In: J.R. Krebs & N.B. Davies (eds), *Behavioural ecology. An evolutionary approach,* 282-309. Oxford.

LEONARD, R.W. & ROBERTSON, M.L. (1992): Nutritional requirements and human evolution: A bioenergetics model. *American Journal of Human Biology* 4, 179-195.

LEWIS-WILLIAMS, J.D. (1981): *Believing and seeing. Symbolic meanings in Southern San rock paintings.* London.

LEWIS-WILLIAMS, J.D. & BIESELE, M. (1978): Eland hunting rituals among northern and southern San groups: Striking similarities. *Africa* 48, 117-134.

LLOYD, L. C. (n.d.): *Unpublished notebooks.* Jagger Library, University of Cape Town.

MARTIN, R.D. (1990): *Primate origins and evolution.* London.

MARTIN, R.D. (1996): Scaling of the mammalian brain: The maternal energy hypothesis. *News in Physiological Sciences* 11, 149-156.

MAYER, P.J. (1982): Evolutionary advantage of the menopause. *Human Ecology* 10, 477-494.

MAYNARD SMITH, J. & SZATHMÁRY, E. (1995): *The major transitions in evolution.* Oxford.

McHENRY, H.M. (1996): Sexual dimorphism in fossil hominids and its socioecological implications. In: J. Steele & S. Shennan (eds), *The archaeology of human ancestry. Power, sex and tradition,* 91-109. London.

MELLARS, P. (1991): Cognitive changes and the emergence of modern humans in Europe. *Cambridge Archaeological Journal* 1, 63-76.

MELLARS, P. & STRINGER, C. (eds) (1989): *The Human Revolution. Behavioural and biological perspectives in the origins of modern humans.* Edinburgh.

MELNICK, D. & HOELZER, G. (1993): What is mtDNA good for in the study of primate evolution? *Evolutionary Anthropology* 2, 2-10.

MENOCAL, P.B. de (1995): Plio-Pleistocene African climate. *Science* 270, 53-59.

NETTLE, D. (in press): Linguistic diversity and the evolution of societies. In: C.D. Knight. C. Power & R.I.M. Dunbar (eds), *The evolution of culture.* Edinburgh.

PECCEI, J.S. (1995): The origin and evolution of menopause: The altriciality-lifespan hypothesis. *Ethology and Sociobiology* 16, 425-449.

PINKER, S. (1994): *The Language instinct.* London.

PINKER, S. & BLOOM, P. (1990): Natural language and natural selection. *Behavioral and Brain Sciences* 13, 707-784.

PITTS, M., & ROBERTS, M. (1997) *Fairweather Eden. Life in Britain half a million years ago as revealed by the excavations at Boxgrove.* London.

POWER, C. (in press a): Old wives' tales: the gossip hypothesis and the reliability of cheap signals. In: J.R. Hurford, M. Studdert-Kennedy & C.D. Knight (eds), *Approaches to the evolution of language: Social and cognitive bases.* Cambridge.

POWER, C. (in press b): A sexual selection model for the origins of art. In: C.D. Knight, C. Power & R.I.M. Dunbar (eds), *The evolution of culture.* Edinburgh.

POWER, C. & AIELLO, L.C. (1997): Female proto-symbolic strategies. In: L.D. Hager (ed.), *Women in human evolution,* 153-171. London.

POWER, C. & WATTS, I. (1996): Female strategies and collective behaviour. The archaeology of earliest *Homo sapiens sapiens.* In: J. Steele & S. Shennan (eds), *The archaeology of human ancestry. Power, sex and tradition,* 306-330. London.

POWER, C. & WATTS, I. (1997): The woman with the zebra's penis. Gender, mutability and performance. *Journal of the Royal Anthropological Institute (N.S.)* 3, 537-560.

ROGERS, A.R. (1993): Why menopause? *Evolutionary Ecology* 7, 406-420.

RUDNER, I. (1982): Khoisan pigments and paints and their relationship to rock paintings. *Annals of the South African Museum* 87, 1-280.

RUFF, C.B., TRINKAUS, E. & HOLLIDAY, T.W. (1997): Body mass and encephalization in Pleistocene *Homo. Nature* 387, 173-176.

SAVAGE-RUMBAUGH, E.S. & RUMBAUGH, D.M. (1993): The emergence of language. In: K. Gibson & T. Ingold (eds), *Tools, language and cognition in human evolution,* 86-108. Cambridge.

SMITH, E.A. (1991): *Inujjuamiut foraging strategies. Evolutionary ecology of an Arctic hunting economy.* New York.

SOLOMON, A. (1992): Gender, representation and power in San ethnography and rock art. *Journal of Anthropological Archaeology* 11, 291-329.

STRINGER, C. & GAMBLE, C. (1993): *In search of the Neanderthals: Solving the puzzle of human origins.* London.

TOOBY, J., & DeVORE, I. (1987): The reconstruction of hominid behavioral evolution through strategic modeling. In: W.G. Kinzey (ed.), *The evolution of human behavior: Primate models,* 183-237. Albany.

TRIVERS, R.L. (1972) Parental investment and sexual selection. In: B. Campbell (ed.), *Sexual selection and the descent of man 1871-1971,* 136-179. Chicago.

ULBAEK, I. (in press): The origin of language and cognition. In: J.R. Hurford, M. Studdert-Kennedy & C.D. Knight (eds), *Approaches to the evolution of language: social and cognitive bases.* Cambridge.

VALIENTE-NOAILLES, C. (1993): *The Kua. Life and soul of the Central Kalahari Bushmen.* Rotterdam.

VIGILANT, L., PENNINGTON, R., HARPENDING, H., KOCHER, T.D. & WILSON, A.C. (1989): Mitochondrial DNA sequences in single hairs from a southern African population. *Proceedings of the National Academy of Sciences, (USA)* 86, 9350-9354.

WAAL, F.B.M. de (1982): *Chimpanzee politics.* London.

WATTS, I. (n.d.): *The origins of symbolic culture: The southern African Middle Stone Age and Khoisan ethnography.* Ph.D. Thesis, University of London, in prep.

WILHELM, J. H. (1954): Die !Kung-Buschleute. *Jahrbuch des Museums für Völkerkunde zu Leipzig* 12, 91-188.

WOODBURN, J. (1964): *The social organization of the Hadza of North Tanganyika.* Unpublished Ph.D. thesis, University of Cambridge.

WRANGHAM, R.W. (1980): An ecological model of female-bonded primate groups. *Behaviour* 75, 269-299.

WRANGHAM, R.W. (1987): The significance of African apes for reconstructing human social evolution. In: W.G. Kinzey (ed.), *The evolution of human behavior: Primate models,* 51-71. Albany.

ZAHAVI, A. (1993): The fallacy of conventional signalling. *Philosophical Transactions of the Royal Society of London.* 340, 227-230.

Life and Death, Mortuary Practices and Survival Strategies in the Palaeolithic

Herbert Ullrich

Abstract

Human bones and their archaeological context are the most informative remains telling about reflections on life and death in Palaeolithic times. The fact that all known Lower Palaeolithic, 94 % of the Middle Palaeolithic and 84 % of the Upper Palaeolithic individuals are represented only by few disarticulated bones, very often deliberately modified after death by man with tools (e.g. cut marks, disarticulation patterns, bone breakage patterns, scraping marks), are not the result of pre- and/or post-depositional disturbance of formerly complete skeletons of buried dead by natural processes or animal activity, but have to be interpreted as the result of intentional human activity in connection with mortuary practices celebrated with disarticulated bones of obviously „favoured" dead.

In the European Palaeolithic two different strategies were manifested in mortuary practices and rites: (1) Mortuary practices based on disarticulated human bones resulting from manipulations (defleshing, dismemberment) on corpses of the deceased; (2) Mortuary practices and rites with the entire intact corpse of the dead.

Homo erectus about 500,000 to 300,000 years ago was the first to celebrate mortuary practices with disarticulated human bones, independently in Europe, Africa and Asia. From that time death has been an immanent part of human life and mortuary practices might have originated in close connection with intensive hunting activities. Reflections on life and death may have their origins in reflections on animal life and death and their importance to the daily life of Palaeolithic humans.

Burial of the entire intact corpse of the deceased, which was seldom practised with „highly favoured" dead in Middle and Upper Palaeolithic times and coexisted with mortuary practices based on defleshing and dismemberment of the corpse and bone fragmentation, reflects a new way of thinking by humans and a new form of confrontation with human life and death during the daily life of Palaeolithic man. Burials of the entire intact corpse of the dead reflect a very close connection between the deceased and the living community and a rapid change in reflections on human death. Human beings were no longer part of the animal life and death in the thoughts of Palaeolithic human groups celebrating burials, but became separated from animals and were given a higher status. This new role of the human being in the thoughts and daily life of Palaeolithic man was documented in completely new mortuary practices.

Reflections on life and death in the Palaeolithic necessarily initiated reflections on the world in which humans were living and on the afterworld. The first evidence of initial reflections on the afterworld might be the intentional deposition/burial of disarticulated bones in the Middle Palaeolithic, directed towards a temporary or permanent keeping the relics, towards the time after death. Fundamental changes in reflections on the afterworld were connected with burials of the entire intact corpse by some human groups in the Middle Palaeolithic, which may be interpreted as a way of keeping alive the memory of the dead in the living community as well as belief in a life after death. This was also documented in the grave-furnishings of the Upper Palaeolithic, which, in our opinion, do reflect an attribute to the afterworld thus enabling a quick return of the dead into the living community.

Mortuary practices and ritual in the Palaeolithic reflect the many unsolved problems and contradictions between life and death as well as the struggle for a universal conception of life and death and influenced the lifestyles and survival strategies of Palaeolithic humans intensively.

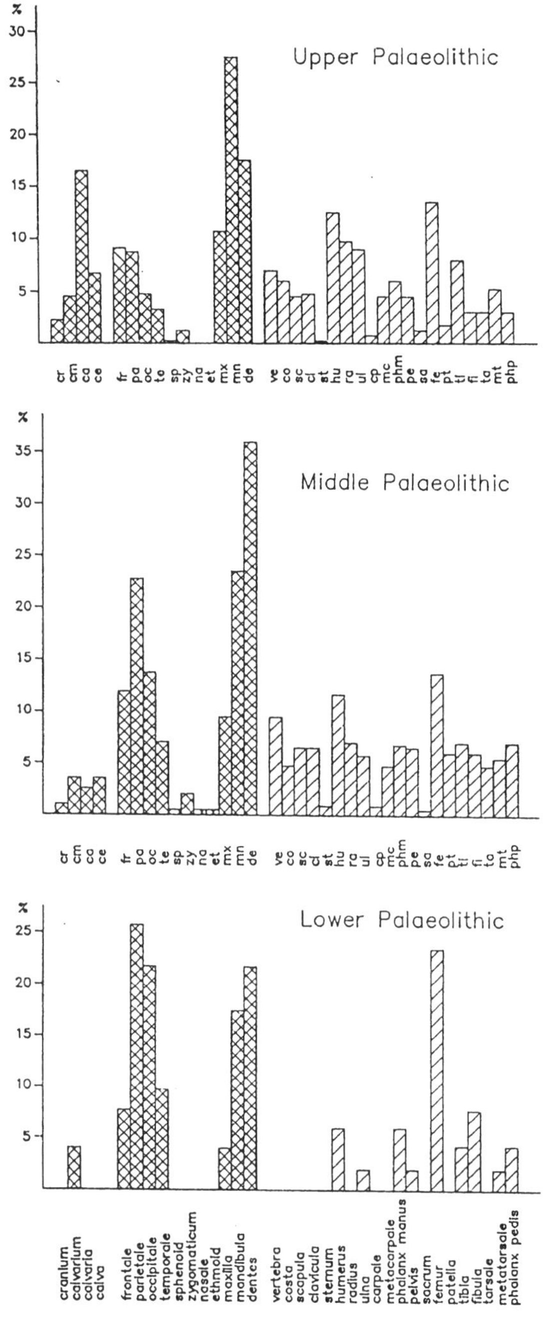

Figure 1 - Patterns of skeletal part representation of the disarticulated human bones for the European Palaeolithic individuals.

Introduction

Mortuary practices and reflections on life and death have their origins in Palaeolithic times. Very often expressed and widely accepted, even in textbooks, is the opinion that Neanderthals and Upper Palaeolithic humans in Europe buried most of their dead. Burial in the Palaeolithic is generally considered to be burial of the entire intact corpse of the deceased. The fact that usually only few bones are found at Palaeolithic sites is interpreted as being the result of natural processes and/or post-depositional disturbance of the initially complete skeleton of the buried dead by animal or human activity.

Mortuary practices in the Palaeolithic have recently again been of special interest to archaeologists reconstructing the ritual and burial of Palaeolithic humans (e.g. Harrold, 1980; May, 1986; Mussi, 1986; Gargett, 1989; Smirnov, 1989, 1991; Holtkamp, 1990; Binant, 1991; Belfer Cohen & Holtkamp, 1992; Defleur, 1993). But archaeologists have only focused on the archaeological background and have not taken into consideration the anthropological aspect of the fossil human remains. Only Smirnow (1991) took into account also aspects of human remains.

An anthropological approach to Palaeolithic mortuary practices is based primarily on the excavated human remains - patterns of skeletal representation, the presence or absence of bone modifications and their interpretation - but is also considering taphonomic processes and the archaeological context of Palaeolithic human remains. The results of this anthropological approach are contributing to a more profound, detailed and more specified comprehension of mortuary practices in the Palaeolithic. They give evidence that mortuary practices in the Palaeolithic were widespread and celebrated in various ways. Burial of the entire intact corpse was a very rare practised ritual, but not a common ritual in Palaeolithic times.

Patterns of skeletal representation

Human remains of about 900 individuals from nearly 350 Palaeolithic sites in Europe were known up to 1990 (Ullrich, 1991a). A detailed analysis of the skeletal part representation for 826 individuals from 320 Palaeolithic sites in Europe (43 individuals from the Lower Palaeolithic, 258 from the Middle and 525 from the Upper Palaeolithic) has clearly shown two representation groups (Ullrich, 1992, in prep.):

(1) All known Lower Palaeolithic, 93.9 % of the Middle Palaeolithic and 84.1 % of the Upper Palaeolithic individuals - that are about 730 Palaeolithic individuals ! - are represented by few *disarticulated bones* (or teeth) only, mainly just one bone or two per individual. The individual representation of the skeletal parts is very low in this representation group: Skull bones are predominant, although represented in less than 20 % of the individuals; parts of the postcranium are represented in hardly more than 10 %, mainly in less than 5 % of the individuals (Figure 1). There are no differences in the patterns of skeletal representation either between adults (more robust bones) and infants (more fragile bones) or between males and females.

(2) Complete or nearly complete *skeletons* are relatively seldom found at European Palaeolithic sites. They are unknown from Lower Palaeolithic sites and only at most 6.1 % of the Middle Palaeolithic and 15.9 % of the Upper Palaeolithic individuals are represented by skeletons. The individual representation of the skeletal parts within this representation group is very high: 70-80 % for the postcranial and 80-90 % for the skull remains.

Taphonomical processes *versus* intentional human activity

Considering the reasons for the different patterns of skeletal part representation of Palaeolithic humans we can conclude that missing skeletal parts in individuals represented by complete/nearly complete skeletons are usually the result of weathering, animal activities or other processes of disturbance and destruction. But what about the reason for missing bones in individuals represented by disarticulated bones only? Taphonomic processes as well as intentional human activities or both can be the reason for the skeletal part representation at Palaeolithic sites.

Geological, physical and chemical processes (sediment pressure, sediment movements, wind and water erosion, weathering, collapses of cave roofs etc.) may result in post-depositonal disturbances of a complete skeleton and in the partly or complete destruction of (mainly fragile) bones. Usually however such processes result in fragmentations of bones and in modifications of the surface of bones. Pre-depositional disturbance and destruction of bones may be caused also by animals (especially carnivores) and post-depositional e.g. by rodents and roots [for the extensive publications on taphonomic processes see only Shipman (1981) and Bonichsen & Sorg (1989)]. The extent of effects of taphonomic processes on Palaeolithic human bones may be very different and depends on the special situation of the relevant Palaeolithic site. The more than 600 human bones from the Neanderthal site Krapina (Croatia) e.g. show very few taphonomically caused changes (Smith, 1976; Ullrich, 1978). The more than 1,600 human remains from Atapuerca (Spain) dated 320,000-200,000 BP, however, have many fractures and bone modifications caused by taphonomic processes (more than half of the remains with carnivore impacts) (Andrews & Fernandez Jalvo, 1997).

Considering the patterns of skeletal part representation of the fossil human remains and the impact of taphonomical processes on skeletons it is impossible to believe that exclusively taphonomical processes are the reason for the fact that about 730 individuals at nearly 300 Palaeolithic sites are represented by few disarticulated bones only. The bones of the initially complete skeletons of these individuals (one human skeleton consists of 217 bones) cannot be destructed completely - with the exception of few bones - by taphonomical processes. From 40

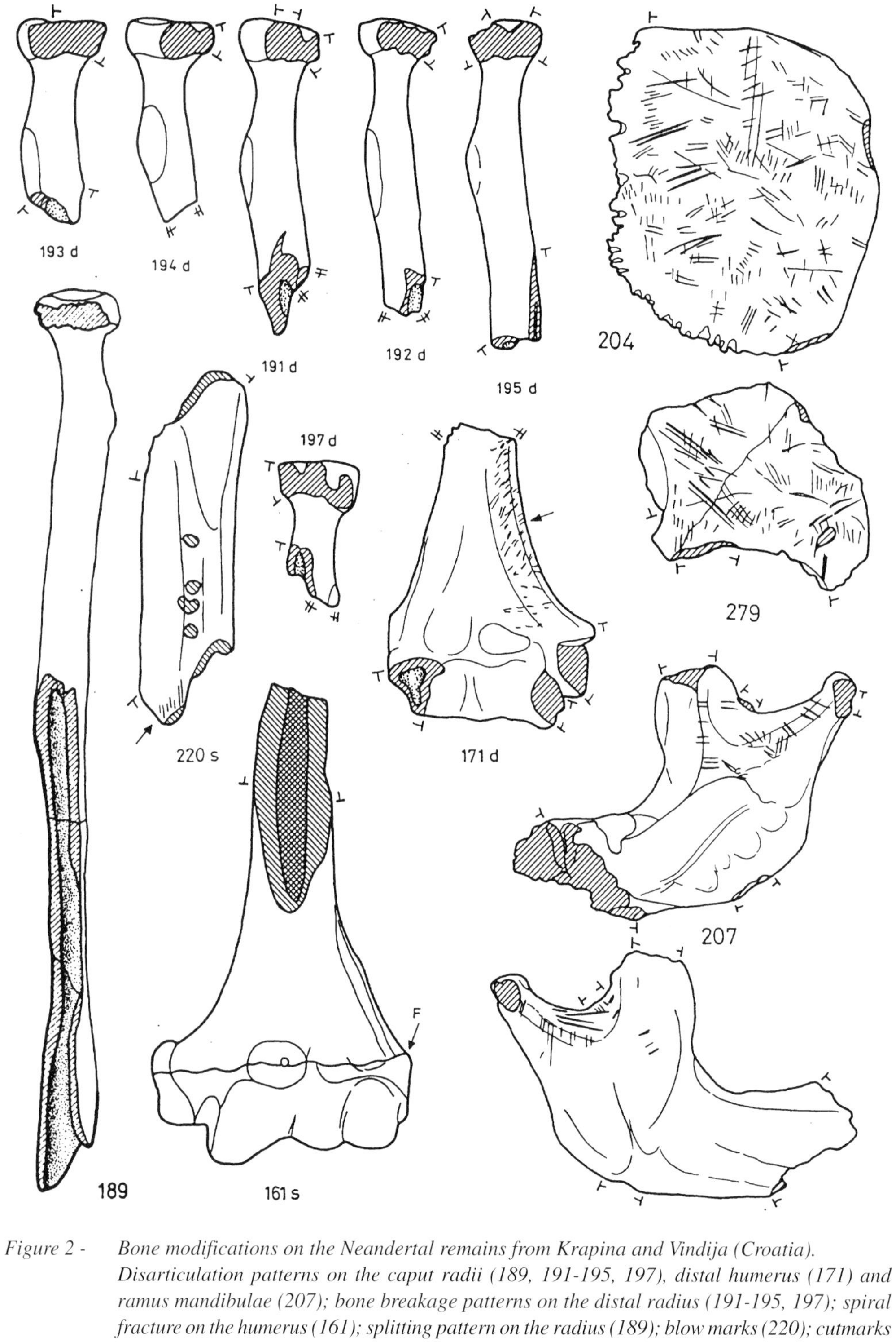

Figure 2 - *Bone modifications on the Neandertal remains from Krapina and Vindija (Croatia). Disarticulation patterns on the caput radii (189, 191-195, 197), distal humerus (171) and ramus mandibulae (207); bone breakage patterns on the distal radius (191-195, 197); spiral fracture on the humerus (161); splitting pattern on the radius (189); blow marks (220); cutmarks and scraping marks (171, 207, 204, 279).*

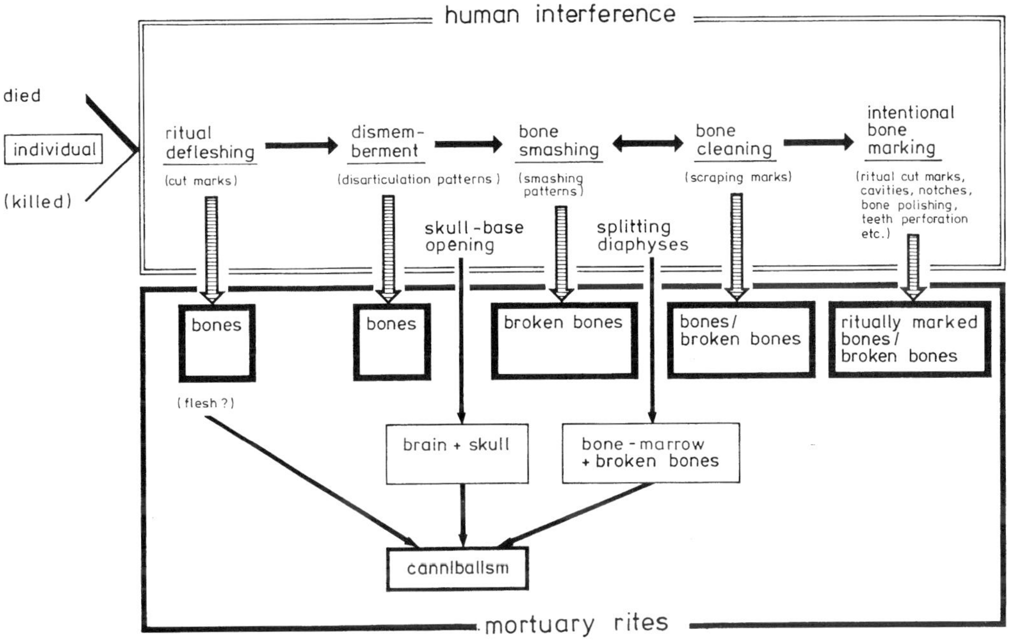

Figure 3 - *Manipulations on human corpses in Palaeolithic times resulted in bones and broken bones for celebrating mortuary ceremonies.*

Lower Palaeolithic (94.3 %), 185 Middle Palaeolithic (71.7 %) and 322 Upper Palaeolithic individuals (61.3 %) - that are 547 individuals - only one bone/tooth or two are preserved.

The only possible explanation given for such a pattern of skeletal part representation is the assumption that intentional human activities are the reasons for the fact that usually only very few bones per individual are found at Palaeolithic sites. There is clear evidence that the intact entire corpses of these individuals were never left nor deposited or buried at Palaeolithic sites (caves, rock-shelters, open-air sites as temporary home bases) where some of their bones have been excavated. Only few bones of the deceased have intentionally been brought by Palaeolithic humans to the home bases and became embedded there.

Bone modifications caused by humans

Very often bone modifications sustained deliberately after death by humans with tools have been found on disarticulated human bones from Palaeolithic sites: cut marks, disarticulation patterns, bone breakage patterns (more than 90 % of the disarticulated bones are broken!), splitting patterns, chopping marks, percussion marks, scraping marks etc. (e.g. Cook, 1991; Czarnetzki, 1977; Gieseler, 1977; Grimm & Ullrich, 1965; Le Mort, 1981, 1986, 1988, 1989; Le Mort & Gambier, 1991; Orschiedt, 1997; Russell, 1987a; Ullrich, 1978, 1979a,b,c, 1982b, 1984, 1989b, 1991b, 1994; White 1986a,b, 1987). These bone modifications (Figure 2) were caused peri-mortem, that means on the fresh, fat and elastic bone, not post-mortem on the dry and brittle bone. Bone modifications caused by intentional human activities indicate defleshing and dismemberment of the corpse, resulting in disarticulated human bones, and manipulations with human bones such as breakage, splitting, scraping off and cleaning.

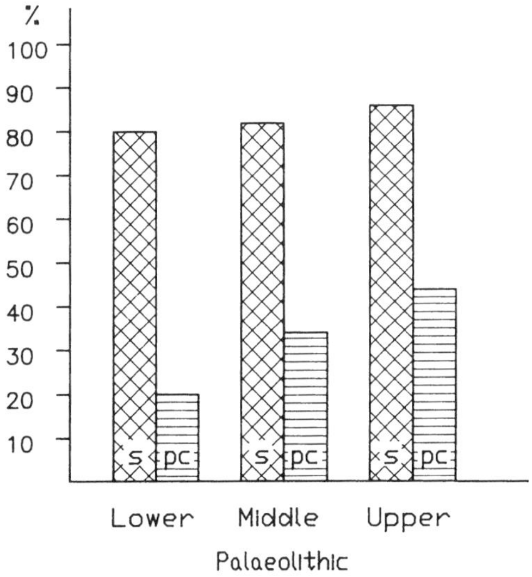

Figure 4 - Individual representation of disarticulated skull (s) and postcranial remains (pc) in the European Palaeolithic record.

Manipulations on human corpses were carried out obviously on individuals who died a natural dead. We have only little evidence, that humans had been intentionally killed before their bodies were defleshed and dismembered (e.g. Middle Palaeolithic child Engis 2/Belgium; Upper Palaeolithic male from Cioclovina/Romania, child from Balla/Hungary, male Mladeč 5/Czech Republic).

Mortuary practices with disarticulated human bones

Burials of entire intact corpses are, of course, the most informative remains telling about reflections on life and death in Palaeolithic times. But mortuary practices with disarticulated human bones resulting from manipulations on human corpses were celebrated much more earlier and widespread in the Palaeolithic, not only in Europe, but also in Africa and Asia. *Homo erectus* about 500,000 to 300,000 years ago was obviously the first to have practised mortuary ceremonies with disarticulated human bones (Ullrich, 1994, 1996a).

We do not know what Palaeolithic humans normally did with their dead and where their corpses were left. It is very likely that most of the individuals of the hunter-gatherers died away from the temporarily or seasonally occupied home bases and that their corpses were left at the place of the death. Bones of these individuals have never been found and will never be found at Palaeolithic sites.

We only have disarticulated human bones from those „favoured" individuals whose corpses were the object of special manipulations. Manipulations on human corpses resulting in defleshing, dismemberment and bone breakage were carried out obviously at the place of the death and resulted in bones and broken bones of the deceased (Figure 3), which were of great importance to Palaeolithic humans for celebrating further mortuary ceremonies within the whole group at the home base. We do not have evidence that manipulations on human corpses were carried out in caves and rock shelters or at open air sites occupied by Palaeolithic humans. But we have clear evidence that most of the disarticulated human bones have been carefully selected and brought intentionally by Palaeolithic men as disarticulated human bones to the sites occupied by them, where further mortuary practices were celebrated. After having completed and finished the celebration of mortuary ceremonies for the deceased, the disarticulated human bones were, according to the archaeological context, either simply thrown away and became mixed with animal bones as food remains or were intentionally deposited or buried with definite ritual motivations at special places within the occupied home base.

We do not believe in the interpretation that disarticulated human bones resulted from primary burials of the entire intact corpse of the deceased outside the home bases in the Palaeolithic (e.g. see Orschiedt, 1997). There is no archaeological evidence for such a ritual in Palaeolithic times. Palaeolithic humans did not stay, so far as we know, for longer periods at the same home base -

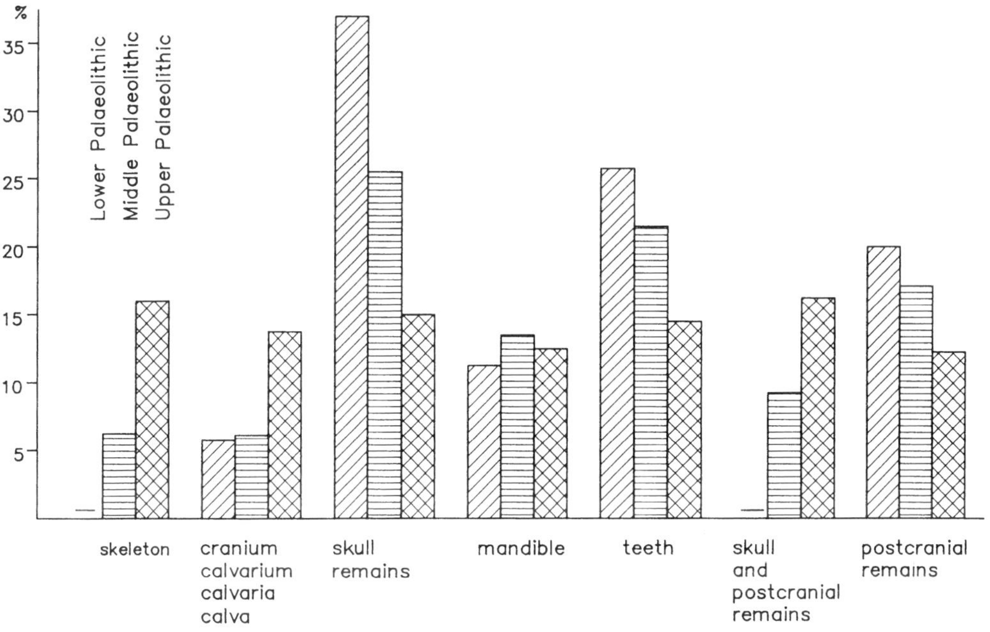

Figure 5 - Patterns of skeletal part representation for disarticulated human bones from the European Palaeolithic (triple columns 2-7).

periods that would have been necessary for the process of decay of the corpse of a primary buried dead in order to get bones of the deceased for further mortuary practices.

Skulls and skull bones of the dead were of special interest to Palaeolithic humans. They are known from about 80 % of the Palaeolithic individuals represented by disarticulated bones only. Postcranial remains of these individuals are represented only in 20-45 % (Figure 4). Only skulls (cranium, calvarium, calvaria, calva) - and no postcranial remains - are known from 8 Middle Palaeolithic and 23 Upper Palaeolithic sites in Europe; only skull bones were discovered at 54 Palaeolithic sites. Very often individuals are represented by mandibles only (Figure 5). This is the case for 96 Lower, Middle and Upper Palaeolithic individuals. From 129 Palaeolithic individuals we have only teeth and no bones; 4 Lower Palaeolithic, 26 Middle Palaeolithic and 29 Upper Palaeolithic sites have yielded only human teeth. These results confirm the often announced assumption that the skull in Palaeolithic times played an important role in ritual ceremonies (cf. also Ullrich 1991b; Smirnov, 1987), which are sometimes described as 'skull cult' - but this is a question of the definition 'cult'.

Throwing away disarticulated human bones

Throwing away disarticulated human bones after having completed mortuary rites is the most frequent and widespread mortuary ceremony in the European Palaeolithic, from the late *Homo erectus* up to the end of the Palaeolithic. Usually it was celebrated without any recognizable ritual motivation and the bones/broken bones became mixed with animal bones at the occupation site.

The Neanderthal site at Krapina (Croatia) is a very instructive example. Intentionally selected human bones (more than 600 skull remains and postcranial bones) of more than 30 individuals (obviously far more) were brought by Palaeolithic humans into the rock-shelter. Nearly all bones

Mortuary practice	Lower Palaeolithic	Middle Palaeolithic	Upper Palaeolithic
Throwing away			
(with ritual motivation)			
Skull bones	Bilzingsleben	Weimar-Ehringsdorf 1-3	---
Mandibles	---	Weimar-Ehringsdorf 6	---
Teeth	---	---	Dolní Věstonice 8; (Bédeilhac; La Combe; Pataud 8-9, 13; Saint-Germain-la-Rivière 5)
Skull/postcranial bones	---	---	Mladeč
(without ritual motivation)			
Skull bones	Castel di Guido 3-6; Pofi 1-2; Vértesszöllös 2	Sakazia; Vindija	Bačo Kiro; Ruse; Tapolca; Pavlov 2-3; Culatovo,
Mandibles	Azych	---	Bačo Kiro; Bervavöl-gy; Pavlov 2; Předmostí 19, 21, 24, 26, 30; Samarkand 1-2
Teeth	Bilzingsleben, Vértesszöllös 2	Taubach 1-2	Dolní Věstonice 7-11
Skull/postcranial bones	---	Barakai; Krapina; Vindija	Předmostí 20,22,23; Maszycka
Postcranial bones	Castel di Guido 1-2; Pofi 3; Sel'ungur 3	Achštyr; Romankovo 1-2; Škurlat; Ohaba-Ponor	Předmostí 27-29; Korman'; Pilisszánto

Table 1 - Mortuary practices with disarticulated human bones: Throwing away bones without and with ritual motivation. (Examples given have been studied by the author; in brackets complements by dates from literature).

were broken, many of them show bone modifications resulting from manipulations on the corpses: cut marks (30.1 % of the postcranial and 14.4 % of the skull fragments), blow marks, bone breakage patterns, splitting patterns etc. The manipulations on human corpses and bone fragmentations were done elsewhere outside the rock-shelter. Only the splitting of some long bones was carried out inside the rock-shelter, obviously in order to extract marrow. After finishing

Mortuary practice	Lower Palaeolithic	Middle Palaeolithic	Upper Palaeolithic
Deposition			
Skeletal parts	---	---	Afontova gora 2
Skull/ Calvarium	---	Saccopastore 1, 2	Veternica 6
Calvaria	---	Engis 2; Gánovce; Weimar-Ehringsdorf 9; Subalyuk 2	Cioclovina
Calva	---	Veternica 1	(Röthekopf); (Le Placard 6, 9-12)
Skull bones	Bilzingsleben A	Kůlna 1-2; Ochoz 2-3	---
Mandibles	---	Ochos 1; Šipka	Ranis
Skull/postcranial bones	---	Weimar-Ehringsdorf 7; Spy 1, 2; Starosel'e; (Neandertal 1)	Balla; Döbritz 2; Brillen-Höhle 2
Postcranial bones	---	---	Döbritz 1;Eliseeviči

Table 2 - Mortuary practices with disarticulated human bones: Deposition of bones. (Examples given have been studied by the author; in brackets complements by dates from literature).

mortuary ceremonies at the occupation site, which included obviously some cannibalistic rites, the human bones were thrown away and became mixed with animal bones. The mortuary practice hypotheses (Ullrich, 1989) is the most probable one interpreting the human remains from Krapina. There is no evidence that the corpses of more than 30 individuals were rapidly buried by their kin or possibly by rock-fall at the rock-shelter and that most of their bones were destructed by post-depositional processes as suggested by Trinkaus (1985). There is also no evidence of a secondary burial - a burial of cleaned bones - at Krapina as concluded by Russell (1989b).

Further examples for throwing away human bones without ritual motivation after having finished mortuary ceremonies are given in Table 1. But throwing away human bones was practiced also with recognizable ritual motivation. At the Lower Palaeolithic site Bilzingsleben (Germany) about 20 skull fragments and some teeth - but no postcranial remains - were scattered on the living floor, especially on the pavement, but had also been intentionally thrown into the nearby former brook (Ullrich, 1994). At the Upper Palaeolithic site Mladeč in Moravia skulls and postcranial bones were thrown through a chimney into the cave.

Intentional deposition of disarticulated human bones

Intentional deposition of disarticulated human bones at special places within the occupation site has been recorded from the Middle and Upper Palaeolithic in Europe. Skeletal parts, skulls,

Mortuary practice	Lower Palaeolithic	Middle Palaeolithic	Upper Palaeolithic
Burial			
Skeleton	---	---	Brno 3; Předmostí 3
Skeletal parts	---	---	Předmostí 4, 9, 10, 14
Skull/ Calvarium	Petralona	Circeo 1; (Pech de l'Aze)	Sungir' 5; Veternica 2-4
Calvaria	---	---	(Maz dÀzil)
Skull bones	---	---	Předmostí 16, 18
Mandibles	---		Předmostí 12-13, 15, 17
Skull/postcranial bones	---	Le Moustier 1; Teshik-Tash	Brno 2; Pavlov 1; Předmostí 1, 2, 5, 6-8, 11

Table 3 - *Mortuary practices with disarticulated human bones: Burial of bones. (Examples given have been studied by the author; in brackets complements by dates from literature).*

skull bones, mandibles, skull and postcranial remains as well as postcranial bones only, peri-mortem human-modified, were deposited along the walls of caves, in wall niches, under stones, nearby fire-places or at other exposed places of the occupation site. In contrast to throwing away disarticulated human bones the intentional deposition includes also more complete skulls (calvarium, calvaria, calva), that means that in many cases the cranium, respectively the neurocranium, had not been fractured completely.

Intentional deposition of disarticulated bones was practised by Middle and Upper Palaeolithic humans at many occupation sites (Table 2). If the occipital fragment from Bilzingsleben discovered in the former brook and covered by the antlers of a deer might be interpreted as intentional deposition then this mortuary practice probably originated at the end of the Lower Palaeolithic.

Burial of disarticulated human bones

Burial in the Palaeolithic indicates in our opinion a stronger relation between life and death and is associated with more definite ritual motivation, often documented in grave-goods, grave-pits, stone markings of a grave, ochre and so on.

Burial of disarticulated cranial and postcranial bones

At many Middle and Upper Palaeolithic sites cranial and postcranial remains representing one individual were discovered in an archaeological context that points to a burial. Archaeologists very often are interpreting those burials as burials of entire intact human corpses where the missing human bones should were destructed by post-depositional processes. Our investigations have shown that such an interpretation very often is incorrect, because only disarticulated human bones have been buried. Bone modifications on those remains caused by human activity clearly

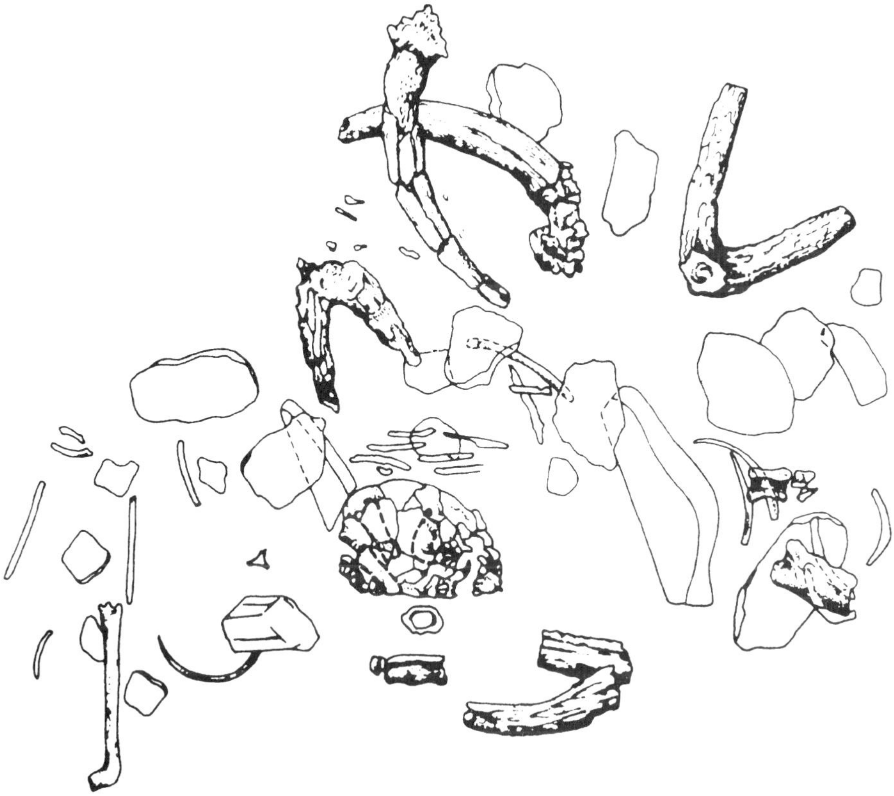

Figure 6 - Burial of disarticulated human bones (skull with mandible and some postcranial remains) inside and outside a circle of five horn pairs of the Siberian mountain goat at the Neandertal site Teshik-Tash in Uzbekistan.

demonstrate that human manipulations on human corpses had been carried out before burial of the often carefully cleaned bones (Table 3).

For example, at the Neanderthal site Teshik-Tash in Uzbekistan a human skull with mandible and some postcranial bones were discovered inside and outside of a circle of five pairs of horns of the Siberian mountain goat (Figure 6). There is no evidence that an entire intact corpse of the boy was buried, but evidence of a burial of disarticulated cleaned bones. The Neanderthal youth Le Moustier 1 (France) is also considered to be a burial of the entire intact corpse, but there is - in contrast to the conventional wisdom - no evidence that either a complete/nearly complete skeleton or postcranial remains in an anatomical connection/position were excavated at the rock-shelter. Only some postcranial bones, scattered and without anatomical position, and the skull with mandible were found. On the skull and mandible distinct cut marks and scraping marks are visible (unpublished results). Therefore the only probable diagnosis for Le Moustier 1 can be: burial of the skull togther with disarticulated human postcranial bones. The same interpretation has to be given to the Upper Palaeolithic males Brno 2 and Pavlov 1 in Moravia. On both sites only the skull and disarticulated postcranial bones were buried, in Pavlov the bones were covered with an mammoth shoulder blade (Figure 7).

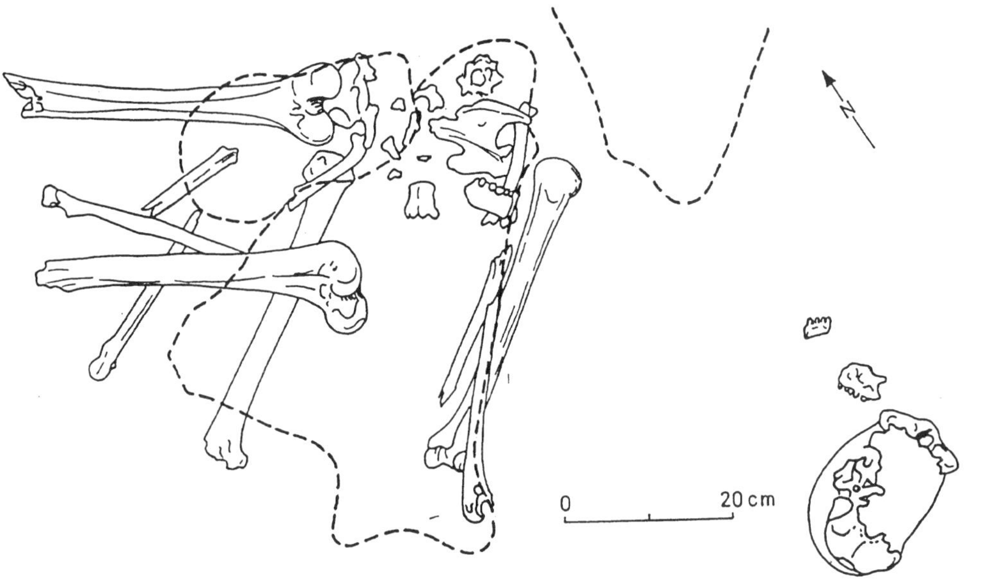

Figure 7 - Skull and disarticulated postcranial bones of the male Pavlov 1 (Moravia), covered with a mammoth shoulder blade.

The Gravettian site at Předmostí in Moravia represents a unique Upper Palaeolithic mass burial (Figure 8). It is generally considered that entire intact corpses of 18 humans were commonly buried in a grave, partly covered with two mammoth shoulder blades and a limestone layer (Klíma, 1991). We have presented an alternative model interpreting the burial rites at Předmostí (Ullrich, 1997). This model is based on the facts that (1) most of the individuals (13) are represented by either one or few disarticulated bones only (cf. Matiegka, 1934), (2) there is clear evidence of artificial modifications on some disarticulated bones of most of the individuals in the mass burial as well as on human bones found outside the grave pointing to modifications on the corpses of the deceased, (3) that with the exception of one male skeleton the bones of the other four almost complete skeletons were found close together, but not in anatomical connection. In our opinion there is no evidence that entire intact corpses of the dead had ever been buried in the mass grave at Předmostí, but only *bones of the dead* (one individual as almost complete *skeleton*). The Upper Palaeolithic hunters from Předmostí obviously buried bones of their dead in the common grave pit before leaving the camp site for the next season or forever. The bones of each individual were put close together, long bones often parallel to each other, and obviously separated from bones of other individuals. There is no evidence that one or all individuals from the mass grave died an unnatural death. There is evidence that the individuals formed a closely related group (Ullrich, 1996b, 1997).

Burial of the skull

Burial of the disarticulated unfractured skull (cranium, calvarium) is a special form of the burial of disarticulated human bones based on activities involving human corpses. This mortuary practice (Table 3) obviously originated at the end of the Lower Palaeolithic, demonstrated by the skull (without mandible, but with disarticulation patterns and cut marks) from Petralona in Greece (Ullrich, 1984), and occurred, although not frequently, through the Middle Palaeolithic up to the

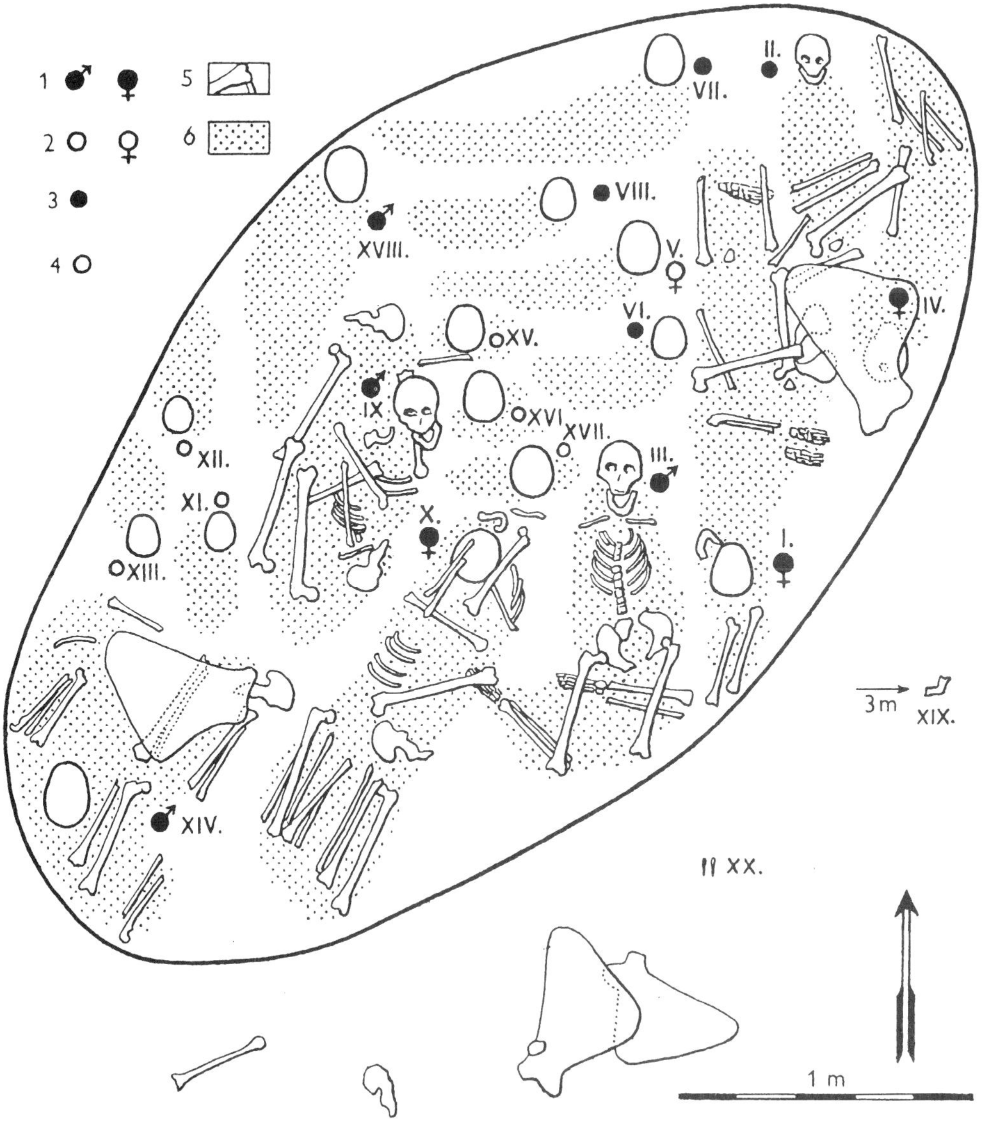

Figure 8 - Reconstruction of the mass burial from Předmostí by Klìma (1991).

end of the Upper Palaeolithic. For instance, in Veternica (Croatia) three calvarium (with disarticulation patterns; two with cut marks and blowing scars) from the Aurignacian were buried under a stone slab at the left wall of the cave (Ullrich, 1991b).

Mortuary practices with entire intact corpses

In debates on burials of entire intact corpses during the Palaeolithic only archaeological criteria (body-position, grave-pit, grave offerings, magical or ritual manipulations etc.) have been considered. Anthropological criteria have completely been disregarded. The main anthropological criteria for entire intact corpse burials are:

- *complete/almost complete human skeletons* (skeletal representation);
- *anatomical connection/position of the bones* (position of the skeleton);
- *absence of human bone modifications* as the result of peri-mortem or post-mortem manipulations on human corpses and bones.

In considering mortuary practices in the Palaeolithic we have also distinguished between deposition and burial of the entire intact corpse in the light of the following criteria:

- *deposition*: complete/almost complete skeleton in an extended, semiflexed or flexed position with more or less strong orientation;
- *burial*: complete/almost complete skeleton with an extended, semiflexed or flexed position with more or less strong orientation; grave-pits and/or other elements of grave-structure (deposited earth, stones, bones etc.), grave-goods (animal bones, artefacts etc.) or other evidence of definite ritual have to be present.

Following the given main anthropological criteria only in a maximum of 5 % of the Middle Palaeolithic and 10-12 % of the Upper Palaeolithic individuals in Europe were buried/deposited as entire intact corpse (Table 4). These results have been obtained on the bases of the published data and should be scrutinized on the relevant original human fossil remains as well as on their archaeological context.

There is no evidence either from Europe and the Near East or from Asia and Africa that Lower Palaeolithic humans ever deposited or buried entire intact corpses of their dead. In the European Middle Palaeolithic only few obviously „highly favoured" dead in limited areas (France, Ukraine) were given such a mortuary practice. Entire intact corpse burial/deposition was first practised by Middle Palaeolithic humans in the Near East (Qafzeh, Skhul) about 100,000-80,000 years BP. In West Europe the oldest burials date to 47,500 BP (La Chapelle, La Ferrassie). In Europe and the Near East burial/deposition of the entire intact corpse was practised in caves and at rock-shelters during the Middle Palaeolithic.

Compared with the data from Middle Palaeolithic sites burial/deposition of the entire intact corpse was more frequent and more widespread in the European Upper Palaeolithic, but as a whole very rare and limited to some areas (France, Italy and Russia mainly, but also Moravia, Germany) in relation to other mortuary rites practised at that time. In France and Italy entire intact corpses were buried/deposited usually in caves and rock-shelters, in Moravia and Russia exclusively at open occupations sites.

Reflections on life and death

Mortuary practices in the Palaeolithic were necessarily closely connected with reflections on life and death. Those reflections began, nearly at the same time independently - as far as we know now -, with late *Homo erectus* about 500,000-300,000 years ago in Europe, Africa and Asia. Prerequisites for such reflections were a highly developed brain with the capacity to develop communication, speech, and abstract thinking as well as a high standard of cultural activities, hunting, co-operation and life in stable units at temporary occupation sites with fire-places. These prerequisites were developed in late *Homo erectus*. From that time death has been an immanent part of the way of human life and has intensively influenced lifestyles, life strategies and survival strategies of Palaeolithic humans - directly or indirectly.

Mortuary practice strategies in the European Palaeolithic

In the European Palaeolithic two different strategies were manifested in mortuary practices and rites:

(1) Mortuary practices based on disarticulated human bones resulting from manipulations (defleshing, dismemberment) on corpses of the deceased;

Mortuary practice	Middel Palaeolithic		Upper Palaeolithic	
	Europe	**Near East**	**Europe**	**Near East**
Deposition	La Quina 5 Starosel'e 1 Mezmaj 1	Amud 1 Qafzeh 15; Skhul 1,6,7; Tabun C1 Shanidar 3,6	(Gough's Cave 1) Le Veyrier 1 Kostenki 7	
Burial	La Chapelle La Ferrassie 1, 2, 4b, 6 (Le Moustier 2) Régourdou Roc-de-Marsal Kiik-Koba (1), 2	Amud 7 Kebera 2 Qafzeh 8, 9, 10, 11 Skhul 4, 5 Shanidar 1, 4, 5, 7 Dederiyeh 1	Dolní Věstonice 3, 13, 14, 15, 16 Bruniquel 24 Cap Blanc Chancelade Combe Capelle (Cro-Magnon 1) Entzheim Le Figuier Les Hoteaux 1 Laugerie-Basse 1 (La Madeleine) (Pataud 6) (Roc-de-Sers 2) Saint-Germain-la-Rivière Neuessing 2 (Oberkassel 1, 2) (Paviland) Arene Candide 1 Continenza Giovane Principé Grimaldi Torre 1, 2, (3) Caviglione 1 Grande 1-5 Enfants 1-2, 4-6 Maritza 1 Ostuni 1-2 Paglicci Ripari Villabruna A Romito 1-2 San Teodoro 1, 3-4 Tagliente 2 Vado all'Arancio 1,2 Veneri 1,2 (Cueva Morin 1, 2) Kostenki 1-4 Mal'ta 1 Sungir' 1-3, 6	(Ein Gev) (Ein Gev I) (Ein Gev X) Neve David Ohala II 2

Table 4 - Mortuary practices with entire intact corpses in the Palaeolithic of Europe and the Near East.

(2) Mortuary practices and rites with the entire intact corpse of the dead.

All mortuary practices of the Lower Palaeolithic and most mortuary rituals during the Middle and Upper Palaeolithic were closely connected with disarticulated human bones and therefore with human manipulations on human corpses of „favoured" dead: defleshing and dismemberment of the corpse, appropriation and fragmentation of bones of the deceased. We have no evidence why Palaeolithic humans defleshed and dismembered (very often completely dismembered) many of their dead. We only know that bones and broken bones of those dead were very important for celebrating mortuary practices. Cannibalism cannot be the reason for those practices, although we can neither deny nor prove that Palaeolithic humans had eaten flesh of their defleshed and dismembered dead. A possible explanation for defleshing and dismemberment of human corpses and the appropriation and fragmentation of bones of the deceased could be given by the main occupation of Palaeolithic man - hunting. Mortuary practices might have originated in close connection with intensive hunting activities. Late *Homo erectus* was very much experienced in defleshing and dismemberment of large animal carcasses and corpses as well as in appropriation and fragmentation of animal bones. Hunted animals, their meat and insides, but also their bones were important and essential for Lower Palaeolithic humans. It is obvious, therefore, that reflections on human life and death originated in reflections on animal life and death and their importance to the daily life of Palaeolithic man. It is obvious furthermore that Lower Palaeolithic humans as well as later on Middle and Upper Palaeolithic humans acted with some of their „favoured" dead in the same way as with hunted animals - with the exception that bones of the deceased had never been manufactured into tools but became the object of ritual mortuary practices and that we do not know, if the flesh of the defleshed dead had been eaten by Palaeolithic humans. Ethnographical parallels of dismemberment of animal carcasses and human corpses tell us that dismemberment means revival of life. Maybe Palaeolithic humans had similar thoughts when celebrating their mortuary ceremonies with defleshing and dismemberment of the corpses of their „favoured" deceased. We only have evidence that very few of these people had been killed intentionally.

Burial/deposition of the entire intact corpse of the deceased, which was seldom practised in Middle and Upper Palaeolithic times and coexisted with other mortuary practices, reflects a complete new way of thinking by some human groups and a new form of confrontation with human life and death during the daily life of Palaeolithic man. In contrast to mortuary practices connected with defleshing, dismemberment and bone fragmentation, entire intact corpses of „highly favoured" dead carefully and intentionally buried at occupation sites demonstrate that it was most important for both the deceased and the living social group to preserve the entire intact body. This reflects a very close connection between the deceased and the living community and a rapid change in reflections on human death. Reflections on human life and death of some Middle and Upper Palaeolithic groups were no longer embedded in reflections on animal life and death, they became independent and attained a much higher meaning. The human being was no longer part of the animal life and death in the thoughts of Palaeolithic humans, but became separated from the animals and was placed higher than animals. This new role of the human being in the thoughts and daily life of Palaeolithic man was documented in completely new mortuary practices - in burials/depositions of the entire intact corpse of the deceased.

This new attitude towards human beings in the world of Palaeolithic man was first documented in the burials/depositions of the entire intact corpse at the sites Qafzeh and Skhul in the near East about 100,000 - 80,000 BP. The hominids from Qafzeh and Skhul are classified as early modern humans. Later on all burials/depositions of the entire intact corpse were connected with Neanderthal populations in the Near East (Kebara, Shanidar, Tabun, Amud - 60,000 - 28,000 BP) as well as in Europe (La Chapelle, La Ferrassie, La Quina, Régourdou, Roc-de-Marsal). We do not know

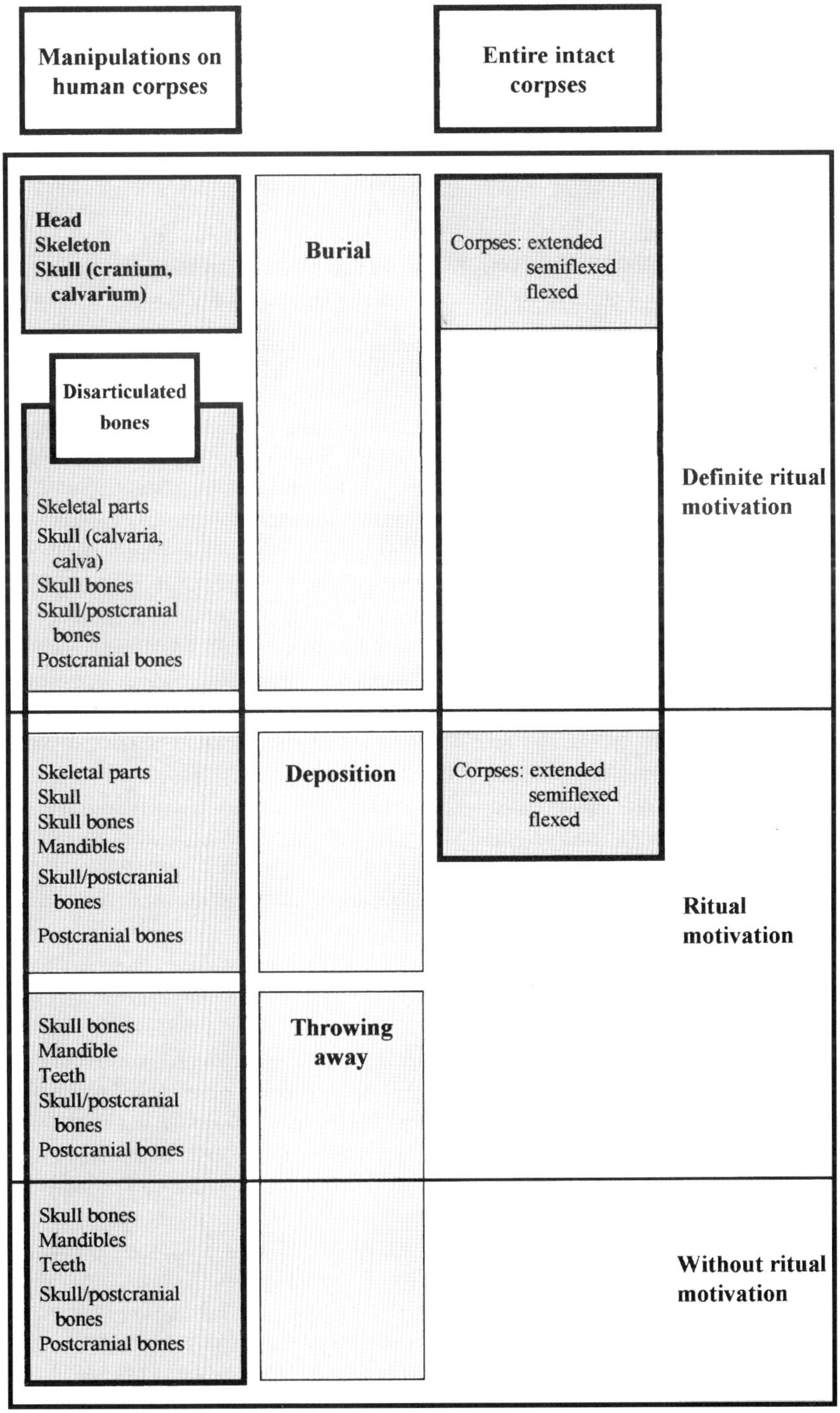

Figure 9 - Various ways of mortuary practices celebrated on disarticulated human bones and entire intact corpses in the European Palaeolithic.

why burials in the Middle and in the Upper Palaeolithic did not become a widespread and universal mortuary practice. We only know that there are no striking differences in mortuary practices connected with defleshing and dismemberment of the corpse, appropriation and fragmentation of bones of the deceased as well as in burials of the entire intact corpse celebrated by Middle Palaeolithic Neanderthals and Upper Palaeolithic populations of modern humans in Europe. The only marked differences in burials are reflected in grave furnishings.

Reflections on the afterworld

Reflections on life and death in the Palaeolithic necessarily initiated reflections on the world in which humans were living and on the afterworld. First evidence of initial reflections on the afterworld might be, in our opinion, the deposition of disarticulated human bones in the Middle Palaeolithic, perhaps at the end of the Lower Palaeolithic. Depositions and burials of human bones are directed towards a temporary or permanent keeping of the relics, towards the time after the death. Deposition and burial of the skull as well as burial/deposition of the defleshed corpse (i.e. a skeleton with ligaments) might mark a progress in the initial obviously very nebulous and very contradictory reflections on the afterworld, because unfractured intact human skulls and defleshed skeletons (e.g. Brno 3 and Předmostí 3) mediate between dismemberment of corpses and fragmentation of bones on the one hand and the entire and intact human body on the other hand.

Fundamental changes in reflections on the afterworld were connected with burial of the entire intact corpse in the Middle Palaeolithic which may be interpreted as a way of keeping the memory of the dead in the living community and as a life after the death. This was documented also in the grave furnitures of the Upper Palaeolithic, which, in our opinion, do not reflect a social differentiation but only an attribute to the afterworld enabling a quick return into the living community. It is suggested that Palaeolithic humans were reflecting on an afterworld that was closely connected to the one in which they were living.

Mortuary practices and survival strategies

The great variety and complexity of mortuary practices and mortuary ritual in the Palaeolithic (Figure 9) reflects the many unsolved problems and contradictions between life and death, between humans and their environments, but also in their social life, which faced the humans daily. Although Palaeolithic humans were intensively engaged in reflecting on life and death they had no universal conception of life and death.

Death became an important and immanent factor of human life at the end of the Lower Palaeolithic that influenced the way of human life intensively. Mortuary practices became more and more part of the social life of Palaeolithic humans and their life strategies. The separation of the human being from the animals in the thoughts of some Palaeolithic human groups marked a rapid change in reflections on human death and resulted in a close connection between the deceased and the living social group in the thoughts of Palaeolithic humans. Reflections on life and death based on thoughts of a life after the death, an afterworld, resulted in mortuary rituals, in burials with grave furniture in the Upper Palaeolithic, that became more and more a social survival strategy and enabled Palaeolithic humans to overcome their social problems and the contradictions between life and death. This was obviously a long lasting process that resulted in the dominance of burials and in cemeteries in connection with the permanent settlements in the Mesolithic.

References

ANDREWS, P. & FERNANDEZ JALVO, Y. (1997): Surface modifications of the Sima de los Huesos fossil humans. *Journal of Human Evolution* 33, 191-217.

BELFER-COHEN, A. & HOVERS, E. (1992): In the eye of the beholders: Mousterian and Natufian burials in the Levant. *Current Anthropology* 33, 463-471.

BINANT, P. (1991): *La préhistoire de la mort. Les premièrs sépultures en Europe.* Paris.

BONICHSEN, R. & SORG, M.H. (eds.) (1989): *Bone modifications.* Orono.

COOK, J. (1991): Preliminary report on marked human bones from the 1986-1987 excavations at Gough's cave, Somerset, England. *Anthropologie* (Brno) 29, 181-187.

CZARNETZKI, A. (1977): Artifizielle Veränderungen an den Skelettresten aus dem Neandertal? In: *Festschrift 75 Jahre Anthropologische Staatssammlung München 1902-1977,* 215-219. München.

DEFLEUR, A. (1993): *Sépultures paléo-moyen.* Paris.

GARGETT, R.H. (1989): Grave shortcommings. The evidence of Neandertal burial. *Current Anthropology* 30, 157-190.

GIESELER, W. (1977): Das jungpaläolithische Skelett von Neuessing. In: *Festschrift 75 Jahre Anthropologische Staatssammlung München 1902-1977,* 39-51. München.

GRIMM, H. & ULLRICH, H. (1965): Ein jungpaläolithischer Schädel und Skelettreste aus Döbritz, Kr. Pößneck, *Alt-Thüringen* 7, 50-89.

HARROLD, F.B. (1980): A comparative analysis of Eurasian Palaeolithic burials. *World Archaeology* 12, 195-211.

HOLTKAMP, E.-V. (1990): *Neandertalerbestattungen und Schädeldeponierungen des Alt- und Mittelpaläolithikums.* Münster.

KLÍMA, B. (1991): Das paläolithische Massengrab von Předmostí. Versuch einer Rekonstruktion. *Quartär* 41/42, 187-194.

Le MORT, F. (1981): *Dégradations artificielle sur des os humains du paléolithique.* Thèse 3e cycle Univ. Paris VI.

Le MORT, F. (1986): Le decharnement du cadavre au paléolithique. *Bulletin de la Societé d'Anthropologie Sud-Ouest* 21, 205-215.

Le MORT, F. (1988): Le décharnement du cadavre chez les néandertaliens: quelches exemples. In: *L'Homme de Néandertal,* vol. 5, 43-55. Liège.

Le MORT, F. (1989): Traces de décharnement sur les ossements néandertaliens de Combe-Grenal (Dordogne). *Bulletin de la Societé Préhistorique Française* 86, 79-87.

Le MORT, F. & GAMBIER, D. (1991): Cutmarks and breakage on the human bones from Placard (France). An example of special mortuary practice during the Upper Palaeolithic. *Anthropologie* (Brno) 29, 189-194.

MALEZ, M. & ULLRICH, H. (1982): Neuere paläanthropologische Untersuchungen am Material aus der Höhle Vindija (Kroatien, Jugoslawien). *Palaeontologia Jugoslavia* 29, 1-44.

MATIEGKA, J. (1934): *Homo předmostensis. Fosilni človek z Předmosti na Morave.* I. *Lebky.* Praha.

MAY, F. (1986): *Les sépultures préhistoriques. Étude critique.* Paris.

MUSSI, M. (1986): Italian Palaeolithic and Mesolithic burials. *Human Evolution* 1, 545-556.

ORSCHIEDT, J. (199/): Der Nachweis einer Sekundärbestattung aus dem Magdaenien der Brillenhöhle, Alb-Donau-Kreis (Baden-Württemberg). *Archäologisches Korrespondenzblatt* 27, 193-206.

RUSSELL, M.D. (1987a): Bone breakage in the Krapina collection. *American Journal of Physical Anthropology* 72, 373-379.

RUSSELL, M.D. (1987b): Mortuary practices at the Krapina Neandertal site. *American Journal of Physical Anthropology* 72, 381-397.

SHIPMAN, P. (1981): *Life history of a fossil. An introduction to taphonomy and paleoecology.* Cambridge/Mass.

SMIRNOV, Yu.A. (1987): K voprosu o kul'te čerepa i nižnej čeljusti v rannem paleolite. In: *Problemy interpretacii archeologičeskich istočnikov,* 50-69. Ordzonikice.

SMIRNOV, Y. (1989): Intentional human burial. Middle Paleolithic (last glaciation) beginnings. *Jurnal of World Prehistory* 3, 2, 199-233.

SMIRNOV, Yu. A. (1991): *Must'erskie pogrebeniya Evrazii. Vozniknovenie pogrebal'noj praktiki i osnovy tafologii.* Moskva.

SMITH, F. (1976): *The Neandertal remains from Krapina. A descriptive and comparative study.* Knoxville.

TRINKAUS, E. (1985): Cannibalism and burial at Krapina. *Journal of Human Evolution* 14, 203-216.

ULLRICH, H. (1978): Kannibalismus und Leichenzerstückelung beim Neandertaler von Krapina. In: M. Malez (red.), *Krapinski pračovjek i evolucija hominida,* 293-318. Zagreb.

ULLRICH, H. (1979a): Bemerkungen zu den Defekten am jungpaläolithischen Kinderunterkiefer von Ranis. *Zeitschrift für Archäologie* 13, 153-161.

ULLRICH, H. (1979b): Artifizielle Veränderungen am jungpaläolithischen Schädel von Cioclovina (SR Rumänien). *Annuaire Roumain d'Anthropologie* 16, 3-12.

ULLRICH, H. (1979c): Artifizielle Veränderungen am Occipitale von Vértesszöllös. *Anthropologiai Közlemenyek* 23, 3-10.

ULLRICH, H. (1982a): Artificial injuries on fossil human bones and the problem of cannibalism, skull-cult, and burial rites. *Anthropos* (Brno) 21, 253-262.

ULLRICH, H. (1982b): Artificial injuries on Upper Paleolithic human fossils found in Czechoslovakia. In: *IInd Anthropological Congress Aleš Hrdlicka,* 451-454. Praha.

ULLRICH, H. (1984): Petralona - eine rituelle Schädelbestattung? *Ethnographisch-Archäologische Zeitschrift* 25, 585-627.

ULLRICH, H. (1986): Manipulations on human corpses, mortuary practice and burial rites in Palaeolithic times. *Anthropos* (Brno) 23, 227-236.

ULLRICH, H. (1989): Neandertal remains from Krapina and Vindija - mortuary practices, burials or cannibalism? In: O.P. Eiben (ed.), *European populations in past, present and future,* 15-19. Budapest.

ULLRICH, H. (1991a): Fossile Hominidenfunde - Übersicht in Tabellenform. In: J. Herrmann & H. Ullrich (Hrsg.), *Menschwerdung. Eine Gesamtdarstellung,* 680-713. Berlin.

ULLRICH, H. (1991b): Bemerkungen zum Schädelkult in der Höhle Veternica (Kroatien). *RAD HAZU* (Zagreb) 458, 173-191.

ULLRICH, H. (1992): Skelettrepräsentation und Totenriten beim archaischen und anatomisch modernen Homo sapiens in Europa. *Wissenschaftliche Zeitschrift der Humboldt-Universität zu Berlin, Reihe Medizin* 41, 2, 135-141.

ULLRICH, H. (1994): Bone modifications in *Homo erectus* remains. *Courier Forschungs-Institut Senckenberg* 171, 257-266.

ULLRICH, H. (1996a): Patterns of skeletal representation and mortuary practices in the African Palaeolithic. In: C.C. Magori, C.B. Saane & F. Schrenk (eds.), *Four million years of hominid evolution in Africa: Papers in honour of Dr. Mary Douglas Leakey's outstanding contribution in palaeoanthropology,* 129-136. Darmstadt (Kaupia: Darmstädter Beiträge zur Naturgeschichte, Heft 6).

ULLRICH, H. (1996b): Reconstruction of close biological relationships in Palaeolithic burials. In: M. Otte (éd.), *Nature et culture,* 765-796. Liège (ERAUL 68).

ULLRICH, H. (1997): Předmostí - An alternative model interpreting burial rites. *Anthropologie* (Brno) 34 (1996), 299-306.

ULLRICH, H. (in prep.): *Totenriten, Schädelkult und Kannibalismus im Paläolithikum Europas.*

WHITE, T.D. (1986a): *Acheulian man in Ethiopia's Middle Awash Valley: the implications of cutmarks on the Bodo cranium.* Haarlem.

WHITE, T.D. (1986b): Cut marks on the Bodo cranium; a case of prehistoric defleshing. *American Journal of Physical Anthropology* 69, 503-509.

WHITE, T.D. (1987): Cannibals at Klasies? *Sagittarius* 2, 2, 6-9.

List of Contributors

ALT, Kurt W., Institut für Humangenetik und Anthropologie der Albert-Ludwigs-Universität, Breisacher Straße 33, D-79106 Freiburg. Germany.

BOCHERENS, Hervé, Laboratoire de Biogéochimie Isotopique, Université Paris 6, 4 Place Jussieu, F-75252 Paris Cedex 05. France.

BRATLUND, Bodil, Archaeological Research Laboratory, Department of Archaeology, Stockholm University, Royal Palace Ulriksdal, S-17071 Solna. Sweden.

BROMAGE, Timothy G., Department of Anthropology, Hunter College, C.U.N.Y., 695 Park Avenue, New York, N.Y. 10021. USA.

BUTOVSKAYA, Marina L., Institute of Ethnology and Anthropology, Leninski prospekt 32a, korp. B, Moscow 117334. Russia.

CACHEL, Susan, Department of Anthropology, Rutgers University, Douglass College, P.O. Box 270, New Brunswick, N.J. 08903-0270. USA.

CLARK, J. Desmond, Department of Anthropology, University of California, Berkeley, CA 94720. USA.

DOMÍNGUEZ-RODRIGO, Manuel, Departamento de Arqueologia y Prehistoria, Universidad Complutense, Facultad de Geografia e Historia, Ciudad Universitaria s/n, E-28040 Madrid. Spain.

FACCHINI, Fiorenzo, Unità Antropologia, Dipartimento di Biologia, Università di Bologna, Via Selmi 3, I-40126 Bologna. Italy.

FOLEY, Robert A., Department of Biological Anthropology and King's College, University of Cambridge, Downing Street, Cambridge CB2 3DZ. U.K.

GAMBLE, Clive, Department of Archaeology, University of Southampton, Highfield, Southampton SO17 1BJ. U.K.

GÜLEÇ, Erksin, D.T.C. Fakultesi, Sihhiye, Ankara Universitesi, Ankara. Turkey.

HARRIS, J.W.K., Department of Anthropology, Rutgers University, Douglass College, P.O. Box 270, New Brunswick, N.J. 08903-0270. USA.

HENKE, Winfried, Institut für Anthropologie, Johannes Gutenberg-Universität, Saarstraße 21, D-55099 Mainz. Germany.

HOWELL, F. Clark, Laboratory for Human Evolutionary Studies, Museum of Vertebrate Zoology, University of California, 3101 Valley Life Science Building, Berkeley, CA 94720-3160. USA.

JÄGER, Klaus-Dieter, Institut für Prähistorische Archäologie der Martin-Luther-Universität, Brandbergweg 23c, D-06099 Halle/Saale. Germany.

JELÍNEK, Jan, Moravian Museum, Anthropos Institute, Zelný trh 6, Brno 65937. Czech Republic.

KORTLANDT, Adriaan, 88 Woodstock Road, Oxford OX2 7ND. U.K.

KOZINTSEV, Alexander G., Museum of Anthropology and Ethnology, Universitetskaya na. 3, St. Petersburg 199034. Russia.

KOZLOWSKI, Janusz K., Institute of Archaeology, Jagellonian University, Golebia 11, PL-31007 Kraków. Poland.

LAITMAN, Jeffrey T., Department of Cell Biology and Anatomy, Mt. Sinai School of Medicine, New York, NY 10029. USA.

LENEY, Mark, New College and Institute of Biological Anthropology, University of Oxford, Oxford OX1 3BN. U.K.

MANIA, Dietrich, Forschungsgrabung Bilzingsleben, Ibrahimstraße 29, D-07744 Jena. Germany.

MANIA, Ursula, Forschungsgrabung Bilzingsleben, Ibrahimstraße 29, D-07744 Jena. Germany.

OTTE, Marcel, Université de Liège, Préhistoire, 7 Place du XX Août, B-4000 Liège. Belgium.

PICHLER, Sandra L., Institut für Ur- und Frühgeschichte der Universität Freiburg, Belfortstraße 22, D-79085 Freiburg. Germany.

PIONTEK, Janusz, Institute of Anthropology, Adam Mickiewicz University, Fredry 10, PL-61 701 Poznan. Poland.

POWER, Camilla, Department of Anthropology, University College London, Gower Street, London WC1E 6BT. U.K.

ROLLAND, Nicolas, Department of Anthropology, University of Victoria, P.O. Box 3050, MS 7046, Victoria BC. Canada V8W 3P5.

RONEN, Avraham, Zinman Institute of Archaeology, University of Haifa, Haifa. Israel.

ROTHE, Hartmut, Institut für Anthropologie der Universität Göttingen, Ethologische Station, Sennickerode 11, D-37130 Gleichen. Germany.

SANTANGELO, Antonio, Piazza Geovanni delle Bande Nere 2, I-20146 Milano. Italy.

SCHÄFER, Dieter, Forschungsinstitut für Alpine Vorzeit, Leopold-Franzens-Universität, Kaiser-Franz-Joseph- Straße 12, A-6020 Innsbruck. Austria.

SCHIEFENHÖVEL, Wulf, Max-Planck-Institut für Verhaltensphysiologie, Von-der-Tann-Straße 3-5, D-82346 Andechs. Germany.

SCHRENK, Friedemann, Hessisches Landesmuseum, Friedensplatz 1, D-64283 Darmstadt. Germany.

SCHUSTER, Sabina, Freiburger Institut für Paläowissenschaftliche Studien F.I.P.S., c/o Universität, Belfortstr.22, D-79085 Freiburg. Germany.

SCHWARTZ, Jeffrey H., Department of Anthropology, University of Pittsburgh, 3HO1 Forbes Quad, Pittsburgh, PA 15260. USA.

SIGMON, Becky A., Department of Anthropology, Erindale College, University of Toronto, 3359 Mississauga Road North, Mississauga, Ontario. Canada L5L 1C6..

SONDAAR, Paul Y., American Museum of Natural History, Central Park West at 79th Street, New York, N.Y. 10024-5192. USA.

STEELE, James, Department of Archaeology, University of Southampton, Highfield, Southampton SO17 1BJ. England.

TATTERSALL, Ian, Department of Anthropology, American Museum of Natural History, Central Park West at 79th St., New York, NY 10024. USA.

TEMBROCK, Günter, Institut für Biologie, Abt. Verhaltensphysiologie, Humboldt-Universität zu Berlin, Invalidenstraße 43, D-10115 Berlin. Germany.

THIEME, Hartmut, Institut für Denkmalpflege, Scharnhorststraße 1, D-30175 Hannover. Germany.

TOBIAS, Phillip V., Department of Anatomical Sciences, Medical School, University of the Witwatersrand, 7 Yorkroad, Parktown, Johannesburg 2193. Republic of South Africa.

TURNER, Elaine, Forschungsbereich Altsteinzeit des Römisch-Germanischen Zentralmuseum, Schloß Monrepos, D-56567 Neuwied. Germany.

ULLRICH, Herbert, Institut für Anthropologie, Medizinische Fakultät der Humboldt-Universität zu Berlin, Tucholskystraße 2, D-10117 Berlin. Germany.

VALOCH, Karel, Moravian Museum, Anthropos Institute, Zelný trh 6, Brno 65937. Czech Republic.

VISHNYATSKY, Leonid B., Institute for the History of Material Culture, Russian Academy of Sciences. Dvortsovaya emb. 18, St. Petersburg 191186. Russia.

VLČEK, Emanuel, National Museum of Prague, U Vojansky 27, Praha 15000 Praha 5. Czech Republic.

VOS, John de, Nationaal Natuurhistorisch Museum, P.O. Box 9517, NL-2300 Leiden. The Netherlands.

WHITE, Tim D., Laboratory for Human Evolutionary Studies, Museum of Vertebrate Zoology, University of California, 3101 Valley Life Science Building, Berkeley, CA 94720-3160. USA.